FIFTY YEARS OF INVASION ECOLOGY

FIFTY YEARS OF INVASION ECOLOGY

THE LEGACY OF CHARLES ELTON

Edited by David M. Richardson

Centre for Invasion Biology
Department of Botany & Zoology
Stellenbosch University

WILEY-BLACKWELL

A John Wiley & Sons, Ltd., Publication

Blackwell Publishing was acquired by John Wiley & Sons in February 2007. Blackwell's publishing program has been merged with Wiley's global Scientific, Technical and Medical business to form Wiley-Blackwell.

Registered office: John Wiley & Sons Ltd, The Atrium, Southern Gate, Chichester, West Sussex, PO19 8SQ, UK

Editorial offices: 9600 Garsington Road, Oxford, OX4 2DQ, UK
The Atrium, Southern Gate, Chichester, West Sussex, PO19 8SQ, UK
111 River Street, Hoboken, NJ 07030-5774, USA

For details of our global editorial offices, for customer services and for information about how to apply for permission to reuse the copyright material in this book please see our website at www.wiley.com/wiley-blackwell.

Library of Congress Cataloging-in-Publication Data

Fifty years of invasion ecology : the legacy of Charles Elton / edited by David M. Richardson.
    p. cm.
  Includes bibliographical references and index.
  ISBN 978-1-4443-3585-9 (hardcover : alk. paper) – ISBN 978-1-4443-3586-6 (pbk. : alk. paper)
1. Biological invasions.   2. Biological invasions–Study and teaching–History..   3. Elton, Charles S.
(Charles Sutherland), 1900–1991.   I. Richardson, D. M. (David M.), 1958-   II. Title: 50 years of invasion ecology.
  QH353.F54 2011
  577'.18–dc22

                                                              2010030974

A catalogue record for this book is available from the British Library.

This book is published in the following electronic formats: ePDF 978-1-4443-2999-5;
Wiley Online Library 978-1-4443-2998-8; ePub 978-1-4443-3000-7

Set in 9/11pt PhotinaMT by Toppan Best-set Premedia Limited
Printed and bound in Malaysia by Vivar Printing Sdn Bhd

1   2011

CONTENTS

Contributors, vii

Foreword, xi

Introduction, xiii

PART 1   HISTORICAL PERSPECTIVES, 1

1 A world of thought: '*The Ecology of Invasions by Animals and Plants*' and Charles Elton's life's work, 3
ROGER L. KITCHING

2 Charles Elton: neither founder nor siren, but prophet, 11
DANIEL SIMBERLOFF

3 The inviolate sea? Charles Elton and biological invasions in the world's oceans, 25
JAMES T. CARLTON

4 The rise and fall of biotic nativeness: a historical perspective, 35
MATTHEW K. CHEW AND ANDREW L. HAMILTON

PART 2   EVOLUTION AND CURRENT DIMENSIONS OF INVASION ECOLOGY, 49

5 Patterns and rate of growth of studies in invasion ecology, 51
HUGH J. MACISAAC, RAHEL A. TEDLA AND ANTHONY RICCIARDI

6 Invasion ecology and restoration ecology: parallel evolution in two fields of endeavour, 61
RICHARD J. HOBBS AND DAVID M. RICHARDSON

PART 3   NEW TAKES ON INVASION PATTERNS, 71

7 Biological invasions in Europe 50 years after Elton: time to sound the ALARM, 73
PETR PYŠEK AND PHILIP E. HULME

8 Fifty years of tree pest and pathogen invasions, increasingly threatening world forests, 89
MICHAEL J. WINGFIELD, BERNARD SLIPPERS, JOLANDA ROUX AND BRENDA D. WINGFIELD

PART 4   THE NUTS AND BOLTS OF INVASION ECOLOGY, 101

9 A movement ecology approach to study seed dispersal and plant invasion: an overview and application of seed dispersal by fruit bats, 103
ASAF TSOAR, DAVID SHOHAMI AND RAN NATHAN

10 Biodiversity as a bulwark against invasion: conceptual threads since Elton, 121
JASON D. FRIDLEY

11 Soil biota and plant invasions: biogeographical effects on plant–microbe interactions, 131
RAGAN M. CALLAWAY AND MARNIE E. ROUT

12 Mutualisms: key drivers of invasions ... key casualties of invasions, 143
ANNA TRAVESET AND DAVID M. RICHARDSON

v

13 Fifty years on: confronting Elton's
    hypotheses about invasion success
    with data from exotic birds, 161
    *TIM M. BLACKBURN, JULIE L. LOCKWOOD
    AND PHILLIP CASSEY*

14 Is rapid adaptive evolution important
    in successful invasions?, 175
    *ELEANOR E. DORMONTT, ANDREW J. LOWE
    AND PETER J. PRENTIS*

15 Why reproductive systems matter
    for the invasion biology of plants, 195
    SPENCER C.H. BARRETT

16 Impacts of biological invasions on
    freshwater ecosystems, 211
    ANTHONY RICCIARDI AND HUGH J. MACISAAC

17 Expanding the propagule pressure concept
    to understand the impact of biological
    invasions, 225
    *ANTHONY RICCIARDI, LISA A. JONES, ÅSA M.
    KESTRUP AND JESSICA M. WARD*

**PART 5 POSTER-CHILD INVADERS,
THEN AND NOW, 237**

18 Elton's insights into the ecology of ant
    invasions: lessons learned and lessons
    still to be learned, 239
    NATHAN J. SANDERS AND ANDREW V. SUAREZ

19 Fifty years of 'Waging war on cheatgrass':
    research advances, while meaningful
    control languishes, 253
    RICHARD N. MACK

**PART 6 NEW DIRECTIONS
AND TECHNOLOGIES, NEW
CHALLENGES, 267**

20 Researching invasive species 50 years after Elton:
    a cautionary tale, 269
    MARK A. DAVIS

21 Invasions and ecosystems: vulnerabilities and
    the contribution of new technologies, 277
    *PETER M. VITOUSEK, CARLA M. D'ANTONIO
    AND GREGORY P. ASNER*

22 DNA barcoding of invasive species, 289
    *HUGH B. CROSS, ANDREW J. LOWE,
    C. FREDERICO D. GURGEL*

23 Biosecurity: the changing face of
    invasion biology, 301
    PHILIP E. HULME

24 Elton and the economics of
    biological invasions, 315
    CHARLES PERRINGS

25 Modelling spread in invasion ecology:
    a synthesis, 329
    *CANG HUI, RAINER M. KRUG
    AND DAVID M. RICHARDSON*

26 Responses of invasive species to a changing
    climate and atmosphere, 345
    JEFFREY S. DUKES

27 Conceptual clarity, scientific rigour and 'The
    Stories We Are': engaging with two challenges
    to the objectivity of invasion biology, 359
    JOHAN HATTINGH

28 Changing perspectives on managing biological
    invasions: insights from South Africa and the
    Working for Water programme, 377
    *BRIAN W. VAN WILGEN, AHMED KHAN
    AND CHRISTO MARAIS*

PART 7 CONCLUSIONS, 395

29 Invasion science: the roads travelled and the
    roads ahead, 397
    DAVID M. RICHARDSON

30 A compendium of essential concepts and
    terminology in invasion ecology, 409
    *DAVID M. RICHARDSON, PETR PYŠEK
    AND JAMES T. CARLTON*

Taxonomic Index, 421

General Index, 425

A companion resources site for this book is available at:

www.wiley.com/go/richardson/invasionecology

CONTRIBUTORS

GREGORY P. ASNER, *Department of Global Ecology, Carnegie Institution of Washington, Stanford, CA 94305, USA.* [gpa@stanford.edu]

SPENCER C.H. BARRETT, *Department of Ecology and Evolutionary Biology, University of Toronto, Toronto, Ontario M5S 3B2, Canada.* [spencer.barrett@utoronto.ca]

TIM M. BLACKBURN, *Institute of Zoology, Zoological Society of London, Regent's Park, London NW1 4RY, UK.* [tim.blackburn@ioz.ac.uk]

RAGAN M. CALLAWAY, *Division of Biological Sciences, University of Montana, Missoula, MT 59812, USA.* [ray.callaway@mso.umt.edu]

JAMES T. CARLTON, *Maritime Studies Program, Williams College-Mystic Seaport, Mystic, CT 06355, USA.* [James.T.Carlton@williams.edu]

PHILLIP CASSEY, *School of Biosciences, Birmingham University, Edgbaston, UK; and School of Earth and Environmental Sciences, University of Adelaide, SA 5005, Australia.* [p.cassey@bham.ac.uk]

MATTHEW K. CHEW, *Arizona State University School of Life Sciences, Tempe, AZ 85287-4501, USA.* [mchew@asu.edu]

HUGH B. CROSS, *State Herbarium of South Australia, Science Resource Centre, Department of Environment and Natural Resources, and Australian Centre for Evolutionary Biology and Biodiversity, School of Earth and Environmental Sciences, University of Adelaide, North Terrace, SA 5005, Australia.* [hugh.cross@adelaide.edu.au]

CARLA M. D'ANTONIO, *Department of Ecology, Evolution and Marine Biology and Program in Environmental Studies, University of California, Santa Barbara, CA 93106, USA.* [dantonio@lifesci.ucsb.edu]

MARK A. DAVIS, *Department of Biology, Macalester College, Saint Paul, MN 55105, USA.* [davis@macalaster.edu]

ELEANOR E. DORMONTT, *Australian Centre for Evolutionary Biology and Biodiversity, School of Earth and Environmental Sciences, University of Adelaide, SA 5005, Australia.* [eleanor.dormontt@adelaide.edu.au]

JEFFREY S. DUKES, *Department of Forestry and Natural Resources and Department of Biological Sciences, Purdue University, West Lafayette, IN 47907-2061, USA.* [jsdukes@purdue.edu]

JASON D. FRIDLEY, *Department of Biology, Syracuse University, Syracuse, NY 13244, USA.* [fridley@syr.edu]

C. FREDERICO D. GURGEL, *School of Earth and Environmental Sciences, University of Adelaide, North Terrace, SA 5005, Australia; and State Herbarium of South Australia, Department of Environment and Natural Resources; and South Australian Research and Development Institute, Aquatic Sciences.* [fred.gurgel@adelaide.edu.au]

ANDREW L. HAMILTON, *Arizona State University School of Life Sciences, Tempe, AZ 85287-4501 USA.* [andrew.l.hamilton@asu.edu]

JOHAN HATTINGH, *Department of Philosophy, Stellenbosch University, South Africa.* [jph2@sun.ac.za]

RICHARD J. HOBBS, *School of Plant Biology, University of Western Australia, Crawley, WA 6009, Australia.* [rhobbs@cyllene.uwa.edu.au]

CANG HUI, *Centre for Invasion Biology, Department of Botany & Zoology, Stellenbosch University, 7602 Matieland, South Africa.* [chui@sun.ac.za]

PHILIP E. HULME, *The Bio-Protection Research Centre, Lincoln University, PO Box 84, Christchurch, New Zealand.* [philip.hulme@lincoln.ac.nz]

LISA A. JONES, *Redpath Museum and Department of Biology, McGill University, Montreal, Quebec H3A 2K6, Canada.* [lisa.jones@mcgill.ca]

AHMED KHAN, *Working for Water Programme, Cape Town, South Africa.* [KhanA@dwa.gov.za]

ÅSA M. KESTRUP, *Redpath Museum and Department of Biology, McGill University, Montreal, Quebec H3A 2K6, Canada.* [asa.kestrup@mcgill.ca]

ROGER L. KITCHING, *Griffith School of the Environment, Griffith University, Brisbane, QLD 4111, Australia.* [r.kitching@griffith.edu.au]

RAINER M. KRUG, *Centre for Invasion Biology, Department of Botany & Zoology, Stellenbosch University, 7602 Matieland, South Africa.* [Rainer@krugs.de]

JULIE L. LOCKWOOD, *Department of Ecology, Evolution and Natural Resources, Rutgers University, New Brunswick, NJ 08901-8551, USA.* [lockwood@aesop.rutgers.edu]

ANDREW J. LOWE, *State Herbarium of South Australia, Science Resource Centre, Department of Environment and Natural Resources, and Australian Centre for Evolutionary Biology and Biodiversity, School of Earth and Environmental Sciences, University of Adelaide, North Terrace, SA 5005, Australia.* [Andrew.Lowe@sa.gov.au]

HUGH J. MACISAAC, *Great Lakes Institute for Environmental Research, University of Windsor, Windsor, Ontario, Canada.* [hughm@uwindsor.ca]

RICHARD N. MACK, *School of Biological Sciences, Washington State University, Pullman, WA 99164, USA.* [rmack@wsu.edu]

CHRISTO MARAIS, *Working for Water Programme, Cape Town, South Africa.* [MaraisC@dwa.gov.za]

HAROLD A. MOONEY, *Department of Biology, Stanford University, Stanford, CA 94305, USA.* [hmooney@stanford.edu]

RAN NATHAN, *Movement Ecology Laboratory, Department of Ecology, Evolution, and Behavior, Alexander Silberman Institute of Life Sciences, The Hebrew University of Jerusalem, Jerusalem 91904, Israel.* [rnathan@cc.huji.ac.il]

CHARLES PERRINGS, *School of Life Sciences, Arizona State University, Tempe, AZ 85287, USA.* [Charles.Perrings@asu.edu]

PETER J. PRENTIS, *School of Land, Crop and Food Sciences, University of Queensland, Brisbane, QLD 4072, Australia.* [p.prentis@uq.edu.au]

PETR PYŠEK, *Institute of Botany, Academy of Sciences of the Czech Republic, CZ-252 43 Průhonice, Czech Republic; and Department of Ecology, Faculty of Science, Charles University, Viničná 7, CZ-128 01 Praha 2, Czech Republic.* [e-mail: pysek@ibot.cas.cz]

ANTHONY RICCIARDI, *Redpath Museum, McGill University, Montreal, Quebec H3A 2K6, Canada.* [tony.ricciardi@mcgill.ca]

DAVID M. RICHARDSON, *Centre for Invasion Biology, Department of Botany & Zoology, Stellenbosch University, 7602 Matieland, South Africa.* [rich@sun.ac.za]

MARNIE E. ROUT, *Division of Biological Sciences, University of Montana, Missoula, MT 59812, USA.* [marnie.rout@mso.umt.edu]

JOLANDA ROUX, *Forestry and Agricultural Biotechnology Institute, University of Pretoria, Pretoria 0002, South Africa.* [e-mail: jolanda.roux@fabi.up.ac.za]

NATHAN J. SANDERS, *Department of Ecology and Evolutionary Biology, University of Tennessee, Knoxville, TN 37996, USA.* [nsanders@utk.edu]

DAVID SHOHAMI, *Movement Ecology Laboratory, Department of Ecology, Evolution, and Behavior, Alexander Silberman Institute of Life Sciences, The Hebrew University of Jerusalem, Jerusalem 91904, Israel.* [david.shohami@mail.huji.ac.il]

DANIEL SIMBERLOFF, *Department of Ecology and Evolutionary Biology, University of Tennessee, Knoxville, TN 37996, USA.* [dsimberloff@utk.edu]

BERNARD SLIPPERS, *Forestry and Agricultural Biotechnology Institute, University of Pretoria, Pretoria 0002, South Africa.* [bernard.slippers@fabi.up.ac.za]

ANDREW V. SUAREZ, *Departments of Entomology and Animal Biology, University of Illinois, Urbana, IL 61801, USA.* [avsuarez@life.uiuc.edu]

RAHEL A. TEDLA, *Great Lakes Institute for Environmental Research, University of Windsor, Windsor, Ontario, Canada.* [tedla1@uwindsor.ca]

ANNA TRAVESET, *Mediterranean Institute of Advanced Studies (CSIC-UIB), 07190 Esporles, Mallorca, Balearic Islands, Spain.* [atraveset@uib.es]

ASAF TSOAR, *Movement Ecology Laboratory, Department of Ecology, Evolution, and Behavior, Alexander Silberman Institute of Life Sciences, The Hebrew University of Jerusalem, Jerusalem 91904, Israel.* [asaf.tsoar@mail.huji.ac.il]

BRIAN W. VAN WILGEN, *Centre for Invasion Biology, CSIR Natural Resources and the Environment, P.O. Box 320, Stellenbosch 7599, South Africa.* [bvwilgen@csir.co.za]

PETER M. VITOUSEK, *Department of Biology, Stanford University, Stanford, CA 94305, USA.* [vitousek@stanford.edu]

JESSICA M. WARD, *Redpath Museum and Department of Biology, McGill University, Montreal, Quebec H3A 2K6, Canada.* [jess.m.ward@gmail.com]

BRENDA D. WINGFIELD, *Department of Genetics, Forestry and Agricultural Biotechnology Institute, University of Pretoria, Pretoria 0002, South Africa.* [brenda.wingfield@fabi.up.ac.za]

MICHAEL J. WINGFIELD, *Forestry and Agricultural Biotechnology Institute, University of Pretoria, Pretoria 0002, South Africa.* [e-mail: mike.wingfield@fabi.up.ac.za]

FOREWORD

Every once in a while a gathering of scholars discuss and debate a particular topic that stands out as a guidepost and significant turning point for an entire field. I believe that this book chronicles the results of just such an event. I deeply regret that I was not present to listen to the obviously stimulating and provocative presentations at the November 2008 symposium in Stellenbosch that gave rise to this book. Fortunately, they have been captured in a comprehensive manner for those of us who were not there. The contributors to this volume are certainly the leaders in this field and have laid its foundations, including the organizer of this remarkable event, David Richardson.

The theme that held this meeting together was a retrospective examination of the field of invasion biology, using the 50th anniversary of Charles Elton's landmark book *The Ecology of Invasions by Animals and Plants* as an anchor for the discussions. What makes the results of this meeting so special to me are that they celebrate Elton's seminal contributions to the field. I read Elton's book as a graduate student. Its messages certainly resonated with me because it called to attention an issue of great significance to society as whole and one where ecological science could contribute to understanding the nature and consequences of what was happening. He underscored the issue in penetrating language: 'We must make no mistake, we are seeing one of the great historical convulsions in the world's fauna and flora'. Although I did not get swept into studying this major phenomenon in my early career, Elton's striking words and global assessment of the issue stayed with me, which brings me to the second reason I regretted missing the meeting. It was held in the lovely town of Stellenbosch, South Africa. In the surrounding countryside of Stellenbosch, which had undergone a major transformation due to invading species, my colleague, Fred Kruger, and I were prompted to initiate, in 1980, a post-Elton global assessment of the status of invasive species under the auspices of the Scientific Committee for Problems of the Environment (SCOPE). This project initially updated the current status of biological invasions globally. Subsequently, with new partners, IUCN and CABI, it took on the larger issue of what factors were driving this phenomenon, and the options for stemming the tide (Global Invasive Species Programme, GISP). This third stage, with the GISP name continuing, focuses on control strategies and methods.

The explosive development of the science of invasion biology in recent years is the subject of this book. It is a critical examination of the rapid progress that has been made. It reveals the extraordinarily richness and complexity of the field, where we have come to a common understanding and where there are still major disagreements. Can we say the field is now maturing because it is attracting the attention of philosophers? Two contributions in this book offer their interesting and debatable viewpoints, and highlight that invasion biology has major social dimensions, including ethical values, and hence warrants deep analysis and scrutiny. Given the social dimensions of the field it is evident that more branches of the social sciences need to become engaged in this area, as noted by several contributors to the volume. At the same time, more of the players of the major driving forces of biotic change, such as globalized trade (for example, the World Trade Organization), have yet to become engaged in a serious manner in discussions, rather than, as at present, often being distant observers.

In addition to the conceptual advances in the field, which so are well outlined in this book, the power of some new investigative tools are discussed, such as the development of comprehensive databases developed by Pyšek and his colleagues, where they are able to document the patterns of introductions and spread through time. These data indicate that the rates of new introductions are accelerating, although not as fast as the current publication rate on invasive species! Then, the remote sensing inventories of invasive species, as chronicled by Vitousek, might help shift us away from sole concern about the end-points of long processes, such as extinctions of a species, to the dynamics of population-size changes of both native and introduced species as they interact.

The chapters in this book call for the development of new directions for the field, building on the founding platform of community ecology, which it so well summarizes. They look to the many other relevant disciplines – such as biosecurity, economics, ecosystem ecology, epidemiology, institution building, risk assessment, sociology, to name a few – that are so crucial for calling attention to the issue of invasive species and that can bring about policies that will mitigate against detrimental trends to human well-being and enhance those that are favourable. The book then chronicles the results of the initial dramatic explosion of knowledge of the past three decades and the beginnings and even bigger burst of new knowledge on one of the great drivers of change, biotic homogenization, which ranks alongside global warming as a major transformative agent of the biotic base that supports human society.

In short order, as this field becomes larger, and more dimensions are explored and integrated, the job of tracking the literature will become even greater and will call for just the kind of overall synthesis that David Richardson has so ably orchestrated in this volume. Let us hope he is ready for another round, which certainly will be called for in a shorter time than the 50-year landmark that stimulated this book.

Harold Mooney
Stanford University, California, USA

INTRODUCTION

'In the 1950s, the planet still had isolated islands, in both geographical and cultural terms—lands of unique mysteries, societies, and resources. By the end of the 20th century, expanding numbers of people, powerful technology, and economic demands had linked Earth's formerly isolated, relatively nonindustrialized places with highly developed ones into an expansive and complex network of ideas, materials, and wealth'.
—*Lutz Warren and Kieffer (2010)*

This book grew from a symposium hosted by the Centre for Invasion Biology at Stellenbosch University in South Africa in November 2008. The meeting, entitled '*Fifty years of invasion ecology – the legacy of Charles Elton*', was attended by 137 delegates from at least 14 countries (Fig. Intro 1). It set out to explore advances in the study of biological invasions in the half-century since the publication of Charles Elton's book *The Ecology of Invasions by Animals and Plants*, and to identify challenges for the future (Garciá-Berthou 2010).

Elton's (1958) book is an important milestone in the history of invasion ecology (Intro Box 1). There are, however, many other good reasons that make it interesting, important and exciting to examine changes in the extent of, and scientific interest in, biological invasions over the past half-century. The quotation cited above neatly summarizes the context for this book. Many processes were set in motion in the middle of the 20th century that were highly influential in shaping the state of our environment today, including the extent, magnitude and trajectories of biological invasions. Although the seeds – literally and figuratively – of many invasions were sown much earlier, the fact that the global human population has grown two and a half times and the global economy has grown eight times since the 1950s (Perrings, this volume) has radically altered the course of biological invasions and

other environmental problems. The way in which people perceive non-native species, including those that become invasive, has also changed radically in this period, but in distinct ways in different sectors and societies and in different parts of the world. The interactions between escalating invasions, other facets of global change, and changing paradigms in ecology, environmental management and ethics, and conservation form the arena in which invasion ecology has evolved, and is still evolving. Invasions are of interest because of the damage they bring to invaded ecosystems and the need to understand their causes and driving forces in order to reduce impacts. They are also of considerable interest to biogeographers, ecologists and evolutionary biologists because they provide natural experiments at temporal and spatial scales that could never be achieved by intentional manipulation for testing theories about the factors that structure biodiversity patterns and affect ecosystem functioning. The study of invasions has borrowed concepts, paradigms and terminology from numerous other fields of enquiry along the way, but has emerged as a discrete field of study, with a growing number of practitioners. It has its own journals, academic centres and a vast and rapidly growing literature. It also has its antagonists and naysayers: those who believe, for instance, that labelling an organism as non-native, trying to keep such organisms out and seeking to eradicate them once they have arrived (even when they have been shown to cause damage) is akin to xenophobia and that this has no place in our homogenized world where diversity of people, cultures, cuisines, etc. is often seen as desirable. The number of stakeholders in decisions about the management of invasive species has grown enormously, creating the need for formal protocols for integrating perspectives among disciplines and domains that traditionally have had little contact. Such developments form the complex backdrop against which the 'game rules' for studying and managing biological invasions are taking shape.

Intro Box 1

British ecologist Charles Sutherland Elton (1900–1991, shown here) was by no means the first author to write about non-native species displaying invasive tendencies. Several pioneering naturalists of the 19th century, notably Charles Darwin, Alphonse de Candolle, Joseph Hooker and Charles Lyell, mentioned invasive species in their writings. In the first half of the 20th century, Joseph Grinnell, Frank Egler, Herbert Baker, Carl Huffaker and other ecologists also published important contributions on introduced species. Many people do, however, recognize Elton's book as the starting point for *focused scientific attention* on biological invasions. It has been described as: 'an accessible and enduring classic', the 'bible of invasion biology', a 'classic book', 'the cornerstone work in [invasion ecology]', an 'invasion classic', a 'magisterial book', 'one of the most forward-looking publications in ecology', a 'pioneering work' and a 'seminal work'. Whether it deserves such hefty accolades is debatable, and this question is indeed discussed at length in this book. Its citation history shows that it has been extremely influential: it has been referred to more than 1500 times in the international literature (more than any other single publication in the field), and continues to be cited more than 100 times a year. It is also held in high regard by the most active researchers in the field today. The book brought together previously disparate themes, including biogeography, conservation biology, epidemiology, human history and population ecology, to show the true global scale and the severe and escalating implications of biological invasions for life on Earth. It placed the phenomenon of invasions in the context of ecological understanding of the time, and provided a map for new research directions (Richardson & Pyšek 2007, 2008). Photograph courtesy of the Department of Zoology, University of Oxford.

Fig. Intro 1 Delegates at the symposium on '*Fifty years of invasion ecology – the legacy of Charles Elton*', Stellenbosch Institute for Advanced Study, South Africa, in November 2008 (photograph: Anton Jordaan).

There is no shortage of excellent books dealing with biological invasions and invasion ecology. Among the most significant titles that deal with general topics in invasion ecology with a more or less global reach in the past decade are the following: Mooney and Hobbs (2000); Groves et al. (2001); Pimentel (2002); Booth et al. (2003); Myers and Bazely (2003); Ruiz and Carlton (2003); Mooney et al. (2005); Sax et al. (2005); Cadotte et al. (2006); Coates (2006); Lockwood et al. (2007); Nentwig (2007); Blackburn et al. (2009); Clout and Williams (2009); Davis (2009); Keller et al. (2009); and Perrings et al. (2010). This list excludes treatments that focus on particular regions, realms or taxa, of which there have been many excellent contributions.

This volume differs from those listed above and all other invasion-related texts in that it aims to provide a collection of thought-provoking essays by leading researchers and thinkers on the evolution of approaches, concepts and paradigms in the study (and management) of biological invasions, with special reference to advances over the past 50 years. The book does not provide comprehensive coverage of all facets of invasion ecology; rather, it explores selected advances, innovations and challenges. The list of contributors to this volume includes many authors of the best-cited and most influential papers in the field of invasion ecology (Pyšek et al. 2006) as well as younger authors and those whose work is less well known in the mainstream literature on invasion ecology. The assembled authors hold a wide range of views, and several authors disagree with others on various issues. Contributors were given wide latitude but were asked to trace the growth and development of their particular themes over the past five decades in particular. They were asked to consider what was known (or what could have been known) about their subject in the 1950s. Was it discussed in Elton's book, and if so, how? If not, why not? What prevented Elton and his contemporaries from addressing certain issues that are now well known or that have become widely accepted as fundamental drivers of invasions? How (in what directions and using which technologies and scientific constructs and in partnership with which other disciplines) has knowledge accumulated in some key areas that were already known in the 1950s? What are the challenges and opportunities for the future?

The volume is divided into seven parts. The first deals with historical perspectives and includes four very different contributions. The first is by Roger Kitching, a student in Charles Elton's group at Oxford University in the 1960s. This chapter provides an engaging account of the academic milieu in which Elton operated and gives important insights on Charles Elton, the man who essentially formalized and presented animal ecology to the English-speaking world in the first half of the 20th century. It describes Elton's contributions to ecology in general and how his interest in invasions developed. In the next chapter, Daniel Simberloff examines the influence of Elton's book, and other publications and events that have shaped the study of biological invasions. James Carlton's chapter emphasizes the importance of Elton's book in focussing attention on the sea, until then virtually ignored in writings about biological invasions. The section closes with an essay by Matthew Chew and Andrew Hamilton that chronicles the history of notions of 'nativeness': the extent to which a given organism can be considered to 'belong' more in one place than in another.

Part two of the book comprises two chapters under the heading Evolution and current dimensions of invasion ecology. In the first, Hugh MacIsaac et al. discuss patterns and the rate of growth of studies in invasion ecology. They explore trends in research on animals compared with plants, terrestrial compared with aquatic habitats, and, for aquatic systems, vertebrates compared with invertebrates. The second chapter examines similarities and contrasts in two fields of enquiry with their roots in the mid-20th century, but with much longer historical precedents: invasion ecology and restoration ecology. Richard Hobbs and Dave Richardson show that both fields are, in some senses, strongly 'mission oriented' in that they address issues of real conservation and management significance. However, they suggest that many aspects of invasion biology have a stronger academic focus (dealing with the evolutionary consequences of invasions, community assembly, limiting similarity, etc.) than is the case for restoration ecology. An interesting distinction is that much (but not all) of invasion biology originated with, and focuses on, problems, whereas restoration ecology grew out of, and focuses largely on, solutions. With the two fields set to interact and intersect more frequently, as rapid environmental change and increased biotic exchange act synergistically to change biophysical envelopes, species' distributions and biotic assemblages, closer alignment is needed in these two fields of endeavour.

New takes on invasion patterns (part 3) explores the current level of knowledge and understanding of biological invasions in one important region (Europe) and for one large group of organisms (tree pests and pathogens). Petr Pyšek and Philip Hulme review the changes in the extent of invasions across Europe in the past half-century. Using several examples cited by Elton, they highlight the massive increases in the number of invaders and the overall extent and impact of invasions in this region. They describe results of a massive data capturing exercise in the past decade that has revolutionized knowledge of the extent and magnitude of invasions in Europe. Michael Wingfield and co-authors discuss trends relating to invasive pathogens that threaten the world's forests and forestry. For microbes, including fungal pathogens, determining whether an organism is native or introduced is a significant challenge.

Part 4 deals with the 'nuts and bolts' of invasion ecology. It has nine chapters, each of which deals with a theme selected to illustrate how advances in technology, changes in scientific methods and paradigms, and momentous changes in the extent and pervasiveness of invasions, have driven changes in focus in research on invasive species.

Asaf Tsoar et al. review the history of understanding of seed dispersal as a fundamental factor in plant invasions. They discuss the emergence of the field of 'movement ecology', and apply these insights to introduce and illustrate a general framework for elucidating the role of dispersal mechanisms as a major driving force in invasion processes.

The reigning paradigm over much of the history of the study of biological invasions has been that communities have 'biotic resistance' to invaders, a notion that was central to Charles Elton's understanding of invasions. Jason Fridley gives a detailed review of Elton's view of biodiversity as a bulwark against invasion and the history of the diversity–invasibility hypothesis, one of Elton's longest-lasting legacies in invasion ecology. The chapter explores whether perspectives on the diversity–invasibility hypothesis remain a useful component of a framework for studying biological invasions.

The role of soil biota as mediators of alien plant invasions was poorly understood in Elton's time. Ray Callaway and Marnie Rout explore the exciting advances in this area. Studies that examined the effects of soil pathogens in the context of invasions have paved the way for groundbreaking work on plant–soil feedbacks, and biogeographical approaches have been applied in exploring shifts between invaded and native ranges in these feedbacks. They then discuss current research initiatives on various microbial mechanisms underlying successful invasions in light of the different effects invaders exhibit on soil biota and the biogeographical nature of these impacts.

Recent research has shown that positive (facilitative) interactions are as important as negative interactions in structuring communities and ecosystems. The past decade has seen a flurry of research on the role of mutualisms in facilitating invasions, especially for plants. Anna Traveset and Dave Richardson review the results of this work and discuss a framework of ecological networks for elucidating how invasive species are integrated into communities and the many ways in which such integration can impact on the functioning of invaded communities. They describe how invasive species often disrupt prevailing mutualisms, and how insights from invasion ecology are shedding new light on the role of mutualisms in structuring communities and the fragility of many interactions.

Blackburn et al. provide an assessment of the current level of understanding of birds as invasive species. They tackle this by considering six key statements in Elton's book that touch on a wide range of potential determinants of invasiveness and invasibility. They describe major advances in our understanding of avian invasions, and point to some important challenges that lie ahead.

Developments in molecular techniques, especially in the past decade, have shown that several genomic processes can drive adaptation to novel environments within 20 generations or less, providing viable alternative or synergistic evolutionary explanations for successful invasions. Such advances are revolutionizing invasion ecology, providing insights that researchers in the 1950s, and even much more recently, could never have dreamt of. Elly Dormonnt et al. discuss emerging insights on adaptation as a mediator of biological invasions, focusing on two of the main ecological explanations for invasion success: propagule pressure and enemy release. Spencer Barrett highlights recent theoretical and empirical work on the reproductive biology of invasive plants. The chapter covers evolutionary transitions in mating strategies associated with migration, the occurrence of pollen limitation and role of reproductive assurance in invasive populations, and strategies in outbreeding species for overcoming the constraints imposed by mate limitation during invasion. The use of genetic markers has provided radical new insights into reproductive diversity in plants, and accurate information on reproductive systems is crucial for understanding the biology of plant invasions.

Lakes and rivers are among the ecosystems that have been most invaded and extensively altered by alien species. Freshwater studies comprise only about 15% of the invasion literature. Although they have had a disproportionately small influence on our evaluation of classical Eltonian concepts, they have been very informative on the community- and ecosystem-level impacts of invasions, particularly involving the structure and function of food webs, the potential for synergistic and cascading effects, and the role of ecological naiveté in declines of native species. The chapter by Tony Ricciardi and Hugh MacIsaac reviews this information.

The concept of propagule pressure was developed long after Elton's book and is largely independent of his influence. Propagule pressure has been shown to be a strong predictor of the establishment of non-indigenous species, but its effects on the ecological impacts of invasions have hardly been examined. Tony Ricciardi et al. review evidence that variation in propagule pressure can alter the magnitude, direction and scope of impacts through its influence on the abundance, functional ecology and range size of invaders.

The aim of part 5 (poster-child invaders – then and now) is to consider some prominent invasive species that were well known half a century ago, to explore how our knowledge of the species has changed and whether the accumulated knowledge has helped to formulate effective management strategies. Nathan Sanders and Andrew Suarez discuss Charles Elton's insights on the ecology of ant invasions and examine how these invasions have changed, and the many ways in which our knowledge has grown. The other case study deals with *Bromus tectorum*, a notorious invasive grass that was well known in the 1950s and has been well studied ever since. Richard Mack chronicles the story of research and management efforts for cheatgrass.

New directions and technologies, new challenges (part 6) includes seven chapters. In the first, Mark Davis considers two areas where he believes invasion ecology has stumbled in its short history. First, he argues that the field has relied for too long on a niche-based approach to understanding invasions. His second concern is that researchers have sometimes overstated and misrepresented certain conclusions and claims, and that this behaviour compromises the

scientific integrity of the field. In this respect he examines claims that invasive species pose the second greatest threat to the survival of species in peril after direct habitat transformation. His message is that the nascent field of invasion ecology should guard against transforming preliminary conclusions and tentative statements into 'invasion gospel' and that all tenets in the field must be subjected to rigorous scrutiny.

Radical advances have been made in our ability to detect, identify and map invasive species. Two chapters review the state of the art in technologies at opposite ends of the spectrum in terms of spatial scale. Peter Vitousek et al. discuss the evolution of methods for remote sensing that are providing exciting opportunities for detecting, understanding and managing ecosystem-transforming invasive species. They review a variety of remote-sensing approaches that have proved useful in evaluating invasions, and summarize key results from work using light detection and ranging (LiDAR) to detect structural features of the plant canopy and substrate in Hawaii. Such technology, with the ability to 'see' beneath forest canopies, is opening new doors for understanding the dynamics of invasions and the impacts they cause. At the opposite end of the spatial spectrum, Hugh Cross et al. review developments towards a standardized method for species identification through the comparative analysis of short DNA sequences, a technique called 'DNA barcoding'. There is huge interest in this approach for assisting identification and confirming the provenance of species, especially microorganisms and for a wide range of other applications in invasion ecology. Philip Hulme's chapter describes the emergence of a specialized field within the broad field of invasion ecology: biosecurity. Although ecologists and biogeographers continue to study the 'nuts and bolts' of biological invasions, research on invasive species is increasingly becoming a multidisciplinary endeavour involving taxonomists and population biologists, statisticians and modellers, economists and social scientists, with its agenda increasingly being shaped by politics, legislation and public perceptions. One area that has enjoyed much research effort in the past decade has been the economics of biological invasions. Charles Perrings reviews developments in this field. Among the many growth areas in this field is the development of approaches for calculating the financial costs of introduction, establishment and spread of potentially harmful alien species, an increasingly significant externality of international trade. Another is the identification of efficient control strategies or policies, which is complicated by the fact that costs of managing invasive species are a 'public good' at several different levels – national, regional and global – which, if left to the market, will be undersupplied. The chapter discusses the many facets of this field.

Biological invasions are a spatial phenomenon. Much research has been directed at explaining and predicting the spatial dynamics of invasive spread. Cang Hui et al. review advances in spatial modelling of invasions. They discuss different categories of spread models and elaborate on advances in parameterizing and simulating dynamics at all stages of the naturalization–invasion continuum. Jeffrey Dukes' chapter examines what is known about responses of invasive species to a changing climate and atmosphere.

The widespread occurrence and increasing abundance of invasive species worldwide and the increasing levels of conflicts of interest where different sectors of societies have radically divergent views on the costs and benefits of certain introduced organisms raises many philosophical and ethical issues. Johan Hattingh examines some prominent challenges to mainstream approaches to dealing with biological invasions.

A wide range of management approaches for dealing with invasive species are underway in different parts of the world. South Africa is selected as a case study to show the evolution of strategies over time. The chapter by Brian van Wilgen et al. reviews the history of dealing with invasive species in the region, and describes the rationale for, and successes and failures in, the 15-year history of the widely lauded Working for Water Programme, which seeks to integrate invasive species management with socio-political priorities.

Part 7 (conclusions). In the penultimate chapter, Dave Richardson explores the dimensions of the current research agenda in invasion science and proposes some profitable avenues for further work. The book ends with a detailed compendium of essential concepts and terminology in invasion ecology, which is the first attempt to provide a systematic listing of fundamental concepts in the field that applies to terrestrial, freshwater and marine ecosystems and across all taxonomic groups.

ACKNOWLEDGEMENTS

All chapters were reviewed by at least two referees. These included most of the chapter authors and the following: James Aronson, Bethany A. Bradley, Jane A. Catford, Rob I. Colautti, Wolfgang Cramer, Melania E. Cristescu, Mitchell B. Cruzan, Paul Debarro, Saara DeWalt, Andre Denth, Marie-Laure Desprez-Loustau, Katrina M. Dlugosch, Paul Downey, Frank N. Egerton, Jennifer Firn, John R. Flowerdew, Mirijam Gaertner, Charles L. Griffiths, Richard H. Groves, Denise Hardesty, Steven I. Higgins, Ben D. Hoffmann, Michael A. Huston, James Justus, Cynthia S. Kolar, Paul Kardol, Salit Kark, Reuben P. Keller, Christian A. Kull, Berndt Janse van Rensburg, Bettine Jansen van Vuuren, Fabien Leprieur, Lloyd L. Loope, Michael P. Marchetti, Laura A. Meyerson, Unai Pascual, Núria Roura-Pascual, Jennifer L. Ruesink, Nanako Shigesada, Paul H. Skelton, Daniel Sol, Wim van der Putten, Charles R. Warren, David A. Westcott, Russel Wise, Peter A. Williams and John R. Wilson. I sincerely thank all these friends and colleagues for providing helpful reviews that improved the chapters.

I also thank the many people who assisted in many ways in ensuring that this book reached production. Firstly, Steven Chown, Sarah Davies and all the staff and students at the Centre for Invasion Biology at Stellenbosch University enthusiastically supported the 'Elton symposium' from which this volume grew. My personal assistant, Christy Momberg, worked tirelessly on the organization of the symposium. Thanks are also due to the main sponsors of the symposium: the Department of Science and Technology–National Research Foundation Centre of Excellence for Invasion Biology; Stellenbosch University (especially Vice Rector: Research, Professor Arnold van Zyl, and Senior Director: Research, Dr Therina Theron, for their interest and support); the Working for Water Programme; and the Oppenheimer Memorial Trust.

Thanks are also due to Ward Cooper, Camille Poire and Kelvin Matthews at Wiley-Blackwell and Chris Purdon for driving the production of the book so efficiently.

I am most grateful to my family, Corlia and Sean, for enduring my preoccupation with things related to 'the Elton meeting' and then 'the Elton book' for so long. I will try not to mention 'the E word' in their presence for a while!

David M. Richardson
Stellenbosch, South Africa
April 2010

REFERENCES

Blackburn, T.M., Lockwood, J.L. & Cassey, P. (2009) *Avian Invasions. The Ecology and Evolution of Exotic Birds.* Oxford University Press, Oxford.

Booth, B.D., Murphy, S.D. & Swanton, C.J. (2003) *Weed Ecology in Natural and Agricultural Systems.* CABI Publishing, Wallingford.

Cadotte, M.W., McMahon, S.M. & Fukami, T. (eds) (2006) *Conceptual Ecology and Invasion Biology: Reciprocal Approaches to Nature.* Invading Nature: Springer Series in Invasion Ecology 1. Springer, Dordrecht.

Clout, M.N. & Williams, P.A. (eds) (2009) *Invasive Species Management. A Handbook of Principles and Techniques.* Oxford University Press, Oxford.

Coates, P. (2006) *American Perceptions of Immigrant and Invasive Species. Strangers on the Land.* University of California Press, Berkeley.

Elton, C.S. (1958) *The Ecology of Invasions by Animals and Plants.* Methuen, London.

Davis, M. (2009) *Invasion Biology.* Oxford University Press, Oxford.

García-Berthou, E. (2010) Invasion ecology fifty years after Elton's book. *Biological Invasions*, **12**, 1941–1942.

Groves, R.H., Panetta, F.D. & Virtue, J.G. (eds) (2001) *Weed Risk Assessment.* CSIRO Publishing, Collingwood.

Keller, R.P., Lodge, D.M., Lewis, M.A. & Shogren, J.F. (eds) (2009) *Bioeconomics of Invasive Species. Integrating Ecology, Economics, Policy, and Management.* Oxford University Press, Oxford.

Lockwood, J.L., Hoopes, M.F. & Marchetti, M.P. (2007) *Invasion Ecology.* Blackwell, Oxford.

Lutz Warren, J. & Kieffer, S. (2010) Risk management and the wisdom of Aldo Leopold. *Risk Analysis*, **30**, 165–174.

Mooney, H.A. & Hobbs, R.J. (eds) (2000) *Invasive Species in a Changing World.* Island Press, Washington, DC.

Mooney, H.A., Mack, R.N., McNeely, J.A., Neville, L.E., Schei, P.J. & Waage, J. (eds) (2005) *Invasive Species. A New Synthesis.* Island Press, Washington, DC.

Myers, J.H. & Bazely, D.R. (2003) *Ecology and Control of Introduced Plants.* Cambridge University Press, Cambridge.

Nentwig, W. (ed.) (2007) *Biological Invasions.* Springer, Berlin.

Perrings, C., Mooney, H.A. & Williamson, M. (eds) (2010) *Bioinvasions & Globalization. Ecology, Economics, Management, and Policy.* Oxford University Press, Oxford.

Pimentel, D. (ed.) (2002) *Biological Invasions. Economic and Environmental Costs of Alien Plant, Animal and Microbe Species.* CRC Press, Boca Raton.

Pyšek, P., Richardson, D.M. & Jarošík, V. (2006) Who cites who in the invasion zoo: insights from an analysis of the most highly cited papers in invasion ecology. *Preslia*, **78**, 437–468.

Richardson, D.M. & Pyšek, P. (2007) Classics in physical geography revisited: Elton, C.S. 1958: The ecology of invasions by animals and plants. Methuen: London. *Progress in Physical Geography*, **31**, 659–666.

Richardson, D.M. & Pyšek, P. (2008) Fifty years of invasion ecology – the legacy of Charles Elton. *Diversity and Distributions*, **14**, 161–168.

Ruiz, G.M. & Carlton, J.T. (eds) (2003) *Invasive Species. Vectors and Management Strategies.* Island Press, Washington, DC.

Sax, D.F., Stachowicz, J.J. & Gaines, S.D. (eds) (2005) *Species Invasions. Insights into Ecology, Evolution, and Biogeography.* Sinauer Associates, Sunderland, Massachusetts.

Part 1

Historical Perspectives

Chapter 1

A WORLD OF THOUGHT: '*THE ECOLOGY OF INVASIONS BY ANIMALS AND PLANTS*' AND CHARLES ELTON'S LIFE'S WORK

Roger L. Kitching

Griffith School of the Environment, Griffith University, Brisbane, QLD 4111, Australia

Fifty Years of Invasion Ecology: The Legacy of Charles Elton, 1st edition. Edited by David M. Richardson

1.1 INTRODUCTION

In this chapter, I present a personal view of the Elton canon: the body of work created by Charles Elton during his whole working life. *The Ecology of Invasions by Animals and Plants* (hereafter abridged as '*Ecology of Invasions*' or '*EIAP*') was produced something like two-thirds of the way through this long period of productivity. I suggest its origins and impacts are best appreciated when viewed as part of Elton's overall intellectual contribution. That *EIAP* may be regarded as foundational to a whole subsequent field of study is indisputable (Richardson & Pyšek, 2007, 2008; but see Simberloff, this volume, for a different view) and yet, I shall show that this is only one of several highly productive and important areas of work that originated within the body of work for which Elton was responsible during his long life.

1.2 THE ECOLOGIST AND THE MAN

Charles Sutherland Elton (1900–1991) did not invent the discipline of animal ecology: that evolved from the many musings of earlier naturalists, beginning to precipitate into modern scientific form courtesy of Charles Darwin, Alfred Russel Wallace and Victor Shelford, among others. Indeed many of the concepts usually associated with Elton's ideas had existed in more or less nascent form in earlier years. The idea of a food chain (although not the phrase) had been described well over 100 years earlier (Bradley 1718; Egerton 2007). The general notion of a pyramid of numbers or at least the underlying 'rule of ten' had been stated clearly and generally by Karl Semper in 1881. The broadening of these simplifications of trophic interactions into food webs began with a series of specific diagrams as early as 1912 (Pierce et al. 1912) and these early efforts are described in detail by Egerton (2007). Even the term 'niche', so often now eponymic with Elton, had been used by Grinnell as early as 1904 and developed significantly in his famous article on the California thrasher (Grinnell 1917). It was Charles Elton, though, who gathered up, clarified and connected these ideas into a cogent whole in his 1927 book *Animal Ecology* and, to my way of thinking, so set an agenda for the entire emerging field of animal ecology. It also established ecological ground rules that subsequently underpinned the emergence of formal conservation agencies in both the UK and USA.

This short text remains vital reading for all ecologists and is a model of clear thinking, pithy writing and penetrating insight. As well as incorporating the concepts already mentioned he also built substantially on Shelford's ideas on succession within natural communities. The insights presented so well in *Animal Ecology* had derived from Elton's years as a general naturalist during his youth and early adulthood in England. In addition, he had spent three very formative seasons before and after graduating from Oxford on expedition in Spitzbergen and Norwegian Lapland observing the ecological communities and animal populations in that almost canonical landscape where, perhaps, the grand patterns can be perceived more clearly because the component parts, so confusing and diverting in less extreme environments, are relatively few in number.

Elton's career has been described at length by his thorough and sympathetic obituarists (Macfadyen 1992; Southwood & Clarke 1999). I repeat here only the observations that Elton's scientific contributions during a long scientific career were marked by a series of contrasting but perhaps surprisingly coherent set of major books, each in itself a landmark for the developing subject. *Animal Ecology* (1927) was succeeded by *Voles, Mice and Lemmings* (1942), *The Ecology of Invasions by Animals and Plants* (1958) and, finally, *The Pattern of Animal Communities* (1966). These undoubtedly seminal contributions should, in my view, be joined by his swansong paper on tropical rainforest biodiversity published in the *Journal of Animal Ecology* in 1967. I shall return to this last major publication towards the end of this account.

From 1932 until his retirement in 1967, Elton worked with a small group of other ecologists and graduate students in the Bureau of Animal Population (the BAP), within but not physically part of the Department of Zoology of the University of Oxford (Crowcroft 1991). It was my privilege to join that group as a doctoral student in October 1966: one of two 'final' students of the BAP. I was supervised, formally, by H.N. Southern but my entry to the Bureau and progress within it were closely directed by Elton himself. Elton's influence within the Bureau was all pervasive. A quiet, even unprepossessing, man, Elton nevertheless imposed his style and philosophy on the life of the Bureau and those of us who were part of the enterprise were willing participants in what we took to be a noble endeavour. I for one have not deviated from that view in the ensuing 40 years.

Life in the Bureau revolved around afternoon tea. Coffee was indeed taken in the mornings but was a low-key affair, usually huddled within the small kitchen. Coffee was an event of small meetings: for us students it was full of surprises. 'Roger, I wonder if I could introduce you to our visitor ... this is Julian Huxley' was just one of several unexpected encounters, later to be treasured as an era in biology gradually passed away. But afternoon tea was of greater splendour altogether. The library of the BAP, normally open to questing undergraduates, would be closed. Elton would preside at the head of the long scrubbed table and the rest of us – staff, students, technicians, distinguished visitors and assistants – would range down either side. If there was some event to be marked, the 'boss' (Elton) would shyly slide a large bag of cream buns onto the table alongside the giant teapot. It was at one such event that I arrived early and discovered that, briefly at least, Elton and I were the only ones at table. 'Tell me', he said, 'what do you do for exercise?' I diffidently said that I played a little squash but his retort took me not a little by surprise. 'I used to box, you know. I knocked a man out once'. All this presented in a quiet, near whisper and emanating from this small balding man renowned, among we postgraduates, for routinely wearing at least one of his several, concurrent, sleeveless pullovers, inside out.

Perhaps this anecdote, though, captures Elton's intellectual impact as well as the man himself. Intellectually speaking he produced knockout blows from what some thought of as an unlikely source. Indeed his impact is, in my view, still not fully appreciated. Much later, while I was a Bullard Fellow at Harvard in 1998, Ernst Mayr, after grilling me about *my* intellectual antecedents actually said, 'Ah yes, Elton ... we expected so much more of him!' It was not exactly clear in this conversation exactly who 'we' were but I can only say that most of the current trends in animal ecology now owe much to Elton: so pervasive are these debts that most do not question their actual foundations.

1.3 *ECOLOGY OF INVASIONS* IN CONTEXT

Ecology of Invasions, as much of this book testifies, is a work of lasting impact (see also Richardson & Pyšek 2008). I have heard it said that the book was 'ahead of its time', but this is misleading. It was, in fact, very

much *of* its time, harking back to the very beginnings of what was then seen as the modern renaissance in biology, reviewing and critiquing the current state of play and then setting a research agenda which is, only now, receiving the attention it demands. Elton presents his thoughts in *EIAP* very much as part of the ecological sub-science of biogeography. It is significant that Elton begins his thesis in *EIAP* with a reprise of Wallace's views of biogeography. Bear in mind that, in 1954, the prevailing paradigm in biogeography was an entirely dispersalist one. Wegener's (1915) ideas of continental drift were still considered by most as part of the lunatic fringe (indeed, so I was taught during my undergraduate years at Imperial College as late as 1965, admittedly by a very conservative and elderly teacher). It was not until 1959 that Heezen, Tharp and Ewing first published their findings confirming the existence of the mid-Atlantic ridge which finally began the process which eventually led to mainstream acceptance of the ideas of a dynamic Earth with vicariant continents (see Miller 1983, for a full, popular account of this process). Of course, this is not to say that Elton was unaware of Wegener's 'crazy' idea. According to Macfadyen (1992) he was actually a keen proponent and advocate for the ideas of vicariance.

However, in 1954, Elton stated that the set of continents was to be regarded as always having been an archipelago. Accordingly his preoccupation with natural and anthropogenic animal movement was to be seen as the very core of biogeography: spatial patterns were to be understood only by examination of past and present movements of organisms (or their ancestors) across the face of the Earth. Davis et al. (2001) draw a long bow (in my view) when claiming that the 1958 book was an entirely new direction in Elton's work. Notwithstanding the fact that Elton's choice for a prize before his (unsuccessful) school graduation, was a set of the works of A.R. Wallace (Southwood & Clarke 1999), his earlier works are redolent with ideas of animal movement across landscapes, and the process of what we would now call community assembly. Indeed, as pointed out by Sir Alister Hardy (1968), Elton compared the processes of dispersal among locations (Elton 1930) with the Mendelian rearrangements of genes that take place within organisms. This was part of a set of ideas in which Elton suggested that animals selecting habitats through re-location should be regarded as a complement to the environmental selection of individuals that is at the heart of ideas of natural selection. Certainly in *EIAP* he emphasizes the subset of species that have been particularly effective at invading new territory, especially if given a helping hand by humans. However, I suggest this is simply the spin he chose to put on this particular phase of his ongoing synthetic work rather than a new direction of thought.

Preparation for this chapter has brought me into contact with the (to me) more or less arcane activities of historians of science. Two doctoral dissertations have concerned themselves largely with the prelude to, genesis of and consequences arising from *EIAP* and Elton's associated work (Cox 1979; Chew 2006). I have had access only to Chew's work. There has been a, perhaps inevitable, hagiographic tone to most of the recent writings about Elton. Chew, though, presents his analyses as a sort of 'anti-hagiography' (to coin a word) belittling Elton's achievements, originality, even the world-view of those who have created the field of invasion biology subsequent to *EIAP*. Of course Chew is entitled to his opinion and his accounts of the surviving private and professional correspondence of Elton, particularly with the American proto-conservationist Aldo Leopold, are both insightful and useful – even to a hagiographer!

It true to say that Elton was a man of his time and, in the English sense, of his class. He emerged from an intellectual middle-class background, grew up and lived in a time of global conflicts, and was deeply moved and changed by early personal tragedies. That all of this could be true without affecting his work is not to be imagined. Yet to suggest that a preoccupation with biotic invasions and clear contradistinctions between 'native' and 'exotic' biotas reflected both an inbuilt militarism and, even, an incipient xenophobia is, to my mind, overegging the pudding. Chew describes *EIAP* as 'Elton's idiosyncratic jeremiad' and an 'alarmist book'.

One other 'yes, but ...' comment comes to my mind from Chew's writing. He suggests that Elton 'seldom played the public intellectual'. I suggest that this is not the case but that the public nature of Elton's contribution – in popular writing, broadcasting and committee work – was of a different kind from that we associate with 'public intellectuals' currently. The social structure of intellectual life in early to mid-century Britain was both well established and formal. There remained a tendency still to speak of 'the Universities' – meaning Oxford and Cambridge (only) – and those who occupied senior positions within them commanded both public respect yet, themselves, followed an unwritten

code of behaviour both within and beyond academe. Elton's position as a reader within the Oxford system was senior indeed – probably most closely to be compared to a research chair currently. Accordingly, his 'public' impact was subtle and political rather than highly visible and vocal. I return to this point when discussing his role in the establishment of the Nature Conservancy within the UK.

Ecology of Invasions was the first of Elton's major works that concerns itself entirely with an ecological process rather than classifying, describing and hypothesizing about the patterns he observed on the landscape. Perhaps the other outstanding and out-of-time innovation in the book is its conclusions about conservation. Conservation as an activity devoted to the preservation of natural landscapes (in contrast to the maintenance of populations of game animals for hunting) was not a mainstream activity in the early 1950s. Although advocated by Wallace as early as 1910, the British Nature Conservancy had been established only nine years before the publication of *EIAP*. In no small part the establishment of this body, now known as English Nature, resulted from the report of a committee of which Elton was a key member (Macfadyen 1992). So, to find the final two chapters of the book devoted to conservation in a very modern way was a major innovation for its time (Richardson & Pyšek 2007). I happened on the following on page 145 of the book: '... only this [conservation] is concerned with reducing direct power over nature, not increasing it; of letting nature do some of the jobs that engineers and chemists and applied biologists are frantically attempting now.' I think we would currently rephrase that as conserving to maximize ecosystem services!

Elton had engaged in an extended (if sometimes one-sided) correspondence with the American conservation advocate, Aldo Leopold, after their first meeting at the Matamuk Conference on Biological Cycles in 1931 (see also Hobbs & Richardson, this volume). Chew (2006) presents an extended account of their correspondence and the interplay of ideas which, in both instances, contributed substantially to the subsequent development of formal conservation efforts in both their nations. In Britain this took the form of a powerful and pervasive government bureaucracy with relatively little private investment and involvement. In the USA a much more 'mixed' model was adopted. As early as 1942 Elton had set out principles for the establishment of government-run nature reserves in a memorandum to A.G. Tansley who was running a committee of the British Ecological Society examining such matters. This memorandum, analysed at length by Chew (2006), ranged over many issues including that of invasive species. It also raised the ideas of what we would now call biophilia (*sensu* Wilson 1984) as well as the more mundane aspects of reserve design. Much of this manifesto was included in the final report of the British Ecological Society Committee (Chew 2006). Elton subsequently became a member of the Science Policy Committee of the nascent Nature Conservancy and remained a member until 1956 (Macfadyen 1992). During this period the highly influential Nature Conservancy Act (1949) was formulated and voted into law. According to Eric Duffey (quoted by Macfadyen 1992) it was through Elton's influence that a research branch was added to the provisions of the Act. He promoted ecological survey (with associated taxonomic services) and detailed ecological work on species (both native and exotic) of applied significance. The system of field stations established as part of the Nature Conservancy (later transmuted into the Natural Environment Research Council) was the result and an ongoing stream of influential ecological reports and actions followed.

Elton was concerned about the conservation of communities and, indeed, *EIAP* is primarily about community ecology: the breakdown of Wallace's realms, the impact of invasives upon native assemblages of animal species, and the way in which exotics insert themselves into existing food-chains within their recipient communities. This overarching concern with communities brings me to the final set of comments I make in this attempt to put Elton into the context of the history of ecology.

1.4 THE CONCERN WITH COMMUNITIES

Like Wallace, Elton was as much concerned with the community ecology of animals as he was with understanding individual populations. Of course, most of the practical work undertaken at the BAP was about populations of mammals and birds, and most of the examples that are described so clearly in *EIAP* are of single, often exotic, pest species for which adequate data had been collected primarily by those concerned with impacts or potential impacts on economic productivity. This did not, however, undermine Elton's persist-

ent preoccupation with ecological communities. Indeed yet another percipient aspect of *EIAP* was its presentation (among the first) of the ongoing conundrum of the relationships between complexity and stability. To say that Elton's comments in *EIAP* (together with those of MacArthur (1955)) set running a robust and muscular 'hare' would be an understatement (although the invasive species metaphor amuses me). Recent overviews of this ongoing debate include that of Lehman and Tilman (2000).

This concern with animals within communities was a unifying theme of Elton's entire body of work, diverted only during World War II into pest biology of rats and mice (Chitty & Southern 1947). Since the 1930s there had been, more or less, a split in ecology along taxonomic and thematic lines. Population ecology was regarded as the very stuff of animal ecology (at least until John Harper's seminal book in 1977) whereas community ecology was principally about associations of plants. I hasten to add ecology did not start out that way but that is how things developed. The demands and funding for pest control, fisheries management and game conservation (for hunting) drove the single-species approach so evident in animal studies. In this binary world of the 1940s, 1950s and 1960s, Elton's approach ran counter to the mainstream.

Animal ecology as it blossomed in the first half of the 20th century was part of the diversification of biology that stemmed from the earlier general acceptance of Darwinian ideas of evolution, especially once these were incorporated with Mendelism as part of the 'new synthesis'. It had been early appreciated that comparative anatomy could only go so far in elucidation of the mechanisms of evolution. Living organisms interacting with their environments held the key to further progress. Along with the ecology of animals (and plants) the disciplines of animal behaviour (ethology) and ecological and population genetics emerged. These were the fields that looked explicitly at whole organisms and complimented the advances in physiological, cellular and biochemical biology that took place concurrently. Not until the much more recent emergence of molecular phylogenetics would we have more effective tools for examining the mechanisms and outcomes of evolution.

Within ecology there was a tendency to regard population ecology and, later, behavioural ecology, as the 'real' fields of study within the evolutionary paradigm. After all, selection works by modifying the genetic heritages of individuals, and we measure this by examining changes in gene frequencies at the level of the population. So how does community ecology – the 'tangled bank' of Darwin's remarks in *Origin of Species* – justify itself under the bright lights of evolutionary thought? The answer, of course, is that individuals and populations do not live in isolation. They evolved and continue to exist (and evolve) in more or less complex webs of interactions in a landscape of locations which present a mosaic of physico-chemical profiles among which organisms move to greater or lesser extents. In a way, community ecology is the top-down approach to understanding evolution in action incorporating as it does both organism and environment as a dynamic whole. The ecology of individuals and populations represents a bottom-up approach in which we study the component parts of communities in a reductionist fashion. These two approaches each feed from the other – and no one was more aware of this than Elton. His capabilities and propensities as a naturalist – like my own – kept his enthusiasm at the level of syntheses. His 1930 vision of animal populations and, by inference, the communities of which they were part, responding to unfavourable conditions by moving across a landscape of diverse selective environments, was perhaps the clearest statement of his awareness of this community/evolution nexus.

The publication of *EIAP* really marked a transition in Elton's career. The University of Oxford had acquired the Wytham Woods estate in 1942 and, from 1945 onwards, Elton had organized the work of the Bureau and its students around a wide-ranging ecological survey of the many habitat types occurring within the estate (Grayson & Jones 1955; Elton 1966). This survey was structured around a database system in which species-specific ecological information was cross-referenced with information on the different habitat components. Nested within this quintessential community-approach were many more or less independent studies of the population dynamics of selected (principally vertebrate) species.

This ground-breaking ecological survey began to come to an end (although much ecological research continued, and continues, on the Wytham Estate) with the publication of Elton's 1966 book *The Pattern of Animal Communities*, and with the forced disbanding of the BAP upon Elton's retirement in 1967 (Crowcroft 1991). This book was Elton's *magnum opus*, yet it never received the prominence of some of his earlier works. From my perspective there are two reasons for this. First, the merely technical. Elton and his co-workers on

the Wytham survey were attempting to construct and interrogate a database that was complex and multidimensional. Modern electronic databases handle such structures with ease: they were not available in the 1950s and 1960s. Elton erected procedures based on record cards and kalamazoo slips, which were cumbersome to use and somewhat opaque to the casual user. Further, and of greater intellectual moment, was the fact that the survey was based upon an insightful but essentially static classification of habitats (Elton & Miller 1954). For me, it was not until Southwood (1977) published his marvellous 'habitat templet', as part of his presidential address to the British Ecological Society in 1976, that the synthesis between life-history strategies, ecological processes, habitat type and community structure became clear.

Southwood, too, was part of the natural historical school of ecology of which Elton was, perhaps, the pre-eminent member and which had dominated the science in Britain since its inception. Southwood, though, was of a later generation with more quantitative skills than Elton. (Much as Elton respected mathematical approaches he was never, by his own admission, a skilled numerical analyst.) It was no surprise, though, when Southwood inherited the mantle of the premier ecological synthesist on the UK scene after Elton's retirement and, in 1991, his death.

Upon retirement, Elton continued the day-to-day maintenance of the Wytham Survey and its vast body of records, collections, literature and supporting information. Neither he nor anyone else published extensively on the survey as a whole after that time. He had, though, one more ace up his sleeve. In 1973 he published 'The structure of invertebrate populations inside Neotropical rain forest' in the *Journal of Animal Ecology*. At the time I remember distinct if mild controversy over whether the approach taken was appropriate or the insights justified: there was an unspoken notion that it was the author's name that got this manuscript through the journal's processes rather than its content. I would say, in fact, that the paper's 'problem' was that it was more than a decade ahead of its time. It is only in the light of the burst of ecological activity and commentary that followed Erwin's paper on tropical forest diversity published in 1982 (see, for example, Erwin & Scott 1980; Erwin 1982; May 1986; Stork 1988) that Elton's true prescience becomes apparent.

Elton's 1973 paper is written in narrative style reminiscent more of the approach adopted in his books than the dry and dull text typical of most scientific papers of the era. Nevertheless it is full of insights that subsequently have been addressed by whole schools of research (most apparently unaware of the Elton paper). In the 1973 paper he pioneered the use of a 'morpho-species' approach to deal with a highly diverse, taxonomically challenging tropical fauna. In addition he promoted the idea of multi-method surveys of tropical diversity for circumventing the biases inherent in any one method. He noted the great dominance of tropical invertebrate faunas by singletons. Elton measured and commented on the apparent imbalance between predatory and non-predatory species in his invertebrate samples and compared available explanations for this. He contrasted top-down explanations (where the plethora of predators was responsible for the dearth of non-predators) with bottom-up explanations (where the low levels of non-herbivores reflected an evolutionary product driven by the scarcity of available resources beneath the forest understorey). He recognized the value of size/abundance analyses as a way of examining species packing. Finally he noted that the abundance of invertebrates in what he called the field layer (up to 2 m from the ground) was low and he speculated on the relative importance and richness of the canopy (which he had no means of accessing). In an Appendix he added comments on the roles of ecological engineering species such as army ants. All of these insights foresaw an agenda for tropical biodiversity studies that has been realized over the past 30 years (for a summary of this development see the papers in Basset et al. (2003)).

Of course, I am not suggesting that those engaged in tropical biodiversity research over that period pursued the topics they did because of Elton's paper: some did, others did not. I simply make the point that the 1973 paper, like virtually all of the Elton canon, was extraordinarily original, perceptive and trend-setting. *Ecology of Invasions* was and remains an extraordinary, important work. It stands out among Elton's post-war products as having been noticed and appreciated yet it is perhaps no more perceptive and insightful than most of his other works.

Perhaps the social moral from this reflection on the life and work of Charles Elton is that there is huge intellectual and, in consequence, practical gain to be had from allowing brilliant scientists to follow the maze of their own imaginings. The only necessity is that from time to time they produce lucid, accessible accounts of their thoughts. As Elton wrote to Leopold in 1945:

'Don't you think we must all resist the deluge of ad hoc work and just sit and think?'

Elton did this to perfection. Yet the tediousness of centralist bureaucracies eventually caught up with him, undervaluing, even degrading, his life's work. This bureaucratic disease has become an epidemic since the 1970s. It is unlikely we shall see Elton's like again, and that will to our great disadvantage.

REFERENCES

Basset, Y., Novotny, V., Miller, S.E. & Kitching, R.L. (eds) (2003) *Arthropods of Tropical Forests: Spatio-Temporal Dynamics and Resource Use in the Canopy*. Cambridge University Press, Cambridge.

Bradley, R. (1718) *New Improvements of Planting and Gardening, Both Philosophical and Practical*, edition 2, part 3. W. Mears, London (quoted in Egerton 2007).

Chew, M.K. (2006) *Ending with Elton: Preludes to Invasion Biology*. PhD thesis, Arizona State University, Tempe.

Chitty, D. & Southern, H.N. (eds) (1947) *Control of Rats and Mice*. Clarendon Press, Oxford.

Cox, D.L. (1979) *Charles Elton and the Emergence of Modern Ecology*. PhD thesis, Washington University, St. Louis.

Crowcroft, P. (1991) *Elton's Ecologists: A History of the Bureau of Animal Population*. University of Chicago Press, Chicago.

Davis, M.A., Thompson, K. & Grime, J.P. (2001) Charles S. Elton and the dissociation of invasion ecology from the rest of ecology. *Diversity and Distributions*, **7**, 97–102.

Egerton, F.N. (2007) Understanding food chains and food webs, 1700–1970. *Bulletin of the Ecological Society of America*, **88**, 50–69.

Elton, C.S. (1927) *Animal Ecology*. Methuen, London.

Elton, C.S. (1930) *Animal Ecology and Evolution*. Clarendon Press, Oxford.

Elton, C.S. (1942) *Voles, Mice and Lemmings*. Clarendon Press, Oxford.

Elton, C.S. (1958) *The Ecology of Invasions by Animals and Plants*. Methuen, London.

Elton, C.S. (1966) *The Pattern of Animal Communities*. Methuen, London.

Elton, C.S. (1973) The structure of invertebrate populations inside Neotropical rain forest. *Journal of Animal Ecology*, **42**, 55–104.

Elton, C.S. & Miller, R.S. (1954) The ecological survey of animal communities: with a practical system of classifying habitats by structural characters. *Journal of Ecology*, **42**, 460–496.

Erwin, T.L. (1982) Tropical forests: their richness in Coleoptera and other arthropod species. *Coleopterists' Bulletin*, **36**, 74–75.

Erwin, T.L. & Scott, J.C. (1980) Seasonal and size patterns, structure and richness of Coleoptera in the tropical arboreal ecosystem: the fauna of the tree *Luehia seemannii* Triana and Planch in the Canal Zone of Panama. *Coleopterists' Bulletin*, **34**, 305–322.

Grayson, A.J. & Jones, E.W. (1955) *Notes on the History of the Wytham Estate with Special Reference to the Woodlands*. Imperial Forestry Institute, Oxford.

Grinnell, J. (1917) The niche relationships of the California Thrasher. *Auk*, **34**, 427–433.

Hardy, A. (1968) Charles Elton's influence in ecology. *Journal of Animal Ecology*, **37**, 3–8.

Harper, J. (1977) *The Population Biology of Plants*. Academic Press, London.

Heezen, B.C., Tharp, M. & Ewing, M. (1959) The floors of the ocean. I. The North Atlantic. *Geological Society of America, Special Paper*, **65**.

Lehman, C.L. & Tilman, D. (2000) Biodiversity, stability, and productivity in competitive communities. *American Naturalist*, **156**, 534–552.

MacArthur, R.H. (1955) Fluctuations of animal populations and a measure of community stability. *Ecology*, **36**, 533–536.

Macfadyen, A. (1992) Obituary: Charles Sutherland Elton. *Journal of Animal Ecology*, **61**, 499–202.

May, R.M. (1986) How many species are there? *Nature*, **324**, 514–515.

Miller, R. (1983) *Continents in Collision*. Time-Life Books, Amsterdam.

Pierce, W.D., Cushman, R.A. & Hood, C.E. (1912) The insect enemies of the cotton boll weevil. *US Department of Agriculture, Bureau of Entomology Bulletin*, **100**, 1–99.

Richardson, D.M. & Pyšek, P. (2007) Classics in physical geography revisited: Elton, C.S. 1958: The ecology of invasions by animals and plants. Methuen: London. *Progress in Physical Geography* **31**, 659–666.

Richardson, D.M. & Pyšek, P. (2008) Fifty years of invasion ecology – the legacy of Charles Elton. *Diversity and Distributions*, **14**, 161–168.

Semper, K.G. (1881) *Animal Life as Affected by the Natural Conditions Of Existence*. Appleton, New York.

Southwood, T.R.E. (1977) Habitat, the templet for ecological strategies? *Journal of Animal Ecology*, **46**, 337–365.

Southwood, T.R.E. & Clarke, J.R. (1999) Charles Sutherland Elton, *Biographical Memoirs of Fellows of the Royal Society of London*, **45**, 129–146.

Stork, N.E. (1988) Insect diversity: facts, fiction and speculation. *Biological Journal of the Linnean Society*, **35**, 321–337.

Wallace, A.R. (1910) *The World of Life: A Manifestation of Creative Power, Directive Mind and Ultimate Purpose*. Chapman & Hall, London.

Wegener, A.L. (1915) *Die Entstehung der Kontinente und Ozeane*. Vieweg und Oshn, Wiesbaden.

Wilson, E.O. (1984) *Biophilia: The Human Bond with Other Species*. Harvard University Press, Cambridge, Massachusetts.

Chapter 2

CHARLES ELTON: NEITHER FOUNDER NOR SIREN, BUT PROPHET

Daniel Simberloff

Department of Ecology and Evolutionary Biology, University of Tennessee, Knoxville, TN 37996, USA

Fifty Years of Invasion Ecology: The Legacy of Charles Elton, 1st edition. Edited by David M. Richardson
© 2011 by Blackwell Publishing Ltd

2.1 INTRODUCTION

Charles Elton is seen as having founded invasion biology as a new, distinct discipline (see, for example, Rejmánek et al. 2002; Ricciardi & MacIsaac 2008; Richardson & Pyšek 2008). This view rests on his remarkable 1958 monograph, *The Ecology of Invasions by Animals and Plants*. Indeed, the announcement for the 2008 symposium in Stellenbosch leading to the present volume says the monograph 'is generally accepted as the foundation for the scientific study of biological invasions' (Anonymous 2007, p. 39). However, Elton is indicted by Davis et al. (2001) for the very act of having, at its founding, separated invasion biology as a discipline distinct from ecology, particularly succession ecology, a persistent dissociation they say has ill-served invasion biology. They speculate that Elton went astray because experiences in the Second World War led him to change his views on invasions. Here I attempt to show that Elton's views on invasions began early, evolved gradually and did not change radically during the war. His longstanding interest in invasions did indeed culminate in the 1958 monograph that is widely read by invasion biologists (Richardson & Pyšek 2007) and adumbrates most of the ecological (but not evolutionary) current invasion research programme. However, the monograph was published too early to have founded the field. Contrary to claims by Ricciardi and MacIsaac (2008) and Richardson and Pyšek (2008), Elton and his monograph, in fact, had limited influence on the development of invasion biology – he was a prophet, not a founder.

2.2 THE EVOLUTION OF ELTON'S VIEW OF INTRODUCED SPECIES

Elton's interest in introduced species began in his pre-First World War childhood in Liverpool, where he haunted a shop displaying exotic animals (Elton 1955 in Coates 2003). His first substantial publication on the topic was his 1927 book *Animal Ecology*, in which both Elton and editor Julian Huxley argued that progress in animal ecology had been hindered by its adoption of methods of plant ecology and that animal ecology required its own approaches. Nevertheless, Elton (1927) accorded plant community succession a key role (cf. Davis et al. 2001). He argued that animal ecologists must understand plant succession to study animal populations and communities, but, by the same token, plant ecologists must study animals, especially because animals often control succession. Where Elton discussed introduced animals in this book, he emphasized how they damage systems formerly undergoing gradual succession: 'Then there are sudden disasters, like fires, floods, droughts, avalanches, the introduction of civilised Europeans and of rabbits, any one of which may destroy much of the existing vegetation' (p. 20).

Davis et al. (2001) attribute significance to the facts that Elton (1927) did not list 'invasion' in the index, and 'invasion,' when used, seemed not to apply particularly to introduced species. These facts, they argue, suggest he had not begun the dissociation between ecology and invasion biology that they feel pervades the monograph. However, Elton's second publication on introduced species (Elton 1933a) was a newspaper article headlined 'Animal invaders' beginning with a list of introduced animals, repeatedly termed 'alien,' causing myriad problems for native species. The muskrat in Britain was termed an 'enemy' (p. 13) and the English sparrow in North America an 'invader' (p. 13). The same year, Elton published a short introductory text, *The Ecology of Animals* (1933b), repeatedly using the word 'pest,' stating explicitly that it can apply to both native and introduced species, but focusing heavily on the latter. He also wrote a popular book that year, *Exploring the Animal World* (Elton 1933c), based (as was his invasion monograph) on BBC radio broadcasts. It closed with a chapter, 'Plagues of animals,' almost wholly about introduced species, both animals and plants. While stressing that some introductions benefit humans, he repeatedly termed them 'plagues' and 'pests' that 'invade' (pp. 103, 112) their new homes. In 1936, Elton entitled a review of a monograph on the Chinese mitten crab 'A new invader', beginning with the statement that the species had just arrived in Great Britain. It was an ominous review, suggesting the crab might soon occupy British river systems and that it caused erosion of banks and was a fisheries pest. He closed by lamenting the 'growing band of invaders' (p. 192) of Britain, citing the grey squirrel, muskrat, French partridge, little owl, willow grouse, rainbow trout, black bass, continental crayfish, American slipper-limpet, oyster-tingles (drills), Colorado potato beetle and hybrid cord grass *Spartina townsendii*.

Further evidence that Elton's views of introduced species and martial metaphors for invasions evolved well before the Second World War comes from projects of the Oxford University Bureau of Animal Population, which he headed. In connection with a 1932 eradication campaign against the muskrat, the British Ministry of Agriculture and Fisheries funded a study, directed by Elton, of muskrat ecology in Britain (Warwick 1934 1940). Elton hired Tom Warwick to assist the project (Crowcroft 1991). Warwick (1935) also studied escapes of introduced nutria, which was eradicated much later. Elton also directed an early study by A.D. Middleton on the potential impacts of the North American grey squirrel in Britain on the native red squirrel and on tree damage (Middleton 1930 1932 1935). The study concluded that the latter impact was already severe but, erroneously, that the former might not be. Interestingly, at the Matamek Conference on Biological Cycles in Labrador in 1931, Elton mentioned both the rapid, tremendous growth of the European muskrat population and the grey squirrel introduction to Great Britain (Leopold 1931).

In short, Elton's great interest in invasions and his pejorative references to them are manifest in his earliest writings and not a product of his activities in the Second World War. During the war, Elton turned the Bureau almost wholly toward aiding the war effort by studying the natural history of four damaging species introduced to Britain – the Norway rat, black rat, house mouse and European rabbit – to aid control (Crowcroft 1991; Southwood & Clarke 1999). Davis et al. (2001) suggest these mammals – introduced pests – were very different from those Elton had studied in his earliest research in the Arctic. However, they did not differ greatly from two the Bureau had researched intensively for years (muskrat and grey squirrel), and Elton had evinced continuing interest in introduced pests at least as early as his 1927 book.

These rodent studies were not Elton's only wartime scholarship on introduced species. In 1942, he submitted to *Polish Science and Learning* a remarkable short essay (Elton 1943) foreshadowing his 1958 classic on biological invasions, featuring the breakdown of the biogeographical realms and including a short summary of many examples from his earlier writings, focusing on animal introductions to Great Britain. He meant this contribution 'to contribute something now towards maintaining the concept of international co-operation in science after the war is over' and as a respite from 'working almost entirely upon practical and immediate problems connected with war needs' (p. 7). In 1944, Elton reviewed a monograph on eradi-

cation of the malaria mosquito *Anopheles gambiae* from Brazil, ending with a brief description of other eradications of introduced species (including the muskrat in Britain), describing them as 'major engagements in a violent struggle against the spread of undesirable plants and animals that is affecting every country' (p. 88), referring to invasions as a 'zoological catastrophe' (p. 88), and pointing readers to his publication in the Polish journal.

Until the 1958 monograph, these were Elton's last publications referring to invasions, except for several on rodent control plus a note on bees attracted to *Ailanthus altissima* that did not mention its non-native status (Elton 1945). The monograph was an expansion of three BBC radio broadcasts (Southwood & Clarke 1999; Simberloff 2000) and closely followed the outlines of the Polish publication, but with many more examples, including plants. After the monograph, Elton published his last major work, which he viewed as the culmination of his research but which has had little impact (Southwood & Clarke 1999): *The Pattern of Animal Communities* (Elton 1966). A chapter on 'Dispersal and invaders' summarized the main monograph themes and also discussed invasions, with particular reference to Elton's long-term studies of Wytham Woods, noting, for example, the paucity there of introduced birds and insects. The conservation section used a martial metaphor: 'a steady bombardment of alien species from other lands and waters, with consequent explosions after the arrival of some of them' (p. 382). On the other hand, 'invader' and 'invasion' are used throughout the book in several contexts unrelated to introduced species: see, for example, 'bark-beetle invasion stage' (p. 289), 'invasion [of animal remains] by carrion insects' (p. 323), 'annual invasions of clothes moths, meal-worm beetles and carpet beetles into houses' (p. 344). One of Elton's last publications was on an introduced insect in Wytham Woods (Elton 1971).

In summary, Elton did not abruptly shift his views on or descriptors for introduced species during the Second World War or at any other time. He often used martial metaphors in referring to biological invasions – 'invasion' itself might be construed as such. Davis et al. (2001) and Davis (2006) see Elton's martial metaphors as part of the dissociation they feel he wrought between invasion biology and the rest of ecology, and as perpetuating a misleading distinction between introduced and native species. However, the analogy between the spread of introduced species (including pathogens) and an invading enemy in warfare may be so patent that it is unsurprising that martial metaphors crop up repeatedly in both invasion biology and public health (Simberloff 2006). Wallace (1890, p. 34) referred to the Norway rat as 'invading ... all over the world,' while Darwin (1996 [1859]) in *The Origin of Species* described natives 'conquered' by introduced species (p. 69) and 'yielding before advancing legions of plants and animals introduced from Europe' (p. 164) and referred to 'intruders' (pp. 259, 314) having 'invaded' (p. 263) territories of other species. It did not take the Second World War for Elton to use such terminology.

2.3 THE DISSOCIATION OF INVASION BIOLOGY FROM SUCCESSION ECOLOGY

As for the claimed dissociation by Elton of biological invasions from ecological succession, it must first be said that Elton was an animal ecologist and skewed his monograph towards animal invasions. Of 183 introduced species specifically discussed in the monograph, 162 were animals, 15 were plants and 6 were pathogens. As noted above, although animal and plant succession are related, Elton (1927) observed that they are not identical and need some different types of study. Perhaps the key point Elton raised repeatedly in his monograph but did not pursue is that most introduced species either do not survive or, if they survive, do not become invasive. Nowadays, this fact has become a dominant research theme: why do certain introduced species become problematic, and how can we predict which these will be? Part of the answer to these questions, particularly for introduced plants, requires research in the tradition of succession ecology.

However, it is unfair to indict Elton for having misled the field on the relevance of plant succession. In the next section, I will show that he did not greatly influence the development of the modern field, so he could not have misled it. However, it is an unfair charge on other grounds. The field of succession ecology was not nearly as advanced in its understanding of mechanisms when Elton wrote as it is today. And Elton (1958), though not using the specific term, clearly called for research of exactly this sort as the most urgent scientific need. Addressing the fact that the white dead-nettle (*Lamium album*) is common in species-rich woodlands in its native range in the

Caucasus, yet after centuries in Britain is only on wasteland, roadsides and field edges, never invading forests, he argued that this was perhaps the 'single most important problem lying underneath all the facts of the present book,' along with the question of, if it had invaded forests, 'would it have replaced some native species, or just added one more to the list?' (Elton 1958, p. 118). To my knowledge, these questions have still not been answered, but surely they are within the ambit of modern plant succession ecology.[1]

2.4 ELTON'S INFLUENCE OF THE DEVELOPMENT OF MODERN INVASION BIOLOGY

Elton's 1958 monograph was favourably but not widely reviewed (see, for example, Cohn 1959; Pearsall 1959; Taylor 1959), and perusal of figures 1 and 2 of Richardson and Pyšek (2008) and the figure in Ricciardi and MacIsaac (2008) shows that the study of invasions did not explode after Elton's monograph appeared. In fact, the figures show few invasion publications until 1991, when there was a rise to approximately 100, with major acceleration after 1995 to 1000 to 3000 by 2006, depending on tallying criteria. Further, the trajectory of citations of Elton's monograph tracks that of invasion publications, numbering fewer than 10 annually until 1991, when the tally began a climb that jumped in 1995 and reached the hundreds annually by 2006. What happened in about 1991 that accounts for these data, and does the fact that Elton suddenly began to be frequently cited in 1991 mean a new field was being established based on his insights?

Elton founded the Bureau of Animal Population in 1932 and directed it until its closure in 1967, when he retired (Crowcroft 1991; Kitching, this volume). He guided Bureau staff members Middleton and Warwick in research on particular introduced species, as noted above. However, neither influenced contemporary ecologists to study invasions or left an important legacy in invasion biology.

Elton directed 25 doctoral theses at the Bureau. Although several degree recipients became prominent biologists, no thesis was on introduced species, and only two of these scientists later studied introduced species. The first was William Murdoch, who received his doctorate in 1963 on the population ecology of several carabid beetles. Murdoch quickly turned to predator–prey interactions, publishing a major paper on this subject in 1969 and many after that. Although he read the monograph, it did not lead him to further research (W.W. Murdoch, personal communication 2008). In a 1975 paper on the relationship of diversity to stability (Murdoch 1975), he cited the monograph as one of four sources for the idea that more species enhance stability. His connection with invasion biology arose from another source: his research on biological control, beginning in 1982 and including many papers through the present. This focus, in turn, stemmed from his interest in predator–prey interactions as a window into population dynamics (W.W. Murdoch, personal communication 2008). Murdoch was inspired to study population dynamics by Elton's 1927 ecology book and discussions at the Bureau by Elton and his earlier student Dennis Chitty about small mammal cycles (W.W. Murdoch, personal communication 2008).

Elton's penultimate student also conducted doctoral research on a different subject (tree hole fauna) but subsequently studied invasions: Roger L. Kitching, awarded his doctorate in 1969. Although Kitching read Elton's monograph while he was at the Bureau, he was more inspired at that time by Elton's recently published *The Pattern of Animal Communities* (R.L. Kitching, personal communication 2008; see Kitching, this volume). He did not begin seriously thinking about and conducting research on invasions until later, when he studied the introduced sheep blowfly in Australia (R.L. Kitching, personal communication 2008). The work, in turn, led Kitching to compile and edit *The Ecology of Pests. Some Australian Case Histories* (Kitching & Jones 1981). This book was not specifically about introduced species, and none of the 12 chapters cited Elton's monograph. However, eight chapters treated introduced species, a fact not lost on Kitching, who subsequently edited *The Ecology of Exotic Animals and Plants – Some Australian Case Histories* (Kitching 1986a). Only then did Kitching begin to consider emerging generalities about invasions. An Englishman, Kitching feels the stimulus of being in a new country (Australia) heavily beset by daily problems with introduced species, combined with the experience of compiling *Pests*, led him to consider general questions about invasions (R.L. Kitching, personal communication 2008). The introduced species Kitching had encountered in his earlier years in England were mostly either ancient or of only local concern, a far different situation than in Australia. Having engaged, then, in the

1980s in invasions as a research concern, Kitching thought back to insights in Elton's monograph that he had read years previously (R.L. Kitching, personal communication 2008).

Many scientists were long-term Bureau visitors. Of these, only David Pimentel worked on introduced species, spending part of 1961 there. Although Pimentel had read Elton's monograph when it appeared (D. Pimentel, personal communication 2008), he was already an established researcher on biological control, having received his doctorate in 1951. During his first position, as Chief of the Tropical Research Laboratory of the U.S. Public Health Service in Puerto Rico, in addition to traditional biological control research, Pimentel studied impacts and management of a species introduced for this purpose that had become a damaging invader, the small Indian mongoose (Pimentel 1955a,b), before publication of Elton's monograph. Along with a research programme expanding into several environmentally important subjects, Pimentel continued to work on biocontrol, including biocontrol introductions that become problematic (see, for example, Pimentel et al. 1984). Subsequently he studied introduced species in general, not just those released for biological control; his estimates of the economic cost of invasions (see, for example, Pimentel et al. 2000) are well known. He was particularly moved in this direction by an Office of Technology Assessment report (U.S. Congress Office of Technology Assessment 1993) estimating costs of a subset of invasions of the United States (D. Pimentel, personal communication 2008).

Elton's influence on the invasion research programmes of his students and colleagues, then, was not great. All cite other factors as the main motivating force. His monograph was widely read but not often cited until 1991. It is difficult to assess how an uncited work influences its readers' thoughts, but it is striking how few references to the monograph appear even in subsequent works devoted to subjects closely allied to Elton's conception of invasions. Three edited volumes are instructive.

In 1964, a symposium was held at Asilomar among biologists of many disciplines to address evolution in introduced organisms (Baker & Stebbins 1965). In 27 papers, Elton's monograph is cited just three times: by Wilson (1965) in a list of authors who have shown that few generalizations about introduced species are possible, by Mayr (1965), who noted that Elton had described several bird introductions, and by Birch

(1965), who observed that Elton had documented many invasions but did not mention evolutionary changes accompanying them (this was not quite true; Elton briefly mentioned the evolution of resistance to both pathogens and pesticides). Birch's observation is key here; Elton's monograph is almost wholly about ecology and not evolution (Simberloff 2000), and the conference was about evolution.[2] However, several ecologists contributed, and papers and discussions often delved into ecology, yet Elton was barely mentioned.

In 1969, a symposium at Brookhaven National Laboratory explored the diversity-stability relationship in ecological systems (Woodwell & Smith 1969). The main generalization Elton had drawn in the monograph (and reiterated in *The Pattern of Animal Communities*) about why some introduced species invaded and spread while others died out or remained restricted was that diversity conferred resistance and species poverty rendered systems susceptible to invasion. Yet, among 19 contributors, almost all addressing ecology, none cited Elton's monograph.

Finally, in 1984, the British Ecological Society and the Linnean Society of London jointly convened a conference at Southampton on the relationship of colonization, succession and stability (Gray et al. 1987). Davis et al. (2001) see this symposium as an explicit attempt to bridge the gap, which they attribute to Elton, between invasion ecology and succession ecology. To the extent that the gap existed, the symposium shows Elton cannot bear all the blame: half the participants were students of plant succession, including J. Philip Grime, co-author of Davis et al. (2001), yet none of the 21 contributions cited Elton's monograph.

2.5 THE RISE OF MODERN INVASION BIOLOGY

If Elton's monograph did not found modern invasion biology or greatly influence its development, what did? The key impetus was a programme on the ecology of biological invasions of the Scientific Committee on Problems of the Environment (SCOPE, an arm of the International Council of Scientific Unions), initiated in mid-1982, and the key figure was Harold A. Mooney. In 1980, many ecologists attended the Third International Conference on Mediterranean-Type Ecosystems in Stellenbosch, South Africa. Invasions were not a major focus in ecology in 1980, but both Richardson et al. (1997) and Mooney (personal communication 2008) note that, to the extent scientists worried about introductions at all then, the prevailing view was that disturbance was a prerequisite for invasion. Elton probably influenced this perspective, which was the thrust of his monograph. Mooney (personal communication 2008) relates that the crystallizing moment came on a field trip when participants could see pines marching over a hill into undisturbed fynbos vegetation, a realization confirmed by Richardson et al. (1997). While at Stellenbosch, Mooney and South African ecologist Fred Kruger began to think about another topic that could be fruitfully explored through the comparative approach of the conferences on Mediterranean-type ecosystems. With invaded fynbos nearby and knowing of similar invasions elsewhere, Mooney and Kruger decided to ask SCOPE to sponsor a programme on plant invasions into the five Mediterranean-type systems (H.A. Mooney, personal communication 2008).

However, at the SCOPE General Assembly in Ottawa in 1982, Ralph Slatyer suggested a global project, in keeping with SCOPE's mandate to advance knowledge on globally significant environmental problems, and one on all habitats, not just Mediterranean-type vegetation (Drake et al. 1989; H.A. Mooney, personal communication 2008). The enthusiasm with which the SCOPE assembly endorsed the invasions programme reflects both observations and interests of several participants in the Stellenbosch meeting and the success of an earlier SCOPE effort on fire ecology (R.H. Groves, personal communication 2009). Thus the project was termed the Scope Programme on the Ecology of Biological Invasions, Mooney was named chair, and the main questions specified by SCOPE were (Williamson et al. 1986; Drake et al. 1989) as follows:
1 What factors determine whether a species becomes an invader or not?
2 What site properties determine whether an ecological system will resist or be prone to invasion?
3 How should management systems be developed to best advantage, given the knowledge gained from studying questions 1 and 2?

The SCOPE project generated great interest and many publications, with national programmes in Great Britain (Kornberg & Williamson 1986; Williamson et al. 1986), Australia (Groves & Burdon 1986), the United States (North America and Hawaii; Mooney & Drake 1986), the Netherlands (Joenje 1987;

Joenje et al. 1987) and South Africa (Macdonald et al. 1986), regional syntheses for Europe and the Mediterranean basin (di Castri et al. 1990) and the tropics (Ramakrishnan 1991) and a global synthesis of the national efforts (Drake et al. 1989).[3] In addition, several other efforts were either begun under the aegis of the SCOPE programme (see, for example, Drake & Williamson 1986; Usher et al. 1988), inspired by the programme (see, for example, Groves & di Castri 1991) or associated themselves with particular parts of the programme (see, for example, Gray et al. 1987; Groves & di Castri 1991). Just the national reports and global synthesis engaged 145 editors and authors, and the associated efforts added many more. Further, many participants in SCOPE workshops continued to conduct invasion research but did not publish in the SCOPE volumes. Another important feature of the SCOPE project is that, although it included most recognized invasion aficionados of the 1980s, it also enlisted many senior figures in related fields (especially ecology and mathematical modelling) who had not previously published on invasions but contributed to one or more of the SCOPE volumes and continued invasion research (Davis 2006).

The impact of the SCOPE programme and SCOPE-inspired publications was enormous. Finding that a Web of Science search of citations of the chapters in the global synthesis book seriously undercounted the citations, I used Google Scholar to tally citations only in journals and published books (not grey literature reports, theses or dissertations) for both the global synthesis volume (Drake et al. 1989) and the North America/Hawaii volume (Mooney & Drake 1986), eliminating duplicate citations of chapters within the same book (Figs 2.1 and 2.2). Both books began to be heavily cited in 1995, with an abrupt spurt in 1999 leading to approximately 100 citations of Drake et al. (1989) and 80 citations of Mooney and Drake (1986) each year between 2001 and 2006. Particularly noteworthy is that citations of the two SCOPE volumes jumped exactly when research citations on invasion biology (figure 2 in Richardson & Pyšek 2008) and citations of Elton's monograph (figure 3 in Richardson & Pyšek 2008) jumped, in about 1995, and jumped to about the same level by about 2005, by which time invasion biology was a firmly established discipline. This coincidence suggests that intellectual forces and/or societal issues made invasion biology an exciting, attractive research topic around the time of the SCOPE programme, not a rediscovery and consideration of Elton's classic; I will return to this thesis shortly.[4]

The timing of the SCOPE project and of the book edited by Kitching (1986a) – *The Ecology of Exotic Animals and Plants – Some Australian Case Histories* – merits further comment. The Australian contribution to SCOPE (Groves & Burdon 1986) – *Ecology of Biological Invasions* – was published in the same year as Kitching's book. The SCOPE contributions tended to ask general questions, or, if they treated a particular group of species, it was a large group: mammals, say, or aquatic

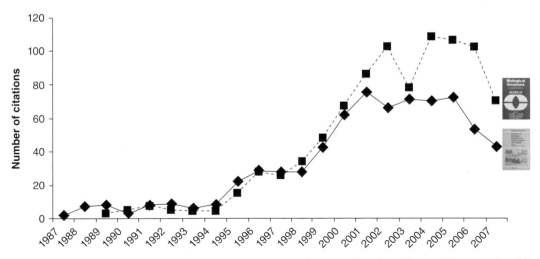

Fig. 2.1 Annual citations of Mooney and Drake (1986) (diamonds, solid line) and Drake et al. (1989) (squares, dotted line).

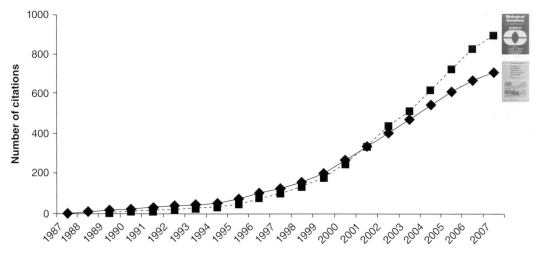

Fig. 2.2 Cumulative citations of Mooney and Drake (1986) (diamonds, solid line) and Drake et al. (1989) (squares, dotted line).

species. However, many of the chapters presented details of invasions by particular species. Further, in addition to the 12 chapters, the book summarized 11 posters; several focused on single species. All the chapters in Kitching (1986a), by contrast, were cases histories, and almost all were on single species. However, several chapters delved into more general discussions of invasions, and the closing chapter (Kitching 1986b) attempted both to seek generalizations about invasion and to relate the biology of the species covered to recent theoretical ideas in population and community ecology. Furthermore, most chapters were on pest species, as may be expected, given Kitching's inspiration (discussed above) for his earlier volume (Kitching & Jones 1981). Kitching's efforts as well as the SCOPE programme were both preceded by a 1977 symposium in Adelaide of the Ecological Society of Australia entitled, 'Exotic species – their establishment and success' (Anderson 1977). Of the 14 published papers, only two cited Elton's monograph. Several contributors sought generalizations about invasions, whereas others detailed invasions of specific introduced pests.

Interestingly, when he compiled his book in 1984, approximately two years before publication, Kitching was unaware of the SCOPE programme (Kitching, personal communication 2008), so this was truly an independent effort. In fact, of the 20 chapter authors in Groves and Burdon (1986) and the 16 chapter authors

in Kitching (1986a), only one is in both groups. The SCOPE volume shares only two authors with the earlier Ecological Society of Australia symposium book. The time was ripe for both the SCOPE programme and the separate Australian efforts for two related reasons. First, the ever-increasing number of problematic invasions had reached a critical mass by the early 1980s. After all, the *raison d'être* of SCOPE was to deal with problems of the environment, and biological invasions were finally perceived as a huge global environmental problem. The public was sensitized to introduced species problems somewhat earlier in Australia than elsewhere (see, for example, Rolls 1969), and agricultural scientists there were heavily invested in studying introduced species, but, as noted above, the fact that most agricultural pests were introduced was not explicitly stated even by Kitching and Jones (1981).

The second reason why the time was ripe for the rise of modern invasion biology is more academic. Any scientist conducting invasion research points to the potential benefit of his/her research to humankind when submitting grant proposals. However, it also became evident in the early 1980s that introduced species, and the increasing amount of data on some of them, suggested avenues for interesting, 'ivory-tower' science. I can cite my own experience. My first paper explicitly on introduced species, in 1981, used data amassed on introduced species to test two community ecology theories – the dynamic equilibrium model and models of limiting similarity (Simberloff 1981). In the national SCOPE volumes and the global synthesis volume, several contributors used introduced species data, as I did, to test theories on population and community structure and dynamics and others generated novel theory on invasions (cf. Davis 2006). By the time of the SCOPE programme, not only were there many more publicized problematic introductions than in 1958, but many more data were available on both previously studied introductions (see, for example, muskrat in Europe) and new or newly studied introductions. Invasions were thus amenable to quantitative analysis and quantitative modelling in the 1980s to a far greater extent than in 1958, and quantitative treatment and modelling have been a hallmark of the field ever since. Elton (1958) was the antithesis of quantitative; although he supported the mathematical theoretician P.H. Leslie as a member of the Bureau, Elton himself was never very interested in mathematical modelling. He was not atheortical (Hardy 1968): much of his invasion monograph is devoted to buttressing a theory that damaging invasions occur because certain communities, especially depauperate ones, offer insufficient resistance. However, he did not test this theory by formulating a testable model.

So, both by the increasing urgency of invasive species problems and increased availability of invasion data, the time was ripe for a burst of interest in introduced species in the 1980s as it was not in 1958. Most scientists participating in the development of modern invasion biology knew Elton's monograph, often from having read it years previously, and admired his perspicacity in recognizing the global scope of the problem (Richardson & Pyšek 2007). Many cited him; for instance, in the SCOPE national volumes and synthesis volume, 33 of 100 chapters cite the monograph, always as a reference for the fact that invasions are important. However, the field did not begin with the monograph, and it would probably have developed in the 1980s much as it did even had Elton not written the monograph. Although he may have planted the seed of the problem in the minds of some researchers who later participated in the field, his role in invasion biology resembled that of Mendel in genetics (cf. Fernberger 1937).

Interestingly, several authors wrote major books on introduced species before the monograph (cf. Chew 2006). Two shed light on why Elton's monograph was the first to draw widespread attention to invasions.

Carl Lindroth (1957) published a lengthy book the year before Elton's book detailing animals introduced between Europe and North America, and George Thomson published an even larger tome in 1922 attempting to summarize everything known about introductions to New Zealand. Except to committed invasions aficionados, both are dry reading, with long lists of species and descriptions of details of their introduction. Both show which species were introduced and when, but neither emphasizes impacts. Lindroth (1957) dealt primarily with biogeography, treating introduced species as impeding understanding of biogeographical patterns. One chapter (94 of 344 pages) is on 'The human transport of animals across the North Atlantic'. The emphasis is on means of transport, and though five pages are on 'The economic importance of introduced animals' – treating agricultural impact – there is no discussion of ecological impact and only cursory discussion of agricultural impact. Thomson (1922) has chapters on animal and plant impacts: 32 pages out of 607. He was partly hamstrung by the fact that he wrote so early that there had been little study of impacts:[5] by the time of Elton's monograph, he could say of New Zealand, 'No place on Earth has received for such a long time such a steady stream of aggressive invaders ...' (p. 89). And Elton detailed the impacts; in fact, most of the monograph was about impacts, and Elton wrote not about one region but about the entire Earth, not about one taxon but about plants, animals and pathogens, not about one habitat but about land, sea and freshwater. In addition, he was an excellent writer, chose his examples well, and constructed a powerful, concise statement. Small wonder his monograph was widely read!

It is also worth considering the meagre impact of two works published before Elton's monograph by high-profile ecologists with very modern views of biological invasions. One scientist was Henry J. Oosting, president in 1955 of the Ecological Society of America and author of an important plant ecology textbook (Oosting 1948). He described how introduced Japanese honeysuckle, chestnut blight, tumble mustard and water hyacinth all change entire native communities, and he presented striking photographs of all four invasions, plus descriptions of impacts of other introduced plants. Even though the book is about plant ecology, he also described the catastrophic impacts of the small Indian mongoose and gypsy moth. In his second edition, Oosting (1956) added descriptions of impacts of kudzu (*Pueraria montana* variety *lobata*) in the South and

Eucalyptus in California, as well as brief discussions of pestiferous activities of the English sparrow and starling in the United States, the muskrat in Europe, and the rabbit in Australia. Yet no one recalls Oosting as having recognized early the pervasive impacts of invaders. In fact, both Harold Mooney (personal communication 2008) and I read this book many years ago and neither of us remembered that it contained anything about invasions. Probably this was because Oosting embedded his discussion of invasions in a text about another, though related, subject, and, unlike Elton, he did not write an entire book about invasions with dozens of examples and with a thesis for why they are so damaging in certain situations but not in others.

In 1942, Frank Egler, a leading plant ecologist and author on vegetation science and conservation (Dritschilo 2008), published a remarkable article in *Ecology* on introduced plants in Hawaii, with discussion of New Zealand and of plant invasions generally. His thesis, supported by Hawaiian examples, was a hybrid of the main theses of Elton (1958) and of the criticism of Elton by Davis et al. (2001), yet neither work cited this paper; nor did Oosting (1956). Egler's key first point is that the most plant invasions in Hawaii are fostered by anthropogenic disturbance, and many would dissipate if human activity ceased. Here he was explicitly following work by Allan (1936) on interactions between introduced and native plants in New Zealand and anticipating the main message of Elton's monograph. However, his second point, which he said also follows Allan but which Allan barely addressed, is that one cannot understand the trajectories of particular invasions by drawing a categorical distinction between introduced and native species. Rather, each species must be studied in its own right by examining how it interacts with other species during succession, almost a paraphrase of the contention of Davis et al. (2001). Egler's paper probably remained obscure because, even more than Elton, he wrote well before problems associated with invasions were recognized as a widespread phenomenon, and, like Oosting, he did not write an entire book with dozens of examples. There was already concern about introduced species in Hawaii, New Zealand and elsewhere, but each invasion was considered, if not *sui generis*, at least a local problem, and the global extent of invasion problems was unrecognized. Further, although he presented a theory of sorts, Egler resembled Elton in buttressing the theory by many natural history examples and did not suggest quantitative tests or criteria for rejection.

2.6 CONCLUSIONS

Charles Elton was a towering figure in ecology, best known for his concepts of the pyramid of numbers and for his contributions to the ideas of the food chain, food web and flow of energy as mechanisms linking populations of different species together into an integrated whole (Hardy 1968). He was one of the first ecologists to focus on animal communities and the first to represent animal communities as dynamic entities rather than lists of co-occurring species (Cox 1979). The 2001 reissue of his 1927 *Animal Ecology* has helped remind us of his breadth, his vision and his influence in all these areas. His contributions on animal population cycles, and those of his student and colleague Dennis Chitty, remain influential. In all of these areas, Elton's writings led rather quickly and directly to the growth of an ongoing research paradigm and programme.

His contribution to invasion biology is quite different, more that of a prophet. When the field arose abruptly in the 1980s, many practitioners, aware of his monograph, cited it; however, the explosion of research in invasion biology in the 1980s, and the directions research has taken since then, are not products of the monograph. Importantly, Elton's role in sensitizing the conservation-minded public to biological invasions as a conservation issue was probably substantial, at least in Great Britain and North America, even independently of his 1958 monograph. In Great Britain, Elton heavily influenced a committee on nature conservation and nature reserves established during the Second World War by the British Ecological Society, both by his service on the committee and by his 1942 memorandum to the committee (Simberloff 2010). The memorandum had a large section on invasion impacts, and the main recommendations and much of the verbiage of the memorandum, including the invasion section, were adopted in the committee report (Committee of the British Ecological Society 1944) and a popular version of the main thrust of the report (Tansley 1945); see Simberloff (2010). In North America, Aldo Leopold is widely viewed as the most influential conservationist and conservation biologist of the 20th century (see, for example, Callicott & Freyfogle 1999) and patron saint of conservation biology (Temple 1999), and his *Sand County Almanac* (Leopold 1949), frequently reprinted, is the bible of the conservation movement (Callicott & Freyfogle 1999). Alone among conservation writers of the early and mid-20th century, Leopold repeatedly and increasingly stressed the myriad problems caused by introduced plants and animals, culminating in his famous essay 'Cheat takes over' in *Sand County Almanac*. Elsewhere I will explore the complex evolution of Leopold's thinking on introduced species; suffice it to say here that Elton's influence was crucial and continuous until Leopold's death in 1948.

The elegance and power of Elton's writing ensure that his 1958 monograph will continue to be cited; it will play a role in enlisting future generations of students into the ranks of invasion biologists. It does not greatly diminish the importance of all of Elton's contributions on introduced species to say that the monograph did not found the field of invasion biology.

ACKNOWLEDGEMENTS

I thank Roger Kitching, Hal Mooney, Bill Murdoch and David Pimentel for important, informative discussion; Jim Drake, Curt Meine and Julianne Lutz Warren for insights and encouragement; Bernard Schermetzler for access to and assistance in using the Aldo Leopold Archives at the University of Wisconsin; and Susan Greenwood and Veronique Plocq Fichelet for facilitating my use of the SCOPE–ICSU archives and for information on the SCOPE invasions programme. Nathan Sanders, Martin Nuñez, Louise Robbins, Mary Tebo, Lloyd Loope and Richard H. Groves constructively criticized early drafts of this manuscript.

ENDNOTES

1. Elton did not restrict his call for such research to plants. For instance, he also lamented that animal ecologists still, in 1958, did not know why the North American grey squirrel had largely replaced the native red squirrel over much of England. It took approximately 35 more years for the reasons to become clear (Williamson 1996).
2. When modern invasion biology did flower subsequently, evolution of invaders and of natives affected by them became a hot research topic (see, for example, Cox 2004).
3. I am indebted to Richard H. Groves for calling my attention to the odd near-absence of New Zealand from the SCOPE programme. This lacuna deserves study by historians. New Zealand was invaded early by a plethora of animal and plant species, some with impacts so evident that Elton (1958) devoted a large part of his chapter 'The fate of remote islands' to New Zealand. The invaders and

their impacts were noted and often detailed well before Elton by several workers, notably James M. Drummond (1907) on birds, William Herbert Guthrie-Smith (1921) on many animal and plant species, George M. Thomson (1922; see description below, and note 5) on the entire gamut of animal and plant introductions, Harry H. Allan (1936; see description below) on plants, and Kazimierz A. Wodzicki (1950) on mammals. Yet among many SCOPE projects, only Robert E. Brockie represented New Zealand; he and three co-authors (Brockie et al. 1988) published part of a report from a working group associated with the SCOPE programme studying invasions in nature reserves (Usher et al. 1988).
4. The one major area of modern invasion biology that was not inspired by the SCOPE programme was the single large area of modern invasion biology that Elton did not treat in his monograph (Simberloff 2000): the evolution of introduced species or natives interacting with them. This area became a focus in the late 1990s. The first comprehensive monograph was by Cox (2004). The three SCOPE questions (listed above) did not refer to evolution, and SCOPE is, after all, concerned with problems of the environment, whereas evolution would seem, at least superficially, rather distantly related to environmental issues.

   Evolution became a key focus within invasion biology about a decade after the SCOPE problems precisely when it became obvious even to casual readers of the literature that the introduction of species to totally new regions, usually in very small numbers, presented obvious possibilities for addressing interesting evolutionary and genetic questions, including the relative strengths of a founder effect, drift and selection, and the genetic distance between different conspecific populations. The advent of molecular tools to detect gene flow probably contributed to the excitement of evolutionists and geneticists about invasions. However, several authors in the SCOPE volumes were primarily evolutionists and geneticists, and it would be interesting for a historian to study the lag in the rise of evolution as an invasion biology focus.
5. Data provided by Thomson (1922) were so comprehensive that their analysis was crucial in the 1990s to the development of the still-growing interest in propagule pressure as a major force determining survival of introduced species (see, for example, Veltman et al. 1996; Duncan 1997; Green 1997) and more recently were key to critical examination of the role of competition in determining invasion success (Duncan & Blackburn 2002).

REFERENCES

Allan, H.H. (1936) Indigene versus alien in the New Zealand plant world. *Ecology*, **17**, 187–193.

Anderson, D. (ed.) (1977) Exotic species in Australia – their establishment and success. *Proceedings of the Ecological Society of Australia*, **10**.

Anonymous (2007) Fifty years of invasion ecology – the legacy of Charles Elton. Symposium: 12–14 November 2008, Stellenbosch, South Africa.

Baker, H.G. & Stebbins, G.L. (1965) *The Genetics of Colonizing Species*. Academic Press, New York.

Birch, L.C. (1965) Evolutionary opportunity for insects and mammals in Australia. *The genetics of colonizing species* (ed. H.G. Baker and G.L. Stebbins), pp. 197–211. Academic Press, New York.

Brockie, R.E., Loope, L.L., Usher, M.B. & Hamann, O. (1988) Biological invasions of nature reserves. *Biological Conservation*, **44**, 9–36.

Callicott, J.B. & Freyfogle, E.T. (1999) Introduction. *For the Health of the Land* (ed. A. Leopold, J.B. Callicott and E.T. Freyfogle), pp. 3–26. Island Press, Washington, DC.

Chew, M.K. (2006) *Ending with Elton: Preludes to Invasion Biology*. Ph.D. dissertation, Arizona State University, Tempe, Arizona.

Coates, P. (2003) Editorial postscript: the naming of strangers in the landscape. *Landscape Research*, **28**, 131–137.

Cohn, M.F. (1959) Review of *The Ecology of Invasions by Animals and Plants*. *Quarterly Review of Biology*, **34**, 303–304.

Committee of the British Ecological Society (1944) Nature conservation and nature reserves. *Journal of Ecology*, **32**, 45–82; *Journal of Animal Ecology*, **13**, 1–25.

Cox, D.L. (1979) *Charles Elton and the Emergence of Modern Ecology*. Ph.D. dissertation, Washington University, St. Louis, Missouri.

Cox, G.W. (2004) *Alien Species and Evolution*. Island Press, Washington, DC.

Crowcroft, P. (1991) *Elton's Ecologists. A History of the Bureau of Animal Population*. University of Chicago Press, Chicago.

di Castri, F., Hansen, A.J. & Debussche, M. (eds.) (1990) *Biological Invasions in Europe and the Mediterranean Basin*. Kluwer, Dordrecht, The Netherlands.

Darwin, C. (1996) *The Origin of Species*, 2nd edn (1859). Oxford University Press, Oxford.

Davis, M.A. (2006) Invasion biology 1958–2005: the pursuit of science and conservation. In *Conceptual Ecology and Invasion Biology* (ed. M.W. Cadotte, S.M. McMahon and T. Fukami), pp. 35–64. Springer, Dordrecht, The Netherlands.

Davis, M.A., Thompson, K. & Grime, J.P. (2001) Charles S. Elton and the dissociation of invasion ecology from the rest of ecology. *Diversity and Distributions*, **7**, 97–102.

Drake, J.A., Mooney, H.A., diCastri, F., et al. (1989) Preface. In *Biological Invasions. A Global Perspective* (ed. Drake, J.A., H.A. Mooney, F. diCastri, R.H. Groves, F.J. Kruger, M. Rejmánek, and M. Williamson), pp. xxiii–xxiv. Wiley, Chichester.

Drake, J.A. & Williamson, M.H. (1986) Invasions of natural communities. *Nature*, **319**, 718–719.

Dritschilo, W. (2008) *Magnificent Failure: Frank Egler and the Greening of Ecology*. Manuscript.

Drummond, J. (1907) On introduced birds. *Transactions and Proceedings of the New Zealand Institute*, **39**, 227–252.

Duncan, R.P. (1997) The role of competition and introduction effort in the success of passeriform birds introduced to New Zealand. *American Naturalist*, **149**, 903–915.

Duncan, R.P. & Blackburn, T.M. (2002) Morphological over-dispersion in game birds (Aves: Galliformes) successfully introduced to New Zealand was not caused by interspecific competition. *Evolutionary Ecology Research*, **4**, 551–561.

Egler, F.E. (1942) Indigene versus alien in the development of arid Hawaiian vegetation. *Ecology*, **23**, 14–23.

Elton, C.S. (1927) *Animal Ecology*. Macmillan, New York. (Republished 2001 by University of Chicago Press, Chicago.)

Elton, C.S. (1933a) Alien invaders. An animal census in Britain. Letter to *The Times* (London), 6 May.

Elton, C.S. (1933b) *The Ecology of Animals*. Methuen, London.

Elton, C.S. (1933c) *Exploring the Animal World*. Allen & Unwin, London.

Elton, C.S. (1936) A new invader. *Journal of Animal Ecology*, **5**, 188–192.

Elton, C.S. (1943) The changing realms of animal life. *Polish Science and Learning*, **2**, 7–11.

Elton, C.S. (1944) The biological cost of modern transport. *Journal of Animal Ecology*, **13**, 87–88.

Elton, C.S. (1945) Honey from *Ailanthus*. *Nature*, **55**, 81.

Elton, C.S. (1958) *The Ecology of Invasions by Animals and Plants*. Methuen, London.

Elton, C.S. (1966) *The Pattern of Animal Communities*. Methuen, London.

Elton, C.S. (1971) *Aulonium trisulcum* Fourc. (Col. Colydiidae) in Wytham Woods, Berkshire, with remarks on its status as an invader. *Entomologist's Monthly Magazine*, **106**, 190–192.

Fernberger, S.W. (1937) Mendel and his place in the development of genetics. *Journal of the Franklin Institute*, **223**, 147–172.

Gray, A.J., Crawley, M.J. & Edwards, P.J. (eds.) (1987) *Colonization, Succession and Stability*. Blackwell Scientific Publications, Oxford.

Green, R.E. (1997) The influence of numbers released on the outcome of attempts to introduce exotic bird species to New Zealand. *Journal of Animal Ecology*, **66**, 25–35.

Groves, R.H. & Burdon, J.J. (eds.) (1986) *Ecology of Biological Invasions*. Cambridge, UK, Cambridge University Press.

Groves, R.H. & di Castri, F. (eds) (1991) *Biogeography of Mediterranean Invasions*. Cambridge University Press, Cambridge, UK.

Guthrie-Smith, H. (1921) *Tutira, the Story of a New Zealand Sheep Station*. William Blackwood and Sons, Edinburgh.

Hardy, A. (1968) Charles Elton's influence in ecology. *Journal of Animal Ecology*, **37**, 3–8.

Joenje, W. (1987) The SCOPE programme on the ecology of biological invasions: an account of the Dutch contribution. *Proceedings of the Koninklijke Nederlandse Akademie van Wetenschappen C*, **90**, 3–13.

Joenje, W., Bakker, K. & and Vlijm, L. (eds.) (1987) The ecology of biological invasions. *Proceedings of the Koninklijke Nederlandse Akademie van Wetenschappen C*, **90**, no. 1.

Kitching, R.L. (ed.) (1986a) *The Ecology of Exotic Animals and Plants. Some Australian Case Histories*. Wiley, Brisbane.

Kitching, R.L. (1986b) Exotics in Australia – synopsis and strategies. In *The Ecology of Exotic Animals and Plants. Some Australian Case Histories* (ed. R.L. Kitching), pp. 262–269. Wiley, Brisbane.

Kitching, R.L. & Jones, R.E. (eds.) (1981) *The Ecology of Pests. Some Australian Case Histories*. CSIRO Australia, Melbourne.

Kornberg, H. & Williamson, M.H. (eds.) (1986) Quantitative aspects of the ecology of biological invasions. *Philosophical Transactions of the Royal Society of London B*, **314**.

Leopold, A. (1931) <http://digital.library.wisc.edu/1711.dl/AldoLeopold>, Matamek Conference on Biological Cycles,m9/25/10-2, p. 38, accessed 10 August 2008.

Leopold, A. (1949) *A Sand County Almanac and Sketches Here and There*. Oxford University Press, New York.

Lindroth, C.H. (1957) *The Faunal Connections between Europe and North America*. Wiley, New York.

Macdonald, I.A.W., Kruger, F.J. & Ferrar, A.A. (eds.) (1986) *The Ecology and Management of Biological Invasions in Southern Africa*. Oxford University Press, Cape Town.

Mayr, E. (1965) The nature of colonization in birds. In *The Genetics of Colonizing Species* (ed. H.G. Baker and G.L. Stebbins), pp. 29–43. Academic Press, New York.

Middleton, A.D. (1930) The ecology of the American grey squirrel (*Sciurus carolinensis* Gmelin) in the British Isles. *Proceedings of the Zoological Society of London*, **100**, 809–843.

Middleton, A.D. (1932) The grey squirrel (*Sciurus carolinensis*) in the British Isles 1930–1932. *Journal of Animal Ecology*, **1**, 166–167.

Middleton, A.D. (1935) The distribution of the grey squirrel (*Sciurus carolinensis*) in Great Britain in 1935. *Journal of Animal Ecology*, **4**, 274–276.

Mooney, H.A. & Drake, J.A. (eds.) (1986) *Ecology of Biological Invasions of North America and Hawaii*. Ecological Studies 58. Springer, New York.

Murdoch, W.W. (1975) Diversity, complexity, stability and pest control. *Journal of Applied Ecology*, **12**, 795–807.

Oosting, H.J. (1948) *The Study of Plant Communities. An Introduction to Plant Ecology*. W.H. Freeman, San Francisco.

Oosting, H.J. (1956) *The Study of Plant Communities. An Introduction to Plant Ecology*, 2nd edn. W.H. Freeman, San Francisco.

Pearsall, W.H. (1959) The ecology of invasion: ecological stability and instability. *New Biology*, **29**, 95–101.

Pimentel, D. (1955a) Biology of the Indian mongoose in Puerto Rico. *Journal of Mammalogy*, **36**, 62–68.

Pimentel, D. (1955b) The control of the mongoose in Puerto Rico. *American Journal of Tropical Medicine and Hygiene*, **41**, 147–151.

Pimentel, D., Glenister, C., Fast, S. & Gallahan, D. (1984) Environmental risks of biological pest controls. *Oikos*, **42**, 283–290.

Pimentel, D., Lach, L., Zuniga, R. & Morrison, D. (2000) Environmental and economic costs of non-indigenous species in the United States. *BioScience*, **50**, 53–65.

Ramakrishnan, P.S. (ed.) (1991) *Ecology of Biological Invasion in the Tropics*. International Scientific Publications, New Delhi.

Rejmánek, M., Richardson, D.M., Barbour, M.G., et al. (2002) Biological invasions: politics and the discontinuity of ecological terminology. *Bulletin of the Ecological Society of America*, **83**, 131–133.

Ricciardi, A. & MacIsaac, H.J. (2008) The book that began invasion biology. *Nature*, **452**, 34.

Richardson, D.M., Macdonald, I.A.W., Hoffmann, J.H. & Henderson, L. (1997) Alien plant invasions. *Vegetation of Southern Africa* (ed. R.M. Cowling, D.M. Richardson & S.M. Pierce), pp. 535–570. Cambridge University Press, Cambridge.

Richardson, D.M. & Pyšek, P. (2007) Classics in physical geography revisited: Elton, C.S. 1958: The ecology of invasions by animals and plants. Methuen: London. *Progress in Physical Geography*, **31**, 659–666.

Richardson, D.M. & Pyšek, P. (2008) Fifty years of invasion ecology – the legacy of Charles Elton. *Diversity and Distributions*, **14**, 161–168.

Rolls, E.C. (1969) *They All Ran Wild. The Story of Pests on the Land in Australia*. Angus and Robertson, Sydney.

Simberloff, D. (1981) Community effects of introduced species. *Biotic crises in ecological and evolutionary time* (ed. M. Nitecki), pp. 53–81. Academic Press, New York.

Simberloff, D. (2000) Foreword in Elton, *The Ecology of Invasions by Animals and Plants*, pp. vii–xiv. University of Chicago Press, Chicago.

Simberloff, D. (2006) Invasional meltdown six years later – important phenomenon, unfortunate metaphor, or both? *Ecology Letters*, **9**, 912–919.

Simberloff, D. (2010) Charles Elton as conservationist and conservation biologist. (Manuscript.)

Southwood, R. & Clarke, J.R. (1999) Charles Sutherland Elton. *Biographical Memoirs of Fellows of the Royal Society of London*, **45**, 129–140.

Tansley, A.G. (1945) *Our Heritage of Wild Nature: A Plea for Organised Nature Conservation*. Cambridge University Press, Cambridge.

Taylor, W.P. (1959) The ecology of invasions. *Ecology*, **40**, 168–169.

Temple, S.A. (1999) Afterword. In *A. Leopold, For the Health of the Land* (ed. J.B. Callicott & E.T. Freyfogle), pp. 227–238. Island Press, Washington, DC.

Thomson, G.M. (1922) *The Naturalisation of Animals and Plants in New Zealand.* Cambridge University Press, Cambridge.

Usher, M.B., Kruger, F.J., Macdonald, I.A.W., Loope, L.L. & Brockie, R.E. (1988) The ecology of biological invasions into nature reserves: an introduction. *Biological Conservation,* **44**, 1–8.

U.S. Congress, Office of Technology Assessment. (1993) *Harmful non-indigenous species in the United States.* OTA-F-565. U.S. Government Printing Office, Washington, DC.

Veltman, C.J., Nee, S. & Crawley, M.J. (1996) Correlates of introduction success in exotic New Zealand birds. *American Naturalist,* **147**, 542–557.

Wallace, A.R. (1890) *Darwinism,* 2nd edn. Macmillan, London.

Warwick, T. (1934) The distribution of the muskrat (*Fiber zibethicus*) in the British Isles. *Journal of Animal Ecology,* **3**, 250–267.

Warwick, T. (1935) Some escapes of coypus (*Myopotamus coypu*) from nutria farms in Great Britain. *Journal of Animal Ecology,* **4**, 146–147.

Warwick, T. (1940) A contribution to the ecology of the musk-rat (*Ondatra zibethica*) in the British Isles. *Proceedings of the Zoological Society of London A,* **110**, 165–201.

Williamson, M. (1996) *Biological Invasions.* Chapman & Hall, London.

Williamson, M.H., H. Kornberg, M.W. Holdgate, A.J. Gray & Conway, G.R. (1986) Preface: the British contribution to the SCOPE Programme on the Ecology of Biological Invasions. *Philosophical Transactions of the Royal Society of London B,* **314**, 1–2.

Wilson, E.O. (1965) The challenge from related species. In *The Genetics of Colonizing Species* (ed. H.G. Baker and G.L. Stebbins), pp. 7–24. Academic Press, New York.

Wodzicki, K.A. (1950) *Introduced mammals of New Zealand: an ecological and economic survey.* Bulletin of the Department of Scientific and Industrial Research, New Zealand, no. 98, Wellington, New Zealand.

Woodwell, G.M. & Smith, H.H. (1969) *Diversity and stability in ecological systems. Brookhaven Symposia in Biology no. 22.* Brookhaven National Laboratory, Upton, New York.

Chapter 3

THE INVIOLATE SEA? CHARLES ELTON AND BIOLOGICAL INVASIONS IN THE WORLD'S OCEANS

James T. Carlton

Maritime Studies Program, Williams College-Mystic Seaport, Mystic, CT 06355, USA

Fifty Years of Invasion Ecology: The Legacy of Charles Elton, 1st edition. Edited by David M. Richardson
© 2011 by Blackwell Publishing Ltd

3.1 INTRODUCTION: SETTING ELTON IN CONTEXT

One of Charles Elton's many unique contributions was the publication of the first global overview of invasions in the sea. That it took an ecologist (and not a biogeographer) – and a primarily terrestrial ecologist at that – until the 1950s to paint the first picture on introduced species in the oceans seems remarkable, the more so given that intensive human traffic across the seas had commenced some 400 years earlier (Carlton 2009).

Chapter 5 of *The Ecology of Invasions by Animals and Plants* (hereafter called *EIAP*) (Elton 1958) is 'Changes in the sea.' Fifteen text pages long, it occupies only 9% of this iconic book, and mentions relatively few species. In this chapter I examine how Elton's work rests within the context of the genre of invasion books and monographs in the 19th and 20th centuries, I ask what Elton may have overlooked (and why) and I consider what Elton could not have foreseen. I also argue that it is not surprising that it took until the mid-20th century, and a worker peering in from another discipline, to point out that the seas, like the land, were not inviolate, on a global scale, to invasions mediated by the hand of man.

As Carlton (1989) noted, Elton's attraction to invasions had begun more than two decades before the publication of his famous volume in 1958. It is instructive to observe that his interest was coupled from the beginning with a desire to engage the public and broader scientific community in what he regarded as one of the great transformations and alterations of life on Earth. One of Elton's early contributions was a letter on 'alien invaders' in *The Times* in 1933 (Carlton 1989); he followed this in 1943 with a semi-popular essay (Elton 1943). He revitalized his interests after the war with BBC radio broadcasts in 1957 (which then directly gave rise to *EIAP*). In 1936, Elton wrote a long review of Peters and Panning's (1933) monograph on 'die Chinesische Wollhandkrabbe' *Eriocheir sinensis*, a catadromous species that was introduced from China to Europe in the early 1900s in ballast water. He concluded that 'The mitten crab will no doubt join the growing band of invaders to this country [Britain] ... there seems no reason to suppose that this list will not continue to grow. ...' Curiously, in a rare and early reference to the potential control of marine invasions, Elton appears to advocate the creation of a 'chemical pollution' (!) barrier in British estuaries to limit the

establishment of *Eriocheir* in England: 'Whether this will ever be deliberately planned depends upon the extent to which fresh-water fishermen are able to influence Government action'.

Elton's goal in writing *EIAP* was clearly not to attempt a summary of the literature on the invasions of animals and plants, nor to provide an in-depth review. Ninety-one per cent of the literature cited by Elton (1958) is from the period 1920–1957. Instead, Elton's book represents carefully selected case histories to make his central point: the extent to which natural communities had become disrupted. Those turning to Elton as a summary of what was known up to the 1950s will find much missing. This is perhaps made no clearer by the fact that he simply did not cite most of the major works on invasions and biotic dispersal of the late 1800s and early 1900s (Table 3.1). This said, what makes Elton's chapter unique is that none of the works listed in Table 3.1 specifically called out the oceans relative to invasions: his was the first book to address marine invasions as a separate topic. Before Elton's 'Changes in the sea,' there was rare mention of marine invasions as part of the 'big picture' of human impacts in the sea. Exceptions are, generally, exceedingly obscure papers. Charles Atwood Kofoid (1915), a well-known zoologist at the University of California at Berkeley, wrote a popular chapter entitled 'Marine biology on the Pacific coast,' for a once better-known book called '*Nature and Science on the Pacific Coast*,' written to coincide with the 1915 World's Fair in San Francisco (the 'Panama-Pacific International Exposition'). In it Kofoid makes the compelling, and all too brief, comment that in southern California, 'The harbor fauna [of San Diego Bay], as elsewhere on the coast, has been contaminated by the cosmopolitan forms brought by shipping such as barnacles, tubularian and campanularian hydroids, mussels (*Mytilus*), clams (*Mya*) and anemones (*Metridium*).' This casual and almost innocuous comment in a semi-popular book would pass without notice were it not for the fact that no other observer or writer had ever previously mentioned that the harbours or bays of the North American Pacific coast had been invaded by a cosmopolitan fauna: no worker had mentioned that any of the barnacles, hydroids, mussels or sea anemones were not native to the American Pacific coast (the North Atlantic softshell clam *Mya arenaria* had been recognized by the 1870s as an introduction to the Pacific coast, having been transported by the commercial oyster industry from New York to San Francisco Bay,

Table 3.1 Selected monographs and books on biotic dispersal or invasions from 1885 to 1957. Those in bold are cited in Elton (1958). Full citations are in the reference list.

Year	Author	Title	Size
1885	Hehn	*The Wanderings of Plants and Animals from their First Home*	523 pp.
1893	Kew	*The Dispersal of Shells*	291 pp.
1895	Warming	*Plantesamfund-Grundtræk af den økologiske Plantegeografi*	335 pp.
1896	Drude	*Manuel de Geographie Botanique*	552 pp.
1909	Spalding	*Distribution and Movement of Desert Plants*	231 pp.
1920	Ritchie	*The Influence of Man on Animal Life in Scotland*	550 pp.
1921	Guthrie-Smith	*Tutira. The Story of a New Zealand Sheep Station*	464 pp.
1922	Thomson	*The Naturalisation of Animals and Plants in New Zealand*	607 pp.
1930	**Ridley**	***The Dispersal of Plants throughout the World***	**701 pp.**
1940	Allan	*A Handbook of the Naturalized Flora of New Zealand*	344 pp.
1949	Clark	*The Invasion of New Zealand by People, Plants, and Animals*	465 pp.
1950	**Wodzicki**	***Introduced Mammals of New Zealand***	**255 pp.**
1957	Lindroth	*The Faunal Connections between Europe and North America*	344 pp.

although ironically, after it was introduced, it was first described as a new species). It is intriguing that Kofoid believed that the large white sea anemone *Metridium senile* was not native, and he may have been privy to information, now lost or obscured, that it had arrived within the time of human memory in Eastern Pacific harbours. That the passing of only a few decades can lead to the loss of historical knowledge – a classic example of the shifting baseline – is illustrated by the Second World War-era arrival of the jellyfish *Aurelia* in the Hawaiian Islands, an appearance not noted as an introduction until 2009 (Carlton & Eldredge 2009). In other times, Kofoid's comment might have inspired a series, if not a field, of investigations, but as with other early seeds in marine ecology, the planting was not watered.

Similarly, Fraser (1926), in an equally obscure piece, and in a rare early 20th century exploration of 'Modifications due to human agencies in the marine life,' listed 'the introduction of new species' as one of five drivers of human change on the American Pacific coast. Throughout the 19th century and early decades of the 20th century, occasional accounts of the intentional or accidental introductions of individual marine and estuarine species were published (for example, Smith 1896; Fulton & Grant 1900; Chilton 1910; Orton 1912). However, the 'thread that binds' continued to elude those who wrote of changes in the sea, despite the growing evidence by the 1940s and 1950s that hundreds if not thousands of species had been

moved around the world in only the past few centuries. A sobering example of this is the massive mid-20th century volume, *Man's Role in Changing the Face of the Earth* (Thomas 1956). There are 54 chapters in this 1200 page tome: three are about the oceans (which we note make up most of the Earth, in terms of both surface area and habitable space). One chapter, 'Harvests in the sea,' includes one paragraph on introductions. A second chapter, 'Influences of man upon coast lines,' does not mention introductions as an influence. A third chapter, 'Man's ports and channels,' does not touch upon the roles of transportation or shipping in altering port and harbour biota. It remained for a general chapter by Marston Bates, entitled 'Man as an agent in the spread of organisms,' to mention invasions in the sea in one paragraph. To his credit, Bates (1956) did note that 'a great many marine invertebrates must have been moved about with the shipping of modern man, but I have come across no general study of this but only isolated mention.'

And thus the stage was set for 1958.

3.2 WHAT ELTON COVERED

Elton mentioned 33 species as examples of invasions in the sea, most in chapter 5. Interestingly, Elton chose (in chapter 1, 'The invaders') two marine invasions (of seven case histories) as stage-setters for the entire book, 'which were brought from one country and exploded into another': these were the mitten crab *Eriocheir* and the salt marsh plant *Spartina*. I list the marine species covered by Elton in Table 3.2, with their current Latin names (evidently crab and fish workers have not been as busy as other taxonomists). Not in Table 3.2 are two categories of species Elton also mentioned: a few species that were transplanted but failed to establish, and several species no longer regarded as invasions; although a generally rare occurrence, native species have at times been mistaken for introductions (Carlton 2009). One-quarter of Elton's examples were invasions either through the Suez Canal into the Mediterranean Sea or invasions in the Caspian Sea.

Elton recognized three vectors of human-mediated dispersal for marine organisms: canals, shipping, and deliberate introductions. For canals, Elton chose to focus on the Suez Canal (giving examples of some of the well-known invaders (Table 3.2)) and the Panama Canal (concluding that its freshwater Gatun Lake would mean that the Canal would not be a 'serious gap' between the Atlantic and Pacific, 'nor much of a transport line for marine life from one ocean to another' (but see below, 'What Elton missed')). Relative to shipping, Elton knew of both ballast water and ships' hull fouling, noting examples of each. For ballast water in particular, Elton noted or implied that *Eriocheir* and the Asian diatom *Odontella* were dispersed by this mechanism (he cited a third example, the Red Sea shrimp *Processa aequimana*, relative to its detection in the North Sea in 1946, but the northern European *Processa* was later shown to be a native species).

For deliberate introductions, Elton felt that 'the greatest agency of all that spreads marine animals to new quarters of the world must be the business of

Table 3.2 Marine biological invasions mentioned in Elton (1958) SC, Suez Canal; C, Caspian Sea.

Name used by Elton	Current (2010) name	Name used by Elton	Current (2010) name
Diatoms		*Mya arenaria*	Same
Biddulphia sinensis	*Odontella sinensis*	*Mytilaster lineatus* (C)	Same
Algae		Crustacea	
Rhodophyta (red algae)		Cirripedia (barnacles)	
Asparagopsis armata	Same	*Elminius modestus*	*Austrominius modestus*
Plants		*Balanus eburneus*	*Amphibalanus eburneus*
Poaceaa (grasses)		*Balanus improvisus*	*Amphibalanus improvisus*
Spartina townsendii	*Spartina x townsendii*		
Platyhelminthes		Decapoda	
Monogenea (ectoparasitic flatworms)		Brachyura (crabs) and Achelata (slipper lobsters)	
Nitzschia sturionis (C)	Same	*Thenus orientalis* (SC)	Same
Annelida		*Neptunus pelagicus* (SC)	Same
Polychaeta (polychaete worms)		*Neptunus sanguinolentus* (SC)	Same
Mercierella enigmatica	*Ficopomatus enigmaticus*	*Myrax fugax* (SC)	Same
		Rhithropanopeus harrisii	Same
Nereis succinea (C)	*Neanthes succinea*	*Eriocheir sinensis*	Same
Mollusca		Caridea (shrimp)	
Gastropoda (snails)		*Leander adspersus* (C)	*Palaemon adspersus*
Crepidula fornicata	Same	Fish	
Urosalpinx cinerea	Same	*Mugil cephalus* (C)	Same
Tritonalia japonica	*Ocinebrina inornata*	*Onchorhynchus kisutch*	Same
Bivalvia (mussels, clams, oysters)		*Onchorhynchus nerka*	Same
		Onchorhynchus tschawytscha	Same
Pinctada vulgaris (SC)	*Pinctada radiata*	*Salmo salar*	Same
Ostrea edulis	Same	*Alosa sapidissima*	Same
Ostrea gigas	*Crassostrea gigas*	*Roccus saxatilis*	*Morone saxatilis*
Paphia philippinarum	*Ruditapes philippinarum*	*Tarpon atlanticus*	Same

oyster culture'. He specifically invoked the movement of Eastern American oysters to Europe, European oysters to Eastern America, Japanese oysters to Western America, and both Japanese and Eastern American oysters to the Hawaiian Islands: 'Oysters', Elton wrote, 'are therefore a kind of sessile sheep, that are moved from pasture to pasture in the sea'. Noting the scale of the epibiota – including fouling organisms and oyster predators – accidentally transported with all of these oyster movements, Elton remarked that,

> 'If a large corporation had been set up just to distribute about the world a selection of organisms living around or just below low-water mark on the shores of the world, it could not have been more efficient at the job, considering that the process has only been going full blast for a hundred years or less!'

Elton then added that certain clams, crabs and tubeworms had been part of what he called this 'chess play', summarizing the role of canals, shipping and oysters, as follows:

> 'In the midst of this rather complex tangle of species and dates and places we can discern the setting in of a very strong historical move, the interchange of the shore fauna of continents, and also sometimes the plankton of different seas. *It is only an advance guard*, yet some of the species have already taken up prominent posts in the new communities they have joined ...'

Italicized here is perhaps Elton's most insightful comment, *It is only an advance guard*. Although he did not extend this thought in any specific predictive nature (such as the number of species that might be involved in the future, or the geography of invasions), it conveys a clear sense that Elton suspected that the invasion roulette wheel had hardly begun to be spun for the oceans of the world. I comment on this further below.

Elton wrapped up his 'deliberate introductions' with the intentional introductions of worms (as fish food) and finfish into the Caspian Sea, the worldwide translocations of North Pacific (*Onchorhynchus*) and North Atlantic (*Salmo*) salmonids (Table 3.2), and the introduction of two Atlantic fish (the shad *Alosa* and the

striped bass *Morone*) to the Pacific coast of North America.

Elton did not spare his editorial thoughts in *EIAP*, and 'Changes in the sea' provides some particularly compelling examples. He noted that the commercial oyster industry operated 'without particularly stringent precaution'. Stronger were his thoughts on moving fish:

> 'From 1872 onwards until 1930 the United States Bureau of Fisheries, with benevolent intent, supplied over 100 million eggs of Pacific salmon to people in other countries, with the idea of establishing new salmon runs there – a considerable attempt to bring in the New World to right the Rest.'

And, noting that the fishery for the Atlantic striped bass on the American Pacific coast was calculated in the early 1950s as providing '2,000,000 man-hours of recreation per annum,' Elton remarked,

> 'A world that begins to assess its recreation in man-hours probably cares fairly little about the breakdown of Wallace's Realms.'

Elton concluded chapter 5 noting that it would be of interest to determine if the introductions of large, dominant predatory fish could indeed be undertaken 'without ill results', particularly relative to the unintentional impact on other, and by implication, native, fisheries.

3.3   WHAT ELTON MISSED

The antiquity of invasions

Perhaps the greatest lacuna in Elton's treatment was his lack of reference to the antiquity of marine invasions. However, here we may ask too much, in two regards. First, and in reference to the earlier observation that it was not Elton's intent to review the history of invasions, he rarely mentions invasions before the 1800s, whether terrestrial, freshwater or marine. Second, the marine literature of Elton's era, and particularly of the decades he focused on (1920s to 1950s), rarely referred to the interplay of maritime history and marine biology. Although the historical record was clear, and certainly well known, of European vessels invading the Pacific theatre on a regular basis by the 1500s, the link between centuries of exploration, colonization and commercialization was not made to the global movement of marine organisms.

Indeed, by the time marine biologists began to appear on the scene in the mid-19th century – as what we might think of as more or less modern manifestations of marine scientists – so many shallow-water species had been introduced in the previous centuries around the world that instead of recognizing these as invasions, the marine biogeographic and systematic community instead adopted a theory of natural cosmopolitanism (Carlton 2009). Carlton (1998) has argued that this was, in part, due to a long-enduring sense (despite the evidence from fisheries!) that the seas were inviolate, noting that Byron (1818) had argued that 'Man marks the earth with ruin, his control Stops with the shore'. Gould (1991) had similarly noted that Lamarck in 1809 had written that the life 'in ocean waters' was not susceptible to the hand-of-man, particularly, in this case, relative to their destruction. Thus Elton's perhaps inevitable sense of invasions in the world's oceans was that it was a comparatively recent phenomenon: 'In contrast to land and fresh waters the sea seems still almost inviolate. Yet big changes in the distribution of species have already begun as a result of human action *during the last hundred years*.' His (italicized here) emphasis on 'during the last hundred years' did not help dissuade or dissolve a general but curious sense among most marine biologists that invasions had somehow only started, and occurred, largely within their lifetimes: yet another example of the shifting baseline (Carlton 2003, 2009).

Vectors

Although Elton was aware of the existence of ships' ballast water (as noted above), he did not connect this mechanism with the ability of ships to easily bridge the freshwater barrier of the Panama Canal (Carlton 1985; Cohen 2006). The Panama Canal reduced the transit time between the Pacific and Atlantic Oceans considerably by avoiding the necessity of sailing around South America: concomitantly, such reduced voyages would mean that living organisms in ballast water would end up having a greater chance of survival between the oceans.

Finally, in terms of what Elton emphasized, it is probable that ships, not oysters, have been 'the greatest agency of all that spreads marine animals to new quarters of the world'. Although certain bays and estuaries with a long history of intensive importation of exotic oysters will appear to have derived more of their exotic biota through the oyster trade (Wonham & Carlton 2005), complicating this conclusion is that many species are polyvectic, and thus fall into a 'multiple vector' column that combines species that may have been introduced by oysters, ship fouling or ballast water. More important, however, is that the intensive global movement of adult oysters with a rich epibiota lasted only about 100 years (1860s to 1970s), and involved selected regions of the world, whereas global shipping started centuries earlier, vessels touched every continent and every island of the world, and continues to expand unabated. Elton's emphasis on the movement of commercial oysters versus the much greater role of ships over time and space is likely linked, again, to the available literature of the time, which emphasized the invasions of the day, rather than the much broader picture of the role of global transportation in modifying the distribution of potentially thousands of species.

Carl Hildebrand Lindroth

Elton finished *EIAP* in July 1957, and thus just missed capturing one of the most important books of the era on invasion science. Had Elton had a chance to incorporate it in his own book, Lindroth (1957) may have been much better known to this day. Carl Hildebrand Lindroth (1911–1975) was a Professor of Entomology at Lund University, Sweden, when he published *The Faunal Connections between Europe and North America*; a more compelling title would also have helped to propel his work. Lindroth was concerned with the history of carabid beetles common to Europe and Newfoundland. While paying due attention to 'natural' dispersal vectors and dispersal history (be it geological or historical), Lindroth turned himself, atypically for most biogeographers, to devoting a good deal of his treatment to the history of human traffic across the North Atlantic Ocean, particularly that traffic that could have accidentally transported beetles. His research led him to the history of ships' ballast, in this case shore ballast consisting of shingle rocks, beach debris, sand and any other such materials that provided not only weight for a ship but also carabid habitat. Lindroth's work, cast in the light of modern invasion science, has many themes that were hardly to be revisited for nearly 50 more years. In one of the first attempts to identify methods by which to distinguish native from non-native species, Lindroth erected what he referred to as 'the five criteria of an introduced species'. For Lindroth these criteria were historical, geographical, ecological, biological and taxonomic. For historical evidence, Lindroth included not only direct evidence of an introduction, but also indications of recent expansion of a species (Lindroth noted that not all such expansions indicate that a species is non-native). For the geographical criterion, he considered as evidence disjunct distributions as well as distributions that did not correlate with seemingly natural boundaries. Lindroth's ecological criterion focused on species restricted to human-modified habitats, such as 'around ports and other settlements'. Biological criteria included species that relied solely on known introduced species, such as 'plant-feeders and parasites monophagously bound to a single host'. Finally, Lindroth felt that taxonomic (morphological) distinctions would aid in determining if certain populations of a species were introduced, where other populations of the same species might be native to a region. Elton himself did not specifically lay out such criteria, but rather focused largely on well-known species whose introduction history was generally without question.

Lindroth further speculated about 'the minimum size of a viable population' as it applied to invasion ecology and population biology, and proposed seven factors that might mediate successful invasion, including substrate, moisture requirements, soil type, food, level of food specialization, means of dispersal and means of reproduction. He identified 174 species of beetles that could have been transported by ballast from Europe, of which 68 were transported and became introduced to North America. 'From the above it can be concluded that the selection of animal species which managed to cross the Atlantic by ships, first and foremost in [shore] ballast, has not been a random one' (Lindroth 1957).

3.4 WHAT ELTON COULD NOT HAVE FORESEEN

Although Elton made the prophetic comment that the invasions he reviewed were 'only an advance guard',

and even emphasized (p. 29) the vast changes that occurred in the scale of global shipping between the 1700s and 1900s, I argue that he could not have foreseen the sheer scale of what was to come.

The Unites States Commission on Ocean Policy (2004) and the Pew Oceans Commission (2003) concluded that the major drivers of ocean change throughout the 20th century have been fisheries (both overextraction and the methods of fishing), water quality alteration (including direct chemical pollution and eutrophication), habitat destruction, invasions and climate change. A critical aspect of recognizing these drivers is that they are inextricably intertwined, although frequently studied (and analysed) as if they were not. Thus the removal of most large predatory fish has substantial impacts on community trophodynamics and structure, which could, in turn, influence the success of new invasions. Both increased and decreased water quality may alter colonization ability. In turn, invasions can alter water quality and influence fisheries. Climate change can influence water chemistry, temperature and the distribution of species (Hellmann et al. 2008; Rahel & Olden 2008; Sorte et al. 2010). The potential linkages and feedbacks are almost uncountable. The global scale of environmental change that would mediate invasions in the oceans was not, and could not have been, known to Elton in the 1950s.

At the same time as change in the ocean has become nearly kaleidoscopic, the vastly increased world human population and the similarly expanded nature of global economies have led to a profound increase in the rate and speed of transportation of people and goods. For marine environments, this has been manifested in a proliferation of the number of vectors capable of moving animals and plants around the world. In contrast to Elton's seemingly simple trichotomy of canals, ships and deliberate introductions, the last half of the 20th century saw a staggering increase in the movement of living organisms. Aquaculture in the sea (mariculture), the live seafood industry, the saltwater aquarium industry and the marine bait industry have all expanded enormously (Carlton 2001; Carlton & Ruiz 2005, table 8.1), facilitated in no small part by the availability of many species through the Internet, an electronic 'bioweb'. Conservation and restoration efforts led to the extensive transplantation of eelgrasses, marsh and other wetland plants, and dune grasses, as well as the intentional release of threatened, endangered or depleted species in new regions where they did not previously occur (Carlton

& Ruiz 2005). Global shipping – the size and speed of ships and the number of voyages – was similarly to increase massively (Carlton 2001). Charles Elton would have taken due note of this increase in global bioflow mediated by the inventiveness of human endeavours to move more species with increasing speed.

The collision of vastly more species flowing on global conveyor belts with highly altered and modified marine ecosystems has led to thousands of species being in daily motion around the world, and a scale of invasions of the world's estuaries, bays, coasts and offshore waters that would astound even Elton (Ruiz et al. 2000; Ruiz & Carlton 2003; Rilov & Crooks 2009).

3.5 CONCLUSIONS

Charles Elton recognized that *EIAP* did not lead to a flood of work on invasions (despite its subsequent popularity). In 1988 Elton remarked that, 'It is only in recent years that people have taken much interest in the break-down of "realms", though I was lecturing to students here about it before the last war [Second World War]' (C. Elton *in litt.* to J. Carlton, 15 May 1988). There were continued contributions on individual marine and estuarine non-native species, and a few summaries for certain regions or estuaries, over the next 20 years, but between 96.4% and 99.7% of the marine bioinvasions literature has appeared since 1980 (J.T. Carlton, unpublished data, based upon a JSTOR search in November 2008 of all combinations of the terms 'ocean' + 'marine' + 'introduced' + 'non-native' + 'invasive'). Rather than inspire work on invasions per se, it appears that Elton's book simply and coincidentally 'hit' the environmental bubble of the 1960s, which saw a flood of works on the human modification of the planet (Carson 1962; Ehrlich 1968; Hardin 1968).

We end where we began: that few workers before Elton's time had devoted any synthetic thoughts to invasions in the sea, and that Elton himself devoted only a small – but important – part of his book to the ocean. More than 50 years later, curiously, the proportional treatment of invasions remains roughly the same, but now on a broader stage, as they did in Elton's time: invasions are best recognized, in terms of diversity and impact, on land, somewhat less so in the world's freshwater habitats, and least in the ocean. The reasons for this cascade in awareness and invasion

science are many, and await exploration elsewhere, but certainly reflect in no small part a greater natural interest in where *Homo sapiens* lives and breathes.

ACKNOWLEDGEMENTS

I am most grateful to David Richardson both for inviting me to the compelling Elton Symposium in Stellenbosch in November 2008 and for his enduring patience in the generation of this contribution. I place special value on my correspondence with Charles Elton discussing his early interests in introduced species. Two reviewers offered helpful changes and additions.

REFERENCES

Allan, H.H. (1940) *A Handbook of the Naturalized Flora of New Zealand.* New Zealand Department of Scientific and Industrial Research, Bulletin 83, Government Printer, Wellington.

Bates, M. (1956) Man as an agent in the spread of organisms. In *Man's Role in Changing the Face of the Earth* (ed. W.L. Thomas), pp. 788–804. University of Chicago Press, Chicago, Illinois.

Byron, G.G. (1818) Apostrophe to the ocean. In *Childe Harold's Pilgrimage*. London.

Carlton, J.T. (1989) Man's role in changing the face of the ocean: biological invasions and implications for conservation of near-shore environments. *Conservation Biology*, **3**, 265–273.

Carlton, J.T. (1998) Apostrophe to the ocean. *Conservation Biology*, **12**, 1165–1167.

Carlton, J.T. (2001) *Introduced Species in U.S. Coastal Waters: Environmental Impacts and Management Priorities.* Pew Oceans Commission, Arlington, Virginia, 28 pp.

Carlton, J.T. (2003) Community assembly and historical biogeography in the North Atlantic Ocean: the potential role of human-mediated dispersal vectors. *Hydrobiologia*, **503**, 1–8.

Carlton, J.T. (1985) Transoceanic and interoceanic dispersal of coastal marine organisms: the biology of ballast water. *Oceanography and Marine Biology*, **23**, 313–371.

Carlton, J.T. (2009) Deep invasion ecology and the assembly of communities in historical time. In *Biological Invasions in Marine Ecosystems* (ed. G. Rilov and J.A. Crooks), pp. 13–56. Springer-Verlag, Berlin and Heidelberg.

Carlton, J.T. & Eldredge, L.G. (2009) *Marine Bioinvasions of Hawai'i. The Introduced and Cryptogenic Marine and Estuarine Animals and Plants of the Hawaiian Archipelago.* Bishop Museum Bulletins in Cultural and Environmental Studies 4, Bishop Museum Press, Honolulu, 202 pp.

Carlton, J. & Ruiz, G. (2005) The magnitude and consequences of bioinvasions in marine ecosystems: implications for conservation biology. In *Marine Conservation Biology* (ed. E. Norse and L. Crowder), pp. 123–148. Island Press, Washington, DC.

Carson, R. (1962) *Silent Spring*. Houghton Mifflin, New York.

Chilton, C. (1910) Notes on the dispersal of marine Crustacea by means of ships. *Transactions of the New Zealand Institute of Science*, **44**, 128–135.

Clark, A.H. (1949) *The Invasion of New Zealand By People, Plants, and Animals*. Rutgers University Press, Brunswick, New Jersey.

Cohen, A.N. (2006) Species introductions and the Panama Canal. In *Bridging Divides. Maritime Canals as Invasion Corridors* (ed. S. Gollasch, B.S. Galil and A.N. Cohen), pp. 127–206. Springer, Dordrecht, the Netherlands.

Drude, O. (1896) *Manuel de Geographie Botanique*. P. Klincksieck, Paris.

Elton, C.S. (1936) A new invader. *Journal of Animal Ecology*, **5**, 188–192.

Elton, C.S. (1943) The changing realms of animal life. *Polish Science and Learning*, **2**, 7–11.

Elton, C.S. (1958) *The Ecology of Invasions by Animals and Plants*. Methuen, London.

Ehrlich, P. (1968) *The Population Bomb*. Ballantine Books, New York.

Fraser, C.M. (1926) Modifications due to human agencies in the marine life of the Pacific coast. *Scientific Monthly*, **22**, 151–157.

Fulton, S.W. & Grant, F.E. (1900) Note on the occurrence of the European crab, *Carcinus maenas*, Leach, in Port Phillip. *Victorian Naturalist*, **27**, 147.

Gould, S.J. (1991) On the loss of a limpet. *Natural History*, **100**, 22–27.

Guthrie-Smith, H. (1921) *Tutira. The Story of a New Zealand Sheep Station*. Blackwood, Edinburgh.

Hardin, G. (1968) Tragedy of the commons. *Science*, **162**, 1243–1248.

Hehn, V. (1885) *The Wanderings of Plants and Animals from their First Home*. Swan Sonnenschein, London.

Hellmann, J.J., Byers, J.E., Bierwagen, B.G. & Dukes, J.S. (2008) Five potential consequences of climate change for invasive species. *Conservation Biology*, **22**, 534–543.

Kew, H.W. (1893) *The Dispersal of Shells. An Inquiry into the Means of Dispersal Possessed by Fresh-Water and Land Mollusca*. Paul, Trench, Trubner & Co., London.

Kofoid, C.A. (1915) Marine biology on the Pacific coast. In *Nature and Science on the Pacific Coast*, pp. 124–132. Paul Elder, San Francisco.

Lindroth, C.H. (1957) *The Faunal Connections between Europe and North America*. John Wiley, New York.

Orton, J.H. (1912) An account of the natural history of the slipper-limpet (*Crepidula fornicata*), with some remarks on its occurrence on the oyster grounds of the Essex coast. *Journal of the Marine Biological Association of the United Kingdom*, **9**, 437–443.

Peters, N. & Panning, A. (1933) Die Chinesische Wollhandkrabbe (*Eriocheir sinensis* H. Milne-Edwards) in Deutschland. *Zoologischer Anzeiger*, **104**, Supplement, 180 pp.

Pew Oceans Commission (2003) *America's Living Oceans. Charting a Course for Sea Change*. Pew Oceans Commission, Arlington, Virginia.

Rahel, F.J. & Olden, J.D. (2008) Assessing the effects of climate change on aquatic invasive species. *Conservation Biology*, **22**, 521–533.

Ridley, H.N. (1930) *The Dispersal of Plants throughout the World*. L. Reeve, Ashford, Kent.

Rilov, G. & Crooks, J. (eds) (2009) *Biological Invasions in Marine Ecosystems*. Springer, Berlin and Heidelberg.

Ritchie, J. (1920) *The Influence of Man on Animal Life in Scotland*. Cambridge University Press, Cambridge, UK.

Ruiz, G. & Carlton, J.T. (eds) (2003) *Invasive Species: Vectors and Management strategies*. Island Press, Washington, DC.

Ruiz, G.M., Fofonoff, P.W., Carlton, J.T., Wonham, M.J. & Hines, A.H. (2000) Invasion of coastal marine communities in North America: apparent patterns, processes, and biases. *Annual Review of Ecology and Systematics*, **31**, 481–531.

Smith, H.M. (1896) A review of the history and results of the attempts to acclimatize fish and other water animals in the Pacific states. *Bulletin of the United States Fisheries Commission for* **1895**, **15**, 379–472.

Sorte, C.J.B., Williams, S.L. & Carlton, J.T. (2010) Marine range shifts and species introductions: comparative spread rates and community impacts. *Global Ecology and Biogeography*, **19**, 303–316.

Spalding, V.M. (1909) *Distribution and Movement of Desert Plants*. Carnegie Institution of Washington, Publication 113.

Thomas, W.L. (ed.) (1956) *Man's Role in Changing the Face of the Earth*. University of Chicago Press, Chicago, Illinois.

Thomson, G.M. (1922) *The Naturalisation of Animals and Plants in New Zealand*. Cambridge University Press, Cambridge, UK.

United States Commission on Ocean Policy (2004) *An Ocean Blueprint for the 21st Century*. Final Report. Washington, DC.

Warming, E. (1895) *Plantesamfund – Grundtræk af den økologiske Plantegeografi*. P.G. Philipsens Forlag, Copenhagen (Republished in 1909 as *Oecology of Plants: An Introduction to the Study of Plant Communities*, by E. Warming and M. Vahl).

Wodzicki, K.A. (1950) *Introduced Mammals of New Zealand*. Department of Scientific and Industrial Research, New Zealand.

Wonham, M.J. & Carlton, J.T. (2005) Trends in marine biological invasions at local and regional scales: The Northeast Pacific Ocean as a model system. *Biological Invasions*, **7**, 369–392.

Chapter 4

THE RISE AND FALL OF BIOTIC NATIVENESS: A HISTORICAL PERSPECTIVE

Matthew K. Chew and Andrew L. Hamilton

Arizona State University School of Life Sciences, Tempe, AZ 85287-4501, USA

Fifty Years of Invasion Ecology: The Legacy of Charles Elton, 1st edition. Edited by David M. Richardson

4.1 INTRODUCTION: THE NATIVENESS PROBLEM

Nativeness is an organizing principle of numerous scientific studies and findings, and the *sine qua non* invoked by many management policies, plans, and actions to justify intervening on prevailing ecosystem processes. In recent years, leading invasion biologists (for example Richardson et al. 2000; Pyšek et al. 2004, 2008) have revisited and subtly revised categories, concepts and definitions related to nativeness to promote increased taxonomic rigor and improve the field's data collection and analysis. Others (for example Klein 2002; Bean 2007) have relied on, applied and extended these revisions. Critiques have emerged from within and without examining invasion biology's concepts and practices (for example Milton 2000; Subramaniam 2001; Sagoff 2002, 2005; Theodoropoulos 2003; Colautti & MacIsaac 2004; Brown & Sax 2005; Gobster 2005; Larson 2007; Warren 2007; Davis 2009; Stromberg et al. 2009). Most of these questioned the appropriateness of the native–alien dichotomy to some degree and some have argued against its continued use (see especially Coates 2003; Aitken 2004; Townsend 2005) whereas others were content to explore its cultural influence (for example Trigger et al. 2008). Given the significance attributed to the distinction between native and alien biota and the growing concern over its quality, it is important to be clear about what these concepts mean. Is nativeness conceptually defensible? Does it accomplish any theoretical work?

Pyšek et al. (2004) argued that 'The search for a precise lexicon of terms and concepts in invasion ecology is not driven by concerns of just semantics'. In that spirit, this chapter reviews the categories underpinning science and policy from historical and conceptual perspectives, not the labels that ecologists and policymakers use. Nevertheless, when scientists describe categories, we must credit their choice of words with meaning, and they must allow us to evaluate their categories by the descriptions they provide.

We address several interpenetrating questions:
1 How did the conception of biotic nativeness develop in historical context?
2 How is nativeness diagnosed and applied?
3 What theoretical considerations does nativeness embody?
4 What rights or privileges does biotic nativeness confer?

In answering, we conclude that its categorical meaning and significance both dissolve under scrutiny. Biotic nativeness is theoretically weak and internally inconsistent, allowing familiar human desires and expectations to be misconstrued as essential *belonging* relationships between biota, places and eras. We believe much well-intended effort is wasted on research contrasting 'native' and 'alien' taxa, and by conservation projects focused primarily on preserving or restoring natives.

4.2 NATIVENESS CODIFIED

In recent discussions, human dispersal is said to render populations, and indeed any successor populations, non-native (Klein 2002; Pyšek et al. 2004; Bean 2007). Nativeness is therefore revocable, but non-nativeness is permanent. Being once human-dispersed accomplishes a mutagenic denaturing. Simberloff's (2005) declaration that 'Non-native species DO threaten the natural environment!' (emphasis original) makes sense only if a human act of dispersal renders nature unnatural.

Should dispersal history determine present belonging? Bean (2007) approvingly cites Ross and Walsh's (2003) reason for distinguishing between native and alien plant taxa: 'Recognition of a taxon as a native, especially if it has a restricted distribution or is known from only a few populations, may result in the expenditure of considerable resources to try and ensure the survival of populations. Conversely, classification as an introduced species may result in efforts being made to eradicate populations'. Ross and Walsh moved seamlessly from elucidating dispersal histories to intervening on those that evidence human participation. They gave no *reason* for intervening, but it must be that 'native' valorizes what 'dispersal history' describes. Fortunately, the conception has a history that provides some insight into its present meaning.

Nativeness is an ancient notion in human social, legal and political contexts from neighbourhood to nation. The idea is codified in modern civil discourse in two basic forms: *jus solis*: right of the soil, or birthplace; and *jus sanguinis*: right of the blood, or inheritance (Alonso 1995). Commonly, citizens by 'native' right of blood, or soil, or both are legally superior in some sense to 'naturalized' citizens. The US Constitutional requirement that Presidents be 'natural born citizens' is a familiar example.

Identifying specific biota with specific places has a deep history, extending to Sumerian epics of about 2500 BCE. Partitioning taxa into *native* and *alien* populations is a relatively modern practice, but it significantly predates invasion biology, which coalesced in the late 1980s (Davis 2006). Before the late 18th century, *native* was a catchall conception for uncultivated or undomesticated biota, i.e. the free-living products of a local landscape. Encounters with unknown taxa and peoples in far-flung locations allowed European civil and biotic applications of *native* to cross-pollinate in new ways.

As biogeographical studies accumulated and floras and faunas were documented, it became common practice to signify additions to existing lists with an asterisk (*) (for example Curtis, 1783, p. 24). Asterisks were applied to inventories for other reasons; notably, when a taxon seemed doubtfully identified or likely introduced through human agency (for example Halleri 1742, p. 374). Asterisks identified doubtful botanical claims the way they now identify sports records achieved by 'performance enhanced' athletes. Asterisks increasingly denoted suspicion of human dispersal, and were routinely applied to agricultural weeds and relict cultivars.

Well into the 19th century, botany was practised largely by physicians, apothecaries and amateurs. Increasing professionalization produced increasing nuance. A stark dichotomy denoted by a single symbol was inadequately informative. Cambridge botany professor John Henslow proposed adding two more standard symbols: the degree (°) denoting obviously introduced plants, and the dagger (†) for uncertain cases (Henslow 1835, p. 84). Almost immediately, Hewett (H.C.) Watson, a well-heeled and sedulous amateur botanist who was apprenticed in law and educated in medicine, adopted Henslow's notation for *The New Botanist's Guide to the Localities of the Rarer Plants of Britain* (Watson 1835a). Later that year, Watson elaborated on the topic in *Remarks on the Geographical Distribution of British Plants*. Ellipses in the quotations below generally indicate that we have elided species listed by authors as exemplars in order to focus on conceptual issues.

> 'Species originally introduced by human agency now exist in a wild state; some ... continued by unintentional sowings ... while several keep their acquired hold of the soil unaided, and often despite our

efforts to dispossess them. Both these classes certainly now constitute a part of the British flora, with just as much claim as the descendants of Saxons or Normans have to be considered a part of the British nation. But there is a third class ... plants which have yet acquired a very uncertain right to be incorporated with the proper spontaneous flora of the island ... species springing up occasionally from seeds or roots thrown out of gardens, and maintaining themselves a few years; and ... those designedly planted for ornamental or economical purposes. Such are no more entitled to be called Britons, than are the Frenchmen or Germans who occasionally make their homes in England.'
> (Watson, 1835b, p. 38)

Subsequently in *Remarks*, Watson cited, refined and expanded on Henslow's notational scheme, applying asterisks to 'species generally supposed to have been introduced, but now to some extent established', daggers to 'species more or less strongly suspected to be in the like circumstance, although now occurring spontaneously', and another mark, the double dagger (‡) to distinguish 'such as may possibly have been introduced, being weeds of cultivated ground or inhabited places' (Watson 1835b, p. 185).

A dozen years later, dissatisfied with daggers and asterisks, Watson published his intention of establishing 'the civil claims and local situation of [British plant] species in accordance with a scale of terms' (Watson 1847, p. 62). In an apparent first, he named his categories, producing an *ad hoc* botanical redefinition of *native*, *alien* and three additional, fully codified categories of intermediate establishment and/or uncertainty:

> '**Native**: Apparently an aboriginal British species; there being little or no reason for supposing it to have been introduced by human agency.
> **Denizen**: At present maintaining its habitats, as if a native, without the aid of man, yet liable to some suspicion of having been originally introduced.
> **Colonist**: A weed of cultivated land, or about houses, and seldom found except in places where the ground has been adapted

for its production by the operations of man; with some tendency, however, to appear also on the shores, landslips, &c.
Alien: Now more or less established, but either presumed or certainly known to have been originally introduced from other countries.
Incognita: Reported as British, but requiring confirmation as such. Some ... through mistakes of the species ... others may have been really seen [as] temporary stragglers from gardens ... others cannot now be found in the localities published for them ... some of these may yet be found again. A few may have existed for a time, and become extinct.' (Watson, 1847: 63–64)

Three of Watson's terms came from English common law regarding human citizenship rights, as is clear from jurist Sir William Blackstone's *Commentaries on the Laws of England 1765–1769* (1922). Blackstone wrote, 'the first and most obvious division of the people is into aliens and natural-born subjects. Natural-born subjects are such as are born within the dominions of the crown of England; that is, within the ... allegiance, of the king; and aliens, such as are born out of it' (Blackstone 1922, p. 365). By contrast, 'a denizen is an alien born, but who has obtained ... [documents] to make him an English subject ... [he] is in a kind of middle state between an alien and natural-born subject, and partakes of both' (Blackstone 1922, p. 374). These civil concepts were discussed in parliament contemporaneously with Watson's work (Anonymous 1843). The son of a lawyer, for a time apprenticed as one, Watson's familiarity with such terms is unsurprising.

Common law regarding colonies was complicated. The 'civil claims' Watson meant to apply to botanical 'colonists' remain uncertain. 'If an uninhabited country be discovered and planted by English subjects, all the English laws then in being, which are the birthright of every subject, are immediately there in force'. (Blackstone 1922, p. 107). The law differed for any inhabited country or ceded territory where a legal system (recognizable to Englishmen) existed. *Incognita* had no specific civil application; it meant *disguised* or *unknown*.

Watson was open to the idea that species were inconstant (Egerton 2003, pp. 147–162), but he had

no better grasp of the mechanics of evolution than his predecessors. He plainly saw plant species as usefully analogous to human individuals. They had places of origin, and by extension, places of belonging. Watson identified a place (the island of Great Britain) and specified that 'native' plant species belonged there by virtue of having appeared, arisen or arrived there without the aid of 'human agency'. Others exhibited lesser, artificial attachments.

Second-generation Zurich phytogeographer Alphonse de Candolle devoted himself to reconciling Genesis with empiricism by expanding his father Augustin's conception of botanical 'centers of creation' (Chew 2006). In 1855, Candolle responded to Watson by publishing his own suite of categories. However, aside from a special emphasis on human intention evidenced by dedicated categories for crops and crop weeds, his conception resembled the Englishman's:

'Cultivated
1. Voluntarily [undefined as if self-evident].
2. Involuntarily: Species which absolutely exist only in the fields, gardens, etc., without being in open country in a spontaneous state.

Spontaneous
1. Adventitious: Of foreign origin, but badly established, being able to disappear from one year to another.
2. Naturalized: Well established in the country, but there is positive evidence of a foreign origin.
3. Probably foreign: Well established ... but according to strong indications, there are more reasons to believe them of origin foreign than primitive in the country ... the odds favoring a foreign origin are better than even.
4. Perhaps foreign: Some indications of a foreign origin, though the species are long and well established in the country. For one reason or another, one can raise some doubts about their indigenousness.
5. Indigenous: Aboriginals, natives ... spontaneous species whose origins are not doubtful, which appear to have existed in the country before the influence

of man, probably for geological rather than historical time.' (Candolle 1855, pp. 642–644; translation from Chew 2006)

Pyšek et al. (2004) traced the native/alien dichotomy to Candolle, but did not discuss the potential implications of basing modern theory and practice on a pre-Darwinian conception. Neither did they note Candolle's citation, discussion and incorporation of Watson's ideas.

In the fourth (mid-1859) volume of his *Cybele Britannica*, H.C. Watson critiqued Candolle's effort and refined his own, providing additional insights into his thinking:

'To the category of *Native Species* we must unavoidably assign all those in regard to which no grounds are now seen for supposing that they were first brought into Britain by human agency. The application of the term is thus simply negative. It can rarely or never be known, whether the species existed in Britain before the advent of mankind, or have immigrated into this country more recently, and if the latter, whether their immigration has been effected by natural means of transport only, as distinguished from those afforded to them by human agency. It is possible that none of these species were aboriginal natives on the present surface of Britain. It may be that all of them were immigrants into the British islands, at different dates, from other lands; those lands, or some of them, having subsequently ceased to be. Such uncertainties belong at present rather to geological, than to geographical botany; and they cannot be here discussed. The broad line of distinction is here to be drawn between natural and human agency; – natural agency being assumed, where human agency is not obvious or suspected.

'In the second category, that of *Introduced Species*, are placed all those which are supposed to have been brought into Britain through the instrumentality of mankind. In some few instances they are known to have been so introduced; notably in the case of some American species, which were unknown in Europe before the discovery and settlement of the Western Continent. In far the majority of instances botanists only infer or suppose that a species has been introduced, because they can detect some remaining indications of human agency in the conditions under which it is still observed to exist here; the inferences suggested by the present conditions being occasionally corroborated by historical or traditional evidences also. But it seems quite within possibility, and even within reasonable probability, that the indications of human agency may have become obliterated in various instances. And if such instances do occur, the plants so situate, although only naturalised aliens, are now unavoidably placed in the same category with the aboriginal natives' (Watson 1859, pp. 65–66).

Watson went on to discuss the relative merits of Candolle's and his own intermediate categories, asserting unequivocally: 'The distinction between native and introduced species is absolute and real; the only difficulty or uncertainty being in a verdict on the matter of fact. There are not degrees of nativity, or degrees of introduction; though there are differences of opinion regarding the evidences in support of either view in reference to individual species' (Watson 1859, p. 68).

By the publication of his fourth *Cybele* volume, Watson was aware of the July 1858 co-debuts of Darwin's and A.R. Wallace's papers on natural selection. Indeed, Darwin's *Origin*, published in November 1859, acknowledged a 'deep obligation' to Watson for 'assistance of all kinds' (Darwin, 1859, p. 48). Watson's book was finalized first; only in a postscript did he allude to the *Origin*: 'If the views of Darwin ... had been earlier explained in print, some change might have been made in ... this volume, where remarks occur on the inequality and permanence of species' (Watson 1859, p. 525).

Darwin sent Watson an advance copy of the *Origin*, and Watson was a quick study. A day before the book went on sale and three before its official publication, Watson wrote 'an enthusiastic letter' to Darwin, speculating on natural selection's implications for human evolution (Burkhardt & Smith 1991, p. 385; Egerton

2003, pp. 191–192). However, the suggestion of human evolution by natural selection, which undermined the nature/artifice divide, failed to dim Watson's enthusiasm for the 'absolute' distinction between his botanical natives and aliens. In 1868, he revisited the categories in detail. 'Native' survived unedited from 1847. To 'denizen' Watson appended 'by human agency, whether by design or by accident'. He elaborated 'colonist' with new examples. 'Casual' replaced 'incognita' but remained a category of ephemeral 'chance stragglers', a subset (like colonists and denizens) of 'alien species ... certainly or very probably of foreign origin; though several ... are now well established amid the indigenous flora of this island; others less perfectly so' (Watson 1868, p. 62).

The practical aspects of Watson's civil model of biotic nativeness remain to be discussed, for it was a model, not an explanation, proposed with specific, limited purposes in mind. Both Watson and Henslow sought to distinguish natural productions from artificial ones, a defining, 'boundary' issue for any natural historian. Each also hoped to discourage what Watson termed 'vainglorious' collectors from artificially establishing discoverable populations (Chew 2006). As an emulator of Humboldt, Watson wanted a 'British' flora whose distributions reliably indicated the natural effects of latitude, altitude, and exposure. He believed only natives were susceptible to such correlation. Undocumented aliens threatened to cloud the picture.

Even though he acknowledged the possibility of extinction (still considered a rare, unlikely phenomenon), Watson (1870, pp. 467–468) did not extend his concerns to conservation, and seems never to have considered suppressing aliens to accommodate natives, much less extirpating aliens as an end in itself. His model of botanical 'civil claims' appreciated natives but did not criminalize aliens.

From these modest, unfamiliar-seeming ambitions, the concept of biotic nativeness has been transmogrified into an obsession of conservationists (Smout 2003) and a pillar of modern ecology. A simple accounting demonstrates that nativeness now stands prominently among ecology's 'title' themes. At mid-20th century, only a handful of articles classified as 'ecology', 'biodiversity conservation' or 'botany' included the term 'native' in their titles. The number did not regularly exceed 10 articles per year until 1980 or 20 per year until 1990, ramping up by nearly an order of magnitude (to 175 per year) by 2007 (ISI Web of Science, http://apps.isiknowledge.com, accessed 3 June 2008).

We extracted another indicator of the prevalence of nativeness in current ecological thinking from the programme of the 10,000+ member Ecological Society of America's 93rd Annual Meeting (Ecological Society of America 2008). An electronic text search one month before the meeting revealed that the combined terms 'nativ*' and 'indig*' (capturing most variants of native and indigenous) occurred 603 times. For comparison, 'ecol*' (ecology, -ogical, etc.) occurred 1685 times; 'effect*' and 'affect*' combined occurred 2016 times, and 'universi*' (university, -ität, -idad, etc., mostly identifying presenter affiliations) occurred 2449 times.

In summary, an olio of ideas from pre-Darwinian botany and pre-Victorian English common law still underpins even the most recent, expert conceptions of biotic nativeness. To the (wide) extent that biotic nativeness is considered actionable and presumed to rest on scientific findings, it is important for scientists, journalists, lawmakers, conservationists and other citizens to understand that those findings express some common beliefs about humans, but nothing about the essences of biota or of particular taxa. How does all this show itself in current science and conservation?

4.3 DIAGNOSING AND APPLYING NATIVENESS

Biotic nativeness is generally diagnosed by time and location indexes that vary with place and purpose. Richardson et al. (2000), Pyšek et al. (2004) and Bean (2007) summarize several such schemes. They share the tradition of distinguishing natives from non-natives by evidence of human intervention and a resulting range expansion. However, what 'human intervention' means to each depends on cultural context. European scientists generally regard human intervention to denote the scale and timing of human impact rather than the fact: 'it is common for an indigenous plant to be regarded as one that was present before the beginning of the Neolithic period (when the widespread growing of crops commenced)' (Bean 2007). In Britain, some make a pre-Roman distinction or identify other temporal thresholds (see, for example, Preston et al. 2004). In the Americas, 'native' usually identifies taxa believed present before 1492, even though there is good evidence of widespread, long-distance trade and plant domestication by pre-Columbian Americans. In Australia, taxa mentioned in the 1770 Banks and Solander flora are considered natives, again despite evidence that plant materials were transported around the Western Pacific and Southeast Asia for millennia before Banks and Solander arrived with Captain Cook aboard *Endeavour*.

Invasion biology has adjusted natural history's strict distinction between nature and artifice to exempt dispersal by pre-commercial, subsistence-level societies, i.e. it is not human agency per se that engenders aliens, but 'civilized' agency; more specifically, European agency wielding complex technologies to exploit far-flung resources. We follow that lead for the remainder of the chapter. Even under this presumption, taxa are still designated *native* purely by default, absent evidence of dispersal by human agency. This is what Watson meant by a 'simply negative' conception. Positive evidence that human agency has never affected a taxon's distribution is hard to come by.

Even thus clarified, the nativeness standard relies on two tacit conceptual transformations. The first takes nativeness to mean a taxon *belongs* where it occurs, geographically, temporally and ecologically. The second takes *belonging* to signify a morally superior claim to existence, making human dispersal tantamount to trespassing. Because human agency is a geologically recent development, this verdict invokes civil rights of prior occupation described earlier. However, neither transformation is deducible from an absence of evidence. Judging *what ought to be* based on *what cannot be demonstrated* seems problematic at best for the purposes of scientific inquiry, recalling the old saw: 'Absence of evidence is not evidence of absence'.

Candolle's primary categorization is likewise tacitly, routinely echoed by invasion biologists who suspend the nativeness criterion for livestock and crops, extending to them rights of occupancy. Here, fulfilled human intentions are treated as essential traits of biota, trumping a defining history of human agency, and even of landscape-scale displacements of natives. (Given the qualitative distinction between subsistence and commercial cultures already described, it seems noteworthy that this 'fulfilled intentions' exemption extends even to international agribusiness.) In contrast, thwarted human intentions obviate any exemption for intentionally introduced species such as common carp (*Cyprinus carpio*) in North America, rabbits (*Oryctolagus cuniculus*) in Australia and muskrats (*Ondatra zibethicus*) in Europe. Iconic instances of recent dispersal *sans* human intention, like the trans-Atlantic segment of the 'remarkable worldwide range expansion' of cattle egrets (*Bulbulcus ibis*) (Telfair 2006) are also exempted

(see, for example, Davis 2009) even though human agency clearly underlies this species' subsequent pattern of American establishment (Telfair 2006).

'Simply negative' native belonging is deployed as a moral imperative for augmenting dwindling populations or re-establishing extirpated ones, to the detriment of (sometimes through detriment to) robust alien populations. That paraphrases Ross and Walsh (2003, quoted earlier) and returns us to evaluating their statement. Favouring natives over aliens as they propose would justify returning Europe to a pre-Neolithic condition, and the Americas to a pre-Columbian one. Such proposals (for example Popper & Popper 1987; Donlan et al. 2006) have achieved notoriety but little traction. However, there are situations where conservation biologists have espoused a more limited version of such a view and attempted its implementation.

The fascinating case of pool frogs (*Pelophylax* (formerly *Rana*) *lessonae*) in Britain represents such an effort. *Pelophylax* species (collectively known as 'water frogs') occur across Europe, and it was long assumed that all pool frogs in Britain were descended from central European animals introduced in the 1800s (Williams & Griffiths 2004). They fulfilled a human desire and thwarted no other, and thus enjoyed the Candolle exemption until the advent of nativeness-based conservation genetics. Genetic analysis indicates that two disjunct, localized Scandinavian pool frog populations, morphologically unremarkable but characterized by a minor vocal 'inflection' are distinguishable at the molecular level from central European populations. This finding was conceptually transformed to signify that different clades of pool frogs belong in different regions of Europe (Beebee et al. 2005).

In 2005 Beebee and coworkers examined a lone male *P. lessonae* specimen that was captured in Norfolk and kept captive until its death in about 1998 (Jim Foster, English Nature, personal communication). This frog, they concluded, sprang not from central European stock, but from the northern (Scandinavian) clade. Archaeologists contributed two pre-12th century bone fragments (from a pool of some 10,000 unhelpful frog bones) morphologically attributable to *P. lessonae*. Lacking evidence for introductions from Scandinavia, Beebee et al. (2005) concluded that 'northern clade' frogs were British natives, and had been extirpated with the death of their test specimen.

After much investigation and planning, conservation ecologists were authorized to release Swedish frogs in Norfolk. Citing a lack of appropriate habitat, they undertook site alterations (tree clearing, pond dredging, artificial water supplies, even pond creation) to favour the return of the 'natives'. In addition, British legal protections were extended to the ex-Scandinavian frogs but specifically denied to the other populations of pool frogs already on the island (Buckley & Foster 2005). Frogs from different *P. lessonae* 'clades' look very much alike, and as far as anyone knows, are ecologically interchangeable. They freely interbreed, and no subgroup has been called uglier, identified as a pest or a disease vector, or as in any way threatening to any other interest. The only reason for preferring the northern clade is a determination of nativeness, but *how* does it provide that reason?

In this case nativeness itself is being treated as both an essential trait and an inherently valuable characteristic of a taxon (but not of its members, which could be experimented with.) This ordinary case embodies an extraordinary logic that requires examining: just what are native taxa, and why should we prefer them? What is it about a finding of nativeness that would move British conservationists to accept the costs of (oxymoronically) creating a new 'native' population from Scandinavian imports when the move confers no further advantages? What hope, expectation or inherent value motivates us to intervene in the name of nativeness?

4.4 BELONGING IN PLACE

This section examines whether and how biotic nativeness accomplishes the theoretical work demanded of it. Nativeness is applied as though a characteristic or trait of a taxon or population, but (as we have seen) it is not diagnosed that way. Once diagnosed, it catalyses conversion of 'rights' of prior (even former) territorial occupation into rights of future occupation. In civil contexts this works tolerably well, if all parties respect the rule of law. It works less well where claims and authorities compete, long-displaced peoples assert long-disputed rights, or current and prior possessors alike claim moral authority. The question germane to biotic nativeness is whether putative natives are contextually (ecologically and geographically) more correct, and thus due some superior consideration. Do plants and animals belong in some places and not others? How can we tell? In any given case, nativeness might be used to assert one or more of the following:

1 A connection between the origin of a species or other taxon and a place. For example, 'This speciation centre [the southeastern part of the Indo-Turanian Centre] harbours nearly 50% of the species of *Tamarix* ... One of its southeastern corners may have been the cradle of the genus' (Baum 1978).
2 Long-standing occurrence (i.e. tenure) in place. For example, '*Castanea sativa* is probably a native species of central and northern Italy' (Tinner et al. 1999).
3 Evolutionary or ecological relationships between biota. Rosen (2000) suggested a co-evolutionary relationship between taxa of some dinoflagellate algae and scleractinian (broadly, reef-forming) corals might have prevailed since the Triassic.

It should go without saying that the biological world differs from the civil world in important respects. Two notorious cases illustrate this point. The northern snakehead (*Channa argus*) is a target of extirpation efforts in the USA, mainly because it is considered alien (Chew & Laubichler 2003). Officials posted a notice asking 'Have you seen this fish?' reading, in part: 'the Northern Snakehead from China is not native to Maryland and could cause serious problems if introduced into our ecosystem' (Derr & McNamara 2003, p. 127). Presumably, a similar objection could have been voiced when channids arrived in China from India or Southeast Asia during the Pleistocene. Episodes accrue through several iterations: there are fossil channids in Switzerland and France dating from the Oligocene (Reichenbacher & Weidmann 1992). Even at the species level, it seems likely that the most common channids in China migrated (or were derived) from species that originated further south (Courtenay & Williams 2004; but see Banerjee et al. 1988). We are not postulating an Oligocene France or a Pleistocene China. We are arguing that with or without human agency, ranges are dynamic. Claims about nativeness like that in the USA 'unwanted poster' are presented as though they adequately capture simple facts, but they mask important complexities. 'Native' does not mean much without a great deal of interpretation. Do snakeheads *belong* in China? How can a fish demonstrate belonging other than by being, surviving and persisting *here*, *now*, any of which probably exceed its awareness of the issues at hand?

One might similarly ask whether any species of the habitually maligned genus *Tamarix* belongs in the American west. Several species of the shrubs were introduced to the USA from Old World locations during the 19th and early 20th centuries (Chew 2009).

Widely planted, they thrived and spread along riparian corridors and reservoir edges. Three species (*T. chinensis, T. ramosissima, T. aphylla*) not known to hybridize where their Old World ranges overlap are doing so in the USA, producing unprecedented, fertile hybrids (Gaskin & Schaal 2002; Gaskin & Shafroth 2005). These lines may constitute new species; hypothetical *Tamarix americana** (a taxonomist's asterisk) paradigmatically native because they evolved within their current ranges and endemic because they exist nowhere else.

We doubt that any of the interests currently devoted to eradicating *Tamarix* in the USA will quail at the suggestion that they are attacking a native species. They can reply, *sensu* Pyšek et al., that the progeny of non-natives are themselves non-native, endemic or otherwise. However, that undermines the distinction licensing eradication in the first place. Where does any *Tamarix* belong? Is it native only to the spot where it first evolved, or where the genus arose? Palaeobiogeography is rife with redistributions and speciations that generated competition for space and resources, the stuff of natural selection (Hall 2003).

Arguments relying on place suggest that places are reliable. Most ecologists likely accept that in the very long term of geological processes, places are unreliable. To appropriate a sentiment from Gertrude Stein, it is also sometimes catastrophically the case that 'there is no *there*, there'. Suddenly or gradually, globally or locally, owing to asteroid impacts, axis wobbles, earthquakes or ice sheets, places with one suite of characteristics are 're-placed' by the advent of another suite. Some 10,000 years ago, iconic elements of the Sonoran Desert flora and fauna including saguaro cacti (*Carnegiea gigantea*), creosote bush (*Larrea tridentata*) and presumably desert tortoises (*Gopherus agassizi*) arrived in what we now call Arizona and supplanted the region's former occupants (see Anderson & Van Devender 1991). Does that sanction an attempt to restore piñon–juniper woodland? To what end?

It is widely asserted that ongoing anthropogenic replacement is a more precipitous process than 'natural' change, as if that adequately explained the native–alien distinction, but it does not. It suggests something about human capacities to generate, understand and deplore particular aspects of change, but reveals nothing essential about biota. Biota persist or not under prevailing conditions, regardless (and presumably ignorant) of how they arose and whether they will change further.

These problems can be solved for certain purposes by stipulating which taxa are native to certain places at certain times and then stipulating what is meant by *place*. This merely shifts the problem to defending the stipulations. Nativeness seems to be unreadable from purely biological or geographical details, undermining any argument that particular taxa *belong* in particular places. Today's native may have displaced some previous native when it arrived. Should we extirpate more recent natives and replace them with less recent ones? Where does the regression stop? Is it simply a matter of expedience? How do we know when we have got back to 'nature's original plan'? (International Association of Fish and Wildlife Agencies 2005).

4.5 BELONGING TOGETHER

We have argued that places and taxa change in ways that make ascriptions of nativeness unstable and uninformative. One might reply that such a view is artificially synchronic. Perhaps events in particular places at particular times matter less than the history of biotic relationships expressed as co-evolution. One obvious rejoinder is that system boundaries are products of particular theories and empirical approaches and therefore not uniquely determinable. Ever-longer views exempt ever more changes leading to ever more indefensible lines. However, this is to miss the heart of the objection: is it not ongoing interdependencies – symbiosis in the broadest sense – that at least some of those who privilege native species are concerned with? Is this what motivates accusing *Tamarix* of 'stealing' water (for example Robinson 1952) or degrading wildlife habitats (for example De Loach et al. 1999)?

Tamarix establishment follows and generates changes to ecological processes, but even the most precise measurement of the most unprecedented changes cannot stand as scientific evidence that those changes should not have occurred. In the most inclusive view, large-scale replacements of some taxa by others are commonplace. How inclusive should ecology's view be?

Taking the least inclusive view, scientists can declare that *Tamarix* is, for any expressible reason, less desirable in an ecosystem than some alternative. We can then promote replacing tamarisk with an alternative. Neither calls for an assertion of nativeness; from scientists it requires primarily 'a lever, and a place to stand' that will facilitate the desired changes.

In a middling, more inclusive view, we can tell the story of taxa that suffered unjustly and declined when human agency inflicted *Tamarix* on them, usurping their places and resources. We can then argue that the prior occupants should be reinstated regardless of their own undocumented (but pre-modern) origins, perhaps even at the high cost of perpetual intervention to suppress the interloper. Science needs to enter this equation first forensically, to sort perpetrators from victims; then, as before, to identify methods for accomplishing our design.

In the most inclusive view, we can see *Tamarix americana** as a taxon demonstrating high fitness under prevailing conditions by replacing and perhaps even displacing a less fit flora that happened to arrive first. Perhaps that earlier flora out-competed another, and then another. Such are the fates of less fit taxa. After all, the American west has not always been American, or even a west. Under this assumption, scientists can read the landscape, describe the processes underway, and hope to elucidate how and why they happened and where they may lead. In doing so, they adhere to science's traditional ideal of maximizing objective knowledge production, rather than facilitating a preferred outcome.

In asking which of these approaches is 'correct', we clearly have a challenge. The first invokes proximate desires, privileging whatever serves best for now, regardless of origins or displacements. This is the basic story of agriculture, horticulture, urbanization and even strip-mining, all of which produce economically rational short-term goods. It dominates the debate, seeming 'to hang over the whole world' as the imperative to meet basic human needs (Elton 1958, p. 144).

The second view deplores *Tamarix* because it harms 'natives' without concern for how natives became native. It rests on a *mis*-anthropic principle of denaturing taxa associated with denatured humans, a struggle of purity against contamination. It asks scientists only for endorsement, but any scientific rationalization broadens its effectiveness by reinforcing sympathetic feeling with intellectual authority. It occupies a qualitatively comfortable middle space between seemingly stark economic rationality (pure need) and seemingly stark intellectual disinterest (no need at all). It is more likely to ally itself with the wholly pragmatic former. In the case of *Tamarix*, those who deplore it as an alien and those who believe it harms their economic interests have formed the 'Tamarisk Coalition', which would be more aptly named the 'anti-Tamarisk Coalition' (Chew 2009).

The third view eschews superficial value judgment and concentrates on ecology. In today's American west, *Tamarix* has evidently landed in extremely favourable habitat. Conditions will not remain so favourable. Over time, and perhaps quite quickly, something will discover, develop or evolve a taste for *Tamarix*. (Coalition interests, led by the US Department of Agriculture, have imported a *Tamarix*-defoliating insect to the region). The iteratively revised or self-revising ecosystem may approach an equilibrium that omits some of the taxa now present. This is theoretically fertile but culturally sterile ground. The story has no monsters, no heroes, no resolution and no moral.

Telling multiple stories demonstrates that ecological relationships, places, taxa and human motivations are all plastic. Places change, taxon boundaries shift, symbioses wax and wane, values and inclinations vary. Better and worse depend on the standard of comparison in the frame. Instead of nativeness we might prefer novelty, diversity, rarity or beauty. If we decide to privilege natives, we must still decide which natives to privilege most. Science can inform, and take a role in writing any of these stories. It must do so if our desire is to impose a particular outcome or condition.

4.6 CONCLUSIONS: BELONGING BELIED

It is remarkably easy to unravel the conception of biotic nativeness. We have argued that the idea is recursive, and that dynamics of places, taxa, ecologies and desires render the label *native* uninformative, even deceptive. Carefully limiting it only reveals that our habit of preferring natives to aliens is poorly founded.

None of the relationships comprising biotic nativeness is an inevitable, permanent or dependable object supporting a conception of belonging. Furthermore, the fact that some organisms thrive 'out of context' mocks rights-based contextual propriety. Yet preference for nativeness permeates ecological thinking, supporting a multi-hundred article-per-year publishing effort. Ecologists can demonstrate that in a relatively small (if quite noticeable) subset of interactions, 'aliens' demonstrate fitness superior to 'natives'; but we cannot explain – in biological terms – how inferior fitness is consistent with superior belonging.

The problem lies not with inadequate terminology or definitions. Many a scientist has attempted and ultimately abandoned categorizing since John Henslow stuck his dagger into the asterisk. Nativeness is a living fossil of an outmoded phytogeography, conceived during the heyday of amateur natural history while a young Darwin explored South America, and fully elaborated before he described natural selection. Nativeness senesced with creationism before ecology or genetics or much else that constitutes modern biology even began, and none of those developments offer to reinvigorate it.

The real crux of the matter is revealed in the publication and presentation statistics discussed in section one, in light of the obvious weakness of biotic nativeness. The dominant 'theoretype' of today's ecological science evolved to reflect the relatively high institutional, organizational, cultural and fiscal fitness of explaining problems in media-friendly terms: misanthropy, misoneism, injustice and displacement. Nativeness is the easy way in. It has been (and will be) fiercely defended and endlessly massaged because it is comfortable and confers advantages; but there is no easy way out. Accommodating nativeness is hampering progress in ecological science. Abandoning nativeness, and with it the hope of belonging, will be costly to those who are overinvested in it. Without nativeness, the ecological past offers us data, but not counsel.

AUTHORS' NOTE

This chapter does not critique civil conceptions of nativeness applied to human individuals, and we consider it inapplicable in that regard.

ACKNOWLEDGEMENTS

We thank participants in the 2005 Southwest Colloquium in the History and Philosophy of the Life Sciences at Arizona State University, participants in this volume's namesake symposium, two anonymous reviewers and many colleagues and students for their valuable feedback and suggestions.

REFERENCES

Aitken, G. (2004) *A New Approach to Conservation: The Importance of the Individual through Wildlife Rehabilitation.* Ashgate, Aldershot.

Alonso, W. (1995) Citizenship, nationality, and other identities. *Journal of International Affairs*, **48**, 585–599.

Anderson, R.S. & Van Devender, T.R. (1991) Comparison of pollen and macrofossils in packrat (*Neotoma*) middens: a chronological sequence from the Waterman Mountains of southern Arizona, USA. *Review of Palaeobotany and Palynology*, **68**, 1–28.

Anonymous (1843) Report on the laws affecting aliens. *The Legal Observer, or Journal of Jurisprudence*, **26**, 385.

Banerjee, S.K., Misra, K.K., Banerjee, S. & Ray-Chaudhuri, S.P. (1988) Chromosome numbers, genome sizes, cell volumes and evolution of snake-head fish (family Channidae). *Journal of Fish Biology*, **33**, 781–789.

Baum, B.R. (1978) *The Genus* Tamarix. Israeli Academy of Sciences and Humanities, Jerusalem.

Bean, A.R. (2007) A new system for determining which plant species are indigenous in Australia. *Australian Systematic Botany*, **20**, 1–43.

Beebee, T.J.C., Buckley, J., Evans, I., et al. (2005) Neglected native or undesirable alien? Resolution of a conservation dilemma concerning the pool frog *Rana lessonae*. *Biodiversity and Conservation*, **14**, 1607–1626.

Blackstone, W. (1922) *Commentaries on the Laws of England.* Bisel, Philadelphia.

Brown, J.H. & Sax, D.F. (2005) Biological invasions and scientific objectivity: reply to Cassey, et al. *Austral Ecology*, **30**, 481–483.

Buckley, J. & Foster, J. (2005) *Reintroduction Strategy for the Pool Frog Rana lessonae in England.* English Nature, Peterborough, UK.

Burkhardt, F. & Smith, S. (1991) *The Correspondence of Charles Darwin: 1858–59*, vol. **7**. Cambridge University Press, Cambridge, UK.

Candolle, A. De (1855) *Géographie Botanique Raisonnée.* Victor Masson, Paris.

Chew, M.K. (2006) *Ending with Elton: Preludes to Invasion Biology.* PhD thesis, Arizona State University, Tempe.

Chew, M.K. (2009) The monstering of tamarisk: how scientists made a plant into a problem. *Journal of the History of Biology*, **42**, 231–266.

Chew, M.K. & Laubichler, M.D. (2003) Natural enemies – metaphor or misconception? *Science*, **301**, 52–53.

Coates, P. (2003) Editorial postscript: naming strangers in the landscape. *Landscape Research*, **28**, 131–137.

Colautti, R.I. & MacIsaac, H.J. (2004) A neutral terminology to define 'invasive' species. *Diversity and Distributions*, **10**, 135–141.

Courtenay, W.R. & Williams, J. D. (2004) *Snakeheads (Pisces, Channidae)–A Biological Synopsis and Risk Assessment.* Circular 1251. US Geological Survey, Denver.

Curtis, W. (1783) *A Catalogue of the British, Medicinal, Culinary and Agricultural Plants Cultivated in the London Botanic Gardens.* B. White, London.

Darwin, C. (1859) *On the Origin of Species by Means of Natural Selection, or the Preservation of Favoured Races in the Struggle for Life.* John Murray, London.

Davis, M.A. (2006) Invasion biology 1958–2005. In *Conceptual Ecology and Invasion Biology: Reciprocal*

Approaches to Nature (ed. M.W. Cadotte, S.M. McMahon and T. Fukami), pp. 35–64. Springer, Dordrecht.

Davis, M.A. (2009) *Invasion Biology*. Oxford University Press, Oxford.

De Loach, C.J., Carruthers, R.I., Lovich, J.E., Dudley, T.L. & Smith, S.D. (1999) Ecological interactions in the biological control of saltcedar (*Tamarix* spp.) in the United States: toward a new understanding. *Proceedings of the X international symposium on biological control of weeds* 4–14 July 1999, Montana State University, Bozeman.

Derr, P.G. & McNamara, E.M. (2003) *Case studies in environmental ethics*. Rowman & Littlefield, Lanham.

Donlan, C.J., Berger, J., Bock C.E., et al. (2006) Pleistocene rewilding: an optimistic agenda for twenty-first century conservation. *American Naturalist*, **168**, 660–681.

Ecological Society of America (2008) *ESA 93rd annual meeting schedule*. http://eco.confex.com/eco/2008. Accessed 8 July 2008.

Egerton, F.N. (2003) *Hewett Cottrell Watson: Victorian Plant Ecologist and Evolutionist*. Ashgate, Aldershot.

Elton, C.S. (1958) *The Ecology of Invasions by Animals and Plants*. Methuen, London.

Gaskin, J. & Schaal, B. (2002) Hybrid tamarix widespread in U.S. invasion and undetected in native Asian range. *Proceedings of the National Academy of Sciences of the USA*, **99**, 11256–11259.

Gaskin, J. & Shafroth, P. (2005) Hybridization of *Tamarix ramosissima* and *T. chinensis* (saltcedars) with *T. aphylla* (Athel) (Tamaricaceae) in the southwestern USA determined from DNA sequence data. *Madroño*, **52**, 1–10.

Gobster, P.H. (2005) Invasive species as ecological threat: is restoration an alternative to fear-based resource management? *Ecological Restoration*, **23**, 261–270.

Hall, M. (2003) Editorial: the native, naturalized and exotic – plants and animals in human history. *Landscape Research*, **28**, 5–9.

Halleri, D.A. (1742) An account of a treatise, 'Archiatri regii & elect. medicin. anatomiae, botan. praelect. &c. enumeratio methodica stirpium helvetiae indigenarum. qua omnium brevis descriptio & synonymia, compendium virium medicarum, dubiarum declaratio, novarum & rariorum uberior historia & icones continentur. Gottingiae' Extracted and Translated from the Latin by William Watson, F.R.S. *Philosophical Transactions of the Royal Society of London*, **42**, 369–380.

Henslow, J.S. (1835) Observations concerning the indigenousness and distinctness of certain species of plants included in the British floras. *The Magazine of Natural History*, **8**, 84–88.

International Association of Fish and Wildlife Agencies (2005) Strange days for invasive species. *Inside IAFWA*, January 2005.

Klein, H. (2002) *Weeds, Alien Plants and Invasive Plants*. PPRI Leaflet Series: Weeds Biocontrol, No.1.1. ARC-Plant Protection Research Institute, Pretoria.

Larson, B.M.H. (2007) An alien approach to invasive species: objectivity and society in invasion biology. *Biological Invasions* **9**, 947–956.

Milton, K. (2000) Ducks out of water: nature conservation as boundary maintenance. *Natural enemies* (ed. J. Knight) pp. 229–248. Routledge, London.

Popper, D.E. & Popper, F.J. (1987) The Great Plains: from dust to dust. *Planning*, **53**(12),12–18.

Preston, C.D., Pearman D.A. & Hall, A.R. (2004) Archaeophytes in Britain. *Botanical Journal of the Linnean Society*, **145**, 257–294.

Pyšek, P., Richardson, D.M., Rejmánek, M., Webster, G.L., Williamson, M. & Kirschner, J. (2004) Alien plants in checklists and floras: towards better communication between taxonomists and ecologists. *Taxon*, **53**, 131–143.

Pyšek, P., Richardson, D.M., Pergl, J., Jaroš, V., Sixtova, Z. & Weber, E. (2008) Geographical and taxonomic biases in invasion ecology. *Trends in Ecology & Evolution*, **23**, 237–244.

Reichenbacher, B. & Weidmann, M. (1992) Fisch-otolithen aus der oligo-miozänen molasse der west-schweiz und der haute-savoie (Frankreich). *Stuttgarter Beitr Naturkunde B*, **184**, 1–83.

Richardson, D.M., Pyšek, P., Rejmanek, M., Barbour, M.G., Panetta, F.D. & West, C.J. (2000) Naturalization and invasion of alien plants: concepts and definitions. *Diversity and Distributions*, **6**, 93–107.

Robinson, T.W. (1952) Water thieves. *Chemurgic Digest*, **11**, 12–15.

Rosen, B.R. (2000) Algal symbiosis and the collapse and recovery of reef communities: Lazarus corals across the K–T boundary. In *Biotic Response to Global Change: The Last 145 Million Years* (ed. S.J. Culver and P.F. Rawson), pp. 164–180. Cambridge University Press, Cambridge, UK.

Ross, J.H. & Walsh, N.G. (2003) *A Census of the Vascular Plants of Victoria*, 7th edn. National Herbarium of Victoria, Royal Botanic Gardens, Melbourne.

Sagoff, M. (2002) What's wrong with exotic species? In *Philosophical Dimensions of Public Policy* (ed. V. Gehring and V.W. Galston), pp. 327–340. Transaction, New Brunswick.

Sagoff, M. (2005) Do non-native species threaten the natural environment? *Journal of Agricultural and Environmental Ethics*, **18**, 215–236.

Simberloff, D. (2005) Non-native species DO threaten the natural environment! *Journal of Agricultural and Environmental Ethics*, **18**, 595–607.

Smout, T.C. (2003) The alien species in 20th-century Britain: constructing a new vermin. *Landscape Research*, **28**, 11–20.

Stromberg, J.C., Chew, M.K., Nagler, P.L. & Glenn, E.P. (2009) Changing perceptions of change: the role of scientists in tamarix and river management. *Restoration Ecology*, **17**, 177–186.

Subramaniam, B. (2001) The aliens have landed! Reflections on the rhetoric of biological invasions. *Meridians: Feminism, Race, Transnationalism*, **2**, 26–40.

Telfair, R.C. (2006) *Cattle egret (Bubulcus ibis). The Birds of North America Online* (ed. A. Poole). Cornell Lab of Ornithology, Ithaca. http://bna.birds.cornell.edu/bna/species/113. Accessed 26 August 2009.

Theodoropoulos, D. (2003) *Invasion Biology: Critique of a Pseudoscience*. Avvar, Blythe.

Tinner, W., Hubschmid, P., Wehrli, M., Ammann, B. & Conedera, M. (1999) Long-term forest fire ecology and dynamics in southern Switzerland. *Journal of Ecology*, **87**, 273–289.

Townsend, M. (2005) Is the social construction of native species a threat to biodiversity? *ECOS*, **26**, 1–9.

Trigger, D., Mulcock, J., Gaynor, A. & Toussaint, Y. (2008) Ecological restoration, cultural preferences and the negotiation of 'nativeness' in Australia. *Geoforum*, **39**, 1273–1283.

Warren, C.R. (2007) Perspectives on the 'alien' and 'native' species debate: a critique of concepts, language, and practice. *Progress in Human Geography*, **31**, 427–446.

Watson, H.C. (1835a) *Remarks on the Geographical Distribution of British Plants*. Longman, London.

Watson, H.C. (1835b) *The New Botanist's Guide to the Localities of the Rarer Plants of Britain*. Longman, London.

Watson, H.C. (1847) *Cybele Britannica*, vol. 1. Longman, London.

Watson, H.C. (1859) *Cybele Britannica*, vol. 4. Longman, London.

Watson, H.C. (1868) *Compendium of the Cybele Britannica*, vol. 1. Privately distributed, Thames Ditton.

Watson, H.C. (1870) *Compendium of the Cybele Britannica*, vol. 3. Privately distributed, Thames Ditton.

Williams, C. & Griffiths, R.A. (2004) A population viability analysis for the reintroduction of the pool frog (*Rana lessonae*) in Britain. English Nature Research Report no. 585.

Part 2

Evolution and Current Dimensions of Invasion Ecology

Chapter 5

PATTERNS AND RATE OF GROWTH OF STUDIES IN INVASION ECOLOGY

Hugh J. MacIsaac[1], Rahel A. Tedla[1] and Anthony Ricciardi[2]

[1]Great Lakes Institute for Environmental Research, University of Windsor, Windsor, Ontario, Canada
[2]Redpath Museum, McGill University, Montreal, Quebec, Canada

Fifty Years of Invasion Ecology: The Legacy of Charles Elton, 1st edition. Edited by David M. Richardson
© 2011 by Blackwell Publishing Ltd

5.1 INTRODUCTION

Human activities are resulting in the spread of many non-indigenous species to habitats in which they are non-native. Introductions appear to be widespread and increasing in marine, freshwater and terrestrial ecosystems alike, with even remote Arctic and Antarctic ecosystems affected (Ruiz et al. 2000; Pyšek et al. 2006; Ricciardi 2006; Aronson et al. 2007). The association between climate change and species invasion suggests that the rising invasion rate will not abate anytime soon (Reid et al. 2007; Cheung et al. 2009). In addition, enhanced globalization of trade provides the means by which non-indigenous species may spread (Hulme 2009). The consequences of the increasing rate of invasion will include an array of ecological, economic and health impacts for invaded countries (Levine & D'Antonio 2003; Pimentel et al. 2005; Reaser et al. 2007; Perrings, this volume).

Charles Elton published his famous volume in 1958, yet more than two decades elapsed before biological invasions became a popular academic pursuit (Richardson & Pysek 2008; Simberloff, this volume). For example, we were able to tabulate publication of seven books on biological invasions between 1959 and 1979, and at least 20 more were published during the 1980s (H.J. MacIsaac, unpublished data). Simberloff (this volume) discusses the role of the Scientific Committee on Problems of the Environment (SCOPE) in raising interest in invasions. Journal publications on non-indigenous species began in earnest around 1990 (Davis et al. 2001; Lockwood et al. 2007). Elton's (1958) legacy to the field is a high citation rate (in more than 40% of papers) in current literature on papers addressing topics developed in his book, including the influence on invasion success of disturbance and native species diversity and stability, and on differences between islands and continents in their vulnerability to the occurrence and impacts of invasion (Richardson & Pyšek 2008; Ricciardi & MacIsaac 2008). Journal publications related to non-indigenous species increased dramatically during the early 1990s (Lockwood et al. 2007; Richardson & Pyšek 2008; Ricciardi & MacIsaac 2008), though it is not clear how this growth is distributed across taxonomic lines.

This chapter explores trends in the modern (since Elton 1958) growth of publications that address non-indigenous species, and tests three hypotheses:

1 Animals accumulate studies earlier than plants. Rationale: animals tend to have effects that are more conspicuous, thus they are likely to be studied earlier.

2 Terrestrial species accumulated studies earlier than aquatic ones. Rationale: firstly, terrestrial organisms are more conspicuous and easier to detect than most aquatic NIS. Secondly, the relative importance of vectors transmitting these groups has changed. The rate of increase for aquatic animals is expected to initially lag behind that of terrestrial animals, consistent with the early historical prevalence of introductions of land mammals and birds, whereas the growing importance of ballast water has driven a rapid increase in aquatic invasions in recent years (see, for example, Ricciardi 2006). Finally, the perception of the consequences of intentional introductions has changed.

3 Aquatic invertebrates accumulate studies later than fishes. Rationale: both the relative importance of vectors transmitting these groups and the perception of consequences of intentional introductions have changed, resulting in especially lower numbers of fish introductions. Fishes have been stocked into systems well before invertebrates were introduced to systems in an attempt to enhance fisheries. Furthermore, documentation of widespread introductions of aquatic invertebrates appears to be driven partly by heightened interest in high-profile ballast-water mediated invasions after 1980 by species including the zebra mussel *Dreissena polymorpha* and comb jelly *Mnemiopsis leidyi*.

We address each of these hypotheses using the ISI Web of Science to track publications in the literature.

5.2 METHODS

We used Thomson's ISI (Institute for Science Information) Web of Knowledge 4.0 to search for and track publications in scientific literature for '100 of the World's Worst Invasive Alien Species' as determined by the World Conservation Union (IUCN; Lowe et al. 2004). We chose these 100 species in an attempt to explore objectively publication patterns for different taxonomic groups. As implied by the name, these species represent many of the most invasive (i.e. problematic) non-indigenous species found in a diversity of habitats around the world, and thus we expected that they would draw researchers' attention. It is important to note, however, that many other highly invasive non-indigenous species were not included on this list (e.g. spiny waterflea *Bythotrephes longimanus*).

Species included on the IUCN list were selected based on two criteria: (i) the significance of invader impacts on either humans or biological diversity; and (ii) their illustration of important issues surrounding biological invasions (Lowe et al. 2004). The IUCN list includes a broad representation of biodiversity including three microorganisms, five macro-fungi, four aquatic plants, 30 terrestrial plants, nine aquatic invertebrates, 17 terrestrial invertebrates, three amphibians, eight fishes, three birds, two reptiles and 14 mammals (Lowe et al. 2004).

For all 100 species, we searched using common and scientific names in the topic box for the period 1965 to 2007 inclusive. We used the Scientific Citation Index Expanded (SCI-EXPANDED), and excluded the Social Sciences Citation Index and the Arts and Humanities Citation Index. ISI Web of Knowledge tracks journals that are written in English and certain principally northern hemisphere languages including Spanish, German, French, Portuguese, Czech, Dutch, Russian, Polish, Turkish, Japanese, Chinese, Swedish, Norwegian, Lithuanian, Slovak, Estonian, Finnish, Hungarian, Slovene, Korean and Serbian.

We conducted two sets of searches for all species in May 2008. The first set consisted of the species' scientific name or common name – as per the IUCN report (Lowe et al. 2004) – in the topic box. Search topics were then refined by subject area, and only field topics relevant to ecology, broadly-speaking, were used. These field topics included ecology, biodiversity conservation, environmental sciences, plant sciences, zoology, limnology and entomology; the same set of field topics were used for all species (Appendix 5.1). Biomedical fields that seemed unrelated to ecology – such as veterinary sciences and immunology – were excluded. Results were tabulated by year, publication count and cumulative numbers of publications for each species. Publication trends were analysed only if the species had accumulated at least one paper in each of a minimum of 6 years. We compared cumulative number of publications of animal versus plant species, of terrestrial versus aquatic species, and of aquatic invertebrates versus fishes, using t-tests and $\log(x + 1)$ data for the period 1965 to 2007. Amphibians were excluded from analysis of terrestrial versus aquatic species since they inhabit both habitats, and the red-eared slider turtle was included with the aquatic group.

To assess only papers that explicitly considered the species as non-indigenous, a second set of searches was conducted that combined the above species' scientific or common name with six additional synonyms (non-native species, alien species, exotic species, introduced species, colonizing species, or exotic species). The same procedures and conditions that were applied for the first set of searches were applied to these combined searches.

We modelled the cumulative number of publications between 1965 and 2007 for each using nonlinear regression in Systat 12.0. Early studied species were considered those that accumulated at least 25% of total publications occurred before 1990. Cumulative growth was fitted as an exponential equation (cumulative publications = α year$^{\beta}$) where α and β were fitted variables and year is the Julian year with 1965 as year 1. The coefficient α and exponent β relate to how early the species was studied and how rapidly the species accumulated publications, respectively. Differences in mean $\log(\alpha + 1)$ and in β were assessed for plants versus animals, for terrestrial versus aquatic species, and for aquatic invertebrates versus fishes were assessed using t-tests. Again, we excluded amphibians from analyses, and the red-eared slider was considered aquatic. It is important to note that growth patterns for all species are assessed versus a 1965 starting point. This allowed us to determine temporal differences in accumulations of studies for different taxa. Taxa with the same relative growth of papers (as a function of the total for that species) could have different α and β values depending on when these publications first started appearing in the literature. Thus our analysis assesses comparative growth relative to 1965, and not whether individual species' accumulation curves differ after the species begins to appear in the literature. We also calculated the time from 1965 that was required to achieve the 25th percentiles for the total number of publications.

5.3 RESULTS AND DISCUSSION

The cumulative publication rate was readily fit using an exponential function for most species, which is not unexpected (Wonham & Pachepsky 2006). However, the Kaphra beetle, a pest species of stored grain, accumulated a relatively large number of papers between the late 1960s and mid-1970s followed by a much reduced rate in later years. This resulted in an asymptotic accumulation pattern and fitted function that had an extremely high α and very low β. Because of the unusual nature of this fitted function, we removed this beetle from further statistical analyses. In addition, curves could not be fit for the fishhook waterflea *Cercopagis pengoi*, Asian longhorn beetle *Anoplophora glabripennis* and yellow crazy ant *Anoplolepis gracilipes*. In each case, studies have been published so recently – mainly since 2000 – that exponential curves could not be fitted. Fishhook fleas originated in the Black Sea, but were first reported in the Baltic Sea and in the Great Lakes in the late 1990s (see Cristescu et al. 2001). The Asian longhorn beetle is a wood-boring insect of quarantine significance that has been found at a variety of North American (e.g. New York, Chicago, Toronto) and European (Braunau, Austria; Gien and Sainte-Anne-sur-Brivet, France; Neukirchen am Inn, Germany) cities (e.g. Haack 2006). The yellow crazy ant has been introduced to parts of Africa, Asia and to islands in the Pacific and Indian Oceans and the Caribbean Sea, where it may exert strong negative impacts on resident species (see, for example, O'Dowd et al. 2003).

Fifteen per cent of the species on the '100 of the World's Worst Invasive Alien Species' list were studied very early. These species include an array of notorious invaders including the common carp *Cyprinus carpio*, bullfrog *Rana catesbeiana*, cane toad *Bufo marinus*, ship rat *Rattus rattus*, starling *Sturnus vulgaris*, phytophthora root rot *Phytophthora cinnamomi* and Dutch elm disease *Ophiostoma ulmi*. Indeed, Dutch elm disease accumulated papers very quickly, recording the 25th and 50th percentiles of its total number between 1982 and 1992, respectively. A second group, constituting approximately half of the species on the list, were studied to some extent before 1990 but very well thereafter. This group includes infamous invaders including water hyacinth *Eichhornia crassipes*, Nile perch *Lates niloticus*, comb jelly *Mnemiopsis leidyi*, little fire ant *Wasmannia auropunctata*, gypsy moth *Lymantria dispar*, leucaena *Leucaena leucocephala* and avian malaria *Plasmodium relictum*. A third group of species were studied exclusively or almost so after 1990. These species include the zebra mussel, green alga *Caulerpa taxifolia*, brown tree snake *Boiga irregularis*, rosy wolf snail *Euglandina rosea*, fire tree *Morella faya*, kudzu *Pueraria montana* var. *lobata*, Brazilian pepper tree *Schinus terebinthifolius* and chestnut blight *Cryphonectria parasitica*. Species that have only recently been studied include the aforementioned fishhook water-flea, Asian longhorn beetle, and frog chytrid fungus *Batrachochytrium dendrobatidis*. Poorly studied species

include the yellow crazy ant, Indian myna bird *Acridotheres tristis*, small Indian mongoose *Herpestes javanicus* and mile-a-minute weed *Mikania micrantha* (fewer than 50 papers each).

The nine species (and publication number) that were excluded from our analysis because they contained too few (no more than five) years of ISI records were mainly terrestrial plants. These species include the African tulip tree *Spathodea campanulata* (16), hiptage *Hiptage benghalensis* (12), Kahili ginger *Hedychium gardnerianum* (10), Koster's curse *Clidemia hirta* (24), quinine tree *Cinchona pubescens* (12), shoe-button ardisia *Ardisia elliptica* (12), yellow Himalayan raspberry *Rubus ellipticus* (13), cypress aphid *Cinara cupressi* (11) and red-vented bulbul *Pycnonotus cafer* (15). We were surprised that so many species from the 100 of the World's Worst Invasive Alien Species list were so poorly documented in the formal literature. Pyšek et al. (2008) noted that only 1.6% of 892 invasive species had more than 20 published studies. The dearth of ISI-tracked publications for many of these species may relate in part to their recent history of spread (e.g. fishhook waterflea, Asian longhorn beetle) or to our recent awareness of the species as a major conservation concern (e.g. rosy wolf snail). It is also possible that some of these species may have been problematic in their native region but were poorly documented as such, owing to a lack of publications in journals tracked by the ISI index. This problem would be particularly acute for invaders originating in, or spreading to, countries in the southern hemisphere or Asia, where invasive species generally receive much less attention – with the notable exceptions of New Zealand, Australia and South Africa. Pyšek et al. (2008) determined that the invasion literature tracked by Web of Science is dominated by studies from North America and Europe, whereas Asia and much of Africa appear vastly understudied.

The single best studied species were all animals: house mouse (*Mus musculus*; 92,221 studies), rainbow trout (*Oncorhynchus mykiss*; 21,584 studies), rabbit (*Oryctolagus cuniculus*; 13,480 studies), pig (*Sus scrofa*; 10,567 studies), and carp (8579 studies) (Fig. 5.1). The best studied plants included the fire tree (2214 studies) and leucaena (1311 studies). Because species are studied for many reasons other than their invasiveness, these values fall considerably when assessments are limited to studies in which the species is specifically identified as invasive by the original authors (see below).

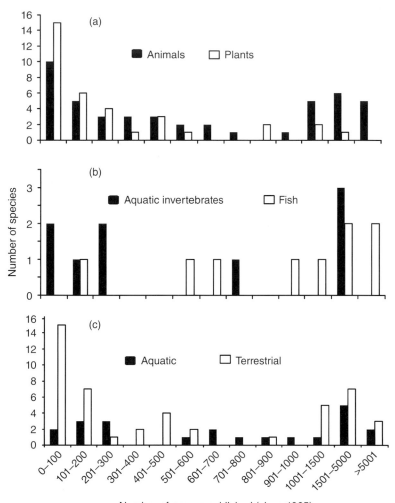

Fig. 5.1 Number of papers published between 1965 and 2007 on non-indigenous species of plants and animals (a), aquatic invertebrates and fishes (b), and aquatic versus terrestrial species (c).

Consistent with our first hypothesis, animals were significantly better studied than plants, in part reflecting the popularity of papers on numerous fishes (*t*-test, $P < 0.001$) (Fig. 5.1). However, this result was also contingent on inclusion of the house mouse, which had more than four times as many studies as any other species. Even with the elimination of obvious biomedical topic fields, some species accumulated large numbers publications that may or may not be related to invasion ecology (e.g. 7852 papers on genetics and heredity); because of our uncertainty on the nature of the individual papers, we retained them studies in our analysis. Overall, animals averaged 3604 studies

(median 658), versus only 374 for plants (median 194). Animals also accumulated studies earlier than plants. For example, the 25th percentile of all publications occurred in 1991 for animals and 1995 for plants.

The hypothesis that terrestrial species tend to be more conspicuous than aquatic species and therefore better studied was not supported. We found no difference (*t*-test, $P = 0.210$) in mean number of publications for aquatic versus terrestrial species (2144 versus 2687, respectively). The time required to achieve the 25th percentile of total publications was also very similar between aquatic (1992) and terrestrial (1993) species.

Among aquatic animals, fishes tended to be better studied than aquatic invertebrates (average: 4782 versus 781 studies, respectively), although variation was pronounced in both groups and the difference was not significant ($P = 0.064$; Fig. 5.1). Consequently, our third hypothesis that the widespread stocking of fishes would result in more publications than for invertebrates was not supported. It should be noted that invertebrates were stocked extensively in the former USSR and Scandinavia to enhance fisheries, although none of the species uses for this purpose are included on the 100 worst invaders list. Furthermore, several of the invertebrate species on the list have been widely dispersed by shipping (e.g. European green crab *Carcinus maenas*, comb jelly, zebra mussel) and/or stocking (Chinese mitten crab *Eriocheir sinensis*, Mediterranean mussel *Mytilus galloprovincialis*), which contributed to enhanced numbers of publications particularly in recent years. Fishes achieved the 25th percentile of all publications slightly earlier than did aquatic invertebrates (1991 versus 1993, respectively).

In addition to rainbow trout and common carp, brown trout *Salmo trutta* (4661 studies) and Mozambique tilapia *Oreochromis mossambicus* (1144 studies) were early and well studied cases. By contrast, Nile perch was poorly studied (174 studies) despite its fundamental and perhaps unprecedented role in the collapse of cichlid fish diversity in Lake Victoria (see, for example, Goudswaard et al. 2008). Among aquatic invertebrates, the European green crab (2089 studies), Mediterranean mussel (1896 studies), and zebra mussel (1596 studies) were the best studied species. The former species has been introduced and is established in coastal marine habitats of all continents except Antarctica (Darling et al. 2008). *Mytilus galloprovincialis* has an extensive and lengthy record of introductions to marine coastal habitats globally (Carlton 1999). *Dreissena polymorpha* is native to fresh

and brackish waters of the Black, Caspian and Azov Seas (see May et al. 2006), though it has spread via canals and shipping beginning in the 18th century throughout much of western and eastern Europe; more recently the species has consolidated its range in Europe, with, for example, initial colonization of Spain and Ireland and enhanced distribution in Italy (Bij de Vaate et al. 2002; Lancioni & Gaino 2006; Rajagopal et al. 2009). The species colonized Lake Erie in the Laurentian Great Lakes no later than 1986 (Carlton 2008), and has since spread widely across eastern North America including most major river systems and hundreds of inland lakes (Drake & Bossenbroek 2004). In 2008, the species was found across the continental divide for the first time, with established populations in Colorado and California.

When names of species from the 100 of the World's Worst Invasive Alien Species were cross-referenced against the seven synonyms for 'invasive species', the number of publications recovered fell dramatically (see, for example, Fig. 5.2a,b). Overall publications for these averaged 2320 papers, though an average of only 55 (2.4%) specifically referenced the species as 'non-indigenous'. When we combined the two searches, 29 of the 91 profiled species had fewer than 20 publications in which they were identified as 'non-indigenous'. This group included all three microorganisms (*Plasmodium relictum*, Banana bunchy top virus, Rinderpest virus), and four of five macro-fungi (*Cryphonectria parasitica*, *Ophiostoma ulmi*, *Batrachochytrium dendrobatidis*, *Phytophthora cinnamomi*). Ten of 25 terrestrial plants were also represented by fewer than 20 papers in the literature in which they were identified as invasive. By contrast, all of the four aquatic plant species, seven of eight fishes and seven of nine aquatic invertebrates (marine and freshwater) were identified in more than 20 studies as 'non-indigenous'. Consistent with Pyšek et al. (2008), the

---

Fig. 5.2 Profiles of growth in number of publications for nine aquatic species (a) and nine terrestrial species (b) listed on the IUCN's 100 of the World's Worst Invasive Alien Species. Aquatic species include carp *Cyprinus carpio*, brown trout *Salmo trutta*, water hyacinth *Eichhornia crassipes*, zebra mussel *Dreissena polymorpha*, marine clam *Corbula amurensis* (previously *Potamocorbula amurensis)*, caulerpa seaweed *Caulerpa taxifolia*, rainbow trout *Oncorhynchus mykiss*, Nile perch *Lates niloticus* and Mozambique tilapia *Oreochromis mossambicus*. Terrestrial species include cogon grass *Imperata cylindrica*, lantana *Lantana camara*, mimosa *Mimosa pigra*, brown tree snake *Boiga irregularis*, pig *Sus scrofa*, ship rat *Rattus rattus*, starling *Sturnus vulgaris*, cane toad *Bufo marinus* and gypsy moth *Lymantria dispar*. In all cases, the dashed line is a fitted exponential curve describing cumulative publications obtained by searching using species' common and scientific names only, whereas solid lines are fitted curves for species names cross-referenced with seven synonyms for non-indigenous species (i.e. only those papers where the species is viewed in a non-indigenous context).

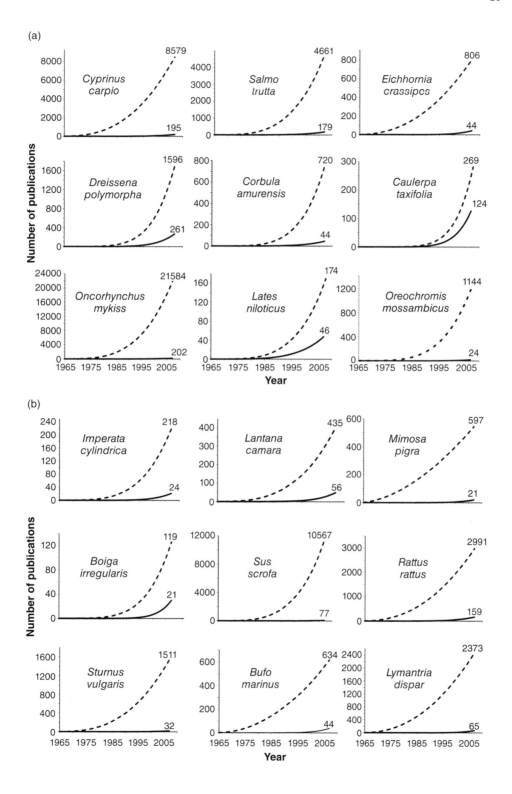

Table 5.1 Mean (standard deviation (SD)) coefficient (α) and exponent (β) describing growth of publications for different taxonomic groups. Values were calculated as: EC = α(Year)$^{\beta}$ where EC is the estimated cumulative number of publications based on nonlinear regression for the period 1965 to 2007. Non-indigenous Ratio is the mean percentage of studies in which the species were considered invasive (see Fig. 5.2 legend). All estimates use a starting point of 1965.

Group	α (SD)	β (SD)	Non-Indigenous ratio	*N*
Microbes	0.0001 (0.0002)	3.7449 (0.4046)	3.4	3
Macro-fungi	0.6797 (0.7885)	2.8023 (1.3229)	9.5	5
Aquatic plants	0.0372 (0.0742)	4.3442 (1.8666)	23.3	4
Land plants	0.0131 (0.0401)	4.2286 (1.3986)	17.8	25
Aquatic invertebrates	0.0070 (0.0188)	4.5096 (1.1442)	15.5	9
Land invertebrates	0.7774 (0.0188)	3.4550 (1.1442)	16.8	15
Amphibians	0.4358 (0.5199)	2.7066 (1.2290)	4.4	3
Fish	0.1288 (0.1893)	3.2059 (0.8317)	6.5	8
Birds	0.0360 (0.0394)	2.4461 (0.3355)	12.7	2
Reptiles	<0.0001 (<0.0001)	4.8710 (0.3099)	10.0	2
Mammals	0.4622 (1.4765)	3.13128 (0.6482)	6.0	14

zebra mussel was identified as 'non-indigenous' in more papers (261) than any other species on the list of 100 of the World's Worst Invasive Alien Species. The fishhook waterflea (62%) and yellow crazy ant (53%) had the highest percentage of studies published in which they were identified as 'non-indigenous', followed by Japanese knotweed *Fallopia japonica* and the green alga *Caulerpa taxifolia* (46%). Among fish, Nile perch had the highest ratio of 'non-indigenous' studies (26%), though surprisingly there is a relative paucity of studies for this invader (Fig. 5.2a). No other fish species had an 'non-indigenous ratio' greater than 10%, illustrating the benign attitude of many researchers to the impacts of introduced fishes as well as the predominantly positive attention given to stocked non-indigenous fishes. As a group, aquatic plants and terrestrial plants were the most likely to be viewed as 'non-indigenous' by the research community (Table 5.1).

We quantified whether growth of publications for the 100 of the World's Worst Invasive Alien Species began relatively early (i.e. around 1965; high α) or at a relatively high rate in recent years (high β). As a group, the highest values of α were observed for terrestrial invertebrates, macro-fungi, mammals and amphibians (Table 5.1). Reptiles, microbes and aquatic invertebrates had very low values of α, reflecting very low publication rates for member species in the 1960s. The greatest growth in publications in recent years occurred among reptiles, aquatic invertebrates, and aquatic and terrestrial plants (Table 5.1). Initial studies

on plants and animals began about the same time, as differences in α were not significant (*t*-test, $P = 0.107$). However, plants apparently accumulated studies more rapidly in recent years (β; *t*-test, $P = 0.022$).

The relatively low rate of identification of the 100 of the World's Worst Invasive Alien Species with seven terms commonly used by the academic community to identify taxa that pose problems may reflect changes in public attitudes or popular terminology in academia. Clearly, terms like 'alien', 'invasive' and the like are more popular (and controversial) today than in the past, whereas others, like 'colonizing' appear to be used less commonly today. However, it seems implausible that the widespread increase in scientific publications on non-indigenous species observed during the 1990s is related to popular terminology. Rather, it is more likely that the increase in publications in scholarly journals follows the pattern of landmark conferences and book publications launched during the 1980s.

Our results have several potential limitations that must be considered. The Web of Knowledge tracks mainly western journals published in 22 languages commonly used in the northern hemisphere. Journals published in other languages are not picked up by the ISI index, and studies on species reported in these outlets would be under-represented in consequence. Publications in the 'grey' literature (e.g. government reports) also are not tracked by the ISI index. The Nile perch has accumulated only 174 papers in our search of the tracked literature and 220 papers if all topics are considered (Fig. 5.2a). Given the importance of this species to the unfolding biodiversity crisis and to African societies dependent on it as a fishery resource (Matsuishi et al. 2006), our analysis almost certainly underestimates the total research effort devoted to this species. Wilson et al. (2007) found that species with high societal importance tended to receive more scientific attention than species with little societal value, thus the comparatively small number of scientific publications uncovered in our search is puzzling. Finally, we assessed growth patterns (α and β) of publications using a common reference date (1965) for all species. Different patterns could be uncovered had we used the publication date when each species was first described as non-indigenous in scientific publications. Nevertheless, our analysis is consistent with our stated purpose of identifying temporal patterns of attention devoted to different introduced taxa.

Publication histories indicate that studies on most of the species that we investigated have increased dramatically since 1990. This can be due to a general surge of interest in invasion ecology because of increased funding available to sponsor these studies, or it may simply reflect a general increase in the volume of scientific literature published. Some of the species on the 100 World's Worst Invasive Alien Species list have been studied for a long time (e.g. water hyacinth, common carp), though for many others (e.g. fishhook waterflea, comb jelly, zebra mussel) studies increased dramatically during the 1990s or later as the species executed high profile and high impact invasions in new localities. Interestingly, the relative growth rate of publications on biological invasions is nearly identical to that for climate change, although the latter is a much larger field with many more studies published overall.

5.4   CONCLUSIONS

Authors have discussed how Charles Elton's views on invasions strongly reflected his particular experiences and interests (Davis et al. 2001; Richardson & Pyšek 2008). Similarly, publication patterns for different non-indigenous species reflect differences in their importance to, interest in or utility of these species to humans. Changes in human interest alter the frequency and locations at which species are intentionally moved globally (Wilson et al. 2009), just as changes in the operation or pathways of principal vectors affect the global movement and unintentional

introduction of non-indigenous species (Carlton 1985). Despite an enormous diversity of species that have been introduced and established in ecosystems throughout the world, and a burgeoning number of case studies that have accrued since Elton (1958), much of invasion biology has developed from studies of a relatively small group of taxa. It remains to be determined how this bias may have influenced key concepts in the field. The surge in interest in invasion biology provides an opportunity to explore concepts using a wider array of taxa and habitats studied.

ACKNOWLEDGMENTS

We are grateful for the invitation from Dave Richardson to participate in the Charles Elton Symposium and for helpful comments on the paper from Dave Richardson and two reviewers. We acknowledge financial support from NSERC discovery grants to H.J.M. and A.R., and from the Canadian Aquatic Invasive Species Network (CAISN).

REFERENCES

Aronson, R.B., Thatje, S., Clarke, A., et al. (2007) Climate change and invasibility of the Antarctic benthos. *Annual Review of Ecology, Evolution and Systematics*, **38**, 129–154.

Bij de Vaate, A., Jażdżewski, K., Ketelaars, H.A.M., Gollasch, S. & Van der Velde, G. (2002) Geographical patterns in range extension of Ponto-Caspian macroinvertebrate species in Europe. *Canadian Journal of Fisheries and Aquatic Sciences*, **59**, 1159–1174.

Carlton, J.T. (1985) Transoceanic and interoceanic dispersal of coastal marine organisms: the biology of ballast water. *Oceanography and Marine Biology*, **23**, 313–371.

Carlton, J.T. (1999) Molluscan invasions in marine and estuarine communities. *Malacologia*, **41**, 439–454.

Carlton, J.T. (2008) The zebra mussel *Dreissena polymorpha* found in North America in 1986 and 1987. *Journal of Great Lakes Research*, **34**, 770–773.

Cheung, W.W.L., Lam, V.W.Y., Sarmiento, J.L., Kearney, K., Watson, R. & Pauly, D. (2009) Projecting global marine biodiversity impacts under climate change scenarios. *Fish and Fisheries*, **10**, 235–251.

Cristescu, M.E., Witt, J., Hebert, P.D.N., Grigorvich, I.A. & MacIsaac, H.J. (2001) An invasion history for *Cercopagis pengoi* based on mitochondrial gene sequences. 2001. *Limnology and Oceanography*, **46**, 224–229.

Darling, J.A., Bagley, M.J., Roman, J., Tepolt, C.K. & Geller, J. (2008) Genetic patterns across multiple introductions of the globally invasive crab genus *Carcinus*. *Molecular Ecology*, **17**, 4992–5007.

Davis, M.A., Thompson, K. & Grime, P. (2001) Charles S. Elton and the dissociation of invasion ecology from the rest of ecology. *Diversity and Distributions*, **7**, 97–102.

Drake, J.M. & Bossenbroek, J.M. (2004) The potential distribution of zebra mussels in the United States. *Bioscience*, **54**, 931–941.

Elton, C.S. (1958) *The Ecology of Invasions by Animals and Plants*. Methuen, London.

Goudswaard, K.P.C., Witte, F. & Katunzi, E.F.B. (2008) The invasion of an introduced predator, Nile perch (*Lates niloticus*, L.) in Lake Victoria (East Africa): chronology and causes. *Environmental Biology of Fishes*, **81**, 127–139.

Haack, R.A. (2006) Exotic bark- and wood-boring Coleoptera in the United States: recent establishments and interceptions. *Canadian Journal of Forest Research*, **36**, 269–288.

Hulme, P.E. (2009) Trade, transport and trouble: managing invasive species pathways in an era of globalization. *Journal of Applied Ecology*, **46**, 10–18.

Lancioni, T. & Gaino, E. (2006) The invasive zebra mussel *Dreissena polymorpha* in Lake Trasimeno (Central Italy): Distribution and reproduction. *Italian Journal of Zoology*, **73**, 335–346.

Levine, J.M. & D'Antonio, C.M. (2003) Forecasting biological invasions with increasing international trade. *Conservation Biology*, **17**, 322–326.

Lockwood, J.L., Hoopes, M.F. & Marchetti, M.P. (2007) *Invasion Ecology*. Blackwell, Oxford.

Lowe, S., Browne, M., Boudjelas, S. & De Poorter, M. (2004) 100 of the world's worst invasive alien species a selection from the global invasive species database. 6–7. Invasive Species Specialist Group and Group of the Species Survival Commission, Auckland, New Zealand.

Matsuishi, T., Muhoozi, L., Mkumbo, O., et al. (2006) Are the exploitation pressures on the Nile perch fisheries resources of Lake Victoria a cause for concern? *Fisheries Management and Ecology*, **13**, 53–71.

May, G.E., Gelembiuk, G.W., Panov, V.E., Orlova, M.I. & Lee, C.E. (2006) Molecular ecology of zebra mussel invasions. *Molecular Ecology*, **15**,1021–1031.

O'Dowd, D.J., Green, P.T. & Lake, P.S. (2003) Invasional 'meltdown' on an oceanic island. *Ecology Letters*, **6**, 812–817.

Pimentel, D., Zuniga, R. & Morrison, D. (2005) Update on the environmental and economic costs associated with alien-invasive species in the United States. *Ecological Economics*, **52**, 273–288.

Pyšek, P., Richardson, D.M. & Jarošík, V. (2006) Who cites who in the invasion zoo: insights from an analysis of the most highly cited papers in invasion ecology. *Preslia*, **78**, 437–468.

Pyšek, P., Richardson, D.M., Pergl, J., Jarošík, V., Sixtová, Z. & Weber, E. (2008) Geogrpahical and taxonomic biases in invasion ecology. *Trends in Ecology & Evolution*, **23**, 237–244.

Rajagopal, S., Pollux, B.J.A., Peters, J.L., et al. (2009) Origin of Spanish invasion by the zebra mussel, *Dreissena polymorpha* (Pallas, 1771) revealed by amplified fragment length polymorphism (AFLP) fingerprinting. *Biological Invasions*, **11**, 2147–2159.

Reaser, J.K., Meyerson, L.A., Cronk, Q., et al. (2007) Economical and socioeconomic impacts of invasive alien species in island ecosystems. *Environmental Conservation*, **34**, 98–111.

Reid, P.C., Johns, D.G., Edwards, M., Starr, M., Poulin, M. & Snoejs, P. (2007) A biological consequence of reducing Arctic ice cover: arrival of the Pacific diatom *Neodenticula seminae* in the North Atlantic for the first time in 800000 years. *Global Change Biology*, **13** 1910–1921.

Ricciardi, A. (2006) Patterns of invasion of the Laurentian Great Lakes in relation to changes in vector activity. *Diversity and Distributions*, **12**, 425–433.

Ricciardi, A. & MacIsaac, H.J. (2008) The book that began invasion ecology. *Nature*, **452**, 34.

Richardson, D.M. & Pyšek, P. (2008) Fifty years of invasion ecology – the legacy of Charles Elton. *Diversity and Distributions*, **14**, 161–168.

Ruiz, G.M., Rawlings, T.K., Dobbs, F.C., et al. (2000) Worldwide transfer of microorganisms by ships. *Nature*, **408**, 49–50.

Wilson, J.R.U., Dormontt, E.E., Prentis, P.J., Lowe, A.J. & Richardson, D.M. (2007) The (bio)diversity of science reflects the interests of society. *Frontiers in Ecology and the Environment*, **5**, 409–414.

Wilson, J.R.U., Proches, S., Braschler, B., Dixon, E.S. & Richardson, D.M. (2009) Something in the way you move: dispersal pathways affect invasion success. *Trends in Ecology & Evolution*, **24**, 136–144.

Wonham, M.J. & Pachepsky, E. (2006) A null model of temporal trends in biological invasion records. *Ecology Letters*, **9**, 663–672.

APPENDIX 5.1

Fields from the ISI searches which were retained in our survey included the following: ecology, evolutionary biology, biology, oceanography, limnology, mycology, multidisciplinary sciences, genetics and heredity, biodiversity conservation, entomology, environmental sciences, marine and freshwater biology, ornithology, plant sciences, agriculture: multidisciplinary, agronomy, forestry, horticulture, geography, physical water resources, agriculture, dairy and animal science, microbiology, toxicology, remote sensing, and virology. Forty-six additional fields tracked by the ISI Web of Science were excluded.

Chapter 6

INVASION ECOLOGY AND RESTORATION ECOLOGY: PARALLEL EVOLUTION IN TWO FIELDS OF ENDEAVOUR

Richard J. Hobbs[1] and David M. Richardson[2]

[1]School of Plant Biology, University of Western Australia, Crawley, WA 6009, Australia
[2]Centre for Invasion Biology, Department of Botany & Zoology, Stellenbosch University, 7602 Matieland, South Africa

Fifty Years of Invasion Ecology: The Legacy of Charles Elton, 1st edition. Edited by David M. Richardson
© 2011 by Blackwell Publishing Ltd

6.1 INTRODUCTION

This book focuses on invasion ecology, but several commentators have recently emphasized the need to consider this topic in a broader interdisciplinary context which draws on other fields of endeavour (see, for example, Davis et al. 2005; Davis 2009). Charles Elton's (1958) book *The Ecology of Invasions by Animals and Plants* has been lauded for bringing together previously disparate themes in considering the dimensions and implications of biological invasions. Richardson and Pyšek (2008), in reviewing Elton's legacy in invasion ecology and suggesting priorities for the future, argued that 'Invasion ecology needs to continue building bridges with other disciplines, following the course charted by Elton. Key areas where improved links with invasion ecology would be mutually beneficial include conservation biology/biogeography, global change biology, restoration ecology, weed science, resource economics, human geography and policy studies'.

In this chapter, we compare invasion ecology with another relatively young discipline: restoration ecology. We examine the similarities and differences between the two fields and consider what lessons might be drawn from the development of restoration ecology, that could benefit invasion ecology, and vice versa. We also consider the recent intermingling of the two disciplines at the practical level where managers are dealing with rapid environmental and biotic change.

6.2 SIMILARITIES AND DIFFERENCES

Both invasion ecology and restoration ecology are relatively new disciplines with their roots in the mid-20th century, but with much longer historical precedents. The need for restoring ecosystems and communities has been recognized for some time. Some see the origin of the activity in the attempts to restore native prairies in the mid-west of America during the middle of the 20th century (Jordan et al. 1987b). However, it is clear that restoration in various forms was practised for much longer than that in Europe and elsewhere, although it was not necessarily labelled as such (Hall 2005). Restoration ecology, the science behind the practice of ecological restoration, has its roots firmly associated with the practice of ecological restoration. A prominent early exponent of meshing science and practice was Aldo Leopold (1887–1948),

who wrote in 1934: 'The time has come for science to busy itself with the earth itself. The first step is to reconstruct a sample of what we had to begin with' (Leopold 1934) (Box 6.1). The first book mentioning restoration ecology specifically (Jordan et al. 1987a) contained a range of chapters discussing the value of using restoration activities as test beds for ecological ideas and theories, and described restored areas as useful field laboratories. Similarly, Bradshaw (1987) saw restoration as an 'acid test for ecology', suggesting that we know how an ecological system works only when we are able to put it back together again effectively. Ewel (1987) suggested that restoration provides 'the ultimate test of ecological theory'.

Despite these suggestions, restoration ecology remained largely focused on individual case studies and did not develop a clear theoretical underpinning, somewhat in contrast to invasion ecology. Even now, a distinct difference between invasion ecology and restoration ecology (at least as reflected in the recent literature) is that the former has devoted more recent research effort towards theoretical issues (towards all-embracing general theories; e.g. Catford et al. 2009). In contrast, in restoration ecology much more attention has been given to solving practical issues. Starting in the early 1990s there were, however, increasing efforts to develop a conceptual basis for restoration ecology, drawing influences from both practice and theoretical ecology (see, for example, Hobbs & Norton 1996; Palmer et al. 1997; and other papers in the same journal volume). Concepts such as ecological succession (Luken 1990; Walker et al. 2007), assembly rules (Temperton et al. 2004) and threshold dynamics (Hobbs & Suding 2009; Suding & Hobbs 2009) have found obvious applicability in restoration ecology. However, the application of these concepts has been mired in the practicality of having to understand the physical and biological basis for ecosystem degradation and renewal across a wide range of disparate ecosystem types (see, for example, Whisenant 1999; Perrow & Davy 2002a,b; King & Hobbs 2006). In addition, many books on restoration continue to focus on particular ecosystem types or locales because of the need to tailor general principles to the details of the situation confronting practitioners. Moving from the general to the specific and from concept to application (and vice versa) remains an ongoing challenge for the discipline (Hobbs & Harris 2001; Hobbs 2007).

Restoration ecology is undergoing a remarkable period of growth and development, mirroring similar

Box 6.1 Elton and Leopold: twin sons of different mothers

Charles Elton (1900–1991) and Aldo Leopold (1887–1948) arc towcring figures in the history of invasion ecology and restoration ecology, respectively. Both are important personalities in 20th century ecology in general, with influence extending well beyond invasions and restoration. Despite their very different backgrounds, nationalities and milieus, the two met once (at the Matamek conference on biological cycles in Labrador, Canada, in 1931), formed a close friendship and corresponded regularly. Much has been written about the two men (especially Leopold: see Meine 1988; Newton 2006), but little emphasis has been placed on their personal interactions, correspondence and 'cross-pollination' on key issues of importance to both invasion ecology and restoration ecology.

Elton is widely recognized as one of the founders of modern population biology and community ecology, but is also widely seen as launching the systematic study of biological invasions (Richardson & Pyšek 2008; but see Simberloff, this volume). Leopold has been hailed as 'the father of wildlife ecology', and, unlike Elton, has become a cult figure in the environmental movement.

Both men were academics who produced important textbooks that were highly influential in their respective fields, but both are also famous (probably more famous, certainly in the case of Leopold) for their less technical writings. Aldo Leopold's *Sand County Almanac* was published in 1949, a year after his death. It describes, in poetic prose, the land around Leopold's home in Sauk County, Wisconsin, and his thoughts on a 'land ethic'. The collection of essays is a landmark in the American conservation movement and is often hailed as one of the most influential environmental books of the 20th century (and is often mentioned in the same sentence as Rachel Carson's 1962 *Silent Spring*). The book is most often cited for the quote which defines 'land ethic': 'A thing is right when it tends to preserve the integrity, stability, and beauty of the biotic community. It is wrong when it tends otherwise.' Elton's (1958) book *The Ecology of Invasions by Animals and Plants* grew from a series of radio talks on the BBC (Simberloff, this volume). Like Leopold's book, Elton's

was written in popular prose. In parts, notably chapters 8 and 9, Elton's style is strongly reminiscent of Leopold's in *Sand County Almanac* and the book's strong conservation theme is almost certainly a result of Elton's interactions with Leopold. Like Leopold's book, Elton's volume told of a rapidly looming crisis, but unlike that of his American friend, Elton's book had negligible impact on public perceptions and launched no environmental movement. Reasons for this discrepancy have not been adequately explored. The fact that Elton's book was published in post-Second World War Europe, where writings of pending threats to biodiversity due to marauding alien species of animals and plants were not given attention they deserved, is probably one reason.

There was much cross-pollination between Elton and Leopold, and many of the ideas that permeate their correspondence are still being debated today. Elton's ideas on naturalness and ecological stability and Leopold's notion of the biotic land pyramid clearly emerged from similar thinking and shared ideas, although the dynamics of what Chew (2006) has termed their 'competitive collaboration' have yet to be fully explored. Elton pioneered the concept of the biotic community in his 1927 book *Animal Ecology* in which he recognized the importance of food chains in structuring biotic communities. These ideas clearly influenced Leopold who in turn seems to have influenced Elton's later formulations of the notion. Elton developed these ideas in chapter 8 of *The Ecology of Invasions by Animals and Plants* where he formulated the 'diversity-invasibility hypothesis' which has been his most lasting legacy in invasion ecology (Richardson & Pyšek 2007; Fridley, this volume).

Matt Chew (2006) wrote: '… within the uncertain boundaries of its infancy, ecology allowed a Yale forester [Leopold] and an Oxford zoologist [Elton] to begin a conversation that still echoes today'. The ideas and concepts that featured most prominently in discussions and correspondence between the two men still feature strongly in research agendas in invasion ecology and restoration ecology today.

trends in invasion ecology. As with invasion ecology, restoration ecology is now covered by several specialist journals, including *Restoration Ecology*, *Ecological Restoration* and the regional Australasian journal *Ecological Management and Restoration*, and there has

been considerable growth in the number of articles in other ecological journals which focus on applied ecology. In addition, there has been a recent proliferation in textbooks covering various aspects of restoration ecology (Falk et al. 2006; van Andel & Aronson

2006; Clewell & Aronson 2007). There is also a growing and active international society, the Society for Ecological Restoration International (www.ser.org), which runs regular conferences and meetings focusing on both science and practice, sometimes in conjunction with other major ecological meetings such as the Ecological Society of America annual conference. The Society has a Science and Policy Working Group, which focuses on both scientific and policy aspects of restoration and has produced several publications aiming to synthesize current ideas and approaches (see, for example, Society for Ecological Restoration International Science & Policy Working Group 2004).

6.3 CONTRASTS IN FOCUS AND APPROACH

Both invasion ecology and restoration ecology are, on the face of it, heavily 'mission oriented' and are focused on issues of real conservation/management significance. That said, many aspects of invasion ecology have a more academic focus, tackling the evolutionary consequences of invasions, community assembly, limiting similarity and the like. Although restoration ecology is also beginning to consider such issues, this arises from the opportunity to use these ideas to guide restoration activities rather than as a purely intellectual pursuit.

A clear difference between the two fields is that much (but not all) of invasion ecology originated with, and focuses on a *problem*, whereas restoration ecology grew out of, and focuses on *solutions*. This subtle difference has likely influenced the development and trajectories of the two disciplines. Invasion ecology begins with the threat or problem, the introduced species. Hence, there has been a large emphasis on understanding what causes some species to become invasive, on predicting invasibility and on characterizing the invasion process and the like. A large part of the literature on invasive species deals with the description of patterns, at various spatial scales, of invasions, i.e. describing the 'macroecology' of biological invasions. This has, to some extent, been divorced from the practicalities of dealing with invasive species or finding solutions, which instead have been largely the domain of weed science and other applied areas (Richardson et al. 2004). Several authors have bemoaned the fact that, despite considerable advances in the conceptualization and understanding of invasion processes, little

progress has been made in reducing the negative impacts of invasions on biodiversity and ecosystem function (see, for example, Hulme 2003). Invasion ecology has also been criticized for its narrow focus on the invader as the problem. Recent studies have found the invader to be only a symptom of other problems such as changes to natural disturbance regimes or limitations in seed production by native species (Seabloom et al. 2003; Didham et al. 2005; MacDougall & Turkington 2005). This narrow focus has masked, until recently, the need for measuring the real impact of these species on an ecosystem (Levine et al. 2003). Invasion ecology research has recently started to take on a wider focus of developing strategies that also maintain the integrity of native dominated ecosystems (Buckley et al. 2007; Firn et al. 2008).

In contrast, the site-specific nature of much restoration work has limited the development of a macroecological perspective. Restoration ecology has focused much more on evaluating the efficacy of management activities than on the elucidating the problem behind ecosystem damage. This has placed more emphasis on understanding the effects of different management treatments on species composition, and ecosystem functioning, often with very effective results. Of course, a lack of emphasis on understanding how or why an ecosystem has become degraded or 'the problem' has resulted in wrong assumptions and consequently ineffective restoration strategies. Although ecosystem degradation has been studied for some time (see, for example, Barrow 1991), attempts to develop a generalized understanding of degradation and restoration processes in tandem have been relatively slow to develop (see, for example, Milton et al. 1994; Harris et al. 1996; Whisenant 1999; King & Hobbs 2006). Increasing emphasis is now being placed on coupling correct diagnosis of problems with setting clear and achievable restoration goals (Hobbs 2007).

6.4 PHILOSOPHICAL AND ETHICAL ISSUES

Recently, in both invasion ecology and restoration ecology, there has been a trend of questioning the basic tenets and assumptions underpinning both the science and the practice. Invasion ecology has been criticized for a simplistic characterization of non-native species and their perceived threat to conservation and other values. There has been increased questioning of the degree of threat actually posed by invasive species in general (see, for example, Gurevitch & Padilla 2004), sometimes linked to the question of whether biological invasions are predominantly drivers or symptoms of degradation (MacDougall & Turkington 2005; Richardson et al. 2007). There has also been much questioning of the use of evocative language and militaristic metaphors to characterize both threats of and response to invasive species (see, for example, Simberloff 2003; Gobster 2005; Davis 2009). Indeed, some commentators have even questioned the value of labelling species as aliens or natives at all, because the concepts could, they argue, be considered to be artificial and elitist (see, for example, Warren 2007; see the discussion of the history of the concept of 'nativeness' in Chew & Hamilton, this volume). At the extreme, invasion ecology has been labelled a 'pseudoscience' (Theodoropoulos 2003).

In restoration ecology, there has been a similar upsurge of questioning basic assumptions, partly because of our changing understanding of ecosystem dynamics. The field has also experienced significant philosophical criticism for some time (see, for example, Katz 1992, 2000; Elliot 1994, 1998). Since its inception, restoration ecology has rubbed shoulders with other disciplines, particularly history, as some practitioners have a strong desire to steer ecosystems to a historical state, and social sciences as restoration activities can impact heavily on the social welfare of local communities and regions. Often, however, the contact between these disciplines is more implicit than explicit, although this is changing quite rapidly (see, for example, Gobster & Hull 2000; Aronson et al. 2007; Jackson & Hobbs 2009). Recently there has been a strong movement away from a purely historically based restoration approach, as we learn more about ecosystem dynamics, the detailed history of human use of ecosystems and the likely changes in ecosystems arising from climate and other environmental changes (Choi 2007; Choi et al. 2008; Davies et al. 2009).

Both fields have to some extent been characterized by a move from simplistic 'black and white' approaches to 'shades of grey'. This has been accompanied by a change in approach from 'just do it' to 'what do we need to do; why, where and when?'. The applied side of invasion ecology has morphed into the domain of 'biosecurity' in which biogeography and ecology are important but where economic and socio-political issues increasingly dominate agendas (Hulme, this volume).

6.5   TERMINOLOGICAL ISSUES

Both fields have been marked by considerable debate over terminology. Davis (2009) points out that the terms to be used when discussing different categories of non-native species have been discussed for well over a century (see also Chew & Hamilton, this volume), and there are ongoing efforts to arrive at a stable and unambiguous terminology (Richardson et al. 2000). Similarly in restoration ecology, a plethora of terms are available describing various types of restoration activity with repeated attempts to derive a stable terminology (see, for example, Clewell & Aronson 2007). Both fields have different types of proponent with, at one extreme, 'terminology police' who emphasize the importance of using a standard terminology based on objective (e.g. biogeographic) criteria, and at the other extreme, those who are less concerned with terminology per se, focusing more on the goals and outcomes of activities (Hodges 2008).

A major issue feeding into terminological debates for both fields is the value-based nature of the subject matter. Although science strives for objectivity, there is increasing recognition that for both invasive species and restoration activities, many decisions are based on values rather than objective science (Colautti & Richardson 2009). For invasion ecology, decisions about which species are classed as 'problems' or which species cause 'harm' are, of necessity, based on underlying questions of 'problems to *what?*' and 'harm to *what?*'. In many cases, the invasive species that have the highest economic impacts are labelled problem species; whereas the species with the worst environmental impacts remain undeclared and difficult to find research funding for. The issue becomes particularly murky when the same species has both 'positive' and 'negative' impacts, typically benefiting and impacting upon different sectors of society. For example, pasture grasses are valuable economically for improving conditions for grazing livestock, but difficult to stop from spreading into conservation areas. For restoration ecology, questions surrounding what a 'good' restoration outcome are also value laden, being dependent on perspective and expectations (Higgs 1997, 2003). The restoration goals based on the latest ecological thinking may not mesh with social expectations based more on historical or nostalgic viewpoints, leading to significant conflict (Gobster 2001; Hobbs 2004). Increasingly there is a much-needed recognition in both fields of the value-based dimensions of the subject matter.

6.6 THE INTERPLAY BETWEEN INVASION ECOLOGY AND RESTORATION ECOLOGY

There are obvious overlaps between the two fields. First, dealing with invasive species is often a key element of restoration (D'Antonio & Meyerson 2002). Many restoration projects focus almost exclusively on the removal of species that are considered to have 'degraded' an ecosystem. In many cases, reducing the abundance of an invader is an important element in achieving other goals such as recovery of endangered species, restoration of historic ecosystems or repair of ecosystem function. However, in some cases, invasive species removal has become a goal in itself, leading to the sort of questioning of underlying assumptions discussed above. Where non-native species are being removed simply because they are non-native, rather than being clearly identified as threats to other values, some sections of society object, leading to social conflict (see, for example, Gobster 2005; Coates 2006; Warren 2007).

Secondly, restoration is increasingly seen as a key element of effectively dealing with invasions. Removal of an invasive species alone may not have lasting and effective outcomes, and indeed may have unforeseen consequences that exacerbate rather than mitigate the 'problem' that triggered the restoration effort. This is because many alien species are readily integrated into networks in the invaded ecosystem, such as food webs, and pollination and seed-dispersal mutualistic interactions (Traveset & Richardson, this volume). Their removal has a very good chance of producing unplanned consequences. In some cases, removal of one problem species simply results in an increase in abundance of another equally or more problematic species (Zavaleta et al. 2001). Such 'secondary invasions' are becoming increasingly common (Cox & Allen 2008; Flory & Clay 2009; Firn et al. 2010). Hence, in many cases, removal of invasive species must be accompanied by measures to restore both the physical environment and the biotic assemblage (for example in riparian habitats; Richardson et al. 2007).

The two fields are likely to interact and intersect more as rapid environmental change and increased biotic exchange act synergistically to change biophysical envelopes, species distributions and biotic assemblages. The ways in which climate change is liable to affect invasive species and restoration activities are only just beginning to be explored in detail (Harris et al. 2006; Hellmann et al. 2008). Similarly, the implications of radically changed biotic assemblages and the increased difficulty in maintaining or restoring historic communities for both conservation and restoration have only recently become prominent (Hobbs et al. 2006, 2009; Hobbs & Cramer 2008). The concept of 'novel ecosystems' – systems where biotic composition and/or abiotic conditions differ radically from historic settings – is one area where invasion ecology and restoration ecology are really coming 'face to face in a dark alley'. This notion jars 'purists' in both fields, but increasingly some degree of pragmatic acceptance is seen as necessary to find practical ways of dealing with rapidly changing environments and species mixes. In any event, the concept is opening all sorts of new doors for research in both fields (see, for example, Firn et al. 2010). These new problems of changing environments and species mixes provide added challenges for relatively young disciplines which are still grappling with fundamental issues.

Based on our comparisons, these two fields can learn much from the other. Invasion ecology can learn from the solution-based focus of restoration ecology and restoration ecology from the problem-based focus of invasion ecology. By integrating a better understanding of causal mechanisms and the efficacy of different solutions into a cohesive framework, both fields could develop a stronger theoretical and practical underpinning. Increased cooperation and interaction among these and other fields can only enhance humanity's ability to respond to the rapidly changing world we share with the rest of the biosphere.

ACKNOWLEDGEMENTS

We thank James Aronson, Jennifer Firn, Mirijam Gaertner and Dan Simberloff for comments on the draft manuscript.

REFERENCES

Aronson, J., Milton, S.J. & Blignaut, J.N. (eds) (2007) *Restoring natural capital: science, business and practice.* Island Press, Washington, DC.

Barrow, C.J. (1991). *Land Degradation. Development and Breakdown of Terrestrial Environments.* Cambridge University Press, Cambridge, UK.

Bradshaw, A.D. (1987) Restoration: an acid test for ecology. In *Restoration Ecology: A Synthetic Approach to Ecological*

Research (ed. W.R. Jordan, M.E. Gilpin and J.D. Aber), pp. 23–30. Cambridge University Press Cambridge, UK.

Buckley, Y.M., Bolker, B.M. & Rees, M. (2007) Disturbance, invasion and re-invasion: managing the weed-shaped hole in disturbed ecosystems. *Ecology Letters*, **10**, 809–817.

Catford, J.A., Jansson, R. & Nilsson, C. (2009) Reducing redundancy in invasion ecology by integrating hypotheses into a single theoretical framework. *Diversity and Distributions*, **15**, 22–40.

Chew, M.K. (2006) *Ending with Elton: Preludes to Invasion Biology*. PhD thesis, Arizona State University, Tempe.

Choi, Y.D. (2007) Restoration ecology to the future: a call for a new paradigm. *Restoration Ecology*, **15**, 351–353.

Choi, Y.D., Temperton, V.M., Allen, E.B., et al. (2008) Ecological restoration for future sustainability in a changing environment. *Ecoscience*, **15**, 53–64.

Clewell, A.F. & Aronson, J. (2007) *Ecological Restoration: Principles, Values and Structure of an Emerging Profession*. Island Press, Washington, DC.

Coates, P. (2006). *American Perceptions of Immigrant and Invasive Species: Strangers on the Land*. University of California Press, Berkeley.

Colautti, R.I. & Richardson, D.M. (2009) Subjectivity and flexibility in invasion terminology: too much of a good thing? *Biological Invasions*, **11**, 1225–1229.

Cox, R.D. & Allen, E.B. (2008) Stability of exotic annual grasses following restoration efforts in southern California coastal sage scrub. *Journal of Applied Ecology*, **45**, 495–504.

D'Antonio, C.M. & Meyerson, L.A. (2002) Exotic plant species as problems and solutions in ecological restoration: a synthesis. *Restoration Ecology*, **10**, 703–713.

Davies, K.W., Svejcar, T.J. & Bates, J.D. (2009) Interaction of historical and nonhistorical disturbances maintains native plant communities. *Ecological Applications*, **19**, 1536–1545.

Davis, M.A. (2009) *Invasion biology*. Oxford University Press, Oxford.

Davis, M.A., Pergl, P., Truscott, A.-M., et al. (2005). Vegetation change – a reunifying concept in plant ecology. *Perspectives in Plant Ecology, Evolution, and Systematics*, **7**, 69–76.

Didham, R.K., Tylianakis, J.M., Hutchison, M.A., Ewers, R.M. & Gemmell, N.J. (2005) Are invasive species the drivers of ecological change? *Trends in Ecology & Evolution* **20**, 470–474.

Elliot, R. (1994) Extinction, restoration, naturalness. *Environmental Ethics*, **16**, 135–144.

Elliot, R. (1998) *Faking Nature*. Routledge, London.

Ewel, J.J. (1987) Restoration is the ultimate test of ecological theory. In *Restoration Ecology: A Synthetic Approach to Ecological Research* (ed. W.R. Jordan, M.E. Gilpin and J.D. Aber), pp. 31–33. Cambridge University Press Cambridge, UK.

Falk, D.A., Palmer, M.A. & Zedler, J.B. (eds.) (2006) *Foundations of restoration ecology*. Island Press, Washington, DC.

Firn, J., House, A.P.N. & Buckley, Y.M. (2010) Alternative states models provide an effective framework for invasive species control and restoration of native communities. *Journal of Applied Ecology*, **47**, 96–105.

Firn, J., Rout T., Possingham, H.P. & Buckley, Y.M. (2008) Managing beyond the invader: manipulating disturbance of natives simplifies control efforts. *Journal of Applied Ecology*, **45**, 1143–1151.

Flory, S.L. & Clay, K. (2009) Invasive plant removal method determines native plant community responses. *Journal of Applied Ecology*, **46**, 434–442.

Gobster, P.H. (2001) Visions of nature: conflict and compatibility in urban park restoration. *Landscape and Urban Planning*, **56**, 35–51.

Gobster, P.H. (2005) Invasive species as ecological threat: is restoration an alternative to fear-based resource management? *Ecological Restoration*, **23**, 261–270.

Gobster, P.H. & Hull, R.B. (eds.) (2000) *Restoring Nature: Perspectives from the Social Sciences and Humanities*. Island Press, Washington, DC.

Gurevitch, J. & Padilla, D.K. (2004) Are invasive species a major cause of extinctions? *Trends in Ecology & Evolution*, **19**, 470–474.

Hall, M. (2005) *Earth Repair: A Transatlantic History of Environmental Restoration*. University of Virginia Press, Charlottesville.

Harris, J.A., Birch, P. & Palmer, J. (1996) *Land Restoration and Reclamation: Principles and Practice*. Addison-Wesley Longman, Harlow, UK.

Harris, J.A., Hobbs, R.J., Higgs, E. & Aronson, J. (2006) Ecological restoration and global climate change. *Restoration Ecology*, **14**, 170–176.

Hellmann, J.J., Byers, J.E., Bierwagen, B.G. & Dukes, J.S. (2008) Five potential consequences of climate change for invasive species. *Conservation Biology*, **22**, 534–543.

Higgs, E. (2003) *Nature by Design: People, Natural Process, and Ecological Restoration*. MIT Press, Cambridge, Massachusetts.

Higgs, E.S. (1997) What is good ecological restoration? *Conservation Biology*, **11**, 338–348.

Hobbs, R.J. (2004) Restoration ecology: the challenge of social values and expectations. *Frontiers in Ecology and the Environment*, **2**, 43–44.

Hobbs, R.J. (2007) Setting effective and realistic restoration goals: key directions for research. *Restoration Ecology*, **15**, 354–357.

Hobbs, R.J., Arico, S., Aronson, J., et al. (2006). Novel ecosystems: theoretical and management aspects of the new ecological world order. *Global Ecology and Biogeography*, **15**, 1–7.

Hobbs, R.J. & Cramer, V.A. (2008) Restoration ecology: interventionist approaches for restoring and maintaining ecosystem function in the face of rapid environmental change. *Annual Review of Environment and Resources* **33**, 39–61.

Hobbs, R.J. & Harris, J.A. (2001) Restoration ecology: repairing the Earth's ecosystems in the new millennium. *Restoration Ecology*, **9**, 239–246.

Hobbs, R.J., Higgs, E. & Harris, J.A. (2009) Novel ecosystems: implications for conservation and restoration. *Trends in Ecology & Evolution*, **24**, 599–605.

Hobbs, R.J. & Norton, D.A. (1996) Towards a conceptual framework for restoration ecology. *Restoration Ecology*, **4**, 93–110.

Hobbs R.J. & Suding K.N. (eds) (2009) *New Models for Ecosystem Dynamics and Restoration*. Island Press, Washington, DC.

Hodges, K.E. (2008) Defining the problem: terminology and progress in ecology. *Frontiers in Ecology and the Environment*, **6**, 35–42.

Hulme, P.E. (2003) Biological invasions: winning the science battles but losing the conservation war? *Oryx*, **37**, 178–193.

Jackson, S.T. & Hobbs, R.J. (2009) Ecological restoration in the light of ecological history. *Science*, **325**, 567–569.

Jordan, W.R.I., Gilpin, M.E. & Aber, J.D. (eds.) (1987a) *Restoration Ecology: A Synthetic Approach to Ecological Research*. Cambridge University Press, Cambridge, UK.

Jordan, W.R.I., Gilpin, M.E. & Aber, J.D. (1987b) Restoration ecology: ecological restoration as a technique for basic research. In *Restoration Ecology: A Synthetic Approach to Ecological Research* (ed. W.R. Jordan, M.E. Gilpin and J.D. Aber) pp. 3–21. Cambridge University Press, Cambridge, UK.

Katz, E. (1992) The big lie: human restoration of nature. *Research in Philosophy and Technology*, **12**, 231–241.

Katz, E. (2000) Another look at restoration: technology and artificial nature. In *Restoring Nature: Perspectives from the Social Sciences and Humanities* (ed. P.H. Gobster and R.B. Hull), pp. 37–48. Island Press, Washington, DC.

King, E.G. & Hobbs, R.J. (2006) Identifying linkages among conceptual models of ecosystem degradation and restoration: towards an integrative framework. *Restoration Ecology*, **14**, 369–378.

Leopold, A.S. (1934) The arboretum and the university. *Parks and Recreation*, **18**, 59–60.

Levine, J.M., Vila, M., D'Antonio, C.M., Dukes, J.S., Grigulis, K. & Lavorel, S. (2003) Mechanisms underlying the impacts of exotic plant invasions. *Proceedings of the Royal Society of London B* **270**, 775–781.

Luken, J.O. (1990) *Directing ecological succession*. Chapman and Hall, New York.

MacDougall A.S. & Turkington R. (2005) Are invasive species the drivers or passengers of change in degraded ecosystems? *Ecology*, **86**, 42–55.

Meine, C.D. (1988) *Aldo Lepold: His Life and Work*. University of Wisconsin Press, Madison.

Milton S.J., Dean W.R.J., du Plessis M.A. & Siegfried W.R. (1994) A conceptual model of arid rangeland degradation: the escalating cost of declining productivity. *BioScience*, **44**, 70–76.

Newton, J. L. (2006) *Aldo Leopold Odyssey. Rediscovering the Author of A Sand County Almanac*. Island Press, Washington, DC.

Palmer, M.A., Ambrose, R.F. & Poff, N.L. (1997) Ecological theory and community restoration ecology. *Restoration Ecology*, **5**, 291–300.

Perrow, M.R. & Davy, A.J. (eds) (2002a) *Handbook of Ecological Restoration*. Volume **1**. Principles of Restoration. Cambridge University Press, Cambridge, UK.

Perrow, M.R. & Davy, A.J. (eds.) (2002b) *Handbook of Ecological Restoration*. Volume **2**. Restoration in Practice. Cambridge University Press, Cambridge, UK.

Richardson, D.M., Holmes, P.M., Esler, K.J., et al. (2007). Riparian zones – degradation, alien plant invasions and restoration prospects. *Diversity and Distributions*, **13**, 126–139.

Richardson, D.M., Moran, V.C., Le Maitre, D.C., Rouget, M. & Foxcroft, L.C. (2004) Recent developments in the science and management of invasive alien plants: Connecting the dots of research knowledge, and linking disciplinary boxes. *South African Journal of Science*, **100**, 126–128.

Richardson, D.M. & Pyšek, P. (2007). Classics in physical geography revisited: Elton, C.S. 1958: The ecology of invasions by animals and plants. Methuen: London. *Progress in Physical Geography*, **31**, 659–666.

Richardson, D.M. & Pyšek, P. (2008) Fifty years of invasion ecology – the legacy of Charles Elton. *Diversity and Distributions*, **14**, 161–168.

Richardson, D.M., Pyšek, P., Rejmánek , M., Barbour, M.G., Panetta, D. and West, C.J. (2000) Naturalization and invasion of alien plants: concepts and definitions. *Diversity and Distributions*, **6**, 93–107.

Seabloom, E.W., Harpole, W.S., Reichman, O.J. & Tilman, D. (2003) Invasion, competitive dominance, and resource use by exotic and native California grassland species. *Proceedings of the National Academy of Sciences of the USA*, **100**, 13384–13389.

Simberloff, D. (2003) Confronting introduced species: a form of xenophobia? *Biological Invasions*, **5**, 179–192.

Society for Ecological Restoration International Science & Policy Working Group (2004). *The SER International Primer on Ecological Restoration*. http://www.ser.org & Society for Ecological Restoration International, Tucson.

Suding, K.N. & Hobbs, R.J. (2009) Threshold models in restoration and conservation: a developing framework. *Trends in Ecology & Evolution*, **24**, 271–279.

Temperton, V.M., Hobbs, R.J., Nuttle, T.J. & Halle, S. (eds) (2004) *Assembly Rules and Restoration Ecology: Bridging the Gap Between Theory and Practice*. Island Press, Washington, DC.

Theodoropoulos, D.I. (2003) *Invasion Biology: Critique of a Pseudoscience*. Avvar Books, Blythe, California.

van Andel, J. & Aronson, J. (eds) (2006) *Restoration Ecology: The New Frontier*. Blackwell, Oxford.

Walker, L.R., Walker, J. & Hobbs, R.J. (eds) (2007) *Linking Restoration and Succession*. Springer, New York.

Warren, C.R. (2007) Perspectives on the 'alien' versus 'native' species debate: a critique of concepts, language and practice. *Progress in Human Geography*, **31**, 427–446.

Whisenant, S.G. (1999) *Repairing Damaged Wildlands: A Process-Orientated, Landscape-Scale Approach*. Cambridge University Press, Cambridge, UK.

Zavaleta, E.S., Hobbs, R.J. & Mooney, H.A. (2001) Viewing invasive species removal in a whole-ecosystem context. *Trends in Ecology & Evolution*, **16**, 454–459.

Part 3

New Takes on Invasion Patterns

Chapter 7

BIOLOGICAL INVASIONS IN EUROPE 50 YEARS AFTER ELTON: TIME TO SOUND THE ALARM

Petr Pyšek[1] and Philip E. Hulme[2]

[1]Institute of Botany, Academy of Sciences of the Czech Republic, CZ-252 43 Průhonice, Czech Republic; and Department of Ecology, Faculty of Science, Charles University, Viničná 7, CZ-128 01 Praha 2, Czech Republic

[2]The Bio-Protection Research Centre, Lincoln University, PO Box 84, Christchurch, New Zealand

Fifty Years of Invasion Ecology: The Legacy of Charles Elton, 1st edition. Edited by David M. Richardson
© 2011 by Blackwell Publishing Ltd

7.1 INTRODUCTION

Although the rates of species' introductions to Europe accelerated significantly during the second half of the 20th century (Hulme et al. 2009c), concerted efforts to understand biological invasions across the continent are relatively recent (Hulme et al. 2009b). Earlier efforts focused mainly on floristic and faunal inventories that led to deliberations on the origins and alien status of species (Chew 2006; Davis 2006). Although such inventories were an important first step, alien species were often viewed as biogeographical curiosities. It was not until the dynamic nature of biological invasions was recognized and awareness of potential impacts increased that the study of alien species began to be incorporated into the field of population ecology. In this respect Europe lagged behind many other continents. For example, concerns about impacts of alien plants in New Zealand were expressed much earlier than in Europe (Allen 1936) but major plant invasions in natural environments in Europe only started being observed in the 1950s in increasingly disturbed landscapes (Williamson et al. 2005).

Invasion ecology has grown enormously in the 50 years since Charles Elton's *The Ecology of Invasions by Animals and Plants* (hereafter 'Elton's book') was published in 1958 (Davis 2006; Pyšek et al. 2006; Richardson & Pyšek 2007, 2008; Ricciardi & MacIsaac 2008; MacIsaac et al., this volume). This chapter reviews the current state of knowledge about biological invasions in Europe and evaluates the dramatic changes that have occurred since Elton's book. Europe, especially its Mediterranean region, has traditionally been considered a donor of invasive species to other parts of the world for historical reasons and the long association of plants and animals with humans since the beginning of agriculture some 10,000 years ago (di Castri 1989). This chapter aims to show that recent systematic research efforts in the past decade in Europe may have changed this long held view: plants and animals of alien origin now form a substantial part of the continent's biodiversity (DAISIE 2009), especially in the Mediterranean (Hulme et al. 2008), and exert huge and diverse impacts on both environment and economy (Vilà et al. 2009). To put the current situation into historical context, we first review how informative Elton's book is in terms of providing an accurate picture of biological invasions in Europe 50 years ago. We then use this as a baseline to explore how the situation and perspectives have changed in the present day.

7.2 STARTING WITH EXAMPLES: ELTON'S ANIMAL-BIASED PERSPECTIVE

Elton's book is very much about changes in species' distributions; biogeography was at the heart of his thinking. He was also a zoologist and, though he dealt with plant invasions in his book, they received far less attention than animals. Of the 195 organisms listed in the book's index, 169 (87%) are animals; of these 51% are arthropods, 27% vertebrates and the rest molluscs (Richardson & Pyšek 2007).

Therefore, for animals, comparison of the distribution maps in his book with current situation provides revealing histories. For example, the Colorado beetle, *Leptinotarsa decemlineata*, has increased its range eastwards and in Scandinavia (compare Fig. 7.1a with b); this species was introduced from South America to France in 1922 and until 1950s expanded throughout the European continent and parts of Asia (Lopez-Vaamonde 2009). The muskrat, *Ondatra zibethicus*, has also expanded its range, having now colonized most of the area encompassed by Elton's maps (compare Fig. 7.1c with d). However, Elton would probably be surprised at the speed and scale of invasions by more recent animal arrivals in Europe: the horse chestnut leaf-miner, *Cameraria ohridella* (Fig. 7.1e), currently a major problem in Europe (see, for example, Girardoz et al. 2006), rose-ringed parakeet, *Psittacula krameri* (Fig. 7.1f), invasion of which in Europe started as late as in the 1960s and 1970s through trading and escape from aviaries (Shwartz & Shirley 2009), or the comb jelly *Mnemiopsis leydii*, first found in Europe in the Black Sea in 1982 (Shiganova & Panov 2009). These species clearly illustrate that the dimensions and complexities of the problem have changed radically since Elton's time.

Only 21 plant taxa feature in Elton's book. A closer look reveals that stories of plant invasions are even less represented in support for the conclusions he drew. For example, the 21 plants mentioned account for 11% of all organisms dealt with in the book, but only 11 refer to invasive plants (others mostly mention plants in their native ranges and/or in relation to animals, mostly invading insects). Only six refer to invasions in

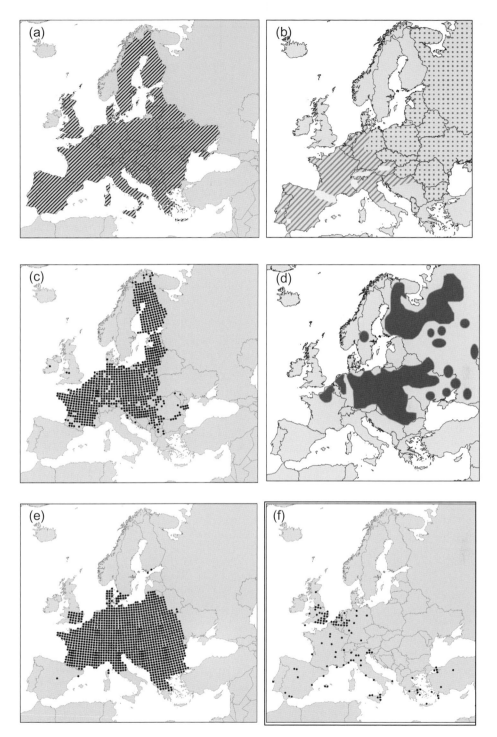

Fig. 7.1 Current distribution of animals mentioned by Elton as invasive in Europe (a, *Leptinotarsa decemlineata*; c, *Ondatra zibethicum*), compared with that given in his book (b, *Leptinotarsa decemlineata*: dots indicate no data; d, *Ondatra zibethicum*), and current distribution of recent invaders (e, *Cameraria ohridella*; f, *Psittacula krameri*). Species' distributions are mapped in 50 km × 50 km grid cells; hatching indicates that the species is reported from the given country, but does not necessarily occur over the whole area. Taken from DAISIE (2009); (b) and (d) redrawn from Elton (1958).

Europe, and only three of these species had by the 1950s undergone dramatic changes in distributions. Not surprisingly, the three primary examples were all from the UK (*Acer pseudoplatanus*, *Rhododendron ponticum* and *Spartina townsendii*) and only these three species are accorded as much detail as some of the animal examples Elton used. Yet, none of the plant species he dealt with was depicted on a map. Of the invasive plants he considered, *Spartina townsendii* (Fig. 7.2a) remains restricted to the UK and western coast of Europe but *Rhododendron ponticum* is now naturalized not only in the British Isles, but also in Belgium, France, the Netherlands and Austria (Fig. 7.2b).

Within Britain and Ireland, it has spread substantially since 1958. It was reported from only 125 hectads (10 km^2 grid cells) in 1970, but in 2248 hectads by 1999 (Preston et al. 2002). A similarly massive increase in Britain and Ireland occurred for sycamore, *Acer pseudoplatanus*, within this period (from 109 to 3400 hectads); the species is nowadays also naturalized in Scandinavia (Fig. 7.2c). *Rosa multiflora*, however, exemplifies a plant species about which Elton was obviously wrong; he considered it beneficial and admired it planted in hedgerows, but this species increased in distribution in the British Isles from 5 hectads to 102 in the last three decades of the 20th

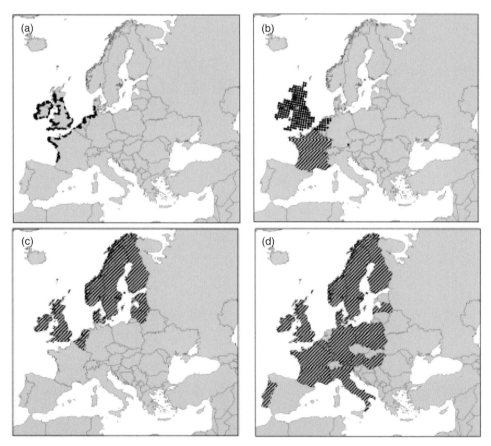

Fig. 7.2 Current distribution of plant species mentioned by Elton (1958) as invasive in Europe. (a) *Spartina townsendii*; (b) *Rhododendron ponticum*; (c) *Acer pseudoplatanus*; (d) *Rosa multiflora*, which he considered harmless, but this species has invaded not only Europe but also the USA since then. Species' distributions are mapped in 50 km × 50 km grid cells; hatching indicates that the species is reported from the country, but does not necessarily occur over the whole area. Taken from DAISIE (2009).

century (Preston et al. 2002) and it is now naturalized over a large part of Europe (Fig. 7.2d). In the eastern United States too, it became a noxious invader and has been the subject of numerous control campaigns since the 1960s (Simberloff 2000).

However, Elton's book is about examples. It does not provide quantitative insights into continent-wide patterns half a century ago. To be more quantitative, we can use current knowledge on alien plants in Europe for which the information is probably the most complete. Of the total number of naturalized neophytes (species introduced after 1500 AD; Pyšek et al. 2004) now present in Europe, over a quarter arrived after 1962, with 10% being even more recent with introductions occurring after 1989 (Lambdon et al. 2008). Using the estimated dates of introduction of alien plants in Europe (Lambdon et al. 2008), three-quarters were probably naturalized somewhere in Europe by 1958. Thus, there were approximately 1300 naturalized neophytes from other continents in Europe when Elton wrote his book; we might therefore have expected more than six examples to be addressed. The main reason for the minor attention given to plant invasions by Elton may be their perceived minor impact since, as in other parts of the world (Allen 1936), most alien plant species (and certainly major 'invasions') were in disturbed habitats (Pyšek et al. 2010), a feature that is acknowledged in his book. Thus the high species numbers of neophytes might not have been automatically translated into a great impact. The publication in 1961 of Sir Edward Salisbury's *Weeds and Aliens*, the first overview of this topic in a single volume, suggests that there was a much wider appreciation of plant invasions than emerges from reading Elton's book. Indeed, Salisbury (1961) addresses some of the very same plant species, though in more detail. He too focuses largely on the British Isles, and hardly addressed patterns for the rest of Europe.

Importantly, the most likely explanation for the dearth of plant examples in Elton's book is that some spectacular plant invasions in Europe only started in the late 1950s or were yet to explode (e.g. *Heracleum mantegazzianum*, see Pyšek et al. (2007a); *Caulerpa taxifolia*, *Carpobrotus edulis*). The indication of the magnitude of changes in the status of plant invasions in Europe over the past four decades can be documented by using the UK as an example. For this country, mapping data systematically collected in several periods are available and show that before 1970 there were 887 neophytes recorded, which together occurred in 50,655 hectads. By 1999 the number of neophyte species increased to 1438 and number of species–hectad records to 283,469, representing an increase by 460% (calculated from data in Preston et al. (2002)). This is indicative of the huge changes in the status of plant invasions in Europe.

7.3 FROM EXAMPLES TO PATTERNS: NUMBERS OF ALIEN SPECIES IN EUROPE

Starting with the big picture, what introduced species are found in Europe? The solid current position of Europe on the global map of research in biological invasions is largely due to two pan-European projects performed under the European Union 6th Framework Programme. The Delivering Alien Invasive Species Inventories for Europe (DAISIE) project established the European Alien Species Database, the European Expertise Registry and the European Invasive Alien Species Information System (Hulme et al. 2009a). Together they provide an online 'one-stop-shop' (www.europe-aliens.org) for information on biological invasions in Europe (Hulme et al. 2009a,b).

Based on data from 71 terrestrial and nine marine regions, DAISIE revealed that 10,771 alien species are known to occur in Europe (see DAISIE 2009 for the checklist of all alien species recorded, ranked taxonomically, that can be used as a reference for future assessment of trends in biological invasions in Europe). The taxa belong to 4492 genera and 1267 families. Plants are best represented and account for 55% of taxa, which is 5789 species; of this number, about a half are invasions within the continent, but the 2846 species of extra-European origin (Lambdon et al. 2008) add to 10,928 native plant species (Winter et al. 2010) and represent thus 20.3% of the total plant diversity in Europe. Terrestrial invertebrates account for 23% (2477 species), followed by vertebrates (6%), fungi (5%; Desprez-Lousteau et al. 2007), molluscs (4%), annelids (1%) and red algae (1%). DAISIE produced the first ever checklists of alien biota for some countries and substantially improved the accuracy of estimates of alien species numbers derived from previous datasets (Hulme et al. 2009d). For example, from the dynamics of introduction of alien plants it can be inferred that approximately 38% of the species present in Europe in the 1980s were not captured by the then completed authoritative Flora Europaea (Tutin et al.

1964–1980). The increase in the numbers of species is thus partly due to the quality of data collated by DAISIE, partly it also reflects new introductions.

The data collated by DAISIE made it possible to assess regional levels of plant and animal invasions in European countries, with large industrialized countries in the western part of the continent harbouring the highest numbers of alien species (Pyšek et al. 2011). There is a significant relationship between the total number of naturalized plant and animal species and the country GDP (Fig. 7.3; Hulme 2007). Of the marine basins, the Mediterranean is most species-rich in alien biota; of the total 737 alien multicellular alien species, 569 are recorded in this basin, whereas the Atlantic coast harbours 200 and Baltic Sea 69 alien species (Galil et al. 2009).

The detailed information about alien species from a wide range of groups makes it possible to evaluate the dynamics of introductions of alien species to Europe over the past century. The numbers of newly recorded naturalized taxa on the continent are generally increasing, in both terrestrial and aquatic environments, and this increase exhibits an accelerating trend (Fig. 7.4). In the past century, the number of naturalized plants increased from 4.1 per annum in 1900s–1920s to 6.5 in the past two decades. Per annum rates

of introductions of amphibians and reptiles tripled (from 0.6 to 1.8 species) and birds exhibited similar dynamics over this period (from 1.4 to 4.8). The only exception to this trend is mammals, where it seems that most suitable introductions arrived in earlier decades, resulting in rather stable rates of introduction over time (Fig. 7.4). The same accelerating dynamics are found in marine environments, with more than 10 and four new species in the Mediterranean basin and Atlantic coast, respectively, recorded each year in the last decade of the 20th century, and one new species in the Baltic Sea (Galil et al. 2009; their figure 7.1). Overall, the rate of new introductions to Europe has increased sharply throughout the past century and is showing little sign of slowing down (Hulme et al. 2009c).

7.4   FROM PATTERNS TO MECHANISMS: WHAT DO WE KNOW?

Since 2000, the European Union has supported a variety of research initiatives that address different aspects of biological invasions (see Hulme et al. 2009b and their table 1). Yet, it was not until the completion of DAISIE that the accumulation of considerable data

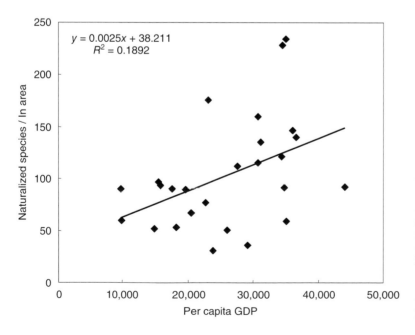

Fig. 7.3 Relationship between the number of naturalized alien species (fungi, bryophytes, plants, terrestrial invertebrates, fishes, amphibians, reptiles, birds and mammals) introduced after 1500 AD and per capita GDP (in US dollars) shown for European Union member countries (excluding one outlier: Luxembourg). $F = 5.60$, df = 1, 24, $P < 0.05$. Data taken from DAISIE database and Lambdon et al. (2008).

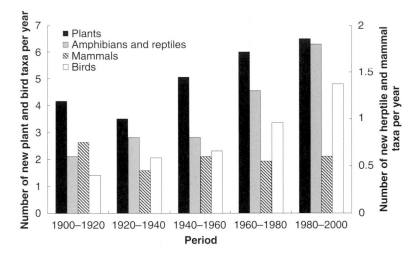

Fig. 7.4 Dynamics of introduction of alien plants and vertebrates in terrestrial environments to Europe. Alien taxa newly recorded as naturalized are shown per annum for the time periods indicated. Based on Hulme et al. (2009c) and reproduced with permission.

made it possible to aim at large-scale syntheses of biological invasions at the continental level, and search for general principles, some of them valid across a wide range of taxonomic groups and environments, as well as to assess the risk from invasions. This work was performed within the framework of the Assessing Large Scale Risks for Biodiversity using Tested Methods (ALARM) project (www.alarmproject.net; Settele et al. 2005). For the first time, it brought together many of the researchers who had previously worked on some of the European projects described above. At the end of 2009, at least 142 journal papers and book chapters dealing with biological invasions have appeared from ALARM (for references, see www.alarmproject. net).

Through DAISIE and ALARM, a conceptual approach emerged which addressed crucial steps or elements of the invasion process at the European-wide, continental scale. Table 7.1 summarizes Europe's most recent key contributions to the knowledge of invasion mechanism and patterns at a large, macroecological scale. The topics studied included pathways of invasion, the role of habitats, the role of species' traits in determining invasiveness, interaction of alien plants with local pollinators, drivers of invasion, ecological and economic impact of invasions, and risk-assessment. Some of these topics were discussed by Elton; for example, he could hardly have missed the crucial importance of how animals and plants are being introduced from one region to another by using various

pathways, and he gave serious attention to this issue. Only recently have invasion pathways been conceptualized, described and parameterized based on European data (Hulme et al. 2008; Hulme this volume) Elton also recognized the key role of disturbed habitats in plant invasions and the great richness of aliens these habitats harbour. Thanks to current focus on habitat invasibility, these patterns have been recently described in detail and analysed (Vilà et al. 2007; Chytrý et al. 2008b; Hejda et al. 2009b; Pyšek et al. 2010). Distribution and spread, as basic features of biological invasions, were the focus of Elton's interest, although from comparison of his data with recent knowledge it clearly appears that periods of massive spread of most invaders were only about to come. It has been estimated that at present, 50 years after Elton, most European neophytes are still filling their potential ranges, a process that is estimated to take between 150 and 300 years after introduction (Williamson et al. 2009).

The substantial progress in creating inventories of invasions in Europe allows for detailed cross-taxonomic comparisons for many taxonomic groups from fungi, plants and invertebrates to vertebrates, in both terrestrial and aquatic environments; several studies aimed at cross-taxonomic analyses of patterns and identifying mechanisms valid for invasive biota in general. The data made it possible to perform these analyses at the continental scale, hence capturing a long-enough gradient of environmental settings, including climate,

Table 7.1 Key findings of recent studies on biological invasions in Europe, performed within the ALARM, DAISIE and other European projects. Only studies based on analyses of original data and addressing the continental or sufficiently large sub-continental scale, and/or a range of taxonomic groups are outlined. Environment: T, terrestrial; F, freshwater; M, marine.

Topic	Scale	Taxa and environment
Pathways of introduction	Continental	Plants, vertebrates, invertebrates, pathogens: T, F, M
Introduction dynamics	Continental	Plants, amphibians and reptiles, birds, mammals: T
Invasion dynamics	Regional	Plants: T
Evolution of invasiveness	Regional	Plants: T
Species' traits and invasiveness	Regional	Plants: T
Habitat affinities	Continental	Plants, insects, amphibians, reptiles, birds, mammals: T, F
Habitat affinities	Continental, regional	Plants: T
Drivers of invasion	Continental	Fungi, bryophytes, plants, invertebrates, fishes, amphibians, reptiles, birds, mammals: T, F
Pathways of introduction	Continental	Aquatic invaders: F
Drivers of invasion: propagule pressure	Continental, regional	Birds: T
Drivers of invasion: climate	Continental	Plants: T

Outcome	Reference
Alien species arrive using six principal pathways: release, escape, contaminant, stowaway, corridor and unaided. Vertebrate pathways tend to be characterized as deliberate releases, invertebrates as contaminants and plants as escapes. Pathogenic micro-organisms and fungi are generally introduced as contaminants of their hosts. The new framework enables these trends to be monitored and develop regulations to stem the number of future introductions.	Hulme et al. (2008)
Dynamics of invasion of alien species from all taxonomic groups and environments are increasing and show no sign of deceleration.	Hulme et al. (2009d)
Naturalized neophytes have smaller range sizes than natives. Historical dynamics indicates that it takes at least 150 years for an alien species to reach its full potential distribution. Most naturalized neophytes are still expanding their ranges in Europe.	Williamson et al. (2009)
Invasiveness does not appear to have a strong phylogenetic component. The presence or absence of native congeners therefore has limited influence on whether an introduced alien plant becomes naturalized.	Lambdon & Hulme (2006a); Lambdon (2008)
A range of traits may influence the likelihood of naturalization but it appears that species origin and the pathways of introduction are as important, if not more so, than the life-history traits of individual species.	Lloret et al. (2004, 2005); Lambdon & Hulme (2006b)
There are two ecologically distinct groups of alien species (plants and insects versus vertebrates) with strikingly different habitat affinities. Invasions by these two contrasting groups are complementary in terms of habitat use. Diversity of alien plants and insects concentrates in riparian and urban habitats, vertebrates are more evenly distributed and also invade aquatic habitats and woodland.	Pyšek et al. (2010)
Habitats are the more important predictors of the local level of invasion than climate and propagule pressure and the patterns of habitat invasions are consistent among biogeographical regions.	Chytrý et al. (2008a,b, 2009a,b); Vilà et al. (2007)
Macroeconomic factors are the most important predictors of the level of invasion in European regions, when analysed jointly with geographical and climatic variables.	Pyšek et al. (2010c)
The introduction and establishment transitions are independent of each other, and species that became widely established did so because their introduction was attempted in many countries, not because of a better establishment capability. The level of invasion of European countries is determined by their area and human population density, but not per capita GDP.	García-Berthou et al. (2005)
Community-level propagule pressure is the major driver shaping the distribution of alien birds in Europe.	Chiron et al. (2009)
Low precipitation constrains alien species richness in warm regions. A rather complex response of alien species to climate change in Europe may be expected, with drought becoming possibly important limiting factor in the future.	Lambdon et al. (2008)

Continued

Table 7.1 *Continued*

Topic	Scale	Taxa and environment
Ecological and economic impact	Continental	Plants, invertebrates, vertebrates: T, F, M
Impact on native pollinators	Continental	Plants: T
Homogenization	Continental	Plants: T
Prediction of future trends	Continental	Plants: T
Species' traits and impact	Continental	Birds, mammals: T

latitudinal and altitudinal trends, habitat heterogeneity and different levels of economies reflecting historical differences in the development of European nations (Table 7.1).

To summarize the most novel results from recent European studies, it appears that the six principal pathways used by alien species – release, escape, contaminant, stowaway, corridor and unaided – differ in importance according to the taxonomic group introduced and environment invaded (Hulme et al. 2008). The importance of propagule pressure associated with individual pathways differs with respect to the type of the pathways and organism in question; although it is of primary importance in, for example, fish and birds, where the pool of invasive species largely results from release (Garcia-Berthou et al. 2005; Chiron et al. 2009), it is less important in determining the level of invasion of plant communities, to which alien species arrive by other pathways, mostly escape from cultivation, contaminant and stowaway. For plants, the type of habitats in which the invasion occurs is a more important predictor of the local level of invasion than

climate and propagule pressure (Chytrý et al. 2008a,b, 2009a,b; Vilà et al. 2007). Assessing the level of invasion of European habitats across taxonomic groups indicates that plants and insects have similarly close habitat affinities to riparian and urban habitat, which is strikingly different from vertebrates that invade in aquatic habitats and woodland; invasions by these two contrasting groups of biota are therefore complementary in terms of habitat use (Pyšek et al. 2010). Species invasiveness does not appear to have a strong phylogenetic component (Lambdon & Hulme 2006a; Lambdon 2008; Pyšek et al. 2009a); a range of traits affect the likelihood of invasion success but these traits interact with each other (Küster et al. 2008) and act in concert with other factors such as species origin, pathways of introduction (Lloret et al. 2004, 2005; Lambdon & Hulme 2006b) and propagule pressure and residence time (Pyšek et al. 2009b). These factors are generally more important than the life-history traits of individual species and the role of traits depends on the stage of invasion process, increasing as species reach more advanced stages of invasion (Pyšek et al.

Outcome	Reference
Ecological and economic impacts are only documented for approximately 10% of alien species in Europe, but many invaders cause multiple impacts over a large area in Europe. Terrestrial vertebrates and aquatic inland invaders are most efficient in causing negative ecological impacts, terrestrial invertebrates and vertebrates an economic impact. Ecological and economic impacts of alien organisms are correlated.	Vilà et al. (2010); Nentwig et al. (2010); Kenis et al. (2009)
Native pollinators depend upon alien plants more than on native plants, but the networks of native pollinators are very permeable and robust to the introduction of invasive alien species into the network.	Vilà et al. (2009)
Invasions of alien and extinctions of native species over the past centuries resulted in increased taxonomic and phylogenetic similarity among European regions that are losing part of their uniqueness due to this homogenization effect.	Winter et al. (2010)
European map of plant invasions based on land-use is a convenient tool for predicting future invasions under contrasting socioeconomic scenarios; the one focused on sustainability may not necessarily result in decreased level of plant invasions in the next 80 years.	Chytrý et al. (2010)
Habitat generalist birds and mammals have greater impact than habitat specialists.	Shirley & Kark (2009); Nentwig et al. (2010)

2009a,b). Since invasions are human-induced processes, macroeconomic factors, as a suitable surrogate of underlying factors, appear to be the most important predictors of the level of invasion in European regions, more so than geographical factors and climate (Pyšek et al. 2010c).

The macroecological analyses highlighted in Table 7.1 are being complemented by numerous studies at smaller scales, focused on individual taxonomic groups, and addressing specific topics, for example impact of invasive species on the diversity of invaded communities (Hejda & Pyšek 2006; Hulme & Bremner 2006; Truscott et al. 2008; Hejda et al. 2009a), competition between native and alien species (Fabre et al. 2004; Kohn et al. 2009) and biotic resistance (Paavola et al. 2005; Vilà et al. 2008), interaction across multiple trophic levels (Girardoz et al. 2006), invasibility of islands (Gimeno et al. 2006) or the role of climate in invasions (Truscott et al. 2006; Broenniman et al. 2007; Walther et al. 2007; Ross et al. 2008). Several case studies of individual invasive species addressing a wide array of methodological approaches were also

produced in recent years, resulting from the above-mentioned European-Union-funded projects.

7.5 FROM UNDERSTANDING TO ACTION: WHERE TO NOW?

The field of invasion ecology suffers from the lack of effective translation of academically gratifying research results to management (Hulme 2003). Stakeholders feel disconnected from the science (Andreu et al. 2009) and theoretical advances need to be more effectively translated into improved management, including objective means for conflict resolution (Richardson & Pyšek 2008; Pyšek & Richardson 2010). We believe that European efforts in the past few years have forged some progress towards a more applied perspective and that a start is being made at adopting successful approaches for putting science into practice that were pioneered in Australia and New Zealand. European scientists have assisted managers by placing their results within a more applied framework such as the giant hogweed manual (Nielsen et al. 2005), have led attempt to raise awareness among the public (DAISIE 2009) and where possible have addressed policy gaps (Hulme et al. 2008).

Among the most important achievements of recent research efforts in Europe was a thorough evaluation of impact, leading to the first continental-wide inventory of the magnitude and variety of ecological and economic impacts of invasive alien species in global terms (Vilà et al. 2010). This assessment concerned negative impacts of alien plants, vertebrates and invertebrates on ecosystem services in terrestrial, freshwater and marine environments. There are currently 1094 species with documented ecological impacts and 1347 with economic impacts in Europe; this points to a serious gap in knowledge because the impact has not been assessed for 90% of the total number of alien species in Europe. Alien species from all taxonomic groups affect supporting, provisioning, regulating and cultural ecosystem services (Binimelis et al. 2007) and interfere with human well-being. Terrestrial vertebrates are responsible for the greatest range of impacts, and these are widely distributed across Europe. Terrestrial invertebrates lead to greater economic impacts than ecological impacts, while the reverse is true for terrestrial plants, where only a small proportion of the total number of species recorded in Europe have impact, and terrestrial vertebrates and aquatic inland invaders are most efficient in causing negative ecological impacts. In economic terms, it is terrestrial invertebrates and particularly terrestrial vertebrates where almost 40% of invaders have an economic impact. Based on this and other studies, the total annual costs of invasive alien species in Europe are estimated at €12.5 billion (Kettunen et al. 2009). Thanks to the study of Vilà et al. (2010), Europe has the most up-to-date information on numbers of aliens and their impacts but lags behind North America, a continent where biological invasions are studied most intensively (Pyšek et al. 2008), in the knowledge of mechanisms underlying impacts (Hulme et al. 2009b). This difference in focus and nature of information on impact is where the two continents can profit from each other's experiences and work towards reliable and comparable estimates of costs from alien species invasions (Vilà et al. 2010).

Remarkable progress has been made towards improving risk-assessment of biological invasions in Europe. For marine environments, an index has been developed that classifies the impacts of alien species on native species, communities, habitats and ecosystem functioning. The method can be used to evaluate impact at five different levels of 'biopollution' and is compatible within the existing schemes for water quality assessment (Olenin et al. 2007). A generic scoring system has been developed for invasive mammals that takes into account both environmental (competition, predation, hybridization, transmission of disease and herbivory) and economic (on agriculture, livestock, forestry, human health and infrastructure) impacts and distinguishes between 'actual' (determined by actual distribution of the invasive species assessed) and 'potential' impacts (Nentwig et al. 2010). For plants, existing risk-assessment schemes were tested to identify those most appropriate for Europe (Křivánek & Pyšek 2006) and the role of deliberate planting for forestry purposes on the invasion process evaluated (Křivánek et al. 2006). Consistent results on habitat invasibility by plants from different European biogeographical regions (Chytrý et al. 2008b) provided an excellent opportunity for mapping plant invasions not only at the regional level (Chytrý et al. 2009b) but for the whole of Europe (Chytrý et al. 2009a). Within the framework of linking the ecological research with socio-economic aspects (see, for example, Binimelis et al. 2007; Kobelt & Nentwig 2008; Andreu et al. 2009), the European map of plant invasions was used to project current levels of invasion under integrated scenarios of future socio-economic development in Europe, based on land-use patterns in Europe projected for 2020, 2050 and 2080 (Spangenberg 2007). This research indicates that an implementation of sustainability policies will not automatically restrict the spread of alien plants, but such policies might rather increase invasions by supporting agriculture and associated invasion-prone land use in less productive areas. This suggests that proactive strategies to manage invasive alien plants will be needed no matter how environmental friendly policies would be adopted in the future (Chytrý et al. 2010).

These examples illustrate that some progress towards developing tool of assessing risk from invasive species is being made, but where does Europe stand in terms of research and management of biological invasions? Recent efforts yielded results and data that will provide more results in the near future. These results represent a great potential to contribute to better understanding biological invasions by testing hypotheses and searching for general patterns valid across a range of taxa and environments. This knowledge can be used to improve scientific-based risk assessments; invasions are complex phenomena and the more key elements of the invasion process (pathways, habitats, species' traits, impact, etc.) are considered in an integrative risk-assessment scheme, the more effective such schemes would be. Last but not least, DAISIE and ALARM received much attention from the European Union administration and their results are used for developing the European strategy against invasive alien species (Kettunen et al. 2009; see http://ec.europa.eu/environment/nature/pdf/council_concl_0609.pdf).

For Europe to address biological invasions at a continental scale, there must be an end to the fragmented legislative and regulatory requirements addressing invasive species and the piecemeal approaches to tackling invasive species across Europe that fail to coordinate pre- and post-border actions (Hulme et al. 2009a). DAISIE established a database on expertise addressing biological invasions in Europe and it is clear that such expertise is heterogeneously distributed across the continent resulting in variable efficiencies in national monitoring and surveillance. As a result, Europe's borders can be easily penetrated by alien species. Furthermore, relative to understanding of the ecology, distribution and taxonomy of alien species in Europe, expertise in management and mitigation of impacts is the principal activity of little more than 10% of invasion scientists. The disparate nature of expertise in Europe also means that as assessing the risks of alien species has become increasingly complex specialist expertise and access to appropriate databases is required. At the same time, the current knowledge base provides an excellent foundation for concerted management action. Europe is nowadays a continent with the most integrated and comprehensive information on its alien biota particularly in terms of distribution patterns, invasion history and impacts. The comprehensiveness of the information is probably better than for any other continent that includes a clear picture of how fast species from other continents invade Europe, by which pathways, how they are distributed in habitats, what is their impact on biodiversity, ecosystems and economy and what can be expected in terms of future development. Although much remains to be done and the improving and completing the data is a never-ending process, the information needed for developing an effective strategy at the continental level is basically available. Our belief is that more than ever before, a single European coordinating

centre with a specific remit to manage biological invasions is needed and should be developed with a mission to identify, assess and communicate current and emerging threats to the economy and environment posed by invasive species (Hulme et al. 2009c). A similar call for a single coordinating centre has been made for the USA (Lodge et al. 2006). Though invasive species can impose considerable impacts on economy and ecosystems, only 2% of Europeans feel that invasions are a significant threat to biodiversity. A single European coordinating body would build public awareness of the problem of invasions, involve the public in finding alternatives and solution, build long-term partnerships with concerned sectors and users, and encourage voluntary approaches and best practices where feasible.

ACKNOWLEDGEMENTS

We thank Mark Davis, David Richardson and an anonymous reviewer for helpful comments on the manuscript, members of DAISIE and ALARM consortia for cooperation, and Jan Pergl and Zuzana Sixtová for technical assistance. This work was supported by the European Union's FP6 projects ALARM (GOCE-CT-2003-506675; Settele et al. 2005), DAISIE (SSPICT-2003–511202), and FP7 project PRATIQUE (212459; Baker et al. 2009). P.P. was further supported by the Ministry of Education of the Czech Republic (MSM0021620828 and LC06073) and the Academy of Sciences of the Czech Republic (AV0Z60050516 and Praemium Academiae award).

REFERENCES

Allen, H.H. (1936) Indigene versus alien in the New Zealand plant world. *Ecology*, **17**, 187–193.

Andreu, J., Vilà, M. & Hulme, P.E. (2009) An assessment of stakeholder perceptions and management of noxious alien plants in Spain. *Environmental Management*, **43**, 1244–1255.

Baker, R.H.A., Battisti, A., Bremmer, J., et al. (2009) PRATIQUE: a research project to enhance pest risk analysis techniques in the European Union. *EPPO Bulletin*, **39**, 87–93.

Binimelis, R., Born, W., Monterroso, I. & Rodrigeuz-Labajos, B. (2007) Socio-economic impact and assessment of biological invasions. In *Biological Invasions* (ed. W. Nentwig), pp. 331–350. Springer, Berlin and Heidelberg.

Broennimann, O., Treier, U.A., Müller-Schärer, H., Thuiller, W., Peterson, A.T. & Guisan, A. (2007) Evidence of climatic niche shift during biological invasion. *Ecology Letters*, **10**, 701–709.

Chew, M.K. (2006) *Ending with Elton: Preludes to Invasion History*. PhD thesis, Arizona State University, Tempe.

Chiron, F., Shirley, S. & Kark, S. (2009) Human-related processes drive the richness of exotic birds in Europe. *Proceedings of the Royal Society B* **276**, 47–53.

Chytrý, M., Jarošík, V., Pyšek, P., et al. (2008a) Separating habitat invasibility by alien plants from the actual level of invasion. *Ecology*, **89**, 1541–1553.

Chytrý, M., Maskell, L., Pino, J., et al. (2008b) Habitat invasions by alien plants: a quantitative comparison between Mediterranean, subcontinental and oceanic regions of Europe. *Journal of Applied Ecology*, **45**, 448–458.

Chytrý, M., Pyšek, P., Wild, J., Maskell, L.C., Pino, J. & Vilà, M. (2009a) European map of alien plant invasions, based on the quantitative assessment across habitats. *Diversity and Distributions*, **15**, 98–107.

Chytrý, M., Wild, J., Pyšek, P., Tichý, L., Danihelka, J. & Knollová, I. (2009b) Maps of the level of invasion of the Czech Republic by alien plants. *Preslia*, **81**, 187–207.

Chytrý, M., Wild, J., Pyšek, P., et al. (2010) Projecting trends in plant invasions in Europe under different scenarios of future land-use change. *Global Ecology and Biogeography*, doi:10.1111/j.1466-8238.2010.00573.x.

DAISIE (ed.) (2009) *Handbook of Alien Species in Europe*. Springer, Berlin.

Davis, M.A. (2006) Invasion biology 1958–2005: the pursuit of science and conservation. In *Conceptual Ecology and Invasion Biology: Reciprocal Approaches to Nature* (ed. M.W. Cadotte, S.M. McMahon and T. Fukami), pp. 35–64. Springer, Berlin.

Desprez-Loustau, M.L., Robin, C., Buée, M., et al. (2007) The fungal dimension of biological invasions. *Trends in Ecology & Evolution*, **22**, 472–480.

di Castri, F. (1989) History of biological invasions with special emphasis on the Old World. In *Biological Invasions: A Global Perspective* (ed. J.A. Drake, H.A. Mooney, F. di Castri, et al.), pp. 1–30. John Wiley, Chichester, UK.

Elton, C.S. (1958) *The Ecology of Invasions by Animals and Plants*. Methuen, London.

Fabre, J.P., Auger-Rozenberg, M.A., Chalon, A., Boivin, S. & Roques, A. (2004) Competition between exotic and native insects for seed resources in trees of a Mediterranean forest ecosystem. *Biological Invasions*, **6**, 11–22.

Galil, B., Gollasch, S., Minchin, D. & Olenin, O. (2009) Alien marine biota of Europe. In *The Handbook of Alien Species in Europe* (ed. DAISIE), pp. 93–104. Springer, Dordrecht.

García-Berthou, E., Alcaraz, C., Pou-Rovira, Q., Zamora, L., Coenders, G. & Feo, C. (2005) Introduction pathways and establishment rates of invasive aquatic species in Europe. *Canadian Journal of Fisheries and Aquatic Sciences*, **62**, 453–463.

Gimeno, I., Vilà, M. & Hulme, P.E. (2006) Are islands more susceptible to plant invasion than continents? A test using *Oxalis pes-caprae* L. in the western Mediterranean. *Journal of Biogeography*, **33**, 1559–1565.

Girardoz, S., Kenis, M. & Quicke, D.L.J. (2006) Recruitment of native parasitoids by an exotic leaf miner, *Cameraria ohridella*: host–parasitoid synchronisation and influence of the environment. *Agricultural and Forest Entomology*, **8**, 48–56.

Hejda, M. & Pyšek, P. (2006) What is the impact of *Impatiens glandulifera* on species diversity of invaded riparian vegetation? *Biological Conservation*, **132**, 143–152.

Hejda, M., Pyšek, P. & Jarošík, V. (2009a) Impact of invasive plants on the species richness, diversity and composition of invaded communities. *Journal of Ecology*, **97**, 393–403.

Hejda, M., Pyšek, P., Pergl, J., Sádlo, J., Chytrý, M. & Jarošík, V. (2009b) Invasion success of alien plants: do habitats affinities in the native distribution range matter? *Global Ecology and Biogeography*, **18**, 372–382.

Hulme, P.E. (2003) Biological invasions: winning the science battles but losing the conservation war? *Oryx*, **37**, 178–193.

Hulme, P.E. (2007) Biological invasions in Europe: drivers, pressures, states, impacts and responses. In *Biodiversity under Threat* (ed. R. Hester and R.M. Harrison), pp 56–80. Royal Society of Chemistry, Cambridge.

Hulme, P.E., Bacher, S., Kenis, M., et al. (2008) Grasping at the routes of biological invasions: a framework for integrating pathways into policy. *Journal of Applied Ecology*, **45**, 403–414.

Hulme, P.E. & Bremner, E.T. (2006) Assessing the impact of *Impatiens glandulifera* on riparian habitats: partitioning diversity components following species removal. *Journal of Applied Ecology*, **43**, 43–50.

Hulme, P.E., Nentwig, W., Pyšek, P. & Vilà, M. (2009a) DAISIE : Delivering Alien Invasive Species Inventories for Europe. In *Atlas of Biodiversity Risk* (ed. J. Settele, L. Penev, T. Georgiev, et al.), pp. 130–131. Pensoft, Sofia and Moscow.

Hulme, P.E., Nentwig, W., Pyšek, P. & Vilà, M. (2009b) Common market, shared problems: time for a coordinated response to biological invasions in Europe? *Neobiota*, **8**, 3–19.

Hulme, P.E., Pyšek, P., Nentwig, W. & Vilà, M. (2009c) Will threat of biological invasions unite the European Union? *Science*, **324**, 40–41.

Hulme, P.E., Roy, D.B., Cunha, T. & Larsson, T.-B. (2009d) A pan-European inventory of alien species: rationale, implementation and implications for managing biological invasions. In *The Handbook of Alien Species in Europe* (ed. DAISIE), pp. 1–18. Springer, Berlin.

Kenis, M., Auger-Rozenberg, M.A., Roques, A., et al. (2009) Ecological effects of invasive alien insects. *Biological Invasions*, **11**, 21–45.

Kettunen, M., Genovesi, P., Gollasch, S., et al. (2009) *Technical Support to EU strategy on Invasive Species (IAS): Assessment of the Impacts of IAS in Europe and the EU.* Final module report for the European commission, Institute for European Environmental Policy, Brussels.

Kobelt, M. & Nentwig, W. (2008) Alien spider introductions to Europe supported by global trade. *Diversity and Distributions*, **14**, 273–280.

Kohn, D., Hulme P.E., Hollingsworth, P. & Butler, A. (2009) Are native bluebells (*Hyacinthoides non-scripta*) at risk from alien congenerics? Evidence from distributions and co-occurrence in Scotland. *Biological Conservation*, **142**, 61–74.

Křivánek, M. & Pyšek, P. (2006) Predicting invasions by woody species in a temperate zone: a test of three risk assessment schemes in the Czech Republic (Central Europe). *Diversity and Distributions*, **12**, 319–327.

Křivánek, M., Pyšek, P. & Jarošík, V. (2006) Planting history and propagule pressure as predictors of invasion by woody species in a temperate region. *Conservation Biology*, **20**, 1487–1498.

Küster, E.C., Kühn, I., Bruelheide, H. & Klotz, S. (2008) Trait interactions help explain plant invasion success in the German flora. *Journal of Ecology* **96**, 860–868.

Lambdon, P.W. (2008) Is invasiveness a legacy of evolution? Phylogenetic patterns in the alien flora of Mediterranean islands. *Journal of Ecology*, **96**, 46–57.

Lambdon, P.W. & Hulme, P.E. (2006a) How strongly do interactions with closely-related native species influence plant invasions? Darwin's naturalization hypothesis assessed on Mediterranean islands. *Journal of Biogeography*, **33**, 1116–1125.

Lambdon, P.W. & Hulme, P.E. (2006b) Predicting the invasion success of Mediterranean alien plants from their introduction characteristics. *Ecography*, **29**, 853–865.

Lambdon, P.W., Pyšek, P., Basnou, C., et al. (2008) Alien flora of Europe: species diversity, temporal trends, geographical patterns and research needs. *Preslia*, **80**, 101–149.

Lloret, F., Medail, F., Brundu, G., et al. (2005) Species attributes and invasion success by alien plants on Mediterranean islands. *Journal of Ecology*, **93**, 512–520.

Lloret, F., Medail, F., Brundu, G. & Hulme, P.E. (2004) Local and regional abundance of exotic plant species on Mediterranean islands: are species traits important? *Global Ecology and Biogeography*, **13**, 37–45.

Lodge, D.M., Williams, S., MacIsaac, H.J., et al. (2006) Biological invasions: recommendations for US policy and management. *Ecological Applications*, **6**, 2035–2054.

Lopez-Vaamonde, C. (2009) *Leptinotarsa decemlineata* Say, Colorado beetle (Chrysomelidae, Coleoptera). In *Handbook of Alien Species in Europe* (ed. DAISIE), p. 336. Springer, Berlin.

Nentwig, W., Kühnel, E. & Bacher, S. (2010) A generic impact-scoring system applied to alien mammals in Europe. *Conservation Biology*, **24**, 302–311.

Nielsen, C., Ravn, H.P., Cock, M. & Nentwig W. (eds) (2005) *The Giant Hogweed Best Practice Manual. Guidelines for the Management and Control of an Invasive Alien Weed in Europe.* Forest and Landscape Denmark, Hoersholm, Denmark.

Olenin, S., Minchin, D. & Daunys, D. (2007) Assessment of biopollution in aquatic ecosystems. *Marine Pollution Bulletin*, **55**, 379–394.

Paavola, M., Olenin, S. & Leppakoski, E. (2005) Are invasive species most successful in habitats of low native species richness across European brackish water seas? *Estuarine Coastal and Shelf Science*, **64**, 738–750.

Preston, C.D., Pearman, D.A. & Dines, T.D. (2002) *New Atlas of the British and Irish Flora.* Oxford University Press, Oxford.

Pyšek, P., Bacher, S., Chytrý, M., et al. (2010) Contrasting patterns in the invasions of European terrestrial and freshwater habitats by alien plants, insects and vertebrates. *Global Ecology and Biogeography*, **19**, 317–331.

Pyšek P., Cock M.J.W., Nentwig W. & Ravn H.P. (eds) (2007a) *Ecology and Management of Giant Hogweed (Heracleum mantegazzianum).* CAB International, Wallingford, UK.

Pyšek, P., Hulme, P.E., Nentwig, W. & Vilà, M. (2011) DAISIE project. In *Encyclopaedia of Biological Invasions* (ed. D. Simberloff D. and M. Rejmánek). University of California Press, Berkeley (in press).

Pyšek, P., Jarošík, V., Hulme, P.E., et al. (2010c) Disentangling the role of environmental and human pressures on biological invasions. *Proceedings of the National Academy of Sciences of the USA*, **107**, 12157–12162.

Pyšek, P., Jarošík, V., Pergl, J., et al. (2009a) The global invasion success of Central European plants is related to distribution characteristics in their native range and species traits. *Diversity and Distributions*, **15**, 891–903.

Pyšek, P., Křivánek, M. & Jarošík, V. (2009b) Planting intensity, residence time, and species traits determine invasion success of alien woody species. *Ecology*, **90**, 2734–2744.

Pyšek, P. & Richardson, D.M. (2010) Invasive species, environmental change and management, and ecosystem health. *Annual Review of Environment and Resources*, **35**, doi:10.1146/annurev-environ-033009-095548.

Pyšek, P., Richardson, D.M. & Jarošík, V. (2006) Who cites who in the invasion zoo: insights from an analysis of the most highly cited papers in invasion ecology. *Preslia*, **78**, 437–468.

Pyšek, P., Richardson, D.M., Pergl, J., Jarošík, V., Sixtová, Z. & Weber, E. (2008) Geographical and taxonomic biases in invasion ecology. *Trends in Ecology & Evolution*, **23**, 237–244.

Pyšek, P., Richardson, D.M., Rejmánek, M., Webster, G., Williamson, M. & Kirschner, J. (2004) Alien plants in checklists and floras: towards better communication between taxonomists and ecologists. *Taxon*, **53**, 131–143.

Ricciardi, A. & MacIsaac, J. (2008) The book that began invasion ecology. *Nature*, **452**, 34.

Richardson, D.M. & Pyšek, P. (2007) Classics in ecology revisited: Elton, C.S. 1958: The ecology of invasions by animals and plants. London: Methuen. *Progress in Physical Geography*, **31**, 659–666.

Richardson, D.M. & Pyšek, P. (2008) Fifty years of invasion ecology: the legacy of Charles Elton. *Diversity and Distributions*, **14**, 161–168.

Ross, L.C., Lambdon, P.W. & Hulme, P.E. (2008) Disentangling the roles of climate, propagule pressure and land use on the current and potential elevational distribution of the invasive weed *Oxalis pes-caprae* L. on Crete. *Perspectives in Plant Ecology, Evolution & Systematics*, **10**, 251–258.

Salisbury, E.J. (1961) *Weeds and aliens*. Collins, London.

Settele, J., Hammen, V., Hulme, P., et al. (2005) ALARM: Assessing LArge-scale environmental Risks for biodiversity with tested Methods. *GAIA – Ecological Perspectives For Science and Society*, **14**, 69–72.

Shiganova, T.A. & Panov V.E. (2009) *Mnemiopsis leidyi* Agassiz, sea walnut, comb jelly (Bolinopsidae, Ctenophora). In *Handbook of Alien Species in Europe* (ed. DAISIE), p. 314. Springer, Berlin.

Shirley, S.M. & Kark, S. (2009) The role of species traits and taxonomic patterns in alien bird impacts. *Global Ecology and Biogeography*, **18**, 450–459.

Shwartz, A. & Shirley, S. (2009) *Psittacula krameri* (Scopoli), rose-ringed parakeet (Psittacidae, Aves). In *Handbook of Alien Species in Europe* (ed. DAISIE), p. 339. Springer, Berlin.

Simberloff, D. (2000) Foreword. In *Elton C.S., The Ecology of Invasions by Animals and Plants*, 2nd edn, pp. vii–xiv. University of Chicago Press, Chicago.

Spangenberg, J.H. (2007) Integrated scenarios for assessing biodiversity risks. *Sustainable Development*, **15**, 343–356.

Truscott, A.-M., Palmer, S.C.F., Soulsby, C., Westaway, S. & Hulme, P.E. (2008) Consequences of invasion by the alien plant *Mimulus guttatus* on the species composition and soil properties of riparian plant communities in Scotland. *Perspectives in Plant Ecology, Evolution & Systematics*, **10**, 231–240.

Truscott, A.-M., Soulsby, C., Palmer, S.C.F., Newell, L. & Hulme, P.E. (2006) The dispersal characteristics of the invasive plant *Mimulus guttatus* and the ecological significance of increased occurrence of high flow events. *Journal of Ecology*, **94**, 1080–1091.

Tutin, T. G., Heywood, V. H., Burges, N. A., et al. (eds) (1964–1980) *Flora Europaea*, vols. 1–5. Cambridge University Press, Cambridge, UK.

Vilà, M., Bartomeus, I., Dietzsch, A.C., et al. (2009) Invasive plant integration into native plant-pollinator networks across Europe. *Proceedings of the Royal Society B*, **276**, 3887–3893.

Vilà, M., Basnou, C., Pyšek, P., et al. (2010) How well do we understand the impacts of alien species on ecological services? A pan-European cross-taxa assessment. *Frontiers in Ecology and the Environment*, **8**, 135–144.

Vilà, M., Pino, J. & Font, X. (2007) Regional assessment of plant invasions across different habitat types. *Journal of Vegetation Science*, **18**, 35–42.

Vilà, M., Siamantziouras, A., Brundu, G., et al. (2008) Widespread resistance of Mediterranean island ecosystems to the establishment of three alien species. *Diversity and Distributions*, **14**, 839–851.

Walther, G.R., Gritti, E.S., Berger, S., Hickler, T., Tang, Z.Y. & Sykes, M.T. (2007) Palms tracking climate change. *Global Ecology and Biogeography*, **16**, 801–809.

Williamson, M., Dehnen-Schmutz, K., Kühn, I., et al. (2009) The distribution of range sizes of native and alien plants in four European countries and the effects of residence time. *Diversity and Distributions*, **15**, 158–166.

Williamson, M., Pyšek, P., Jarošík, V. & Prach, K. (2005) On the rates and patterns of spread of alien plants in the Czech Republic, Britain and Ireland. *Ecoscience*, **12**, 424–433.

Winter, M., Schweiger, O., Nentwig, W., et al. (2010) Losing of uniqueness: plant extinctions and introductions lead to phylogenetic and taxonomic homogenization of the European flora. *Proceedings of the National Academy of Sciences of the USA*, **106**, 21721–21725.

Chapter 8

FIFTY YEARS OF TREE PEST AND PATHOGEN INVASIONS, INCREASINGLY THREATENING WORLD FORESTS

Michael J. Wingfield, Bernard Slippers, Jolanda Roux and Brenda D. Wingfield

Forestry and Agricultural Biotechnology Institute, University of Pretoria, Pretoria 0002, South Africa

Fifty Years of Invasion Ecology: The Legacy of Charles Elton, 1st edition. Edited by David M. Richardson
© 2011 by Blackwell Publishing Ltd

8.1   INTRODUCTION

Charles Elton's 1958 book *The Ecology of Invasions by Animals and Plants* provided the first comprehensive treatment of the topic and is widely regarded as marking the birth of the field of invasion biology. Elton's book included two examples that relate to the field of tree health. These were Chestnut blight caused by the fungus *Cryphonectria parasitica* and the Gypsy moth *Lymantria dispar*. The alien invasive insect had intentionally been introduced into the USA around 1868 or 1869 in the hope of developing a silk industry in that country (Liebhold et al. 1989, 1992). The chestnut blight pathogen was accidentally introduced into the USA, probably through the movement of timber from Asia, and was first recognized in 1904 (Anagnostakis 1987).

Even though chestnut blight and the gypsy moth were recognized as invasive alien species more than 100 years ago, and despite their importance being highlighted by Elton (1958) and many other authors, they remain among the most important scourges to the world's forests. It is even more unfortunate that the accidental introduction of alien invasive pests increased dramatically post the emergence of gypsy moth and chestnut blight as non-native invasives. New arrivals of damaging pests and pathogens have continued to be recorded in natural forests and plantations worldwide (Mireku & Simpson 2002; Desprez-Loustau et al. 2007a; Jones & Baker 2007; Brasier 2008; Hansen 2008). Although great efforts have been made, particularly over the past 50 years, to reduce the movement of tree pests and pathogens, there is little evidence to show a great improvement in the situation. There are also worrying trends, such as the 'plants for planting' craze that suggest deterioration rather than improvement in the future.

It is not the aim of this chapter to provide lists of forest pests and pathogens that are non-native and threatening to forest ecosystems worldwide. Such lists exist for few countries, and even where they have been produced, they commonly underestimate the complexity of the situation. Beyond the process of listing, baseline studies providing data on alien invasive tree pests and pathogens for countries are few in number. Particularly for the microbes, experimental evidence to show that a pathogen is native or introduced has been achieved in only a minimal number of cases. Rather than producing lists, this chapter, following the spirit of Elton, focuses on trends relating to particular pathogens that threaten the world's forests and forestry.

8.2   NATIVE OR INTRODUCED

Knowing whether an organism is native or not appears on the surface to be a relatively simple question. This is surely because organisms such as mammals and plants are large, most have been identified, and their distributions are relatively comprehensively determined. In the case of microbes, including fungal pathogens, the situation is very different (Taylor et al. 2006). For the fungi, it has been estimated that only 7–10% of the species have been identified (Hawksworth 2001; Crous & Groenewald 2005). Although we believe that this is an underestimation (Crous et al. 2006), whether underestimated or overestimated, the fact remains that a very small percentage of the total number have been described. Thus knowing whether newly discovered fungi are native or not is difficult to determine and the reality is that this has been achieved for a relatively small number of tree pathogens.

When a new tree disease is discovered, one of the first questions asked is where it might have originated. Where the causal agent has been studied elsewhere in the world, the situation is relatively simple, particularly if knowledge already exists relating to its origin. Where the disease occurs on a native tree in a relatively undisturbed environment, the causal agent is usually thought to be native, although this might not necessarily be the case (Gilbert 2002; Loo 2009). In situations where a new disease breaks out on a non-native tree species, and where the pathogen is not known elsewhere, it is extremely difficult to be sure of its origin, and it might equally be native or introduced.

A pertinent recent example that illustrates the difficulty of determining the origin of a tree pathogen is found in the case of the new disease of *Pinus radiata* in Chile caused by *Phytophthora pinifolia* (Duran et al. 2008). This disease appeared relatively suddenly and spread rapidly in plantations of a non-native tree, causing dramatic damage. The pathogen is unusual in being the only aerial *Phytophthora* species that infects the above-ground parts of trees and it is not known anywhere else in the world. The fact that it occurs only on the non-native *P. radiata* and not on other *Pinus* species or conifers that grow in the most severely affected areas suggests that it is pine specific (Duran et al. 2008). In this case, the most likely explanation is that it is a non-native pathogen, probably introduced from some area of the world where pines are native. However, it is also possible that it is native on some native plant and that it has adapted to infect *P. radiata*. Alternatively, it could represent a hybrid with this capacity, as was suggested for the *Phytophthora* disease of alder in Europe (Brasier et al. 1999; Brasier 2000; Gibbs et al. 2003), but later found not to be the case (Man in't Veld et al. 2007). A possible resolution to the question lies in conducting population genetic studies on the pathogen. Thus if an appropriately assembled population of *P. pinifolia* isolates are found to have a limited genetic diversity in Chile, then assuming a founder effect (Clegg et al. 2002; Linde et al. 2009), it has most likely been introduced. Conversely, if the population is genetically diverse, there would be a reasonable chance that the pathogen is native. Clearly, complex and expensive studies are needed and this is the reason why it is only in relatively recent years that experimental evidence has been provided to demonstrate the origin of some of the world's best known alien invasive pathogens in native forests (Anagnostakis 1987, 2001; Brasier 1990; Milgroom & Cortesi 1999; Liu & Milgroom 2007).

The origin of host-specific tree pathogens is typically easier to determine than those that have broad host ranges. For example, when chestnut blight first began to spread in the USA, it was logically assumed that it was an alien invader, probably originating on native chestnuts elsewhere in the world (Anagnostakis 1987; Milgroom & Cortesi 1999; Liu & Milgroom 2007). The same would have been true for Dutch elm disease, white pine blister rust and other similar host-specific pathogens (Boyce 1961; Sinclair and Lyon 2005). However, even in the case of host specific pathogens, it can take some time for a likely site of origin to emerge. This was for example true for pine wilt caused by the pine wood nematode *Bursaphelenchus xylophilus* (Mamiya 1983; Wingfield 1987). Large numbers of pine trees had been dying in Japan since the early part of the 20th century, but it was only in 1971 that *B. xylophilus* was shown to be the cause of the disease (Tokushige & Kiyohara 1969; Mamiya 1983). When the nematode was first found associated with dying pines in the USA, it was believed that it was an invasive alien that might cause substantial damage to native North American forests (Dropkin et al. 1981; Wingfield et al. 1984). It took some years to recognize that the nematode had been described in the USA many years

earlier under a different name and that it was likely native in North America where it commonly occurs on pines dying of causes other than nematode infestation (Wingfield et al. 1984; Wingfield 1987).

The origins of tree pathogens with broad host ranges are typically more difficult to determine than host-specific pathogens. Perhaps one of the best examples is found in *Phytophthora cinnamomi* that has devastated native woody ecosystems in Australia (Zentmyer 1988; Shearer & Tippet 1989). The invasive nature of the pathogen in native forests provided strong clues to suggest that the pathogen was alien, but the debates about its origin raged on for many decades (Arentz & Simpson 1986, Shepherd 1975; Weste 1994). Interestingly, *P. cinnamomi* is the only tree pathogen for which there is experimental evidence to show invasion of natural woody ecosystems in Africa (Von Broembsen 1984; Von Broembsen & Kruger 1985; Linde et al. 1999). This pathogen is common and causes serious damage to forestry, fruit and flower crops in South Africa, but it has also invaded fynbos vegetation in the Cape Floristic Region where it is particularly devastating on species of *Leucadendron* and *Leucospermum* (Von Broembsen 1984; Von Broembsen & Kruger 1985).

8.3 PROBLEMS ASSOCIATED WITH IDENTIFICATION AND DETECTION

One of the most complicated problems in dealing with new tree diseases resides in accurately identifying the causal agents. The problem arises largely from the fact that microbes are small and they lack easily distinguishable morphological features. Convergent evolution, particularly in the fungi, has resulted in even distantly related groups looking very similar. Many important and different tree pathogens have in the past been treated as single taxa. An excellent example is found in the *Eucalyptus* canker pathogens that were treated as *Cryphonectria cubensis* until only recently (Gryzenhout et al. 2004, 2009). Molecular tools, particularly phylogenetic inference based on DNA sequence comparisons, have provided clear evidence that this species is unrelated to *Cryphonectria sensu stricto* and that at least three species previously treated as *C. cubensis* and causing serious *Eucalyptus* canker diseases now reside in the genus *Chrysoporthe* (Gryzenhout et al. 2004, 2009). Many more similar examples of cryptic species can be found in virtually every genus of tree pathogen known. For example, the *Eucalyptus* tree pathogens in the Botryosphaeriaceae resided in only a few species in 2004, but are now recognized as representing at least 23 species on this host (Slippers et al. 2009). Such examples have high-lighted a major shortcoming in diagnosis of tree disease until very recently, and quarantine measures based on names of fungi are highly questionable.

DNA barcoding is set to become a most important tool to be used in the identification of tree pathogens (Cross et al., this volume). Although this is a technol-ogy that is already well advanced for insect pests (Armstrong & Ball 2005; Ball & Armstrong 2006) it is likely to require a substantially longer time to develop for the fungi. Fungi represent a kingdom that is dra-matically more diverse than, for example, animals. Thus the phylogenetic distance between even species of the single fungal genus *Saccharomyces* is greater than that between all mammals or the most different insects (Nishida & Sugiyama 1993; Dujon 2006). Although it is possible to barcode insects using a single gene region (Ball & Armstrong 2006), it is now widely accepted that different gene regions will be required for the fungi and their closest relatives (Seifert 2009). Nonetheless, substantial advances in barcoding fungi have been made and this technology will ultimately become very useful for fungal tree pathogens (Cross et al., this volume).

8.4 FORESTS AND FORESTRY

Trees represent an interesting dichotomy in terms of the environments in which they occur. They are found in natural forests, as natives in intensively managed forests and in plantations or orchards of either natives or non-natives. Although, it is commonly overlooked, these different environments, particularly the native versus non-native environments, have had very sub-stantial implications relating to diseases (see Gilbert (2002), Wingfield (2003) and Drenth (2004) for dis-cussions of different systems). In forest ecosystems, trees are often keystone or foundational species. Where alien invasive pests and pathogens affect native trees in natural stands or forests, and functionally remove a particular foundational tree species, it can fundamentally change these ecosystems through dra-matic cascades of effects on the other organisms that depend on these trees and their ecosystem services (Loo 2009).

Native trees grown in intensively managed planta-tions can be very seriously affected by diseases. Typically, these trees represent a substantially more uniform genetic base than those that occur in the surrounding native forests, which can have a signi-ficant effect on epidemic development (as explored by Drenth (2004) for crop plants). Pathogens that are not particularly important in nearby natural forests can cause very serious damage to them. Many exam-ples are available for pines, eucalypts, acacias and poplars where these trees have been established in plantations in areas where they also grow as natives. Thus plantations of *P. elliottii* in the southeastern USA can be very severely affected by fusiform rust caused by *Cronartium quercuum* (Sinclair & Lyon 2005). Likewise, plantations of *Eucalyptus* in Australia have been seriously damaged by species of the leaf and shoot blight pathogens *Quambalaria eucalypti* and *Q. pitereka* (Pegg et al. 2005, 2008) and *Neofusicoccum australe* (=*Botryosphaeria australis*) (Burgess et al. 2006). What is relevant here is that the trees are easily exposed to pathogens that occur naturally in the region and opportunities for the pathogens to increase to epidemic levels are great. Management practices such as hybrid-ization of species and genetic modification are also less acceptable in areas where native trees are grown in close proximity to those that are being modified.

Plantation forestry based on non-native species has been highly successful, particularly in the tropics and Southern hemisphere where acacias, eucalypts and pines have been widely planted for timber and pulp production. The rapid growth and vigour of trees in these plantations has been widely attributed to the ability to grow trees in the absence of their natural enemies (enemy escape) (Bright 1998; Wingfield et al. 2000, 2001b; Wingfield 2003). Yet, in all situations, diseases have gradually appeared and in many areas, these have come to represent a significant constraint to plantation forestry. Some of the diseases have resulted from the accidental introduction of pathogens as is the case with *Dothistroma* needle blight of pines caused by *Dothistroma septosporum* (Barnes et al. 2004; Gibson 1972) or *Mycosphaerella* leaf blotch of Eucalypts caused by *Teratosphaeria nubilosa* (Hunter et al. 2008, 2009). Other diseases are caused by native pathogens with wide host ranges such as *Armillaria* spp. causing root rot (Coetzee et al. 2000, 2005; Sinclair & Lyon 2005) and *Cylindrocladium* spp. causing leaf blight (Crous 2002), which appear to have adapted to be able to infect the non-native tree species. An example of such a pathogen in South and Eastern Africa is that of the *Acacia mearnsii* pathogen *Ceratocystis albifundus*, which results in wilt and death of these Australian trees (Roux et al. 2007; Roux & Wingfield 2009).

Among the most intriguing and worrying examples of diseases of non-native plantation-grown trees are those caused by relatively host-specific pathogens that have undergone very obvious host shifts (Slippers et al. 2005). Perhaps the best example is that of *Eucalyptus* rust caused by *Puccinia psidii*. This pathogen is native on native Myrtaceae in south and Central America and it gained notoriety when it began to infect plantation grown *Eucalyptus* (Coutinho et al. 1998; Glen et al. 2007). There are no known rust diseases of *Eucalyptus* spp. in the native range of these trees and this new *Eucalyptus* disease seriously threatens these trees and other Myrtaceae particularly in Australasia and Africa (Alfenas et al. 2005; Glen et al. 2007). Similar examples are emerging in the Cryphonectriaceae where, for example, species of *Chrysoporthe* native on indigenous Myrtaceae and Melastomataceae in the South/Central America, Africa and Asia have adapted to infect *Eucalyptus* (Wingfield et al. 2001a; Wingfield 2003; Heath et al. 2006; Gryzenhout et al. 2006, 2009).

Very little is known about host jumps in tree pathogens (Parker & Gilbert 2004; Slippers et al. 2005) resulting in new and relatively host specific pathogens. They appear to originate where non-native trees, relatively closely related to native trees are grown in close proximity. Little is known about the level of relatedness of the hosts that allow these shifts to occur. For example, it would have been difficult to predict the apparent jump in *Chr. cubensis* from native *Tibouchina* (Melastomataceae) to *Eucalytpus* in South America (Wingfield et al. 2001a; Rodas et al. 2005). Yet, *Eucalyptus* (Myrtaceae) and *Tibouchina* both reside in the same plant order (Myrtales) and this is obviously sufficiently close to allow adaptation of the pathogen. One might expect additional examples of such host shifts in the future and the implication of the new pathogens that arise from them, at least in terms of their threat the new tree hosts in their areas of origin through 'back introductions' (see Guo 2005), will be important.

8.6 PATHWAYS OF MOVEMENT

When the accidental introduction of pathogens to new environments was recognized as a threat to forests,

quarantine measures were introduced to deal with this problem. For a long time, the focus of these abatement strategies has been on assembling lists of threatening pathogens. Although this approach has had some value, it has suffered the weakness of relying on incomplete records and names that are not necessarily accurate, as discussed above. Furthermore, reliance on lists of names entirely overlooks the fact that species represent populations. Thus the presence of an organism in a country should not detract from the fact that additional introductions of the same pathogen could extend its genetic adaptability and fitness, exacerbating an already existing problem (McDonald & Linde 2002; Suarez & Tsutsui 2008). For these reasons, a contemporary approach is to shift the focus of quarantine efforts from lists of threatening organisms to understanding and managing pathways of introduction.

Knowledge of the pathways of movement of pathogens is fundamental to reducing their introduction into new environments and to the development of effective management strategies (Wilson et al. 2008). During the past century, one of the most important sources of tree pests and pathogens in new environments has been wood and wood packaging material. For many years, waste wood and logs were used as dunnage in ships, and untreated wood was used as packaging for products shipped internationally. It is certain that this practice resulted in the accidental introduction of many tree pests and pathogens (Ridley et al. 2000; Palm & Rossman 2003; Haack 2006; Desprez-Loustau et al. 2007a; Brasier 2008). It is worrying that it has taken so many years since the understanding of the associated threats to implement strict regulations for the effective treatment of wood packaging material (www.ippc.int). The challenge of implementing and enforcing these regulations will be especially difficult in developing countries that lack the capacity to do so.

The international nursery trade represents one of the most serious sources of pathogens accidentally introduced into new environments (Davison et al. 2006; Ivors et al. 2006; Brasier 2008). The relatively recent appearance of the devastating sudden oak death on the east coast of the USA, caused by *Phytophthora ramorum*, provides a sobering example of an alien pathogen that most likely moved via the nursery trade. It is now well recognized that *P. ramorum* is a serious pathogen of *Rhododendron* plants that are commonly traded between countries (Rizzo et al. 2005; Ivors et al. 2006), and it is unfortunate that it required the onset of a devastating disease to recognize this threat.

Sudden oak death has contributed substantially to a global re-examination of regulations relating to the trade in plants and planting material (Brasier 2008). Yet this is a global business of huge proportions and it will be extremely difficult to manage effectively. So called 'plants for planting' are highly desired internationally and there are many examples of plants, sometimes with roots and associated soil attached, which are shipped between countries. The risks of such shipping are poorly understood, as discussed extensively above, relating to identification and recognition of pathogens. Given the magnitude of the risks, there seems little justification for moving whole plants between countries, yet regulations to restrict movement will require significant effort.

Although it might be possible to safely manage legal imports of some categories of plants and planting stock, managing illegal imports presents an enormous challenge. In developed economies, quarantine procedures can be relatively strictly enforced to reduce illegal importation of plants and planting stock. This is, however, almost impossible in developing world countries were borders are in many cases managed very informally. In many parts of the world, people move across borders virtually without control and they often carry their plants with them. Such movement increases the global distribution of pathogens and it likewise increases the risks of movement to additional new environments.

Globalization has contributed to the movement of tree pathogens and it will most likely continue to do so in the future. International forestry and fruit tree growing companies increasingly have a global presence with holdings in many different countries. Where planting stock performs well in one country, the temptation to move this material to other areas of the world where the companies also operate is great. This can often be achieved without sufficient attention being given to the risks of introducing pests and pathogens into new environments. One might thus predict that diseases such as *Eucalyptus* rust that occurs in plantations of forestry companies in South America will emerge in Asia where the same companies own land.

Once a pest or pathogen becomes established in a new area, subsequent introductions into other geographic areas appear to occur increasingly rapidly. Thus a first introduction appears to magnify the risks of subsequent spread. This is probably connected to the fact that the pest or pathogen is one that has the ability to spread easily and also substantial population growth that is typical of alien invasive pests or pathogens. The global movement of the *Sirex* wood wasp *Sirex noctilio* and its associated symbiotic fungus *Amylostereum areolatum* during the course of the past 100 years illustrates how a pest or pathogen continues to move to new environments increasingly rapidly, once it has first become established in a new environment (Slippers et al. 2003; Wingfield et al. 2000, 2008). Similar patterns of distribution could be presented for pine pathogens such as *Dothistroma septosporum* (Gibson 1972; Groenewald et al. 2007) and *Diplodia pinea* (Smith et al. 2000; Burgess et al. 2004), as well as *Eucalyptus* pathogens such as *Mycosphaerella nubilosa* (Hunter et al. 2008).

Various important tree pathogens live in apparently asymptomatic plant tissue as endophytes. In their native environment, these fungi are typically unimportant unless the trees are stressed. However, there is growing evidence that tree endophytes, unimportant in their areas of origin, can be important pathogens where they are introduced (Burgess & Wingfield 2002). *Diplodia pinea* and other members of the Botryosphaeriaceae provide good examples of this category of pathogen (Burgess et al. 2004; Slippers & Wingfield 2007). Thus the movement of apparently disease-free and asymptomatic planting material and even fruit could be the source of pathogens in new environments. Likewise, plant material traded as part of the floral trade bears risks, particularly through the movement of endophytes.

The floral trade could also result in the movement of pathogens other than endophytes. Thus the recent appearance of *Puccinia psidii* in Hawaii where it infects the native Ohia (*Meterosideros polymorpha*) is believed to have possibly occurred as the result of *Eucalyptus* foliage that is used in floral arrangements (Uchida et al. 2006).

8.7 CHANGING SYSTEMS

It is clear that forests and forestry have been increasingly damaged by diseases during the course of the past 50 years. These problems have primarily emerged from the introduction of pathogens into new environments. In terms of plantation forestry, new host–pathogen associations (Wingfield et al. 2010) have also resulted in severe damage. These trends are likely to continue although it is hoped that an improved awareness of the dangers will provide some opportunities to reduce losses. However, a changing world environment is likely to also substantially influence the manifestation and impact of tree diseases in the future.

Invasive alien organisms are strongly influenced by environmental conditions and global climate change will inevitably have a substantial impact on biological invasions. Walther et al. (2009) have provided examples of likely effect of climate change on all stages of the invasion process including introduction, colonization establishment and spread. It is clear that tree pathogens and the biological invasions associated with them will also be strongly influenced by changing weather patterns, particularly those associated with temperature and precipitation (Desprez-Loustau et al. 2006, 2007b).

The effect of climate change on tree diseases will be multifaceted, complex and often difficult to characterize. Rising temperatures will result in range expansion for some pathogens, as has been seen for *Phytophthora cinnamomi* in Europe (Bergot et al. 2008). The devastation caused by the mountain pine beetle with its fungal associates in North America provides a sobering example of climate change allowing populations of a native forest pest to explode (Kurz et al. 2008; Walther et al. 2009). It is, however, important to recognize that climate change will not necessarily result in a worsening of the impact of tree diseases and that some pathogens will be negatively influenced by changing climatic conditions.

Novel associations between pathogens and the trees they infect through unexpected host shifts (Parker & Gilbert 2004; Slippers et al. 2005) and the evolution of hybrid fungi (Brasier 2000, 2001) will likely result in unexpected tree diseases in the future, as discussed above. Likewise, novel associations between tree infesting insects, such as bark beetles, and fungi are already emerging as potentially new threats to forest ecosystems (Wingfield et al. 2010). This is as a result of either direct or indirect human influence, through introductions or expanding ranges of the organisms due to climate change. In this regard, Six and Benz (2007) have shown that temperature influences bark beetle–fungus symbioses and it seems likely that symbioses between forest insect pests and fungi will be influenced by climate change in the future (Six 2009).

8.8 CONCLUSIONS

Tree diseases caused by accidentally introduced pathogens have devastated natural woody ecosystems for more than 100 years. These pathogens have been transported around the world by humans, mainly through the movement of wood and wood products and the trade in plant material. Great effort has been expended and continues to be made to reducing the movement of tree pathogens. This is increasingly through identifying and limiting the most important pathways of introduction. Yet new and important tree diseases continue to appear in forests and plantations and this vividly illustrates the difficulty in preventing the global movement of tree pathogens.

Efforts to reduce the movement of tree pathogens internationally are negatively counterbalanced by globalization. Bridging of countries and continents by increasingly seamless borders, prompted by trade and tourism, makes quarantine increasingly difficult. Although knowledge of the risks associated with the movement of plants and plant products has clearly increased, the difficulty in managing the associated risks has increased and it is likely to continue to do so.

There are many challenges to reducing the movement of tree pathogens globally. Not least is the need for knowledge about these pathogens and the diseases that they cause. Ironically, forest protection is an area of science that has not been well supported internationally, and we see little evidence that the numbers of forest pathologists and forest entomologists is growing concomitantly with the importance of tree diseases. This is a challenge that needs to be presented to governments globally.

Technologies that make it possible to identify tree pathogens more accurately and more rapidly are emerging (Cross et al., this volume). These will certainly enhance the quality of quarantine. However, their value will depend strongly on investments in the characterization of tree pathogens and in the study of the diseases that they cause. Here, it is only possible to identify those organisms that are known to science, and although many (most) remain to be discovered, new technologies will fail to deliver the depth of their promise.

There is growing evidence that global climate change is influencing the biology of tree pathogens in many parts of the world. Range shifts will occur with concomitant changes (both positive and negative) in the impact of pathogens. Novel associations between pathogens, tree hosts, as well as between important insect pests of trees and their symbionts, are also likely to be important. These new associations are only beginning to emerge as important and their outcomes are difficult to predict, but they will certainly strongly influence the future health of global forests and forestry.

Tree diseases will continue to pose one of the greatest threats to forests and forestry worldwide, for the foreseeable future. This threat is especially serious given the foundational role of trees in forest ecosystems and the consequent effect of their removal on the survival of numerous other species.Optimism about our capacity to reduce this threat is desirable and every effort must be made to do so. Yet few of the suggestions and warnings of Charles Elton in his keystone 1958 treatise have been seriously attended to. We would do well to revisit his sage predictions and suggestions regularly and to continue to learn from them.

ACKNOWLEDGEMENTS

We are grateful to the members of the Tree Protection Co-operative Programme (TPCP), the THRIP initiative of the Department of Trade and Industry and the DST/NRF Centre of Excellence in Tree Health Biotechnology, South Africa, for funding many of the studies that provided the foundation for this paper.

REFERENCES

Alfenas, A.C., Zauza, E.A.V., Wingfield, M.J., Roux, J. & Glen, M. (2005) *Heteropyxis natalensis*, a new host of *Puccinia psidii* rust. *Australian Plant Pathology*, **34**, 285–286.

Anagnostakis, S.L. (1987) Chestnut blight, the classical problem of an introduced pathogen. *Mycologia*, **79**, 23–37.

Anagnostakis, S.L. (2001) The effect of multiple importations of pests and pathogens on a native tree. *Biological Invasions*, **3**, 245–254.

Arentz, F. & Simpson, J.A. (1986) Distribution of *Phytophthora cinnamomi* in Papua New Guinea and notes on its origin. *Transactions of the British Mycological Society*, **88**, 289–295.

Armstrong, K.F. & Ball, S.L. (2005) DNA barcodes for biosecurity: invasive species identification. *Philosophical Transactions of the Royal Society B*, **360**, 1813–1823.

Ball, S.L. & Armstrong, K.F. (2006) DNA barcodes for insect pest identifications. A test case with tussock moths (Lepidoptera: Lymantriidae). *Canadian Journal of Forest Research*, **36**, 337–350.

Barnes, I., Crous, P.W., Wingfield, B.D. & Wingfield, M.J. (2004) Multigene phylogene reveal that red band needle blight of *Pinus* is caused by two distinct species of *Dothistroma*, *D. septosporum* and *D. pini*. *Studies in Mycology*, **50**, 551–565.

Boyce, J.S. (1961) *Forest Pathology*, 3rd edn. McGraw-Hill, New York.

Bergot, M., Cloppet, E., Perarnaud, V., Dequet, M., Marcais, B. & Desprez-Loustau, M.-L. (2008) Simulation of the potential range expansion of oak disease caused by *Phytophthora cinnamomi* under climate change. *Global Change Biology*, **10**, 1539–1552.

Brasier, C.M. (1990) China and the origins of Dutch elm disease: an appraisal. *Plant Pathology*, **39**, 5–16.

Brasier, C.M. (2001) Rapid evolution of introduced plant pathogens via interspecific hybridization. *Bioscience*, **52**, 123–133.

Brasier, C.M. (2000) The rise of the hybrid fungi. *Nature*, **405**, 134–135.

Brasier, C.M., Cooke, D.L. & Duncan, J. (1999) Origins of a new *Phytophthora* pathogen through interspecific hybridization. *Proceedings of the National Academy of Sciences of the USA*, **96**, 5878–5883.

Brasier, C.M. (2008) The biosecurity threat to the UK and global environment from international trade in plants. *Plant Pathology*, **57**, 792–808.

Bright, C. (1998) *Life Out of Bounds. Bioinvasion in a Borderless World*. WW Norten, New York.

Burgess, T. & Wingfield, M.J. (2002) Impact of fungal pathogens in natural forest ecosystems: a focus on *Eucalyptus*. In *Microorganisms in Plant Conservation and Biodiversity* (ed. K. Sivasithamparam and K.W. Dixon), pp. 285–306. Kluwer Academic Press, Dordrecht.

Burgess, T., Wingfield, M.J. & Wingfield, B.D. (2004) Global distribution of the pine pathogen *Sphaeropsis sapinea* revealed by SSR markers. *Australasian Plant Pathology*, **33**, 513–519.

Burgess, T.I., Sakalidis, M.L. & Hardy, G.E.S. (2006) Gene flow of the canker pathogen *Botryosphaeria australis* between *Eucalyptus globulus* plantations and native eucalypt forests in Western Australia. *Austral Ecology*, **31**, 559–566.

Clegg, S.M., Degnan, S.M., Kikkawa, J., Moritz, C., Estoup, A. & Owens, I.P.F. (2002) Genetic consequences of sequential founder events by an island-colonizing bird. *Proceedings of the National Academy of Sciences of the USA*, **99**, 8127–8132.

Coetzee, M.P.A., Wingfield, B.D., Coutinho, T.A. & Wingfield, M.J. (2000) Identification of the causal agent of *Armillaria* root rot of *Pinus* species in South Africa. *Mycologia*, **92**, 777–785.

Coetzee, M.P.A., Wingfield, B.D., Bloomer, P. & Wingfield, M.J. (2005) Phylogenetic analyses of DNA sequences reveal species partitions amongst isolates of *Armillaria* from Africa. *Mycological Research*, **109**, 1223–1234.

Coutinho, T.A., Wingfield, M.J., Alfenas, A.C. & Crous, P.W. (1998) Eucalyptus rust: a disease with the potential for serious international implications. *Plant Disease*, **82**, 819–825.

Crous, P.W. (2002) *Taxonomy and Pathology of Cylindrocladium (Calonectria) and Allied Genera*. American Phytopathological Society Press, St Paul, Minnesota.

Crous, P.W. & Groenewald, J.Z. (2005) Hosts, species and genotypes: opinions versus data. *Australasian Plant Pathology*, **34**, 463–470.

Crous, P.W., Rong, I.H., Wood, A., et al. (2006) How many species of fungi are there at the tip of Africa? *Studies in Mycology*, **55**, 13–33.

Davison, E.A., Drenth, A., Kumar, S., Mack, S., Mackie, A.E. & McKirdy, S. (2006) Pathogens associated with nursery plants imported into Western Australia. *Australasian Plant Pathology*, **35**, 473–475.

Desprez-Loustau, M.L., Marcais, B., Nageleisen, L.M., Piou, D. & Vannini, A. (2006) Interactive effects of drought and pathogens in forest trees. *Annals of Forest Science*, **63**, 597–612.

Desprez-Loustau, M., Robin, C., Bue'e, M., et al. (2007a) The fungal dimension of biological invasions. *Trends in Ecology & Evolution*, **22**, 472–480.

Desprez-Loustau, M.L., Robin, C., Reynaud, G., et al. (2007b) Simulating the effect of a climate change scenario on the geographical range and activity of forest-pathogenic fungi. *Canadian Journal of Plant Pathology*, **29**, 101–120.

Drenth, A. (2004) Fungal epidemics – does spatial structure matter? *New Phytologist*, **163**, 4–7.

Dropkin, V.H., Foudin, A.S., Kondo, E., Linit, M., Smith, M.T. & Robbins. (1981) Pinewood nematode: a threat to U.S. Forests? *Plant Disease*, **65**, 1022–1027.

Dujon, B. (2006) Yeasts illustrate the molecular mechanisms of eukaryotic genome evolution. *Trends in Genetics*, **22**, 375–387.

Duran, A., Gryzenhout, M., Slippers, B., et al. (2008) *Phytophthora pinifolia* sp. nov. associated with a serious needle disease of *Pinus radiata* in Chile. *Plant Pathology*, **57**, 715–727.

Elton, C. (1958) *The Ecology of Invasions by Animals and Plants*. Methuen, London.

Gibbs, J.N., van Dijk, C. & Webber, J.F. (2003) *Phytophthora disease of alder in Europe*. Forestry Commission: Forestry Commission Bulletin 126, Edinburgh, UK.

Gibson, I.A.S. (1972). Dothistroma blight of *Pinus radiata*. *Annual Review of Phytopathology*, **10**, 51–72.

Gilbert, G.S. (2002) Evolutionary ecology of plant diseases in natural ecosystems. *Annual Review of Phytopathology*, **40**, 13–43.

Glen, M., Alfenas, A.C., Zauza, E.A.V., Wingfield, M.J. & Mohammed, C. (2007) *Puccinia psidii*: a threat to the Australian environment and economy. *Australasian Plant Pathology*, **36**, 1–16.

Groenewald, M., Barnes I., Bradshaw, R.E., et al. (2007) Characterisation and distribution of mating type genes in the *Dothistroma* needle blight pathogens. *Phytopathology*, **97**, 825–834.

Gryzenhout, M., Myburg, H., van der Merwe, N.A., Wingfield, B.D. & Wingfield, M.J. (2004) *Chrysoporthe*, a new genus to accommodate *Cryphonectria cubensis*. *Studies in Mycology*, **50**, 119–142.

Gryzenhout, M., Rodas, C.A., Mena Portales, J., Clegg, P., Wingfield, B.D. & Wingfield, M.J. (2006) Novel hosts of the Eucalyptus canker pathogen *Chrysoporthe cubensis* and a new *Chrysoporthe* species from Colombia. *Mycological Research*, **110**, 833–845.

Gryzenhout, M., Wingfield, B.D. & Wingfield, M.J. (2009) *Taxonomy, Ecology and Phylogeny of Bark Inhabiting and Tree Pathogenic Fungi in the Cryphonectriaceae*. APS Press, St. Paul, Minnesota.

Guo, Q. (2005) Possible cryptic invasion through 'back introduction'? *Frontiers in Ecology and the Environment*, **3**, 470–471.

Haack, R.A. (2006) Exotic bark- and wood-boring Coleoptera in the United States: recent establishments. *Canadian Journal of Forest Research*, **36**, 269–288.

Hansen, E.M. (2008) Alien forest pathogens: *Phytophthora* species are changing world forests. *Boreal Environment Research* **13**, 33–41.

Hawksworth, D.L. (2001) The magnitude of fungal diversity: the 1.5 million species estimate revisited. *Mycological Research*, **105**, 1422–1432.

Heath, R.N., Gryzenhout, M., Roux, J. & Wingfield, M.J. (2006) Discovery of the *Cryphonectria* canker pathogen on native *Syzygium* species in South Africa. *Plant Disease*, **90**, 433–438.

Hunter, G.C., van der Merwe, N.A., Burgess, T.I., et al. (2008) Global movement and population biology of *Mycosphaerella nubilosa* infecting leaves of cold-tolerant *Eucalyptus globulus* and *E. nitens*. *Plant Pathology*, **57**, 235–242.

Hunter, G.C., Crous, P.W., Carnegie, A.J. & Wingfield, M.J. (2009) *Teratosphaeria nubilosa*, a serious leaf disease pathogen of *Eucalyptus* spp. in native and introduced areas. *Molecular Plant Pathology*, **10**, 1–14.

Ivors, K.L., Garbelotto, M., Vries, I.D.E., et al. (2006) Microsatellite markers identify three lineages of *Phytophthora ramorum* in US nurseries, yet single lineages in US forest and European nursery populations. *Molecular Ecology*, **15**, 1493–505.

Jones, D.R. & Baker, R.H.A. (2007) Introductions of non-native pathogens into Great Britain 1970–2004. *Plant Pathology*, **56**, 891–910.

Kurz, W.A., Dymond, C.C., Stinson, G., Rampley, G.J., Nielsen, E.T., Caroll, A.L., Ebata, T. & Safranyik, L. (2008). Mountain pine beetle and forest carbon feedback to climate change. *Nature*, **452**, 978–990.

Liebhold, A., Mastro, V. & Schaefer, P.W. (1989) Learning from the legacy of Leopold Trouvelot. *Bulletin of the Entomological Society of America*, **35**, 20–22.

Liebhold, A.M., Halverson, J.A. & Elmes, G.A. (1992) Gypsy moth invasion in North America: a quantitative analysis. *Journal of Biogeography*,**19**, 513–520.

Linde, C., Drenth, A. & Wingfield, M.J. (1999) Gene and genotypic diversity of *Phytophthora cinnamomi* in South Africa and Australia revealed by DNA polymorphisms. *European Journal of Plant Pathology*, **105**, 667–680.

Linde, C., Mzala, M. & McDonald, B.A. (2009) Molecular evidence for recent founder populations and human-mediated migration in the barley scald pathogen *Rhynchosporium secalis*. *Molecular Phylogenetics and Evolution*, **51**, 454–464.

Liu, Y. & Milgroom, M.G. (2007) High diversity of vegetative compatibility types in *Cryphonectria parasitica* in Japan and China. *Mycologia*, **99**, 279–284.

Loo, J. (2009) Ecological impacts of non-indigenous invasive fungi as forest pathogens. *Biological Invasions*, **11**, 81–96.

Mamiya, Y. (1983). Pathology of the pine wilt disease caused by *Bursaphelenchus xylophilus*. *Annual Review of Phytopathology*, **21**, 201–220.

Man in't Veld, W.A., Cock, A. & Summerbell, R. (2007) Natural hybrids of resident and introduced *Phytophthora* species proliferating on multiple new hosts. *European Journal of Plant Pathology*, **117**, 25–33.

McDonald, B.A. & Linde, C. (2002) Pathogen population genetics, evolutionary potential, and durable resistance. *Annual Review of Phytopathology*, **40**, 349–379.

Milgroom, M.G. & Cortesi, P. (1999) Analysis of population structure of the chestnut blight fungus based on vegetative incompatibility genotypes. *Proceedings of the National Academy of Sciences of the USA*, **96**, 10518–10523.

Mireku, E. & Simpson, J.A. (2002) Fungal and nematode threats to Australian forests and amenity trees from importation of wood and wood products. *Canadian Journal of Plant Pathology*, **24**, 117–124.

Nishida, H. & Sugiyama, J. (1993) Phylogenetic relationships among *Taphrina*, *Saitoella* and other fungi. *Molecular Biology and Evolution*, **10**, 431–436.

Palm, M.E. & Rossman, A.Y. (2003) Invasion pathways of terrestrial plant-inhabiting fungi. In *Invasive species* (ed. G.M. Ruiz and J.T. Carlson), pp. 31–43, Island Press, Washington, DC.

Parker, I.M. & Gilbert, G.S. (2004) The evolutionary ecology of novel plant-pathogen interactions. *Annual Review of Ecology, Evolution, and Systematics*, **35**, 675–700.

Pegg, G.S., Drenth, A. & Wingfield, M.J. (2005) *Quambalaria pitereka* on spotted gum plantations in Queensland and northern New South Wales, *Australia. International Forestry Review*, **7**, 337.

Pegg, G.S., O'Dwyer, C., Carnegie, A.J., Burgess, T.I., Wingfield, M.J. & Drenth, A. (2008) *Quambalaria* species associated with plantation and native eucalypts in Australia. *Plant Pathology*, **57**, 702–714.

Ridley, G., Bain, J., Bulman, L., Dick, M. & Kay, M. (2000) Threats to New Zealand's indigenous forests from exotic pathogens and pests. In *Science for Conservation*, p. 1–68. Department of Conservation, Wellington, New Zealand.

Rizzo, D.M., Garbelotto, M. & Hansen, E.M. (2005) *Phytophthora ramorum*: integrative research and management of an emerging pathogen in California and Oregon forests. *Annual Review of Phytopathology*, **43**, 309–335.

Rodas, C.A., Gryzenhout, H., Wingfield, B.D. & Wingfield, M.J. (2005) Discovery of Eucalyptus canker pathogen *Chrysoporthe cubensis* on native *Miconia* (Melastomataceae) in Colombia. *Plant Pathology*, **54**, 460–470.

Roux, J., Heath, R.N., Labuschagne, L., Kamgan Nkuekam, G. & Wingfield, M.J. (2007) Occurrence of the wattle wilt pathogen, *Ceratocystis albifundus* on native South African trees. *Forest Pathology*, **37**, 292–302.

Roux, J. & Wingfield, M.J. (2009) *Ceratocystis* species: emerging pathogens of non-native plantation *Eucalyptus* and *Acacia* species. *Southern Forests*, **71**, 115–120.

Seifert, K.A. (2009) Progress towards DNA barcoding of fungi. *Molecular Ecology Resources*, **9**, 83–89.

Shearer, B.L. & Tippett, J.T. (1989) Jarrah dieback: the dynamics and management of *Phytophthora cinnamomi* in the jarrah (*Eucalyptus marginata*) forest of south-western Australia. Department of Conservation and Land Management, Perth, Western Australia.

Shepherd, C.J. (1975) *Phytophthora cinnamomi* – an ancient immigrant to Australia. *Search*, **6**, 484–490.

Sinclair, W.A. & Lyon, H.H. (2005) *Diseases of Trees and Shrubs*, 2nd edn. Cornell University Press, Ithaca.

Six, D.L. (2009). Climate change and mutualism. *Nature Reviews Microbiology*, **7**, 680.

Six, D.L. & Benz, B.J. (2007). Temperature determines symbiont abundance in a multipartite bark beetle–fungus ectosymbiosis. *Microbial Ecology*, **54**, 112–118.

Slippers, B., Coutinho, T., Wingfield, B.D. & Wingfield, M.J. (2003) A review of the genus *Amylostereum* and its association with woodwasps. *South African Journal of Science*, **99**, 70–74.

Slippers, B., Stenlid, J. & Wingfield, M.J. (2005) Emerging pathogens: fungal host jumps following anthropogenic introduction. *Trends in Ecology & Evolution*, **20**, 420–421.

Slippers, B. & Wingfield, M.J. (2007) Botryosphaeriaceae as endophytes and latent pathogens of woody plants: diversity, ecology and impact. *Fungal Biology Reviews*, **21**, 90–106.

Slippers, B., Burgess, T., Pavlic, D., et al. (2009) A diverse assemblage of Botryosphaeriaceae infect *Eucalyptus* in native and non-native environments. *Southern Forests*, **71**, 101–110.

Smith, H., Wingfield, M.J., De Wet, J. & Coutinho, T.A. (2000) Genotypic diversity of *Sphaeropsis sapinea* from South Africa and Northern Sumatera. *Plant Disease*, **84**, 139–142.

Suarez, A.V. & Tsutsui, N.D. (2008) The evolutionary consequences of biological invasions. *Molecular Ecology*, **17**, 351–360.

Taylor, J.W., Turner, E., Townsend, J.P., Dettman, J.R. & Jacobson, D. (2006) Eukaryotic microbes, species recognition and the geographic limits of species: examples from the kingdom Fungi. *Philosophical Transactions of the Royal Society B*, **361**, 1947 1963.

Tokushige, Y. & Kiyohara, T. (1969) *Bursaphelenchus* sp. in the wood of dead pine trees. *Journal of the Japanese Forestry Society*, **51**, 93–95.

Uchida, J., Zhong, S. & Killgore, E. (2006) First report of a rust disease on 'Ohi'a caused by *Puccinia psidii* in Hawaii. *Plant Disease*, **90**, 524.

Von Broembsen, S.L. (1984) Occurrence of *Phytophthora cinnamomi* on indigenous and exotic hosts in South Africa with special reference to the South Western Cape Province. *Phytophylactica*, **16**, 221–225.

Von Broembsen, S.L. & Kruger, F.J. (1985) *Phytophthora cinnamomi* associated with mortality of native vegetation in South Africa. *Plant Disease*, **69**, 715–717.

Walther, G.-R., Roques, A., Hulme, P.E., et al. (2009) Alien species in a warmer world: risks and opportunities. *Trends in Ecology & Evolution*, **24**, 686–693.

Weste, G. (1994) Impact of *Phytophthora* species on native vegetation of Australia and Papua New Guinea. *Australasian Plant Pathology*, **23**, 190–209.

Wilson, J.R.U., Dormontt, E.E., Prentis, P.J., Lowe, A.J. & Richardson, D.M. (2008) Something in the way you move: dispersal pathways affect invasion success. *Trends in Ecology & Evolution*, **24**, 136–144.

Wingfield, M.J. (1987) *Pathogencity of the Pine Wood Nematode*. American Phytopathological Society Press, St. Paul, Minnesota.

Wingfield, M.J. (2003) Daniel McAlpine memorial lecture. Increasing threat of disease to exotic plantation forests in the southern hemisphere: lessons from *Cryphonectria* canker. *Australasian Plant Pathology*, **32**, 133–139.

Wingfield, M.J., Slippers, B., Roux J. & Wingfield, B.D. (2000) World-wide movement of exotic forest fungi, especially in the tropics and Southern Hemisphere. *Bioscience*, **51**, 134–140.

Wingfield, M.J., Slippers, B. & Wingfield B.D. (2010) Novel associations between pathogens, insects and tree species threaten world forests. *New Zealand Journal of Forestry Science*, **40** (Supplement), S95–S103.

Wingfield, M.J., Blanchette, R.A. & Nicholls, T.H. (1984) Is the pine wood nematode an important pathogen in the United States? *Journal of Forestry*, **82**, 232–235.

Wingfield, M.J., Rodas, C., Wright, J., Myburg, H., Venter, M. & Wingfield, B.D. (2001a) First report of *Cryphonectria* canker on *Tibouchina* in Colombia. *Forest Pathology*, **31**, 297–306.

Wingfield, M.J., Roux, J., Coutinho, T.A., Govender, P. & Wingfield, B.D. (2001b) Plantation disease and pest management in the next century. *South African Forestry Journal* **190**, 67–72.

Wingfield, M.J., Slippers, B., Hurley, B.P., Coutinho, T.A., Wingfield, B.D. & Roux, J. (2008) Eucalypt pests and diseases: growing threats to plantation productivity. *Southern Forests*, **70**, 139–144.

Zentmyer, G.A. (1988) Origin and distribution of four species of *Phytophthora*. *Transactions of the British Mycological Society*, **91**, 367–378.

Part 4

The Nuts and Bolts of Invasion Ecology

Chapter 9

A MOVEMENT ECOLOGY APPROACH TO STUDY SEED DISPERSAL AND PLANT INVASION: AN OVERVIEW AND APPLICATION OF SEED DISPERSAL BY FRUIT BATS

Asaf Tsoar, David Shohami and Ran Nathan

Movement Ecology Laboratory, Department of Ecology, Evolution, and Behavior, Alexander Silberman Institute of Life Sciences, The Hebrew University of Jerusalem, Jerusalem 91904, Israel

9.1 BIOLOGICAL INVASIONS AND DISPERSAL PROCESSES

Biological invasions – the entry, establishment and spread of non-native species – are a major cause of human-induced environmental change (Vitousek et al. 1997; Ricciardi 2007). Beyond their substantial economic impact and human health hazards (Vitousek et al. 1997; Pimentel et al. 2001), biological invasions threaten global biodiversity by altering the structure and function of ecosystems and disrupting key biological interactions (Levine et al. 2003; Traveset & Richardson 2006). Consequently, they also constitute a major cause of recent extinctions (Clavero & García-Berthou 2005; but see Didham et al. 2005).

Dispersal, or the unidirectional movement of an organism away from its home or place of birth, is a key process in an organism's life cycle, operating at multiple scales and levels of organization from the single organism through population, metapopulation and community dynamics (Harper 1977; Clobert et al. 2001; Bullock et al. 2002; Cousens et al. 2008; Nathan et al. 2009). For many plants seed dispersal is the primary mobile stage, typically mediated by vectors that disperse seeds over short distances, affecting local-scale plant population and community persistence, structure and dynamics (for recent reviews, see Cousens et al. 2008, Nathan et al. 2009 and references therein). Relatively few seeds are dispersed over long distances (long-distance dispersal (LDD)), affecting large, landscape-scale dynamics of plant populations and communities (Nathan 2006). Defining which dispersal events account for LDD typically involves setting arbitrary or system-specific thresholds, taking either a proportional approach (e.g. all the seeds that travelled the upper 1% of the distance distribution) or an absolute approach (e.g. all the seeds that travelled more than 1000 m, as adopted in section 9.4 of this chapter); the latter is generally preferable, mostly for practical reasons (see Nathan et al. 2008b for discussion).

The vector-mediated seed dispersal process consists of three main phases, each characterized by a basic, key parameter (Nathan et al. 2008b; Table 9.1): (i) the initiation phase, in which seeds are picked up by the vector, characterized by the vector's 'seed load' parameter, or the number of seeds taken per time unit; (ii) the transport phase, in which the vector transports the seeds away from the source, characterized by the vector's 'displacement velocity' parameter after seed uptake; and (iii) the termination phase, in which the seeds are deposited, characterized by the vector's 'seed passage time' parameter, or the duration of seed transport by the vector. The contribution of different vectors to local dispersal versus LDD depends mostly on the two former parameters (Nathan et al. 2008b). LDD vectors, for example, should at least occasionally have high displacement velocity, and are especially efficient when combined with long seed passage time (Nathan et al. 2008b; Schurr et al. 2009). Human transportation is presumably the only mechanism which has a high relative effect on all three parameters, making mankind the most important LDD vector nowadays.

In 1958 Charles Elton launched the systematic scientific study of biological invasions with the publication of his book *The Ecology of Invasions by Animals and Plants* (Elton 1958; reviewed in Richardson & Pyšek 2008). Though Elton argued that the major cause for spread of species was the increased extent of human travel around the globe, he noted that even without human intervention, 'exceptionally good powers of dispersal' (Elton 1958, p. 33) have enabled many species to spread and achieve a wide distribution, hence acknowledging the importance of LDD for the dynamics of population spread. It is the arrival of humans, Elton wrote, that has made 'this process of dispersal so much easier and faster' (Elton 1958, p. 79). Dispersal is now recognized as a major and essential component in the dynamics of invasions (see also Hui et al., this volume). However, although dispersal is necessary it is insufficient to generate continuing spatial spread: dispersed seeds must germinate, survive and grow to become reproductive plants that produce and disperse seeds, and so forth. Thus plant dispersal occurs within one generation (usually lasting a very short time), whereas continuing spread is a multi-generation process.

The process of biological invasion can generally be divided into three dynamic stages – entry, establishment and spread – with barriers, or filters, hindering or preventing transition from one stage to the next (Richardson et al. 2000b; Colautti & MacIsaac 2004). Although dispersal is not the only filter plants must pass through, it is an important one that has key impacts on the survival and success of the invading plant. The entry stage often results from human-mediated extreme LDD (Mack & Lonsdale 2001; Ricciardi 2007; Hulme et al. 2008), but might also result from other LDD mechanisms if the invading species is already present as an alien in a neighbouring region (Hulme et al. 2008). Natural dispersal by man-made infrastructures connecting otherwise unlinked biogeographical regions (e.g. Lessepsian migration through the Suez Canal (Por 1978; Ben-Eliahu & ten Hove 1992)) may also bring about the initial introduction of alien species (Hulme et al. 2008). The establishment stage involves mainly local dispersal, whereas the spread stage involves rapid expansion that is mostly dominated by LDD (Kot et al. 1996; Clark 1998). In these two post-entry stages, humans play an increasingly important role (see, for example, Von der Lippe & Kowarik 2007), yet natural vectors are probably still the key dispersers (Debussche & Isenmann 1990; Richardson et al. 2000a; Murphy et al. 2008; Westcott et al. 2008). Dispersal at the post-entry stage is crucially important to the extent that successful dispersal away from the initial point of introduction marks the transition from 'alien' through 'naturalized' to 'invasive' (*sensu* Richardson et al. 2000b). Therefore, dispersal is a necessary step at several stages of the plant's invasion; understanding dispersal processes not only to, but also within, the invaded region is therefore crucial for understanding and predicting invasion success (Richardson et al. 2000a; Higgins et al. 2003a; Buckley et al. 2006).

The aims of this chapter are to review the progress in seed dispersal research especially in the context of plant invasion (section 9.2), and to introduce (section 9.3) and illustrate (section 9.4) a general framework for elucidating the role of dispersal mechanisms as a major driving force in invasion processes.

9.2 OVERVIEW OF THE CONCEPTS AND METHODS APPLIED TO QUANTIFY SEED DISPERSAL PATTERNS AND UNDERSTAND THEIR UNDERLYING MECHANISMS

Advances in understanding dispersal

Research during the past 50 years has yielded innovative insights into the ecological and evolutionary processes underlying dispersal in general (Bullock et al. 2006; Nathan et al. 2009) with recent advances in LDD in particular (Cain et al. 2000; Nathan 2006; Nathan et al. 2008b). Seed dispersal research has seen an important shift in focus in recent years, from the traditional 'seed-centred' approach focusing on seed attributes and asking by which mechanisms and over which distances these seeds are dispersed, to a 'vector-centred' approach now focusing on a dispersal vector and asking how many seeds this vector disperses over which distances (Nathan et al. 2008b; Schurr et al. 2009).

Studies have shown that 'standard' vectors, those inferred directly from seed morphology, have low impact on LDD and spread rate, compared with other, 'non-standard' vectors (Higgins et al. 2003b; see examples in Nathan et al. 2008b). Nevertheless, and despite the common consensus that LDD events are rare and largely unpredictable under most circumstances, LDD is strongly associated with a limited and identifiable set of environmental conditions and dispersal vectors (Nathan et al. 2008b). Six major generalizations of mechanisms that likely promote plant LDD have been identified (Nathan et al. 2008b), such as open terrestrial landscapes that are free of obstacles to seed and vector movement and thus have a relatively long seed passage time; migratory animals that move in a fast and directional manner and thus have a relatively high displacement velocity; extreme meteorological events that can result in exceptionally high displacement velocity and seed load; and human transportation, presumably the mechanism most likely to move seeds the longest possible distances (Nathan et al. 2008b).

A parallel, highly relevant shift in seed dispersal research has been the relatively recent recognition that animal seed dispersers tend to be generalists rather than specialists (Richardson et al. 2000a; Herrera 2002) and that the coevolutionary vector–seed interactions are not as tight and common as was previously thought (Richardson et al. 2000a; Bascompte et al. 2006). Overall, dispersal systems are complex assemblages of multiple dispersers operating at various scales to generate jointly the 'total dispersal kernel' (Nathan et al. 2008b). Therefore, taking the vector-based approach of seed dispersal in investigating invasion processes requires identifying the key players facilitating passage through the dispersal-related invasion filters, such as in the initial introduction stage (Hulme et al. 2008).

Advances in data collection

Tracking seed movement away from the source plant has always been a challenging, and often limiting, part of studying seed dispersal patterns and mechanisms (Wheelwright & Orians 1982; Nathan &

Muller-Landau 2000; Wang & Smith 2002). This has been especially true when attempting to quantify and identify LDD processes in the field (Cain et al. 2000; Nathan 2006). Seed traps have made an important contribution to dispersal research (see, for example, Clark et al. 1998, Bullock & Clarke 2000) and are still being used today, despite several inherent problems given the difficulty in identifying the source of the seeds (Nathan & Muller-Landau 2000), though this can be resolved using genetic methods (see, for example, Jones et al. 2005). Artificially marking seeds at the source and finding their deposition sites (Levey & Sargent 2000; Xiao et al. 2006) has become increasingly used, and novel methods are still being developed (Carlo et al. 2009; Lemke et al. 2009). Controlled manual seed release is also advancing current wind dispersal research (Tackenberg 2003; Soons et al. 2004).

Research on animal-dispersed plants has progressed from directly observing animal movement, to tracking the animals with radio-telemetry (see, for example, Murray 1988; Westcott & Graham 2000; Spiegel & Nathan 2007). Recent technological advancements such as satellite-tracking using the Argos system or tracking units based on global positioning systems (GPS) have revolutionized the quality, quantity and scale of animal tracking data in the wild; this, in turn, also improved the input parameters inserted into models predicting the animal's seed dispersal ability (Campos-Arceiz et al. 2008). However, two main drawbacks – high costs and heavy power supply – currently limit their use and make most small-sized animals unapproachable by such technologies. Future miniaturization will enable tracking of many additional animal vectors and even seeds at large scales with the high spatio-temporal resolution required to revolutionize this field of research.

The study of genetic variation and molecular ecology has seen tremendous technological and analytical advancements in the past few decades (Ouborg et al. 1999; Cain et al. 2000; Jones & Ardren 2003). This has provided extremely useful data collection and novel analysis methods, that have enabled us to track relatedness of individual seeds and their dispersing plant (Godoy & Jordano 2001; Jones et al. 2005; Jordano et al. 2007; Robledo-Arnuncio & García 2007) or relatedness of individual plants and their dispersing parent (i.e. effective seed dispersal) (Meagher & Thompson 1987; Burczyk et al. 2006; González-Martínez et al. 2006; Hardesty et al. 2006). Valuable inferences on historical gene flow can also be gained by using molecular methods, revealing important ecological and evolutionary consequences of dispersal (Cain et al. 2000; Broquet & Petit 2009).

Advances in modelling and statistical analysis

Models of seed dispersal have played a fundamental role in representing patterns, investigating processes, elucidating the consequences of dispersal, and explaining dispersal evolution for populations and communities (Levin et al. 2003). In addition, modelling is often applied to predict dispersal rates, directions and intensity, which is of prime importance in assessing invasion dynamics (Higgins & Richardson 1999; Neubert & Caswell 2000; Higgins et al. 2003a; Skarpaas & Shea 2007; Jongejans et al. 2008; Soons & Bullock 2008). Modelling studies elucidating the potential role of spatial heterogeneity in determining invasion speed (With 2002) facilitated the development of models predicting dispersal in a spatially explicit and realistic environment (Russo et al. 2006; Levey et al. 2008; Schurr et al. 2008).

A common goal in modelling seed dispersal is estimating the dispersal kernel, the probability density function describing the number (or density) of dispersal units as a function of the distance from the source. In general, we can distinguish between two types of models for seed dispersal: phenomenological and mechanistic (Nathan & Muller-Landau 2000). Phenomenological models have been frequently used to estimate dispersal kernels for plant species (Kot et al. 1996; Clark 1998; Higgins & Richardson 1999; Bullock & Clarke 2000). These models use some functional forms, calibrated against observed data, to describe the distribution of distances of progeny from the seed source. Because model parameters are fitted from observed data, the identity of the dispersal agents is unimportant, thus relaxing the need to identify and quantify the role of different dispersal vectors. Phenomenological models enable us to deduce the spread potential of the plant simply by analysing the kernel tail 'fatness', which largely determines the speed and pattern of colonization (Kot et al. 1996; Clark 1998; Clark et al. 1998; Higgins & Richardson 1999; Higgins et al. 2003a; reviewed in Klein et al. 2006). However, this approach entails several disadvantages, including the high sensitivity of the fitted functions to variation not only in dispersal data but also in data

Table 9.1 Specific parameters of the three key components of a general model for passive dispersal (Nathan et al. 2008b) for four major dispersal systems.

	Anemochory	Hydrochory	Endozoochory	Epizoochory
Vector seed load (seeds time^{-1})	Seed abscission rate (potentially wind-induced)	Seed abscission rate (potentially wind- or water-induced)	Seed intake rate	Seed adhesion rate
Displacement velocity (distance time^{-1})	Flow speed		Animal movement speed	
Seed passage time (time)	Seed release height divided by seed terminal velocity	Seed buoyancy time	Gut retention time	Adhesion time

collection procedures (Hastings et al. 2005). The variation in dispersal processes between species, sites and times implies that this modelling approach is best used for a posteriori analysis of invasions (Higgins & Richardson 1999), which can also be achieved by models that correlate the observed patterns of spatial spread of invasive species with climatic, edaphic or other environmental variables (Peterson & Vieglais 2001; Foxcroft et al. 2004).

Compared with phenomenological models that calibrate dispersal kernels, mechanistic models use data on factors influencing dispersal processes to predict dispersal kernels. The general model for vector-mediated dispersal (Table 9.1) disentangles three basic components, from which further modelling can be carried out on specific cases of vectors and systems. A great deal of work in mechanistic modelling of seed dispersal by wind has been done since the publication of Elton's book, especially in recent years (Okubo & Levin 1989; Nathan et al. 2002; Tackenberg 2003; Soons et al. 2004; Bohrer et al. 2008; Wright et al. 2008; reviewed in Kuparinen 2006). An important advance has been made in fitting mechanistic models to LDD by wind, which earlier models often underestimated and could not explain (Nathan et al. 2002; Tackenberg 2003). Recent studies have shown that wind-speed-induced non-random seed release promotes LDD and increases spread rates (Soons & Bullock 2008), and that canopy structure height affects vertical winds and turbulence structure which in turn affect LDD (Bohrer et al. 2008). Mechanistic models of seed dispersal by animals, in their simplest form, calculate dispersal distances as the product of the vector seed load, displacement velocity and seed passage time (Table 9.1; Murray 1988; Sun et al. 1997; Holbrook &

Smith 2000; Westcott & Graham 2000; Vellend et al. 2003; Wehncke et al. 2003). These models are amenable for incorporating the effects of dispersal by multiple vectors (Dennis & Westcott 2007; Spiegel & Nathan 2007), including 'non-standard' vectors such as cassowaries dispersing the invasive 'water-dispersed' pond apple (Westcott et al. 2008).

9.3 SEED DISPERSAL AND MOVEMENT ECOLOGY OF INVASIVE SPECIES

Seed dispersal research can have important contributions to the field of plant invasion. By applying methodologies developed for seed dispersal research, researchers may improve their understanding of the relationships between invasive plants and their vectors, and better understand the dynamics of species spread within the landscape. Seed dispersal can be generalized within the movement ecology framework, enabling researchers from different disciplines to standardize the study of seed dispersal processes, and to identify the key life history traits, behaviours and external factors determining seed movement (Nathan et al. 2008a).

Movement ecology mechanistically defines the movement of organisms by four basic components and their interaction: internal state (why move?), motion capacity (how to move?), navigation capacity (where and when to move?) and external factors influencing all of the above (Nathan et al. 2008a). Although plants differ considerably from animals in their movement ability, their spatial movement can be conveniently implemented within the movement ecology framework. The external factors of the plant movement ecology framework are proximate extrinsic drivers determined at ecological time spans (Damschen et al. 2008; Nathan et al. 2008a), including environmental characteristics in general and the dispersal vectors in particular. The internal state, motion capacity and navigation capacity are plant attributes selected by ultimate drivers operating mostly at evolutionary time spans (Damschen et al. 2008; Nathan et al. 2008a). In the evolutionary sense, the internal drivers of dispersal, as identified in basic dispersal theory, include bet-hedging in unpredictable ecological conditions, avoiding kin competition and distance- or density-dependent mortality, and promoting outbreeding (Howe & Smallwood 1982; Levin et al. 2003; Nathan et al. 2009). Motion capacity, defined as the traits enabling seed movement, includes plant, fruit and seed characteristics that facilitate transport by external vectors such as plant height, fruit colour and scent, seed size and shape, and fruit and seed structural and chemical composition (Jordano 2000; Herrera 2002). The navigation capacity of plants primarily refers to the traits that synchronize the timing of fruiting and seed release with favourable dispersal conditions (Wright et al. 2008).

Mechanistic models for seed dispersal by wind effectively fit into this conceptual framework, incorporating atmospheric conditions (external factors) and plant traits such as wings and hairs enabling transport by wind (motion capacity) and seed abscission tissue determining timing of seed release (navigation capacity) as input parameters, outputting seed dispersal trajectories (movement); these, in turn, enable predictions of post-dispersal patterns to assess potential consequences for fitness (internal state), such as the probability of hitting a non-occupied establishment site away from sibling seeds (Wright et al. 2008).

Applying the movement ecology framework to the movement of animal-dispersed plants necessitates a twofold nested design (Fig. 9.1). In the inner loop, the dispersed seed is the focal individual and the vector is a major external factor affecting its movement. In the outer loop, the animal serving as the dispersal vector is the focal individual. In other words, modelling seed dispersal by animals requires considering not only the movement ecology of the plant but also the movement ecology of its vectors. The interplay between the two will ultimately determine the movement path of the plant. From the animal disperser's point of view, the internal state includes the need to obtain food (the properties of which are determined by the plant), to

avoid predators, to seek shelter, etc. The motion capacity relates to the internal machineries enabling animals to fly, walk, swim, climb, etc. The navigation capacity comprises the animal's ability to sense and respond to environmental cues related to movement by using, for example, the visual, olfactory or auditory systems, echolocation and magnetic field detection. Among the many biotic and abiotic external factors affecting these three internal components of animal movement are landscape structure, atmospheric or meteorological conditions, movement of predators and location or movement of food sources (Nathan et al. 2008a).

Identifying the relevant crossroads of interactions between the plant and the vector is where the movement ecology framework can greatly assist in identifying traits and mechanisms that could, at least partly, explain and predict plant dispersal processes. Fruit and seed characteristics interact with the set of frugivores the plant attracts, which in turn may differ in their navigation and motion capacities resulting in different movement paths of the seeds. For example, synchronization of fruiting with the passage of long-distance migrating animals could favour LDD and population spread, whereas attracting dispersers that consume fruit and rest on the source plant would favour dispersal over shorter distances. This also depends on the seed passage time and for some animals on the timing of fruiting (e.g. during the breeding season animals may carry seeds back to their offspring or mate).

In the following part we will apply movement ecology to examine the potential of a flying frugivore to disperse seeds of native and potentially invasive species. Studies of plant–vector interactions of an invasive species need not wait until empirical data on the invasion process itself becomes available, but can reasonably assume that, at least in early stages of invasions, animal movements are not significantly affected by the presence and distribution of the invading plant itself (Richardson et al. 2000a). This can be explained by the fact that at early stages the invading plant species is relatively rare and unfamiliar to the foraging animal. Thus, a priori predictions of the spread of potentially invasive animal-dispersed plants can be based on existing data on the foraging movements of local animal species capable of serving as dispersal vectors. Furthermore, comparing the properties of the different framework components between sympatric native and alien plants, or a potentially invasive plant in its native versus invaded range, could facilitate understanding of invasion dynamics and success, and

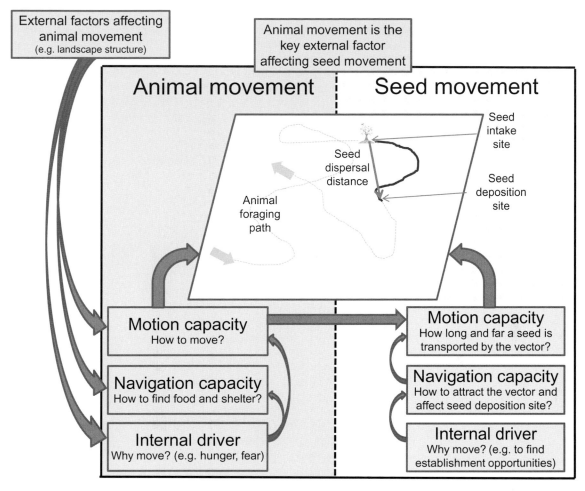

Fig. 9.1 A general conceptual framework for movement ecology of animal-dispersed plants. The framework has a twofold nested design (see main text). In the inner loop, the dispersed seed is the focal individual and the animal (the dispersal vector) is the major external factor affecting its movement. In the outer loop, the dispersal vector is the focal individual.

could assist in identifying invasion filters and pointing out candidate elements for management plans (Buckley et al. 2006).

9.4 FRUIT BATS AS LONG-DISTANCE SEED DISPERSERS OF BOTH NATIVE AND ALIEN SPECIES

Our illustration of the movement ecology approach focuses on seed dispersal of native and alien (including naturalized and potentially invasive) plant species (*sensu* Richardson et al. 2000a) by a common generalist dispersal vector, the Egyptian fruit bat (*Rousettus aegyptiacus*). We apply the twofold nested design of movement ecology (Fig. 9.1) to predict bat-mediated dispersal of the two groups of species, combining vector movement, foraging behaviour and seed passage time. In addition to this specific case study, we will show how simple allometric relationships can predict the seed dispersal distance for frugivorous birds and mammals (Box 9.1).

Box 9.1 Allometric relationships as a generic model for animal seed dispersal

Allometric relationships between body mass and various other characteristics of organisms have been well studied (Calder 1996). Animals with a larger body mass are predicted to have a larger home range, higher travel velocity and longer seed retention time, compared with smaller animals within the same taxonomic

(a)

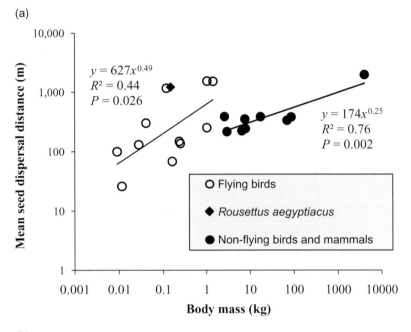

(b)

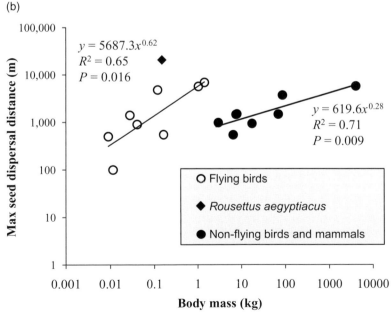

group (Calder 1996). Larger animals are therefore expected, by allometric relations alone, to disperse seeds to greater distances (Westcott & Graham 2000). Moreover, large animals often take up seeds of a wide variety of plant species, irrespective of the plant's dispersal morphology (see, for example, Westcott et al. 2005). Schurr et al. (2009) presented a meta-analysis of endozoochorous dispersal by birds, showing that seed dispersal distance increases with the body mass of avian dispersers as predicted from allometric relationships. They presented a simple general model that relates the body mass of animals to the mean dispersal distance of the seeds they disperse endozoochorously. Here we tested Schurr's et al. (2009) prediction with additional data, as well as with our own data of the Egyptian fruit bat (see section 9.4), and compared it with allometric predictions for flying birds, mammals and non-flying birds (see Rowell & Mitchell 1991; Mack 1995; Zhang & Wang 1995; Julliot 1996; Sun & Moermond 1997; Sun et al. 1997; Holbrook & Smith 2000; Stevenson 2000; Westcott & Graham 2000; Mack & Druliner 2003; Vellend et al. 2003; Wehncke et al. 2003; Westcott et al. 2005; Russo et al. 2006; Pons & Pausas 2007; Spiegel & Nathan 2007; Ward & Paton 2007; Weir & Corlett 2007; Campos-Arceiz et al. 2008).

Mean gut retention time (GRT) and mean speed of movement (SM) can both be expressed allometrically as a function of animal body mass (BM) (Robbins 1993; Calder 1996). For birds, these relationships were estimated as

$$GRT(h) = 1.6BM_{(kg)}^{0.33} \quad \text{(Robbins 1993)} \qquad (1)$$

and

$$SM(m/s) = 15.7BM_{(kg)}^{0.17} \quad \text{(Calder 1996)} \qquad (2)$$

For the Egyptian fruit bat's mean body mass measured in our study (147.5 ± 11.1 g), mean GRT from equation (1) is 51.1 minutes, very close to our measured value (52.82 ± 26.5 minutes). The mean SM from equation (2) is 11.34 m/s, higher than our measured value (9.1 ± 0.86 m/s). Indeed, owing to their general wing shape and flight mode, bats are expected to fly more slowly than birds of similar mass (Hedenström et al. 2009). Theoretical modelling of Egyptian fruit bat power flight (Flight 1.21 software (Pennycuick 2008)) predicts a minimum power speed of 9.3 m/s, in agreement with our empirical results.

We fitted a power curve to literature data of mean and maximum dispersal distances against mean disperser body mass, and compared it with our own predictions of the Egyptian fruit bat's dispersal kernel (Box Fig. 1). The mean and maximum seed dispersal distances by the Egyptian fruit bat are obviously much larger than expected for non-flying birds and mammals, and fairly similar to those of flying birds, though higher than expected, implying the large contribution fruit bats may have for LDD. In general, this multi-species analysis provides the means to approximate the dispersal potential of different vectors from body mass alone, or serve as a generic model for the expected dispersal distances of species differing in their body mass.

Box Fig. 1 The allometric relationships between body mass and the mean (a) and maximum (b) dispersal distances, divided into flying birds (*Onychognathus tristramii, Pycnonotus xanthopygos, Mionectes oleagineus, Ceratogymna atrata, Ceratogymna cylindricus, Dicaeum hirundinaceum, Garrulus glandarius, Pycnonotus jocosus, Corythaeola cristata, Tauraco schuetti* and *Ruwenzorornis johnstoni*) and non-flying birds and mammals (*Casuarius bennetti, Casuarius casuarius, Odocoileus virginianus, Cebus capucinus, Ateles paniscus, Lagothrix lagotricha, Alouatta seniculus, Cebus apella* and *Elephas maximus*). The figure includes data presented in this chapter for *Rousettus aegyptiacus*.

Most fruit bats of the Pteropodidae family are generalist consumers of a high variety of fruit species (Marshall 1983; Muscarella & Fleming 2007). They are common within the Old World tropical region and are claimed to be one of the major seed dispersers of tropical ecosystems (Mickleburgh et al. 1992), yet have been studied mainly for the type of fruit they consume and their qualitative potential contribution to dispersal (see, for example, Shilton et al. 1999; Muscarella & Fleming 2007; Nakamoto et al. 2009).

One of the most widely distributed bats within the Pteropodidae family is the Egyptian fruit bat (*Rousettus aegyptiacus*), a medium sized bat (100–200 g) that is considered a generalist forager, feeding on almost all fleshy fruited trees within its range including native, alien, naturalized and invasive species (Izhaki et al.

1995; Korine et al. 1999; Kwiecinski & Griffiths 1999). The Egyptian fruit bat exhibits commensalism with humans, commonly foraging in rural and urban habitats (Korine et al. 1999). Thus, as a human commensal and generalist feeder, the species has a large potential to disperse alien plants at the post-entry stage and in areas neighbouring human-dominated environments, where it has a higher probability of encountering a rich assortment of alien plants (Reichard & White 2001; Smith et al. 2006).

We captured fruit bats as they exited the roost cave in the Judean lowlands of central Israel (31° 40′ 58″ N 34° 54′ 34″ E), and equipped them with a tracking device combining a radiotelemetry unit (BD-2, Holohil Systems, Canada) and a lightweight GPS datalogger (GiPSy2, TechnoSmArt, Italy), together weighing 9.66 ± 2.3 g (mean $\pm$ SD; range 6.9–12.8 g) including batteries, protective casing and glue used to attach the device to the bat's back, approximately 4% to 9% of the tracked bat's total body mass (147.5 ± 11.1 g). From preliminary experiments, GPS accuracy was estimated to be lower than 5 metres 95% of the time, enabling us to track the exact route of the bat to a specific tree. All bat captures and tracking were approved by the Hebrew University of Jerusalem ethics committee and the Israel Nature and National Parks Protection Authority (licence 33060 given to A.T.). Ten bats were tracked throughout the entire nightly foraging excursion in a high spatiotemporal resolution of 0.1–1 Hz ($n = 9$) or once every 3 minutes ($n = 1$). The tracking device fell off the bats within 1–5 weeks from the time of attachment and was collected for data retrieval.

Tracked fruit bats exhibited long ($14,491 \pm 4,160$ m), straight (straightness index: 0.95 ± 0.04) and fast (33.4 ± 3.1 km/h) continuous commuting flight in relatively high altitudes above ground level (130.7 ± 50.3 m) upon departing from their roost after sunset and while flying back from the foraging site to the roost before sunrise, and showed a consistent foraging pattern where they feed mainly during the start and end of the night.

The bats' foraging site was found to constitute a relatively small area with a median convex hull of 0.052 km^2 per bat. The fruit bats showed a strong preference for foraging near human settlements (Monte-Carlo, $P < 0.001$. The test was conducted by averaging the distances of a set of random points from their nearest settlement centre within the potential foraging area of the bats (a circle of 21 km radius around the roost), repeating this 10^6 times and comparing the distribu-

tion created by the simulations with the measured mean distance from the bats' foraging sites to their nearest settlement centre (795 ± 490 m)). Each tree visited by the tracked bats was identified to the species level and was assessed for fruit fecundity and ripening (see text of Fig. 9.2 for a list of fruit tree species visited by the bats). A foraging event was defined only if the bat had landed for longer than 1 minute on a tree with ripe fruits. The fruit bats were generalist feeders, showing no preference for native or alien species (t-test; $t = 0.686$, $P = 0.515$). Gut retention time (GRT) was tested in a set of standard laboratory experiments (Sun et al. 1997; Holbrook & Smith 2000) on 13 individual wild bats from a recently established captive colony. They were offered two different fruits, selected to represent common plants endozoochorously dispersed by bats (Izhaki et al. 1995): the native common fig (*Ficus carica*) and the naturalized white mulberry (*Morus alba*), which is considered invasive in other parts of the world (Global Invasive Species Team, The Nature Conservancy: www.nature.org). We assumed both fruit species are consumed similarly by the bats, as indicated by field observations. GRT, representing all swallowed and defaecated seeds, was calculated and a gamma function was fitted for each of the two fruit species separately and for all data pooled together. Mean GRT was 55 minutes (range 16–414 minutes) for *F. carica* and 47 minutes (range 18–105 minutes) for *M. alba*. The fitted GRT gamma distributions were significantly different between the two species (Kolmogorov–Smirnov two-sample test; $Z = 3.902$, $P < 0.001$). Mean GRT for all data pooled together was 53 minutes.

Bat-generated dispersal distance kernels were calculated by multiplying the probability that the bat is located at a certain distance from the source tree at a certain time after feeding (based on the tracking data) and the defecation probability of a seed at that time (estimated from the fitted GRT gamma function). We separated the movement data of the tracked bats into two groups, according to the fruit trees they visited, and calculated four dispersal distance kernels from the different GRT distributions (Fig. 9.2): (i) alien, naturalized and invasive tree species with the pooled (*F. carica* and *M. alba*) GRT distribution; (ii) native tree species with the pooled GRT distribution; (iii) all trees with the GRT distribution for *F. carica* alone; and (iv) all trees with the GRT distribution for *M. alba* alone. Although the fitted GRT functions differed between the naturalized *M. alba* and the native *F. carica*, the dispersal dis-

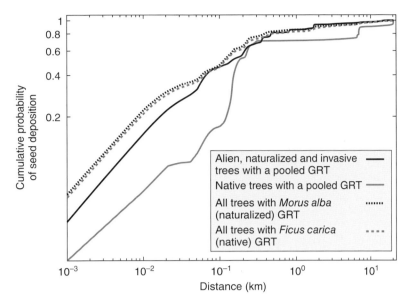

Fig. 9.2 Dispersal distance kernels of bat-dispersed seeds: the cumulative probability of seed deposition, calculated separately for the alien, naturalized and invasive tree species group (*Ficus sycomorus*, *Morus alba*, *Melia azedarach*, *Phoenix dactylifera* and *Washingtonia* sp., solid black line) and for the native tree species group (*Ficus carica*, *Ceratonia siliiqua* and *Olea europaea*, solid grey line) both using the pooled GRT distribution (see main text). The dotted black and grey lines represent the cumulative probability of seed deposition calculated for all trees using the GRT distribution of *M. alba* and *F. carica*, respectively.

tance kernels representing groups (iii) and (iv) were almost identical (Fig. 9.2, dotted grey and black lines), indicating that, controlling for the spatial movement of the vector, the difference in GRT had very little effect on the dispersal kernel. This result indicates that, in this case, seed dispersal distance is not as sensitive to the measured variation in GRT as was expected. In contrast, differences in the spatial distribution of the trees between group (i) and (ii) had a considerable effect on the kernels (Fig. 9.2, solid grey and black lines), which differed significantly (Kolmogorov–Smirnov two-sample test; $Z = 5.233$, $P < 0.001$). A spatially explicit simulation of bat-dispersed seeds revealed that what might be conceived as a simple seed shadow (the spatial distribution of seeds originated from a single source) around each tree is actually a complex mixture of overlapping seed shadows generated by the foraging bat. That is, the seed rain around a fruiting tree commonly encompasses seeds taken from other trees in the neighbourhood, and multiple-source 'seed shadows' are generated even in the vicinity of roosts and non-fruiting resting trees (Fig. 9.3). Although bats are predicted to disperse many (43.9%) seeds near (0–100 metres) the source plant, a high portion (17.2%) of seeds are dispersed long (more than 1 km) distances of up to 20 km (Fig. 9.2), owing to their fast and long commuting flights.

In summary, the Egyptian fruit bats act as both local and LDD vectors of both native and alien seed species. The local dispersal generates seed aggregations around source trees, whereas LDD tends to generate remote seed aggregations elsewhere. The surprising prediction that bats generate aggregations of long-distance dispersed seeds, rather than isolated individual events, can be attributed to the substantial proportion of LDD events. This is facilitated by several characteristics of the fruit bat, such as its fast and straight commuting flights to foraging sites far away from the main roost and its tendency to rest for a long time on non-fruiting trees and outside its main roost, providing new establishment opportunities for dispersed seeds away from the source plant. The landscape structure, or more specifically the spatial distribution of the fruit trees, rather than gut retention time, had the strongest effect on the dispersal kernel. Fruit bats exhibited a generalist habit to eat fruits from a wide range of plant species, readily feeding on alien plants. Their role as dispersers of potentially invasive species is further emphasized by their tendency to forage near human settlements where the initial introduction of invasive species is most expected. Altogether, our findings illustrate that understanding the movement ecology of the dispersal vector is mandatory for understanding and predicting the spatial dynamics of invasive, or potentially

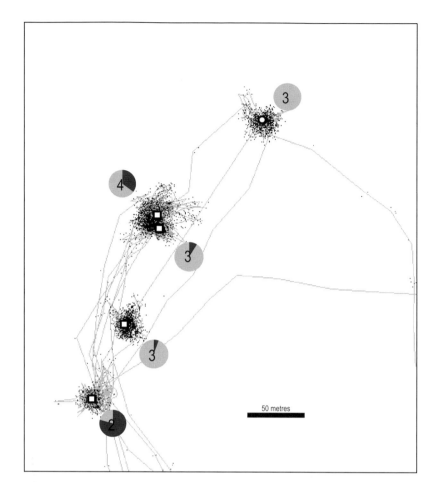

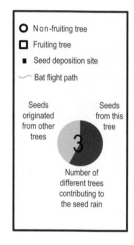

Fig. 9.3 An example of overlapping seed shadows (black dots) predicted for a full nightly path of a foraging bat (grey lines). The pie charts portray the proportions of seeds deposited in a radius of 10 m around a tree that have originated from this tree (dark grey) or from other trees (light grey). The number within the pie chart represents the total number of trees that contributed to the seed rain around each tree. White squares and circles represent fruit and non-fruit trees, respectively. Notice the upper tree is not a fruit tree but has a seed shadow similar to that of the fruit trees.

invasive, plant species (see also Murphy et al. 2008; Westcott et al. 2008). We note, however, that these findings might be specific to our study system and generalizations about the role of fruit bats in driving invasive spread should await data from different plant species and other systems as well. We emphasize again that invasion success strongly depends on post-dispersal processes that determine the survival and establishment of dispersed seeds, an important phase in a plant life cycle (Nathan & Muller-Landau 2000; Wang & Smith 2002), not elaborated in this chapter.

9.5 CONCLUSIONS

Five decades since the publication of Elton's book have witnessed new tools and concepts developed to study seed dispersal. Elton has identified humans' overriding role as the most pronounced dispersal vector responsible for the entry stage of current invasions; yet dispersal is also critically important in the establishment and spread stages of successful invasions, being a major factor in determining the spatial dynamics of plant populations. The dispersal ability of plant species strongly relies on the movement properties of the dispersal vector. Thus, to advance our understanding of the factors and mechanisms influencing seed dispersal and invasion processes, a vector-based approach should be promoted. Here we illustrate the application of a twofold nested design of the movement ecology framework to study dispersal of native and alien fleshy-fruited plant species dispersed by a generalist frugivore, the Egyptian fruit bat. We found that bats fly long distances to restricted foraging sites, generate complex seed shadows with peaks at the vicinity of both fruiting and non-fruiting trees, and are likely to play a key role in dispersing potentially invasive species as LDD vectors exhibiting strong preference to forage near human settlements. We have also shown in this case that the dispersal distance kernel is more strongly affected by the spatial distribution of the fruiting trees, than by the differences in gut retention times among the native or alien plant species examined. The mean and maximum distances of seed dispersal by the fruit bats are much higher than the corresponding dispersal distances expected from allometric relationships, even though their flight speed and gut retention time are relatively similar; the Egyptian fruit bat (presumably like many other fruit bats) is thus exceptional among mammals in its mean seed dispersal distance, even compared with flying frugivorous birds.

Our take-home message emphasizes the need to elucidate the movement ecology of any potentially invasive organism for understanding invasion processes and reducing associated hazards. Understanding the interactions between the plant and its vector should improve our ability to manage and prevent the establishment and spread of invasive species. In our case study, the habit of fruit bats to aggregate in large roosts opens opportunities for monitoring invasion processes by identifying new alien species of seeds in the bat guano dropped within the roosts, while reducing fruit bat activity within settlements could reduce their ability to disperse seeds of plants that have just passed the preliminary entry stage.

REFERENCES

Bascompte, J., Jordano, P. & Olesen, J.M. (2006) Asymmetric coevolutionary networks facilitate biodiversity maintenance. *Science*, **312**, 431–433.

Ben-Eliahu, M.N. & ten Hove, H.A. (1992) Serpulids (Annelida: Polychaeta) along the Mediterranean coast of Israel: new population build-ups of Lessepsian migrants. *Israel Journal of Zoology*, **38**, 35–53.

Bohrer, G., Katul, G.G., Nathan, R., Walko, R.L. & Avissar, R. (2008) Effects of canopy heterogeneity, seed abscission and inertia on wind-driven dispersal kernels of tree seeds. *Journal of Ecology*, **96**, 569–580.

Broquet, T. & Petit, E.J. (2009) Molecular estimation of dispersal for ecology and population genetics. *Annual Review of Ecology Evolution and Systematics*, **40**, 193–216.

Buckley, Y.M., Anderson, S., Catterall, C.P., et al. (2006) Management of plant invasions mediated by frugivore interactions. *Journal of Applied Ecology*, **43**, 848–857.

Bullock, J.M. & Clarke, R.T. (2000) Long distance seed dispersal by wind: measuring and modelling the tail of the curve. *Oecologia*, **124**, 506–521.

Bullock, J.M., Kenward, R.E. & Hails, R. (eds) (2002) *Dispersal Ecology: The 42nd Symposium of the British Ecological Society*. Blackwell, Malden.

Bullock, J.M., Shea, K. & Skarpaas, O. (2006) Measuring plant dispersal: an introduction to field methods and experimental design. *Plant Ecology*, **186**, 217–234.

Burczyk, J., Adams, W.T., Birkes, D.S. & Chybicki, I.J. (2006) Using genetic markers to directly estimate gene flow and reproductive success parameters in plants on the basis of naturally regenerated seedlings. *Genetics*, **173**, 363–372.

Cain, M.L., Milligan, B.G. & Strand, A.E. (2000) Long-distance seed dispersal in plant populations. *American Journal of Botany*, **87**, 1217–1227.

Calder, W.A. (1996) *Size, Function and Life History*, 2nd edn. Dover Publications, New York.

Campos-Arceiz, A., Larrinaga, A.R., Weerasinghe, U.R., et al. (2008) Behavior rather than diet mediates seasonal differences in seed dispersal by Asian elephants. *Ecology*, **89**, 2684–2691.

Carlo, T.A., Tewksbury, J.J. & Martinez del Rio, C. (2009) A new method to track seed dispersal and recruitment using ^{15}N isotope enrichment. *Ecology*, **90**, 3516–3525.

Clark, J.S. (1998) Why trees migrate so fast: confronting theory with dispersal biology and the paleorecord. *American Naturalist*, **152**, 204–224.

Clark, J.S., Macklin, E. & Wood, L. (1998) Stages and spatial scales of recruitment limitation in southern Appalachian forests. *Ecological Monographs*, **68**, 213–235.

Clavero, M. & García-Berthou, E. (2005) Invasive species are a leading cause of animal extinctions. *Trends in Ecology & Evolution*, **20**, 110.

Clobert, J., Danchin, E., Dhondt, A.A. & Nichols, J.D. (eds) (2001) *Dispersal*. Oxford University Press, Oxford.

Colautti, R.I. & MacIsaac, H.J. (2004) A neutral terminology to define 'invasive' species. *Diversity and Distributions*, **10**, 135–141.

Cousens, R., Dytham, C. & Law, R. (2008) *Dispersal in Plants: A Population Perspective*. Oxford University Press, New York.

Damschen, E.I., Brudvig, L.A., Haddad, N.M., Levey, D.J., Orrock, J.L. & Tewksbury, J.J. (2008) The movement ecology and dynamics of plant communities in fragmented landscapes. *Proceedings of the National Academy of Sciences of the USA*, **105**, 19078–19083.

Debussche, M. & Isenmann, P. (1990) Introduced and cultivated fleshy-fruited plants: consequences for a mutualistic Mediterranean plant-bird system. *Biological Invasions in Europe and the Mediterranean Basin* (ed. F. Di Castri, A.J. Hansen & M. Debussche), pp. 399–416. Kluwer Academic Publishers, Dordrecht.

Dennis, A.J. & Westcott, D.A. (2007) Estimating dispersal kernels produced by a diverse community of vertebrates. *Seed Dispersal: Theory and its Application in a Changing World* (ed. A.J. Dennis, R.J. Green, E.W. Schupp & D.A. Westcott), pp. 201–228. CAB International, Wallingford, UK.

Didham, R.K., Tylianakis, J.M., Hutchison, M.A., Ewers, R.M. & Gemmell, N.J. (2005) Are invasive species the drivers of ecological change? *Trends in Ecology & Evolution*, **20**, 470–474.

Elton, C.S. (1958) *The Ecology of Invasions by Animals and Plants*. Methuen, London.

Foxcroft, L.C., Rouget, M., Richardson, D.M. & MacFadyen, S. (2004) Reconstructing 50 years of *Opuntia stricta* invasion in the Kruger National Park, South Africa: environmental determinants and propagule pressure. *Diversity and Distributions*, **10**, 427–437.

Godoy, J.A. & Jordano, P. (2001) Seed dispersal by animals: exact identification of source trees with endocarp DNA microsatellites. *Molecular Ecology*, **10**, 2275–2283.

González-Martínez, S.C., Burczyk, J., Nathan, R., Nanos, N., Gil, L. & Alía, R. (2006) Effective gene dispersal and female reproductive success in Mediterranean maritime pine (*Pinus pinaster* Aiton). *Molecular Ecology*, **15**, 4577–4588.

Hardesty, B.D., Hubbell, S.P. & Bermingham, E. (2006) Genetic evidence of frequent long-distance recruitment in a vertebrate-dispersed tree. *Ecology Letters*, **9**, 516–525.

Harper, J.L. (1977) *Population Biology of Plants*. Academic Press, London.

Hastings, A., Cuddington, K., Davies, K.F., et al. (2005) The spatial spread of invasions: new developments in theory and evidence. *Ecology Letters*, **8**, 91–101.

Hedenström, A., Johansson, L.C. & Spedding, G.R. (2009) Bird or bat: comparing airframe design and flight performance. *Bioinspiration & Biomimetics*, **4**, 015001, doi:10.1088/1748-3182/4/1/015001.

Herrera, C.M. (2002) Seed dispersal by vertebrates. *Plant–Animal Interactions: An Evolutionary Approach* (ed. C.M. Herrera & O. Pellmyr), pp. 185–208. Blackwell, Oxford.

Higgins, S.I. & Richardson, D.M. (1999) Predicting plant migration rates in a changing world: the role of long-distance dispersal. *American Naturalist*, **153**, 464–475.

Higgins, S.I., Clark, J.S., Nathan, R., et al. (2003a) Forecasting plant migration rates: managing uncertainty for risk assessment. *Journal of Ecology*, **91**, 341–347.

Higgins, S.I., Nathan, R. & Cain, M.L. (2003b) Are long-distance dispersal events in plants usually caused by nonstandard means of dispersal? *Ecology*, **84**, 1945–1956.

Holbrook, K.M. & Smith, T.B. (2000) Seed dispersal and movement patterns in two species of *Ceratogymna* hornbills in a West African tropical lowland forest. *Oecologia*, **125**, 249–257.

Howe, H.F. & Smallwood, J. (1982) Ecology of seed dispersal. *Annual Review of Ecology and Systematics*, **13**, 201–228.

Hulme, P.E., Bacher, S., Kenis, M., et al. (2008) Grasping at the routes of biological invasions: a framework for integrating pathways into policy. *Journal of Applied Ecology*, **45**, 403–414.

Izhaki, I., Korine, C. & Arad, Z. (1995) The effect of bat (*Rousettus aegyptiacus*) dispersal on seed-germination in eastern Mediterranean habitats. *Oecologia*, **101**, 335–342.

Jones, A.G. & Ardren, W.R. (2003) Methods of parentage analysis in natural populations. *Molecular Ecology*, **12**, 2511–2523.

Jones, F.A., Chen, J., Weng, G.J. & Hubbell, S.P. (2005) A genetic evaluation of seed dispersal in the Neotropical tree *Jacaranda copaia* (Bignoniaceae). *American Naturalist*, **166**, 543–555.

Jongejans, E., Shea, K., Skarpaas, O., Kelly, D., Sheppard, A.W. & Woodburn, T.L. (2008) Dispersal and demography contributions to population spread of *Carduus nutans* in its native and invaded ranges. *Journal of Ecology*, **96**, 687–697.

Jordano, P. (2000) Fruits and frugivory. *Seeds: The Ecology of Regeneration in Plant Communities* (ed. M. Fenner), pp. 125–165. CAB International, Wallingford, UK.

Jordano, P., García, C., Godoy, J.A. & García-Castaño, J.L. (2007) Differential contribution of frugivores to complex seed dispersal patterns. *Proceedings of the National Academy of Sciences of the USA*, **104**, 3278–3282.

Julliot, C. (1996) Seed dispersal by red howling monkeys (*Alouatta seniculus*) in the tropical rain forest of French Guiana. *International Journal of Primatology*, **17**, 239–258.

Klein, E.K., Lavigne, C. & Gouyon, P.-H. (2006) Mixing of propagules from discrete sources at long distance: comparing a dispersal tail to an exponential. *BMC Ecology*, **6**, 3.

Korine, C., Izhaki, I. & Arad, Z. (1999) Is the Egyptian fruit-bat *Rousettus aegyptiacus* a pest in Israel? An analysis of the bat's diet and implications for its conservation. *Biological Conservation*, **88**, 301–306.

Kot, M., Lewis, M.A. & van den Driessche, P. (1996) Dispersal data and the spread of invading organisms. *Ecology*, **77**, 2027–2042.

Kuparinen, A. (2006) Mechanistic models for wind dispersal. *Trends in Plant Science*, **11**, 296–301.

Kwiecinski, G.G. & Griffiths, T.A. (1999) *Rousettus egyptiacus*. *Mammalian Species*, **611**, 1–9.

Lemke, A., von der Lippe, M. & Kowarik, I. (2009) New opportunities for an old method: using fluorescent colours to measure seed dispersal. *Journal of Applied Ecology*, **46**, 1122–1128.

Levey, D.J. & Sargent, S. (2000) A simple method for tracking vertebrate-dispersed seeds. *Ecology*, **81**, 267–274.

Levey, D.J., Tewksbury, J.J. & Bolker, B.M. (2008) Modelling long-distance seed dispersal in heterogeneous landscapes. *Journal of Ecology*, **96**, 599–608.

Levin, S.A., Muller-Landau, H.C., Nathan, R. & Chave, J. (2003) The ecology and evolution of seed dispersal: a theoretical perspective. *Annual Review of Ecology Evolution and Systematics*, **34**, 575–604.

Levine, J.M., Vilà, M., D'Antonio, C.M., Dukes, J.S., Grigulis, K. & Lavorel, S. (2003) Mechanisms underlying the impacts of exotic plant invasions. *Proceedings of the Royal Society of London B*, **270**, 775–781.

Mack, A.L. (1995) Distance and nonrandomness of seed dispersal by the dwarf cassowary *Casuarius bennetti*. *Ecography*, **18**, 286–295.

Mack, A.L. & Druliner, G. (2003) A non-intrusive method for measuring movements and seed dispersal in cassowaries. *Journal of Field Ornithology*, **74**, 193–196.

Mack, R.N. & Lonsdale, W.M. (2001) Humans as global plant dispersers: getting more than we bargained for. *Bioscience*, **51**, 95–102.

Marshall, A.G. (1983) Bats, flowers and fruit: evolutionary relationships in the Old World. *Biological Journal of the Linnean Society*, **20**, 115–135.

Meagher, T.R. & Thompson, E. (1987) Analysis of parentage for naturally established seedlings of *Chamaelirium luteum* (Liliaceae). *Ecology*, **68**, 803–812.

Mickleburgh, S.P., Hutson, A.M. & Racey, P.A. (1992) *Old World Fruit Bats: An Action Plan For Their Conservation*. International Union for the Conservation of Nature and Natural Resources (IUCN), Gland.

Murphy, H.T., Hardesty, B.D., Fletcher, C.S., Metcalfe, D.J., Westcott, D.A. & Brooks, S.J. (2008) Predicting dispersal and recruitment of *Miconia calvescens* (Melastomaceae) in Australian tropical rainforests. *Biological Invasions*, **10**, 925–936.

Murray, K.G. (1988) Avian seed dispersal of 3 Neotropical gap-dependent plants. *Ecological Monographs*, **58**, 271–298.

Muscarella, R. & Fleming, T.H. (2007) The role of frugivorous bats in tropical forest succession. *Biological Reviews*, **82**, 573–590.

Nakamoto, A., Kinjo, K. & Izawa, M. (2009) The role of Orii's flying-fox (*Pteropus dasymallus inopinatus*) as a pollinator and a seed disperser on Okinawa-jima Island, the Ryukyu Archipelago, Japan. *Ecological Research*, **24**, 405–414.

Nathan, R. (2006) Long-distance dispersal of plants. *Science*, **313**, 786–788.

Nathan, R. & Muller-Landau, H.C. (2000) Spatial patterns of seed dispersal, their determinants and consequences for recruitment. *Trends in Ecology & Evolution*, **15**, 278–285.

Nathan, R., Katul, G.G., Horn, H.S., et al. (2002) Mechanisms of long-distance dispersal of seeds by wind. *Nature*, **418**, 409–413.

Nathan, R., Getz, W.M., Revilla, E., et al. (2008a) A movement ecology paradigm for unifying organismal movement research. *Proceedings of the National Academy of Sciences of the USA*, **105**, 19052–19059.

Nathan, R., Schurr, F.M., Spiegel, O., Steinitz, O., Trakhtenbrot, A. & Tsoar, A. (2008b) Mechanisms of long-distance seed dispersal. *Trends in Ecology & Evolution*, **23**, 638–647.

Nathan, R., Bullock, J.M., Ronce, O. & Schurr, F.M. (2009) *Seed dispersal. Encyclopedia of Life Sciences*. John Wiley, Chichester, UK.

Neubert, M.G. & Caswell, H. (2000) Demography and dispersal: calculation and sensitivity analysis of invasion speed for structured populations. *Ecology*, **81**, 1613–1628.

Okubo, A. & Levin, S.A. (1989) A theoretical framework for data-analysis of wind dispersal of seeds and pollen. *Ecology*, **70**, 329–338.

Ouborg, N.J., Piquot, Y. & van Groenendael, J.M. (1999) Population genetics, molecular markers and the study of dispersal in plants. *Journal of Ecology*, **87**, 551–568.

Pennycuick, C.J. (2008) *Modelling the Flying Bird*. Academic Press, London.

Peterson, A.T. & Vieglais, D.A. (2001) Predicting species invasions using ecological niche modeling: new approaches from bioinformatics attack a pressing problem. *Bioscience*, **51**, 363–371.

Pimentel, D., McNair, S., Janecka, J., et al. (2001) Economic and environmental threats of alien plant, animal, and microbe invasions. *Agriculture Ecosystems & Environment*, **84**, 1–20.

Pons, J. & Pausas, J.G. (2007) Acorn dispersal estimated by radio-tracking. *Oecologia*, **153**, 903–911.

Por, F.D. (1978) *Lessepsian Migration*. Springer-Verlag, Berlin.

Reichard, S.H. & White, P. (2001) Horticulture as a pathway of invasive plant introductions in the United States. *Bioscience*, **51**, 103–113.

Ricciardi, A. (2007) Are modern biological invasions an unprecedented form of global change? *Conservation Biology*, **21**, 329–336.

Richardson, D.M. & Pyšek, P. (2008) Fifty years of invasion ecology – the legacy of Charles Elton. *Diversity and Distributions*, **14**, 161–168.

Richardson, D.M., Allsopp, N., D'Antonio, C.M., Milton, S.J. & Rejmánek, M. (2000a) Plant invasions – the role of mutualisms. *Biological Reviews*, **75**, 65–93.

Richardson, D.M., Pyšek, P., Rejmánek, M., Barbour, M.G., Panetta, F.D. & West, C.J. (2000b) Naturalization and invasion of alien plants: concepts and definitions. *Diversity and Distributions*, **6**, 93–107.

Robbins, C.T. (1993) *Wildlife Feeding and Nutrition*, 2nd edn. Academic Press, San Diego.

Robledo-Arnuncio, J.J. & García, C. (2007) Estimation of the seed dispersal kernel from exact identification of source plants. *Molecular Ecology*, **16**, 5098–5109.

Rowell, T.E. & Mitchell, B.J. (1991) Comparison of seed dispersal by guenons in Kenya and capuchins in Panama. *Journal of Tropical Ecology*, **7**, 269–274.

Russo, S.E., Portnoy, S. & Augspurger, C.K. (2006) Incorporating animal behavior into seed dispersal models: implications for seed shadows. *Ecology*, **87**, 3160–3174.

Schurr, F.M., Steinitz, O. & Nathan, R. (2008) Plant fecundity and seed dispersal in spatially heterogeneous environments: models, mechanisms and estimation. *Journal of Ecology*, **96**, 628–641.

Schurr, F.M., Spiegel, O., Steinitz, O., Trakhtenbrot, A., Tsoar, A. & Nathan, R. (2009) Long-distance seed dispersal. In *Fruit Development and Seed Dispersal* (ed. L. Østergaard), pp. 204–237. Blackwell, Oxford.

Shilton, L.A., Altringham, J.D., Compton, S.G. & Whittaker, R.J. (1999) Old World fruit bats can be long-distance seed dispersers through extended retention of viable seeds in the gut. *Proceedings of the Royal Society of London B*, **266**, 219–223.

Skarpaas, O. & Shea, K. (2007) Dispersal patterns, dispersal mechanisms, and invasion wave speeds for invasive thistles. *American Naturalist*, **170**, 421–430.

Smith, R.M., Thompson, K., Hodgson, J.G., Warren, P.H. & Gaston, K.J. (2006) Urban domestic gardens (IX): Composition and richness of the vascular plant flora, and implications for native biodiversity. *Biological Conservation*, **129**, 312–322.

Soons, M.B. & Bullock, J.M. (2008) Non-random seed abscission, long–distance wind dispersal and plant migration rates. *Journal of Ecology*, **96**, 581–590.

Soons, M.B., Heil, G.W., Nathan, R. & Katul, G.G. (2004) Determinants of long-distance seed dispersal by wind in grasslands. *Ecology*, **85**, 3056–3068.

Spiegel, O. & Nathan, R. (2007) Incorporating dispersal distance into the disperser effectiveness framework: frugivorous birds provide complementary dispersal to plants in a patchy environment. *Ecology Letters*, **10**, 718–728.

Stevenson, P.R. (2000) Seed dispersal by woolly monkeys (*Lagothrix lagothricha*) at Tinigua National Park, Colombia: dispersal distance, germination rates, and dispersal quantity. *American Journal of Primatology*, **50**, 275–289.

Sun, C. & Moermond, T.C. (1997) Foraging ecology of three sympatric turacos in a montane forest in Rwanda. *Auk*, **114**, 396–404.

Sun, C., Ives, A.R., Kraeuter, H.J. & Moermond, T.C. (1997) Effectiveness of three turacos as seed dispersers in a tropical montane forest. *Oecologia*, **112**, 94–103.

Tackenberg, O. (2003) Modeling long-distance dispersal of plant diaspores by wind. *Ecological Monographs*, **73**, 173–189.

Traveset, A. & Richardson, D.M. (2006) Biological invasions as disruptors of plant reproductive mutualisms. *Trends in Ecology & Evolution*, **21**, 208–216.

Vellend, M., Myers, J.A., Gardescu, S. & Marks, P.L. (2003) Dispersal of *Trillium* seeds by deer: implications for long-distance migration of forest herbs. *Ecology*, **84**, 1067–1072.

Vitousek, P.M., Mooney, H.A., Lubchenco, J. & Melillo, J.M. (1997) Human domination of Earth's ecosystems. *Science*, **277**, 494–499.

Von der Lippe, M. & Kowarik, I. (2007) Long-distance dispersal of plants by vehicles as a driver of plant invasions. *Conservation Biology*, **21**, 986–996.

Wang, B.C. & Smith, T.B. (2002) Closing the seed dispersal loop. *Trends in Ecology & Evolution*, **17**, 379–385.

Ward, M.J. & Paton, D.C. (2007) Predicting mistletoe seed shadow and patterns of seed rain from movements of the mistletoebird, *Dicaeum hirundinaceum*. *Austral Ecology*, **32**, 113–121.

Wehncke, E.V., Hubbell, S.P., Foster, R.B. & Dalling, J.W. (2003) Seed dispersal patterns produced by white-faced monkeys: implications for the dispersal limitation of neotropical tree species. *Journal of Ecology*, **91**, 677–685.

Weir, J.E.S. & Corlett, R.T. (2007) How far do birds disperse seeds in the degraded tropical landscape of Hong Kong, China? *Landscape Ecology*, **22**, 131–140.

Westcott, D.A. & Graham, D.L. (2000) Patterns of movement and seed dispersal of a tropical frugivore. *Oecologia*, **122**, 249–257.

Westcott, D.A., Bentrupperbäumer, J., Bradford, M.G. & McKeown, A. (2005) Incorporating patterns of disperser behaviour into models of seed dispersal and its effects on estimated dispersal curves. *Oecologia*, **146**, 57–67.

Westcott, D.A., Setter, M., Bradford, M.G., McKeown, A. & Setter, S. (2008) Cassowary dispersal of the invasive pond apple in a tropical rainforest: the contribution of subordinate dispersal modes in invasion. *Diversity and Distributions*, **14**, 432–439.

Wheelwright, N.T. & Orians, G.H. (1982) Seed dispersal by animals: contrasts with pollen dispersal, problems of terminology, and constraints on coevolution. *American Naturalist*, **119**, 402–413.

With, K.A. (2002) The landscape ecology of invasive spread. *Conservation Biology*, **16**, 1192–1203.

Wright, S.J., Trakhtenbrot, A., Bohrer, G., et al. (2008) Understanding strategies for seed dispersal by wind under contrasting atmospheric conditions. *Proceedings of the National Academy of Sciences of the USA*, **105**, 19084–19089.

Xiao, Z.S., Jansen, P.A. & Zhang, Z.B. (2006) Using seed-tagging methods for assessing post-dispersal seed fate in rodent-dispersed trees. *Forest Ecology and Management*, **223**, 18–23.

Zhang, S.Y. & Wang, L.X. (1995) Fruit consumption and seed dispersal of *Ziziphus cinnamomum* (Rhamnaceae) by two sympatric primates (*Cebus apella* and *Ateles paniscus*) in French Guiana. *Biotropica*, **27**, 397–401.

Chapter 10

BIODIVERSITY AS A BULWARK AGAINST INVASION: CONCEPTUAL THREADS SINCE ELTON

Jason D. Fridley

Department of Biology, Syracuse University, Syracuse, NY 13244, USA

Fifty Years of Invasion Ecology: The Legacy of Charles Elton, 1st edition. Edited by David M. Richardson

10.1   INTRODUCTION

The diversity–invasibility hypothesis (DIH) is the proposition that more biologically diverse communities are more difficult to invade by novel species or genotypes. The measurement of biodiversity in DIH studies is often species richness or indices involving species abundances. However, the same general arguments can apply at the level of populations (genetic diversity) or involve distinctions between species using trait attributes (functional diversity) or more basal taxonomic levels (phylogenetic diversity). The measurement of community invasibility typically involves performance measures of one or a few potential invaders, or the total abundance or species richness of all invaders. The meaning of 'invader' depends on the objectives of the study; many researchers address DIH only in reference to modern biogeographic notions of 'native' communities and 'alien' invaders (*sensu* Pyšek et al. 2004), whereas others are most interested in the process of new species establishment itself, regardless of whether the community and invader share a coevolutionary history. Theoretical approaches related to DIH have been common since the 1970s (Justus 2008); only in the past decade or so have experimental and observational DIH studies come to rival the influence of theory in our understanding of diversity and invasions (Levine & D'Antonio 1999, Fridley et al. 2007).

The renowned British animal ecologist Charles Elton was the first to argue that the diversity and complexity of natural ecosystems may play a role in mediating patterns of biological invasions. Ever since, Elton has, perhaps unfairly, been the poster child for the idea that biodiversity has important functional consequences for species invasions (Richardson & Pyšek 2008). As pointed out by Levine and D'Antonio (1999), Elton's arguments were only one of several conceptual threads that have given rise to modern diversity–invasibility theory. Elton's particular views revolved around poorly developed ideas about relations between ecosystem complexity (e.g. food chain length) and population-level stability (McCann 2009). Although he did invoke what little theory existed at the time, modern understanding of the diversity–stability hypothesis has advanced considerably since Elton's time, although still without substantial consensus (Justus 2008).

Even if Elton's arguments lack modern consensus, to what extent does modern diversity–invasibility theory depend on Elton's insights? The question is of more than historical interest, because many DIH studies, both in invasion ecology and theoretical ecology, cite Elton's work generally and disregard the fact that Elton was referring largely to trophic interactions and population stability. Morevoer, DIH is one of the major lasting contributions of Elton's work, both for invasion ecology and more broadly in theoretical and community ecology (Richardson & Pyšek 2007). In this chapter I argue that Elton's imprint on DIH studies, although significant, is on the whole less than concepts of species packing and complementary resource-use introduced by Robert MacArthur and others in the decades after Elton's work. The concept of limiting similarity in particular (MacArthur & Levins 1967) is perhaps the strongest conceptual thread that unites modern studies of community invasibility, both in recent theoretical work (for example Tilman 2004) and among experimental studies (Davis 2009). There is also an emerging viewpoint that biodiversity may have evolutionary significance in understanding invasion patterns, in that more diverse biotic regions may be global source areas for highly competitive species that spread into other regions, and thus be less invasible than less diverse regions. This view can be traced back to Darwin (1859) and has been invoked to explain historical biotic interchanges in addition to modern invasion patterns (Vermeij 1996, 2005). Although this view has not been typically seen as a component of DIH, it offers a new perspective that could shift DIH focus from small-scale communities to whole biotic regions (Stachowicz & Tilman 2005).

The significance of formulating a conceptual road map for DIH studies (Fig. 10.1) is in how it helps to clarify DIH mechanisms; in this way insights derived from different DIH approaches can be applied to the management and restoration of ecological communities. In what follows, I use Elton's (1958) seminal work as a starting point for describing the conceptual history of DIH studies, as practised in the contemporary invasion literature (see also Fridley et al. 2007). An excellent starting point for any such synthesis is the review by Levine and D'Antonio (1999); here I add detail to their assessment along with a more subjective discussion of where DIH studies stand in the recent opinions of many invasion researchers. My assessment is by no means comprehensive: readers are referred to a recent review of the diversity–stability debate by Justus (2008) and syntheses of empirical and theoretical DIH studies by Fridley et al. (2007) and Davis (2009).

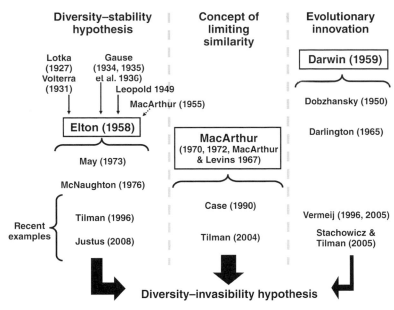

Fig. 10.1 Conceptual threads of the modern diversity–invasibility hypothesis (DIH), the proposition that biologically diverse communities are better able to resist invasions by novel species. Three major components of DIH involve population or community stability, highlighted by Elton; limiting similarity of competing species, as theoretically framed by MacArthur; and whether more diverse regional assemblages may represent greater evolutionary advancement or innovation, as originally suggested by Darwin. Arrows indicate significant intellectual antecedents of Elton's (1958) synthesis; the dotted arrow notes that MacArthur (1955) was not actually cited by Elton. Recent examples of each conceptual thread are selections from a large literature.

10.2 ELTON'S ARGUMENT: COMPLEXITY, STABILITY AND ECOLOGICAL RESISTANCE

Elton's *The Ecology of Invasions by Animals and Plants*, published in 1958, is widely considered the first formal synthesis of invasion biology (Richardson & Pyšek 2007). In it, Charles Elton made the first detailed argument that more diverse ecosystems should be more resistant to invasion by introduced alien species, what he called *ecological resistance*. In the context of the book, his argument for ecological resistance was brief, laid out in about six pages out of 170, and the theoretical, experimental and observational research he pulled from the literature was only weakly organized into several 'lines of evidence' to support his assertion that 'the balance of relatively simple communities of plants and animals is more easily upset than that of richer ones; that is, more subject to destructive oscillations in populations, especially of animals, and more

vulnerable to invasions' (p. 145). In the order that Elton presented them, these arguments included the following:

1 *Populations in 'simpler' communities are more prone to population fluctuations.* Elton cited Lotka–Volterra predator–prey models (Lotka 1927; Volterra 1931 (cited in Elton 1958)) and Gause's experiments on paramecia (Gause 1934, 1935; Gause et al. 1936), noting that 'it is very difficult to keep small populations of this simple mixture in balance'. Notably, he did not offer evidence that it was any easier to maintain a balance in more complex communities, nor did he state how population fluctuations themselves relate to invasibility.

2 *Habitat complexity can modulate population fluctuations due to source-sink dynamics and rescue effects.* Here he again references Gause's experiments, suggesting that the addition of environmental heterogeneity can supply prey refuges from predation and limit predator-induced population cycling.

3 *Oceanic islands are more invaded than mainland assemblages,* presumably because they tend to have fewer species than mainland areas of similar size, although this was not specified.

4 *Pest outbreaks occur more often in anthropogenically disturbed habitats.* Elton references the pest problems associated with crop monocultures, and also notes the issue of artificially shortened food chains that reduce top-down regulation of species abundances.

5 *Tropical ecosystems are less invaded than more species-poor temperate and boreal ecosystems.* Elton noted the few existing records of insect outbreaks in the tropics, and asserted 'the ecological stability of tropical rainforest seems to be a fact'.

6 *Experimental research in orchard pest management suggests that as food chain complexity increases, the frequency of pest outbreaks declines.* Elton's examples involve unintended consequences of pesticide use on non-target insect predators, whose declines presumably explained subsequent pest explosions with continued chemical use.

As is evident from these examples, Elton's focus was on community complexity, largely in the form of food webs, and the notion that population fluctuations could be modulated by the richer array of trophic interactions associated with more diverse and complex ecosystems (Fig. 10.1). Here and elsewhere in the book it's clear that Elton had deeply considered the influence of human activities on (especially) animal populations ('invasions most often come to cultivated land, or to land much modified by human practice' (p. 63)), and viewed naturalness, or lack of anthropogenic disturbance, as a prerequisite for stability and 'balance' in coexisting populations. (Earlier in the book he argues that natural forces can produce 'some kind of permanent balance' between populations (p. 115), and that 'the complexity of natural balance in populations is evident enough to anyone who cares to recognize it' (p. 137).) Elton's views on naturalness and ecological stability were shaped significantly by the pioneering conservationist Aldo Leopold, who introduced concepts of ecological integrity and ecosystem health in his chapter on *The Land Ethic* in his 1949 *Sand County Almanac* (Elton quotes a passage that links community stability with 'integrity' in his chapter on the anthropogenic disruption of natural food chains). Like Leopold, however, Elton's treatment of these concepts was vague, and he and Leopold's views of the integrity and steadiness of natural ecosystems look quaint in view of modern research in trophic and ecosystem

ecology. This perspective has been labelled the 'balance of nature camp', which Cooper (2001) argues is 'guided by the image of nature as full – of populations with boundless reproductive potential being held in check by density sensitive governors'.

It is unfair, however, to typify Elton's views on the subject of diversity and invasibility as strictly a 'balance of nature' perspective (*sensu* Cronk & Fuller 1995), at least to the extent that invasions from a 'balance of nature' perspective require imbalances brought on by anthropogenic disturbances. For one thing, Elton claimed that differences in diversity among ecosystems even in the absence of anthropogenic disturbance should influence invasion rates, as is evident in his suggestion that mainlands (as opposed to islands) and diverse tropical ecosystems should be less invasible. He also referenced historical (non-anthropogenic) invasion patterns in this way; for example, in reference to the late Pleistocene palaeorecord, he wondered 'whether it is just a coincidence that the erasure of earlier [i.e. Tertiary] differences by mutual migrations shows most of all in the simpler ecosystems of the Arctic tundra' (p. 41). Elton was also aware of many examples of invasions in relatively complex systems. In these cases where diversity or disturbance failed to provide an easy explanation of invasibility, Elton variously invoked enemy release (e.g. in the spread of *Rhododendron ponticum* in Britain), empty niches (e.g. ship rats on Cyprus) or invaders of unusual aggressiveness or competitive ability (e.g. the replacement of the 'gentler' Pacific island rat by more aggressive ship and Norway rats). These other mechanisms were not well integrated into his synthesis, however, and the lasting impression from Elton's work is that diverse ecosystems are less invasible because they contain more stable populations (McCann 2009). Elton's argument thus rests on two questions. Are in fact populations in more diverse ecosystems less prone to fluctuation? And how are population fluctuations related to invasibility?

10.3 HAVE ELTON'S CLAIMS ABOUT STABILITY IN DIVERSE ECOSYSTEMS HELD UP?

By resting his argument about diversity and invasibility largely on the premise that population fluctuations are modulated in more diverse or complex ecosystems, Elton placed invasibility research on a contentious foundation that would soon come to dominate theoretical research in community ecology. Indeed, shortly before Elton's work, Robert MacArthur (1955) published the first direct theoretical treatment of the relationship between trophic complexity and population stability. Like Elton, MacArthur argued that communities of more complex trophic structure, in this case measured as number of links in a food web, tend to exhibit smaller fluctuations in the abundances of component species. Unlike Elton, MacArthur's reasoning was presented as a formal proof involving food web and information theory (although it did not directly involve population or community dynamics (McCann 2009)), which facilitated further refinement by later theoreticians. Interestingly, although MacArthur's work was perhaps the most pertinent evidence in support of Elton's argument at the time, it does not appear in Elton's 1958 book. This may stem from Elton's underlying wariness of the scientific utility of mathematical modelling in ecology (Justus 2008), one of MacArthur's enduring legacies.

Evidence counter to the diversity–stability hypothesis began to accumulate in the 1970s. Within two decades of Elton's book there was no consensus that more diverse communities were more stable, and indeed several studies had found the opposite. As Justus (2008) describes in a recent review of the diversity–stability debate, the turning point was May's (1973) theoretical demonstration that more complex communities were often less stable over a wide array of different community structures. In particular, May countered Elton by demonstrating that moving from simple to more species-rich Lotka–Volterra predator–prey systems actually increased population fluctuations. Goodman (1975) reviewed Elton's arguments and noted a lack of empirical support for many, including new data about outbreaks of tropical insects. Studies of the past several decades, including more empirical research field studies and laboratory microcosms, have failed to provide further resolution (Levine & D'Antonio 1999; Justus 2008). One contentious issue is whether diversity stabilizes population- or community-level processes. For example, McNaughton (1977) argued that diversity was more likely to stabilize ecosystem processes like productivity than the abundances of individual populations, a perspective also shared in more recent work by Tilman (1996). Justus (2008) concluded that the diversity-stability debate remains unresolved, in part because 'diversity' and 'stability' can be manifest in myriad ways, relatively few of which have been subject to detailed theoretical or empirical treatment. To the extent that Elton's view of the DIH rests on the diversity–stability hypothesis, then, it is by no means clear that more diverse ecosystems are more stable.

Even if Elton was correct that populations fluctuate less in more diverse systems, how do population fluctuations relate to invasibility? Surprisingly few researchers have made an explicit connection between stability – particularly Elton's concept of stability that involved constancy in abundance over time (Justus 2008) – and invasibility. Davis (2009) suggests that population stability should relate to the temporal frequency at which communities are open to new species, assuming such fluctuations occur independently of resource availability. Davis cites two recent empirical studies involving grassland (Bezemer & van der Putten 2007) and sessile marine invertebrate communities (Dunstan & Johnson 2006) that demonstrate how such 'windows of opportunity' are correlated with increased invasion events, and thus link population constancy with invasibility. For plant communities or others involving species within particular guilds, population-level stability seems less relevant than community-level stability (such as total productivity), in that community stability may be a better indicator of the availability of free resources (Tilman 1999). This may be particularly true in systems where species abundances are seasonally complementary, such that reductions in diversity occur in parallel with the opening up of resources during particular seasons (see, for example, Stachowicz et al. 2002). However, this invokes a perspective of community organization involving niche differences and the partitioning of resources, which has its origins not in Elton's writings on invasion (although Elton did much to develop the niche concept (Elton 1927)), but in MacArthur's concept of limiting similarity.

10.4 FROM ELTON TO MACARTHUR: AN INVASION FRAMEWORK BUILT ON LIMITING SIMILARITY

In a series of contributions beginning with his paper (with Richard Levins) in 1967 on the concept of *limiting similarity*, Robert MacArthur promoted a body of theory based on species' differences in resource acquisition that has come to define conceptually a large area of community ecology (MacArthur 1970, 1972). This view, referred to as niche-driven community assembly (Hubbell 2001) or species packing (Slobodkin 2001), asserts that patterns of biodiversity, such as species richness, are determined by differences between species in how effectively they acquire particular resources in an environment where the quality of one or more resources varies (e.g. food size or soil nutrient concentrations). Because species maximize their carrying capacities and competitive abilities at different positions in resource space, they can stably coexist at equilibrium in such an environment, in accord with standard Lotka–Volterra rules that describe how interspecific coexistence is dictated by intraspecific population regulation (MacArthur 1972). The appeal of the theory lies not in demonstrating that species differences can promote coexistence, which had been demonstrated decades before. Rather, the key insight MacArthur and Levins produced, later elaborated by May (1973), was that there was a finite limit to how similar competing species could be if they were to coexist at equilibrium. Although such limits depend on many factors related to how species consume resources and use them for population growth (which have been difficult to parameterize and thus test in empirical systems), the central idea that biodiversity patterns are tied to resource-use differences between species has profound ramifications for diversity–invasibility theory (middle of Fig. 10.1).

One consequence of limiting similarity, as MacArthur (1970) noted, is that the same rules that dictate whether a new species is different enough to successfully invade an established community also require that successful invasions be accompanied by greater community productivity. Thus, under the standard equilibrium assumptions required by limiting similarity, more diverse assemblages should be more productive because species are optimized to best exploit different resource conditions (Tilman et al. 1997; Tilman 1999). Put another way, limiting similarity requires coexisting species to be *complementary* in their use of resources, which in turn enhances total resource acquisition, and thus productivity, in communities containing more species. In this view diversity (particularly species richness), productivity and community invasibility are all intimately connected and tied together by the requirement of species to differ in resource use, a view espoused perhaps most strongly by Tilman (1999). This view is relatively distinct from Elton's argument about population fluctuations; in particular there need be no temporal dimension to

limiting similarity, and indeed much of the theory built on limiting similarity presupposes equilibrium conditions (see, for example, Tilman 1982; Tilman et al. 1997; Case 1990; cf. Tilman (2004) for a stochastic example). MacArthur may not have felt Elton's arguments were particularly relevant, either: although Elton (1958) is listed in the bibliography of MacArthur's great synthesis, *Geographical Ecology* (1972), it is never referred to in the text. Davis (2009) also notes that the suggestion that successful invasions depend in part on the similarity in resource requirements between potential invaders and resident species is older than species coexistence theory. Darwin (1859) suggested introduced plant species will have a difficult time invading if they have to compete with close relatives, assuming their shared phylogeny indicates similar niches or pest pressures (Darwin's naturalization hypothesis (Daehler 2001)).

The hypothesis that diversity enhances total resource uptake, and, as a consequence leads to reduced community invasibility, has recently faced fierce challenges on theoretical and empirical grounds. Theoretically, the universality of limiting similarity as the driver of biodiversity patterns has been questioned repeatedly, particularly by proponents of nonequilibrium views of community assembly (Grime 1973; Huston 1979, 1994) including the more recent concept of neutral theory (Hubbell 2001; note the paradoxical importance of MacArthur & Wilson's (1967) *Theory of Island Biogeography* to this view). Communities assembled by neutral dynamics do not exhibit diversity–invasibility patterns, except in the trivial case where overall species membership is limited by a paucity of total individuals (i.e. sampling or rarefaction effects (Fridley et al. 2004; Herben et al. 2004, Palmer et al. 2008)). Empirically, evidence for the negative associations of diversity and invasibility expected from limiting similarity theory has been highly inconsistent. Experimental studies conducted at small scales tend to find support for DIH, largely through limiting similarity mechanisms, but observational studies tend to find the opposite pattern (Fridley et al. 2007). Many researchers have suggested that such conflicting patterns indicate that the role of diversity itself in invasibility is weak compared to other factors like propagule pressure (Brown & Peet 2003), variation in resource availability (Stohlgren et al. 1999, 2003) or disturbance levels (Byers & Noonburg 2003; see also Shea & Chesson 2002; Huston 2004; Tilman 2004; Fridley et al. 2007; Davis 2009). The current state of the hypothesis seems to be that some researchers would prefer to emphasize the role that community diversity plays, even if weak (see, for example, Stachowicz & Tilman 2005), whereas others view the poor ability of diversity in natural systems to predict invasibility as damning evidence against the DIH paradigm (Rejmánek 1996; Williamson 1996; Huston 2004; Davis 2009). Still another perspective is the possibility that high levels of diversity imply not only many potential competitors, but also many potential mutualists (Richardson et al. 2000). Several studies in marine systems in particular have demonstrated links between diversity and facilitative interactions in invasions (see, for example, Stachowicz & Byrnes 2006). Nonetheless, to the extent that DIH is true at all for the spatial and temporal scales Elton envisaged, its predictive power is generally weak (Fridley et al. 2007; Davis 2009), thus suggesting that this most-cited aspect of Elton's work is not a robust framework for invasion research despite its continued popularity (Richardson & Pyšek 2007).

It is interesting to note that Elton was primarily an animal ecologist, and there is a stronger historical tendency for animal ecologists to emphasize the role of limiting similarity in structuring communities than for plant ecologists (Grubb 1977). Should DIH be more relevant for animal communities? Huston (1994) argued that equilibrium-based mechanisms of species coexistence are more relevant for mobile organisms that, owing to their greater capacity for selective patch foraging, are less influenced by spatial and temporal resource heterogeneity and thus more likely to approach the equilibrium conditions necessary for competitive exclusion. If communities of mobile organisms are more strongly regulated by equilibrium-dependent mechanisms like limiting similarity, then DIH may be a more powerful framework for animal invasions. Indeed, much of the observational evidence against DIH comes from plant communities (Fridley et al. 2007), and much theory and discussion against DIH comes from researchers who primarily study plants (see, for example, Rejmánek 1996; Davis 2009; Huston 2004; Stohlgren et al. 2008). Elton himself discussed plants only rarely in his 1958 book, despite the title (plants represent less than 10% of the book's examples (Richardson & Pyšek 2007)). As a general rule, researchers who emphasize the importance of diversity in invasibility studies are those who place high significance on limiting similarity as a driver of community assembly in natural systems, and there

seem to be proportionally fewer plant ecologists in this group.

10.5 CONSIDERING DIH MORE BROADLY: WERE ELTON AND MACARTHUR WRONG AT THE LOCAL SCALE, BUT CORRECT AT BIOGEOGRAPHIC SCALE?

At the conclusion of his 1972 book, MacArthur briefly considered whether the rules of limiting similarity could be applied to the case where two historically isolated assemblages, each with their own niche-based structure, meet: would species from one assemblage invade the other, and vice-versa? In the same section, MacArthur remarked on Darlington's (1965) principle that 'colonists from [larger continental] species-rich areas are more likely to succeed than those from smaller continents in less favourable climates'. In particular, MacArthur observed that, 'the adage "practice makes perfect" should apply to competition as well as to other activities, and the emigrants from species-rich continents in tropical climates have had much practice in competing. They certainly should be good at invading a new community of competitors'. MacArthur stopped short of demonstrating this phenomenon theoretically; indeed his example of two biotas meeting was used to demonstrate how niche structures could be arranged such that the biotas would not subsequently invade one another. It would be easy to show that one simple change in MacArthur's logic would lead to a much different outcome: instead of having the resource utilization curves of species from each assemblage perfectly match available production (a case precluding any invasions at equilibrium), let species of one community track production levels only loosely. In this scenario, such a non-saturated community could be invaded by species from the other assemblage, whereas the saturated assemblage would remain uninvaded. Would such an occurrence happen in nature, and is it related to differences in diversity between biotic assemblages?

Darlington's (1965) arguments were first articulated by Darwin. In the last chapter of *On the Origin of Species* (1859), Darwin wrote that 'As natural selection acts by competition, it adapts the inhabitants of each country only in relation to the degree of perfection of their associates; so that we need feel no surprise at the inhabitants of any one country, although on the ordinary view supposed to have been specially created and adapted for that country, being beaten and supplanted by the naturalized productions from another land'. Like Darlington, Darwin suggested species from larger land masses would be 'advanced through natural selection and competition to a higher stage of perfection or dominating power'. Dobzhansky (1950) made a similar argument to explain why so many plant and animal lineages stem from the tropics and have repeatedly invaded temperate areas, stating that 'the environment provides "challenges" to which organisms "respond" by adaptive changes ... tropical environments [containing more species] provide more evolutionary challenges than do the environments of temperate and cold lands'. To Darwin, Dobzhansky and Darlington, the biological diversity of a whole biotic assemblage, such as a continental fauna, is a representation of evolutionary potential, in that species of unusually diverse regions are better equipped by a greater intensity of competition to match both abiotic and biotic challenges (right side of Fig. 10.1). In MacArthurian terms, this could be interpreted to mean that evolution has better equipped some communities to better match ambient production levels, i.e. exhibit higher carrying capacities and/or have higher competitive abilities under a given resource condition. When whole biotic assemblages meet, whether historically through continental drift or through modern human transport of propagules around the world, some floras and faunas – either those more diverse, or from larger land masses, or both – should be less invasible because of their greater evolutionary advancement.

Among modern researchers, Vermeij (1991, 1996, 2005) has perhaps made the strongest connection between regional species diversity and biological invasions, both as an explanation for historical biotic interchanges and as a guiding principle for modern invasions, based loosely on the adaptation-based arguments of Darwin, Dobzhansky, Darlington and others. More recently, Stachowicz and Tilman (2005) have put this general argument in terms of resource-use trade-off surfaces. They suggest greater evolutionary innovation in more diverse biogeographic realms allows those species to maximize resource use efficiency; species from regions of larger area and greater diversity may be able to tolerate lower resource levels, and thus competitively replace less advantaged species from other regions.

This argument, a biogeographic version of Elton–MacArthur expectations about diversity and invasibil-

ity, may explain many of the major patterns in global species invasions, such as why there is a large imbalance in rates of modern species exchanges between continental floras and faunas (see, for example, Por 1978; Di Castri 1989; Lonsdale 1999; Fridley 2008), and why, paradoxically, some invaders appear better adapted to foreign environments than resident species (Sax & Brown 2000). However, as far as I am aware, no one has yet demonstrated the metabolic superiority of species from particular biotic regions given similar environmental conditions (but see Lillegraven et al. (1987) concerning eutherian and marsupial mammals), nor shown that such differences in resource utilization stem from predictable differences in regional diversity, size or history. This would seem to be an issue of great potential value in invasion biology.

10.6   CONCLUSIONS

Elton set the stage for half a century of theoretical and empirical research into whether more diverse ecosystems are more able to withstand invasions from novel species. Although Elton's specific arguments have generally not held up to theoretical scrutiny, there is room for more empirical study of how ecosystem diversity and complexity relate to the stability of component populations, and the extent to which such fluctuations promote invasion by novel species. Most modern empirical research into Elton's general hypothesis stems more directly from the concepts of limiting similarity and species packing put forth by MacArthur. However, like Elton's notion of diversity and stability, the linking of invasibility to diversity by way of limiting similarity has also faced significant challenges, both from more recent theory of how communities are assembled and from empirical studies that lack a consistent diversity–invasibility pattern. Thus, although Elton was the first major proponent of biodiversity as a driver of species invasion patterns, his contributions to modern invasion studies have been weakened in light of the notable lack of consensus as to whether diversity itself plays a significant role in community invasibility.

A revised framework of the relationship between biodiversity and community invasibility may involve a broader, larger-scale consideration of the consequences of biodiversity for community assembly and how species evolve to meet biotic and abiotic challenges. Historical patterns of biotic interchange suggest differences in diversity between whole regions may indicate the susceptibility of those regions to invasion, but biologists are a long way from understanding how to translate such patterns into predictive frameworks for understanding which species will invade which ecosystems.

ACKNOWLEDGEMENTS

I am grateful to Mark Davis, Michael Huston, James Justus and Dave Richardson for their critical comments on the manuscript. Dov Sax provided stimulating discussion of invasion mechanisms in the context of global biogeography.

REFERENCES

Bezemer, T.M. & van der Putten, W.H. (2007) Ecology: diversity and stability in plant communities. *Nature*, **446**, E6–7.

Brown, R.L. & R.K. Peet. (2003) Diversity and invasibility of Southern Appalachian plant communities. *Ecology* **84**, 32–39.

Byers, J.E. & Noonburg, E.G. (2003) Scale dependent effects of biotic resistance to biological invasion. *Ecology*, **84**, 1428–1433.

Case, T.J. (1990) Invasion resistance arises in strongly interacting species-rich model competition communities. *Proceedings of the National Academy of Sciences of the USA*, **87**, 9610–9614.

Cooper, G. (2001) Must there be a balance of nature? *Biology and Philosophy*, **16**, 481–506.

Cronk, Q.B. & Fuller, J.L. (1995) *Plant invaders*. Chapman and Hall, London, UK.

Daehler, C.C. (2001) Darwin's naturalization hypothesis revisited. *The American Naturalist*, **158**, 324–330.

Darlington, P.J. (1965) *Biogeography of the Southern End of the World*. Harvard University Press, Cambridge, Massachusetts.

Darwin, C. (1859) *On the Origin of Species*. Facsimile reproduction of the First Edition by Atheneum Press, New York.

Davis, M.A. (2009) *Invasion Biology*. Oxford University Press, Oxford.

Di Castri, F. (1989) History of biology invasions with special emphasis on the Old World. *Biology Invasions: A Global Perspective* (ed. J.A. Drake et al.), pp. 1–29. Wiley, Chichester, UK.

Dobzhansky, T. (1950) Evolution in the tropics. *American Scientist*, **38**, 209–221.

Dunstan, P.K. & Johnson, C.R. (2006) Linking richness, community variability and invasion resistance with patch size a model marine community. *Ecology*, **87**, 2842–2850.

Elton, C. (1927) *Animal Ecology*. Sidgwick and Jackson, London.

Elton, C. (1958) *The Ecology of Invasions by Animals and Plants*. Methuen, London.

Fridley, J.D., Brown, R.L. & Bruno, J.F. (2004) Null models of exotic invasion and scale-dependent patterns of native and exotic species richness. *Ecology*, **85**, 3215–3222.

Fridley, J.D., Stachowicz, J.J., Naeem, S., et al. (2007) The invasion paradox: reconciling pattern and process in species invasions. *Ecology*, **88**, 3–17.

Fridley, J.D. (2008) Of Asian forests and European fields: Eastern U.S. plant invasions in a global floristic context. *PLoS ONE*, **3**, e3630.

Gause, G.F. (1934) *The Struggle for Existence*. Williams & Wilkins, Baltimore.

Gause, G.F. (1935) *Vérifications Experimentales de la Théorie Mathématique de la Lutte pour la Vie*. Hermann, Paris.

Gause, G.F., Smaragdova, N.P. & Witt, A.A. (1936) Further studies of interaction between predators and prey. *Journal of Animal Ecology*, **5**, 1–18.

Goodman, D. (1975) The theory of diversity–stability relationships in ecology. *Quarterly Review of Biology*, **50**, 237–266.

Grime, J.P. (1973) Control of species density in herbaceous vegetation. *Journal of Environmental Management*, **1**, 151–167.

Grubb, P.J (1977) The maintenance of species-richness in plant communities: the importance of the regeneration niche. *Biological Reviews*, **52**, 107–145.

Herben, T., Mandák, B., Bímova, K. & and Münzbergova, Z. (2004) Invasibility and species richness of a community: a neutral model and a survey of published data. *Ecology*, **85**, 3223–3233.

Hubbell, S.P. (2001) *The Unified Neutral Theory of Biodiversity and Biogeography*. Princeton University Press, Princeton.

Huston, M.A. (1979) A general hypothesis of species diversity. *The American Naturalist*, **113**, 81–101.

Huston, M.A. (1994) *Biological Diversity: The Coexistence of Species on Changing Landscapes*. Cambridge University Press, Cambridge, UK.

Huston, M.A. (2004) Management strategies for plant invasions: manipulating productivity, disturbance, and competition. *Diversity and Distributions*, **10**, 167–178.

Justus, J. (2008) Complexity, diversity, stability. In *A companion to the philosophy of biology* (ed. S. Sarkar & A. Plutynski), pp. 321–350. Blackwell, Oxford.

Leopold, A. (1949) *A Sand County Almanac and Sketches Here and There*. Oxford University Press, New York.

Levine, J.M. & C.M. D'Antonio. (1999) Elton revisited: a review of evidence linking diversity and invasibility. *Oikos*, **87**, 15–26.

Lillegraven, J.A., Thompson, S.D., McNab, B.K. & Patton, J.L. (1987) The origin of eutherian mammals. *Biological Journal of the Linnean Society*, **32**, 281–336.

Lonsdale, W.M. (1999) Global patterns of plant invasions and the concept of invasibility. *Ecology*, **80**, 1522–1536.

Lotka, A.J. (1927) Fluctuations in the abundance of species considered mathematically. *Nature*, **119**, 12–13.

MacArthur, R.H. (1955) Fluctuations of animal populations and a measure of stability. *Ecology*, **36**, 533–536.

MacArthur, R.H. (1970) Species-packing and competitive equilibrium for many species. *Theoretical Population Biology*, **1**, 1–11.

MacArthur, R.H. (1972) *Geographical Ecology: Patterns in the Distribution of Species*. Harper & Row, New York.

MacArthur, R.H. & Levins, R. (1967) The limiting similarity, convergence, and divergence of coexisting species. *The American Naturalist*, **101**, 377–385.

MacArthur, R.H. & Wilson, E.O. (1967) *The Theory of Island Biogeography*. Princeton University Press, Princeton.

May, R.M. (1973) *Stability and Complexity in Model Ecosystems*. Princeton University Press, Princeton.

McCann, K. (2009) The structure and stability of food webs. In *The Princeton Guide to Ecology* (ed. S.A. Levin), pp. 305–311. Princeton University Press, Princeton.

McNaughton, S.J. (1977) Diversity and stability of ecological communities: a comment on the role of empiricism in ecology. *The American Naturalist*, **111**, 515–525.

Palmer, M., McGlinn, D. & Fridley, J.D. (2008) Artifacts and artifictions in biodiversity research. *Folia Geobotanica*, **43**, 245–257.

Por, F.D. (1978) *Lessepsian Migration : The Influx of Red Sea Biota into the Mediterranean by Way of the Suez Canal*. Springer, Berlin.

Pyšek, P., Richardson, D.M., Rejmánek, M., Webster, G.L., Williamson, M. & Kirschner, J. (2004) Alien plants in checklists and floras: towards better communication between taxonomists and ecologists. *Taxon*, **53**, 131–143.

Rejmánek, M. (1996) Species richness and resistance to invasions. In *Biodiversity and Ecosystem Processes in Tropical Forests* (ed. G.H. Orians, R. Dirzo and J.H. Cushman), pp. 153–172. Springer, Berlin.

Richardson, D.M., Allsopp, N., D'Antonio, C.M., Milton, S.J. & Rejmánek, M. (2000) Plant invasions: the role of mutualisms. *Biological Reviews*, **75**, 65–93.

Richardson, D.M. & Pyšek, P. (2007) Classics in physical geography revisited: Elton, C.S. 1958: The ecology of invasions by animals and plants. Methuen: London. *Progress in Physical Geography*, **31**, 659–666.

Richardson, D.M. & Pyšek, P. (2008) Fifty years of invasion ecology – the legacy of Charles Elton. *Diversity and Distributions*, **14**, 161–168.

Sax, D.F. & Brown, J.H. (2000) The paradox of invasion. *Global Ecology & Biogeography*, **9**, 363–372.

Shea, K. & P. Chesson. (2002) Community ecology theory as a framework for biological invasions. *Trends in Ecology & Evolution*, **17**, 170–176.

Slobodkin, L.B. (2001) Limits to biodiversity (species packing). In *Encyclopedia of Biodiversity*, volume **3** (ed. S.A. Levin), pp. 729–738. Academic Press, New York.

Stachowicz, J.J. & Tilman, D. (2005) What species invasions tell us about the relationship between community saturation, species diversity and ecosystem functioning. In *Species Invasions: Insights into Ecology, Evolution and Biogeography* (ed. D. Sax, J. Stachowicz and S. Gaines), pp. 41–64. Sinauer, Sunderland, Massachusetts.

Stachowicz, J.J. & Byrnes, J.E. (2006) Species diversity, invasion success, and ecosystem functioning: disentangling the influence of resource competition, facilitation, and extrinsic factors. *Marine Ecology Progress Series*, **311**, 251–262.

Stachowicz, J.J., Fried, H., Osman, R.W. & Whitlatch, R.B. (2002) Biodiversity, invasion resistance, and marine ecosystem function: reconciling pattern and process. *Ecology*, **83**, 2575–2590.

Stohlgren, T.J., Binkley, D., Chong, G.W., et al. (1999) Exotic plant species invade hot spots of native plant diversity. *Ecological Monographs*, **69**, 25–46.

Stohlgren, T.J., Barnett, D.T. & Kartesz, J.T. (2003) The rich get richer: patterns of plant invasions in the Unites States. *Frontiers in Ecology and the Environment*, **1**, 11–14.

Stohlgren, T.J., Barnett, D.T., Jarnevich, C.S., Flather, C. & Kartesz, J. (2008) The myth of plant species saturation. *Ecology Letters*, **11**, 313–326.

Tilman, D. (1982) *Resource Competition and Community Structure*. Princeton University Press, Princeton.

Tilman, D. (1996) Biodiversity: population versus ecosystem stability. *Ecology*, **77**, 350–363.

Tilman, D. (1999) The ecological consequences of changes in biodiversity: a search for general principles. *Ecology*, **80**, 1455–1474.

Tilman, D. (2004) Niche tradeoffs, neutrality, and community structure: a stochastic theory of resource competition, invasion, and community assembly. *Proceedings of the National Academy of Sciences of the USA*, **101**, 10854–10861.

Tilman, D., Lehman, C.L. & Thomson, K.T. (1997) Plant diversity and ecosystem productivity: theoretical considerations. *Proceedings of the National Academy of Sciences of the USA*, **94**, 1857–1861.

Vermeij, G.J. (1991) When biotas meet: understanding biotic interchange. *Science*, **253**, 1099–1104.

Vermeij, G.J. (1996) An agenda for invasion biology. *Biological Conservation*, **78**, 3–9.

Vermeij, G.J. (2005) Invasion as expectation: a historical fact of life. In *Species Invasions: Insights into Ecology, Evolution and Biogeography* (ed. D. Sax, J. Stachowicz & S. Gaines), pp. 315–339. Sinauer, Sunderland, Massachusetts.

Volterra, V. (1931) *Lecons sur la theorie mathematique de la lutte pour la vie*. Gautheir-Villars, Paris.

Williamson, M. (1996) *Biological Invasions*. Chapman & Hall, London.

Chapter 11

SOIL BIOTA AND PLANT INVASIONS: BIOGEOGRAPHICAL EFFECTS ON PLANT–MICROBE INTERACTIONS

Ragan M. Callaway and Marnie E. Rout

Division of Biological Sciences, The University of Montana, Missoula, MT 59812, USA

'Thou shalt suffer in alternate years the new reaped fields to rest ... thus by rotation like repose is gained nor earth meanwhile uneared and thankless left.'

Virgil (29 BC)

Fifty Years of Invasion Ecology: The Legacy of Charles Elton, 1st edition. Edited by David M. Richardson
© 2011 by Blackwell Publishing Ltd

11.1   INTRODUCTION

Ancient agriculturists realized that crops, if left in the same place for long enough, would begin to perform poorly (White 1970; Howard 1996; Altieri 2004). They also appeared to recognize that by removing crops from an area for a season or more, rotating different crops among years in the same place, or by planting other species with their crops, they could trade high short-term productivity for long-term sustainability of crop production at rates of lower productivity. There are several reasons why agricultural monocultures lose productivity over time, but among the most important causes are host-specific soil-borne diseases, which proliferate and have disproportionally strong effects when the densities of their hosts are unu-sually high. In the past century, modern agronomists have clearly shown that the accumulation of host-specific pathogenic soil organisms is a major cause of damage, productivity loss, and mortality in crop mono-cultures (Hoestra 1975; Magarey 1999). Charles Elton (1958) did not address soil biota as important natural enemies in his landmark book but since then many studies have clearly shown that pathogenic soil organisms, or the lack of them, play a major role in the devel-opment of other kinds of monocultures: those caused by exotic plant invasions. Furthermore, the study of soil biota in the context of exotic plant invasions has led to an explosion in our understanding of the role of these organisms in community ecology in general (see Fig. 11.1). The primary aim of this chapter is to explore how our understanding of the role of soil biota in exotic

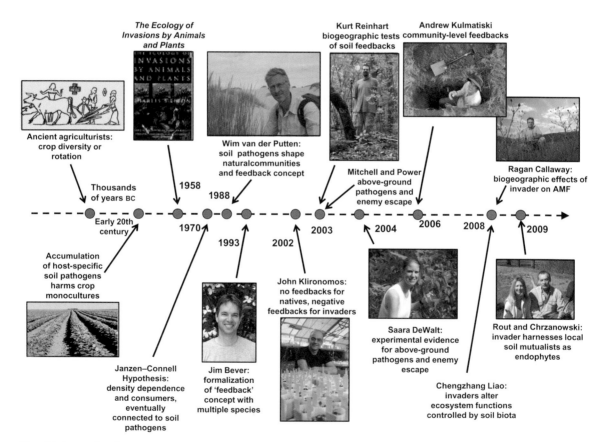

Fig. 11.1 Timeline for some conceptual developments involving the role of soil biota in structuring plant communities and in exotic plant invasions.

plant invasions has increased since the time of Elton. First, we examine studies looking at the effects of soil pathogens in the context of exotic plant invasions which paved the way for the groundbreaking work on plant–soil feedbacks. We then discuss how biogeographical approaches have been used to explore shifts between invaded and native ranges in these feedbacks. Lastly, we turn to current research on various microbial mechanisms underlying successful invasions in light of the different effects invaders exhibit on soil biota and the biogeographical nature of these impacts.

11.2 BIOLOGICAL INVASIONS AND SOIL PATHOGENS

Since Elton, we have found that soil pathogens have powerful effects in natural plant communities, but with the exception of studies of soil biota in the context of the Janzen-Connell hypothesis for tropical tree diversity (Janzen 1970; Connell 1971), soil pathogens are just beginning to find their place in community theory. Inclusion of pathogens in community theory has been helped by efforts to understand the causes of exotic plant invasions as studies of pathogens and exotic invaders have provided new insights into the role of pathogens as important drivers of classic patterns and processes in plant communities (Mitchell & Power 2003). For example, Saara DeWalt et al. (2004) (Fig. 11.1) examined the role of fungi in the invasion of the Neotropical shrub *Clidemia hirta* on Hawaii. They planted *C. hirta* into understorey and open habitats in its native range of Costa Rica and its non-native range in Hawaii and applied insecticide and fungicide to some treatments. They found that the survival of the invader in its native range increased with protection from these two groups of natural enemies. In contrast, suppression of pathogenic fungi and herbivores had no effect in the non-native range. It is important to note that this study inhibited fungi and insects via foliar leaf applications. Although soil fungal and insect populations were not directly manipulated, these novel experiments indicate that pathogens were important for invasive success. Additionally, this study indicated that pathogens had important effects on the local distribution and abundance of *C. hirta* in its native range. One of the most important early breakthroughs for understanding the importance of pathogens in natural communities, and specifically soil pathogens, was made by Wim van der Putten et al. (1988) (Fig. 11.1)

who found that *Ammophila arenaria*, a European beachgrass, grew more poorly in 'rhizosphere sand' (sand in which the plant had been growing) than in fresh 'sea sand' (sand in which no *Ammophila* had grown). They sterilized the soil with gamma-irradiation and found that the biomass of *Ammophila* seedlings increased. They concluded that sand in which the grass had been growing accumulated biota harmful to *Ammophila*, which was reflected in the condition of the roots. In a later paper (Van der Putten et al. 1993) the role of soil biota on *Ammophila* was further investigated in the context of succession. *Ammophila* was suppressed by soil biota that accumulated in its rhizospheres, but plant species that followed *Ammophila* in succession were not. Van der Putten and others established that *Ammophila* was strongly inhibited by soil biota in its European native range, where it is naturally succeeded by other species.

In an extension of this work, others explored the role of soil biota on *Ammophila* in parts of the world where the grass has become highly invasive. In California, where *Ammophila* has invaded coastal dunes, Beckstead and Parker (2003) measured the effects of belowground pathogens on *Ammophila*. They manipulated soil biota by sterilizing soils to eliminate all soil-dwelling organisms and found that seed germination, seedling survival, and seedling growth was reduced in nonsterilized soil compared with sterilized soil, thus failing to find a release from natural enemies. They also found several common fungal pathogens, but no pathogenic nematodes. The absence of nematodes could be an important factor contributing to successful invasions in California because they can have strong additive pathogenic effects on plant growth when combined with fungi from soils in the native range (De Rooij-van der Goes 1995; but see Van der Putten and Troelstra 1990). In a global survey of nematodes on *Ammophila*, Van der Putten et al. (2005) found that in the non-native ranges of *Ammophila* more specialist nematode taxa attacked the local native plant species than attacked the invasive *Ammophila*. Knevel et al. (2004) compared the effects of the soil biota from the native European range versus the non-native range in South Africa where *Ammophila* has also invaded. As shown before, biota in the soil from the native range had strong inhibitory effects on *Ammophila* grown in native soils, but unlike what was found in California, soil biota from the non-native range were far less damaging to *Ammophila* growing in non-native soils. Importantly, van der Putten's initial research on soil biota in the context of plant *Ammophila* invasions (*Ammophila*) stimulated a growing body of work on plant–soil feedbacks that has yielded insights into the nature of invasions and the forces that structure native communities.

11.3 PLANT–SOIL FEEDBACKS

Plant rhizospheres contain many organisms that inhibit plant growth (parasites) and organisms that promote plant growth (mutualists), and many in each group are bacteria and fungi. The species-specific biochemistry of root exudates arguably plays a key role in determining the composition of the microbial community in the rhizosphere and suggests a 'degree of co-evolution' among plant–microbe interactions (Hoestra 1975). The relative accumulation of parasitic or mutualistic bacteria over time in rhizospheres can elicit 'feedback' responses in which a species declines or increases in growth or fitness as it affects its own rhizosphere community. Again, biological invasions have allowed comparisons of soil feedbacks in ways that have contributed tremendously to our overall understanding of the role of soil biota in ecology.

Jim Bever (1994) (Fig. 11.1) explored plant–soil feedbacks with four old-field plant species and biota that developed in the soil occupied by each plant species. He found that the survival of *Krigia* dandelion was lower when the plant was grown in soil previously occupied by conspecifics and the other three species had lower growth and root–shoot ratios when exposed to inoculum from soil previously occupied by conspecifics compared with the inoculum derived from other species. These effects were tentatively attributed to root pathogens. Mills and Bever (1998) then followed up the earlier study by testing the effects of the soil pathogen *Pythium* on several different old-field plant species. They found that some plant species were more susceptible to *Pythium* in general than other species and that some plant species accumulated *Pythium* at faster rates than other species. They made the case that this sort of species-specific accumulation of soil pathogens could explain negative plant–soil feedbacks.

In 2002, John Klironomos connected the effects of soil feedbacks to the relative abundance of both exotic and native plant species in communities (Fig. 11.1). The relative strength of feedback interactions between plants and soil biota was highly correlated with the relative abundance of species in old-field communities.

In other words, species that developed strong negative feedbacks over time with soil biota were less common and species that developed strong positive feedbacks with soil biota were more common in the community. In an explicit comparison of 'rare plants' to 'invasive plants' Klironomos found that all rare species tested experienced strong negative feedbacks, whereas four of five invasive species experienced significant positive feedbacks. He noted that these invasive species may have simply possessed inherent physiological traits that prevented large build ups of soil pathogens in their rhizospheres, but importantly that the invaders may have 'escaped their harmful pathogens by invading foreign territory'. To understand whether invaders possess inherent resistance to soil pathogens or have escaped host-specific pathogens in their home ranges, several researchers have conducted explicit biogeographical comparisons of interactions between exotic species, pathogens and soil biota (Hierro et al. 2005).

11.4 BIOGEOGRAPHY OF INVASIONS

The most fascinating attribute of exotic invaders is that they become far more abundant or dominant in their new non-native ranges than they were in their native ranges; and this occurs regardless of the continent of origin or the continent of destination. Unless by some coincidence the abiotic conditions of non-native ranges happen to dramatically favour some non-native species over locally adapted native species, it would seem that powerful *biotic* forces are at work (Elton 1958). Of a suite of potential biotic forces from which invaders may benefit, the overwhelming focus has been on escape from specialist insect herbivores, enemies that were fully recognized by Elton (Keane & Crawley 2002). Taking a novel tack on enemy escape, Mitchell and Power (2003) (Fig. 11.1) searched the literature and databases for evidence of infection by viruses, rust, smut and powdery mildew fungi on the above-ground parts of 473 plant species that had been introduced from Europe into the USA. They found that on average, 84% fewer fungi and 24% fewer viruses infected plant species in the non-native North American range than infected the same species in the native range. They also found that species ranked as 'invasive' were more highly infected in their native ranges than species not ranked as invasive. These results suggest that many relatively host-specific leaf and flower pathogens occurred on plant species in their native ranges and that many of these were escaped when the species were introduced by humans into new ranges. Mark Van Kleunen and Markus Fischer (2009) tackled the same question but in the opposite biogeographical direction. They used global data for 140 North American plant species that had naturalized in Europe, for which fungus–plant host distributions were known. In the invaded European range, they found 58% fewer floral and foliar pathogen loads on these invasive North American plant species. However, when they also looked at the known distributions of the fungal pathogens, only 10.3% were not found in the European range, which suggests the enemy release hypothesis does not account for the successful distributions of these North American species. Most interestingly, the geographic spread of these non-native species in Europe was *negatively* correlated with decreased loads of fungal pathogens. They noted that North American plants may have escaped specific fungal species that control them in their native range, but based on total abundances of fungal species reported in the database, release from foliar and floral fungal pathogens does not explain the spread of North American plant species in Europe. It is important to note that van Kleunen and Fischer equated 'geographic spread' with 'invasiveness' owing to the nature of the data set, but species that have broad distributions are not necessarily invasive (Hierro et al. 2005). Very high densities and high negative impacts on other species are traits more connected to invasiveness. Nevertheless, these kinds of biogeographical comparisons are crucial for understanding invasions. These studies focused on pathogens that target above-ground plant components. However, soils are reservoirs for many above-ground pathogens. For example, virtually all plant pathogenic fungi spend a relatively small fraction of their life cycle on their host plants, with most the life cycle spent in the soil or in plant detritus within soils (Pepper 2000). Thus survival and effects of even above-ground pathogenic fungi are primarily regulated by biotic and abiotic soil factors (Pepper 2000).

11.5 PLANT–SOIL FEEDBACKS AND THE BIOGEOGRAPHY OF INVASIONS

To our knowledge, Kurt Reinhart et al. (2003) (Fig. 11.1) were the first to undertake intercontinental biogeographical comparisons of soil biota feedback effects in the context of invasion. They studies *Prunus serotina*,

a native tree in temperate North America, which is typically early successional and rarely abundant in its native range. In the native range, Packer and Clay (2000) found that the soil pathogen *Pythium* had stronger inhibitory effects on *P. serotina* seedlings near mature trees, where the fungus appeared to accumulate, than on seedlings occurring far from mature trees. When seedlings were grown in soil collected either 0–5 m or 25–30 m from *P. serotina* trees, sterilization of the soil collected near mature trees improved seedling survival, but sterilization of soil collected far from trees had no effect. In sharp contrast to the typical low densities of *P. serotina* in its native range, it has become very dense and highly invasive in parts of Europe. Reinhart et al. (2003) examined populations of *P. serotina* in the Netherlands and in Indiana and found that the densities of saplings and trees were roughly six to nine times higher in the non-native range than in the native range. This dramatic difference in tree density among ranges correlated with less negative soil feedbacks in the non-native range than in the native range where soil pathogens appeared to strongly limit seedling growth and tree density. Later, Reinhart et al. (2005) studied the generality of the negative feedbacks in the native range by sampling 22 populations over a substantial area of the eastern USA. In soils collected under *P. serotina* trees, sterilization and fungicide treatments specific to oomycetes (including *Pythium*) improved seedling survival, suggesting that oomycetes were causing the death of seedlings. They also compared the effects of soil biota collected under other tree species and found that biota collected under *P. serotina* were more pathogenic to conspecifics than biota collected under other species. Considered together, these results indicate that relatively host-specific soil biota have important regulatory effects on *P. serotina* over large areas in its home range, but that this biotic constraint is escaped in Europe where the tree has successfully invaded.

In sterilization experiments similar to those conducted by Reinhart et al. (2003), Callaway et al. (2004a) found that soil microbes from several sites in the home range of the invasive exotic forb *Centaurea stoebe* (née *maculosa*) L. had stronger inhibitory effects on the this species than soil microbes from several sites in the North American invasive range. They then tested feedback effects using soils from one site in each range. The microbial community in the soil from Europe was trained with either *C. stoebe* or *Festuca ovina*, a small perennial bunchgrass native to Europe.

The microbial community in the soil from North America was pre-cultured by planting *C. stoebe* or *Festuca idahoensis*, a bunchgrass similar to *F. ovina* but native to western North America. The trained soils were used to inoculate substrate in which *C. stoebe* was planted alone or in competition with one of the two grass species. In the native European soil *C. stoebe* cultivated strong negative feedbacks with soil biota either alone or in competition. However, in soils from North America, *C. stoebe* cultivated strong positive feedbacks. The same treatments in sterilized soil eliminated feedback effects in soils from both ranges indicating the crucial role of soil biota, but they did not identify soil taxa that might have been involved. In related experiments, Callaway et al. (2004b) investigated the effects of soil fungi on interactions between *C. stoebe* and six species native to the grassland invaded by *C. stoebe*. Fungicide reduced arbuscular mycorrhizal fungi (AMF) colonization of *C. stoebe* roots, but did not reduce non-AMF. In soils without fungicide *C. stoebe* grew larger when grown with *F. idahoensis* or *Koeleria cristata* than when grown alone. However, when fungicide was applied to the soil the positive effects of *Festuca* and *Koeleria* were not present. Fungicide reduced the competitive effects of the native bunchgrass *Pseudoroegneria* on *C. stoebe* and did not affect the way two native North American forbs, *Achillea millefolium* and *Linum lewisii*, competed with *C. stoebe*. These results suggested that *C. stoebe* invasion can be affected by complex and often beneficial effects of fungal communities in the non-native range. Interestingly the effects of soil fungi were not manifest as simple direct effects. Only when native plants, invasive plants, and soil microbial communities were interacting at the same time were the effects observed, suggesting indirect interactions among plants and soil microbes.

Positive effects of soil biota from the non-native range were also observed by Reinhart and Callaway (2004) for *Acer platanoides*, a widespread tree in Europe that has become invasive in North America, and *A. negundo*, a widespread tree in North America that has become invasive in Europe. They compared the relative importance of soil biota in the context of the 'enemy release hypothesis' and biogeographical differences in the effects of mutualisms. As found by Reinhart et al. (2003) for *P. serotina*, distances from focal trees to the nearest *Acer* conspecifics were 56–77% less in their non-native ranges than in their native ranges demonstrating higher densities in the non-native range. Unlike the experiments with *P. serotina* and *C.*

stoebe, soil collected under *Acer* species in their non-native ranges decreased the growth of conspecific seedlings by 64–112%, but the soil associated with native tree species in the non-native ranges *increased* the growth of *Acer* seedlings. These results suggest that native soil biota initially boost invasion by the two *Acer* species, but over time soil biota communities become more pathogenic to *Acers* as they establish over time. Like for *C. stoebe*, mutualistic interactions with soil biota may be relatively more beneficial to *Acer* species in their non-native ranges than in their native ranges. Similarly, Cui and He (2009) found that the native shrub *Saussurea deltoidea* altered soil biota in ways that increased the growth of the invader *Bidens pilosa* in China.

As noted above, some invaders benefit from soils trained by certain native species early in the invasion process, over time shifting the soil biota. These shifts in plant–soil feedbacks can not only enhance the initial invader, but can also be beneficial to other invasive species. For example, in the same intermountain grasslands invaded by *C. stoebe*, Jordan et al. (2008) measured the feedback effects of three other aggressive invaders: *Euphorbia esula*, *Bromus inermis* and *Agropyron cristatum*. They found that *B. inermis* and *A. cristatum* demonstrated strong positive plant–soil feedbacks. However, in a novel twist, *B. inermis* and *A. cristatum* also had facilitative effects on other invasive species through soil feedbacks. Jordan et al. (2008) argued that their results suggested that the alteration of native soil biota by invasives could promote self-invasion or 'cross-facilitation' of other invasive species. Studies of *Ageratina adenophora*, a species native to Mexico that has become highly invasive throughout the world in semi-tropical ecosystems, have provided a conceptually similar picture. In China, where *A. adenophora* has become an exceptionally aggressive invader, Niu et al. (2007) found that the soil biota collected from heavily invaded sites suppressed native plant species, but did not suppress *A. adenophora*. They also found that soil biota in the invaded site had greater positive effects on *A. adenophora* than soil biota in the non-invaded site, suggesting that the invader altered soil microbial communities in ways that favoured itself.

As noted above, soil biota include complex mixtures of deleterious and beneficial taxa. Van der Putten et al. (2007) explored the contribution of soil pathogens and AMF in the feedback effects of an invasive and two native grasses in the Kalahari savanna in Botswana.

The exotic grass *Cenchrus biflorus*, established neutral to positive soil feedbacks, whereas the two native grass species, *Eragrostis lehmanniana* and *Aristida meridionalis*, showed neutral to negative feedback effects. This study found two host-specific pathogens on the roots of *E. lehmanniana* that did not affect the invasive *C. biflorus*. Fungi were also isolated from the roots of the *C. biflorus*, but these fungi had no effects on any of the grasses. Native grasses showed higher diversities of AMF taxa in their roots than *C. biflorus*, but it was not clear what role these different species played in the general feedbacks generated by the different grasses.

Plant–soil feedbacks can impact native communities in ways that suggest community-level alterations are occurring, including those that directly inhibit native species as well as those that directly enhance invasives. One example of a native species experiencing direct inhibition by an invader was conducted by Batten et al. (2008) in a novel experiment on feedbacks in the Western USA. They tested the effects of the soil microbial biota found in the soil rhizosphere of the invasive grass *Aegilops triuncialis* and biota found in the rhizospheres of the natives *Lasthenia californica* and *Plantago erecta*, on the growth of the three species. *Lasthenia californica* experienced a delayed flowering date and reduced above-ground biomass when exposed to soil microbes that accumulated in the rhizospheres of *A. triuncialis*. They proposed that changes in soil microbes elicited by invaders might cause community-scale positive feedbacks (not necessarily positive feedbacks on the invader themselves) that ultimately lead to dominance by the invader. In other words, these observed changes in the native species contributed directly to their decreased abundance which could indirectly contribute to increased invader dominance. In a long-term field experiment on northern shrub steppe communities Andrew Kulmatiski et al. (2006) examined 'soil history', a proxy for feedbacks, and found that exotics appeared to facilitate their own growth by maintaining small beneficial fungal populations. Whether or not soils were occupied by exotics or native was a crucial factor for determining exotic and native plant distributions. Following this with a major synthetic analysis, Kulmatiski et al. (2008) (Fig. 11.1) gathered the literature on these kinds of feedback studies with invasive species and used meta-analytical models to examine generality in the effects of plant–soil feedbacks on plant diversity and invasions. After including the literature on the subject to date, they found that plant–soil feedbacks were larger and more negative for native species than for invasive species.

The strong biogeographical patterns noted in several of the studies discussed above suggest that plants in a region affect the evolution of soil pathogens, and vice versa, in a way that leads to some degree of difference in host preference, host response, or host effect. Human-facilitated dispersal beyond natural geographic limits may then lead to escape from these pathogens. There is other evidence that plants can influence the evolutionary trajectories of their neighbours through the effects of microbial symbionts (Aarssen & Turkington 1985; Chanway et al. 1989). These studies showed that the legume *Trifolium repens* had increased biomass when grown with genotypes of the grass *Lolium perenne* with which they had coexisted (Aarssen & Turkington 1985). This increased biomass was due to increased effectiveness of nodule induction by nitrogen-fixing *Rhizobium* strains collected from co-existing *Trifolium–Lolium* communities compared with *Rhizobium* strains collected from *Trifolium*-only populations (Chanway et al. 1989). In sum, these studies showed that the plant–microbe symbiosis may not only be influenced by coevolution between the microbial partner and the respective host plant, but also by other neighbouring plant species. Evidence that invasive plants disrupt microbially mediated ecological functions supports the notion that coevolution among plants and microbes may occur within communities.

11.6 EFFECTS OF INVASIVE PLANTS ON SOIL BIOTA

Invasive plants can alter soil biota in ways that do not explicitly lead to feedback relationships, and these effects can be positive, neutral or negative (Wolfe & Klironomos 2005). There are far more examples of this sort of invader–soil biota interaction in the literature than examples of feedbacks, thus we focus on only a few of these to illustrate the ranges of variation observed for these effects.

Some experimental studies have found no relationship between plant species diversity and microbial diversity (Gastine et al. 2003; Porazinska et al. 2003; Niklaus et al. 2007). Other experiments have found that plant diversity correlates positively with bacterial composition or diversity (Grüter et al. 2006; Wardle et al. 1999; see Wardle 2002). Yet others have found no correlation between plant diversity and bacterial diversity but strong correlations with fungal diversity (Millard & Singh 2009). Despite the lack of a consistent relationship between plant diversity per se and the diversity of soil microbes in general, the presence or absence of particular plant species or functional groups is often correlated with the composition or diversity of soil microbial communities (Grüter et al. 2006; Bremer et al. 2009; Zhang et al. 2010). The strong effects of different plant species on soil microbial composition and function may derive from differences in the carbon: nitrogen ratios of leaves (or other basic stoichiometric differences), changes in disturbance regimes or species-specific differences in secondary metabolites (Wolfe & Klironomos 2005). Regardless of the mechanism, the arrival of many new plant species and the decline of native species has important implications for microbial communities, particularly their functions in ecosystem processes.

Biological invasions commonly drive marked decreases in the diversity of native plants (Rout & Callaway 2009), and whether caused by this decrease in diversity per se or the presence of a new plant species, soil biota often change in composition or decrease in diversity (Wolfe & Klironomos 2005 and references therein). In this context, the system that has been the most thoroughly studied is arid shrub-steppe invaded by the annual grass *Bromus tectorum*. During the past decade Jayne Belnap and colleagues have conducted a series of studies yielding a great deal of insight into the effects of this invader on soil microbes and their functions. Belnap et al. (2005) showed that sites historically invaded with *Bromus tectorum* had lower diversity measures of both plants and soil biota, while non-invaded sites had the greatest plant diversity and soil biota diversity. *Halogeton glomeratus* is an annual in the Chenopodiaceae from western Asia that has invaded grassland and shrubland throughout the western USA. Duda et al. (2003) found that the abundance of *H. glomeratus* correlated with increases in soil fertility (see also Liao et al. 2008, Rout & Callaway 2009) and increases in the diversity of soil bacteria. Whether invaders alter soil biota through their effects on the stoichiometry of nutrient cycling or through secondary metabolites is not known.

Taking a different approach to studying the effects of invaders on soil microbes, in 2006 Maarten Eppinga and colleagues in the Netherlands integrated experimental data with models to advance the theory that invasive species might accumulate local native gener-alist pathogens to a degree where native species are harmed more than the invasive species. They based their models on *Ammophila arenaria* arguing that in the non-native range 'accumulation of local soil pathogens could enhance dominance and rate of spread of *A. arenaria* as a result of negative specific soil community effects on native plant species'. Consistent with this idea, Mangla et al. (2007) found that soils from the rhizospheres of *Chromolaena odorata*, one of the world's most invasive tropical species, accumulated high concentrations of *Fusarium* (apparently *semitectum*), a generalist soil fungi, which was harmful to native plant species. Soils collected beneath *Chromolaena* in India inhibited growth of native species and contained over 25 times more spores of the pathogenic fungi than soils collected beneath neighbouring native species. Sterilization of the *Chromolaena* soils eliminated the inhibitory effect. In an experiment linking root exudates to the effect of the invader on soil biota *Chromolaena* root leachate increased *Fusarium* in spore density by over an order of magnitude. This exacerbation of the effects of native soil biota on native plants suggests a novel mechanism that might contribute to exotic plant invasion.

11.7 MICROBIAL MECHANISMS UNDERLYING PLANT INVASIONS

Plant invasions provide biogeographical contexts in which microbial ecological functions can be explored, and invasions demonstrate biogeographical patterns in ecological functions that might help to unify biogeographical concepts across all life forms (Green et al. 2004; Horner-Devine et al. 2004). Although recent studies suggest that regional or local evolutionary trajectories exist for microbes (McInerney et al. 2008), until recently, ecologists have focused on broad pathogenic or mutualistic functions of microbes and the broad functional role of microbes in nutrient cycling; processes suggesting that soil microbes rarely have restricted biogeographical distributions. However, invasive plants often interact differently with microbial functional groups in invaded ranges than they do with the same functional groups in native ranges, causing substantial shifts in the ecological nature of microbial mutualistic relationships (Parker et al. 2006), escape from fungal pathogens (Reinhart et al. 2005) and functional changes in soil bacterial communities (Liao et al. 2008). These biogeographical effects of invasive plants soil microbes suggest a new paradigm for microbial biogeography – biogeographical patterns exist – which is the case for all other taxa on Earth.

Support for the idea that invasive plants affect soil biota differently in native and non-native ranges has been found in studies of bacteria and fungi. For example, biological invasions commonly disrupt ecosystem processes mediated by soil bacteria (Liao et al. 2008; Rout & Callaway 2009) (Fig. 11.1). In the context of biogeographical patterns for soil microbes, plant invasions reduce local plant species richness, but typically correlate with *increased* net primary productivity; a correlation that conflicts conceptually with the current diversity–ecosystem function paradigm; i.e. species richness increases net primary productivity (Tilman et al. 2001). Even more puzzling, invasions commonly increase net primary productivity but they generally do not deplete soil resources such as nitrogen; instead they generally correlate with increased soil nitrogen pools and total ecosystem nitrogen stocks (Liao et al. 2008; Rodgers et al. 2008; Rout & Chrzanowski 2009; Rout & Callaway 2009), all processes mediated by soil bacteria. Rout & Chrzanowski (2009) (Fig. 11.1) found that the ecosystem effects of *Sorghum halepense* may be caused through mutualistic endophytic bacteria, including nitrogen-fixers, housed in the rhizomes of the invader. The effects of invaders on these processes in their home ranges are unknown, but since many invaders are far less abundant in their home ranges similar effects are unlikely. The dramatic and consistent alterations of soil nutrient cycles associated with invaders in introduced ranges (Liao et al. 2008) suggests that they exert novel effects, relative to the effects of native plants, on soil bacteria. As mentioned, this differential effect exists for soil fungi as well. Callaway et al. (2008) showed that one of North America's most aggressive invaders of undisturbed forest understories, *Alliaria petiolata*, had much stronger inhibitory effects on AMF in invaded North American soils than on AMF in European soils where *A. petiolata* is native. This biogeographical effect was also found for specific flavonoid fractions in extracts from *A. petiolata*. Together, these patterns suggest that ecological functions of soil microorganisms (as measured by various microbially mediated parameters like nutrient cycling, litter decomposition, etc.) have biogeographical patterns.

These patterns may provide a conceptual link with the differences in plant–soil feedbacks between native and invaded ranges (less negative in the invaded range) described above. If invaders affect microbially mediated soil processes in predictably different ways than natives (see Rout & Callaway 2009) two equally interesting hypotheses exist. Either plants with 'invasive' traits, as a group, also happen to affect soil microbes differently than plants without such invasive traits (but somehow do not display these traits in their native range), or microbes in various parts of the world are functionally similar but subtly different with respect to the plants with which they interact.

Future research on plant invasions requires attention how they alter the structure and function of soil biota. Invaders can have dramatic effects on aboveground community structure, most dramatically as decreases in species richness and functional diversity, and these decreases are likely to strongly soil biota. Disrupted ecosystem processes that are microbially mediated often appear to shift the balance of particular cycles in favour of the invaders. Recent research has explored how interactions between plants and their symbionts (both parasitic and mutualistic associations) respond to shifts in plant diversity (Nuismer & Doebeli 2004; Thrall et al. 2007).

There is growing interest in how invaders might affect mutualistic soil biota. This interest is associated with recent sea changes in plant ecology in which the effects of positive interactions (i.e. facilitation and mutualisms) are emphasized in shaping natural and invaded communities (Bruno et al. 2003; Richardson et al. 2000). Perhaps successful invasions are due in part to the lack of coevolved interactions with soil biota, thus invaders are interacting with microbial symbionts in ways that are more beneficial through positive feedbacks (Klironomos 2002; Packer & Clay 2000; Reinhart et al. 2003, 2005; Callaway et al. 2004a,b), increased soil nutrient pools (Liao et al. 2008; Rodgers et al. 2008; Rout & Chrzanowski 2009), and perhaps the accumulation of more beneficial mutualisms.

11.8 CONCLUSIONS

As discussed, when humans disperse invasive species in new areas they can escape soil-borne natural enemies. Incorporating biogeographical approaches into research on plant invasions has contributed not only to our understanding of plant performance in invaded ecosystems, but also to our understanding of microbial processes underlying invasions. Clearly, interactions among plants and microbes are bidirectional: plant communities affect microbial communities and vice versa. Biogeographical differences in the pathogenic effects of soil biota can be due to different pathogen densities, the composition of pathogen communities, and differences in the virulence of individual species or genotypes in the soil biota. Interestingly, the phylogenetic relatedness of the invader relative to resident species might affect how susceptible exotic species are to resident pathogens (Gilbert & Webb 2007) further supporting the necessity for using biogeographical approaches to study. The effects of soil mutualists on invasions are understood less than the effects of pathogens, but mutualists also appear to play important roles. Invaders might be limited by the absence of appropriate mutualists in their new ranges (Parker 2001) or benefit from mutualists they encounter in the soil of invaded ranges (Marler et al. 1999; Parker et al. 2007). Invaders can also suppress soil mutualists of other plant species in invaded ranges more aggressively than mutualists in their original range (Callaway et al. 2008) or have effects that are similar to the effects of nitrogen-fertilization on decreased mutualisms among coevolved organisms (see Denison & Kiers 2004; Thrall et al. 2007). Whether or not the benefits of new mutualistic relationships with soil biota in invaded regions are generally stronger, weaker or similar to mutualistic interactions in native regions remains an important unanswered question. Focusing on the role of microbial mutualisms in the invasion process might also advance our general understanding of how these interactions coevolve in general. Furthermore, whether invaders will in time acquire more effective parasites (see Parker & Gilbert 2004) remains a crucial yet unexplored area. Testing these and other processes in the context of invasion, and specifically in a biogeographical context, will provide insight into the function of natural communities and ecosystems.

REFERENCES

Aarssen, L.W. & Turkington, R. (1985) Vegetation dynamics and neighbour associations in pasture-community evolution. *Journal of Ecology*, **73**, 585–603.

Altieri, M.A. (2004) Linking ecologists and traditional farmers in the search for sustainable agriculture. *Frontiers in Ecology and the Environment*, **2**, 35–42.

Batten, K.M., Scow, K.M. & Espeland, E.K. (2008) Soil microbial community associated with an invasive grass differentially impacts native plant performance. *Microbial Ecology*, **55**, 220–228.

Beckstead, J. & Parker, I.M. (2003) Invasiveness of *Ammophila arenaria*: release from soil-borne pathogens? *Ecology*, **84**, 2824–2831.

Bever, J.D. (1994) Feedback between plants and their soil communities in an old field community. *Ecology*, **75**, 1965–1977.

Belnap, J., Phillips, S.L., Sherrod, S.K. & Moldenke, A. (2005) Soil biota can change after exotic plant invasion: does this affect ecosystem processes? *Ecology*, **86**, 3007–3017.

Bremer, C., Braker, G., Matthies, D., Beierkuhnlein, C. & Conrad, R. (2009) Plant presence and species combination, but not diversity, influence denitrifier activity and the composition of nirK-type denitrifier communities in grassland soil. *FEMS Microbiology Ecology*, **70**, 45–55.

Bruno, J.F., Stachowitcz, J.J. & Bertness, M.E. (2003) Inclusion of facilitation into general ecological theory. *Trends in Ecology and Evolution* **18**, 119–125.

Callaway, R.M., Cipollini, D., Barto, K., et al. (2008) Novel weapons: invasive plant suppresses fungal mutualists in America but not in its native Europe. *Ecology*, **89**, 1043–1055.

Callaway, R.M., Thelen, G.C., Rodriguez, A. & Holben, W.E. (2004a) Release from inhibitory soil biota in Europe may promote exotic plant invasion in North America. *Nature*, **427**, 731–733.

Callaway, R.M., Thelen, G.C., Barth, S., Ramsey, P.W. & Gannon, J.E. (2004b) Soil fungi alter interactions between North American plant species and the exotic invader *Centaurea maculosa* in the field. *Ecology*, **85**, 1062–1071.

Chanway, C.P., Holl, F.B. & Turkington, R. (1989) Effect of *Rhizobium leguminosarum* biovar *trifolii* genotype on specificity between *Trifolium repens* and *Lolium perenne*. *Journal of Ecology*, **77**, 1150–1160.

Connell, J.H. (1971) On the role of natural enemies in preventing competitive excusion. In *Dynamics in Populations* (ed. P.J. den Boer and G.R. Gradwell), pp. 298–312. Center for Agricultural Publishing and Documentation, Wageningen, the Netherlands.

Cui, Q. & He, W. (2009) Soil biota, but not soil nutrients, facilitate the invasion of *Bidens pilosa* relative to a native species *Saussurea deltoidea*. *Weed Research*, **49**, 201–206.

Denison, R.F. & Kiers, E.T. (2004) Why are most *Rhizobium* beneficial to their plant hosts, rather than parasitic? *Microbes and Infection*, **6**, 1235–1239.

De Rooij-van der Goes, P.C.E.M. (1995) The role of plant-parasitic nematodes and soil-borne fungi in the decline of *Ammophila arenaria* L. Link. *New Phytologist*, **129**, 661–669.

DeWalt, S.J., Denslow, J.S. & Ickes, K. (2004) Natural enemy release facilitates habitat expansion of the invasive tropical shrub *Clidemia hirta*. *Ecology*, **85**, 471–483.

Duda, J.J., Freeman, D.C., Emlen, J.M., et al. (2003) Differences in native soil ecology associated with invasion of the exotic annual chenopod, *Halogeton glomeratus. Biology and Fertility of Soils*, **38**, 72–77.

Elton, C.S. (1958) *The Ecology of Invasions by Plants and Animals.* Methuen, London.

Eppinga, M.B., Rietkerk, M., Dekker, S.C., De Ruiter, P.C. & van der Putten, W.H. (2006) Accumulation of local pathogens: a new hypothesis to explain exotic plant invasions. *Oikos*, **114**, 168–176.

Gastine, A., Scherer-Lorenzenb M. & Leadley, P.W. (2003) No consistent effects of plant diversity on root biomass, soil biota and soil abiotic conditions in temperate grassland communities. *Applied Soil Ecology*, **24**, 101–111.

Gilbert, G.S. & Webb, C.O. (2007) Phylogenetic signal in plant pathogen-host range. *Proceedings of the National Academy of Sciences of the USA*, **104**, 4979–4983.

Green, J.L., Holmes, A.J., Westoby, M., et al. (2004) Spatial scaling of microbial eukaryotic diversity. *Nature*, **432**, 747–750.

Grüter, D., Schmid, B. & Brandl, H. (2006) Influence of plant diversity and elevated atmospheric carbon dioxide levels on belowground bacterial diversity. *BMC Microbiology*, **6**, 68, doi:10.1186/1471-2180-6-68.

Hierro, J.L., Maron, J.L. & Callaway, R.M. (2005) A biogeographical approach to plant invasions: the importance of studying exotics in their introduced and native range. *Journal of Ecology*, **93**, 5–15.

Hoestra, H. (1975) Crop rotation, monoculture and soil ecology. *European and Mediterranean Plant Protection Organization Bulletin*, **5**, 173–180.

Horner-Devine, M., Lage, C., Hughes, J. & Bohannan, B.J.M. (2004) A taxa–area relationship for bacteria. *Nature*, **432**, 750–753.

Howard, R.J. (1996) Cultural control of plant diseases: a historical perspective. *Canadian Journal of Plant Pathology*, **18**, 145–150.

Janzen, D.H. (1970) Herbivores and the number of tree species in tropical forests. *American Naturalist*, **104**, 501–508.

Jordan, N.R., Larson, D.L. & Huerd, S.C. (2008) Soil modification by invasive plants: effects on native and invasive species of mixed-grass prairies. *Biological Invasions*, **10**, 177–190.

Keane, R.M. & Crawley, M.J. (2002) Exotic plant invasions and the enemy release hypothesis. *Trends in Ecology & Evolution*, **17**, 164–170.

Klironomos, J.N. (2002) Feedback with soil biota contributes to plant rarity and invasiveness in communities. *Nature*, **417**, 67–70.

Knevel, I.C., Lans, T., Menting, F.B.J., Hertling, U.M. & van der Putten,W.H. (2004) Release from native root herbivores and biotic resistance by soil pathogens in a new habitat both affect the alien *Ammophila arenaria* in South Africa. *Oecologia*, **141**, 502–510.

Kulmatiski, A., Beard, K.H. & Stark, J.M. (2006) Soil history as a primary control on plant invasion in abandoned agricultural fields. *Journal of Applied Ecology*, **43**, 868–876.

Kulmatiski, A., Beard, K.H., Stevens, J.R. & Cobbold, S.M. (2008) Plant–soil feedbacks: a meta-analytical review. *Ecology Letters*, **11**, 980–992.

Liao, C., Peng, R., Luo, Y., et al. (2008) Altered ecosystem carbon and nitrogen cycles by plant invasion: a meta-analysis. *New Phytologist*, **177**, 706–714.

Magarey, R.C. (1999) Reduced productivity in long term monoculture: where are we placed? *Australasian Plant Pathology*, **28**, 11–20.

Mangla, S., Inderjit & Callaway, R.M. (2007) Exotic invasive plant accumulates native soil pathogens which inhibit native plants. *Journal of Ecology*, **96**, 58–67.

Marler M., Zabinski, C.A. & Callaway, R.M. (1999) Mycorrhizae indirectly enhance competitive effects of an invasive forb on a native bunchgrass. *Ecology*, **80**, 1180–1186.

McInerney, J.O., Cotton J.A. & Pisani, D. (2008) The prokaryotic tree of life: past, present ... and future? *Trends in Ecology & Evolution*, **23**, 276–281.

Millard, P. & Singh, B.K. (2009) Does grassland vegetation drive soil microbial diversity? *Nutrient Cycling in Agroecosystems*, doi:10.1007/s10705-009-9314-3.

Mills, K.E. & Bever, J.D. (1998) Maintenance of diversity within plant communities: soil pathogens as agents of negative feedback. *Ecology*, **79**, 1595–1601.

Mitchell, C.E. & Power, A.G. (2003) Release of invasive plants from fungal and viral pathogens. *Nature*, **421**, 625–627.

Niklaus, P.A., Alphei, J., Kampichler, C., et al. (2007) Interactive effects of plant species diversity and elevated CO_2 on soil biota and nutrient cycling. *Ecology*, **88**, 3153–3163.

Niu, H., Liu, W., Wan, F. & Liu, B. (2007) An invasive aster (*Ageratina adenophora*) invades and dominates forest understories in China: altered soil microbial communities facilitate the invader and inhibit natives. *Plant and Soil*, **294**, 73–85.

Nuismer, S.L. & Doebeli, M. (2004) Genetic correlations and coevolutionary dynamics of three-species systems. *Evolution*, **58**, 1165–1177.

Packer, A. & Clay, K. (2000) Soil pathogens and spatial patterns of seedling mortality in a temperate tree. *Nature*, **404**, 278–281.

Parker, I.M. & Gilbert, G.S. (2004) The evolutionary ecology of novel plant-pathogen interactions. *Annual Review Ecology Evolution Systematics*, **35**, 675–700.

Parker, I.M., Wurtz, A. & Paynter, Q. (2007) Nodule symbiosis of invasive *Mimosa pigra* in Australia and in ancestral habitats: a comparative analysis. *Biological Invasions*, **9**, 127–138.

Parker, M.A. (2001) Mutualism as a constraint on invasion success for legumes and rhizobia. *Diversity and Distributions*, **7**, 125–131.

Parker, M.A., Malek, W. & Parker, I.M. (2006) Growth of an invasive legume is symbiont limited in newly occupied habitats. *Diversity and Distributions*, **12**, 563–571.

Pepper, I.L. (2000) Beneficial and pathogenic microbes in agriculture. *Environmental Microbiology* (ed. R.M. Maier, I.L. Pepper and C.P. Gerba), pp. 425–426. Academic Press, San Diego.

Porazinska, D.L., Bardgett, R.D., Blaauw, M.B., et al. (2003) Relationships at the aboveground–belowground interface: plants, soil biota, and soil processes. *Ecological Monographs*, **73**, 377–390.

Reinhart, K.O., Packer, A., van der Putten, W.H. & Clay, K. (2003) Plant–soil biota interactions and spatial distribution of black cherry in its native and invasive ranges. *Ecology Letters*, **6**, 1046–1050.

Reinhart, K.O. & Callaway, R.M. (2004) Soil biota facilitate exotic *Acer* invasion in Europe and North America. *Ecological Applications*, **14**, 1737–1745.

Reinhart, K.O., Royo, A.A., van der Putten, W.H. & Clay, K. (2005) Soil feedback and pathogen activity in *Prunus serotina* throughout its native range. *Journal of Ecology*, **93**, 890–898.

Richardson D.M., Allsopp N., D'Antonio C.M., Milton S.J., Rejmanek M. (2000) Plant invasions – the role of mutualisms. *Biological Reviews of the Cambridge Philosophical Society* **75**, 65–93.

Rodgers, V.L., Wolfe, B.E., Werden, L.K. & Fenzi, A.C. (2008) The invasive species *Alliaria petiolata* (garlic mustard) increases soil nutrient availability in northern hardwood-conifer forests. *Oecologia*, **157**, 459–471.

Rout, M.E. & Callaway, R.M. (2009) An invasive plant paradox. *Science*, **324**, 734–735.

Rout, M.E. & Chrzanowski, T.H. (2009) The invasive *Sorghum halepense* harbors endophytic N_2-fixing bacteria and alters soil biogeochemistry. *Plant and Soil*, **315**, 163–172.

Tilman, D., Reich, P.B., Knops, J., Wedin, D., Mielke, T. & Lehman, C. (2001) Diversity and productivity in a long-term grassland experiment. *Science*, **294**, 843–845.

Thrall, P.H., Hochberg, M.E., Burdon, J.L. & Bever, J.D. (2007) Coevolution of symbiotic mutualists and parasites in a community context. *Trends in Ecology & Evolution*, **22**, 120–126.

Van der Putten, W.H., Van Dijk, C. & Troelstra, S.R. (1988) Biotic soil factors effecting growth and development of *Ammophila arenaria*. *Oecologia*, **76**, 313–320.

Van der Putten, W.H. & Troelstra, S.R. (1990) Harmful soil organisms in coastal foredunes involved in degeneration of *Ammophila arenaria* and *Calammophila baltica*. *Canadian Journal of Botany*, **68**, 1560–1568.

Van der Putten, W.H., Van Dijk, C. & Peters, B.A.M. (1993) Plant specific soil-borne diseases contribute to succession in foredune communities. *Nature*, **362**, 53–56.

Van der Putten, W.H., Yeates, G.W., Duyts, H., Schreck Reis, C. & Karssen, G. (2005) Invasive plants and their escape from root herbivory: a worldwide comparison of the root-feeding nematode communities of the dune grass *Ammophila arenaria* in natural and introduced ranges. *Biological Invasions*, **7**, 733–746.

Van der Putten, W.H., Kowalchuk, G.A. & Brinkman, E.P. (2007) Soil feedback of exotic savanna grass relates to pathogen absence and mycorrhizal selectivity. *Ecology*, **88**, 978–988.

Van Kleunen, M. & Fischer, M. (2009) Release from foliar and floral fungal pathogen species does not explain the geographic spread of naturalized North American plants in Europe. *Journal of Ecology*, **97**, 385–392.

Virgil (1900) *Bucolics, Aeneid, and Georgics of Virgil*, volume **1** (ed. J.B. Greenough), pp. 71–117. Ginn and Company, Boston, Massachusetts.

Wardle, D.A., Bonner, K.I., Barker, G.M., et al. (1999) Plant removals in perennial grassland: vegetation dynamics, decomposers, soil biodiversity and ecosystem properties. *Ecological Monographs*, **69**, 535–568.

Wardle, D.A. (2002) *Communities and Ecosystems: Linking the Aboveground and Belowground Components*. Princeton University Press, Princeton.

White, K.D. (1970) *Roman Farming*. Cornell University Press, Ithaca, New York.

Wolfe, B.E. & Klironomos, J.N. (2005) Breaking new ground: soil communities and exotic plant invasion. *BioScience*, **55**, 477–487.

Zhang, C., Wang, J., Liu, W., et al. (2010) Effects of plant diversity on microbial biomass and community metabolic profiles in a full-scale constructed wetland. *Ecological Engineering*, **36**, 62–68.

Chapter 12

MUTUALISMS: KEY DRIVERS OF INVASIONS … KEY CASUALTIES OF INVASIONS

Anna Traveset and David M. Richardson

Fifty Years of Invasion Ecology: The Legacy of Charles Elton, 1st edition. Edited by David M. Richardson

12.1 INTRODUCTION

The reigning paradigm over much of the history of the study of biological invasions has been that communities have 'biotic resistance' to invaders, a notion that was central to Charles Elton's (1958) understanding of invasions. This view is based on the assumption that natural communities are mainly structured by negative interactions; it thus emphasizes the biotic relationships between native and invasive alien species mediated through competition, herbivory, parasitism, etc. It predicts (i) the risk of invasions decreases when resource capture by the native community increases, for instance when species diversity in the community is higher, and (ii) the establishment of invasive species is favoured by the absence of natural enemies (herbivores, predators, pathogens) (Simberloff 1986; Rejmánek 1998); the enemy-release hypothesis (see, for example, Keane & Crawley 2002) proposes that introduced species have better opportunities for establishment when freed from the negative effects of natural enemies that, in their native range, lead to high mortality rates and reduced productivity.

This longstanding paradigm has, however, been increasingly challenged recently as many studies have shown that positive (facilitative) interactions are as important, or even more so, than negative interactions in structuring communities and ecosystems (Bertness & Callaway 1994; Callaway 1995; Bruno et al. 2003, 2005; Valiente-Banuet et al. 2006; Brooker et al. 2008). Facilitation can have strong effects at the level of individuals (on fitness), populations (on growth and distribution), communities (on species composition and diversity) and even landscapes (see, for example, Valiente-Banuet et al. 2006; Brooker et al. 2008).

When positive interactions among species are incorporated in population and community models, they change many fundamental assumptions and predictions (Bruno et al. 2005; Bulleri et al. 2008). Clearly, a robust predictive framework for invasion biology demands an improved understanding of the role of facilitation in mediating biotic resistance of communities to the incursion of introduced species. In particular, it is important to consider (i) the effect that the establishment of such positive interactions between the invasive alien species and already-present biota (native or alien) can have from overcoming such biotic

resistance, and (ii) that such positive interactions can bring about significant changes to invaded ecosystems. Much evidence has accumulated in the last decade from terrestrial and aquatic ecosystems to show that native species frequently promote (facilitate) the colonization and establishment of introduced species through a variety of mechanisms (Richardson et al. 2000a; Bruno et al. 2005; Badano et al. 2007; Milton et al. 2007; Olyarnik et al. 2008). Simberloff and von Holle (1999) were the first to incorporate facilitation explicitly in an invasion biology framework; they coined the term 'invasional meltdown' for the process whereby two or more introduced species facilitate establishment and/or spread of each other (and potentially other species). This contributes to increased invasibility and accelerated invasion rates and to a synergic amplification of the disruptive effects of invasive species.

Mutualisms are a type of facilitative interaction in which the two (or more) species involved both benefit. Pollinator and seed dispersal mutualisms are especially important for plant invasions, as the production and dispersal of propagules are usually fundamental requirements for invasion (see reviews in Davis 2009; Simberloff 2009). Native and alien animals clearly assist the spread of alien plants by pollinating their flowers or by dispersing their seeds (reviewed in Richardson et al. 2000a). Native and alien plants also facilitate the spread of alien animals (pollinators and seed dispersers) by providing them with important food resources (pollen, nectar, resins, fruit pulp, etc.). In this chapter we deal mostly with plant–animal mutualistic interactions involving pollination (for which most data are available) and seed dispersal. Plant–fungal mutualistic interactions (see, for example, Callaway et al. 2004, Kottke et al. 2008, Collier & Bidartondo 2009) and plant–plant interactions (e.g. Badano et al. 2007) are also crucial for the success of many plant invasions, as has been documented in several systems.

Another established concept in ecology is that invasion success is influenced by the phylogenetic relationships between biological invaders and residents of the target community. In *The Origin of Species*, Darwin (1859) explored whether species with a common evolutionary history interact with each other more closely than unrelated species. He predicted that introduced species with close relatives were less likely to succeed owing to fiercer competition resulting from their similarity to residents. Using data from eastern North America, Darwin found that most naturalized tree genera had no native counterparts, suggesting that aliens may be handicapped by more intense competition from established congenerics. A century later, Elton (1958) supported this view by arguing that unique traits allow invaders to exploit 'empty niches' in species-poor island communities. There has been a lack of consensus among studies that have tested 'Darwin's naturalization hypothesis', some finding support for it and others not (reviewed in Proches et al. 2008; see also Thuiller et al. 2010).

Considering mutualistic interactions, we might predict that plant invaders similar to natives in morphological traits (flower/fruit colour, size, shape, etc.) and physiological traits (chemical composition of nectar, fruit pulp, etc.) are more likely to share pollinators/seed dispersers with native plants. This could lead to successful establishment in the receptive community. Similarly, an invader pollinator/disperser might 'fit better' in the new environment if its requirements are similar to those of the resident/native pollinators. Therefore, when considering positive interactions, predictions about invasive success based on the phylogenetic relatedness between invaders and residents might differ from those made when considering negative interactions. Recent developments in coexistence theory demonstrate that invasion success can result either from fitness differences between invader and residents that favour the former, or from niche differences that allow the establishment of the invader despite having a lower fitness (MacDougall et al. 2009).

Mutualisms have received increasing attention recently, and are now widely accepted to be important mediators of ecosystem functioning (Bruno et al. 2003, 2005; Agrawal et al. 2007; Brooker et al. 2008; Bronstein 2009). There has been a rapid increase in the number of published papers linking mutualisms and invasions. A literature search in the ISI Web of Knowledge including the terms 'mutualis' and 'invasi' showed only three papers dealing with both topics in 1999, but 50 in 2009. In this chapter we review studies that have examined the importance of mutualistic interactions in determining the success and impact of invasive species. Our goal is to identify general patterns as well as topics that need more research.

12.2   MUTUALISMS AS DRIVERS OF INVASIONS

The role of mutualists in the naturalization–invasion continuum

To colonize, survive, regenerate and disperse, a species introduced to a new area must negotiate several biotic and abiotic filters/barriers (Richardson et al. 2000b; Mitchell et al. 2006). This, and the fact that the introduction of species and their mutualistic or antagonistic partners often do not take place simultaneously (see, for example, Richardson et al. 2000b; Grosholz 2005), greatly reduces the probability of an alien species interacting with the same mutualistic and antagonistic species in the new environment as in their native range. An increasing body of literature demonstrates that positive interactions between species, specifically those established among plants and animals, are crucial for the integration of invasive species into native communities, and that these can mediate the impacts of introduced species.

Mutualisms are important at all stages of the invasion process (Fig. 12.1). An alien plant introduced by humans can be transported to new areas far from the original site through dispersal by animals; such a plant may in turn establish in that area owing to symbiotic microorganisms in the soil and/or because pollinators mediate seed production. A plant species can spread and become invasive because frugivorous animals disperse its seeds far from where the plant was originally established. The same dispersal vector may be implicated at more than one stage. Human activities, for instance, are by definition the vector of arrival for alien species, but can also disseminate the invader within the introduced region. The rate of spread is influenced by mean dispersal distance and more importantly by unpredictable, rare long-distance dispersal events that have a disproportionate effect on population growth and aerial spread (Trakhtenbrot et al. 2005). Moreover, disturbances (whether natural or anthropogenic) can also influence the initial establishment of invasive mutualists.

The importance of mutualistic interactions during the invasion process depends on different traits/requirements of the invader. In the case of invasive plants, we expect these interactions to determine invasion success when the plant:

1 Is an obligate outcrosser (e.g. owing to self-incompatibility or dioecy, or if self-compatible has no

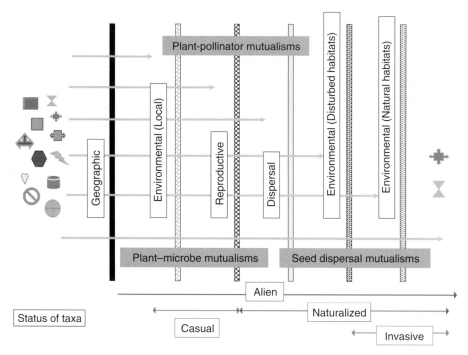

Fig. 12.1 Schematic representation of the naturalization–invasion continuum, depicting various barriers that an introduced plant must negotiate to be become 'casual', 'naturalized' or 'invasive' (based on Richardson et al. 2000). Bars show the main phases and stages at which different categories of mutualisms are influential in invasion dynamics. Once species become 'invasive' they often become widespread and abundant and may affect a wide range of naturally occurring mutualisms.

capacity for autonomous self-pollination), and therefore requires pollinators to set seeds. Examples include the obligate outcrossers *Centaurea diffusa* and *C. maculosa* (Harrod & Taylor 1995) and purple loosestrife, *Lythrum salicaria* (Mal et al. 1992), both invasive in North America.

2 Needs animals to disperse its seeds. For instance, *Crataegus monogyna* produces larger fruit displays of higher quality than a native congener, and this has contributed to the rapid spread of this shrub in western North America (Sallabanks 1993).

3 Needs specific microorganisms (nitrogen-fixing bacteria or mycorrhizal fungi) to establish and grow. Examples include many introduced conifers in the Southern hemisphere (which failed until appropriate fungal symbionts were introduced; Richardson et al. 2000a) and many herbs (Reinhart & Callaway 2006).

Because mutualistic interactions, specifically plant–animal mutualisms, are more prevalent in the tropics than in the temperate zones (Schemske et al. 2009), they are more likely to influence invasion success at lower latitudes. Unfortunately, this hypothesis has yet to be tested, as most data elucidating the role of mutualisms on invasions are from temperate areas. Likewise, we would expect a stronger influence of pollinators and seed dispersers on plant invasions at lower altitudes owing to the declining diversity of such animals with increasing elevation.

Several studies have shown that systems rich in native species often support large numbers of alien species (see, for example, Rejmánek 1996; Stohlgren et al. 1999; Richardson et al. 2005). In fact, if plant invasions are facilitated by a diverse array of pollinators, dispersers, fungi and bacteria. So, if disturbance and/or fluctuations in resource availability create windows of opportunity, we should expect invasibility to be positively correlated with native species richness (Richardson et al. 2000a).

The importance of mutualisms as mediators of invasion success

Baker's rule states that plants capable of uniparental reproduction, especially self-compatible species, are more likely to be successful colonists than are self-incompatible or dioecious species (Baker 1955). Unfortunately, we still have rather little information on the reproductive systems of most invasive alien plants (but see Barrett, this volume). A study of 17 invasive alien woody and herb species in South Africa using controlled pollination experiments revealed that all were either self-compatible or apomictic, and that 72% of species were capable of autonomous self pollination (Rambuda & Johnson 2004). A survey in Missouri, USA, found a similar pollination ecology and degree of autogamy between 10 closely related pairs of native and introduced plant species, although of those that differed, the introduced species were more autogamous than their native congeners (Harmon-Threatt et al. 2009). Other surveys in different areas, however, show that many introduced plant species, particularly woody ones, require pollinator mutualisms to become invasive (Richardson et al. 2000a).

Pollen limitation does not seem to represent a major barrier to the success of introduced plants (Richardson et al. 2000a; Rambuda & Johnson 2004). This may be because of the high level of generalization of pollinators (see section 'Mechanisms whereby mutualisms can drive invasions'). Among the few exceptions reported so far are *Trifolium pratense* (red clover), which did not set seed in New Zealand before bumblebees were introduced, and *Cytisus scoparius*, *Ficus* spp. and *Melilotus* sp. (details in Richardson et al. 2000a). Figs are the best-studied case of pollinator-mediated constraints on invasion. Several species have spread in alien habitats only after their specific wasp pollinators arrived, either through accidental introduction by humans or by long-distance dispersal (Gardner & Early 1996). Assuming that pollinator specificity is greater in tropical than in temperate forests (Bawa 1990), and given the significant positive relationship between pollen limitation and plant species richness (Vamosi et al. 2006), we predict that invasive plants should be more pollen limited in the former, but more data are needed to test this hypothesis.

An example of how a plant-seed dispersal interaction may drive an invasion is provided by Milton et al. (2007). They report that birds drive the invasion of arid savannas in South Africa when alien fleshy-fruited plants infiltrate prevailing seed-dispersal networks. Once infiltrated, the natural dispersal network is disrupted because some invasive plants transform the savannas by overtopping and suppressing native trees that act as crucial perch sites and foci for directed dispersal (Iponga et al. 2008).

The importance of mutualistic interactions is also emphasized in the invasion ecology of ectomycorrhizal plants, such as Northern hemisphere conifers introduced to the Southern hemisphere. For instance, the establishment and spread of *Pinus* species in many parts of the Southern hemisphere was initially thwarted by the absence of appropriate fungal symbionts (Richardson et al. 2000a; Nuñez et al. 2009). In the Galápagos Islands, the invasion by the obligately arbuscular mycorrhizal *Psidium guajava* was only possible because arbuscular mycorrhizas were already naturally present on the islands (Schmidt & Scow 1986).

Towards a framework of ecological networks to study integration (and impact) of invasive species at the community level

Understanding the evolution and diversification of ecological interactions requires more than an elucidation of interactions between pairs of species. Mutualisms between plants and pollinators or plants and seed dispersers must be considered as networks of interactions involving many species, some with high degrees of generalization (Jordano 1987; Waser et al. 1996). The formalization of interactions between plant communities and their mutualistic animals, using analytical methods developed for the ecology of trophic webs, provides an appropriate conceptual framework for studying the dynamics of such interactions. Such an approach helps us to understand how new species are incorporated into the community and how the community responds to new members (see, for example, Memmot & Waser 2002; Olesen et al. 2002, Olesen & Jordano 2002; Bascompte et al. 2003, Aizen et al. 2008; Padrón et al. 2009). Nonetheless, most information derived using this network approach to date is based on qualitative data (presence/absence of interactions) and assumes that all interactions between plants and animals have similar weights, at least regarding the quality of the interaction (i.e., assumes all pollinator/disperser visits are equally effective from the plant's viewpoint). A functional approach is needed to make more robust predictions about the impact of a

Fig. 12.2 Example of a plant–flower visitor network. The nodes in the lower part represent the insect species in the web; nodes in the upper part represent plant species that are visited by them. Lines between nodes symbolize links between species pairs. The yellow node is the invasive species *Opuntia maxima*, which is integrated in the community and which interacts with a high number of flower-visiting insects (lines indicated in green). Data from Padrón et al. (2009). The image was produced using FoodWeb3D, written by R.J. Williams and provided by the Pacific Ecoinformatics and Computational Ecology Laboratory (www.foodwebs.org; Yoon et al. 2004), courtesy of B. Padrón.

disturbance (the entrance of an invader, for instance) on the entire community. We must obtain good quantitative data to assess whether the patterns that have emerged so far hold when the different effectiveness of mutualists are considered.

Other mutualistic interactions, such as those established between plants and soil biota (mycorrhizas or nitrogen-fixing bacteria) should also be studied at a community level (see, for example, Kottke et al. 2008; Collier & Bidartondo 2009; van der Heijden & Horton 2009). Recent research has highlighted their important role, either facilitating or inhibiting plant invasions (Callaway & Rout, this volume). These underground interactions can in turn affect the aboveground mutualisms. For instance, arbuscular mycorrhizas in some plants can promote an increase in floral display and/or in the quantity and quality of nectar which directly affects the pollination visitation rates to flowers (Wolfe et al. 2005). Similarly, underground mutualisms can alter plant–herbivore interactions, as observed in the invasive *Ammophila arenaria*, a dominant species in many dune systems around the world whose mycorrhizas improve resistance to the attack of parasitic nematodes to its roots (Beckstead & Parker 2003).

An alien species is incorporated in a mutualistic plant–animal network when it establishes a link with another species, native or alien, forming a new node in the web. The new interactions between alien and native species affect the demographic success of the former, and influence interactions among natives with subsequent consequences for them. Figure 12.2 shows a network of plant-flower visitor interactions in a plant community in the Balearic Islands that has been invaded by the alien cactus *Opuntia maxima*. This species has become integrated in the community and interacts with many native insects (Padrón et al. 2009).

Pollination and seed dispersal systems tend to be dominated by generalists. Consequently, it has been predicted that alien species are easily accommodated in such networks (Waser et al. 1996). Studies in invaded natural communities seem to confirm this prediction in the case of native pollinators, because their level of generalization is positively correlated with the probability of servicing a given invasive plant (Lopezaraiza-Mikel et al. 2007), or with the number of alien species included in their diet (Memmott & Waser 2002). In turn, most successful animal-pollinated invasive plants are pollinated by generalist species (Richardson et al. 2000a). Less information is available on seed dispersal assemblages of alien species.

Mechanisms whereby mutualisms can drive invasions

Species niche width may be one of the traits that determines the success of an invader (Vázquez 2005). If we apply this hypothesis to mutualistic interactions, the prediction is that generalist alien species have a higher probability of receiving more visits, in the case of plants, or of acquiring more resources in the case of animals, than specialist species (Richardson et al. 2000a). As result, generalist aliens are more likely to integrate into local networks. Validation of this hypothesis requires the comparison, from a biogeographical perspective, of the generalization level of invasive alien species with those of non-invasive aliens (Vázquez 2005). Available evidence suggests that even if reproduction of some highly invasive plant species is limited by pollen availability (Parker et al. 2002), only a small proportion of potentially invasive introduced plants seem to have failed because of the absence of pollinators (Richardson et al. 2000a). This is the case of many species with highly specialized flowers, such as orchids or *Ficus*, emblematic examples that represent examples of mutual specialization.

We may also ask whether, in a given community, invasive species are more generalist than native species. Intra-community comparisons have not shed much light on this so far. Whereas some communities show similar levels of generalization between native and alien mutualists, either for plants or pollinators (see Morales & Aizen 2002, 2006; Olesen et al. 2002), others show differences between them. For instance, Memmott and Waser (2002) found alien flower visitors to be the most generalist species in the community, including super-generalist species such as *Apis mellifera*, whereas alien plants were on average less generalist than natives.

In mutualistic networks, the distribution of the generalization level (or degree) is highly asymmetrical (Vázquez & Aizen 2004), leading to a nested pattern that in turn confers a high coherence to the network (Bascompte et al. 2003). This implies that specialist species tend to interact exclusively with generalist species, whereas generalist species interact both with specialists and generalists. The former might raise the possibility of integration of specialist alien species into mutualistic networks more than would be expected by random. This might indeed explain the similar generalization levels in native and alien species.

Given that the number of links that an alien plant species can establish is partly explained by the phylogenetic affinity of that species with the native flora (Memmott & Waser 2002), it seems likely that specialist alien species connect to those generalist natives that have some taxonomic affinity with their original mutualists. For example, the European native *Cytisus scoparius*, with zygomorphic flowers specialized for bumblebee pollination, is visited almost exclusively by native and generalist bumblebees in the invaded range in South America (Morales & Aizen 2002). The ornithophilous *Nicotiana glauca* is pollinated by hummingbirds in its native range in central and northern Argentina (Nattero & Cocucci 2007), and receives visits of other hummingbird species in Venezuela (Grases & Ramírez 1998) and of sunbirds in South Africa (Geerts & Pauw 2009). By contrast, *Kalanchoe* sp., another ornithophilous species introduced to Venezuela, pollinated by passeriforms in its native region (South Africa), is not visited by hummingbirds in Venezuela (N. Ramírez, personal communication).

Facilitation among invasive species and evidence for invasional meltdown

A preferential interaction between alien mutualists in which species interact with a higher frequency than expected by chance can lead to a core of alien generalists, or what has been termed an invasion complex (D'Antonio 1990). Positive interactions among invasive species are relatively frequent, especially plant–pollinator and plant–seed disperser interactions. However, there is no evidence to date that such a core of alien generalists within networks is the rule. Indeed, Bruno et al., (2005) showed that alien species are not more likely to benefit alien species than natives. More recently, Aizen et al. (2008) have suggested that the existence of differential interactions among invaders might take place in the most advanced stages of invasion; at the early stages, invasives are integrated in the webs by interacting with natives.

Examples of such invader complexes have been reported from many different types of ecosystems, but especially islands (Traveset & Richardson 2006). The honeybee is an important pollinator of many invasive plants on islands where it has been introduced, such as the Bonin Islands, New Zealand, Tasmania, Azores, Santa Cruz and Tenerife. Some species of bumblebees and *Megachile rotundata* show a preference for alien plants in New Zealand and Australia, respectively. Alien wasps have favoured the spread of introduced *Ficus* species in continental areas of North America, in Hawaii and New Zealand (references in Richardson et al. 2000a). The same intrinsic plant traits (for example a large floral display, high nectar and/or pollen production) can promote more frequent interactions with invasive pollinators (usually social insects, mainly owing to their high energetic demand to maintain their colonies) than with native pollinators. Likewise, invasive plants that produce large fruit crops can also be mainly dispersed by alien invasive animals. For instance, on Mediterranean islands, rats and rabbits, introduced thousands of years ago, are important dispersers of invasive plants such as *Carpobrotus* spp. In the Canary Islands, the alien squirrel *Atlantoxerus getulus* is dispersing the also invasive *Opuntia maxima* (López-Darias & Nogales 2008). In island ecosystems specifically, mainly because of the lower species richness, we may expect a greater effect of invasion complexes on native biodiversity than in the mainland. More data are, however, needed for a robust test of this hypothesis.

Invader complexes can lead to a process whereby the species involved in the interaction reciprocally improve conditions for survival and spread (invasional meltdown; Simberloff & Von Holle 1999). Many authors have suggested that such processes may be operating in a wide range of ecosystems (references in Simberloff & Von Holle 1999; Traveset & Richardson 2006). We need more research to determine how widespread such invasion complexes are, and how they impact on pollination and seed dispersal networks. Much research is currently underway in this area and a clearer picture should emerge soon.

12.3 THE OTHER SIDE OF THE COIN: MUTUALIMS DISRUPTED – CASUALTIES OF INVASIONS

The dynamics of disruption

Invasive alien species can bring about substantial changes to prevailing mutualistic interactions (Traveset & Richardson 2006), and such alterations can, in turn, mediate subsequent invasion dynamics (Mitchell et al. 2006). The most dramatic changes have been documented in island ecosystems such as those on Hawaii (Waring et al. 1993) and New Zealand

(Kelly et al. 2006), but also in continental situations, such as in South African fynbos (Christian 2001), the dry forests of Thailand (Ghazoul 2004) or the rainforest of Mexico (Roubik & Villanueva-Gutierrez 2009). Such mutualistic disruptions may be caused by plants, animals or pathogens. An example of the last-mentioned was recently reported by McKinney et al. (2009) from the northern Rocky Mountains, where an invasive alien fungus is disrupting the obligate seed-dispersal mutualism involving *Pinus albicaulis*, a keystone subalpine tree species, and the only bird capable of dispersing its large, wingless seeds, the Clark's Nutckacker (*Nucifraga columbiana*). The fungus kills tree branches and significantly reduces cone production, which influences the nutcracker's occurrence and seed dispersal success.

One of the best documented categories of mutualistic disruptions caused by invasions has been widespread changes to plant–animal mutualisms that affect pollination and reproductive success of native plant species. Competitive interactions between native and alien species can involve different, not mutually exclusive mechanisms, which mainly imply changes in the frequency of visits and in interspecific pollen transfer (reviewed in Morales & Traveset 2008). In the presence of more attractive alien species, natives can experience fewer pollinator visits (see, for example, Brown et al. 2002), and/or a reduction in the quality (Ghazoul 2004; Lopezaraiza-Mikel et al. 2007) of visits of some pollinator species owing to changes in their abundance or behaviour (Ghazoul 2002, 2004). In either case, this may lead to a subsequent decrease in pollination levels and seed production of native plants. Research is needed to investigate whether such reductions in seed set have demographic consequences.

Alternatively, the presence of a highly attractive invasive species may facilitate visits to the less attractive native species by means of an 'overall attraction' of pollinators (see Rathcke 1983; Moeller 2005). In the past decade, an increasing number of studies have experimentally evaluated changes in pollination levels and in reproductive success of natives in response to the presence of aliens. A recent meta-analysis found an overall significantly negative effect of alien plants on visitation to and reproduction of native species (Morales & Traveset 2009). The negative effect increased at high relative densities of alien plants and, interestingly, their effect on visitation and reproductive success was most detrimental when alien and native plants had similar flower symmetry or colour, thus highlighting the importance of phenotypic similarity between aliens and natives in determining the outcome of the interaction. Such a finding is indeed consistent with a prediction of niche theory: that functionally similar invaders should impose the greatest harm on native communities (MacDougall et al. 2009).

Once an alien species is integrated into a network of mutualistic interactions, it can modify key parameters that describe network structure or topology. In natural systems with a nested interaction structure (Bascompte et al. 2003) the impact of an alien species may rapidly cascade through the entire network because all species are closely interlinked (Lopezaraiza-Mikel et al. 2007). Thus, the importance of aliens at the level of network is expected to be pronounced. However, research in this area of invasion biology is in its infancy, both for alien plants and alien pollinators (see Lopezaraiza-Mikel et al. 2007; Aizen et al. 2008; Padrón et al. 2009; Valdovinos et al. 2009; Vilà et al. 2009).

One of the consequences of the asymmetry of mutualistic networks is a low reciprocal dependence between mutualists: if a pollinator, for instance, is strongly reliant on a given plant species, this plant species typically depends only weakly on the services of that pollinator species, and vice versa. Such asymmetry confers stability and robustness to the networks against the loss of species (Vázquez & Aizen 2004; Bascompte et al. 2006). In a study of the impact of invasive plants and pollinators on network architecture, Aizen et al., (2008) analysed the connectivity of 10 networks characterized by contrasting levels in the incidence of invasive species and in their mutual dependence among the interacting species. There were no differences in connectivity (proportion of links relative to all those possible) between invaded and uninvaded networks (as also found by Memmott & Waser (2002) and Olesen et al. (2002)), but the invasive species were found to promote a 'redistribution' of links between plants and animals in the network: a high number of links were transferred from generalist native species to invasive super-generalist species, and therefore the entire network topology was modified. In other words, the invasive species usurp interactions among native mutualists in the process of invasion. In this way, at least at advanced stages of the invasion, supergeneralist invasives can alter the 'foundations' of the network architecture itself, becoming central nodes in it (Aizen et al. 2008), and their removal might significantly alter network topology (Valdovinos et al. 2009). Moreover, because interactions of a low reciprocal dependence are the most robust against disturbances promoting extinctions (Ollerton et al. 2003), such results suggest that the invasive species and its mutualists might be more resistant to disturbances, which would increase the probability of permanence and survival of such invasive in the network. This, added to a preferential interaction among invasive species, might lead to invasional meltdown, precipitating an even higher impact of the invasion on the native community, similar to those reported for other systems and interactions (see, for example, Grosholz 2005; Griffen et al. 2008; Belote & Jones 2009).

The presence of an invasive species can affect the patterns of pollen flow if pollinators transfer pollen interspecifically, depositing pollen of the invasive on the stigmas of the native, or vice versa, generating a decrease in the quality of pollination. Such a decrease may be due to the deposition of heterospecific pollen on their stigmas, which can interfere with the deposition and/or germination of conspecific pollen and/or to a loss of conspecific pollen on flowers of other species (Morales & Traveset 2008). Such changes can occur independently of changes in the absolute frequency of visits, although sometimes both phenomena occur simultaneously (see, for example, Brown et al. 2002; Ghazoul 2002; Lopezaraiza-Mikel et al. 2007). Only a few studies have evaluated changes in pollen deposition in the presence of an invasive species. Some have found reductions in the deposition of conspecific pollen (Ghazoul 2002; Larson et al. 2006), others have found increases in the deposition of heterospecific pollen (Ghazoul 2002), whereas yet others reported no consistent changes (Grabas & Laverty 1999; Moragues & Traveset 2005; Larson et al. 2006). Although the amount of heterospecific pollen on native stigmas is generally too low to interfere with the deposition of conspecific pollen (Morales & Traveset 2008), it might affect the reproduction of closely related native species if it produces hybrid seeds (Wolf et al. 2001; Burgess et al. 2008). So far, these mechanisms (changes in the frequency of visits and interspecific pollen transfer) have been evaluated separately, and the difference in the approaches used has hampered the evaluation of the relative importance of both mechanisms and their interaction.

Although individual invasive alien species may have a negative impact on particular mutualistic interactions, their effect on the overall community may be neutral or even positive for mutualistic interactions. It is also important to consider that the spatial scale of investigation affects the estimated strength of competition for pollinators between invasive and native plant species. For instance, the alien geophyte *Oxalis pes-caprae* appears to compete for pollinators with the native *Diplotaxis erucoides* at a scale of a few metres, but at larger scales the presence of *Oxalis* flowers in invaded fields attracts pollinators, facilitating visits to *Diplotaxis* (Jakobsson et al. 2009). Consistent with the arguments of Bulleri et al. (2008), it might well be that competition could operate inherently at a smaller spatial scale than facilitation and thus be more likely to produce results observed in fine-scale studies.

Can invasive alien species replace extinct or declining native mutualists?

Native island pollinators and seed dispersers, which tend to be more prone to extinction than plants (at least at the time scale of years or decades), have been occasionally and unintentionally, replaced by 'ecologically similar' alien species. For instance, in Hawaii, the widespread alien bird *Zosterops japonica* has replaced the extinct native honeycreepers (see Cox & Elmqvist 2000), whereas *Z. lateralis* has replaced native bird pollinators in New Zealand (Kelly et al. 2006). On Mauritius, the ornithophilous flowers of *Nesocodon mauritianus* (Campanulaceae) are currently visited almost exclusively by the introduced red-whiskered bulbul (*Pycnonotus jocosus*) (Linnebjerg et al. 2009). In Mallorca (Balearic Islands, western Mediterranean), the main disperser of the native shrub *Cneorum tricoccon* is the alien pine marten *Martes martes*, which appears to have replaced the extinct endemic lizards that performed this role before carnivores arrived on these islands (Traveset 1995). Given that the decline in native species and the arrival of alien species is frequently associated with habitat disturbance, it is often difficult to discern the relation of causality between a native species regression and an alien invasion.

An illustrative case of a functional replacement is the invasion of the alien bumblebee *Bombus ruderatus* in the Andean forests of Patagonia, after their deliberate introduction to southern Chile in 1982. This species was detected in native communities in 1994, and since then its abundance and distribution range have increased (Morales 2007). In 1996, a census of flower visitors to the native *Alstroemeria aurea* along a gradient of anthropogenic disturbance showed a decrease in the relative frequency of its main pollina-

tor, the native bumblebee *B. dahlbomii*, parallel to an increase in the frequency of the invasive pollinator, suggesting that such disturbance would favour the invasive species and hamper the native one. Later, a long-term study on the same plant species in a non-disturbed area confirmed that the invasive species is favoured by habitat alteration but also that it has indeed displaced the native species (Madjidian et al. 2008). This study highlights the necessity of long-term studies that consider different factors that can co-vary with changes in abundance of invasive and native species at the time that they evaluate the impact of the former on the latter (Stout & Morales 2009). This will allow distinguishing between passive replacements and competitive displacements of native by the invasive alien species. The study also highlights that these replacement processes can occur very quickly, in just a few years.

In general, we have rather little information on whether alien species act as functional surrogates of, and occupy the same niches as, extinct native species. Regarding pollinators, several studies have shown that introduced pollinators are not as effective as natives they have replaced (see Traveset & Richardson 2006). For instance, they may promote reduced outcrossing rates which can result in a reduced gene flow and/or promote hybridization between native plants (England et al. 2001; Dick et al. 2003). Introduced bees can affect plant fitness by actively reducing pollination of native plants (physical interference with native pollinators on the flowers; Gross & Mackay 1998) or by altering pollen dispersal (see, for example, Westerkamp 1991; Paton 1993; Celebrezze & Paton 2004).

Hansen et al. (2002) showed experimentally that the exclusion of bird pollinators reduced seed set in two Mauritian trees that were otherwise visited primarily by alien honeybees, possibly owing to higher levels of within-plant foraging behaviour of honeybees. Likewise, the introduced *Bombus ruderatus* is not as effective for *Alstroemeria* as is *B. dahlbomi* (Madjidian et al. 2008). For seed dispersers, we do not know to what extent they have similar foraging behaviours and move seeds to similar sites as extinct native species. Many oceanic islands, for instance, have lost a large proportion of native frugivorous avifauna, although avian species richness has remained fairly constant because extinction has been balanced by colonization and naturalization of alien bird species (see, for example, Sax et al. 2002; Foster & Robinson 2007; Cheke & Hume 2008). Despite this, alien birds often act as legitimate, and highly effective, dispersers of native plants. We do not know exactly how these alien birds influence the dispersal of native plants. In Hawaii, alien birds disperse many native understorey shrubs (Foster & Robinson 2007), but elsewhere alien birds disperse mainly alien plant species, like the red-whiskered bulbul on La Reunion (Mandon-Dalger et al. 2004). In New Zealand, the contribution of alien birds to seed dispersal of native plants is unexpectedly small (Kelly et al. 2006). In the Bonin Islands, Kawakami et al. (2009) found that introduced white-eyes appear to compensate for extinct native seed dispersers.

The magnitude of the current biodiversity crisis calls for radical conservation measures in some cases (e.g. rewilding in North America; 'managed relocation' involving the planned movement of threatened species to areas outside their current range where prospects of long-term survival are improved; habitat restoration using alien birds in Hawaii). Native species that have recently become extinct or those in which populations are declining might be replaced by functionally equivalent species. This might be especially necessary to buffer ecosystems against the loss of keystone species (*hubs* in the community networks).

Invasions and mutualisms in fragile ecosystems (islands)

The integration of invasive species into receptive communities by means of facilitative interactions with native species is likely to occur frequently in island ecosystems, where many native mutualists have wider trophic niches than in mainland systems. On islands, all possible detrimental effects are magnified exponentially, with reduced population sizes, absence of specialization in the native interactions or the unpredictability of resource production (see review in Kaiser-Bunbury et al. 2010). Many cases have been documented on islands where alien species have strongly negative impacts on native communities; both the magnitude and the mechanisms of the impacts vary depending on the functional group to which the alien species belongs and its abundance (Traveset & Richardson 2006; Traveset et al. 2009a). In any case, the invasive species has the capacity to alter significantly the reproductive success of the natives at the same time as altering the structure of pollination or disperser networks.

The introduction of *Apis mellifera* and *Bombus terrestres* to many islands around the world has had devastating effects on native bees (Goulson 2003) as a direct result of competition for floral resources (see, for example, Kato et al. 1999) or competition for nest sites (see, for example, Wenner & Thorp 1994), and even on the reproductive success of plants that depended upon them. This is because such alien species may reduce seed production, modify gene flow and promote hybridization between closely related species (see Traveset & Richardson 2006). Honeybees are already reported to have displaced native pollinators on many islands. For instance, small endemic solitary bees and white-eyes have been replaced in the Bonin Islands, Japan (Kato et al. 1999) and in Mauritius (Hansen et al. 2002), respectively. Similarly, honeyeaters have decreased in numbers in Australia, which has been attributed to the introduction of honeybees (Paton 1993) or bumblebees (Hingston et al. 2002; Hingston 2005). These two alien species have integrated well into the pollinator networks of many invaded island communities in Japan (Abe 2006), Tasmania (Hingston et al. 2002), the Mascarene Islands (Olesen et al. 2002) and the Canary Islands (Dupont et al. 2004). Invasive ants, such as *Anoplolepis gracilipes*, *Technomyrmex albipes* and *Wasmannia auropunctata* are also having dramatic effects on the biota of different islands where they have been introduced, such as New Caledonia (Jourdan et al. 2001), Mauritius (Hansen & Müller 2009) and Samoa (Savage et al. 2009). The Argentine ant *Linepithema humile* has significantly reduced the abundance of two important pollinators in Hawaii, the moth *Agrotis* sp. and the solitary bee *Hylaeus volcanica*, with potentially severe negative effects on the seed set of many native plant species (Cole et al. 1992). Moreover, a recent review by Rodríguez-Cabal et al., (2009) on the impact of this invasive ant on seed dispersal shows that it displaces native ants in most invaded areas and that it does not act as a legitimate disperser, although this depends upon seed traits such as size and percent reward. This suggests that Argentine ant invasions may drive shifts in community diversity (e.g. Christian 2001) and parallel changes in ecosystem functioning.

The integration of invasives into island communities is usually facilitated by generalist (sometimes supergeneralist) pollinators and dispersers that include nectar and pollen or fleshy fruits in their diets (Olesen et al. 2002; Morales & Aizen 2006; Traveset & Richardson 2006; Aizen et al. 2008;

Linnebjerg et al. 2009). For instance, *Carpobrotus* spp. are highly invasive on many Mediterranean islands, and are pollinated by a large diversity of native insects that are attracted by their abundant and attractive flowers (Moragues & Traveset 2005). A different situation occurs with the invasive *Kalanchoe pinnata* (Crassulaceae) in the Galápagos Islands. This plant, with a great capacity for vegetative growth, has complex flowers, which are not effectively visited by native pollinators in these islands (L. Navarro, personal communication). Kueffer et al. (2009) compared fruit traits between native and invasive alien plants on oceanic islands. They found fruit quality to be more variable in the latter, suggesting that this might represent an advantage for them during the invasion process. They also proposed that island plants produce fruits of lower energy content than invasives, probably because of reduced competition for dispersal. Further work is needed on more oceanic islands to examine this idea.

Introduced vertebrates are the animal group with most detrimental effects on the native island floras and faunas (Sax & Gaines 2008), including indirect effects on native mutualisms (see, for example, Nogales et al. 2004, 2005, 2006; Traveset & Riera 2005; Kelly et al. 2006; Traveset et al. 2009b). The negative impacts of, for example, introduced goats, rats and parrots on plant fitness and dispersal can be multifold. These include the direct consumption of native plants and, more indirectly, the reduction of populations of legitimate seed dispersers (see Riera et al. 2002; Traveset & Richardson 2006). There is also much evidence that rodents, cats and opossums have overwhelming effects on native seed dispersers (see, for example, Jourdan et al. 2001; García 2002; Nogales et al. 2004; Kelly et al. 2006; Towns et al. 2006; Hansen & Müller 2009).

Given the alarming increase in disturbances in island ecosystems, further studies are needed to determine the range of impacts of invasive species on mutualistic interactions at different levels (population, species, community and ecosystem). The loss of some mutualistic interactions often leads to decreases in recruitment rates of plants that depend upon them, even cascading into local or total extinctions. The lack of studies at community level precludes us from making global estimates of the impact of the loss of such mutualisms and from evaluating the degree of resilience of mutualistic networks possess with respect to different types of disturbance.

12.4 CHALLENGES FOR THE FUTURE

Biological invasions provide superb natural experiments at a global scale, offering exciting new insights on many aspects of ecology, including the factors that mediate ecosystem functioning and stability. Ecologists have begun to grasp this opportunity in the past decade, resulting in substantial advances in our understanding of the role of mutualisms in structuring communities, and of the fragility of many interactions.

The impacts of invasive species on naturally occurring mutualisms are still poorly documented, but the picture that is emerging is that they are often profound. Most insights are still from observations, and more manipulative experiments are needed to disentangle the full complexity of species interactions and to unravel the implications of the multiple interactive factors for community and ecosystem stability. Assessments of actual and potential impacts of introduced species have until recently all but ignored effects on mutualisms: a revised framework for incorporating such effects in impact assessments is needed.

Many restoration projects are underway to address degradation of ecosystems due to invasions and other factors. Too few such projects consider positive species interactions explicitly, or pay enough attention to keystone species (e.g. species that interact with large number of pollinators or dispersers; 'hubs' in network terminology) and ecosystem engineers (species that create or modify habitats). The integration of results such as those reviewed in this chapter is crucial for improving the efficiency of restoration efforts in yielding sustainable restored communities and landscapes. The use of ecosystem engineers (either native or alien) can enhance recruitment of native species, either directly or indirectly. For instance, many studies have shown the important effect of plant facilitation for the recruitment of many plant species, especially in stressful environments (Brooker et al. 2008). Such engineers can also create new opportunities for invaders, and there are examples of this both for terrestrial (Badano et al. 2007) and marine systems (Tweedley et al. 2008). Pollination and seed dispersal processes operate on very different spatial and temporal scales, as pollination is mainly dominated by invertebrates whereas seed dispersal is largely carried out by vertebrates. Therefore, restoration programmes that focus mainly on pollination may not necessarily favour the maintenance of seed dispersal interactions, and vice versa.

Insights from invasions can inform decisions on radical conservation strategies, such as managed relocation and 'rewilding' of ecosystems by introducing extant species, taxon substitutions, as functional replacements or ecological analogues for extinct native species. Using alien species as a management tool can in some cases provide beneficial interactions in systems where native mutualists have disappeared. For example, in some heavily degraded island systems, honeybees may act as pollinators of native species, thus contributing positively to the native plant fitness. In Mauritius, honeybees were the major flower visitors of 43 out of 74 plant species (58%) in a weeded conservation management area (Kaiser-Bunbury et al. 2010).

When trying to understand the dynamics of plant–animal interactions in an increasingly fragmented and homogenized world, we also need to consider the role of evolution. Some of the new interactions established between species in invaded communities can evolve quickly, sometimes over a few decades and across complex geographical landscapes (Thompson 2002). We need much better information on the geographical scales at which mutualistic interactions are organized ecologically and evolutionarily.

It is increasingly accepted by conservation managers that what we need to preserve is the 'interaction biodiversity' of communities. Interactions are the glue of biodiversity, linking species locally and over broader geographical scales. Entire community networks can collapse if such interactions weaken or if a fraction of them disappear. A network framework may improve many conservation efforts and help to take better management decisions. We need remnants of relatively pristine, geographically complex landscapes as benchmarks for interaction biodiversity (Thompson 1996). Such landscapes are becoming scarce; maintaining interaction biodiversity must therefore also require a better understanding of how to minimize the increasingly pervasive influence of the different drivers of global change, one of them being invasive species. This comes back to understanding species interactions and evolution of specialization in such interactions.

ACKNOWLEDGEMENTS

We thank Marcelo A. Aizen and Spencer C.H. Barrett for helpful comments on the manuscript and acknowledge the Spanish Ministry of Science (project CGL2007-61165BOS), the DST-NRF Centre of Excellence for Invasion Biology and the Hans Sigrist Foundation for financial support.

REFERENCES

Abe, T. (2006) Threatened pollination systems in native flora of the Ogasawara (Bonin) Islands. *Annals of Botany*, **98**, 317–334.

Agrawal, A.A., Ackerly, D.D., Adler, F., et al. (2007) Filling key gaps in population and community ecology. *Frontiers in Ecology and the Environment*, **5**, 145–152.

Aizen, M.A., Morales, C.L. & Morales, J.M. (2008) Invasive Mutualists Erode Native Pollination Webs. *PLoS Biology*, **6**, 396–403.

Baker, H.G. (1955) Self-compatibility and establishment of long-distance dispersal. *Evolution*, **9**, 337–349.

Badano, E.I., Villaroel, E., Bustamante R.O., Marquet, P.A. & Cavieres, L.A. (2007) Ecosystem engineers facilitate invasions by exotic plants in high-Andean ecosystems. *Journal of Ecology*, **95**, 682–688.

Bascompte, J., Jordano, P., Melian, C.J. & Olesen, J.M. (2003) The nestedness assembly of plant–animal mutualistic networks. *Proceedings of the National Academy of Sciences of the USA*, **100**, 9838–9837.

Bascompte, J., Jordano, P. & Olesen, J.M. (2006) Asymmetric coevolutionary networks facilitate biodiversity maintenance. *Science*, **312**, 431–433.

Bawa, K.S. (1990) Plant–pollinator interactions in tropical rain forests. *Annual Review of Ecology and Systematics*, **21**, 399–422.

Beckstead, J. & Parker, I.M. (2003) Invasiveness of *Ammophila arenaria*: Release from soil-borne pathogens? *Ecology*, **84**, 2824–2831.

Belote, R.T. & Jones, R.H. (2009) Tree leaf litter composition and nonnative earthworms influence plant invasion in experimental forest floor mesocosms. *Biological Invasions*, **11**, 1045–1052.

Bertness, M.D. & Callaway, R.M. (1994) Positive interactions in communities. *Trends Ecology and Evolution*, **9**, 91–193.

Bronstein, J.L. (2009) The evolution of facilitation and mutualism. *Journal of Ecology*, **97**, 1160–1170.

Brooker, R.W., Maestre, F.T., Callaway, R.M., Lortie, C.L., Cavieres, L.A. et al. (2008) Facilitation in plant communities: the past, the present, and the future. *Journal of Ecology*, **96**, 18–34.

Brown, B.J., Mitchell, R.J. & Graham, S.A. (2002) Competition for pollination between an invasive species (purple loosestrife) and a native congener. *Ecology*, **83**, 2328–2336.

Bruno, J.F., Stachowicz, J.J. & Bertness, M.D. (2003) Inclusion of facilitation into ecological theory. *Trends Ecology & Evolution*, **18**, 119–125.

Bruno, J.F., Fridley, J.D., Bromberg, K. & Bertness, M.D. (2005) Insights into ecology, evolution and biogeography. In *Species Invasions. Insights into Ecology, Evolution, and Biogeography* (ed. D.F. Sax, J.J. Stachowicz and S.D. Gaines), pp. 13–40, Sinauer Associates, Sunderland, Massachusetts.

Bulleri, F., Bruno, J.F. & Benedetti-Cecchi, L. (2008) Beyond competition: incorporating positive interactions between species to predict ecosystem invasibility. *PLoS Biology*, **6**, 1136–1140.

Burgess, K.S., Morgan, M. & Husband, B.C. (2008) Interspecific seed discounting and the fertility cost of hybridization in an endangered species. *New Phytologist*, **177**, 276–284.

Callaway, R. (1995) Positive interactions among plants. *Botanical Review*, **61**, 306–349.

Callaway, R.M., Thelen, G.C., Rodriguez, A. & Holben, W.E. (2004) Soil biota and alien plant invasion. *Nature*, **427**, 731–733.

Celebrezze, T. & Paton, D.C. (2004) Do introduced honeybees (*Apis mellifera*, Hymenoptera) provide full pollination service to bird-adapted Australian plants with small flowers? An experimental study of *Brachyloma ericoides* (Epacridaceae). *Austral Ecology*, **29**, 129–136.

Cheke, A. & Hume, J. (2008) *Lost Land of the Dodo. An Ecological History of Mauritius, Réunion & Rodrigues*. Yale University Press, New Haven, Connecticut.

Christian, C.E. (2001) Consequences of biological invasion reveal the importance of mutualism for plant communities. *Nature*, **413**, 635–639.

Cole, F.R., Medeiros, A.C., Loope, L.L. & Zuehlke, W.W. (1992) Effects of the argentine ant on arthropod fauna of Hawaiian high-elevation shrubland. *Ecology*, **73**, 1313–1322.

Collier, F.A. & Bidartondo, M.I. (2009) Waiting for fungi: the ectomycorrhizal invasion of lowland heathlands. *Journal of Ecology*, **97**, 950–963.

Cox, P.A. & Elmqvist, T. (2000) Pollinator extinction in the Pacific islands. *Conservation Biology*, **14**, 1237–1239.

D'Antonio, C.M. (1990) Seed production and dispersal in the non-native, invasive succulent *Carpobrotus edulis* (Aizoaceae) in coastal strand communities of central California. *Journal of Applied Ecology*, **27**, 693–702.

Darwin, C. (1859) *The Origin of Species*. John Murray, London.

Davis, M.A. (2009) *Invasion Biology*. Oxford University Press, Oxford.

Dick, C.W., Etchelecu, G., Austerlitz, F. (2003) Pollen dispersal of tropical trees (*Dinizia excelsa*: Fabaceae) by native insects and African honeybees in pristine and fragmented Amazonian rainforest. *Molecular Ecology*, **12**, 753–764.

Dupont, Y.L., Hansen, D.M., Valido, A. & Olesen, J.M. (2004) Impact of introduced honey bees on native pollination interactions of the endemic *Echium wildpretii* (Boraginaceae) on Tenerife, Canary Islands. *Biological Conservation*, **118**, 301–311.

Elton, C.S. (1958) *The Ecology of Invasions by Animals and Plants*. Methuen, London.

England, P.R., Beynon, F., Ayre, D.J. & Whelan, R.J. (2001) A molecular genetic assessment of mating system variation in a naturally bird-pollinated shrub: contributions from birds and introduced honeybees. *Conservation Biology*, **15**, 1645–1655.

Foster, J.T. & Robinson, S.K. (2007) Introduced birds and the fate of Hawaiian rainforests. *Conservation Biology*, **21**, 1248–1257.

Gardner, R.O. & Early, J.W. (1996) The naturalisation of banyan figs (*Ficus* spp. Moraceae) and their pollinating wasps (Hymenoptera: Agaonidae) in New Zealand. *New Zealand Journal of Botany*, **34**, 103–110.

García, J.D.D. (2002) Interaction between introduced rats and a frugivore bird–plant system in a relict island forest. *Journal of Natural History*, **36**, 1247–1258.

Geerts, S. & Pauw, A. (2009) African sunbirds hover to pollinate an invasive hummingbird-pollinated plant. *Oikos*, **118**, 573–579.

Ghazoul, J. (2002) Flowers at the front line of invasion? *Ecological Entomology*, **27**, 638–640.

Ghazoul, J. (2004) Alien abduction: disruption of native plant–pollinator interactions by invasive species. *Biotropica*, **36**, 156–164.

Goulson, D. (2003) Effects of introduced bees on native ecosystems. *Annual Review of Ecology and Systematics*, **34**, 1–26.

Grabas, G.P. & Laverty, T.M. (1999) The effect of purple loosestrife (*Lythrum salicaria* L.; Lythraceae) on the pollination and reproductive success of sympatric co-flowering wetland plants. *Ecoscience*, **6**, 230–242.

Grases, C. & Ramírez, N. (1998) Biología reproductiva de cinco especies ornitófilas en un fragmento de bosque caducifolio secundario en Venezuela. *Revista de Biología Tropical*, **46**, 1095–1108.

Griffen, B.D., Guy, T. & Buck, J.C. (2008) Inhibition between invasives: a newly introduced predator moderates the impacts of a previously established invasive predator. *Journal of Animal Ecology*, **77**, 32–40.

Gross, C.L. & Mackay, D. (1998) Honeybees reduce fitness in the pioneer shrub *Melastoma affine* (Melastomataceae). *Biological Conservation*, **86**, 169–178.

Grosholz, E.D. (2005) Recent biological invasion may hasten invasional meltdown by accelerating historical introductions. *Proceedings of the National Academy of Sciences of the USA*, **102**, 1088–1091.

Hansen, D.M., Olesen, J.M. & Jones, C.G. (2002) Trees, birds and bees in Mauritius: exploitative competition between introduced honey bees and endemic nectarivorous birds? *Journal of Biogeography*, **29**, 721–734.

Hansen, D.M. & Müller, C.B. (2009) Invasive ants disrupt gecko pollination and seed dispersal of the endangered plant *Roussea simplex* in Mauritius. *Biotropica*, **41**, 202–208.

Harmon-Threatt, A.N., Burns, J.H., Shemyakina, L.A. & Knight, T.M. (2009) Breeding system and pollination ecology of introduced plants compared to their native relatives. *American Journal of Botany*, **96**, 1544–1550.

Harrod, R.J. & Taylor, R.J. (1995) Reproduction and pollination biology of *Centaurea* and *Acroptilon* species, with emphasis on *Centaurea diffusa*. *Northwest Science*, **69**, 97–105.

Hingston, A.B. (2005) Does the introduced bumblebee, *Bombus terrestris* (Apidae), prefer flowers of introduced or native plants in Australia? *Australian Journal of Zoology*, **53**, 29–34.

Hingston, A.B., Marsden-Smedley, J. & Driscoll, D.A., et al. (2002) Extent of invasion of Tasmanian native vegetation by the exotic bumblebee *Bombus terrestris* (Apoidea: Apidae). *Austral Ecology*, **27**, 162–172.

Iponga, D.M., Milton, S.J. & Richardson, D.M. (2008) Superiority in competition for light: A crucial attribute defining the impact of the invasive alien tree *Schinus molle* (Anacardiaceae) in South African savanna. *Journal of Arid Environments*, **72**, 612–623.

Jakobsson, A., Padrón, B. & Traveset, A. (2009) Competition for pollinators between invasive and native plants – the importance of spatial scale of investigation. *Ecoscience*, **16**, 138–141.

Jordano, P. (1987) Patterns of mutualistic interactions in pollination and dispersal: connectance, dependence asymmetries, and coevolution. *American Naturalist*, **129**, 657–677.

Jourdan, H., Sadlier, R.A. & Bauer, A.M. (2001) Little fire ant invasion (*Wasmannia auropunctata*) as a threat to New Caledonian lizards: evidences from a sclerophyll forest (Hymenoptera: Formicidae). *Sociobiology*, **38**, 283–301.

Kaiser-Bunbury, C.N., Traveset, A. & Hansen, D.M. (2010) Conservation and restoration of plant–animal mutualisms on oceanic islands. *Perspectives in Plant Ecology, Evolution and Systematics*, **12**, 131–143.

Kato, M., Shibata, A. & Yasui, T. (1999) Impact of introduced honeybees, *Apis mellifera*, upon native bee communities in the Bonin (Ogasawara) Islands. *Researches on Population Ecology*, **41**, 217–228.

Kawakami, K., Mizusawa, L. & Higuchi, H. (2009) Re-established mutualism in a seed-dispersal system consisting of native and introduced birds and plants on the Bonin Islands, Japan. *Ecological Research*, **24**, 741–748.

Keane, R.M. & Crawley, M.J. (2002) Exotic plant invasions and the enemy release hypothesis. *Trends in Ecology & Evolution*, **17**, 164–170.

Kelly, D., Robertson, A.W., Ladly, J.J., Anderson, S.H. & McZenzie, R.J. (2006) Relative (un)importance of introduced animals as pollinators and disperses of native plants. *Biological invasions in New Zealand* (ed. R.B. Allen and W.G. Lee), pp 227–245, Springer, Berlin.

Kottke, I., Haug, I., Setaro, S. et al. (2008) Guilds of mycorrhizal fungi and their relation to trees, ericads, orchids and liverworts in a neotropical mountain rain forest. *Basic and Applied Ecology*, **9**, 13–23.

Kueffer, C., Kronauer, L. & Edwards, P.J. (2009) Wider spectrum of fruit traits in invasive than native floras may increase the vulnerability of oceanic islands to plant invasions. *Oikos*, **118**, 1327–1334.

Larson, D.L., Royer, R.A. & Royer, M.R. (2006) Insect visitation and pollen deposition in an invaded prairie plant community. *Biological Conservation*, **130**, 148–159.

Linnebjerg, J.F., Hansen, D.M. & Olesen, J.M. (2009) Gut passage effect of the introduced red-whiskered bulbul (*Pycnonotus jocosus*) on germination of invasive plant species in Mauritius. *Austral Ecology*, **34**, 272–277.

Lopezaraiza-Mikel, M.E., Hayes, R.B., Whalley, M.R., et al. (2007) The impact of an alien plant on a native plant–pollinator network: An experimental approach. *Ecology Letters*, **10**, 539–550.

López-Darias, M. & Nogales, M. (2008) Effects of the invasive Barbary ground squirrel (*Atlantoxerus getulus*) on seed dispersal systems of insular xeric environments. *Journal of Arid Environments*, **72**, 926–939.

MacDougall, A.S., Gilbert, B. & Levine, J.M. (2009) Plant invasions and the niche. *Journal of Ecology*, **97**, 609–615.

Madjidian, J.A., Morales, C.L. & Smith, H.G. (2008) Displacement of a native by an alien bumblebee: lower pollinator efficiency overcome by overwhelmingly higher visitation frequency. *Oecologia*, **156**, 835–845.

Mal, T.K., Lovettdoust, J. & Lovedoust, L. et al. (1992) The biology of Canadian weeds.100. *Lythrum salicaria*. *Canadian Journal of Plant Science*, **72**, 1305–1330.

Mandon-Dalger, I., Clergeau, P., Tassin, J., Riviere, J.N. & Gatti, S. (2004) Relationships between alien plants and an alien bird species on Réunion Island. *Journal of Tropical Ecology*, **20**, 635–642.

McKinney, S.T., Fiedler, C.E. & Tomback, D.F. (2009) Invasive pathogen threatens bird–pine mutualism: implications for sustaining a high-elevation ecosystem. *Ecological Applications*, **19**, 597–607.

Memmott, J. & Waser, N. (2002) Integration of alien plants into a native flower–pollinator visitation web. *Proceedings of the Royal Society of London B*, **269**, 2395–2399.

Mitchell, C.E. et al. (2006) Biotic interactions and plant invasions. *Ecology Letters*, **9**, 726–740.

Milton, S.J., Wilson, J.R.U., Richardson, D.M., et al. (2007) Invasive alien plants infiltrate bird-mediated shrub nucleation processes in arid savanna. *Journal of Ecology*, **95**, 648–661.

Moragues, E. & Traveset, A. (2005) Effect of *Carpobrotus* spp. on the pollination success of native plant species of the Balearic Islands. *Biological Conservation*, **122**, 611–619.

Morales, C.L. (2007) Introducción de abejorros (*Bombus*) no nativas: causas, consecuencias ecológicas y perspectivas. *Ecología Austral*, **17**, 51–65.

Morales, C.L. & Aizen, M.A. (2002) Does the invasion of alien plants promote invasion of alien flower visitors? A case study from the temperate forests of southern Andes. *Biological Invasions*, **4**, 87–100.

Morales, C.L. & Aizen, M.A. (2006) Invasive mutualisms and the structure of plant–pollinator interactions in the temperate forests of north-west Patagonia, Argentina. *Journal of Ecology*, **94**, 171–180.

Morales, C.L. & Traveset, A. (2008) Interspecific pollen transfer: Magnitude, prevalence and consequences for plant fitness. *Critical Reviews in Plant Sciences*, **27**, 221–238.

Morales, C. & Traveset, A. (2009) A meta-analysis of impacts of alien vs. native plants on pollinator visitation and reproductive success of co-flowering native plants. *Ecology Letters*, **12**, 716–728.

Moeller, D.A. (2005) Pollinator community structure and sources of spatial variation in plant–pollinator interactions in *Clarkia xantiana* ssp *xantiana*. *Oecologia*, **142**, 28–37.

Nattero, J. & Cocucci, A.A. (2007) Geographic variation in floral traits of tree tobacco in relation to its hummingbird pollinator fauna. *Biological Journal of the Linnean Society*, **90**, 657–667.

Nogales, M., Martín, A., Tershy, B.R., et al. (2004) A review of feral cat eradication on islands. *Conservation Biology*, **18**, 310–319.

Nogales, M., Nieves, C., Illera, J.C., Padilla, D.P. & Traveset, A. (2005) Effect of native and alien vertebrate frugivores on seed viability and germination patterns of *Rubia fruticosa* (Rubiaceae) in the eastern Canary Islands. *Functional Ecology*, **19**, 429–436.

Nogales, M., Rodríguez-Luengo, J.L. & Marrero, P. (2006) Ecological effects and distribution of invasive non-native mammals on the Canary Islands. *Mammal Review*, **36**, 49–65.

Nuñez, M.A., Horton, T.R. & Simberloff, D. (2009) Lack of belowground mutualisms hinders Pinaceae invasions. *Ecology*, **90**, 2352–2359.

Olesen, J.M., Eskildsen, L.I. & Venkatasamy, S. (2002) Invasion of pollination networks on oceanic islands: importance of invasive species complexes and endemic super generalists. *Diversity and Distributions*, **8**, 181–192.

Olesen, J.M. & Jordano, P. (2002) Geographic patterns in plant-pollinator mutualistic networks. *Ecology*, **83**, 2416–2424.

Ollerton, J., Johnson S.D., Cramer, L. & Kellie, S. (2003) The pollination ecology of an assemblage of grassland asclepiads in South Africa. *Annals of Botany*, **92**, 807–834.

Olyarnik, S.V., Bracken, M.E.S., Byrnes, J.E., Hughes, A.R., Hulgren, K.M. & Stachowicz, J.J. (2008) Ecological factors affecting community invasibility. In *Biological Invasions of Marine Ecosystems: Ecological, Management, and Geographic Perspectives* (ed. G. Rilov and J.A. Crooks), pp. 215–240. Springer, Berlin.

Padrón, B., Traveset, A., Biedenweg, T., Diaz, D., Olesen, J.M. & Nogales, M. (2009) Impact of invasive species in the pollination networks of two different archipelagos. *PLoS One*, **4**(7), e6275.

Parker, I.M., Engel, A., Haubensak, K.A. & Goodell, K. (2002) Pollination of *Cytisus scoparius* (Fabaceae) and *Genista mon-*

spessulana (Fabaceae), two invasive shrubs in California. *Madroño*, **49**, 25–32.

Paton, D.C. (1993) Honeybees in the Australian environment: does *Apis mellifera* disrupt or benefit the native biota? *Bioscience*, **43**, 95–103.

Procheş, Ş., Wilson, J.R.U., Richardson, D.M. & Rejmánek, M. (2008). Searching for phylogenetic pattern in biological invasions. *Global Ecology and Biogeography*, **17**, 5–10.

Rambuda, T.D. & Johnson, D. (2004) Breeding systems of invasive alien plants in South Africa: does Baker's rule apply? *Diversity and Distributions*, **10**, 409–416.

Rathcke, B. (1983) Competition and facilitation among plants for pollination. In *Pollination Biology* (ed. L. Real), pp. 305–327. Academic Press.

Reinhart, K.O. & Callaway, R.M. (2006) Soil biota and invasive plants. *New Phytologist*, **170**, 445–457.

Rejmánek, M. (1996) A theory of seed plant invasiveness: the first sketch. *Biological Conservation*, **78**, 171–181.

Rejmánek, M. (1998) Invasive plants and invasible ecosystems. In *Invasive Species and Biodiversity Management* (ed. O.T. Sandlund, P.J. Schei and A. Viken), pp. 79–102. Kluwer, Dordrecht.

Richardson, D.M., Allsopp, N., D'Antonio, C., Milton, S.J. & Rejmánek, M. (2000a) Plant invasions – the role of mutualism. *Biological Reviews*, **75**, 65–93.

Richardson, D.M., Pyšek, P., Rejmánek, M., Barbour, M.G., Panetta, D.F. & West, C.J. (2000b) Naturalization and invasion of alien plants: concepts and definitions. *Diversity and Distributions*, **6**, 93–107.

Richardson, D.M., Rouget, M., Ralston, S.J., et al. (2005) Species richness of alien plants in South Africa: Environmental correlates and the relationship with indigenous plant species richness. *Ecoscience*, **12**, 391–402.

Riera, N., Traveset, A. & García, O. (2002) Breakage of mutualisms in islands caused by exotic species: the case of *Cneorum tricoccon* in the Balearics (Western Mediterranean Sea). *Journal of Biogeography*, **29**, 713–719.

Rodríguez-Cabal, M.A., Stuble, K.L., Núñez, M.A. & Sanders, N.J. (2009) Quantitative analysis of the effects of the exotic Argentine ant on seed-dispersal mutualisms. *Biology Letters*, **5**, 499–502.

Roubik, D.W. & Villanueva-Gutierrez, R. (2009) Invasive Africanized honey bee impact on native solitary bees: a pollen resource and trap nest analysis. *Biological Journal of the Linnean Society*, **98**, 152–160.

Sallabanks, R. (1993) Fruiting plant attractiveness to avian seed dispersers: native vs. invasive *Crataegus* in Western Oregon. *Madroño*, **40**, 108–116.

Savage, A.M., Rudgers, J.A. & Whitney, K.D. (2009) Elevated dominance of extra floral nectary-bearing plants is associated with increased abundances of an invasive ant and reduced native ant richness. *Diversity and Distributions*, **15**, 751–761.

Sax, D.F., Gaines, S.D. & Brown, J.H. (2002) Species invasions exceed extinctions on islands worldwide: a comparative study of plants and birds. *American Naturalist*, **160**, 766–783.

Sax, D.F. & Gaines, S.D. (2008) Species invasions and extinction: the future of native biodiversity on islands. *Proceedings of the National Academy of Sciences of the USA*, **105**, 11490–11497.

Schemske, D.W., Mittelbach, G.G., Cornell, H.V., Sobel, J.M. & Roy, K. (2009) Is there a latitudinal gradient in the importance of biotic interactions? *Annual Review of Ecology, Evolution, and Systematics*, **40**, 245–69.

Schmidt, S.K. & Scow, K.M. (1986) Mycorrhizal fungi on the Galapagos Islands. *Biotropica*, **18**, 236–240.

Simberloff, D. (1986) Introduced insects: a biogeographic and systematic perspective. In *Ecology of Biological invasions of North America and Hawaii* (ed. H.A. Mooney and J.A. Drake), pp. 3–26. Springer, New York.

Simberloff, D. (2009) The role of propagule pressure in biological invasions. *Annual Review of Ecology, Evolution, and Systematics*, **40**, 81–102.

Simberloff, D. & von Holle, B. (1999) Positive interactions of non-indigenous species: invasional meltdown? *Biological Invasions*, **1**, 21–32.

Stohlgren, T.J., Binkley, D., Chong, G.W., et al. (1999) Exotic plant species invade hot spots of native plant diversity. *Ecological Monographs*, **69**, 25–46.

Stout, J.C. & Morales, C.L. (2009) Ecological impacts of invasive alien species on bees. *Apidologie*, **40**, 388–409.

Thompson, J.N. (1996) Conserving interaction diversity. In *The Ecological Basis of Conservation: Heterogeneity, ecosystems, and biodiversity* (ed. S.T.A. Pickett, R.S. Ostfeld, M. Shachak and G.E. Likens), pp. 285–293, Chapman & Hall, New York.

Thompson, J.N. (2002) Plant–animal interactions: future directions. In *Plant–Animal Interactions. An Evolutionary Approach* (ed. C.M. Herrera and O. Pellmyr), pp. 236–247, Blackwell, Oxford.

Thuiller, W., Gallien, L., Boulangeat, I., et al. (2010) Resolving Darwin's naturalization conundrum: a quest for evidence. *Diversity and Distributions*, **16**, 461–475.

Towns, D.R., Atkinson, I.A.E. & Daugherty, C.H. (2006) Have the harmful effects of introduced rats on islands been exaggerated? *Biological Invasions*, **8**, 863–891.

Traveset, A. (1995) Reproductive ecology of *Cneorum tricoccon* L. (Cneoraceae) in the Balearic Islands. *Botanical Journal of the Linnean Society*, **117**, 221–232.

Traveset, A. & Riera, N. (2005) Disruption of a plant–lizard seed dispersal system and its ecological consequences for a threatened endemic plant in the Balearic Islands. *Conservation Biology* **19**, 421–431.

Traveset, A. & Richardson, D.M. (2006) Biological invasions as disruptors of plant reproductive mutualisms. *Trends in Ecology & Evolution*, **21**, 208–216.

Traveset, A., Nogales M. & Navarro, L. (2009a) Mutualismos planta–animal en islas: influencia en la evolución y mantenimiento de la biodiversidad. In *Ecología y Evolución de Interacciones Planta–Animal: Conceptos y Aplicaciones* (ed. R. Medel, M.A. Aizen and R. Zamora), pp. 157–180. Editorial Universitaria, Santiago, Chile.

Traveset, A., Nogales, M., Alcover, J.A., et al. (2009b) A review on the effects of alien rodents in the Balearic (W Mediterranean) and Canary Islands (Atlantic Ocean). *Biological Invasions*, **11**, 1653–1670.

Trakhtenbrot, A., Nathan, R., Perry, G. & Richardson, D.M. (2005) The importance of long-distance dispersal in biodiversity conservation. *Diversity and Distributions*, **11**, 173–181.

Tweedley, J.R., Jackson, E.L. & Attrill, M.J. (2008) *Zostera marina* seagrass beds enhance the attachment of the invasive alga *Sargassum muticum* in soft sediments. *Marine Ecology Progress Series*, **354**, 305–309.

Valdovinos, F.S., Ramos-Jiliberto, R., Flores, J.D., Espinoza, C. & Lopez, G. (2009) Structure and dynamics of pollination networks: the role of alien plants. *Oikos*, **118**, 1190–1200.

Valiente-Banuet, A., Rumebe, A.V., Verdú, M. & Callaway, R.M. (2006) Modern quaternary lineages promotes diversity through facilitation of ancient Tertiary lineages. *Proceedings of the National Academy of Sciences of the USA*, **103**, 16812–16817.

van der Heijden, M.G.A. & Horton, T.R. (2009) Socialism in soil? The importance of mycorrhizal fungal networks for facilitation in natural ecosystems. *Journal of Ecology*, **97**, 1139–1150.

Vamosi, J.C., Knight, T.M., Steets, J.A., Mazer, S.J., Burd, M. & Tia-Lynn, A. (2006) Pollination decays in biodiversity hotspots. *Proceedings of the National Academy of Sciences of the USA*, **103**, 956–961.

Vázquez, D.P. (2005) Exploring the relationship between niche breadth and invasion success. In *Conceptual Ecology and Invasions Biology* (ed. M.W. Cadotte, S.M. McMahon and T. Fukami), pp. 317–332. Springer, Berlin.

Vázquez, D.P. & Aizen, M.A. (2004) Asymmetric specializations: a pervasive feature of plant–pollinator interactions. *Ecology*, **85**, 1251–1257.

Vilà, M., Bartomeus, I., Dietzsch, A.C., et al. (2009) Invasive plant integration into native plant–pollinator networks across Europe. *Proceedings of the Royal Society B*, **276**, 3887–3893.

Waring, G.H., Loope, L.L & Medeiros, A.C. (1993) Study on the use of alien versus native plants by nectarivorous forest birds on Maui, Hawaii. *Auk*, **110**, 917–920.

Waser, N.M., Chittka, L., Price, M.V., Williams, N.M. & Ollerton, J. (1996) Generalization in pollination systems and why it matters. *Ecology*, **77**, 1043–1060.

Wenner, A.M. & Thorp, R.W. (1994) Removal of feral honey bee (*Apis mellifera*) colonies from Santa Cruz Island. *Fourth California Islands Symposium: Update on the Status of Resources* (ed. W. L. Halvorson and G.L. Maender), pp. 513–522, Santa Barbara Museum of Natural History, Santa Barbara, California.

Westerkamp, C. (1991) Honeybees are poor pollinators – why? *Plant Systematics and Evolution*, **177**, 71–75.

Wolf, D.E., Takebayashi, N. & Rieseberg, L.H. (2001) Predicting the risk of extinction through hybridization. *Conservation Biology*, **15**, 1039–1053.

Wolfe, B.E., Husband, B.C. & Klironomos, J.N. (2005) Effects of a belowground mutualism on an aboveground mutualism. *Ecology Letters*, **8**, 218–223.

Yoon, I., Williams, R.J., Levine, E., Yoon, S., Dunne, J.A. & Martinez, N.D. (2004) Webs on the web (WoW): 3D visualization of ecological networks on the WWW for collaborative research and education. *Proceedings of the IS&T/SPIE Symposium on electronic Imaging, Visualization and Date Analysis*, **5295**, 124–132.

Chapter 13

FIFTY YEARS ON: CONFRONTING ELTON'S HYPOTHESES ABOUT INVASION SUCCESS WITH DATA FROM EXOTIC BIRDS

Tim M. Blackburn[1], Julie L. Lockwood[2] and Phillip Cassey[3]

[1]Institute of Zoology, Zoological Society of London, Regent's Park, London NW1 4RY, UK
[2]Department of Ecology, Evolution and Natural Resources, Rutgers University, New Brunswick, NJ 08901-8551, USA
[3]School of Biosciences, Birmingham University, Edgbaston, UK

Fifty Years of Invasion Ecology: The Legacy of Charles Elton, 1st edition. Edited by David M. Richardson

13.1   INTRODUCTION

Charles Elton was by no means the first person to recognize the importance of studying invasive species, but has come to be seen as perhaps the most significant of the founding fathers of this field (Cadotte 2006; Richardson & Pyšek 2007, 2008; but see Simberloff, this volume, for a counterpoint). Elton is justly renowned within the natural sciences for his work on animal ecology: his classic book on the subject (Elton 1927) identifies many of the issues that are still of core relevance today (Leibold & Wootton 2001). It is perhaps no surprise then that as the need to understand species invasions has grown in importance, as their distribution and abundance across the environment increase, researchers will have been more likely to come across his monograph on the subject (Elton 1958) than those of other early writers (see, for example, Baker & Stebbins 1965). In doing so they will have encountered an accessible and stimulating introduction to the topic; this combination of serendipity and accessibility may explain Elton's apparent influence. Nevertheless, whatever the cause, Elton (1958) has attained the position as 'one of the central scientific books of the [twentieth] century' (Quammen 1999).

The modern reader may well be struck by the extent to which case studies of specific invasions dominate Elton's monograph. This focus perhaps reflects the volume's genesis in a series of popular radio lectures, as well as the level of knowledge about invasions at the time he was writing (see Pyšek & Hulme, this volume, for discussion). Yet, Elton (1958) is probably most frequently cited as a source of hypotheses for the general mechanisms by which invasions occur (Richardson & Pyšek 2008). For example, of the first 10 references we were able to access from a Google Scholar search (June 2009; http://scholar.google.co.uk/) for 'Elton C 1958', six cited him for the hypothesis that more species-rich communities are harder to invade (Holway 1998; Lonsdale 1999; Rozdilsky & Stone 2001; Hewitt 2002; Byers & Noonburg 2003; Cleland et al. 2004; see Fridley, this volume, for discussion), and one for the related idea that continents are harder to invade than islands (Daehler & Carino 2000). We have ourselves cited Elton (1958) for these two hypotheses (see, for example, Blackburn et al. 2009a). Other hypotheses about invasion success presented in Elton (1958) are that invasions fail because they meet with biotic resistance, that invasions are more likely to occupy cultivated or modified land, and that invasive organisms replace or reduce the numbers of native species. These are all essentially ideas about diversity processes: the influence of native diversity on invasion success, or of successful invaders on native diversity (Fridley, this volume).

Since *The Ecology of Invasions by Animals and Plants* was first published, however, there has been a revolution in our understanding of how the natural world works. Elton (1958) is often argued to be highly influential in the discipline of invasion biology, but 50 years on, how do the ideas he presented about invasions stand up to modern scrutiny? Here, we explore how the core mechanistic hypotheses presented by Elton (1958) fit with our understanding of the invasion process in one specific taxon: birds. We confront a series of five statements taken from Elton (1958), referenced to page number, with our modern understanding of the situation for bird invasions. We focus on birds because this group presents as good information as we have on most aspects of the invasion process (as well as on their ecology and evolution more widely), and for a large and varied set of introduction events (Blackburn et al. 2009a). Birds thus represent an invaluable resource for analysing the invasion process. We conclude the chapter by briefly considering whether bird invasions might be unusual in comparison to other taxa.

13.2   STATEMENT 1 (PAGE 145): 'I WILL NOW TRY TO SET OUT SOME OF THE EVIDENCE THAT THE BALANCE OF RELATIVELY SIMPLE COMMUNITIES OF PLANTS AND ANIMALS IS MORE EASILY UPSET THAN THAT OF RICHER ONES ... AND MORE VULNERABLE TO INVASIONS.'

The predominance of case studies means that it is not until almost the end of the book that Elton (1958) presents the mechanistic hypotheses about the drivers of the invasion process for which he is most commonly cited. These, and indeed almost all of Elton's statements about mechanism, relate to the idea that biotic interactions determine the probability of invasion, and that the fewer species composing a native assemblage, the more likely it is to be invaded. This focus is interesting for at least four reasons.

First, although he does not directly cite it, Elton's (1958) ideas about invasions clearly relate to

MacArthur's (1955) work on the relationship between diversity and stability (see also Levine & D'Antonio 1999; Fridley, this volume), and the idea that simpler food webs are more subject to population fluctuations if one link is disrupted. This idea generated hot debate through the 1970s and 1980s (from May (1973) to Pimm (1991)), and continues to be of interest, especially in relation to the potential consequences of elevated extinction rates (McCann 2000). However, stability does not necessarily correlate with invasibility. Moreover, Elton's language and examples refer to invader impacts rather than establishment probability, and hence do not refer to the effects of diversity on invasibility per se (Levine & D'Antonio 1999). In what follows, we will tend to interpret Elton's ideas in terms of invasibility directly, with invasibility usually meant as the susceptibility of a community to the establishment of an exotic species, although sometimes as the susceptibility to exotic spread. This use seems to us to be more in line with interpretations of Elton's ideas in the recent invasion biology literature.

Second, although the bulk of Elton's book comprises examples of how human activities have facilitated the spread of invasive organisms, Elton makes no mechanistic connection between the magnitude of these activities and vulnerability to invasions. Yet, data from birds increasingly suggest that the richness of exotic species in large measure depends on human activities, through their contribution to the dual influences of colonization pressure and propagule pressure (Blackburn et al. 2008; Chiron et al. 2009).

Colonization pressure is defined as the number of exotic species introduced into a single location, some of which may succeed in establishing and some of which will not (Blackburn et al. 2009a; Lockwood et al. 2009). The likely influence of colonization pressure on patterns of exotic species richness was first clearly stated by Case (1996), who noted that the number of exotic species present in an avifauna cannot exceed the number introduced. In fact, post-establishment spread from other locations of introduction means that the truth of this statement is likely to be scale dependent. Nevertheless, in general we would expect locations with more species introduced to them to be those with higher exotic species richness. Moreover, we would also expect this relationship even if the persistence of exotic introductions were random with respect to location. The correlation between the number of exotic species established y and the number introduced is a relationship between y and $(x + y)$,

where x is the number of species introduced that fail to establish, which produces a spurious positive correlation (Brett 2004; Lockwood et al. 2009).

If human activities may largely explain the number of species established at a location though colonization pressure, they may also largely explain the identities of the species established, through the influence of propagule pressure. Propagule pressure is defined as the total number of individuals of a species introduced at a given location (Williamson 1996). Not all individuals are necessarily introduced in one go, and so propagule pressure can be viewed as having two components: the number of introduction events (propagule number) and the number of individuals per introduction (propagule size) (Pimm 1991; Carlton 1996; Veltman et al. 1996; Lockwood et al. 2005). Establishment success has been shown consistently to be a positive function of propagule pressure or its components, and indeed propagule pressure is usually the strongest predictor of success in any given situation (Cassey et al. 2004a, 2005a; Lockwood et al. 2005; Colautti et al. 2006; Hayes & Barry 2008). Its importance parallels the small-population paradigm for conservation biology, with more individuals presumably buffering introduced populations against the problems of demographic, environmental and genetic stochasticities, and Allee effects (Blackburn et al. 2009a). Colonization pressure and propagule pressure may often be positively related, through well-known relationships between the number of individuals and species in samples of assemblages (Preston 1948; Lockwood et al. 2009), leading to additive effects on exotic species richness.

Third, Elton's (1958) focus is predominantly on the influence of the exotic location on the invasion process. It is now recognized, at least in the context of bird invasions, that there are three types of characteristic that influence success: those relating to location (as emphasized by Elton), to the species, and to the individual introduction event (Duncan et al. 2003). The influence of event-level characteristics, of which propagule pressure is the most important, has only relatively recently been recognized (largely following Williamson (1996)). It is more surprising that Elton gives so little consideration to species-level traits in driving invasion success. It may be that this reflects his view of ecology, which on the basis of his classic textbook (Elton 1927) is much more focused on population- and community-level interactions between organisms (and between organisms and the environment) than it is on the identities of the organisms themselves. Yet, there is now accumulating evidence that these can influence all stages of the invasion process in birds, from determining which species get chosen for transport and introduction (Cassey et al. 2004b), through which species succeed in establishing (Cassey et al. 2004a; Sol et al. 2005a; Blackburn et al. 2009b), to the extent of post-establishment spread (Duncan et al. 1999, 2001).

Fourth, as we will see in subsequent sections, Elton's (1958) views of the effects of the location on invasion success largely relate to how interspecific interactions might retard invasions. In fact, there are three broad aspects of the environment that may influence invasion success: interspecific interactions, the availability of resources and the physical environment (Shea & Chesson 2002). Elton was well aware of the importance of all three aspects to the structure and function of ecological systems (Elton 1927), but the last two barely figure in his treatment of the invasion process. Elton's poor treatment of the last looks like a particularly glaring omission as we write this chapter in 2009, when the response of species distributions to climate change is a major area of research, and the responses of exotic species to different climates a major strand in this work (Duncan et al. 2009).

13.3 STATEMENT 2 (PAGE 147): 'THE NATURAL HABITATS ON SMALL ISLANDS SEEM TO BE MUCH MORE VULNERABLE TO INVADING SPECIES THAN THOSE ON THE CONTINENTS. THIS IS ESPECIALLY SO ON OCEANIC ISLANDS, WHICH HAVE RATHER FEW INDIGENOUS SPECIES.'

An obvious prediction that arises out of the view that simple communities should be more easily invaded than complex ones is that islands should be more easily invaded than continents. Elton (1958) couched this view in terms of the relative invasibility of natural habitats, but this element is often omitted in treatments of the prediction. In fact, a strict test of the proposition as stated by Elton (1958) is difficult, as human impacts mean that any habitat in either type of location will be only more or less natural. The prediction arises both because island habitats tend to contain fewer species than mainland habitats, as Elton noted, but also because island species are often viewed as competitively inferior (Sol 2000). It is certainly true that islands, especially oceanic islands, tend to have more species of exotic bird established than do equivalent areas of the continental mainland (Fig. 13.1).

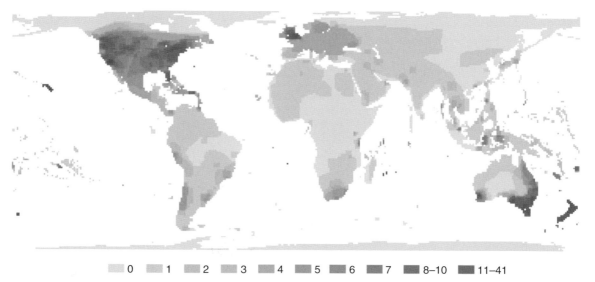

0   1   2   3   4   5   6   7   8–10   11–41

Fig. 13.1 An estimate of the number of exotic bird species found in equal area grid cells across the land surface of the world (reprinted with permission of Oxford University Press, from Blackburn et al. (2009a)). The data were compiled by M. Parnell from the maps in Long (1981), and visualized by N. Pettorelli.

However, differences in the numbers of exotic species across sites do not necessarily indicate that the richer sites are more invasible, for at least three reasons.

First, differences in numbers of exotic species may be a consequence of differences in colonization pressure, rather than invasibility. For example, deliberate introductions have resulted in 26 exotic bird species becoming established on the South Island of New Zealand (Blackburn et al. 2008), versus nine in South Africa (Dean 2000). However, the 26 South Island exotics derive from 74 introduced species, versus 21 such species for South Africa. Thus 43% of the introductions to continental South Africa succeeded in establishing, versus 35% of the introductions to South Island. Accounting for colonization pressure, the island is actually *less* invasible in this case.

Colonization pressure alone is also sufficient to explain patterns of variation in exotic species richness across islands. For example, the relationship between the number of exotic bird species established on islands and island area ($z = 0.18$) simply mirrors the species–area relationship for introduced (i.e. failed plus successful) exotics ($z = 0.20$), with the intercept lowered to reflect the fact that establishment success is less than 100% (Blackburn et al. 2008). Larger islands have more exotic bird species because more were introduced. An interesting unanswered question is why bird species were introduced in a manner that reasonably well mimics natural species–area relationships for islands, which tend to have slopes in the range 0.25–0.35 (Rosenzweig 1995).

Second, differences in numbers of exotic species may be a consequence of differences in the characteristics of the species introduced, rather than invasibility. As noted earlier, species differ in traits that influence the probability that they will establish and spread. Islands might appear more invasible if they have predominantly received species with traits that promote establishment. Sol (2000) addressed this problem by comparing the success of the same bird species when introduced to islands or mainlands, in this case New Zealand versus Australia, and the Hawaiian Islands versus mainland USA. Of the 42 species introduced to both Australia and New Zealand, 18 species succeeded in both locations, and 16 failed in both. Of the eight species with mixed success, five succeeded in New Zealand but not Australia, and three vice versa. Similar results pertained for the comparison between the Hawaiian Islands and mainland USA. Thus Sol (2000) found no evidence that islands were easier to invade.

Third, differences in numbers of exotic species may be a consequence of consistent differences in event-level rather than location-level characteristics. For example, the primary determinant of establishment success is propagule pressure: if this tended to be lower for introductions to continental mainlands, islands could appear more invasible without any necessary location-level differences in invasibility. In fact, Cassey et al. (2004a) found that propagule pressure was significantly *higher* on mainlands than islands, which could have confounded previous analyses that failed to find that islands were more invasible (see, for example, Sol 2000; Blackburn & Duncan 2001; Cassey 2003). Nevertheless, Cassey et al. (2004a) assessed the causes of exotic bird establishment success at the global scale in an analysis that accounted for propagule pressure and a variety of other species- and location-level characteristics, and found no significant effect of island location on success.

There is in fact no good evidence that establishment success, at least, is higher for exotic bird species introduced to islands versus mainlands (Sol 2000; Blackburn & Duncan 2001; Cassey 2003; Cassey et al. 2004a; Sol et al. 2005b). The high numbers of exotic bird species found on islands appear to be a consequence of the many attempts to introduce birds to islands, rather than any inherent feature of islands that makes them easier to invade (Simberloff 1995).

13.4 STATEMENT 3 (PAGES 148–149): 'THE FIFTH LINE OF EVIDENCE COMES FROM THE TROPICS … THERE ARE ALWAYS ENOUGH ENEMIES AND PARASITES AVAILABLE TO TURN ON ANY SPECIES THAT STARTS BEING UNUSUALLY NUMEROUS.'

Another obvious prediction that arises out of the view that simple communities should be more easily invaded than complex ones is that the tropics should be hard to invade. Elton (1958) presents this idea in terms of the dearth of outbreaks of insect pests in tropical forests, making only an implied comparison to forests at other latitudes, and saying nothing about the ability or otherwise of those pests to establish in tropical forest locations. Nevertheless, a more general link is often drawn in the invasion literature between the likelihood of invasion success at any invasion stage and whether the location of introduction is in the temperate or

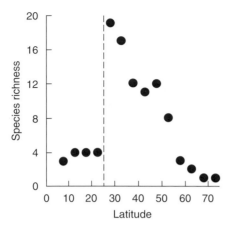

Fig. 13.2 The relationship between species richness and latitude for exotic bird species established in North America. The dashed line indicates the Tropic of Cancer. Reprinted from Sax (2001) with permission of Wiley-Blackwell.

tropical zones, or between invasion success and latitude (see, for example, Holdgate 1986; Rejmánek 1996; Blackburn & Duncan 2001; Sax 2001; Sax & Gaines 2006).

There is a definite dearth of exotic bird species established at tropical relative to temperate latitudes, at least on continental mainlands (see, for example, Fig. 13.1). For North America, Sax (2001) shows that the number of exotic bird species increases as latitude decreases, but only as far as the Tropic of Cancer: few exotic bird species are found within the North American tropics (Fig. 13.2). Once again, however, the influence of colonization pressure imposes itself: the tropics may house few exotic bird species because few have been introduced there. Sax (2001) argues that the pattern illustrated in Fig. 13.2 is not a consequence of the number of species introduced but is instead to be expected if contemporary biotic conditions are excluding exotic species. However, global analyses of bird introductions have failed to find a relationship between establishment success and latitude of introduction (Blackburn & Duncan 2001; Cassey et al. 2004a), implying that either colonization pressure probably does play a role or that our knowledge of establishment success across different latitudes is biased. Moreover, although there is evidence that avian establishment success is higher in some regions than in others (see, for example, Blackburn & Duncan

2001; Cassey et al. 2004a), it is not obvious that tropical regions are harder (or easier) to invade than others.

As noted earlier, locations differ in more than the presence of enemies (encompassing competitors, predators and parasites): resource availability and the physical environment also vary (Shea & Chesson 2002). Yet, Elton (1958) specifically couches the resistance of tropical forests to invaders in relation to biotic resistance. A more refined test of Elton's ideas with respect to tropical biotic resistance would assess the effects of enemies directly, rather than assuming that latitude, quantitatively or qualitatively, acts as a surrogate for their effects. Here, however, one runs into the problem of which are the relevant enemies? It is difficult to predict *a priori* which of all potential competitors, predators and parasites that an introduced species might encounter might be the ones that prevent establishment. Conversely, it is easy to imagine that the presence of one specific enemy might be sufficient to prevent an exotic species from establishing or spreading. The importance of higher tropical species richness may thus result either because it increases the chances that a specific prophylactic enemy is present at a location, or because the additive effects of a larger suite of enemies acting in concert are likely to resist invasion by most species introduced.

More direct tests of the biotic resistance hypothesis for birds can be divided into those that consider species richness in total, and those that try to identify and test the effects of specific groups of natural enemies. There is little evidence from the former that avian establishment success declines as species richness increases. For example, neither Case (1996) nor Cassey et al. (2005b) found significant evidence for a relationship between establishment success and native bird species richness for a sample of islands around the world. Stohlgren et al. (2006) actually found a positive relationship between native and exotic species richness across 3074 counties in 49 of the states of the United States. A lack of information on introduction failure rates means that it is difficult to draw firm conclusions about establishment success from Stohlgren et al.'s results, and there is wider evidence of scale dependence in such comparisons (see, for example, Fridley et al. 2007), but it is difficult to find in them evidence of biotic resistance. There is little evidence relating to the influence of biotic resistance on invasive spread in birds, but what there is, is not supportive. Thus those bird species that have established in tropical North

America tend to have wide latitudinal distributions (Sax 2001).

Tests of the influence of biotic resistance by more specific subsets of species on establishment probability have had more success in producing results consistent with the idea. Cassey et al. (2005b) showed that the number of species of exotic mammal predators on oceanic islands was negatively correlated with the establishment probability of exotic birds. Exotic mammals are the main avian predators in oceanic island systems, and the likely importance of high predation pressure in the early stages of an invasion, when the exotic bird population is small, is obvious. There is also a large literature on the effect on establishment probability of competition between newly introduced and previously established exotic bird species (see, for example, Moulton & Pimm 1983, 1986a,b, 1987; Moulton 1985; Simberloff 1992; Lockwood et al. 1993; Lockwood & Moulton 1994; Brooke et al. 1995; Moulton et al. 1996, 2001). These studies quantify morphological over-dispersion and priority effects as evidence of competitive exclusion of later arrivals by similar but already established exotic bird species. However, whether these findings are truly a consequence of competition has been questioned, and the same data can yield alternative explanations (see, for example, Simberloff & Boecklen 1991; Duncan 1997; Duncan & Blackburn 2002; Blackburn et al. 2009a). Nevertheless, recent work modelling interactions between propagule pressure, the abundance of a competitor species and the strength of competition suggests that competition may help determine the fate of introduced species (Duncan & Forsyth 2006). When a newly arrived exotic is morphologically very different to an already established competitor, or if the abundance of the competitor is low, then the probability of establishment for the new arrival is predicted mainly by propagule pressure. When the abundance of the competitor and the newly released species are both high, there is a pronounced effect of the strength of competition as indexed by morphological similarity. Relationships observed in data for New Zealand bird introductions fit well with these model predictions (Duncan & Forsyth 2006).

The converse of Elton's (1958) idea that higher levels of natural enemies in the tropics will prevent exotic species invasions is that invasions may be facilitated if the species is introduced in a location with fewer of the enemies that keep it in check in its native range. 'Enemy release' (Keane & Crawley 2002) has

most often been assessed in terms of parasite prevalence and diversity, as these are easily compared in native and exotic populations, and the small numbers of individuals typically involved in any introduction event gives plenty of scope for sampling effects to cause native parasites to be left behind. Although conceptually simple, firmly establishing that an exotic species has truly escaped enemies has proven difficult for two reasons. First, the degree to which the exotic is escaping enemy impacts is strongly dependent on the exact identity of its enemies in the native range, and their per capita effect. This information is not easily acquired except through detailed study in the native range (Colautti et al. 2005). Second, even if an exotic left its serious enemies behind, it may have encountered new enemies in its non-native range, and these may impose higher per capita effects than those from which it escaped (Colautti et al. 2004). Given this complexity it is not surprising that there is little consistent evidence that exotic populations of birds support fewer parasite species than do native populations (Steadman et al. 1990; Paterson et al. 1999; Torchin et al. 2003; Colautti et al. 2004; Ishtiaq et al. 2006). Release from enemies in the form of predators has been implicated in the invasion success of exotic ring-necked parakeets *Psittacula krameri* in Israel and the UK, where nest predation is significantly lower than in their native India (Shwartz et al. 2009).

13.5 STATEMENT 4 (PAGE 63): 'IT WILL BE NOTICED THAT INVASIONS MOST OFTEN COME TO CULTIVATED LAND, OR TO LAND MUCH MODIFIED BY HUMAN PRACTICE.'

If 'the balance of relatively simple communities of plants and animals is more easily upset than that of richer ones ... and more vulnerable to invasions' (Elton 1958), then one might predict that the hugely simplified ecological systems covering most cultivated land would be readily invasible. So too should be the much modified areas of towns and cities. It is unclear from his book whether Elton's ideas about biotic resistance led him to the observation heading this section or vice versa. Whichever way round, it is an observation that seems to be broadly supported by data for exotic birds.

Several authors have noted that exotic birds preferentially establish in land much modified by human practice. For example, Diamond and Veitch (1981) asked whether introductions and extinctions in the New Zealand avifauna were cause and effect. Their answer (in the negative) noted that exotic bird species are largely excluded from unbrowsed climax forest communities with intact bird communities, and only enter forests where browsing by exotic herbivores has opened up the habitat, and where predation by exotic predators has decimated the native avifauna. Exotic bird species comprised only 0.1–0.2% of the individual birds found in censuses of forest on the island nature reserve of Little Barrier reported by Diamond and Veitch (1981), and 2% of the individual birds reported by Cassey (unpublished data) from four separate censuses at the same location 25 years later. In contrast, exotic birds tend to attain phenomenally high densities on agricultural land in New Zealand (MacLeod et al. 2009), where they greatly outnumber native bird species (T.M. Blackburn, personal observations).

Exotics species are frequently chosen to be commensal with humans and, unsurprisingly, urban environments often favour exotic species (see Garden et al. 2006; Kark et al. 2007; van Rensburg et al. 2009). Most of the exotic bird species on Mauritius are found predominantly in urban, agricultural or secondary habitats (Cheke & Hume 2008). Sixty-nine percent of all sightings on surveys of resident bird species in urban Melbourne referred to exotics (Green 1984), and some exotic bird species (for example the house crow *Corvus splendens* (Nyári et al. 2006)) seem essentially to be obligate users of anthropogenically modified habitats.

The tendency for exotic bird species to favour urban and agricultural environments, whereas native species prefer native habitats, may explain why it has been hard to find evidence of biotic resistance in this taxon. Invasions by exotic species are unlikely to be regulated by interspecific interactions with natives if the two never meet. In this vein, Case (1996) attributed a negative relationship between the numbers of native and exotic bird species across a range of island and mainland habitats to environmental differences in native versus human modified habitats, rather than to the importance of biotic resistance. Moreover, if exotic birds tended to be collected and transported from human-modified habitats, then the introduction process may tend to introduce species pre-adapted to such habitats (Blackburn et al. 2009a). Of course, it is nevertheless possible that exotics and natives may segregate by habitat because of biotic resistance (Case 1996).

13.6   STATEMENT 5 (PAGE 118): 'INTRODUCED ANIMALS OFTEN DO REPLACE OR REDUCE THE NUMBERS OF NATIVE ONES.'

At one level, this statement is almost a truism. There is an upper limit to the number (or at least the biomass) of animals that can occupy an ecosystem or habitat, which is broadly determined by the amount of productive energy available to those animals through primary productivity. Any introduced animal is likely to reduce the amount of energy available to native animals, through competition, predation or parasitism, and so is likely in turn to 'replace or reduce their numbers'. This is especially true in situations where the exotic species attain high densities, as is the case for introduced birds in New Zealand (see above), or become widespread across the exotic environment (Parker et al. 1999). Introduced predators may also directly reduce the numbers of native species, especially in situations where the natives have no experience of predators.

In practice, however, a direct negative impact of introduced animals on native populations can be difficult to demonstrate, and this seems especially the case for negative impacts of introduced birds. In part, this may arise for the same reason as the difficulty in demonstrating biotic resistance in bird invasions: introduced species often favour the kinds of anthropogenically modified habitats that are simply less likely to support many native species. Introduced animals then are not replacing or reducing the numbers of native ones, but simply using habitats from which native animals have already been excluded by habitat change or destruction. Across South Africa, for example, common mynas are found more frequently than expected by chance in areas with greater human population numbers and land transformation indices (Peacock et al. 2007), and are less likely to be associated with protected natural areas.

A negative impact of introduced birds on native populations can still be difficult to demonstrate even in cases where everything seems to point to the likelihood of an effect. For example, competition between cavity nesting bird species for nest holes is expected to be intense, as these are often a scarce and limiting resource, especially for species that cannot excavate their own holes (Newton 1998). The European starling *Sturnus vulgaris* is one such species, which often aggressively evicts other species from their holes. Nest-site competition with increasing starling populations has been widely implicated with population declines of native cavity nesting birds in the USA (Wiebe 2003), yet Koenig (2003) concluded that European starlings have yet unambiguously and significantly to threaten any species of North American cavity-nesting bird, with the possible exception of sapsuckers. Starlings also compete with native parrot species for nest holes in Australia (Blakers et al. 1984), but quantitative data on population-level effects of this behaviour are lacking here too. It has also proven difficult to demonstrate competition between native and exotic bird species for food resources. For example, native and exotic species of white-eye (*Apalopteron familiare* and *Zosterops japonicus*, respectively) in the Bonin Islands overlap in distribution in secondary forests, and have very similar ecologies (Kawakami & Higuchi 2003). However, although there is evidence that the native species adjusts its foraging behaviour when the exotic species is present, there is no detectable negative effect of the exotic on the native, and densities of the native species have not declined.

Predation by introduced birds has been more strongly implicated than competition in the declines of several native bird species. For example, the common myna *Acridotheres tristis* has frequently been suspected of causing both clutch failure and population reduction of native island species (Holyoak & Thibault 1984; Thibault 1988). Blanvillain et al. (2003) found that interactions between the endangered endemic Tahiti flycatcher *Pomarea nigra* and exotic common mynas were more common during breeding activities than during non-breeding, and that more such encounters were observed in territories that experienced nest failure or fledging death compared with those where nesting was successful. This strongly suggests the existence of nest predation by common myna. The swamp harrier *Circus approximans* was also introduced to Tahiti, and has been implicated as a cause of the imminent extinction of the Tahiti imperial pigeon *Ducula aurorae* on the island (Thibault 1988). Predation by introduced birds seems likely to have led to reductions in the numbers of native animal species other than birds, though evidence for this is lacking (Ebenhard 1988).

13.7 CONCLUSIONS

We draw the following conclusions from the large-scale experiment in nature that bird invasions represent.

First, there is little evidence from birds for broad predictions deriving from Elton's (1958) hypothesis that relatively simple communities of plants and animals are easier to invade. Islands appear to be no easier for birds to invade than continental mainland areas, and neither do low latitude regions appear less susceptible than higher. In both cases, the apparent contraindications represented by the distribution of exotic bird species richness (Fig. 13.1) can be explained by differences in colonization pressure (Blackburn et al. 2009a; Lockwood et al. 2009). There is also little evidence that the total number of enemies (or potential enemies) at a location predicts invasion success, either in terms of initial establishment or subsequent spread: native bird species richness seems not to suppress the richness of exotic bird species. Although arguably the important natural enemies of exotic birds may not be native birds, areas rich in birds also tend broadly to be rich in many other taxa as well (Grenyer et al. 2006; Jetz et al. 2009).

Yet, more nuanced studies seem to suggest that biotic resistance can play a role in preventing avian invasions. There is a substantial body of evidence relating to the competitive effects of already established avian invaders on later arrivals, and although its results have been called into question, competitive effects continue to be found as such models increase in sophistication and take into account the important effect of propagule pressure (Duncan & Forsyth 2006). It has been suggested that predation holds more promise as a force for biotic resistance (see, for example, Blackburn et al. 2009a), and the one study to test for its effect on exotic bird establishment obtained a positive result (Cassey et al. 2005b). Conversely, there is so far little evidence that escape from enemies facilitates invasion success in birds. It remains possible that escape from a specific enemy is a primary determinant of invasion success (and certain biocontrol successes lend support to this idea), but the generality of this possibility will be very difficult to prove, as will be its action in any specific case.

Second, Elton's (1958) prediction about the invasibility of anthropogenically modified habitats is broadly supported. Whether this is for the reason he apparently envisaged – simplified ecological systems should have lower biotic resistance – or simply because of habitat segregation between exotics and natives generated by the invasion process (see, for example, Blackburn et al. 2009a), would benefit from further study. Third, although it seems almost inevitable that exotic species will reduce or replace the numbers of native species, evidence for direct effects is currently sparse, at least for exotic birds. Instead, they more often seem to use space or energy already usurped from native species by human transformation of the environment.

Overall, therefore, one could conclude that Elton's (1958) mechanistic hypotheses are poorly supported when confronted with data from bird invasions. However, is that because Elton was wrong, or is it because Elton was broadly right, but that bird invasions are somehow unusual? Certainly, it is clear that taxa do not all behave the same way in response to the abiotic and/or biotic environments. Animals and plants have different resource requirements and interact differently, the aquatic and terrestrial realms pose different challenges to the organisms that inhabit them, and even within the terrestrial realm, there is far from perfect congruence in the distributions of vascular plant, amphibian, mammal and bird species richness (Grenyer et al. 2006; Jetz et al. 2009). Different taxa may thus differ in their susceptibility to biotic resistance, or in the extent to which human influences determine invasion success (for example, through propagule and colonization pressures). Relatively little consideration has so far been given in the literature to the extent to which different taxa differ in their invasion responses, and the question deserves consideration in its own right. Here, we simply draw attention to three points that we think may be relevant in this regard.

First, we think that propagule and colonization pressures will be general processes influencing invasion success in all taxa. Artefacts (colonization pressure) are not taxon specific, whereas the small population paradigm (propagule pressure) is likely to be universal, albeit that taxa are likely to differ in the population level that is critically small. Most invasion studies have not, and indeed in most cases cannot, take propagule and colonization pressures into account, and this will inevitably lead to problems of interpretation with respect to mechanism, especially for studies that draw conclusions from analyses of the number of established exotic species.

Second, we are not convinced that competition is a major structuring force in bird assemblages. We suspect that predation, being terminal for the victim, is much more so. The failure of biotic resistance as a paradigm in bird assemblages may be because studies have focused on competition. Those studies that do consider predator effects (see, for example, Cassey et al.

2005b; Shwartz et al. 2009) find evidence for them. Similarly, numerous biocontrol successes over plant and insect pests also point to the importance of enemy release.

Finally, most studies on exotic birds are correlational, and the lack of experimental approaches may, in part, explain why the biotic resistance hypothesis is less well-supported in birds than it is in other taxa, such as plants. We suspect that birds may provide a relatively harsh test of Elton's (1958) mechanistic hypotheses for invasion success. Nevertheless, even if the mechanisms that Elton espoused are eventually proved to fall short, which of us would not be happy to think that, in 50 years time, our research was still the subject of such stimulating investigation and discussion?

ACKNOWLEDGEMENTS

T.M.B. thanks Dave Richardson for inviting him to speak at the symposium that gave rise to this book, and Steven Chown, Dave Richardson and all the post-docs and students at the Centre for Invasion Biology for their wonderful hospitality and rewarding interactions over his several visits there.

REFERENCES

Baker, H.G. & Stebbins, G.L. (1965) *The Genetics of Colonizing Species*. Academic Press, New York.

Blackburn, T.M., Cassey, P. & Lockwood, J.L. (2008) The island biogeography of exotic bird species. *Global Ecology and Biogeography*, **17**, 246–251.

Blackburn, T.M., Cassey, P. & Lockwood, J.L. (2009b) The role of species traits in the establishment success of exotic birds. *Global Change Biology*, **15**, 2852–2860.

Blackburn, T.M. & Duncan, R.P. (2001) Determinants of establishment success in introduced birds. *Nature*, **414**, 195–197.

Blackburn, T.M., Lockwood, J.L. & Cassey, P. (2009a) *Avian Invasions. The Ecology and Evolution of Exotic birds*. Oxford University Press, Oxford.

Blakers, N., Davies, S.J.J.F. & Reilly, P.N. (1984) *The Atlas of Australian Birds*. Melbourne University Press, Carlton.

Blanvillain, C., Salducci, J.M., Tutururai, G. & Maeura, M. (2003) Impact of introduced birds on the recovery of the Tahiti Flycatcher (*Pomarea nigra*), a critically endangered forest bird of Tahiti. *Biological Conservation*, **109**, 197–205.

Brett, M.T. (2004) When is correlation between non-independent variables "spurious"? *Oikos*, **105**, 647–656.

Brooke, R.K., Lockwood, J.L. & Moulton, M.P. (1995) Patterns of success in passeriform bird introductions on Saint Helena. *Oecologia*, **103**, 337–342.

Byers, J.E. & Noonburg, E.G. (2003) Scale dependent effects of biotic resistance to biological invasion. *Ecology*, **84**, 1428–1433.

Cadotte, M.W. (2006). Darwin to Elton: early ecology and the problem of invasive species. In *Conceptual ecology and invasions biology: reciprocal approaches to nature*. (ed. M.W. Cadotte, S.M. McMahon and T. Fukami), pp. 15–33. Kluwer, Netherlands.

Carlton, J.T. (1996) Pattern, process, and prediction in marine invasion ecology. *Biological Conservation*, **78**, 97–106.

Case, T.J. (1996) Global patterns in the establishment and distribution of exotic birds. *Biological Conservation*, **78**, 69–96.

Cassey, P. (2003) A comparative analysis of the relative success of introduced landbirds on islands. *Evolutionary Ecology Research*, **5**, 1011–1021.

Cassey, P., Blackburn, T.M., Duncan, R.P. & Gaston, K.J. (2005b) Causes of exotic bird establishment across oceanic islands. *Proceedings of the Royal Society B*, **272**, 2059–2063.

Cassey, P., Blackburn, T.M., Duncan, R.P. & Lockwood, J.L. (2005a) Lessons from the establishment of exotic species: a meta-analytical case study using birds. *Journal of Animal Ecology*, **74**, 250–258.

Cassey, P., Blackburn, T.M., Russell, G.J., Jones, K.E. & Lockwood, J.L. (2004b) Influences on the transport and establishment of exotic bird species: an analysis of the parrots (Psittaciformes) of the world. *Global Change Biology*, **10**, 417–426.

Cassey, P., Blackburn, T.M., Sol, D., Duncan, R.P. & Lockwood, J.L. (2004a) Introduction effort and establishment success in birds. *Proceedings of the Royal Society of London B*, **271** (Suppl.), S405–S408.

Cheke, A. & Hume, J. (2008) *Lost Land of the Dodo. An Ecological History of Mauritius, Réunion and Rodrigues*. T & A D Poyser, London.

Chiron, F., Shirley, S. & Kark, S. (2009) Human-related processes drive the richness of exotic birds in Europe. *Proceedings of the Royal Society B*, **276**, 47–53.

Cleland, E.E., Smith, M.D., Andelman, S.J., et al. (2004) Invasion in space and time: non-native species richness and relative abundance respond to interannual variation in productivity and diversity. *Ecology Letters*, **7**, 947–957.

Colautti, R.I., Grigorovich, I.A. & MacIsaac, H.J. (2006) Propagule pressure: a null model for biological invasions. *Biological Invasions*, **8**, 1023–1037.

Colautti, R.I., Muirhead, J.R., Biswas, R.N. & MacIsaac, H.J. (2005) Realized vs apparent reduction in enemies of the European starling. *Biological Invasions*, **7**, 723–732.

Colautti, R.I., Ricciardi, A., Grigorovich, I.A. & MacIsaac, H.J. (2004) Is invasion success explained by the enemy release hypothesis? *Ecology Letters*, **7**, 721–733.

Daehler, C.C. & Carino, D.A. (2000) Predicting invasive plants: prospects for a general screening system based on current regional models. *Biological Invasions*, **2**, 93–102.

Dean, W.R.J. (2000) Alien birds in southern Africa: what factors determine success? *South African Journal of Science*, **96**, 9–14.

Diamond, J.M. & Veitch, C.R. (1981) Extinctions and introductions in the New Zealand avifauna: cause and effect? *Science*, **211**, 499–501.

Duncan, R.P. (1997) The role of competition and introduction effort in the success of passeriform birds introduced to New Zealand. *American Naturalist*, **149**, 903–915.

Duncan, R.P. & Blackburn, T.M. (2002) Morphological over-dispersion in game birds (Aves: Galliformes) successfully introduced to New Zealand was not caused by interspecific competition. *Evolutionary Ecology Research*, **4**, 551–561.

Duncan, R.P., Blackburn, T.M. & Sol, D. (2003) The ecology of bird introductions. *Annual Review of Ecology, Evolution and Systematics*, **34**, 71–98.

Duncan, R.P., Blackburn, T.M. & Veltman, C.J. (1999) Determinants of geographical range sizes: a test using introduced New Zealand birds. *Journal of Animal Ecology*, **68**, 963–975.

Duncan, R.P., Bomford, M., Forsyth, D.M. & Conibear, L. (2001) High predictability in introduction outcomes and the geographical range size of introduced Australian birds: a role for climate. *Journal of Animal Ecology*, **70**, 621–632.

Duncan, R.P., Cassey, P. & Blackburn, T.M. (2009) Do climate envelope models transfer? A manipulative test using dung beetle introductions. *Proceedings of the Royal Society B*, **276**, 1449–1457.

Duncan, R.P. & Forsyth, D.M. (2006). Competition and the assembly of introduced bird communities. In *Conceptual Ecology and Invasions Biology*. (ed. M.W. Cadotte, S.M. McMahon and T. Fukami), pp. 415–431. Springer, Dordrecht.

Ebenhard, T. (1988) Introduced birds and mammals and their ecological effects. *Swedish Wildlife Research*, **13**, 1–107.

Elton, C. (1927) *Animal Ecology*. Sidgwick & Jackson, London.

Elton, C. (1958) *The Ecology of Invasions by Animals and Plants*. Methuen, London.

Fridley, J.D., Stachowicz, J.J., Naeem, S., et al. (2007) The invasion paradox: reconciling pattern and process in species invasions. *Ecology*, **88**, 3–17.

Garden, J., McAlpine, C., Peterson, A., Jones, D. & Possingham, H. (2006) Review of the ecology of Australian urban fauna: a focus on spatially explicit processes. *Australian Ecology*, **31**, 126–148.

Green, R.J. (1984) Native and exotic birds in a suburban habitat. *Australian Wildlife Research*, **11**, 181–190.

Grenyer, R., Orme, C.D.L., Jackson, S.F., et al. (2006) Global distribution and conservation of rare and threatened vertebrates. *Nature*, **444**, 93–96.

Hayes, K.R. & Barry, S.C. (2008) Are there consistent predictors of invasion success? *Biological Invasions*, **10**, 483–506.

Hewitt, C.L. (2002) Distribution and biodiversity of Australian tropical marine bioinvasions. *Pacific Science*, **56**, 213–222.

Holdgate, M.W. (1986) Summary and conclusions: characteristics and consequences of biological invasions. *Philosophical Transactions of the Royal Society of London B*, **314**, 733–742.

Holway, D.A. (1998) Factors governing rate of invasion: a natural experiment using Argentine ants. *Oecologia*, **115**, 206–212.

Holyoak, D.T. & Thibault, J.-C. (1984) Contribution à l'étude des oiseaux de Polynésie orientale. *Memoirs du Museum National d'Histoire Naturelle (France). Nouvelle Serie. Serie A. Zoologie*, **127**, 1–209.

Ishtiaq, F., Beadell, J.S., Baker, A.J., Rahmani, A.R., Jhala, Y.V. & Fleischer, R.C. (2006) Prevalence and evolutionary relationships of haematozoan parasites in native versus introduced populations of common myna *Acridotheres tristis. Proceedings of the Royal Society B*, **273**, 587–594.

Jetz, W., Kreft, H., Ceballos, G. & Mutke, J. (2009) Global associations between terrestrial producer and vertebrate consumer diversity. *Proceedings of the Royal Society B*, **276**, 269–278.

Kark, S., Iwaniuk, A., Schalimtzek, A. & Banker, E. (2007) Living in the city: can anyone become an 'urban exploiter'? *Journal of Biogeography*, **34**, 638–651.

Kawakami, K. & Higuchi, H. (2003) Interspecific interactions between the native and introduced White-eyes in the Bonin Islands. *Ibis*, **145**, 583–592.

Keane, R.M. & Crawley, M.J. (2002) Exotic plant invasions and the enemy release hypothesis. *Trends in Ecology & Evolution*, **17**, 164–170.

Koenig, W.D. (2003) European starlings and their effect on native cavity-nesting birds. *Conservation Biology*, **17**, 1134–1140.

Leibold, M.A. & Wootton, J.T. (2001) *Introduction to Elton, C.S. (1927) Animal Ecology*. University of Chicago Press, Chicago.

Levine, J.M. & D'Antonio, C.M. (1999) Elton revisited: a review of evidence linking diversity and invasibility. *Oikos*, **87**, 15–26.

Lockwood, J.L., Cassey, P. & Blackburn, T. (2005) The role of propagule pressure in explaining species invasions. *Trends in Ecology & Evolution*, **20**, 223–228.

Lockwood, J.L., Cassey, P. & Blackburn, T.M. (2009) The more you introduce the more you get: the role of colonization and propagule pressure in invasion ecology. *Diversity and Distributions* **15**, 904–910.

Lockwood, J.L. & Moulton, M.P. (1994) Ecomorphological pattern in Bermuda birds: the influence of competition and implications for nature preserves. *Evolutionary Ecology*, **8**, 53–60.

Lockwood, J.L., Moulton, M.P. & Anderson, S.K. (1993) Morphological assortment and the assembly of communities of introduced passeriforms on oceanic islands: Tahiti versus Oahu. *American Naturalist*, **141**, 398–408.

Long, J.L. (1981) *Introduced Birds of the World. The Worldwide History, Distribution and Influence of Birds Introduced to New Environments*. David & Charles, London.

Lonsdale, W.M. (1999) Global patterns of plant invasions and the concept of invasibility. *Ecology*, **80**, 1522–1536.

MacArthur, R.H. (1955) Fluctuations of animal populations and a measure of community stability. *Ecology*, **36**, 533–536.

MacLeod, C.J., Newson, S.E., Blackwell, G. & Duncan, R.P. (2009) Enhanced niche opportunities: can they explain the success of New Zealand's introduced bird species? *Diversity and Distributions*, **15**, 41–49.

May, R.M. (1973) *Stability and Complexity in Model Ecosystems*. Princeton University Press, Princeton, New Jersey.

McCann, K.S. (2000) The diversity–stability debate. *Nature*, **405**, 228–233.

Moulton, M.P. (1985) Morphological similarity and coexistence of congeners: an experimental test with introduced Hawaiian birds. *Oikos*, **44**, 301–305.

Moulton, M.P. & Pimm, S.L. (1983) The introduced Hawaiian avifauna: biogeographic evidence for competition. *American Naturalist*, **121**, 669–690.

Moulton, M.P. & Pimm, S.L. (1986a). The extent of competition in shaping an introduced avifauna. In *Community Ecology* (ed. J. Diamond and T.J. Case), pp. 80–97. Harper & Row, New York.

Moulton, M.P. & Pimm, S.L. (1986b). Species introductions to Hawaii. In *The Ecology of Biological Invasions of North America and Hawaii*. (ed. H.A. Mooney and J.A. Drake), pp. 231–249. Springer, New York.

Moulton, M.P. & Pimm, S.L. (1987) Morphological assortment in introduced Hawaiian passerines. *Evolutionary Ecology*, **1**, 113–124.

Moulton, M.P., Sanderson, J. & Simberloff, D. (1996) Passeriform introductions to the Mascarenes (Indian Ocean): an assessment of the role of competition. *Ecologie*, **27**, 143–152.

Moulton, M.P., Sanderson, J.G. & Labisky, R.F. (2001) Patterns of success in game bird (Aves: Galliformes) introductions to the Hawaiian islands and New Zealand. *Evolutionary Ecology Research*, **3**, 507–519.

Newton, I. (1998) *Population Limitation in Birds*. Academic Press, San Diego.

Nyári, A., Ryall, C. & Peterson, A.T. (2006) Global invasive potential of the house crow *Corvus splendens* based on ecological niche modelling. *Journal of Avian Biology*, **37**, 306–311.

Parker, I.M., Simberloff, D., Lonsdale, W.M., et al. (1999) Impact: towards a framework for understanding the ecological effects of invaders. *Biological Invasions*, **1**, 3–19.

Paterson, A.M., Palma, R.L. & Gray, R.D. (1999) How frequently do avian lice miss the boat? Implications for coevolutionary studies. *Systematic Biology*, **48**, 214–223.

Peacock, D.S., van Rensburg, B.J. & Robertson, M.P. (2007) The distribution and spread of the invasive alien common myna *Acridotheres tristis*, L (Aves: Sturnidae) in southern Africa. *South African Journal of Science*, **103**, 465–473.

Pimm, S.L. (1991) *The Balance of Nature? Ecological Issues in the Conservation of Species And Communities*. University of Chicago Press, Chicago.

Preston, F.W. (1948) The commonness, and rarity, of species. *Ecology*, **29**, 254–283.

Quammen, D. (1999) Foreword. In *Killer Algae: The True Tale of a Biological Invasion* (ed. A. Meinesz), pp. vii–x. University of Chicago Press, Chicago.

Rejmánek, M. (1996) Species richness and resistance to invasions. In *Biodiversity and Ecosystem Processes in Tropical Forests* (ed. G.H. Orians, R. Dirzo and J.H. Cushman), pp. 153–172. Springer, Berlin.

Richardson, D.M. & Pyšek, P. (2007) Classics in physical geography revisited: Elton, C.S. 1958: The ecology of invasions by animals and plants. Methuen: London. *Progress in Physical Geography*, **31**, 659–666.

Richardson, D.M. & Pyšek, P. (2008) Fifty years of invasion ecology – the legacy of Charles Elton. *Diversity and Distributions*, **14**, 161–168.

Rosenzweig, M.L. (1995) *Species Diversity in Space and Time*. Cambridge University Press, Cambridge.

Rozdilsky, I.D. & Stone, L. (2001) Complexity can enhance stability in competitive systems. *Ecology Letters*, **4**, 397–400.

Sax, D.F. (2001) Latitudinal gradients and geographic ranges of exotic species: implications for biogeography. *Journal of Biogeography*, **28**, 139–150.

Sax, D.F. & Gaines, S.D. (2006). The biogeography of naturalised species and the species–area relationship: reciprocal insights to biogeography and invasion biology. In *Conceptual Ecology and Invasions Biology: Reciprocal Approaches to Nature* (ed. M.W. Cadotte, S.M. McMahon and T. Fukami), pp. 449–479. Springer, Dordrecht.

Shea, K. & Chesson, P. (2002) Community ecology as a framework for biological invasions. *Trends in Ecology & Evolution*, **17**, 170–176.

Shwartz, A., Strubbe, D., Butler, C.J., Matthysen, E. & Kark, S. (2009) The effect of enemy-release and climate conditions on invasive birds: a regional test using the rose-ringed parakeet (*Psittacula krameri*) as a case study. *Diversity and Distributions*, **15**, 310–318.

Simberloff, D. (1992) Extinction, survival, and effects of birds introduced to the Mascarenes. *Acta Oecologica*, **13**, 663–678.

Simberloff, D. (1995) Why do introduced species appear to devastate islands more than mainland areas? *Pacific Science*, **49**, 87–97.

Simberloff, D. & Boecklen, W.J. (1991) Patterns of extinction in the introduced Hawaiian avifauna: a reexamination of the role of competition. *American Naturalist*, **138**, 300–327.

Sol, D. (2000) Are islands more susceptible to be invaded than continents? Birds say no. *Ecography*, **23**, 687–692.

Sol, D., Blackburn, T.M., Cassey, P., Duncan, R.P. & Clavell, J. (2005a). The ecology and impact of non-indigenous birds. In *Handbook of the Birds of the World, volume 10, Cuckoo-Shrikes to Thrushes* (ed. J. del Hoyo, A. Elliott and J. Sargatal), pp. 13–35. Lynx Ediçions and BirdLife International, Cambridge, UK.

Sol, D., Duncan, R.P., Blackburn, T.M., Cassey, P. & Lefebvre, L. (2005b) Big brains, enhanced cognition, and response of birds to novel environments. *Proceedings of the National Academy of Sciences of the USA*, **102**, 5460–5465.

Steadman, D.W., Greiner, E.C. & Wood, C.S. (1990) Absence of blood parasites in indigenous and introduced birds from the Cook Islands, South Pacific. *Conservation Biology*, **4**, 398–404.

Stohlgren, T.J., Barnett, D., Flather, C., et al. (2006) Species richness and patterns of invasion in plants, birds, and fishes in the United States. *Biological Invasions* **8**, 443–463.

Thibault, J.-C. (1988). Menacés et conservation des oiseaux de Polynésie Française. In *Livre Rouge des Oiseaux Menacés des Regions Françaises d'outre-Mer* (ed. J.-C. Thibault and I. Guyot), pp. 87–124. Conseil International pour la Protection des Oiseaux, Saint Cloud.

Torchin, M.E., Lafferty, K.D., Dobson, A.P., McKenzie, V.J. & Kuris, A.M. (2003) Introduced species and their missing parasites. *Nature*, **421**, 628–630.

van Rensburg, B.J., Peacock, D.S. & Robertson, M.P. (2009) Biotic homogenization and alien bird species along an urban gradient in South Africa. *Landscape and Urban Planning*, **92**, 233–241.

Veltman, C.J., Nee, S. & Crawley, M.J. (1996) Correlates of introduction success in exotic New Zealand birds. *American Naturalist*, **147**, 542–557.

Wiebe, K.L. (2003) Delayed timing as a strategy to avoid nest-site competition: testing a model using data from starlings and flickers. *Oikos*, **100**, 291–298.

Williamson, M. (1996) *Biological Invasions*. Chapman and Hall, London.

Chapter 14

IS RAPID ADAPTIVE EVOLUTION IMPORTANT IN SUCCESSFUL INVASIONS?

Eleanor E. Dormontt[1], *Andrew J. Lowe*[1,2]
and Peter J. Prentis[3]

[1]Australian Centre for Evolutionary Biology and Biodiversity, School of Earth and Environmental Sciences, University of Adelaide, SA 5005, Australia
[2]State Herbarium of South Australia, Science Resource Centre, Department for Environment and Heritage
[3]School of Land, Crop and Food Sciences, University of Queensland, Brisbane, QLD 4072, Australia

Fifty Years of Invasion Ecology: The Legacy of Charles Elton, 1st edition. Edited by David M. Richardson
© 2011 by Blackwell Publishing Ltd

14.1 INTRODUCTION

Ecological explanations of invasion success have dominated the literature on biological invasions since Elton's seminal publication (Elton 1958). Traditionally this focus has primarily been on traits of the exotic organism or recipient environment that make invasion more likely (e.g. growth rate and leaf area (Grotkopp & Rejmánek 2007); species richness, reviewed in Fridley et al. (2007)). Other theories concentrate on the unique interactions that can arise between environment and invader, such as enemy release (reviewed in Liu & Stiling 2006) and novel weapons (see, for example, Thorpe et al. 2009). Introduction dynamics such as propagule pressure (Simberloff 2009), residence time (Wilson et al. 2007) and human use (Thuiller et al. 2006), have also been identified as important ecological factors facilitating successful invasions. The possible evolutionary determinants of invasion success have received much less attention (Callaway & Maron 2006), but the past few years have seen a surge in interest, with publications considering evolutionary processes now more common than those with a purely ecological perspective (Fig. 14.1).

Part of the reason for this delay in realizing the potential relevance of evolutionary mechanisms to invasions (despite some early recognition, for example Baker & Stebbins (1965)) is that invasion dynamics were typically considered likely to constrain rather than promote adaptive evolution. For example, introduction of a few propagules from a single area would likely result in a genetic bottleneck in the invasive range. The ensuing small population would likely be more inbred, have lower genetic diversity, have a more

uniform susceptibility to pathogens, and have limited (or non-existent) gene flow with native populations. In fact, many of these characteristics are those generally associated with rare, endangered species. Empirical evidence supports one or more of these assumptions in a variety of invasive taxa (see, for example, Amsellem et al. 2000, 2001; Schmid-Hempel et al. 2007; Bailey et al. 2009; Prentis et al. 2009; Zang et al. 2010; reviewed studies in Dlugosch & Parker 2008a), supporting the notion that many species succeed in their new environment without need for adaptive evolution. However, there are a growing number of case studies that do not conform to this pattern (see, for example, Marrs et al. 2008a; Chun et al. 2010; reviewed studies in Roman & Darling 2007 and Dlugosch & Parker 2008a), or exhibit evidence for rapid adaptive evolution despite these constraints (see, for example, Dlugosch & Parker 2008b). This new evidence, combined with a conceptual shift towards viewing biological invasions as both quantitatively and qualitatively different to natural colonizations (Ricciardi 2007; Wilson et al. 2009), has led to greater consideration of the importance of evolutionary processes in facilitating successful invasions.

In this chapter we begin by looking at two of the main ecological explanations for invasion success, propagule pressure and enemy release. Excellent reviews on these hypotheses already exist (see, for example, Simberloff 2009; Orians & Ward 2010) so it is not our intention to cover old ground; rather, it is to summarize some of the recent literature and highlight how these ecological features can promote conditions conducive to adaptive evolution. We then examine what the genetic mechanisms behind rapid adaptive

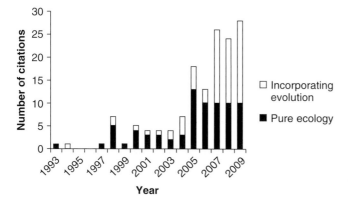

Fig. 14.1 Citations returned from the ISI Web of Science, search topic = invasion-biology AND ecology NOT evolution, (referred to here as 'pure ecology'); and topic = invasion-biology AND evolution (referred to here as 'incorporating evolution').

evolution might be, highlight the empirical regimes required to distinguish between alternate hypotheses and look towards fruitful future work that can help further clarify the role of rapid adaptive evolution in biological invasions.

14.2 PROPAGULE PRESSURE AND RAPID ADAPTIVE EVOLUTION

Support for the importance of propagule pressure in biological invasions

One of the most influential parameters of invasion success appears to be propagule pressure, which is the number and size of introductions to the recipient environment (reviewed in Simberloff 2009). In studies examining the relative impacts of various parameters thought to influence invasion success, propagule pressure has consistently been identified as the primary determinant. For example, propagule pressure exerts a greater influence than source population latitude (Maron 2006), abiotic conditions (Von Holle & Simberloff 2005) and species richness of the recipient environment (Von Holle & Simberloff 2005; Eriksson et al. 2006). As propagule pressure can fluctuate both temporally and geographically for a single species, invasion success can appear unpredictable unless propagule pressure is considered (Lockwood et al. 2005). Models seeking to explain abundance and distribution of invasive species that incorporate propagule pressure have been consistently more successful than those that have not (Richardson & Pyšek 2006), and any attempts to predict future invasions must similarly allow for this variable (Rouget & Richardson 2003). In fact, Colautti et al. (2006) suggest that propagule pressure should be the null model for all biological invasions.

How propagule pressure can promote conditions conducive to rapid adaptive evolution

The relationship between propagule pressure and invasion success is not necessarily linear, i.e. more propagules more often, does not always equate to optimal propagule pressure. For example, patch occupancy modelling of sexual diploid organisms by Travis et al. (2005) revealed a parabolic relationship between propagule pressure and invasion success for species poorly adapted to the environment; an intermediate propagule pressure aided establishment and facilitated rapid adaptive evolution to the local environment.

Optimal propagule pressure can promote conditions conducive to rapid adaptive evolution in several different ways. Theoretically, propagule pressure could reduce mate limitation in outcrossing species, ultimately relaxing Baker's law (Baker 1955; Stebbins 1957) and increasing the opportunities for obligate outcrossers to establish invasive populations. A continual flow of propagules into a novel environment can also relieve inbreeding depression and limit genetic drift, as well as increase genetic diversity (Simberloff 2009). In this way, diversity within introduced populations can reach levels comparable to native populations (Table 14.1), or if propagules arrive from multiple differentiated source populations, can even exceed that which is observed in the native range (see, for example, Lavergne & Molofsky 2007). The prominence of multiple source populations in biological invasions (Table 14.1) (Bossdorf et al. 2005; Roman & Darling 2007; Dlugosch & Parker 2008a; Wilson et al. 2009) has really only come to light since the advent of molecular techniques able to distinguish putative source origin. However, the importance of multiple sources is still under some debate, the increased diversity brought about by multiple sources can certainly theoretically contribute to rapid adaptive evolution (Prentis et al. 2008) but a causal association has not been demonstrated (Dlugosch & Parker 2008a). Simberloff (2009) reviewed 14 studies in which propagule pressure had been positively associated with increased genetic diversity, three of which were also able to demonstrate rapid evolution (Saltonstall 2002; Roman 2006; Lavergne & Molofsky 2007) but were unable to distinguish between neutral and adaptive processes.

14.3 ENEMY RELEASE AND RAPID ADAPTIVE EVOLUTION

Evidence for enemy release in biological invasions

The enemy release hypothesis (ERH) (reviewed by Keane & Crawley 2002; Colautti et al. 2004; Liu & Stiling 2006; Orians & Ward 2010) theorizes that exotic species escape the negative effects of their

Table 14.1 Review of studies examining genetic diversity in the native and introduced range of invasive plant species. In 20 of the 37 species studied, the authors observed genetic variation in the introduced range that was equal to or greater than that observed in the native range. Of the 24 studies that were able to deduce source number, 22 were attributed to multiple sources. These figures support a role for multiple introductions in reducing the severity of genetic bottlenecks in invasive species. Molecular markers used were as follows: nuclear microsatellites (nSSR); nuclear inter-simple sequence repeats (nISSR); allozymes; chloroplast DNA gene sequences (cpDNA); chloroplast microsatellites (cpSSR); chloroplast restriction fragment length polymorphisms (cpRFLP); random amplified polymorphic DNA (RAPD); amplified fragment length polymorphisms (AFLP); internal transcribed spacer sequences (ITS); plastid DNA gene sequences (ptDNA). Within-population genetic diversity estimates were used wherever possible. Genetic diversity was defined as the following: reduction in the invasive range (−); no significant difference between the native and invasive ranges (=); increase in the invasive range (+). All data are reported according to the original authors' conclusions. Refer directly to individual studies for specific units/indices used.

Species	Common name	Family	Native range	Invasive range(s) studied
Aegilops triuncialis	Barbed goatgrass	Poaceae	Eurasia	California
Alliaria petiolata	Garlic mustard	Brassicaceae	Europe	North America
Ambrosia artemisiifolia	Annual ragweed	Asteraceae	North America	France
Apera spica-venti	Silky bentgrass	Poaceae	Europe	Canada
Avena barbata	Slender oat	Poaceae	Mediterranean Basin	California
Brachypodium sylvaticum	False brome	Poaceae	Eurasia	North America
Bromus mollis	Soft brome	Poaceae	England	Australia
Bromus tectorum	Cheatgrass	Poaceae	Eurasia	North America
Bryophyllum delagoense	Mother of millions	Crassulaceae	Madagascar	Australia
Capsella bursa-pastoris	Shepherd's purse	Brassicaceae	Europe	North America
Centaurea diffusa	Diffuse knapweed	Asteraceae	Eurasia	North America
Centaurea stoebe micranthos	Spotted knapweed	Asteraceae	Eurasia	North America
Chondrilla juncea	Skeleton weed	Asteraceae	Turkey	Australia
Clidemia hirta	Soapbush	Melastomataceae	Costa Rica	Hawaii
Cytisus scoparius	Scotch broom	Fabaceae	Europe	Australia; California; New Zealand; Chile
Echium plantagineum	Paterson's curse	Boraginaceae	Europe	New South Wales
Epipactis helleborine	Broad leaved Helleborine	Orchidaceae	Europe	Canada

Molecular marker(s)	Native breeding system	Genetic diversity	Inferred number of sources and admixture	References
nSSR	Predominant inbreeder	–	Multiple without admixture	Meimberg et al. 2006
nISSR; nSSR	Predominant inbreeder	=/–	Multiple without admixture	Meekins et al. 2001; Durka et al. 2005
nSSR	Predominant outcrosser	=	Multiple with admixture	Genton et al. 2005
Allozymes	Obligate outcrosser	=	Multiple	Warwick et al. 1987
Allozymes	Predominant inbreeder	=/–	n/a	Clegg & Allard 1972; Garcia et al. 1989
nSSR; cpDNA	Mixed	–	Multiple with admixture	Rosenthal et al. 2008
Allozymes	Predominant inbreeder	=	n/a	Brown & Marshall 1981
Allozymes	Predominant inbreeder	+	Multiple	Novak et al. 1991; Novak & Mack 1993
nSSR	Mixed? + Vegetative	–	n/a	Hannan-Jones et al. 2005
Allozymes	Predominant inbreeder	=	Multiple	Neuffer, & Hurka 1999
cpDNA; nSSR	Obligate outcrosser	=/–	Multiple with admixture	Hufbauer & Sforza 2008; Marrs et al. 2008b
cpDNA; nSSR	Obligate outcrosser	=	Multiple with admixture	Hufbauer & Sforza 2008; Marrs et al. 2008a
Allozymes	Obligate inbreeder (apomict)	–	n/a	Chaboudez 1994.
Allozymes	Mixed	+	n/a	DeWalt & Hamrick 2004
cpSSR; nSSR	Predominant outcrosser	=	Multiple with admixture	Kang et al. 2007
Allozymes	Predominant outcrosser	=	n/a	Burdon & Brown 1986
Allozymes; cpDNA trnL intron PCR-RFLP	Predominant outcrosser	=/+	n/a	Squirrell et al. 2001a,b

Continued

Table 14.1 *Continued*

Species	Common name	Family	Native range	Invasive range(s) studied
Erigeron annuus	Daisy fleabane	Asteraceae	North America	Europe
Heracleum mantegazzianum	Giant hogweed	Apiaceae	South West Asia	Europe
Heracleum persicum	Persian hogweed	Apiaceae	South West Asia	Europe
Heracleum sosnowskyi	Sosnowskyi's hogweed	Apiaceae	South West Asia	Europe
Hirschfeldia incana	Hoary mustard	Brassicaceae	Southern Europe	UK
Hypericum canariense	Canary Island St. John's wort	Clusiaceae	Europe	North America
Hypericum perforatum	St. John's wort	Clusiaceae	Canary Islands	California; Hawaii
Ligustrum robustum	Tree privet	Oleaceae	Sri Lanka; India	Mascarene Islands
Olea europaea cuspidata	Wild olive	Oleaceae	Africa; Asia	Eastern Australia; Hawaii
Olea europaea europaea	Cultivated olive	Oleaceae	Mediterranean Basin	South Australia
Pennisetum setaceum	Fountain grass	Poaceae	Egypt	Namibia; South Africa; Hawaii; California
Phalaris arundinacea	Reed canarygrass	Poaceae	Europe	North America
Phyla canescens	Lippia	Verbenaceae	South America	Australia; France
Pueraria lobata	Kudzu	Fabaceae	China	USA
Rubus alceifolius	Giant bramble	Rosaceae	South East Asia	La Reunion; Mayotte; Mauritius; Madagascar; Queensland
Schinus terebinthifolius	Brazilian pepper tree	Anacardiaceae	South America	Florida
Silene latifolia	White campion	Caryophyllaceae	Europe	North America
Silene vulgaris	Bladder campion	Caryophyllaceae	Europe	North America
Spartina alterniflora	Smooth cordgrass	Poaceae	North American Atlantic/Gulf coast	North American Pacific coast
Trifolium hirtum	Rose clover	Fabaceae	Turkey	California

Molecular marker(s)	Native breeding system	Genetic diversity	Inferred number of sources and admixture	References
RAPD	Predominant inbreeder (apomict)	−	n/a	Edwards et al. 2006
AFLP	Mixed	n/a	Multiple	Jahodova et al. 2007
AFLP	Mixed	n/a	Multiple	Jahodova et al. 2007
AFLP	Mixed	n/a	Multiple	Jahodova et al. 2007
RAPD	Predominant outcrosser	=	Multiple	Lee et al. 2004.
AFLP; ITS	Mixed	−	Single	Dlugosch & Parker 2008b
AFLP	Predominant inbreeder (apomict)	=	Multiple	Maron et al. 2004
cpRFLP; RAPD	Predominant outcrosser	=	n/a	Milne & Abbott 2004
nSSR; ITS; ptDNA	Obligate outcrosser	−	Single	Besnard et al. 2007
nSSR; ITS; ptDNA	Obligate outcrosser	=	Multiple with admixture	Besnard et al. 2007
ITS; nSSR; nISSR	Obligate inbreeder (apomict)	=	n/a	Le Roux et al. 2007
Allozymes	Predominant outcrosser	+	Multiple with admixture	Lavergne & Molofsky 2007
nISSR; ITS	Predominant outcrosser + Vegetative	n/a	Multiple	Fatemi et al. 2008; Xu et al. 2010
nISSR	Predominant outcrosser + Vegetative	=	Multiple	Sun et al. 2005
AFLP	Mixed	−	n/a	Amsellem et al. 2000, 2001
cpDNA; nSSR	Obligate outcrosser	=	Multiple with admixture	Williams et al. 2005
cpDNA	Obligate outcrosser	−	Multiple with admixture	Taylor & Keller 2007
cpDNA	Mixed	=	n/a	Taylor & Keller 2007
cpDNA; nSSR	Predominant outcrosser + Vegetative	=	Multiple with admixture	Blum et al. 2007
	Mixed	=/+	n/a	Molina-Freaner & Jain 1992

natural enemies when in the introduced range (Darwin 1859; Elton 1958). Although this argument is intuitive, the positive effect of enemy release in the introduced range depends upon the extent to which abundance in the native range is controlled by natural enemies (Hierro et al. 2005). A recent review of studies which compare herbivore damage in native and introduced ranges revealed that roughly half found reduced damage in the introduced range, the reminder of studies showed no significant difference; indicating that for herbivores at least, the assumptions of ERH do not always hold true (Bossdorf et al. 2005). On the other hand, a comparison of invasive and non-invasive naturalized species in Ontario, New York and Massachusetts found 96% more herbivore damage on the non-invasives (Cappuccino & Carpenter 2005). These results suggest that although changes in herbivore load between native and introduced ranges may not always occur, preferential herbivory is likely an important regulator of plant community dynamics. Aside from herbivory, a study on 473 naturalized plant species in the USA found 84% fewer fungal and 24% fewer viral pathogens in the exotic compared with the native range, and that those plants experiencing the greatest levels of pathogen release were reported most widely as noxious weeds (Mitchell & Power 2003).

Enemy release, much like propagule pressure, can fluctuate in both space and time. Siemann et al. (2006) found that herbivore damage and tree performance were lowest in the areas of first introduction of the Chinese tallow tree (*Sapium sebiferum*), suggesting that herbivore release is important in establishment, but that insect herbivory accumulates over time. Conversely, certain habitats appear resistant to colonization in the native range but not in the exotic; for example soapbrush (*Clidemia hirta*) is excluded from native Costa Rican forest areas by herbivores and fungal pathogens not present in the introduced range and hence it has become an invasive forest species in Hawaii (DeWalt et al. 2004; see also Callaway & Rout, this volume).

Enemy release seems to be important in some invasions but not others. For example, differences between soil biota in the native and introduced ranges had no effect on growth and survival of marram grass (*Ammophila arenaria*) (Beckstead & Parker 2003). Instead Eppinga et al. (2006) suggest that *A. arenaria* may dominate by accumulating local pathogens, which in turn have a stronger negative effect on the surrounding native plants than on *A. arenaria* itself. Another possible explanation for ERH's lack of general applicability is that plants adapted to grow in high resource environments are more likely to be limited by herbivores, whereas low resource adapted plants are better defended due to the comparatively higher cost of herbivory (Blumenthal 2006). Therefore, although higher resource plants will experience a greater release from herbivore enemies upon introduction, low resource-adapted, well-defended plants may eventually produce a stronger evolutionary response to the reduction in natural enemy threat (Blumenthal 2006).

Plant enemies can be specialist or generalist and there is good evidence that ERH does not apply equally to both types. A recent study of *Senecio jacobaea* in its introduced and native ranges showed that invasive populations had reduced defence against specialist herbivores as predicted by ERH, but conversely had increased protection against generalist herbivores (Joshi & Vrieling 2005). Similarly, although finding an overall reduction in herbivores in the introduced range, Liu & Stiling (2006) revealed that the reduction was predominantly in specialists and those feeding on reproductive parts. In fact, there is good evidence that exotics are often preferentially chosen by native generalist herbivores (Parker & Hay 2005) supporting the idea of biotic resistance to invasive species (Elton 1958). A fascinating application of this theory has been used to explain the comparative success of Old World versus New World invasive plants: In a meta-analysis of studies covering over 100 invasive species, Parker et al. (2006) concluded that plants were particularly susceptible to novel generalist herbivores. The authors note that during European colonization of the New World, Old World generalist herbivores such as cattle, replaced the New World native generalist fauna. Old world invasive plants may therefore thrive by following their native natural enemies, not escaping them (Parker et al. 2006).

Can enemy release facilitate rapid adaptive evolution?

Based on the assumptions of the ERH, Blossey & Notzold (1995) put forward their theory of 'evolution of increased competitive ability' (EICA), which states that in the absence of herbivores, natural selection will favour those genotypes with increased resource allocation to competitive ability. This hypothesis generates two testable predictions, namely that individual plants from the invasive range will allocate more resources to growth and/or reproduction than those from the native range, and that specialist herbivores will favour plants from the introduced range over those from the native range as they possess fewer defensive mechanisms (Blossey & Notzold 1995).

The original study found supporting evidence for both predictions in the purple loosestrife (*Lythrum salicaria*) (Blossey & Notzold 1995), but work since has provided mixed support (see Bossdorf et al. 2004; Hierro et al. 2005; Orians & Ward 2010 for reviews). In particular studies tend to find support for one assumption but not the other, for example the Tansy Ragwort *Senecio jacobaea* has increased growth and reproduction, and decreased specialist herbivore defences in the invasive range in accordance with EICA, but actually has increased general herbivore defence and greater herbivore tolerance (Joshi & Vrieling 2005; Stastny et al. 2005). Similarly the giant goldenrod *Solidago gigantea* showed reduced herbivory tolerance in the invasive range but this did not translate to better performance in the absence of herbivores (Meyer et al. 2005). *S. gigantea* seems to invest more resources into rhizome production than flowers in the invasive range (Meyer & Hull-Sanders 2008). Although no changes in chemical defences could be identified, a generalist herbivore from the native range performed significantly better when grown on invasive plants, suggesting that some defence has been lost in the invaded range (Hull-Sanders et al. 2007).

A neat study that has provided recent support for EICA looked at phytochemical shifts in the invasive wild parsnip *Pastinaca sativa* by analysing herbarium records (Zangerl & Berenbaum 2005). Levels of various furanocoumarins were initially low in the introduced range but increase dramatically at the same time as their specialist herbivore, the parsnip webworm *Depressaria pastinacella*, was accidentally introduced into the invasive range (Zangerl & Berenbaum 2005). However, the associated assumption of increased vigour was not investigated so does not provide evidence for all aspects of EICA; in fact it could be argued that the observed shifts in resource allocation were merely a plastic rather than evolutionary response.

Several studies have shown increased defence in the invasive range which is contrary to the predictions of EICA (see, for example, Muller & Martens 2005;

Wikstrom et al. 2006) and in *Alliaria petiolata*, evolution of *reduced* competitive ability has been proposed based on the theory that if competition is lower in the invasive range, selection will favour reduced competitive ability if it has an associated fitness cost (Bossdorf et al. 2004). The conflicting evidence for EICA may be illustrative of the complex and variable mechanisms that are associated with plant invasions but a confounding factor may be the lack of control for sampling bias and maternal effects in many common garden studies (Bossdorf et al. 2005; Colautti et al. 2009).

The EICA hypothesis has arguably received most attention and empirical scrutiny, but is by no means the only possible evolutionary outcome of changes in enemy pressures in the invaded range. Orians & Ward (2010) put forward a set of testable evolutionary hypotheses based on the ecological conditions of the novel environment, of which EICA is but one of nine. Predicted evolutionary outcomes include increased or reduced defence allocation; changes in growth rate; allocation of defences more towards generalist or specialist enemies; and shifts between reliance on induced and constitutive defence traits (Orians & Ward 2010). This framework provides an excellent basis for future work on the effects of changing enemy loads on adaptive evolution in invasive species and could potentially be effectively combined with genetic and genomic approaches (as discussed in section 14.5) to reliably detect the genetic bases of any rapid adaptive evolution in plant defences.

14.4 RAPID ADAPTIVE EVOLUTION IN INVASIVE SPECIES

Support for the role of rapid adaptive evolution in biological invasions

Rapid adaptive evolution in introduced populations is proposed to have played a major role in some successful invasions (Maron et al. 2004; Phillips et al. 2006; Prentis et al. 2008; Whitney & Gabler 2008). When species are introduced to biogeographical regions where they did not evolve, they may encounter a suite of novel environmental conditions and selection regimes. Under these new selection regimes, genetically based phenotypic changes might affect individual fitness, the establishment of introduced populations and the spread of invasive populations across the landscape (Maron et al. 2004; Colautti et al. 2010; Xu et al. 2010). Consequently, rapid adaptive evolution in response to altered selection regimes could be an important determinant in the establishment, proliferation and spread of invasive species (Hendry et al. 2007; Prentis et al. 2008). Many studies of invasive species have found evidence of rapid phenotypic change between their native and introduced range, such as increased phenotypic plasticity, changes in body shape and size, and changes in breeding systems (Huey et al. 2000; Lavergne & Molofsky 2007; Barrett et al. 2008; Barrett, this volume), but few studies have documented whether selection was the mechanism responsible for these phenotypic shifts (Barrett et al. 2008; Keller & Taylor 2008).

So what is the evidence for rapid adaptive evolution as a mechanism promoting invasion success? Direct evidence for adaptive evolution promoting biological invasions is currently limited, but some excellent experiments have demonstrated the role of selection on ecologically relevant traits after the colonization of novel environments via human-mediated invasions. As evolutionary change in introduced populations can occur through neutral or adaptive processes, it is important to determine when adaptive evolution is responsible for observed changes. To establish that phenotypic evolution in introduced populations is the result of adaptation, it is necessary to statistically control for neutral processes that can also generate phenotypic change (reviewed in Keller & Taylor 2008). In their seminal paper, Maron et al., (2004) controlled for colonization history and determined that clinal variation observed in the introduced range of St. John's wort (*Hypericum perforatum*) evolved through adaptive processes. Another study controlling for neutral evolution has found that both adaptive evolution and colonization history have influenced the generation of phenotypic clines in two introduced species of *Silene* in North America (Keller et al. 2009). Recently, Xu et al. (2010) demonstrated that adaptive evolution was responsible for phenotypic divergence between the native range of *Phyla canescens* and two different invaded regions. Although these excellent studies highlight that adaptive evolution can drive phenotypic divergence within and between the ranges of invasive species, they do not elucidate the type of genetic variation underlying ecologically relevant traits.

Several different mechanisms have been implicated in generating the genetic variation underlying rapid adaptive evolution and the colonization of new habitats (Colosimo et al. 2005; Whitney et al. 2006; Barrett & Schluter 2008; Prentis et al. 2008). Although the exact source of genetic variation underlying traits important to successful invasions remains uncharacterized, in the following pages we contrast standing genetic variation with genetic novelty resulting from hybridization, and discuss whether these processes might promote adaptive evolution during the invasion process. Understanding whether standing genetic variation, genetic novelty or both contribute to evolution in invasive species has wide implications in the field of invasion biology. This key information is important for predicting future invasions; identifying genes underlying ecologically relevant traits in invasive populations; and designing strategies to better manage the import of species to decrease the potential for rapid evolution in invasive populations.

Adaptation from standing genetic variation or new mutation

Genetically based phenotypic changes that allow species to adapt to new environments can arise from new mutations or standing genetic variation, which is defined as the presence of more than one allele at a locus in a population. Understanding whether standing genetic variation or new mutation is responsible for rapid evolution in invasive species is currently an important question in invasion biology. We predict that most rapid adaptive evolution in invasive species should occur from standing genetic variation because favourable alleles are immediately available for selection to act upon and usually occur at a greater frequency within populations than new mutations (Barrett & Schluter 2008). Furthermore, as invasive populations often face new environments, neutral or even deleterious alleles in the native range may become advantageous in the introduced range. An elegant experimental demonstration supporting this prediction comes from the natural invasion of the Caribbean islands and Central America by the Brazilian water hyacinth (*Eichhornia paniculata*). In this species, recessive modifier genes that promote selfing occur at low frequencies in outcrossing source populations but fail to spread, possibly as reliable pollination services are available or because of the genetic costs associated with inbreeding depression. Selfing has evolved in the Caribbean islands where modifier genes have increased in frequency, possibly as a result of unreliable pollination services in this new environment (reviewed in

Barrett et al. 2008). Rapid morphological change resulting from adaptation from standing genetic variation has also been documented in repeated natural invasions of novel freshwater environments by marine stickleback fishes (Colosimo et al. 2005). Both these examples demonstrate that rapid adaptive evolution can occur from standing genetic variation during the colonization of a novel habitat.

If sufficient genetic variation exists in introduced populations, adaptation from standing genetic variation could dominate rapid evolution during range expansion. This may be a particularly important mechanism of rapid evolution for invasive species that are exposed to ecogeographic variation during range expansion. Clines have been observed in the introduced range of several invasive species, including some striking examples in *Drosophila* and flowering plants (Huey et al. 2000; Keller et al. 2009). However, the relative roles of introduction history, demographic processes and selection on generating clines in invasive species have rarely been tested in most studies (but see Keller et al. 2009; Colautti et al. 2010). In fact, Colautti et al. (2010) present strong evidence for adaptation from standing genetic variation for ecologically relevant traits during the northward expansion of purple loosestrife (*Lythrum salicaria*) in North America. Although this study provides compelling evidence of local adaptation to environmental conditions it also demonstrated that adaptation has been constrained at the northern limits of the introduced range because of a dearth of genetic variation for particular combinations of traits. Therefore adaptation from standing genetic variation in invasive populations may be limited if genetic constraints prevent selection from improving particular combinations of traits simultaneously. In such cases populations occupying marginal habitats at range extremes may suffer from a reduction in population growth, unless alleles that are beneficial in these environments arise by new mutation.

New beneficial mutations may also provide phenotypic variation for selection to act upon during colonization of new environments or range expansion in invasive species. Evidence for their role in the rapid adaptive evolution of invasive species, however, is currently lacking. Some of the best examples of adaptation through new mutation come from studies of microbes (Rainey & Travisano 1998) and the evolution of resistance in pest species (Wootton et al. 2002). A recent study has found that adaptation from new beneficial mutations has enabled a species of bacteria to colonize

a novel fluctuating environment through the evolution of bet-hedging genotypes that persist because of rapid phenotype switching (Beaumont et al. 2009). Although this example provides strong evidence for the colonization of a novel environment by adaptation from new mutation, we lack sufficient data to make confident conclusions about the role of new beneficial mutations in adaptive evolution of invasive species.

Current population genetic theory and data (reviewed in Barrett & Schluter 2008) has made it possible to distinguish between adaptation from standing genetic variation and adaptation from new mutation, because they leave different molecular signatures in the genome (Hermisson & Pennings 2005). By applying this genetic theory to population genomic data, such as restriction-site-associated DNA (RAD) markers (Miller et al. 2007), generated by recently developed sequencing technologies, including 454, Illumina and Solid sequencing, it will become possible to decipher whether rapid adaptive evolution in invasive species occurs mainly through selection on standing genetic variation or new mutation (see section 14.5).

Adaptation from genetic mixing

Genetic mixing resulting from hybridization between different species (interspecific) and between different source populations of a single species (admixture) has been hypothesized to stimulate invasiveness in plants (see Abbott 1992; Ellstrand & Schierenbeck 2000). Evidence for admixture or hybridization as a stimulus for invasion in specific taxa is increasing with many examples of highly invasive intra- and inter-specific hybrids, including lizards, toads and plants (see Abbott 1992; Ellstrand & Schierenbeck 2000; Estoup et al. 2001; Kolbe et al. 2008). However, some recent studies are also showing that the general role of hybridization as a stimulus for invasiveness may be over estimated (Dlugosch & Parker 2008b; Whitney et al. 2009). A recent study (Whitney et al. 2009) examined if the number of hybrids and the number of invasive species in 256 plant families were correlated. This relationship would be expected to develop if hybridization leads to invasiveness, because the number of invasive species should be greater in hybrid prone plant families. This study found that the present data did not support this hypothesis. Hybrids may not always be successful invaders because many hybrids may be largely sterile or unviable (Prentis et al. 2007), and because admixture after multiple introductions can lead to a mosaic of maladaptation, where trait values from one population might be better suited to another (Dlugosch & Parker 2008a). Therefore, we believe it is premature to generalize that hybridization is a stimulus for invasion, but nevertheless there are specific cases where hybridization has been associated with invasiveness which deserve consideration.

Invasive plants of hybrid origin may have increased fitness due to heterosis or later-generation recombination of parental genotypes (Prentis et al. 2008). Excluding asexual and allopolyploid invaders (Ellstrand & Schierenbeck 2000), heterosis is likely to be transient in outcrossing species because of recombination and reductions in heterozygosity in later-generation hybrids (Baack & Rieseberg 2007). Therefore, we predict that recombination will be more important than heterosis in the rapid evolution of invasive species. Recombination in hybrids can generate novel combinations of traits upon which selection can act to produce a phenotype that is better suited to its new environmental conditions (Rieseberg et al. 2003; Kolbe et al. 2008). Specifically, recombination can cause genetically based phenotypic changes in invasive species through two main processes: adaptive trait introgression and transgressive segregation.

Adaptive trait introgression involves the transfer of beneficial alleles that increase fitness, between divergent populations or species. Consequently, adaptive trait introgression could be an important mechanism for evolution in admixed populations resulting from multiple introductions or after hybridization between different species. For example, adaptive introgression has transferred abiotic stress resistance and herbivore-resistant traits from *Helianthus debilis* into *Helianthus annuus* ssp. *annuus*, and formed a stabilized hybrid lineage, *H. annuus* ssp. *texanus* (Whitney et al. 2006, 2010). This new hybrid lineage has higher fitness than *H. annuus* ssp. *annuus* in the new environments it encounters in central and southern Texas (Whitney et al. 2006, 2010). Although this example highlights the potential for adaptive introgression to promote evolution in an invasive species, we need more data before concluding this is an important process for promoting biological invasions.

Transgressive segregation is the formation of traits that are novel or extreme relative to those of either parental line in hybrid progeny (Rieseberg et al. 1999). This phenomenon has been reported to be a common feature of plant hybrids and has been implicated as a key component in the success of some hybrid species (Arnold & Hodges 1995). Recent experiments have demonstrated that extreme phenotypes resulting from transgressive segregation can aid the colonization of novel environments in hybrid sunflower species (Lexer et al. 2003). These studies have also demonstrated selection on the extreme phenotypes, which act to increase fitness of the hybrid species in desert environments compared with either parental species (Rieseberg et al. 2003). These results indicate that transgressive segregation can produce phenotypic novelty that could facilitate biological invasions, but reciprocal transplant experiments are needed in each case to determine whether extreme traits are under selection in the introduced range of invasive species.

Testing whether hybridization is actually a stimulus for invasion will be an important question in future research. To answer this question, researchers will first need to demonstrate whether or not invasive species are of hybrid origin. This will require genetic markers and population genetic analyses to determine that introduced species exhibit mixed ancestry from either genetically divergent populations in the native range or between two different species. Secondly, researchers will need to demonstrate the increased fitness of hybrids in the invasive range compared with their progenitors. Such studies will require either reciprocal transplant experiments; or the use of long-term selection experiments, where different species or multiple introductions are grown in sympatry and allowed to hybridize. After several generations, the resulting populations can be genotyped to determine the ancestry of individuals and their relative fitness in these environments.

14.5 FUTURE DIRECTIONS IN STUDYING ADAPTIVE EVOLUTION IN INVASIVE SPECIES: A GENOMIC APPROACH

Biological invasions provide a unique opportunity to examine how adaptive and neutral processes influence phenotypic evolution over ecological timescales. Although evolutionary change was hypothesized as a process that could increase invasion success over 35 years ago (Baker 1974), most research in this area has been conducted in the past decade. This research has largely concentrated on comparisons of genetic diversity between the native and introduced range of invasive species (reviewed in Dlugosch & Parker 2008a). Recently, there has been a surge of interest in investigating adaptive evolution in invasive species (Maron et al. 2004; Keller & Taylor 2008; Prentis et al. 2008), yet many studies have not adequately controlled for neutral evolutionary processes (but see Maron et al. 2004; Keller et al. 2009; Colautti et al. 2010; Xu et al. 2010). Several methods exist to determine whether adaptive evolution is responsible for rapid phenotypic change in biological invasions, while controlling for neutral evolution, phenotypic plasticity and pre-adaptation. These approaches include phenotypic ancestor–descendent comparisons between native and introduced populations when the source populations are known (Dlugosch & Parker 2008b); comparisons of genetic differentiation at neutral genetic markers (F_{ST}) versus phenotypic traits (Q_{ST}) (Xu et al. 2010); and finally reciprocal transplant experiments between source populations from the native range and introduced populations, using selection gradient analysis (Rundle & Whitlock 2001). Although these three approaches can demonstrate that phenotypic change in introduced populations is adaptive, they will not elucidate the genetic variation underlying this evolution.

Understanding the gene or genes underlying ecologically relevant traits that are involved in adaptive evolution to novel environments is a major goal in evolutionary biology and should be equally important to scientists studying biological invasions. Through the application of such methods the invasion biology community will be able to determine whether rapid adaptive evolution in the invasive range results from standing genetic variation rather than new mutation. Further, we can begin to understand whether the same genes are the targets of selection across multiple independent invasions of the same species; or if similar genes are the target of selection in different species invading the same habitat. Of course to get the most out of these experiments we first need to know which traits have been the target of selection-driven adaptive change in the introduced range.

Once adaptive phenotypic evolution has been established, such as in the case of *Phyla canescens* (Xu et al. 2010), we advocate the approach put forward by Stinchcombe & Hoekstra (2008) of a genome scan in combination with quantitative trait loci (QTL) mapping. These methods will allow identification of chromosomal regions and genes within these regions that underlie ecologically relevant traits important to adaptive evolution in invasive species.

Genome scans involve genotyping numerous loci throughout the genome of many individuals in several populations (Beaumont & Balding 2004) to detect 'outlier' loci that show unusually high levels of differentiation. This technique allows a distinction to be made between evolutionary forces that affect the whole genome (e.g. bottlenecks and drift) and those that affect particular loci (i.e. selection) (Stinchcombe & Hoekstra 2008). Historically, genome scans have used amplified fragment length polymorphism (AFLP) markers or microsatellites (SSRs) obtained from the expressed portion of the genome to detect outlier loci. Although this approach has had some success in finding candidate genes underlying adaptive phenotypic evolution (Stinchcombe & Hoekstra 2008), it can lack power to detect outliers using these marker systems. By applying this analysis technique to RAD markers (Miller et al. 2007), generated with high-throughput next generation sequencing, the power of genome scans can be substantially improved. Hohenlohe et al. (2010) conducted a genome scan using Illumina-sequenced RAD tags in two marine and three freshwater populations of threespine stickleback and were able to identify and genotype over 45,000 single nucleotide polymorphisms (SNPs). Using the statistical power of population genomics, this study identified several genomic regions indicative of divergent selection after the colonization of freshwater. This approach in isolation, however, has a major limitation as it is often unclear which 'outlier' loci underlie phenotypic traits of interest. To overcome this problem, genome scans should be used in combination with QTL mapping.

QTL mapping uses statistical analyses of molecular markers distributed throughout the genome and traits measured in the progeny of controlled crosses to identify stretches of DNA that are closely linked to genes that underlie the trait in question (Stinchcombe & Hoekstra 2008). Consequently, if it is possible to make controlled crosses between native and introduced individuals of the same species differing in ecologically relevant traits of interest, QTL mapping can be used to identify the genomic regions associated with these traits. This approach requires a large amount of genetic markers to saturate the genome, and population genomic markers such as sequenced RAD tags make an ideal marker for QTL mapping. QTL mapping with RAD tags in controlled crosses of freshwater and marine sticklebacks (see Baird et al. 2008) fine mapped the genetic bases conferring reduction of lateral plate armour in freshwater populations of the three-spine stickleback by identifying recombinant breakpoints in F2 individuals. Theoretically, by adding the outlier loci detected by Hohenlohe et al. (2010) to this QTL map, it would be possible to determine if the outlier loci identified in genome scans are found in the chromosomal regions underlying the reduction of lateral plate armour. Experimental evolution research has determined that a reduction in lateral plate number is under selection in freshwater populations of three-spine stickleback (Barrett et al. 2008). Therefore, a population genomics approach using sequenced RAD tags has identified the genetic bases of adaptive evolution in this case.

A combined genome scan and QTL mapping approach can help to identify genes or small chromosomal regions underlying adaptive phenotypic change. A well-illustrated example of this approach comes from maize, where outlier loci identified from genome scans were integrated onto linkage maps to determine if they map to chromosomal regions underlying domestication traits that have a history of artificial selection (Vigouroux et al. 2002). *Zea mays* ssp. *mays* (maize) was crossed to its wild progenitor *Z. mays* ssp. *parviglumis* (Teosinte) and two outlier loci detected from genome scans were found to map close to a chromosomal region controlling two traits that differ significantly between maize and teosinte (Vigouroux et al. 2002). Although this example is not of an invasive species, it does highlight how a combined approach can be applied to study adaptive phenotypic change in any organism including invasive species.

Overall these examples demonstrate that genome scans and QTL mapping, using RAD markers generated by next generation sequencing, can be used to determine the genetic bases of adaptive phenotypic evolution of species. Using these new technological advances has provided unprecedented insights into the genetics underlying adaptive trait evolution and validated previous research with much less effort. It is now time to apply these new technologies to biological invasions. We believe that using this combined population genomics approach can move the field of invasion biology forward by allowing researchers to determine the genetic basis of ecologically relevant traits involved in successful invasions, and potentially to identify candidate genes or mutations involved in rapid adaptive evolution of invasive species.

ACKNOWLEDGEMENTS

We thank David Richardson for the opportunity to contribute this chapter. We are especially indebted to Spencer Barrett and two anonymous reviewers, whose constructive criticisms greatly improved this chapter. P.J.P. thanks Daniel Ortiz-Barrientos for long and thought-provoking discussions on this topic. This work was supported in part by a grant from the Council of Australasian Weed Societies awarded to E.E.D. and an ARC Discovery Grant (DP0664967) awarded to A.J.L.

REFERENCES

Abbott, R.J. (1992) Plant invasions, interspecific hybridization and the evolution of new plant taxa. *Trends in Ecology & Evolution*, **7**, 401–405.

Amsellem, L., Noyer, J.L., Le Bourgeois, T. & Hossaert-McKey, M. (2000) Comparison of genetic diversity of the invasive weed *Rubus alceifolius* Poir. (Rosaceae) in its native range and in areas of introduction, using amplified fragment length polymorphism (AFLP) markers. *Molecular Ecology*, **9**, 443–455.

Amsellem, L., Noyer, J.L. & Hossaert-McKey, M. (2001) Evidence for a switch in the reproductive biology of *Rubus alceifolius* (Rosaceae) towards apomixis, between its native range and its area of introduction. *American Journal of Botany*, **88**, 2243–2251.

Arnold, M.L. & Hodges, S.A. (1995) Are natural hybrids fit or unfit relative to their parents? *Trends in Ecology & Evolution*, **10**, 67–70.

Baack, E.J. and Rieseberg, L.H. (2007) A genomic view of introgression and hybrid speciation. *Current Opinion in Genetics and Development*, **17**, 513–518.

Bailey, J.P., Bimova, K. & Mandak, B. (2009) Asexual spread versus sexual reproduction and evolution in Japanese Knotweeds.1. sets the stage for the 'Battle of the Clones'. *Biological Invasions*, **11**, 1189–1203.

Baird, N.A., Etter, P.D., Atwood, T.S., Currey, M.C., Shiver, A.L., Lewis, Z.A., Sleker, E.U. Cresko, W.A. & Johnson, E.A. (2008) Rapid SNP discovery and genetic mapping using sequenced RAD markers. *PLoS ONE*, **3**, e3376.

Baker, H.G. (1955) Self-compatibility and establishment after 'long-distance' dispersal. *Evolution*, **9**, 347–349.

Baker, H.G. (1974) The evolution of weeds. *Annual Review of Ecology and Systematics*, **5**, 1–24.

Baker, H.G & Stebbins, G.L. (eds.) (1965) *The genetics of colonizing species*. Academic Press, New York.

Barrett, R.D.H., Rogers, S.M. & Schluter, D. (2008) Natural selection on a major armor gene in threespine stickleback. *Science*, **322**, 255–257.

Barrett, R.D.H. & Schluter, D. (2008) Adaptation from standing genetic variation. *Trends in Ecology & Evolution*, **23**, 38–44.

Beaumont, M.A. and Balding, D.J. (2004) Identifying adaptive genetic divergence among populations from genome scans. *Molecular Ecology*, **13**, 969–980.

Beaumont, H.J.E., Gallie, J., Kost, C., Ferguson, G.C. & Rainey, P.B. (2009) Experimental evolution of bet-hedging. *Nature*, **462**, 90–93.

Beckstead, J. & Parker, I.M. (2003) Invasiveness of *Ammophila arenaria*: Release from soil-borne pathogens? *Ecology*, **84**, 2824–2831.

Besnard, G., Henry, P., Wille, L., Cooke, D. & Chapuis, E. (2007) On the origin of the invasive olives (*Olea europaea* L., Oleaceae). *Heredity*, **99**, 608–619.

Blossey, B. & Notzold, R. (1995) Evolution of increased competitive ability in invasive nonindigenous plants – a hypothesis. *Journal of Ecology*, **83**, 887–889.

Blum, M.J., Bando, K.J., Katz, M. & Strong, D.R. (2007) Geographic structure, genetic diversity and source tracking of *Spartina alterniflora*. *Journal of Biogeography*, **34**, 2055–2069.

Blumenthal, D.M. (2006) Interactions between resource availability and enemy release in plant invasion. *Ecology Letters*, **9**, 887–895.

Bossdorf, O., Auge, H., Lafuma, L., Rogers, W.E., Siemann, E. & Prati, D. (2005) Phenotypic and genetic differentiation between native and introduced plant populations. *Oecologia*, **144**, 1–11.

Bossdorf, O., Prati, D., Auge, H. & Schmid, B. (2004) Reduced competitive ability in an invasive plant. *Ecology Letters*, **7**, 346–353.

Brown, A.H.D. & Marshall, D.R. (1981) Evolutionary changes accompanying colonization in plants. In *Evolution Today* (ed. G.E.C. Scudder and J.L. Reveal), pp. 351–363. Carnegie-Mellon University, Pittsburgh.

Burdon, J.J. & Brown, A.H.D. (1986) Population genetics of *Echium plantagineum* L. – target weed for biological control. *Australian Journal of Biological Sciences*, **39**, 369–378.

Callaway, R.M. & Maron, J.L. (2006) What have exotic plant invasions taught us over the past 20 years? *Trends in Ecology & Evolution*, **21**, 369–374.

Cappuccino, N. & Carpenter, D. (2005) Invasive exotic plants suffer less herbivory than non-invasive exotic plants. *Biology Letters*, **1**, 435–438.

Chaboudez, P. (1994) Patterns of clonal variation in skeleton weed (*Chondrilla juncea*), an apomictic species. *Australian Journal of Botany*, **42**, 283–295.

Chun, Y.J., Fumanal, B., Laitungm, B. & Bretagnolle, F. (2010) Gene flow and population admixture as the primary post-invasion processes in common ragweed *Ambrosia artemisiifolia* populations in France. *New Phytologist*, **185**, 1100–1107.

Clegg, M.T. & Allard, R.W. (1972) Patterns of genetic differentiation in the slender wild oat species *Avena barbata*.

Proceedings of the National Academy of Sciences of the USA, **69**, 1820–1824.

Colautti, R.I., Eckert, C.G. & Barrett, S.C.H. (2010) Evolutionary constraints on adaptive evolution during range expansion in an invasive plant. *Proceedings of the Royal Society B*, **277**, 1799–1806.

Colautti, R.I., Grigorovich, I.A. & MacIsaac, H.J. (2006) Propagule pressure: a null model for biological invasions. *Biological Invasions*, **8**, 1023–1037.

Colautti, R.I., Maron, J.L. & Barrett, S.C.H. (2009) Common garden comparisons of native and introduced plant populations: latitudinal clines can obscure evolutionary inferences. *Evolutionary Applications*, **2**, 187–199.

Colautti, R.I., Ricciardi, A., Grigorovich, I.A. & MacIsaac, H.J. (2004) Is invasion success explained by the enemy release hypothesis? *Ecology Letters*, **7**, 721–733.

Colosimo, P.F., Hosemann, K.E., Balabhadra, S., et al. (2005) Widespread parallel evolution in sticklebacks by repeated fixation of ectodysplasin alleles. *Science*, **5717**, 1928–1933.

Darwin, C. (1859) *On the Origin of Species by Means of Natural Selection or the Preservation of Favoured Races in the Struggle for Life*. Murray, London.

DeWalt, S.J., Denslow, J.S. & Ickes, K. (2004) Natural-enemy release facilitates habitat expansion of the invasive tropical shrub *Clidemia hirta*. *Ecology*, **85**, 471–483.

DeWalt, S.J. & Hamrick, J.L. (2004) Genetic variation of introduced Hawaiian and native Costa Rican populations of an invasive tropical shrub, *Clidemia hirta* (Melastomataceae). *American Journal of Botany*, **91**, 1155–1163.

Dlugosch, K.M. & Parker, I.M. (2008a) Founding events in species invasions: genetic variation, adaptive evolution, and the role of multiple introductions. *Molecular Ecology*, **17**, 431–449.

Dlugosch, K.M. & Parker, I.M. (2008b) Invading populations of an ornamental shrub show rapid life history evolution despite genetic bottlenecks. *Ecology Letters*, **11**, 701–709.

Durka, W., Bossdorf, O., Prati, D. & Auge, H. (2005) Molecular evidence for multiple introductions of garlic mustard (*Alliaria petiolata*, Brassicaceae) to North America. *Molecular Ecology*, **14**, 1697–1706.

Edwards, P.J., Frey, D., Bailer, H. & Baltisberger, M. (2006) Genetic variation in native and invasive populations of *Erigeron annuus* as assessed by RAPD markers. *International Journal of Plant Sciences*, **167**, 93–101.

Ellstrand, N.C. & Schierenbeck, K (2000) Hybridization as a stimulus for the evolution of invasiveness in plants? *Proceedings of the National Academy of Sciences of the USA*, **97**, 7043–7050.

Elton, C.S. (1958) *The Ecology of Invasions by Animals and Plants*. Methuen, London.

Eppinga, M.B., Rietkerk, M., Dekker, S.C., De Ruiter, P.C. & Van der Putten, W.H. (2006) Accumulation of local pathogens: a new hypothesis to explain exotic plant invasions. *Oikos*, **114**, 168–176.

Eriksson, O., Wikstrom, S., Eriksson, A. & Lindborg, R. (2006) Species-rich Scandinavian grasslands are inherently open to invasion. *Biological Invasions*, **8**, 355–363.

Estoup, A., Wilson, I.J., Sullivan, C., Cornuet, J.M. & Moritz, C. (2001) Inferring population history from microsatellite and enzyme data in serially introduced cane toads, *Bufo marinus*. *Genetics*, **159**, 1671–1687.

Fatemi, M., Gross, C., Julien, M. & Duggin, J.A. (2008) *Phyla canescens*: multiple introductions into Australia as revealed by ISSR markers and nuclear ribosomal DNA internal transcribed spacers (ITS). In *Proceedings of the 16th Australian Weeds Conference* (ed. R.D. Van Klinken, V.A. Osten, F.D. Panetta and J.C. Scanlan), pp. 247–249. Queensland Weeds Society, Brisbane.

Fridley, J.D., Stachowicz, J.J., Naeem, S., Sax, D.F., Seabloom, E.W., Smith, M.D., Stohlgren, T.J., Tilman, D. & Von Holle, B. (2007) The invasion paradox: reconciling pattern and process in species invasions. *Ecology*, **88**, 3–17.

Garcia, P., Vences, F.J., Delavega, M.P. & Allard, R.W. (1989) Allelic and genotypic composition of ancestral Spanish and colonial Californian gene pools of *Avena barbata* – evolutionary implications. *Genetics*, **122**, 687–694.

Genton, B.J., Shykoff, J.A. & Giraud, T. (2005) High genetic diversity in French invasive populations of common ragweed, *Ambrosia artemisiifolia*, as a result of multiple sources of introduction. *Molecular Ecology*, **14**, 4275–4285.

Grotkopp, E. & Rejmánek, M. (2007) High seedling relative growth rate and specific leaf area are traits of invasive species: phylogenetically independent contrasts of woody angiosperms. *American Journal of Botany*, **94**, 526–532.

Hannan-Jones, M.A., Lowe, A.J., Scott, K.D., Graham, G.C., Playford, J.P. & Zalucki, M.P. (2005) Isolation and characterization of microsatellite loci from mother-of-millions, *Bryophyllum delagoense* (Crassulaceae), and its hybrid with *Bryophyllum daigremontianum*, 'Houghton's hybrid'. *Molecular Ecology Notes*, **5**, 770–773.

Hendry, A.P., Nosil, P. & Rieseberg, L.H. (2007) The speed of ecological speciation. *Functional Ecology*, **21**, 455–464.

Hermisson, J. & Pennings, P.S. (2005) Soft sweeps: molecular population genetics of adaptation from standing genetic variation. *Genetics*, **169**, 2335–2352

Hierro, J.L., Maron, J.L. & Callaway, R.M. (2005) A biogeographical approach to plant invasions: the importance of studying exotics in their introduced and native range. *Journal of Ecology*, **93**, 5–15.

Hohenlohe, P.A., Bassham, S., Etter, P.D., Stiffler, N., Johnson, E.A. & Cresko, W.A. (2010) Population genomics of parallel adaptation in threespine stickleback using sequenced RAD tags. *PLoS Genetics*, **6**, e1000862.

Huey, R.B., Gilchrist, G.W., Carlson, M.L., Berrigan, D. & Serra, L. (2000) Rapid evolution of a geographic cline in size in an introduced fly. *Science*, **287**, 308–309.

Hull-Sanders, H.M., Clare, R., Johnson, R.H. & Meyer, G.A. (2007) Evaluation of the evolution of increased competitive ability (EICA) hypothesis: loss of defense against generalist but not specialist herbivores. *Journal of Chemical Ecology*, **33**, 781–799.

Hufbauer, R.A. & Sforza, R. (2008) Multiple introductions of two invasive *Centaurea* taxa inferred from cpDNA haplotypes. *Diversity and Distributions*, **14**, 252–261.

Jahodova, S., Trybush, S., Pyšek, P., Wade, M. & Karp, A. (2007) Invasive species of *Heracleum* in Europe: an insight into genetic relationships and invasion history. *Diversity and Distributions*, **13**, 99–114.

Joshi, J. & Vrieling, K. (2005) The enemy release and EICA hypothesis revisited: incorporating the fundamental difference between specialist and generalist herbivores. *Ecology Letters*, **8**, 704–714.

Kang, M., Buckley, Y.M. & Lowe, A.J. (2007) Testing the role of genetic factors across multiple independent invasions of the shrub Scotch broom (*Cytisus scoparius*). *Molecular Ecology*, **16**, 4662–4673.

Keane, R.M. & Crawley, M.J. (2002) Exotic plant invasions and the enemy release hypothesis. *Trends in Ecology & Evolution*, **17**, 164–170.

Keller, S.R., Sowell, D.R., Neiman, M., Wolfe, L.M. & Taylor D.R. (2009) Adaptation and colonization history affect the evolution of clines in two introduced species. *New Phytologist*, **183**, 678–690.

Keller, S.R. & Taylor, D.R. (2008) History, chance and adaptation during biological invasion: separating stochastic phenotypic evolution from response to selection. *Ecology Letters*, **11**, 852–866.

Kolbe, J.J., Larson, A., Losos, J.B. & de Queiroz, K. (2008) Admixture determines genetic diversity and population differentiation in the biological invasion of a lizard species. *Biology Letters*, **4**, 434–437.

Lavergne, S. & Molofsky, J. (2007) Increased genetic variation and evolutionary potential drive the success of an invasive grass. *Proceedings of the National Academy of Sciences of the USA*, **104**, 3883–3888.

Lee, P.L.M., Patel, R.M., Conlan, R.S., Wainwright, S.J. & Hipkin, C.R. (2004) Comparison of genetic diversities in native and alien populations of hoary mustard (*Hirschfeldia incana* [L.] Lagreze-Fossat). *International Journal of Plant Sciences*, **165**, 833–843.

Le Roux, J.J., Wieczorek, A.M., Wright, M.G. & Tran, C.T. (2007) Super-genotype: global monoclonality defies the odds of nature. *PLoS ONE*, **2**, e590.

Lexer, C., Welch, M.F., Raymond, O. & Rieseberg, L.H. (2003) The origin of ecological divergence in *Helianthus paradoxus* (Asteraceae): selection on transgressive characters in a novel hybrid habitat. *Evolution*, **57**, 1989–2000.

Liu, H. & Stiling, P. (2006) Testing the enemy release hypothesis: a review and meta-analysis. *Biological Invasions*, **8**, 1535–1545.

Lockwood, J.L., Cassey, P. & Blackburn, T. (2005) The role of propagule pressure in explaining species invasions. *Trends in Ecology & Evolution*, **20**, 223–228.

Maron, J.L. (2006) The relative importance of latitude matching and propagule pressure in the colonization success of an invasive forb. *Ecography*, **29**, 819–826.

Maron, J.L., Vila, M., Bommarco, R., Elmendorf, S. & Beardsley, P. (2004) Rapid evolution of an invasive plant. *Ecological Monographs*, **74**, 261–280.

Marrs, R.A., Sforza, R. & Hufbauer, R.A. (2008a) Evidence for multiple introductions of *Centaurea stoebe micranthos* (spotted knapweed, Asteraceae) to North America. *Molecular Ecology*, **17**, 4197–4208.

Marrs, R.A., Sforza, R. & Hufbauer, R.A. (2008b) When invasion increases population genetic structure: a study with *Centaurea diffusa*. *Biological Invasions*, **10**, 561–572.

Meekins, J.F., Ballard, H.E. & McCarthy, B.C. (2001) Genetic variation and molecular biogeography of a North American invasive plant species (*Alliaria petiolata*, Brassicaceae). *International Journal of Plant Sciences*, **162**, 161–169.

Meimberg, H., Hammond, J.I., Jorgensen, C.M., et al. (2006) Molecular evidence for an extreme genetic bottleneck during introduction of an invading grass to California. *Biological Invasions*, **8**, 1355–1366.

Meyer, G., Clare, R. & Weber, E. (2005) An experimental test of the evolution of increased competitive ability hypothesis in goldenrod, *Solidago gigantea*. *Oecologia*, **144**, 299–307.

Meyer, G.A. & Hull-Sanders, H.M. (2008) Altered patterns of growth, physiology and reproduction in invasive genotypes of *Solidago gigantea* (Asteraceae). *Biological Invasions*, **10**, 303–317.

Miller, M.R., Dunham, J.P., Amores, A., Cresko, W.A. & Johnson, E.A. (2007) Rapid and cost-effective polymorphism identification and genotyping using restriction site associated DNA (RAD) markers. *Genome Research*, **17**, 240–248.

Milne, R.I. & Abbott, R.J. (2004) Geographic origin and taxonomic status of the invasive privet, *Ligustrum robustum* (Oleaceae), in the Mascarene Islands, determined by chloroplast DNA and RAPDs. *Heredity*, **92**, 78–87.

Mitchell, C.E. & Power, A.G. (2003) Release of invasive plants from fungal and viral pathogens. *Nature*, **421**, 625–627.

Molina-Freaner, F. & Jain, S.K. (1992) Isozyme variation in Californian and Turkish populations of the colonizing species *Trifolium hirtum*. *Journal of Heredity*, **83**, 423–430.

Muller, C. & Martens, N. (2005) Testing predictions of the 'evolution of increased competitive ability' hypothesis for an invasive crucifer. *Evolutionary Ecology*, **19**, 533–550.

Neuffer, B. & Hurka, H. (1999) Colonization history and introduction dynamics of *Capsella bursa-pastoris* (Brassicaceae) in North America: isozymes and quantitative traits. *Molecular Ecology*, **8**, 1667–1681.

Novak, S.J. & Mack, R.N. (1993) Genetic variation in *Bromus tectorum* (Poaceae) – comparison between native and introduced populations. *Heredity*, **71**, 167–176.

Novak, S.J., Mack, R.N. & Soltis, D.E. (1991) Genetic variation in *Bromus tectorum* (Poaceae) – population differentiation in its North-American range. *American Journal of Botany*, **78**, 1150–1161.

Orians, C.M. & Ward, D. (2010) Evolution of plant defenses in nonindigenous environments. *Annual Review of Entomology*, **55**, 439–459.

Parker, J.D., Burkepile, D.E. & Hay, M.E. (2006) Opposing effects of native and exotic herbivores on plant invasions. *Science*, **311**, 1459–1461.

Parker, J.D. & Hay, M.E. (2005) Biotic resistance to plant invasions? Native herbivores prefer non-native plants. *Ecology Letters*, **8**, 959–967.

Phillips, B.L., Brown, G.P., Webb, J.K. & Shine, R. (2006) Invasion and the evolution of speed in toads. *Nature*, **439**, 803.

Prentis, P.J., Sigg, D.P., Raghu, S., Dhileepan, K., Pavasovic, A. & Lowe, A.J. (2009) Understanding invasion history: genetic structure and diversity of two globally invasive plants and implications for their management. *Diversity and Distributions*, **15**, 822–830.

Prentis, P.J., White, E.M., Radford, I.J., Lowe, A.J. & Clarke, A.R. (2007) Can hybridization cause local extinction: a case for demographic swamping of the Australian native *Senecio pinnatifolius* by the invasive *Senecio madagascariensis*? *New Phytologist*, **176**, 902–912.

Prentis, P.J., Wilson J.R.U., Dormontt, E.E., Richardson, D.M. & Lowe, A.J. (2008) Adaptive evolution in invasive species. *Trends in Plant Science*, **13**, 288–294.

Rainey, P.B. & Travisano, M. (1998) Adaptive radiation in a heterogeneous environment. *Nature*, **394**, 69–72.

Rieseberg, L.H., Archer, M.A. & Wayne, R.K. (1999) Transgressive segregation, adaptation and speciation. *Heredity*, **83**, 363–372.

Rieseberg, L.H., Raymond, O., Rosenthal, D.M., et al. (2003) Major ecological transitions in wild sunflowers facilitated by hybridization. *Science*, **301**, 1211–1216.

Ricciardi, A. (2007) Are modern biological invasions an unprecedented form of global change? *Conservation Biology*, **21**, 329–336.

Richardson, D.M. & Pyšek, P. (2006) Plant invasions: merging the concepts of species invasiveness and community invasibility. *Progress in Physical Geography*, **30**, 409–431.

Roman, J. (2006) Diluting the founder effect: cryptic invasions expand a marine invader's range. *Proceedings of the Royal Society B*, **273**, 2453–2459.

Roman, J. & Darling, J.A. (2007) Paradox lost: genetic diversity and the success of aquatic invasions. *Trends in Ecology & Evolution*, **22**, 454–464.

Rosenthal, D.M., Ramakrishnan, A.P. & Cruzan, M.B. (2008) Evidence for multiple sources of invasion and intraspecific hybridization in *Brachypodium sylvaticum* (Hudson) Beauv. in North America. *Molecular Ecology*, **17**, 4657–4669.

Rouget, M. & Richardson, D.M. (2003) Inferring process from pattern in plant invasions: A semimechanistic model incorporating propagule pressure and environmental factors. *American Naturalist*, **162**, 713–724.

Rundle, H.D. & Whitlock, M.C. (2001) A genetic interpretation of ecologically dependent isolation. *Evolution*, **55**, 198–201.

Saltonstall, K. (2002) Cryptic invasion by a non-native genotype of the common reed, *Phragmites australis*, into North America. *Proceedings of the National Academy of Sciences of the USA*, **99**, 2445–2449.

Schmid-Hempel, P., Schmid-Hempel, R., Brunner, P.C., Seeman, O.D. & Allen, G.R. (2007). Invasion success of the bumblebee, *Bombus terrestris*, despite a drastic genetic bottleneck. *Heredity*, **99**, 414–422.

Siemann, E., Rogers, W.E. & Dewalt, S.J. (2006) Rapid adaptation of insect herbivores to an invasive plant. *Proceedings of the Royal Society B*, **273**, 2763–2769.

Simberloff, D. (2009) The role of propagule pressure in biological invasions. *Annual Review of Ecology Evolution and Systematics*, **40**, 81–102.

Squirrell, J., Hollingsworth, P.M., Bateman, R.M., et al. (2001a) Partitioning and diversity of nuclear and organelle markers in native and introduced populations of *Epipactis helleborine* (Orchidaceae). *American Journal of Botany*, **88**, 1409–1418.

Squirrell, J., Hollingsworth, P.M., Bateman, R.M., et al. (2001b) Erratum: partitioning and diversity of nuclear and organelle markers in native and introduced populations of *Epipactis helleborine* (Orchidaceae) *American Journal of Botany*, **88**, 1927.

Stastny, M., Schaffner, U. & Elle, E. (2005) Do vigour of introduced populations and escape from specialist herbivores contribute to invasiveness? *Journal of Ecology*, **93**, 27–37.

Stebbins, G.L. (1957) Self fertilization and population variability in the higher plants. *American Naturalist*, **91**, 337–354.

Stinchcombe, J.R. & Hoekstra, H.E. (2008) Combining population genomics and quantitative genetics: finding the genes underlying ecologically important traits. *Heredity*, **100**, 158–170.

Sun, J.H., Li, Z.C., Jewett, D.K., Britton, K.O., Ye, W.H. & Ge, X.J. (2005) Genetic diversity of *Pueraria lobata* (kudzu) and closely related taxa as revealed by inter-simple sequence repeat analysis. *Weed Research*, **45**, 255–260.

Taylor, D.R. & Keller, S.R. (2007) Historical range expansion determines the phylogenetic diversity introduced during contemporary species invasion. *Evolution*, **61**, 334–345.

Thorpe, A.S., Thelen, G.C., Diaconu, A. & Callaway, R.M. (2009) Root exudate is allelopathic in invaded community but not in native community: field evidence for the novel weapons hypothesis. *Journal of Ecology*, **97**, 641–645.

Thuiller, W., Richardson, D.M., Rouget, M., Procheş, S. & Wilson, J.R.U. (2006) Interactions between environment, species traits, and human uses describe patterns of plant invasions. *Ecology*, **87**, 1755–1769.

Travis, J.M.J., Hammershoj, M. & Stephenson, C. (2005) Adaptation and propagule pressure determine invasion dynamics: insights from a spatially explicit model for sexually reproducing species. *Evolutionary Ecology Research*, **7**, 37–51.

Von Holle, B. & Simberloff, D. (2005) Ecological resistance to biological invasion overwhelmed by propagule pressure. *Ecology*, **86**, 3212–3218.

Vigouroux, Y., McMullen, M., Hittinger, C.T., et al. (2002) Identifying genes of agronomic importance in maize by screening microsatellites for evidence of selection during domestication. *Proceedings of the National Academy of Sciences of the USA*, **99**, 9650–9655.

Warwick, S.I., Thompson, B.K. & Black, L.D. (1987) Genetic-variation in Canadian and European populations of the colonizing weed species *Apera spica-venti*. *New Phytologist*, **106**, 301–317.

Whitney, K.D., Ahern, J.R. & Campbell, L.G. (2009) Hybridization-prone plant families do not generate more invasive species. *Biological Invasions*, **11**, 1205–1215.

Whitney, K.D. & Gabler, C.A. (2008) Rapid evolution in introduced species, 'invasive traits' and recipient communities: challenges for predicting invasive potential. *Diversity and Distributions*, **14**, 569–580.

Whitney, K.D., Randell, R.A. & Rieseberg, L.H. (2006) Adaptive introgression of herbivore resistance traits in the weedy sunflower *Helianthus annuus*. *American Naturalist*, **167**, 794–807.

Whitney, K.D., Randell, R.A. & Rieseberg, L.H. (2010) Adaptive introgression of abiotic tolerance traits in the sunflower *Helianthus annuus*. *New Phytologist*, **187**, 230–239.

Wikstrom, S.A., Steinarsdottir, M.B., Kautsky, L. & Pavia, H. (2006) Increased chemical resistance explains low herbivore colonization of introduced seaweed. *Oecologia*, **148**, 593–601.

Williams, D.A., Overholt, W.A., Cuda, J.P. & Hughes, C.R. (2005) Chloroplast and microsatellite DNA diversities reveal the introduction history of Brazilian peppertree (*Schinus terebinthifolius*) in Florida. *Molecular Ecology*, **14**, 3643–3656.

Wilson, J.R.U., Richardson, D.M., Rouget, M., et al. (2007) Residence time and potential range: crucial considerations in modelling plant invasions. *Diversity and Distributions*, **13**, 11–22.

Wilson, J.R.U., Dormontt, E.E., Prentis, P.J., Lowe, A.J. & Richardson, D.M. (2009) Something in the way you move: dispersal pathways affect invasion success. *Trends in Ecology & Evolution*, **24**, 136–144.

Wootton, J.C., Feng, X., Ferdig, M.T., et al. (2002) Genetic diversity and chloroquine selective sweeps in *Plasmodium falciparum*. *Nature*, **418**, 320–323.

Xu, C.Y., Julien, M.H., Fatemi, M., et al. (2010) Phenotypic divergence during the invasion of *Phyla canescens* in Australia and France: evidence for selection-driven evolution. *Ecology Letters*, **13**, 32–44.

Zang, Y.Y., Zang, D.Y. & Barrett, S.C.H. (2010) Genetic uniformity characterizes the invasive spread of water hyacinth (*Eichhornia crassipes*), a clonal aquatic plant. *Molecular Ecology*, **19**, 1774–1786.

Zangerl, A.R. & Berenbaum, M.R. (2005) Increase in toxicity of an invasive weed after reassociation with its coevolved herbivore. *Proceedings of the National Academy of Sciences of the USA*, **102**, 15529–15532.

WHY REPRODUCTIVE SYSTEMS MATTER FOR THE INVASION BIOLOGY OF PLANTS

Spencer C. H. Barrett

Department of Ecology and Evolutionary Biology, University of Toronto, Toronto, Ontario M5S 3B2, Canada

Fifty Years of Invasion Ecology: The Legacy of Charles Elton, 1st edition. Edited by David M. Richardson

15.1   INTRODUCTION

Human-assisted species introductions and the invasion of newly occupied territory have long fascinated naturalists and biogeographers. In the *Origin of Species*, Charles Darwin (1859) discussed introduced species and considered how their taxonomic relationships might influence competitive interactions and community composition ('Darwin's naturalization hypothesis'; see Hill & Kotanen (2009)). Despite this early interest it was not until almost a century later, with the publication of *The Ecology of Invasions by Animals and Plants* by the Oxford zoologist Charles S. Elton in 1958, that biological invasions were recognized as a distinct phenomenon worthy of study. Today, Elton's volume is still the most cited work in the field of invasion biology and many of the topics he discussed are still active areas of research (Richardson & Pyšek 2008).

Much of Elton's book concerned case histories of animal species and genetic and evolutionary issues were only mentioned occasionally. He briefly considered the role of polyploidy in colonization of mudflats by *Spartina townsendii* (p. 26), speculated on the role of genetics in the decline of Canadian Pondweed (*Elodea canadensis*) in the UK (p. 115) and mentioned the evolution of resistance in insect pests and fungi (p. 141). However, he ignored the evolutionary consequences of biological invasions and the possibility that local adaptation may play a role in the spread of invasive species. In essence, Elton treated species as uniform entities despite the explosive ecological and demographic changes that he documented so well. From a historical viewpoint this is surprising; Oxford was the birthplace of 'ecological genetics' (Ford 1964), evolutionary studies were actively pursued there when his book was written, and Elton's own work was cited in Ford's classic volume.

An evolutionary perspective on biological invasions was not to emerge until the appearance of *The Genetics of Colonizing Species*, edited by the plant evolutionists Herbert G. Baker and G. Ledyard Stebbins (Baker & Stebbins 1965). This influential work arose from a conference in Asilomar, California, the preceding year and among the contributors were many leading evolutionary biologists and several prominent ecologists. The volume focused on the evolutionary changes that occur when plants and animals are introduced to novel environments, although contributors also considered traits promoting colonizing success, and the role of stochastic forces in affecting patterns of genetic variation during colonization. Baker and Stebbins' volume initiated a new branch of invasion biology firmly rooted in population and evolutionary genetics.

An evolutionary framework for understanding biological invasions has yet to fully emerge and the field still tends to be dominated by case histories despite some attempts at synthesis (Parsons 1983; Williamson 1996; Cox 2004; Sax et al. 2005). One of the difficulties is that invasive species represent a heterogeneous assortment of taxa, with much greater variety in species attributes than is often appreciated. This is especially the case for flowering plants because of their great diversity in life history, reproductive biology and genetic systems. This variation makes attempts at generality a daunting task, even though plants are subject to many of the same ecological and demographic mechanisms that drive evolutionary change in animal populations. Nevertheless, several of the distinctive features of plants including their immobility, hermaphroditism, modularity and clonality, although not unique to plants, influence the character of the invasion process.

In this chapter I consider why reproductive systems matter for the invasion biology of plants. I focus on reproductive systems for two reasons. First, among plant life history traits, modes of reproduction are the most influential in governing evolutionary response to environmental change. This fundamental role arises because reproduction governs genetic transmission between generations, and this in turn affects the organization of genetic variation within and among populations (population genetic structure). Second, plants display enormous diversity in their modes of reproduction, affecting many features of the invasion process including biogeography, demography, and opportunities for the evolution of local adaptation. In this review I pay particular attention to two key topics: (i) plant reproductive diversity and its influence on genetic variation and the evolution of local adaptation; (ii) the role of demographic forces, particularly low density and how this may affect selection on reproductive systems. I conclude by considering future developments in invasion biology and how recent advances in evolutionary genetics may assist in promoting maturity of the field.

15.2   PLANT REPRODUCTIVE DIVERSITY

The reproductive system of a population encompasses traits that determine (i) the balance between sexual

Fig. 15.1 Variation in reproductive systems in flowering plants; the apices (top, left, right) of the triangle represent asexual reproduction (clonal propagation and apomixis), cross-fertilization and self-fertilization, respectively. Species or populations can be located in or on the triangle, depending on their reproductive modes. Arrows indicate shifts in reproductive systems commonly associated with invasion, as discussed in the text for Brazilian (*EpB*) outcrossing and Caribbean (*EpC*) selfing populations of annual *Eichhornia paniculata*, and *E. crassipes* populations from its native (*EcN*) and introduced (*EcI*) range, which differ in the relative importance of clonal versus sexual reproduction. *Eichhornia crassipes* is self-compatible and produces both cross- and self-fertilized seed.

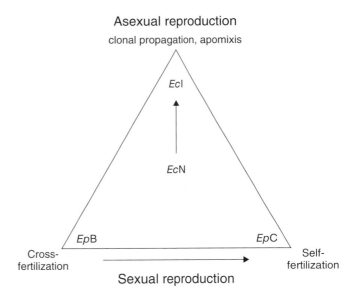

Asexual reproduction
clonal propagation, apomixis

*Ec*I

*Ec*N

*Ep*B   *Ep*C

Cross-fertilization   Self-fertilization

Sexual reproduction

and asexual reproduction, and (ii) the relative amounts of cross-fertilization ('outcrossing') versus self-fertilization ('selfing'). Figure 15.1 illustrates these axes of variation in the form of a triangle in which species or populations can be located, depending on their modes of reproduction (Fryxell 1957). Many angiosperms combine several reproductive modes, and one of the striking features of plant reproductive systems is the diversity that can occur among related species, and among populations within a species. In Fig. 15.1 the location of populations of two *Eichhornia* species featured in this chapter serve to emphasize shifts in reproductive modes that are commonly associated with invasion. Reproductive systems are evolutionarily labile and transitions among different modes are commonplace, especially in herbaceous species with widespread distributions (reviewed in Barrett 2008). In this chapter, I consider how invasion fosters such changes and why transitions to uniparental reproduction are a common trend (as indicated in Fig. 15.1). Also, because plants have little direct control over the environments in which they reproduce, ecological and demographic factors play an important role in determining variation in reproductive systems. Much of the observed diversity in modes of reproduction among populations is therefore context dependent.

The reproductive system is an especially important life-history trait for invasive species because the quantity and genetic quality of propagules influences dispersal and spread to novel environments with both demographic and biogeographical consequences. Reproductive systems also determine opportunities for adaptive evolution because they influence population-genetic parameters including levels of diversity, rates of recombination, effective population size, gene flow and selection response (Schoen & Brown 1991; Charlesworth 1992; Hamrick & Godt 1996). Thus knowledge of the reproductive system is crucial for understanding the demographic and genetic characteristics of populations and for predicting how invasive species are likely to respond to future environmental challenges.

Despite the complexity of plant reproductive systems, a fundamental dichotomy is whether offspring arise from uniparental or biparental reproduction. This distinction is of particular importance for invasive species because mates or pollinators may be limiting during establishment and during subsequent colonizing episodes. Under these circumstances, uniparental reproduction may be favoured over biparental reproduction, especially in species with short life histories. Two axes of quantitative variation are also of significance for invasive populations. First, for species that reproduce both sexually and asexually it is important to establish their relative frequency, as this can influence the clonal diversity of populations (Silvertown 2008). Similarly, determining the fraction of seeds resulting from cross- versus self-fertilization is relevant for understanding the maintenance of heterozygosity and offspring fitness. Selfing is commonly associated with inbreeding depression and this can have both demographic and evolutionary consequences (Charlesworth & Charlesworth 1987). The use of polymorphic allozyme and microsatellite markers allows measurement of these variables and provides a more quantitative assessment of plant reproductive systems than inferences based on morphology. Much research on plant reproduction (reviewed in Barrett & Eckert 1990; Goodwillie et al. 2005) is concerned with measuring mating patterns in plant populations and evaluating the fitness consequences of outcrossing and selfing.

A recent survey of mating patterns reported that populations occurring in open disturbed habitats on average experienced higher selfing rates than those in undisturbed habitats (Eckert et al. 2010). There is now growing interest in whether anthropogenic changes to environments, including habitat fragmentation, competition between native and introduced species, and various facets of global climate change, influence plant reproduction (Aguilar et al. 2006; Aizen and Vázquez 2006; Traveset & Richardson 2006; Bjerknes et al. 2007; Memmott et al. 2007). Shifts from specialized to generalized pollination and from biparental to uniparental reproduction may characterize future trends in the reproductive biology of populations in human-altered environments.

15.3 REPRODUCTIVE SYSTEMS AND GENETIC DIVERSITY

Early studies during the biosystematics era established that mating systems were a strong determinant of the patterns of phenotypic variation within and among plant populations (Baker 1953). Inbreeding species generally contained relatively low levels of variation within populations but displayed high differentiation among populations. In contrast, populations of outbreeders maintained more variation but were less differentiated from one another. These early generalizations have now been corroborated theoretically (Charlesworth & Charlesworth 1995) and are supported by surveys of diverse species using allozyme markers (Hamrick & Godt 1996) and, more recently, microsatellites and DNA nucleotide sequences (Charlesworth & Pannell 1999; Baudry et al. 2001;

Ness et al. 2010). Although the total molecular diversity of inbreeders is usually only slightly lower than outbreeders, differences are more dramatic at the population level with inbreeders maintaining, on average, about half of the diversity found in outcrossers. Population structure is more pronounced in inbreeders and this is reflected by significantly higher values of G_{st} and F_{st}, both measures of population subdivision. The partitioning of genetic diversity within and among populations is important for invasion biologists, as population differentiation is often associated with ecological differences in the behaviour of populations.

A variety of genetic and demographic factors cause selfing populations to have reduced genome-wide diversity compared with outcrossers (Wright et al. 2008). Homozygosity increases with the selfing rate causing reductions in effective population size (N_e) up to twofold with complete selfing (Nordborg 2000). Moreover, because of higher linkage disequilibrium in selfers other processes in the genome, including selective sweeps and background selection, further reduce genetic variation within populations (Charlesworth & Wright 2001). These processes can be augmented by the demographic and life history characteristics of species, especially genetic bottlenecks when a single individual founds a colony. Populations of many annual selfers are characterized by frequent colonization–extinction cycles and these demographic processes can lead to strong population subdivision and erosion of diversity within populations.

The effects of mating patterns and demography on genetic diversity are illustrated by considering the annual *Eichhornia paniculata* (Pontederiaceae), a colonizer of aquatic habitats in the neotropics. This species occurs primarily in northeast Brazil, where populations occupy ephemeral ponds and ditches and are most commonly tristylous, a genetic polymorphism promoting outcrossing. Through long-distance dispersal, *E. paniculata* has also colonized Cuba and Jamaica, where it is a weed of rice fields (see Plate 6a). In contrast to Brazil, populations in the Caribbean are primarily inbreeding because selfing variants capable of setting seed without the requirement of pollinators established after dispersal from South America (Barrett et al. 1989, 2008). Populations in the Caribbean have significantly lower levels of genetic diversity at allozyme loci due to high selfing and bottlenecks associated with island colonization (Barrett & Husband 1990) (Fig. 15.2a). A recent comparison of DNA nucleotide

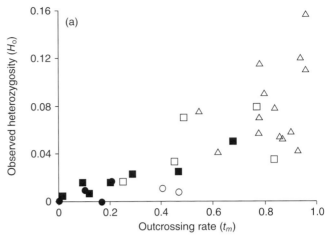

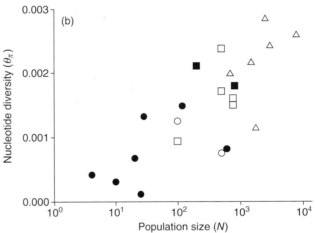

Fig. 15.2 Mating patterns and demography are associated with genetic diversity in *Eichhornia paniculata* populations. (a) Relation between multilocus outcrossing rate (t_m) and observed heterozygosity (H_o) at 19–27 allozyme loci in 32 populations from Brazil and Jamaica. Open and filled symbols are from Brazil and Jamaica, respectively. After Barrett & Husband (1990). (b) Relation between population size (N) and nucleotide diversity (θ_π) at 10 nuclear loci in 25 populations sampled from Brazil, Cuba and Jamaica. Open symbols are from Brazil, filled symbols are from Cuba and Jamaica. After Ness et al. (2010). In both (a) and (b) the triangles, squares and circles refer to trimorphic (outcrossing), dimorphic (mixed mating) and monomorphic (selfing) populations, respectively.

diversity in populations of varying size and style morph structure from the two regions revealed similar patterns (Ness et al. 2010) (Fig. 15.2b). Low levels of diversity characterized the smallest populations, which were largely composed of selfing variants, whereas larger tristylous populations maintained significantly higher levels of diversity. The association between selfing and small population size in *E. paniculata* highlights the important role that demographic history plays in affecting genetic diversity. The demographic and genetic characteristics of populations are not independent properties and disentangling their joint effects on patterns of diversity is a major challenge.

15.4 GENETIC VARIATION AND RAPID EVOLUTION OF LOCAL ADAPTATION

Biological invasions expose populations to novel selection pressures and the potential for rapid evolution of local adaptation (Prentis et al. 2008). Common garden studies are routinely used to detect differentiation among invasive populations in life-history traits of ecological significance (reviewed in Bossdorf et al. 2005). However, demonstrating that population differentiation results from the evolution of local adaptation is not straightforward because a variety of neutral processes associated with immigration history can also give rise to similar patterns (Maron et al. 2004; Colautti et al. 2009; Keller et al. 2009). The amount of genetic variation for ecologically relevant traits will determine how rapidly invasive populations respond to natural selection and the rate of response can directly affect the speed of biological invasion (García-Ramos & Rodríguez 2002). Therefore, understanding whether local adaptation is likely to evolve, and how rapidly it will occur, are important questions for invasion biologists and managers.

Evolution during the contemporary timescales typically investigated in studies of invasive species may be constrained by low standing genetic variation within populations and/or the availability of new mutations (Orr & Bettancourt 2001). If adaptation is initiated by new mutations it will depend on the rate at which these arise and whether they are beneficial. In contrast, if a large component of adaptive substitutions results from alleles already present in a population, the rate of change will depend on contemporary selective pressures, the history of selection and the previous environmental conditions that populations encountered (Barrett & Schluter 2008). Given the relatively short timescales involved in most invasions, it is more probable that adaptation will arise from standing genetic variation because beneficial alleles are more likely to be present at low to moderate frequencies.

In some cases, alleles that may be neutral or even deleterious in the native range may become advantageous after introduction to novel environments. This appears to be the case for genes controlling selfing in *E. paniculata* that have enabled colonization of the Caribbean, as discussed above. Selfing modifiers are not uncommon in outcrossing populations in Brazil, but under most circumstances do not spread because they have no advantage because pollinator service is reliable, unlike in the Caribbean (reviewed in Barrett et al. 2008). Understanding the sources and amounts of genetic variation contributing to the evolution of adaptive traits is important because this determines selective outcomes (Hermisson & Pennings 2005). Unfortunately, with the exception of a few well-studied examples of agricultural weeds, for example the evolution of herbicide resistance (reviewed in Neve et al. 2009), little is known about the origin and genetic basis of adaptations in invasive plants.

Most studies of genetic diversity in invasive populations have measured variation at neutral or near neutral genetic markers, usually allozymes (reviewed in Barrett & Shore 1989). Although genetic markers are of use in reconstructing the immigration history of introduced species (see, for example, Taylor & Keller 2007), they are of less value in predicting the amount of genetic variation for ecologically relevant traits, because the correlation between molecular and quantitative measures of genetic variation is weak at best (Lewontin 1984; Reed & Frankham 2001). The mating system also plays an important role in affecting this relation (Brown & Burdon 1987). In populations of many inbreeding weeds it is not uncommon to find very low levels, or no polymorphism at allozyme loci (see, for example, Schachner et al. 2008). However, genetic uniformity does not necessarily reflect an absence of variation in quantitative characters, and where this has been measured significant amounts of between-family (additive) genetic variation have been revealed in selfing species (Barrett 1988; Warwick 1990). Mutation rates of quantitative characters in selfers are sufficient to generate considerable heritable variation following genetic bottlenecks (Lande 1975, 1977). Adaptive evolution is therefore unlikely to be stalled in most selfing species because of an absence of genetic variation in life-history traits, except where populations have extremely small effective sizes (Schoen & Brown 1991). Indeed, a recent survey of transplant experiments of primarily selfing versus outcrossing species concluded that there was no difference in the likelihood of local adaptation evolving (Hereford 2010).

Reproductive systems mediate the conflicting roles of demographic and historical processes in affecting patterns of genetic variation, with consequences for the evolution of local adaptation. Founder events and genetic bottlenecks result in a loss of diversity, whereas multiple introductions and hybridization can increase diversity. Inbreeding restricts recombination and opportunities for genetic admixture magnifying founder effects and increasing the likelihood that stochastic processes will influence patterns of genetic diversity (Husband & Barrett 1991). In contrast, in outbreeding species hybridization with relatives can boost diversity fostering the evolution of increased invasiveness (Ellstrand & Schierenbeck 2000). Genetic admixture and the origin of genotypic novelty can also occur within invasive species as a result of multiple introductions to the introduced range followed by gene flow between formerly differentiated populations (Kolbe et al. 2004; Lavergne & Molofsky 2007). This process may result in levels of diversity in introduced populations that exceed those generally found in the native range (Novak & Mack 2005; Dlugosch & Parker 2008a). Multiple introductions from source regions are increasingly reported and the extent to which genetic admixture occurs will be strongly influenced by the mating system of populations. Levels of hybridization should be associated with likelihood of outcrossing, although even highly selfing populations are not immune from occasional outcrossing resulting in genetic admixture (Vaillant et al. 2007).

15.5 GENETIC CONSTRAINTS ON THE EVOLUTION OF ADAPTATION

Genetic constraints can restrict opportunities for the evolution of local adaptation, but there is little empirical evidence on how commonly they influence response to selection in natural populations (Antonovics 1976; Blows & Hoffman 2005). Biological invasions provide opportunities for investigating constraints on contemporary evolution because rapid range expansion, especially along climatic gradients, can expose populations to strong selection on traits associated with reproductive phenology. However, fitness trade-offs have the potential to constrain adaptive differentiation across environmental gradients, despite the occurrence of abundant genetic variation for life-history traits within populations (Etterson & Shaw 2001; Griffith & Watson 2006). Genetic constraints may also influence species distributions by preventing local adaptation at range margins, although this has rarely been investigated (reviewed in Eckert et al. 2008). Plant invasions provide experimental systems for understanding the genetic basis of range limits and the role of genetic constraints on responses to selection.

Recent investigations of the invasive spread of the Eurasian wetland plant *Lythrum salicaria* (purple loosestrife, Lythraceae, see Plate 6b) have examined limits to local adaptation associated with northern migration in eastern North America (Montague et al. 2008; Colautti et al. 2010a). This species is highly outcrossing, and molecular evidence suggests that multiple introductions followed by gene flow among introduced populations is responsible for the high levels of diversity within populations (Chun et al. 2009). The showy floral displays of introduced populations attract numerous insect pollinators that mediate pollen transport within and among populations.

Common garden experiments of *L. salicaria* populations sampled along a latitudinal transect of approximately 1200 km in eastern North America detected large amounts of heritable genetic variation in life-

history traits, of which days to first flower and size were of particular significance (Colautti et al. 2010a). Variation in these two traits formed distinct clines: northern populations flowered more rapidly at a smaller size and those from southern latitudes delayed flowering and were larger (Fig. 15.3a,b; and see Montague et al. 2008). This pattern was associated with a strong genetic correlation between time to first flower and vegetative size. Population means for these traits were distributed along the major axis of covariance of families within populations implicating strong constraints on population divergence (Fig. 15.3c). Northward spread of *L. salicaria* was also associated with a decline in genetic variation of quantitative characters and limited variation for small, early-flowering genotypes. These patterns are consistent with strong selection for local adaptation to shorter growing seasons when there is a trade-off between age and size at reproduction. Striking differences in fecundity distinguished northern and southern populations of *L. salicaria*. Northern populations produced only 4% of the fruit matured by southern populations (Colautti et al. 2010a). Reduced seed production should slow population growth and contribute to reducing rates of spread.

These results suggest that natural selection on reproductive phenology has accompanied the invasive spread of *L. salicaria* in eastern North America. However, genetic constraints and a dearth of early flowering genotypes in northern populations probably contribute to establishing the northern limit of the invasive range, at least over short timescales. Studies at the northern margin of the range of *L. salicaria* in Europe would be valuable to investigate whether genetic factors also play a role in establishing boundaries to the species' native range.

15.6 SELECTION ON REPRODUCTIVE SYSTEMS DURING INVASION

Colonization is often associated with low density and/or small population size. These demographic conditions can result in pollen limitation of seed set (Ashman et al. 2004) resulting in 'Allee effects', and a slowing of the invasion process (Allee 1931; Cheptou 2004; Taylor et al. 2004). Empirical efforts to detect Allee effects in invasive species have provided mixed results (Davis et al. 2004; Cheptou & Avendaño 2006; Elam et al. 2007; but see van Kleunen & Johnson 2005; van Kleunen et al. 2007). The vulnerability of populations

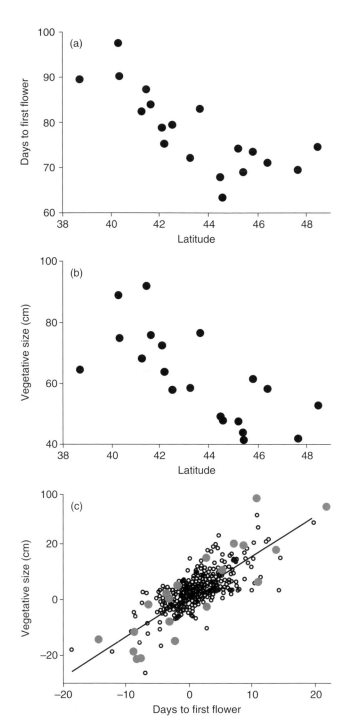

Fig. 15.3 Clinal variation and trade-offs between life-history traits associated with the evolution of local adaptation in invasive populations of *Lythrum salicaria* from eastern North America. Data are from 20 populations, 17 families per population and two individuals per family grown under uniform glasshouse conditions. See Colautti et al. (2010a) for details. (a) Relation between latitude (degrees north) and days to first flower based on population means; (b) relation between latitude (degrees north) and vegetative size based on population means; (c) relation between days to first flower and vegetative size for families (small open symbols) and population means (grey filled symbols). The strong positive relation indicates a genetic trade-off between the two traits.

to Allee effects will be strongly influenced by their mating systems. Self-incompatible species are more susceptible than those that are self-compatible to pollen limitation (Larson & Barrett 2000). Self-compatible plants, particularly those capable of autonomous self-pollination, are more likely to maintain fecundity and establish colonies from a single individual following dispersal. This general principle, known as 'Baker's law' (Baker 1955), was initially used to explain the rarity of self-incompatible species on oceanic islands compared with those that were self-compatible. However, it can be extended to any demographic situation where the reproduction of individuals is limited by an absence of pollinators or mates. This includes conditions at the leading edge of an invasion front or in a metapopulation in which colonization–extinction cycles result in episodes of low density (Pannell & Barrett 1998; Cheptou & Diekmann 2002; Dornier et al. 2008). Uniparental reproduction, by providing reproductive assurance, may allow small populations to grow faster than if they were outcrossing and reduce the duration of post-introduction bottlenecks.

Efforts to investigate the importance of uniparental reproduction for colonization have largely come from surveys of introduced species (Brown & Marshall 1981; Price & Jain 1981; Williamson & Fitter 1996; Rambuda & Johnson 2004). However, interpretation of associations between reproductive systems and colonizing success is complicated by the absence of data on relative rates of introduction (propagule pressure), and by the diverse phylogenetic affinities of species compared. A more powerful method uses paired samples of congeneric introduced species in which one has become invasive, and the other has not. Here, shared ancestry and equivalent opportunity are to some extent controlled. This approach identified the facility for autonomous selfing as an important correlate of invasiveness in South African Iridaceae exported to various regions for horticulture (van Kleunen et al. 2008). More studies of this type using comparative and phylogenetic methods combined with historical information on immigration history would be valuable.

Another way to investigate the reproductive consequences of the invasion process involves looking for direct evidence of evolutionary transitions to uniparental reproduction. Comparisons of conspecific native and introduced populations have been used to explore this question (Brown & Marshall 1981). Shifts to selfing or apomixis are commonly associated with island colonization (see, for example, Barrett et al.

1989; Barrett & Shore 1987; Amsellem et al. 2001). However, at more restricted spatial scales it has proved more difficult to find evidence for genetic changes to reproductive systems, particularly the breakdown in self-incompatibility to self-compatibility (see, for example, Cheptou *et. al.* 2002; Brennan et al. 2005; Cheptou & Avendaño 2006; Lafuma and Maurice 2007; Colautti et al. 2010b). Whether recurrent colonization episodes exert selection for uniparental reproductive systems depends on a range of factors including the spatial scale of colonization, gene flow, inbreeding depression, and the availability of genetic modifiers of mating system traits.

Many obligate outbreeders are successful colonizers, especially among introduced woody species where longevity reduces the risk of reproductive failure if mates arrive later in the invasion process. However, outcrossing enforced by self-incompatibility or dioecy should represent a severe liability for annual species because only a single season is available for mating. Not surprisingly in flowering plants as a whole, mating systems are strongly associated with life history, especially longevity, with annual and short-lived species exhibiting, on average, significantly higher selfing rates than long-lived woody species (Barrett et al. 1996). Nevertheless, some self-incompatible annuals have become successful invaders (see, for example, Sun & Ritland 1998; Brennan et al. 2005), raising the question of how these species maintain fecundity during population establishment.

Recent studies of common ragweed (*Ambrosia artemisiifolia*, Asteraceae) are instructive in this regard (Friedman & Barrett 2008). This species is a wind-pollinated annual native to North America but is now invasive in Europe, Asia and Australia. Ragweed had been assumed to be largely selfing, presumably because many annual invaders are selfers. However, Friedman & Barrett (2008) demonstrated high levels of outcrossing enforced by a strong self-incompatibility system using genetic markers. Ragweed possesses several traits that limit opportunities for Allee effects to occur. These include the production of large quantities of windborne pollen reducing the likelihood of pollen limitation, prolific seed production and dormant seed banks. Metapopulation models indicate that high fecundity and seed banks can offset costs associated with low density in self-incompatible colonizing species (Pannell & Barrett 1998). Not all invasive species are equally subject to Allee effects, as the dispersal biology of species plays a crucial role in establishment and spread. Studies investigating associations between reproductive and dispersal systems of invasive species are long overdue.

15.7 INVASION THROUGH ASEXUAL REPRODUCTION

So far, this review has largely considered sexual reproduction, where recombination and gene flow among populations provide the genetic fuel for the evolution of local adaptation. I now turn to species in which sex is infrequent and asexual reproduction predominates through either clonal propagation or apomixis. How likely is the evolution of local adaptation during invasion in these species?

Populations of predominantly clonal plant species are not necessarily genetically depauperate. On the contrary, surveys using genetic markers have demonstrated that populations can maintain considerable amounts of genetic diversity (reviewed in Silvertown 2008). Nevertheless, species that reproduce by asexual means are vulnerable to the influence of founder events, which can become magnified through predominant asexual reproduction, resulting in extensive areas of genetic uniformity. This is a particularly common feature of aquatic invaders through the dispersal of floating vegetative propagules (Barrett et al. 1993; Kliber & Eckert 2005; B. Wang et al. 2005). In some cases, either local or long-distance dispersal events have disabled sexual systems because of the stochastic loss of sexual morphs preventing seed reproduction (e.g. *Fallopia japonica* (Hollingsworth & Bailey 2000); *Nymphoides peltata* (Y. Wang et al. 2005)). When these situations occur, opportunities for local adaptation are curtailed and phenotypic plasticity plays an important role in enabling populations to respond to environmental heterogeneity.

During the past 150 years, water hyacinth (*Eichhornia crassipes*, Pontederiaceae) (see Plate 6c) has become the world's most serious invader of aquatic environments (Gopal & Sharma 1981). It is native to lowland South America but has spread to more than 50 countries, including tropical, subtropical and temperate zones on five continents. Several attributes of *E. crassipes* contribute to its success, including prolific clonal reproduction, the mobility of its free-floating life form and high rates of growth. Interestingly, these features are not represented among the remaining seven species of *Eichhornia*, none of which have become

serious weeds (Barrett 1992). This represents a rare case where the traits responsible for invasion success are clear.

Water hyacinth is tristylous, and the geographical distribution of floral morphs indicates that founder events have played a prominent role in the species' worldwide spread (Barrett 1989). Tristylous populations are confined to lowland South America, but in the introduced range the mid-styled morph predominates, with the long-styled morph occurring infrequently. A recent global survey of amplified fragment length polymorphism in ramets sampled from 54 populations has recently confirmed that genetic bottlenecks characterize the species invasive spread (Zhang et al. 2010). Although 49 clones were detected, introduced populations exhibited very low genetic diversity and little differentiation compared with those from the native range. Eighty per cent of introduced populations were genetically uniform, with one clone dominating in 74.5% of the populations sampled (Fig. 15.4). These patterns of genetic diversity result from extreme bottlenecks during colonization and prolific clonal propagation. This study clearly demonstrates that significant amounts of genetic diversity are not necessarily a prerequisite for global invasive spread over contemporary timescales.

Continent-wide genetic uniformity also occurs in invasive species that reproduce asexually by apomixis (but see Maron et al. (2004) for a counter-example). A recent survey of amplified fragment length polymorphism diversity in *Hieracium aurantiacum* (Asteraceae) revealed almost no genetic variation over much of its introduced range in North America (Loomis & Fishman 2009). *Hieracium* species are particularly interesting because they frequently hybridize, despite being apomictic, providing opportunities for the creation of novel genotypes. In New Zealand reversions from apomixis to obligate sexuality have originated on at least three occasions in populations of *H. pilosella* as a result of hybridization (Chapman et al. 2003). This demonstrates that transitions from uniparental to biparental reproduction can also occur in invasive populations.

15.8 FUTURE GENETIC STUDIES ON INVASIVE SPECIES

Invasive species provide opportunities for investigating biological processes over historic timescales. Indeed,

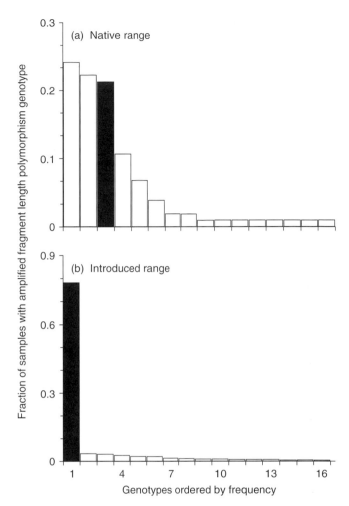

Fig. 15.4 Genetic bottlenecks and founder events characterize the worldwide invasive spread of *Eichhornia crassipes* (water hyacinth). Comparison of the relative frequency of clones in the native and introduced range based on a survey of amplified fragment length polymorphism. The shaded bar identifies the same clone in the two samples; note the dominance of this clone in the introduced range. Samples size were 104 and 1036 ramets sampled from 7 and 47 populations in the native and introduced range, respectively. A total of 49 clones was detected in the survey; only the most frequent are depicted. From data in Zhang et al. (2010).

this was part of what attracted Charles Elton to the study of introduced species. Elton's focus was primarily ecological; however, because biological invasions are demographic processes, they can have profound genetic consequences of significance for predicting the future evolution of species. Although there is good evidence that reproductive systems of invasive plants play a major role in affecting population genetic structure, it is less clear how important the amounts of genetic diversity are for persistence and spread. In addition, although part of the appeal of studying invasive species is the possibility of integrating historical information into predictions concerning contemporary patterns

and processes, in reality we are only just beginning to use the available tools and approaches for the quantitative analysis of immigration history. I conclude by outlining several future avenues that might be profitably explored on these topics.

Although 45 years have passed since publication of Baker & Stebbins' (1965) seminal volume on the genetics of colonizing species, we still know surprisingly little about the ecological consequences of genetic variation (reviewed in Hughes et al. 2008). As a result, the relations between genetic variation and colonizing success are poorly understood. Long-term field studies are needed to determine the significance of genetic variation for invasive species. Ideally, these should involve the establishment across environmental gradients of replicated colonies with different genetic inputs of both marker-based polymorphism and quantitative genetic variation for traits relevant to establishment and persistence. Monitoring of population growth and the estimation of demographic parameters of these colonies would help to define the significance of genetic variation for invasive species. Few studies of this type have been conducted on any plant species (although see Martins & Jain 1979; Polans & Allard 1989; Newman & Pilson 1997). It is important to recognize, however, that demonstrating that genetic diversity is associated with colonizing success provides only limited insight into the mechanisms responsible. Experimental manipulations of colonies in which genotype–phenotype relations are identified, combined with phenotypic selection analyses, should help to provide the mechanistic explanations about how and why evolution occurs in invasive populations.

Finally, invasion biology rests on a firmer foundation when information about contemporary patterns of variation can be interpreted using historical information on the source of introductions. Unfortunately, for most invasions we are largely ignorant of immigration history (Marsico et al. 2010). Knowing when, where and how many introductions are responsible for a particular invasion are essential parts of the puzzle of predicting future ecological and evolutionary responses. Recent developments in evolutionary genetics offer some encouragement for more rigorously addressing issues related to migration and demographic history. Phylogeographic methods and the use of coalescence models can reveal novel insights into the evolutionary history of range expansion and the nature of genetic diversity sampled during invasion (Saltonstall 2002; Schaal et al. 2003; Taylor & Keller 2007; Dlugosch & Parker 2008a,b; Rosenthal et al. 2008). This information is of both basic and applied significance and may inform management strategies, such as the choice of geographical areas in the native range sampled for biological control agents of invasive species.

Two recent examples illustrate how evolutionary genetic approaches can aid in providing historical information on invasive species. I begin by returning to the case of *Eichhornia paniculata* and the colonization of Caribbean islands from Brazil. It could be surmised that the occurrence of *E. paniculata* as a weed of rice fields in Cuba and Jamaica resulted from human assisted migration in historic times, as appears to be the case for the related *E. crassipes*, also native to Brazil (Barrett & Forno 1982). However, coalescent simulations based on nucleotide sequence data firmly reject this hypothesis and show that colonization of the Caribbean by *E. paniculata* probably occurred approximately 125,000 years before present, well before the origins of agriculture (Ness et al. 2010). The second example involves *Arabidopsis thaliana*, an annual selfing weed that is the primary model organism for plant biology. Using Bayesian computations and explicit spatial modelling of molecular variation, François et al. (2008) detected a major wave of migration from east to west about 10,000 years ago in Europe with a westward spread of approximately 0.9 km per year. The spread of *A. thaliana* was associated with a progressive decline in nucleotide diversity probably as a result of repeated bottlenecks. This migration appears to have been associated with spread of agriculture during the Neolithic transition.

These examples demonstrate that gene genealogies and coalescence models can be used to investigate the migration history and demography of species. These approaches have provided unprecedented insights into the global spread of the most successful colonizing species – *Homo sapiens* – (DeGiorgio et al. 2009). It is now time to also apply these approaches to the study of plant invasions so that the value of historical information can be fully realized.

ACKNOWLEDGEMENTS

I thank Dave Richardson for the opportunity to contribute to this volume, Rob Colautti, Rob Ness, Yuan-Ye Zhang and Da-Yong Zhang for discussions and permission to cite unpublished work and Bill Cole for assistance with figures. My work on invasive species is funded by Discovery Grants from NSERC (Canada), the Canada Research Chair's programme, and a Premier's Discovery Award in Life Sciences and Medicine from the Ontario Government.

REFERENCES

Aguilar, R., Ashworth, L., Galetto, L. & Aizen, M.A. (2006) Plant reproductive susceptibility to habitat fragmentation: review and synthesis through a meta-analysis. *Ecology Letters*, **9**, 968–980.

Aizen, M.A. & Vázquez, D.P. (2006) Flower performance in human-altered habitats. In *Ecology and evolution of flowers* (ed. L.D. Harder and S.C.H. Barrett), pp. 159–179. Oxford University Press, Oxford.

Allee, W.C. (1931) *Animal Aggregations: A Study in General Sociology*. University of Chicago Press, Chicago.

Amsellem, L., Noyer, J.L. & Hossaert-McKey, M. (2001) Evidence for a switch in the reproductive biology of *Rubus alceifolius* Poir. (Rosaceae) towards apomixis, between its native range and its area of introduction. *American Journal of Botany*, **88**, 2243–2251.

Antonovics, J. (1976) The nature of limits to natural selection. *Annals of the Missouri Botanic Garden*, **63**, 224–247.

Ashman, T.-L., Knight, T.M., Steets, J.A., et al. (2004) Pollen limitation of plant reproduction: ecological and evolutionary causes and consequences. *Ecology*, **85**, 2408–2421.

Baker, H.G. (1953) Race formation and reproductive method in flowering plants. *Society for Experimental Biology Symposia*, **7**, 114–145.

Baker, H.G. (1955) Self-compatibility and establishment after 'long-distance' dispersal. *Evolution*, **9**, 347–349.

Baker, H.G. & Stebbins, G.L. (eds) (1965) *The Genetics of Colonizing Species*. Academic Press, New York.

Barrett, R.D.H. & Schluter, D. (2008) Adaptation from standing genetic variation. *Trends in Ecology & Evolution*, **22**, 465–471.

Barrett, S.C.H. (1988) Genetics and evolution of agricultural weeds. In *Weed Management in Agroecosystems: Ecological Approaches* (ed. M. Altieri and M.Z. Liebman), pp. 57–75, CRC Press, Boca Raton, Florida.

Barrett, S.C.H. (1989) Waterweed invasions. *Scientific American*, **260**, 90–97.

Barrett, S.C.H. (1992) Genetics of weed invasions. In *Applied Population Biology* (ed. S.K. Jain and L. Botsford), pp. 91–119. Wolfgang Junk, the Netherlands.

Barrett, S.C.H. (ed.) (2008) *Major Evolutionary Transitions in Flowering Plant Reproduction*. University of Chicago Press, Chicago.

Barrett, S.C.H. & Eckert, C. (1990) Variation and evolution of mating systems in seed plants. In *Biological Approaches and Evolutionary Trends in Plants* (ed. S. Kawano), pp. 229–254. Academic Press, Tokyo.

Barrett, S.C.H. & Forno, I.W. (1982) Style morph distribution in New World populations of *Eichhornia crassipes* (water hyacinth). *Aquatic Botany*, **13**, 299–306.

Barrett, S.C.H. & Husband, B.C. (1990) Variation in outcrossing rates in *Eichhornia paniculata*: the role of demographic and reproductive factors. *Plant Species Biology*, **5**, 41–56.

Barrett, S.C.H. & Shore, J.S. (1987) Variation and evolution of breeding systems in the *Turnera ulmifolia* L. complex (Turneraceae). *Evolution*, **41**, 340–354.

Barrett, S.C.H. & Shore, J.S. (1989) Isozyme variation in colonizing plants. In *Isozymes in Plant Biology* (ed. D. Soltis and P. Soltis), pp. 106–126. Dioscorides Press, Portland, Orgeon.

Barrett, S.C.H., Morgan, M.T. & Husband, B.C. (1989) The dissolution of a complex genetic polymorphism: the evolution of self-fertilization in tristylous *Eichhornia paniculata* (Pontederiaceae). *Evolution*, **41**, 1398–1416.

Barrett, S.C.H., Eckert, C.G. & Husband, B.C. (1993) Evolutionary processes in aquatic plant populations. *Aquatic Botany*, **44**, 105–145.

Barrett, S.C.H., Harder, L.D. & Worley, A.C. (1996) The comparative biology of pollination and mating in flowering plants. *Philosophical Transactions of the Royal Society of London B*, **351**, 1271–1280.

Barrett, S.C.H., Colautti, R.I. & Eckert, C.G. (2008) Plant reproductive systems and evolution during biological invasion. *Molecular Ecology*, **17**, 373–383.

Barrett, S.C.H., Ness, R.W. & Vallejo-Marín, M. (2008) Evolutionary pathways to self-fertilization in a tristylous species. *New Phytologist*, **183**, 546–556.

Baudry, E., Kerdelhue, C., Innan, H. & Stephan, W. (2001) Species and recombination effects on DNA variability in the tomato genus. *Genetics*, **158**, 1725–1735.

Bjerknes, A.-L., Totland, O., Hegland, S.J. & Nielsen A. (2007) Do alien plant invasions really affect pollination success in native plant species? *Biological Conservation*, **138**, 1–12.

Blows, M.W. & Hoffman, A. (2005) A reassessment of genetic limits to evolutionary change. *Ecology*, **86**, 1371–1384.

Bossdorf, O., Auge, H., Lafuma, L., Rogers, W.E., Siemann, E. & Prati, D. (2005) Phenotypic and genetic differentiation between native and introduced plant populations. *Oecologia*, **144**, 1–11.

Brennan, A.C., Harris, S.A. & Hiscock, S.J. (2005) Modes and rates of selfing and associated inbreeding depression in the self-incompatible plant *Senecio squalidus* (Asteraceae): a successful colonizing species in the British Isles. *New Phytologist*, **168**, 475–486.

Brown, A.H.D. & Burdon, J.J. (1987) Mating systems and colonizing success in *plants. Colonization, succession and stability* (ed. A.J. Gray, M.J. Crawley and P.J. Edwards), pp 115–131. Blackwell, Oxford.

Brown, A.H.D. & Marshall, D.R. (1981) Evolutionary changes accompanying colonization in plants. In *Evolution Today: Proceedings of the Second International Congress of Systematic and Evolutionary Biology* (ed. G.C.E. Scudder and J.L. Reveal), pp. 351–363. Carnegie-Mellon University, Pittsburgh, Pennsylvania.

Chapman, H., Houliston, G.J., Robson, B. & Iline, I. (2003) A case of reversal: the evolution and maintenance of sexuals from parthenogenetic clones in *Hieracium pilosella*. *International Journal of Plant Sciences*, **164**, 719–728.

Charlesworth, B. (1992) Evolutionary rates in partially self-fertilizing species. *American Naturalist*, **140**, 126–148.

Charlesworth, D. & Charlesworth, B. (1987) Inbreeding depression and its evolutionary consequences. *Annual Review of Ecology and Systematics*, **18**, 237–268.

Charlesworth D. & Charlesworth, B. (1995) The effect of the breeding system on genetic variability. *Evolution*, **49**, 911–920.

Charlesworth, D. & Pannell, J.R. (1999) Mating systems and population genetic structure in the light of coalescent theory. *Integrating Ecology and Evolution in a Spatial Context* (ed. J. Silvertown and J. Antonovics), pp 73–95. Blackwell Science, Oxford.

Charlesworth, D. & Wright, S. (2001) Breeding systems and genome evolution. *Current Opinion in Genetics & Development*, **11**, 685–690.

Cheptou, P.O. (2004) Allee effect and self-fertilization in hermaphrodites: reproductive assurance in demographically stable populations. *Evolution*, **58**, 2613–2621.

Cheptou, P.O. & Avendaño, L.G. (2006) Pollination processes and the Allee effect in highly fragmented populations: consequences for the mating system in urban environments. *New Phytologist*, **172**, 774–783.

Cheptou, P.O. & Diekmann, U. (2002) The evolution of self-fertilization in density-regulated populations. *Proceedings of the Royal Society of London B*, **269**, 1177–1186.

Cheptou, P.O., Lepart, J. & Escarre J. (2002) Mating system variation along a successional gradient in the allogamous colonizing plant *Crepis sancta* (Asteraceae). *Journal of Evolutionary Biology*, **15**, 753–762.

Chun, Y.J., Nason, J.D. & Moloney, K.A. (2009) Comparison of quantitative and molecular genetic variation of native vs. invasive populations of purple loosestrife (*Lythrum salicaria* L. Lythraceae). *Molecular Ecology*, **18**, 3020–3035.

Colautti, R.I., Maron, J.L. & Barrett, S.C.H. (2009) Common garden comparisons of native and introduced plant populations: latitudinal clines can obscure evolutionary inferences. *Evolutionary Applications*, **3**, 187–189.

Colautti, R.I., Eckert, C.G. & Barrett, S.C.H. (2010a) Evolutionary constraints on adaptive evolution during range expansion in an invasive plant. *Proceedings of the Royal Society of London B*, **277**, 1799–1806.

Colautti, R.I., White, N.A. & Barrett, S.C.H. (2010b) Variation of self-incompatibility within invasive populations of purple loosestrife (*Lythrum salicaria* L.) from eastern North America. *International Journal of Plant Sciences*, **171**, 158–166.

Cox, G.W. (2004) *Alien Species and Evolution: The Evolutionary Ecology of Exotic Plants, Animals, Microbes, and Interacting Native Species*. Island Press, Washington, DC.

Darwin, C. (1859) *On the Origin of Species*. London, John Murray.

Davis, H.G., Taylor, C.M., Lambrinos, J.G. & Strong, D.R. (2004) Pollen limitation causes an Allee effect in a wind-pollinated invasive grass (*Spartina alterniflora*). *Proceedings of the National Academy of Sciences of the USA*, **101**, 13804–13807.

DeGiorgio, M., Jakobsson, M. & Rosenberg, N.A. (2009) Explaining worldwide patterns of human genetic variation using a coalescent-based serial founder model of migration outward from Africa. *Proceedings of the National Academy of Sciences of the USA*, **106**, 16057–16062.

Dlugosch, K.L.M. & Parker, I.M. (2008a) Founding events in species invasions: genetic variation, adaptive evolution, and the role of multiple introductions. *Molecular Ecology*, **17**, 431–449.

Dlugosch, K.L.M. & Parker, I.M. (2008b) Invading populations of an ornamental shrub show rapid life history evolution despite genetic bottlenecks. *Ecology Letters*, **11**, 701–709.

Dornier, A., Munoz, F. & Cheptou, P.O. (2008) Allee effect and self-fertilization in hermaphrodites: reproductive assurance in a structured metapopulation. *Evolution*, **62**, 2558–2569.

Eckert, C.G., Samis, K.E. & Lougheed, S.C. (2008) Genetic variation across species' geographical ranges: the central–marginal hypothesis and beyond. *Molecular Ecology*, **17**, 1170–1188.

Eckert, C.G. Kalisz, S., Geber, M.A., et al. (2010) Plant mating systems in a changing world. *Trends in Ecology & Evolution*, **25**, 35–43.

Elam, D.R., Ridley, C.E., Goodell, K. & Ellstrand, N.C. (2007) Population size and relatedness affect plant fitness in a self-incompatible invasive plant. *Proceedings of the National Academy of Sciences of the USA*, **104**, 549–552.

Ellstrand, N.C. & Schierenbeck, K.A. (2000) Hybridization as a stimulus for the evolution of invasiveness in plants? *Proceedings of the National Academy of Sciences of the USA*, **97**, 7043–7050.

Elton, C.S. (1958) *The Ecology of Invasions by Animals and Plants*. Methuen, London.

Etterson, J.R. & Shaw, R.G. (2001) Constraint on adaptive evolution in response to global warming. *Science*, **294**, 151–154.

Ford, E.B. (1964) *Ecological Genetics*. Chapman and Hall, London.

François, O., Blum, M.G.B., Jakobsson, M. & Rosenberg, N.A. (2008) Demographic history of European populations of *Arabidopsis thaliana*. *PLoS Genetics*, **4**, e1000075.

Friedman, J. & Barrett, S.C.H. (2008) High outcrossing in the annual colonizing species *Ambrosia artemisiifolia* (Asteraceae). *Annals of Botany*, **101**, 1303–1309.

Fryxell, P.A. (1957) Mode of reproduction of higher plants. *Botanical Review*, **23**, 135–233.

García-Ramos, G. & Rodríguez, D. (2002) Evolutionary speed of species invasions. *Evolution*, **56**, 661–668.

Goodwillie, C., Kalisz, S. & Eckert, C.G. (2005) The evolutionary enigma of mixed mating systems in plants: occurrence, theoretical explanations, and empirical evidence. *Annual Reviews of Ecology, Evolution and Systematics*, **36**, 47–79.

Gopal, B. & Sharma, K.P. (1981) *Water-Hyacinth (Eichhornia crassipes) the Most Troublesome Weed of the World*. Hindasia, Delhi, India.

Griffith, T.M. & Watson, M.A. (2006) Is evolution necessary for range expansion? Manipulating reproductive timing of a weedy annual transplanted beyond its range. *American Naturalist*, **167**, 153–164.

Hamrick, J.L. & Godt, M.J. (1996) Effects of life history traits on genetic diversity in plant species. *Philosophical Transactions of the Royal Society of London B*, **351**, 1291–1298.

Hereford, J. (2010) Does selfing or outcrossing promote local adaptation. *American Journal of Botany*, **97**, 298–302.

Hermisson, J. & Pennings, P.S. (2005) Soft sweeps: molecular population genetics of adaptation from standing genetic variation. *Genetics*, **169**, 2335–2352.

Hill, S.B. & Kotanen, P.M. (2009) Evidence that phylogenetically novel non-indigenous plants experience less herbivory. *Oecologia*, **161**, 581–590.

Hollingsworth, M.L. & Bailey, J.P. (2000) Evidence for massive clonal growth in the invasive weed *Fallopia japonica* (Japanese knotweed). *Botanical Journal of the Linnean Society*, **133**, 463–472.

Hughes A.R., Inouye, B.D., Johnson, M.T.K., Underwood, N. & Vellend, M. (2008) Ecological consequences of genetic diversity. *Ecology Letters*, **11**, 609–623.

Husband, B.C. & Barrett, S.C.H. (1991) Colonization history and population genetic structure of *Eichhornia paniculata* in Jamaica. *Heredity*, **66**, 287–296.

Keller, S.R., Sowell, D.R., Neiman, M., Wolfe, L.M. & Taylor, D.R. (2009) Adaptation and colonization history affect the evolution of clines in two introduced species. *New Phytologist*, **183**, 678–690.

Kliber, A. & Eckert, C.G. (2005) Interaction between founder effect and selection during biological invasion in an aquatic plant. *Evolution*, **59**, 1900–1913.

Kolbe, J.J., Glor, R.E., Schettino, L.R.G., Lara, A.C., Larson, A. & Losos, J.B. (2004) Genetic variation increases during biological invasion by a Cuban lizard. *Nature*, **431**, 177–181.

Lafuma, L. & Maurice, S. (2007) Increase in mate availability without loss of self-incompatibility in the invasive species *Senecio inaequidens* (Asteraceae). *Oikos*, **116**, 201–208.

Lande, R. (1975) The maintenance of genetic variability by mutation in polygenic characters with linked loci. *Genetical Research*, **26**, 221–235.

Lande, R. (1977) The influence of the mating system on the maintenance of genetic variability in polygenic characters. *Genetics*, **86**, 485–498.

Larson, B.M.H. & Barrett, S.C.H. (2000) A comparative analysis of pollen limitation in flowering plants. *Biological Journal of the Linnean Society*, **69**, 503–520.

Lavergne, S. & Molofsky, J. (2007) Increased genetic variation and evolutionary potential drive the success of an invasive grass. *Proceedings of the National Academy of Science of the USA*, **104**, 3883–3888.

Lewontin, R.C. (1984) Detecting population differences in quantitative characters as opposed to gene frequencies. *American Naturalist*, **123**, 115–124.

Loomis, E.S. & Fishman, L. (2009) A continent-wide clone: population genetic variation of the invasive plant *Hieracium aurantiacum* (orange hawkweed; Asteraceae) in North

America. *International Journal of Plant Science*, **170**, 759–765.

Maron, J.L., Vilà, M., Bommarco, R., Elmendorf, S. & Beardsley, P. (2004) Rapid evolution of an invasive plant. *Ecological Monographs*, **74**, 261–280.

Martins, P.S. & Jain, S.K. (1979) Role of genetic variation in the colonizing ability of rose clover (*Trifolium hirtum* All.). *American Naturalist*, **114**, 591–595.

Marsico, T.D., Burt, J.W., Espeland, E.K., et al. (2010) Underutilized resources for studying the evolution of invasive species during their introduction, establishment, and lag phases. *Evolutionary Applications*, **3**, 203–219.

Memmott, J., Craze, P.G., Waser, N.M. & Price M.V. (2007) Global warming and the disruption of plant–pollinator interactions. *Ecology Letters*, **10**, 710–717.

Montague, J.L., Barrett, S.C.H. & Eckert, C.G. (2008) Re-establishment of clinal variation in flowering time among introduced populations of purple loosestrife (*Lythrum salicaria*, Lythraceae). *Journal of Evolutionary Biology*, **21**, 234–245.

Ness, R.W., Wright, S.I. & Barrett, S.C.H. (2010) Mating-system variation, demographic history and patterns of nucleotide diversity in the tristylous plant *Eichhornia paniculata*. *Genetics*, **184**, 381–392.

Neve, P., Villa-Aiub, M. & Roux, F. (2009) Evolutionary-thinking in agricultural weed management. *New Phytologist*, **184**, 783–793.

Newman, D. & Pilson, D. (1997) Increased probability of extinction due to decreased genetic effective population size: experimental populations of *Clarkia pulchella*. *Evolution*, **51**, 354–362.

Nordborg, M. (2000) Linkage disequilibrium, gene trees and selfing: ancestral recombination graph with partial self-fertilization. *Genetics*, **154**, 923–929.

Novak, S.J. & Mack, R.N. (2005) Genetic bottlenecks in alien plant species. Influence of mating system and introduction dynamics. In *Species Invasions Insights into Ecology, Evolution, and Biogeography* (ed. D.F. Sax, J.J. Stachowicz and D.D. Gaines), pp. 201–228. Sinauer & Associates, Sunderland, Massachusetts.

Orr, H.A. & Bettancourt, A.J. (2001) Haldane's sieve and adaptation from the standing variation. *Genetics*, **157**, 875–884.

Pannell, J.R. & Barrett, S.C.H. (1998) Baker's Law revisited: reproductive assurance in a metapopulation. *Evolution*, **52**, 657–668.

Parsons, P.A. (1983) *The Evolutionary Biology of Colonizing Species*. Cambridge University Press, Cambridge, UK.

Polans, N.O. & Allard, R.W. (1989) An experimental evaluation of the recovery of ryegrass populations from genetic stress resulting from restriction of population size. *Evolution*, **43**, 1320–1324.

Prentis, P.J., Wilson, J.R.U., Dormontt, E.E., Richardson, D.M. & Lowe, A.J. (2008) Adaptive evolution in invasive species. *Trends in Plant Science*, **13**, 288–294.

Price, S.C. & Jain, S.K. (1981) Are inbreeders better colonizers? *Oecologia*, **49**, 283–286.

Rambuda, T.D. & Johnson, S.D. (2004) Breeding systems of invasive alien plants in South Africa: does Baker's rule apply? *Diversity and Distributions*, **10**, 409–416.

Reed, D.N. & Frankham, R. (2001) How closely correlated are molecular and quantitative measures of genetic variation? A meta-analysis. *Evolution*, **55**, 1095–1103.

Richardson, D.M. & Pyšek, P. (2008) Fifty years of invasion ecology – the legacy of Charles Elton. *Diversity and Distributions*, **14**, 161–168.

Rosenthal, D.M., Ramakrishnan, A.P. & Cruzan, M.B. (2008). Evidence for multiple sources of invasion and intraspecific hybridization in *Brachypodium sylvaticum* (Hudson) Beauv. in North America. *Molecular Ecology*, **17**, 4657–4669.

Saltonstall, K. (2002) Cryptic invasion by a non-native genotype of the Common Reed, *Phragmites australis*, into North America. *Proceedings of the National Academy of Sciences of the USA*, **99**, 2445–2449.

Sax, D.F., Stachowicz, J.J. & Gaines, S.D. (eds) (2005) *Species Invasions: Insights into Ecology, Evolution And Biogeography*. Sinauer Associates, Sunderland.

Schaal, B.A., Gaskin, J.F. & Caciedo, A.L. (2003) Phylogeography, haplotype trees, and invasive plant species. *Journal of Heredity*, **94**, 197–204.

Schachner, L.J., Mack, R.N. & Novak, S.J. (2008) *Bromus tectorum* (Poaceae) in midcontinental United States: population genetic analysis of an ongoing invasion. *American Journal of Botany*, **95**, 1584–1595.

Schoen, D.J. & Brown, A.H.D. (1991) Intraspecific variation in population gene diversity and effective population size correlates with the mating system in plants. *Proceedings of the National Academy of Sciences of the USA*, **88**, 4494–4497.

Silvertown, J. (2008) The evolutionary maintenance of sexual reproduction: evidence from the ecological distribution of asexual reproduction in clonal plants. *International Journal of Plant Sciences*, **169**, 157–168.

Sun, M. & Ritland, K. (1998) Mating system of yellow starthistle (*Centaurea solstitialis*), a successful colonizer in North America. *Heredity*, **80**, 225–232.

Taylor, D.R., & Keller, S.R. (2007) Historical range expansion determines the phylogenetic diversity introduced during contemporary species invasion. *Evolution*, **61**, 334–345.

Taylor, C.M., Davis, H.G., Civille, J.C., Grevstad, F.S. & Hastings A. (2004) Consequences of an Allee effect on the invasion of a Pacific estuary by *Spartina alterniflora*. *Ecology*, **85**, 3254–3266.

Traveset, A. & Richardson, D.M. (2006) Biological invasions as disruptors of plant reproductive mutualisms. *Trends in Ecology & Evolution*, **21**, 208–216.

Vaillant, M.T., Mack, R.N. & Novak, S.J. (2007) Introduction history and population genetics of the invasive grass *Bromus tectorum* (Poaceae) in Canada. *American Journal of Botany*, **94**, 1159–1169.

van Kleunen, M. & Johnson, S.D. (2005) Testing for ecological and genetic Allee effects in the invasive shrub *Senna didymobotrya* (Fabaceae). *American Journal of Botany*, **92**, 1124–1130.

van Kleunen, M., Fischer, M. & Johnson, S.D. (2007) Reproductive assurance through self-fertilization does not vary with population size in the alien invasive plant *Datura stramonium*. *Oikos*, **116**, 1400–1412.

van Kleunen, M., Manning, J.C., Pasqualetto, V. & Johnson, S.D. (2008) Phylogenetically independent associations between autonomous self-fertilization and plant invasiveness. *American Naturalist*, **171**,195–201.

Wang, B.R., Li, W.G. & Wang, J.B. (2005) Genetic diversity of *Alternanthera philoxeroides* in China. *Aquatic Botany*, **81**, 277–283.

Wang, Y., Wang, Q-F. Guo, Y-H. & Barrett, S.C.H. (2005) Reproductive consequences of interactions between clonal growth and sexual reproduction in *Nymphoides peltata*: a distylous aquatic plant. *New Phytologist*, **165**, 329–336.

Warwick, S.I. (1990) Allozyme and life history variation in five northwardly colonizing weed species. *Plant Systematics and Evolution*, **169**, 41–54.

Williamson, M.H. (1996) *Biological Invasions*. Chapman and Hall, London.

Williamson, M.H. & Fitter, A. (1996) The characters of successful invaders. *Biological Conservation*, **78**, 163–170.

Wright, S.I., Ness, R.W., Foxe, J.P. & Barrett, S.C.H. (2008) Genomic consequences of outcrossing and selfing in plants. *International Journal of Plant Sciences*, **169**, 105–118.

Zhang, Y.-Y., Zhang, D.-Y. & Barrett, S.C.H. (2010) Genetic uniformity characterizes the invasive spread of water hyacinth (*Eichhornia crassipes*), a clonal aquatic plant. *Molecular Ecology*, **19**, 1774–1786.

Chapter 16

IMPACTS OF BIOLOGICAL INVASIONS ON FRESHWATER ECOSYSTEMS

Anthony Ricciardi[1] and Hugh J. MacIsaac[2]

[1]Redpath Museum, McGill University, Montreal, Quebec, Canada
[2]Great Lakes Institute for Environmental Research, University of Windsor, Windsor, Ontario, Canada

Fifty Years of Invasion Ecology: The Legacy of Charles Elton, 1st edition. Edited by David M. Richardson
© 2011 by Blackwell Publishing Ltd

16.1 INTRODUCTION: A BRIEF HISTORICAL PERSPECTIVE

The human-assisted spread of non-indigenous fishes and aquatic invertebrates, microbes and plants has had strong ecological impacts in lakes and rivers worldwide (see, for example, Nesler & Bergersen 1991; Witte et al. 1992; Flecker & Townsend 1994; Hall & Mills 2000; Latini & Petrere 2004). Cumulative invasions have disproportionately transformed freshwater communities such that they are dominated by non-indigenous species to a greater extent than their terrestrial counterparts (Vitousek et al. 1997). Although some lakes and rivers have documented invasion histories spanning several decades (see, for example, Mills et al. 1996; Hall & Mills 2000; Ricciardi 2006; Bernauer & Jansen 2006), the ecological impacts of freshwater invasions were rarely studied until many years after the publication of Elton's (1958) influential book. Until the late 20th century, concern over freshwater invasions focused almost exclusively on the economic impacts of pest species, particularly those that threatened fisheries (Morton 1997). Earlier reports of impacts on freshwater biodiversity (for example, Rivero 1937; Sebestyen 1938) were rare and largely ignored. However, two dramatic examples were well documented. The first was the spread of a fungal pathogen that destroyed native crayfish populations throughout Europe starting in the 1870s and continuing for several decades (Reynolds 1988). The second was an example highlighted by Elton (1958): the invasion of the upper Great Lakes by the parasitic sea lamprey *Petromyzon marinus*, which contributed to the destruction of commercial fisheries (Mills et al. 1993).

Nevertheless, by the end of the 20th century it became apparent that freshwater fishes had been introduced to virtually everywhere on the planet and that the impacts of most of these introductions were completely unknown (Lever 1996), although cases were accumulating. In a landmark study, Zaret & Paine (1973) demonstrated cascading food-web effects of an introduced piscivore in a Panamanian lake; theirs is among the most highly cited of impact studies. The first syntheses of the impacts of introduced freshwater fishes indicated a wide array of ecological effects arising from predation, competition, hybridization, disease transfer and habitat modification (Moyle 1976; Taylor et al. 1984; Moyle et al. 1986). These negative effects were viewed as surprising at the time because, as Moyle and colleagues (1986) noted, fish introduc-

tions were generally made with the assumption that the new species filled a 'vacant niche' and thus would integrate into their communities without consequence. This assumption was further disputed by the 1990 international symposium 'The Ecological and Genetic Implications of Fish Introductions', which offered the first global perspective of the effects of fish introductions in case studies from Europe, Africa, Australasia, tropical regions of Asia and America, and North America (see Allendorf 1991, and references therein). It concluded that any beneficial effects of fish introductions were immediate, whereas detrimental effects were delayed and often overlooked, but could constitute a threat to the persistence of native populations (Allendorf 1991).

For decades, freshwater invertebrates were transplanted into lakes and rivers to supplement the diets of sport fishes. During the 1970s, the first negative impacts of non-indigenous freshwater mysid shrimp were documented in multiple countries and, as in other cases of introduced predators, revealed far-reaching effects on benthic communities, phytoplankton, zooplankton and the upper trophic levels of food webs in North American and Scandinavian lakes (Nesler & Bergersen 1991; Spencer et al. 1991). During the 1980s, surging interest in conservation provided further impetus for investigating the effects of invasions on freshwater biodiversity, spurred on by dramatic impacts of the Nile perch *Lates niloticus* in Lake Victoria (Witte et al. 1992) and the zebra mussel *Dreissena polymorpha* in the North American Great Lakes (MacIsaac 1996).

Indeed, there has been a rapid increase in studies on freshwater invasions over the past two decades (MacIsaac et al., this volume). Freshwater studies comprise approximately 15% of the entire invasion research literature published over the past 50 years, but, compared with terrestrial studies, they have made a disproportionately smaller contribution to classical Eltonian concepts such as biotic resistance, enemy release and disturbance (Fig. 16.1). They have made more substantive contributions to modern concepts such as propagule pressure and human vector dispersal, but their predominant focus has been on the effects of invasions on recipient communities and ecosystems. In a random sample of 100 journal articles on freshwater invasions published in 2008, over 40% of the articles specifically address ecological impacts of introduced species (A. Ricciardi, unpublished data). The number of published quantitative studies on the com-

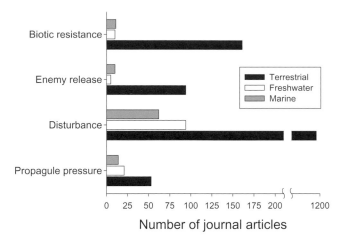

Fig. 16.1 Proportional contribution of published terrestrial, freshwater and marine studies to some key concepts in invasion ecology. Articles published between 1960 and 2008 were located using Web of Knowledge 4.0 (Thomson Institute for Science Information).

munity- and ecosystem-level impacts of freshwater fishes, for example, exceeds those for other individual groups such as terrestrial plants, terrestrial invertebrates and marine invertebrates (see Parker et al. 1999). It is not clear whether this overwhelming focus exists because the impacts of freshwater invasions are more conspicuous or more amenable to study in freshwater systems; in any case, it has enhanced our understanding of the context-dependent forces that structure communities and affect ecosystem processes. Herein, we evaluate the current state of knowledge of the impacts of invasions on freshwater biodiversity and food webs.

16.2 IMPACTS OF INVASIONS ON FRESHWATER BIODIVERSITY

Human activities have extensively altered freshwater ecosystems, whose species are disappearing at rates that rival those found in tropical forests (Ricciardi & Rasmussen 1999). Although biological invasions have been implicated as a principal cause of freshwater extinctions (Miller et al. 1989; Witte et al. 1992), researchers are confronted with the question of whether non-indigenous species are drivers or merely 'passengers' of ecological changes leading to biodiversity loss (Didham et al. 2005). Disentangling the respective roles of invasions and other anthropogenic stressors in causing major ecological impacts is a key challenge to invasion ecology. In freshwater systems,

this is made difficult by myriad stressors interacting with invasions. Habitat alteration facilitates invasions and exacerbates their impacts by increasing the connectedness of watersheds (for example through canalization), reducing native competitors and predators (through disturbance), and increasing habitat homogenization (through impoundment, reservoir construction and shoreline development) to the benefit of opportunistic invaders (see Moyle & Light 1996; Scott & Helfman 2001; Scott 2006; Johnson et al. 2008b). Impoundment and reservoir construction promote the replacement of endemic riverine species by cosmopolitan lentic species that would otherwise be poorly suited to natural river flows (Marchetti et al. 2004a). Further homogenization of fish communities results from the introductions of piscivores that reduce the abundance and diversity of littoral fishes, particularly in lakes where refugia have been removed by shoreline development (MacRae & Jackson 2001). Consequently, the interaction of invasion and land use has homogenized freshwater biota across multiple spatial scales (Duncan & Lockwood 2001; Rahel 2002; Clavero & García-Berthou 2006).

Nevertheless, invasions appear to be a principal contributor to biodiversity loss in some systems. A correlative analysis (Light & Marchetti 2007) suggests that introduced species, rather than habitat alteration, are the primary driver of population declines and extinctions of California freshwater fishes. Experimental studies are needed to confirm this result, but it is consistent with documented impacts of invasions throughout the region (Moyle 1976). Elsewhere, another correlative study showed that zebra mussel invasion has added to the impacts of environmental stressors, resulting in a 10-fold acceleration in rates of local extinction of native mussels in the Great Lakes region (Ricciardi et al. 1998). In this case, there is ample experimental evidence that native mussel mortality is increased by zebra mussel activities (Ricciardi 2004). Recoveries of some native species after experimental removals of non-indigenous species provide further evidence of the significant role of invasions in freshwater biodiversity loss (see, for example, Vredenburg 2004; Lepak et al. 2006; Pope 2008).

Lakes and other insular systems are naïve to the effects of a broad range of invaders, owing to their evolutionary isolation (Cox & Lima 2006). In historically fishless lakes, introduced trout have caused the extirpation of native fauna that have evolved without selection pressures to adapt to large aquatic predators (see Knapp & Matthews 2000; Pope 2008; Schabetsberger et al. 2009). Major community-level impacts are often observed in species-poor systems such as alpine lakes, desert pools, isolated springs and oligotrophic waters (Moyle & Light 1996), but other studies have demonstrated that even species-rich freshwater systems are vulnerable to disruption. For example, North American rivers contain the planet's richest assemblage of freshwater mussels, which have no evolutionary experience with dominant fouling organisms like the zebra mussel. Intense fouling by the zebra mussel on the shells of other molluscs interferes with metabolic activities of native species and has led to severe rapid declines or extirpation of many native mussel populations (Ricciardi et al. 1998; Ricciardi 2004). In contrast, extirpations of native mussels have rarely been reported from invaded European lakes and rivers, whose fauna had a shared evolutionary history with an ancestral form of the zebra mussel in the geological past (Ricciardi et al. 1998). A more extreme example is Lake Victoria, in which the greatest vertebrate mass extinction in modern history followed the introduction of the Nile perch, a piscivore that grew to a maximum body size at least two orders of magnitude larger than any native resident in the lake. Its introduction has been implicated in the loss of nearly 200 endemic cichlid species that had no evolutionary experience with large piscivores (Witte et al. 1992).

These examples support the contention that invasions are more likely to alter communities that lack entire functional groups – such as large piscivores or fouling bivalves – rather than high numbers of native species per se (Simberloff 1995). However, a community may be disrupted by any introduced species possessing sufficiently novel traits, particularly if such species are predators. The sea lamprey invasion of the Great Lakes contributed to the near total extirpation of lake trout *Salvelinus namaycush* from these waters and the extinction of some endemic salmonids (see Miller et al. 1989). The Great Lakes contain a few native species of lamprey, but these are smaller and lack the well developed jaws of the more predaceous sea lamprey; thus, the latter is ecologically distinct within the system. Ecologically distinct species are more likely to encounter naïve prey and less likely to encounter enemies that are adapted to them. Given that phylogenetically distant species tend to be ecologically distinct from each other, we expect that invaders that cause substantial declines in native populations will belong to novel taxa more often than low-impact invaders (Ricciardi & Atkinson 2004). Indeed, this appears to be the case in freshwater systems (Fig. 16.2).

Impacts on biodiversity can also result when invaders are brought into contact with closely related native species. Hybridization with introduced relatives appears to be at least partially responsible for over 30% of the North American freshwater fishes considered extinct in the wild; one example is the Amistad gambusia *Gambusia amistadensis*, which was hybridized to extinction when it interbred with introduced mosquitofish; such events can occur in only a few years (Miller et al. 1989). Although perhaps most widespread in fishes, hybridization and introgression have likely affected a broad range of freshwater taxa, but the global extent of these impacts remain unknown (Perry et al. 2002, and references therein).

16.3 CASCADING IMPACTS ON FOOD WEBS

Experimental studies on introduced aquatic predators have demonstrated that their effects extend beyond the replacement of native species. Such studies have enhanced our understanding of trophic cascades, wherein the biomass of primary producers is altered through indirect food-web effects (Flecker & Townsend 1994; Nyström et al. 2001; Schindler et al. 2001). A classic case involves European brown trout *Salmo trutta* in New Zealand streams, where their predation reduces invertebrate grazers, thereby releasing benthic

Fig. 16.2 Proportions of high-impact invaders (i.e. those implicated in a greater than 50% decline of a native species population) and low-impact invaders that belong to novel genera in independent freshwater systems with extensive invasion histories. The null hypothesis of no difference between proportions was tested with one-tailed Fisher exact tests and rejected for the Hudson River ($P = 0.0007$) and North American Great Lakes ($P = 0.00002$), but not for the Rhine River ($P = 0.49$), Lake Biwa ($P = 0.23$), and the Potomac River ($P = 0.07$). The null hypothesis was rejected by a meta-analysis of the entire data set (Fisher's combined probability test, $\chi^2 = 45.8$, $P < 0.0001$). Primary data sources: Mills et al. 1993, Ricciardi 2001 (Great Lakes), Mills et al. 1996 (Hudson River), Ruiz et al. 1999 (Potomac River), Hall and Mills 2000 (Lake Biwa), Bernauer and Jansen 2006 (Rhine River). Additional data were obtained from literature and online databases.

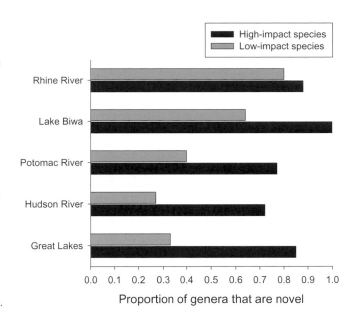

algae from herbivory; as a result, higher algal biomass is achieved in the presence of the brown trout compared with streams containing only native fishes (see Flecker & Townsend 1994). Through top-down effects and nutrient regeneration, trophic cascades generated in ponds, lakes and streams by introduced fish and crayfish have resulted in twofold to sixfold increases in algal biomass (Leavitt et al. 1994; Nyström et al. 2001; Herbst et al. 2009) and can lead to undesirable phytoplankton blooms (Reissig et al. 2006).

Cascading effects may extend beyond ecosystem compartments, particularly when they involve predators or omnivores that are not regulated by higher trophic levels. This was observed after the introduction of a piscivorous South American cichlid to Gatun Lake, Panama (Zaret & Paine 1973). Several native fishes were rapidly decimated through predation, which led to local declines of fish-eating birds such as kingfishers and herons. More significant was the loss of insectivorous fish, an entire guild, which was linked to an increase in the abundance of local mosquito populations in subsequent years and apparently affected the incidence and type of malaria infecting humans in the area. This phenomenon was noted decades earlier by Rivero (1937), who reported the decline of mosquitofish and a subsequent increase in malaria in Cuba

as a result of predation by introduced largemouth bass *Micropterus salmoides*.

In another spectacular case, the opossum shrimp *Mysis diluviana* (formerly *M. relicta*), a glacial relict species, was introduced deliberately by wildlife managers into Flathead Lake (Montana, USA) to supplement the diet of a non-indigenous landlocked salmon (kokanee, *Oncorhynchus nerka*). The shrimp is nocturnal, and during daylight hours it remained at the bottom of the lake while the salmon fed in shallower waters. Flathead Lake lacks upwelling currents that could have made the shrimp accessible to the salmon. Thus the shrimp avoided predation by the salmon and, moreover, outcompeted it for zooplankton prey, which subsequently became scarce. Consequently, the salmon population crashed, followed by the near disappearance of eagle and grizzly bear populations that depended on spawning salmon as a food resource (Spencer et al. 1991).

Cascading effects are also caused by introduced suspension feeders (see below) and aquatic plants. Being both autogenic engineers and primary producers, introduced aquatic plants have a great capacity to transform freshwater systems. Their impacts include altered habitat structure (Valley & Bremigan 2002), altered water quality (Rommens et al. 2003; Perna &

Burrows 2005) and reduced diversity (Boylen et al. 1999; Schooler et al. 2006). The replacement of natives by non-indigenous plants, even those that appear to be ecologically equivalent to the native, can have subtle negative effects on associated communities and food webs (Brown et al. 2006; Wilson & Ricciardi 2009).

Other food-web effects include diet shifts, which have substantive consequences for ecosystem processes such as contaminant cycling and secondary production. In a California lake, competition with an introduced planktivore (threadfin shad, *Dorosoma petenense*) caused three native planktivores to shift their diets from zooplankton to nearshore benthos, resulting in elevated mercury concentrations in each species (Eagles-Smith et al. 2008). In Canadian Shield lakes, introduced bass reduce the abundance and diversity of littoral fishes, the preferred prey of a native piscivore, lake trout (Vander Zanden et al. 1999). To mitigate such competition, lake trout change their foraging behaviour and shift their diets towards pelagic zooplankton in lakes where pelagic fishes are absent, and suffer reduced growth rates as a result (Vander Zanden et al. 1999, 2004; Lepak et al. 2006). In a Japanese woodland stream, introduced rainbow trout *Oncorhynchus mykiss* monopolized terrestrial invertebrate prey that fell into the stream. The loss of this prey subsidy caused native fish to shift their diet to stream insects that graze on benthic algae, thereby increasing algal biomass and reducing the biomass of adult insects that emerge from the stream; the latter effect led to a 65% reduction in the density of riparian spiders (see Baxter et al. 2004).

16.4 COMPLEX INTERACTIONS AMONG SPECIES

A major contributor to spatial variation in impact is the interaction between the invader and the invaded community. Antagonistic interactions, such as predation, can limit the impact of an invader by regulating its abundance or by constraining its behaviour (Robinson & Wellborn 1988; Harvey et al. 2004; Roth et al. 2007). Several case studies from freshwater systems also demonstrate synergistic interactions among invading species (Ricciardi 2001), such as when one invader indirectly enhances the success and impact of another by releasing it from predation. An experimental study in western North America revealed that local invasions of the bullfrog *Rana catesbeiana* are facilitated by the presence of introduced bluegill sunfish *Lepomis macrochirus*. The latter indirectly enhance survival of tadpoles by reducing densities of predatory macroinvertebrates (see Adams et al. 2003). Similarly, the population collapse of the historically dominant piscivore in the Great Lakes as a result of sea lamprey predation that targeted lake trout apparently facilitated the invasion of the alewife *Alosa pseudoharengus*, an introduced planktivorous fish that was previously sparse or absent in Lakes Michigan, Huron and Superior when lake trout populations were intact. Subsequent population explosions of alewife triggered changes in the composition and abundance of zooplankton and abrupt declines of native planktivores (see Kitchell & Crowder 1986).

A bottom-up food-web effect that is rarely studied in freshwater systems is hyperpredation, in which an introduced prey indirectly facilitates the decline of a native species by enabling a shared predator to increase its abundance. This has been observed in lakes of the Klamath Mountains, California, where introduced trout have reduced populations of a native frog both through their own predation and by facilitating predation by an introduced aquatic snake (Pope et al. 2008).

Freshwater studies have also highlighted an important synergistic interaction between introduced suspension-feeding bivalves and both native and non-indigenous plants. Through their filtration activities, dreissenid mussels (the zebra mussel and quagga mussel *D. rostriformis bugensis*) have increased water clarity, thereby stimulating the growth of vegetation that includes non-indigenous species such as Eurasian watermilfoil *Myriophyllum spicatum* and curly pondweed *Potamogeton cripsus* in the Great Lakes region (MacIsaac 1996; Vanderploeg et al. 2002), which in turn provide settlement surfaces for juvenile mussels. In Lake St. Clair, a formerly turbid system, zebra mussels stimulated macrophyte growth to unprecedented levels, which caused a major shift in the fish community by reducing the abundance of species adapted to turbid waters (for example walleye, *Sander vitreus*) and favouring species adapted to foraging in weed beds (smallmouth bass, *Micropterus dolomieu*, and northern pike, *Esox lucius*) (Vanderploeg et al. 2002). Similar cascading effects were observed after the invasion of the Potomac River estuary by the Asiatic clam *Corbicula fluminea*. Water clarity in the estuary tripled within a few years of the clam's discovery, and coincided with the development of submerged macrophyte beds that had previously been absent for 50 years. These beds included the non-indigenous species *Hydrilla verticillata* and *Myriophyllum spicatum*, and supported an increased abundance of fishes, including introduced bass and waterfowl. The overwhelming influence of the clam on these changes was confirmed after it suffered a major decline that precipitated reductions in submerged aquatic vegetation and associated populations of birds and fish (Phelps 1994).

Particular combinations of invaders, especially co-adapted species, can also produce strong synergistic impacts. In western Europe, multiple invasions by Ponto-Caspian species completed the parasitic life cycle of the trematode *Bucephalus polymorphus*. The introductions of the trematode's first intermediate host, the zebra mussel, and its definitive host, the pike-perch *Sander lucioperca*, allowed *B. polymorphus* to become established in inland waters and cause high mortality in local populations of cyprinid fishes that served as secondary intermediate hosts (Combes & Le Brun 1990). Interactions among Ponto-Caspian species introduced to the Great Lakes have also produced synergistic impacts (Ricciardi 2001). A new food-web link composed of the Eurasian round goby *Neogobius melanostomus* and its natural prey, dreissenid mussels, created a contaminant pathway that has increased the heavy-metal burden of piscivorous fish (Hogan et al. 2007). Furthermore, the prolific growth of benthic plants and algae in response to the filtration effects of dreissenid mussels has resulted in the build-up of excessive decaying vegetation that periodically depletes oxygen levels in bottom waters, generating conditions that promote the proliferation of botulism bacteria. Outbreaks of type E botulism have occurred in the Great Lakes every summer since 1999, and are responsible for die-offs of over 90,000 waterfowl, primarily fish-eating and scavenging species. The botulin toxin occurs in the mussels and in their principal predator, the round goby, which itself is commonly found in the stomachs of fish-eating birds. Therefore it is hypothesized that the round goby is transferring toxin from the mussels to higher trophic levels (Yule et al. 2006).

16.5 PREDICTABILITY OF IMPACT

Predictive methods are needed to prioritize invasion threats, but progress in predicting impact has been modest, largely because of the moderating influence of local environmental conditions such as those mentioned above (Parker et al. 1999; Ricciardi 2003). So far, very few general hypotheses have related impact to physical habitat conditions, although some researchers have observed that invaders are more likely to extirpate native species in nutrient-poor aquatic systems with low species diversity and extremely high or extremely low environmental (for example flow, temperature) variability or severity (Moyle & Light 1996).

The context-dependent nature of impact has been explored by studies correlating the impact of introduced species to variation in physicochemical variables. Freshwater studies on amphipod crustaceans and fishes have demonstrated that the magnitude and direction of antagonistic interactions (such as intraguild predation) that determine whether an introduced species replaces or is inhibited by a resident species can vary along environmental gradients such as conductivity, salinity, oxygen and temperature (Taniguchi et al. 1998; Alcaraz et al. 2008; Piscart et al. 2009; Kestrup & Ricciardi 2009). These studies demonstrate that although impacts are highly context dependent, only a few key environmental variables might be important for prediction. Such is the case for the zebra mussel's impact on native mussel populations, which is largely driven by their level of fouling on the shells of native species. Their fouling intensity is positively correlated with calcium concentration and negatively correlated to the mean particle size of surrounding sediments (Jokela & Ricciardi 2008), suggesting that the most vulnerable native mussel populations can be identified from these two habitat variables before invasion.

Only a few introduced organisms in a given region appear to cause severe ecological impacts (Ricciardi & Kipp 2008) and tools are needed to identify these high-impact species. The use of species traits to predict high-impact aquatic invaders has lagged behind terrestrial studies. Predictive traits vary across different stages of invasion (Kolar & Lodge 2002; Marchetti et al. 2004b) and the invasiveness, or colonizing ability, of a species is an inadequate predictor of impact for terrestrial and aquatic organisms (see Ricciardi & Cohen 2007); even poor colonizers (for example Atlantic salmon, *Salmo salar*) may have strong local impacts, whereas widely successful colonizers (for example freshwater jellyfish, *Craspedacusta sowerbyi*) do not necessarily disrupt ecosystems. However, there is intriguing evidence that the relative risk posed by introduced species to their invaded communities may be predicted by a comparison of their functional responses. Comparing native and non-indigenous amphipod crustaceans, Bollache et al. (2008) suggested that high-impact invasive predators have a higher functional response (*sensu* Holling 1959) than more benign species. Moreover, they argue that this comparative analysis can be extended to other trophic groups. A similar approach uses bioenergetics modelling to estimate predator consumptive demand, and thus compares the relative threats of introduced piscivores in a given system (Johnson et al. 2008a).

A burgeoning number of case studies tentatively suggest that the most severe ecological impacts are caused by aquatic invaders with the following characteristics: (i) they tend have a higher fecundity and abundance than related native species (Hall & Mills 2000; Keller et al. 2007); (ii) they are often generalist predators or omnivores (Moyle & Light 1996; Hall & Mills 2000) and, in the case of crustaceans, are highly aggressive (Gamradt et al. 1997; Dick 2008); (iii) they are introduced to systems where functionally similar species do not exist (Ricciardi & Atkinson 2004); (iv) they use resources differently from resident species such that they can alter the availability of critical resources such as light, nutrients, food or habitat space, for example through ecosystem engineering (Phelps 1994; Rodriguez et al. 2005) or by building or breaking trophic links between different ecosystem compartments (Simon & Townsend 2003; Baxter et al. 2004); (5) their physiological requirements are closely matched to abiotic conditions in the invaded environment (the intensity of impact is higher in environments that are optimal to the invader (see, for example, Jokela & Ricciardi 2008)); and (6) they have a history of strong impacts in other invaded regions (Marchetti et al. 2004b). Invasion history, when sufficiently documented, provides a valuable basis for developing quantitative predictions (for example using synthetic tools such as meta-analysis) of the impacts of widespread aquatic invaders such as zebra mussels (Ricciardi 2003; Ward & Ricciardi 2007), common carp (Matsuzaki et al. 2009) and various species of crayfish (McCarthy et al. 2006; Matsuzaki et al. 2009). These studies have shown that although the magnitude of an impact is often quite variable, the types and direction of impacts might exhibit patterns that are consistent, and thus predictable, across a range of invaded habitat types and geographic regions.

On the other hand, some impacts are so idiosyncratic that they defy prediction. For example, the mass die-offs of Great Lakes' waterfowl linked to the mussel–goby–botulism interaction described above could not have been predicted from the literature because similar events involving these species have not been previously recorded, despite the extensive invasion histories of zebra mussels and round gobies in Europe. Such unpredictable synergies are likely to accrue in systems that are heavily invaded (Ricciardi 2001). Furthermore, the effects of introduced species can vary substantially over time, owing to shifts in resident species composition (for example other invasions), changes to abiotic variables and evolutionary processes; such longterm feedbacks between introduced species and their invaded environments have rarely been studied and present a challenge to predicting the chronic effects of an invasion (Strayer et al. 2006).

16.6   ARE FRESHWATER SYSTEMS MORE SENSITIVE TO THE IMPACTS OF INTRODUCED SPECIES?

To our knowledge, no one has explicitly compared the impacts of invasions in freshwater, terrestrial and marine systems. In particular, it would be interesting to determine if differences between freshwater and terrestrial trophic interactions contribute to differential sensitivity to the effects of introduced consumers. Top-down control of plant biomass is stronger in water than on land (Shurin et al. 2006), thus the indirect effects of consumers may be larger in freshwater systems. Furthermore, the insularity of lakes and small rivers may result in a greater naiveté of their biota compared with continental terrestrial and marine biota (Cox & Lima 2006). This is evident in contrasting patterns of impact by terrestrial mammals and freshwater fishes; invasions by mammals are more likely to contribute to extirpations of native species when they involve dispersal between continents than within continents, whereas invasions by fishes are just as likely to cause extirpations whether they occur intra-continentally or inter-continentally (Fig. 16.3). Strong negative impacts of invasions on native species appear to be more frequent in freshwater systems than in marine systems (Ricciardi & Kipp 2008). Freshwater systems tend to have a higher proportion of invaders that are reported to have caused native species declines (Fig. 16.4). On average, high-impact invaders comprise 11% (95% confidence limits 0.069–0.154) of the total number of invaders in a freshwater system and

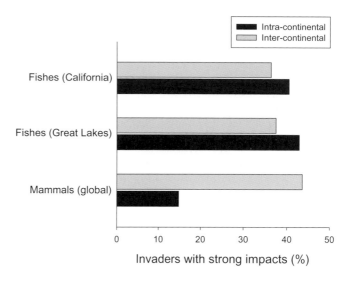

Fig. 16.3 Data on the percentage of invaders with strong negative impacts, i.e. those implicated in the severe (greater than 50%) decline of a native species population. For freshwater fishes in California and the Great Lakes, intra-continental and inter-continental invaders had similar proportions of species with strong negative impacts (Fisher's exact test, $P > 0.05$). For terrestrial mammals, inter-continental invasions had much stronger impacts than intra-continental invasions ($P < 0.05$). After Ricciardi and Simberloff (2009).

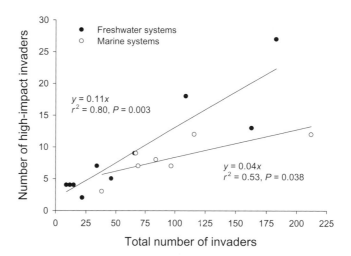

Fig. 16.4 Relationships between the number of high-impact invaders (i.e. those implicated in the severe decline of a native species population) and the total number of invaders in freshwater and marine systems. After Ricciardi and Kipp (2008), with data from Marsden and Hauser (2009).

4% (95% confidence limits 0.004–0.084) in a marine system.

In terms of their well-defined boundaries, their insularity, and the generally limited diversity and size of the populations that inhabit them, lakes and small river basins can be thought of as analogous to oceanic islands. Therefore, it might be more appropriate and useful to compare freshwater systems to islands when testing the generality of concepts across biomes.

16.7   CONCLUSIONS AND FUTURE PROSPECTS

Lakes and river basins offer well-defined, replicated model systems for studying the effects of non-indigenous species (see, for example, Vander Zanden et al. 2004; Rosenthal et al. 2006; Sharma et al. 2009). Freshwater ecosystems are among the most invaded in the world, and have revealed patterns and problems that have added substantially to our knowledge of the population-, community- and ecosystem-level effects of introduced species. Yet research has only glimpsed the extent to which invasions have transformed these ecosystems. There is a dearth of impact studies in regions where introductions of aquatic species are most prevalent (for example Asia (Leprieur et al. 2009)). Even in the North American Great Lakes, which have an extensive history of invasion involving over 180 non-indigenous species (see Mills et al. 1993; Ricciardi 2006), the effects of most invasions remain unknown. The identification of general patterns has been hindered by taxonomic biases in impact studies, which focus more often on fishes than invertebrates (see Parker et al. 1999). Furthermore, although an invader can potentially affect different levels of ecological organization – including individuals (for example behaviour, morphology), populations (abundance, genetic structure), communities (richness, evenness, food-web structure) and ecosystems (habitat structure, primarily production, nutrient dynamics, contaminant cycling), few impact studies have examined multiple levels simultaneously (Simon & Townsend 2003).

Notwithstanding the challenge it poses to prediction, the context-dependent variation of the effects of an invader across a set of lakes or rivers offers valuable opportunities to identify environmental conditions that enhance or inhibit impact. The scarcity of pre-invasion data sets impedes efforts to measure impacts against a background of temporal variation and multiple stressors; but in cases where long-term data exist, the effects of non-indigenous species may be isolated from other anthropogenic perturbations (Yan & Pawson 1997).

The risk of a highly disruptive invasion will increase as more non-indigenous species are added to a system, or as the same species invade more systems (Ruesink 2003; Ricciardi & Kipp 2008). An important insight is that many introduced species, if widely distributed, can become highly disruptive in at least some systems (Ruesink 2003). Given (i) the complexity of direct and indirect interactions between introduced species and other stressors (Didham et al. 2007), (ii) the insularity and acute sensitivity of freshwater ecosystems and (iii) our current level of understanding of impact, we conclude that humans cannot engage safely in deliberate introductions of freshwater species, and that management plans to conserve freshwater biodiversity must include strategies for preventing invasions. Fortunately, this view seems to be better appreciated by biologists and wildlife managers today than it was during Elton's time.

ACKNOWLEDGEMENTS

We thank Dave Richardson for inviting us to participate in the symposium that inspired this volume. Michael Marchetti and two anonymous referees provided valuable comments on the manuscript. Funding was provided to both authors by the Natural Science and Engineering Research Council (NSERC) and the Canadian Aquatic Invasive Species Network (CAISN).

REFERENCES

Adams, M.J., Pearl, C.A. & Bury, R.B. (2003) Indirect facilitation of an anuran invasion by non-native fishes. *Ecology Letters*, **6**, 343–351.

Alcaraz, C., Bisazza, A. & Garcia-Berthou, E. (2008) Salinity mediates the competitive interactions between invasive mosquitofish and an endangered fish. *Oecologia*, **155**, 205–213.

Allendorf, F.W. (1991) Ecological and genetic effects of fish introductions: synthesis and recommendations. *Canadian Journal of Fisheries and Aquatic Sciences*, **48**, 178–181.

Baxter, C.V., Fausch, K.D., Murakami, M. & Chapman, P.L. (2004) Fish invasion restructures stream and forest food webs by interrupting reciprocal prey subsidies. *Ecology*, **85**, 2656–2663.

Bernauer, D. & Jansen, W. (2006) Recent invasions of alien macroinvertebrates and loss of native species in the upper Rhine River, Germany. *Aquatic Invasions*, **1**, 55–71.

Bollache, L., Dick, J.T.A., Farnsworth, K.D. & Montgomery, W.I. (2008) Comparison of the functional responses of invasive and native amphipods. *Biology Letters*, **4**, 166–169.

Boylen, C.W., Eichler, L.W. & Madsen, J.D. (1999) Loss of native aquatic plant species in a community dominated by Eurasian watermilfoil. *Hydrobiologia*, **415**, 207–211.

Brown, C.J., Blossey, B., Maerz, J.C. & Joule, S.J. (2006) Invasive plant and experimental venue affect tadpole performance. *Biological Invasions*, **8**, 327–338.

Clavero, M. & García-Berthou, E. (2006) Homogenization dynamics and introduction routes of invasive freshwater fish in the Iberian Peninsula. *Ecological Applications*, **16**, 2313–2324.

Combes, C. & Le Brun, N. (1990) Invasions by parasites in continental Europe. In *Biological Invasions in Europe and the Mediterranean Basin* (ed. F. Di Castri, A.J. Hansen and M. Debussche), pp. 285–296. Kluwer Academic Publishers, Dordrecht, the Netherlands.

Cox, J.G. & Lima, S.L. (2006) Naiveté and an aquatic–terrestrial dichotomy in the effects of introduced predators. *Trends in Ecology & Evolution*, **21**, 674–680.

Dick, J.T.A. (2008) Role of behaviour in biological invasions and species distributions: lessons from interactions between the invasive *Gammarus pulex* and the native *G. duebeni* (Crustacea: Amphipoda). *Contributions to Zoology*, **77**, 91–98.

Didham, R.K., Tylianakis, J.M., Gemmell, N.J., Rand, T.A. & Ewers, R.M. (2007) Interactive effects of habitat modification and species invasion on native species decline. *Trends in Ecology & Evolution*, **22**, 489–496.

Didham, R.K., Tylianakis, J.M., Hutchison, M.A., Ewers, R.M. & Gemmell, N.J. (2005) Are invasive species the drivers of ecological change? *Trends in Ecology & Evolution*, **20**, 470–474.

Duncan, J.R. & Lockwood, J.L. (2001) Spatial homogenization of the aquatic fauna of Tennessee: extinction and invasion following land use change and habitat alteration. In *Biotic Homogenization* (ed. J.L. Lockwood and M.L. McKinney), pp. 245–257, Kluwer Academic Publishers, New York.

Eagles-Smith, C.A., Suchanek, T.H., Colwell, A.E., Anderson, N.L. & Moyle, P.B. (2008) Changes in fish diets and food web mercury bioaccumulation induced by an invasive planktivorous fish. *Ecological Applications*, **18** (special issue), A213–A226.

Elton, C.S. (1958) *The Ecology of Invasions by Animals and Plants*. Methuen, London.

Flecker, A.S. & Townsend, C.R. (1994) Community-wide consequences of trout introduction in New Zealand streams. *Ecological Applications*, **4**, 798–807.

Gamradt, S.C., Kats, L.B. & Anzalone, C.B. (1997) Aggression by non-native crayfish deters breeding in California newts. *Conservation Biology*, **11**, 793–796.

Hall, S.R. & Mills, E.L. (2000) Exotic species in large lakes of the world. *Aquatic Ecosystem Health and Management*, **3**, 105–135.

Harvey, B.C., White, J.L. & Nakamoto, R.J. (2004) An emergent multiple predator effect may enhance biotic resistance in a stream fish assemblage. *Ecology*, **85**, 127–133.

Herbst, D.B, Silldorff, E.L. & Cooper, S.D. (2009) The influence of introduced trout on the benthic communities of paired headwater streams in the Sierra Nevada of California. *Freshwater Biology*, **54**, 1324–1342.

Hogan, L.S., Marschal, E., Folt, C. & Stein, R.A. (2007) How non-native species in Lake Erie influence trophic transfer of mercury and lead to top predators. *Journal of Great Lakes Research*, **33**, 46–61.

Holling, C.S. (1959) Some characteristics of simple types of predation and parasitism. *Canadian Entomologist*, **91**, 385–398.

Johnson, B.M., Martinez, P.J., Hawkins, J.A. & Bestgen, K.R. (2008a) Ranking predatory threats by nonnative fishes in the Yampa River, Colorado, via bioenergetics modeling. *North American Journal of Fisheries Management*, **28**, 1941–1953.

Johnson, P.T., Olden, J.D. & Vander Zanden, M.J. (2008b) Dam invaders: impoundments facilitate biological invasions in freshwaters. *Frontiers in Ecology and the Environment*, **6**, 357–363.

Jokela, A. & Ricciardi, A. (2008) Predicting zebra mussel fouling on native mussels from physicochemical variables. *Freshwater Biology*, **53**, 1845–1856.

Keller, R.P., Drake, J.M. & Lodge, D.M. (2007) Fecundity as a basis for risk assessment of nonindigenous freshwater molluscs. *Conservation Biology*, **21**, 191–200.

Kestrup, Å. & Ricciardi, A. (2009) Environmental heterogeneity limits the local dominance of an invasive freshwater crustacean. *Biological Invasions*, **11**, 2095–2105.

Kitchell, J.F. & Crowder, L.B. (1986) Predator–prey interactions in Lake Michigan: model predictions and recent dynamics. *Environmental Biology of Fishes*, **16**, 205–211.

Knapp, R.A. & Matthews, K.R. (2000) Non-native fish introductions and the decline of the mountain yellow-legged frog from within protected areas. *Conservation Biology*, **14**, 428–438.

Kolar, C.S. & Lodge, D.M. (2002) Ecological predictions and risk assessment for alien fishes in North America. *Science*, **298**, 1233–1236.

Latini, A.O. & Petrere, Jr. M. (2004) Reduction of a native fish fauna by alien species: an example from Brazilian freshwater tropical lakes. *Fisheries Management and Ecology*, **11**, 71–79.

Leavitt, P.R., Schindler, D.E., Paul, A.J., Hardie, A.K. & Schindler, D.W. (1994) Fossil pigment records of phytoplankton in trout-stocked lakes. *Canadian Journal of Fisheries and Aquatic Sciences*, **51**, 2411–2423.

Lepak, J.M., Kraft, C.E. & Weidel, B.C. (2006) Rapid food web recovery in response to removal of an introduced apex predator. *Canadian Journal of Fisheries and Aquatic Sciences*, **63**, 569–575.

Leprieur, F., Brosse, S., Garcia-Berthou, E., Oberdorff, T., Olden, J.D. & Townsend, C.R. (2009) Scientific uncertainty and the assessment of risks posed by non-native freshwater fishes. *Fish and Fisheries*, **10**, 88–97.

Lever, C. (1996) *Naturalized Fishes of the World*. Academic Press, London.

Light, T. & Marchetti, M.P. (2007) Distinguishing between invasions and habitat changes as drivers of diversity loss among California's freshwater fishes. *Conservation Biology*, **21**, 434–446.

MacIsaac, H.J. (1996) Potential abiotic and biotic impacts of the zebra mussel on the inland waters of North America. *American Zoologist*, **36**, 287–299.

MacRae, P.S.D. & Jackson, D.A. (2001) The influence of smallmouth bass (*Micropterus dolomieu*) predation and habitat complexity on the structure of littoral zone fishes. *Canadian Journal of Fisheries and Aquatic Sciences*, **58**, 342–351.

Marchetti, M.P., Light, T., Moyle, P.B. & Viers, J.H. (2004a) Fish invasions in California watersheds: testing hypotheses using landscape patterns. *Ecological Applications*, **14**, 1507–1525.

Marchetti, M.P., Moyle, P.B. & Levine, R. (2004b) Alien fishes in California watersheds: characteristics of successful and failed invaders. *Ecological Applications*, **14**, 587–596.

Marsden, J.E. & Hauser, M. (2009) Exotic species in Lake Champlain. *Journal of Great Lakes Research*, **35**, 250–265.

Matsuzaki, S.S., Usio, N., Takamura, N. & Washitani, I. (2009) Contrasting impacts of invasive engineers on freshwater ecosystems: an experiment and meta-analysis. *Oecologia*, **158**, 673–686.

McCarthy, J.M., Hein, C.L., Olden, J.D. & Vander Zanden, M.J. (2006) Coupling long-term studies with meta-analysis to investigate impacts of non-native crayfish on zoobenthic communities. *Freshwater Biology*, **51**, 224–235.

Miller, R.R., Williams, J.D. & Williams, J.E. (1989) Extinctions of North American fishes during the past century. *Fisheries*, **14**(6), 22–38.

Mills, E.L., Leach, J.H., Carlton, J.T. & Secor, C.L. (1993) Exotic species in the Great Lakes: a history of biotic crises and anthropogenic introductions. *Journal of Great Lakes Research*, **19**, 1–54.

Mills, E.L., Strayer, D.L., Scheuerell, M.D. & Carlton, J.T. (1996) Exotic species in the Hudson River Basin: A history of invasions and introductions. *Estuaries*, **19**, 814–823.

Morton, B. (1997) The aquatic nuisance species problem: a global perspective and review. In *Zebra Mussels and Aquatic Nuisance Species* (ed. F. D'Itri), pp. 1–54, Ann Arbor Press, Chelsea, Michigan.

Moyle, P.B. (1976) Fish introductions in California: history and impact on native fishes. *Biological Conservation*, **9**, 101–118.

Moyle, P.B., Li, H.W. & Barton, B.A. (1986) The Frankenstein effect: impact of introduced fishes on native fishes in North America. In *Fish Culture in Fisheries Management* (ed. R.H. Stroud), pp. 415–426, American Fisheries Society, Bethesda, Maryland.

Moyle, P.B. & Light, T. (1996) Biological invasions of fresh water: empirical rules and assembly theory. *Biological Conservation*, **78**, 149–161.

Nesler, T.P. & Bergersen, E.P. (1991) *Mysids in Fisheries: Hard Lessons from Headlong Introductions*. American Fisheries Society Symposium No. 9, Bethesda, Maryland.

Nyström, P., Svensson, O., Lardner B., Brönmark, C. & Granéli, W. (2001) The influence of multiple introduced predators on a littoral pond community. *Ecology*, **82**, 1023–1039.

Parker, I.M., Simberloff, D., Lonsdale, W.M., et al. (1999) Impact: toward a framework for understanding the ecological effects of invaders. *Biological Invasions*, **1**, 3–19.

Perna, C. & Burrows, D. (2005) Improved dissolved oxygen status following removal of exotic weed mats in important fish habitat lagoons of the tropical Burdekin River floodplain, Australia. *Marine Pollution Bulletin*, **51**, 138–148.

Perry, W.L., Lodge, D.M. & Feder, J.L. (2002) Importance of hybridization between indigenous and nonindigenous freshwater species: an overlooked threat to North American freshwater biodiversity. *Systematic Biology*, **51**, 255–275.

Phelps, H.L. (1994) The Asiatic clam (*Corbicula fluminea*) invasion and system-level ecological change in the Potomac River Estuary near Washington, D.C. *Estuaries*, **17**, 614–621.

Piscart, C., Dick, J.T.A., McCrisken, D. & MacNeil, C. (2009) Environmental mediation of intraguild predation between the freshwater invader *Gammarus pulex* and the native *Gammarus duebeni celticus*. *Biological Invasions*, **11**, 2141–2145.

Pope, K.L. (2008) Assessing changes in amphibian population dynamics following experimental manipulations of introduced fish. *Conservation Biology*, **22**, 1572–1581.

Pope, K.L., Garwood, J.M., Welsh, Jr. H.H. & Lawler, S.P. (2008) Evidence of indirect impacts of introduced trout on native amphibians via facilitation of a shared predator. *Biological Conservation*, **141**, 1321–1331.

Rahel, F.J. (2002) Homogenization of freshwater faunas. *Annual Reviews in Ecology and Systematics*, **33**, 291–315.

Reissig, M., Trochine, C., Queimaliños, C., Balseiro, E. & Modenutti, B. (2006) Impact of fish introduction on planktonic food webs in lakes of the Patagonian Plateau. *Biological Conservation*, **132**, 437–447.

Reynolds, J.D. (1988) Crayfish extinctions and crayfish plague in central Ireland. *Biological Conservation*, **45**, 279–285.

Ricciardi, A. (2001) Facilitative interactions among aquatic invaders: is an 'invasional meltdown' occurring in the Great Lakes? *Canadian Journal of Fisheries and Aquatic Sciences*, **58**, 2513–2525.

Ricciardi, A. (2003) Predicting the impacts of an introduced species from its invasion history: an empirical approach applied to zebra mussel invasions. *Freshwater Biology*, **48**, 972–981.

Ricciardi, A. (2004) Assessing species invasions as a cause of extinction. *Trends in Ecology & Evolution*, **19**, 619.

Ricciardi, A. (2006) Patterns of invasion in the Laurentian Great Lakes in relation to changes in vector activity. *Diversity and Distributions*, **12**, 425–433.

Ricciardi, A. & Atkinson, S.K. (2004) Distinctiveness magnifies the impact of biological invaders in aquatic ecosystems. *Ecology Letters*, **7**, 781–784.

Ricciardi, A. & Cohen, J. (2007). The invasiveness of an introduced species does not predict its impact. *Biological Invasions*, **9**, 309–315.

Ricciardi, A. & Kipp, R. (2008) Predicting the number of ecologically harmful exotic species in an aquatic system. *Diversity and Distributions*, **14**, 374–380.

Ricciardi, A., Neves, R.J. & Rasmussen, J.B. (1998) Impending extinctions of North American freshwater mussels (Unionidae) following the zebra mussel (*Dreissena polymorpha*) invasion. *Journal of Animal Ecology*, **67**, 613–619.

Ricciardi, A. & Rasmussen, J.B. (1999) Extinction rates of North American freshwater fauna. *Conservation Biology*, **13**, 1220–1222.

Ricciardi, A. & Simberloff, D. (2009) Assisted colonization is not a viable conservation strategy. *Trends in Ecology & Evolution*, **24**, 248–253.

Rivero, L.H. (1937) The introduced largemouth bass, a predator upon native Cuban fishes. *Transactions of the American Fisheries Society*, **66**, 367–368.

Robinson, J.V. & Wellborn, G.A. (1988) Ecological resistance to the invasion of a freshwater clam, *Corbicula fluminea*: fish predation effects. *Oecologia*, **77**, 445–452.

Rodriguez, C.F., Bécares, E., Fernández-Aláez, M. & Fernández-Aláez, C. (2005) Loss of diversity and degradation of wetlands as a result of introducing exotic crayfish. *Biological Invasions*, **7**, 75–85.

Rommens, W., Maes, J., Dekeza, N., Inghelbrecht, P., Nhiwatiwa, T., Holsters, E., Ollevier, F., Marshall, B. & Brendonck, L. (2003) The impact of water hyacinth (*Eichhornia crassipes*) in a eutrophic subtropical impoundment (Lake Chivero, Zimbabwe). I. Water quality. *Archiv für Hydrobiologie*, **158**, 373–388.

Rosenthal, S.K., Stevens S.S. & Lodge, D.M. (2006). Whole-lake effects of invasive crayfish (*Orconectes* spp.) and the potential for restoration. *Canadian Journal of Fisheries and Aquatic Sciences*, **63**, 1276–1285.

Roth, B.M., Tetzlaff, J.C., Alexander, M.L. & Kitchell, J.F. (2007) Reciprocal relationships between exotic rusty crayfish, macrophytes, and *Lepomis* species in northern Wisconsin lakes. *Ecosystems*, **10**, 74–85.

Ruesink, J.L. (2003) One fish, two fish, old fish, new fish: which invasions matter? In *The Importance of Species – Perspectives on Expendibilty and Triage* (ed. P. Kareiva and S.A. Levin), pp. 161–178, Princeton University Press, Princeton, New Jersey.

Ruiz, G.M, Fofonoff, P., Hines, A.H. & Grosholz, E.D. (1999) Non-indigenous species as stressors in estuarine and marine communities: assessing invasion impacts and interactions. *Limnology and Oceanography*, **44**, 950–972.

Schabetsberger, R., Luger, M.S., Drozdowski, G. & Jagsch, A. (2009) Only the small survive: monitoring long-term changes in the zooplankton community of an Alpine lake after fish introduction. *Biological Invasions*, **11**, 1335–1345.

Schindler, D.E., Knapp, K.A. & Leavitt, P.R. (2001) Alteration of nutrient cycles and algal production resulting from fish introductions into mountain lakes. *Ecosystems*, **4**, 308–321.

Schooler, S.S., McEvoy, P.B. & Coombs, E.M. (2006) Negative per capital effects of purple loosestrife and reed canary grass on plant diversity of wetland communities. *Diversity and Distributions*, **12**, 351–363.

Scott, M.C. (2006) Winners and losers among stream fishes in relation to land use legacies and urban development in the southeastern US. *Biological Conservation*, **127**, 301–309.

Scott, M.C. & Helfman, G.S. (2001) Native invasions, homogenization, and the mismeasure of integrity of fish assemblages. *Fisheries*, **26**(11), 6–15.

Sebestyen, O. (1938) Colonization of two new fauna-elements of Pontus-origin (*Dreissensia polymorpha* Pall. and *Corophium curvispinum* G.O. Sars *forma devium* Wundsch) in Lake Balaton. *Verhandlungen des Internationalen Verein Limnologie*, **8**, 169–182.

Sharma, S., Jackson, D.A. & Minns, C.K. (2009) Quantifying the potential effects of climate change and the invasion of smallmouth bass on native lake trout populations across Canadian lakes. *Ecography*, **32**, 517–525.

Shurin, J.B., Gruner, D.S. & Hillebrand, H. (2006) All wet or dried up? Real differences between aquatic and terrestrial food webs. *Proceedings of the Royal Society of London B*, **273**, 1–9.

Simberloff, D. (1995) Why do introduced species appear to devastate islands more than mainland areas? *Pacific Science*, **49**, 87–97.

Simon, K.S. & Townsend, C.R. (2003) Impacts of freshwater invaders at different levels of ecological organisation, with emphasis on salmonids and ecosystem consequences. *Freshwater Biology*, **48**, 982–994.

Spencer, C.N., McClelland, B.R. & Stanford, J.A. (1991) Shrimp stocking, salmon collapse, and eagle displacement. *BioScience*, **41**, 14–21.

Strayer, D.L., Eviner, V.T., Jeschke, J.M. & Pace, M.L. (2006) Understanding the long-term effects of species invasions. *Trends in Ecology & Evolution*, **21**, 645–651.

Taniguchi, Y., Rahel, F.J. Novinger, D.C. & Geron, K.G. (1998) Temperature mediation of competitive interactions among three fish species that replace each other along longitudinal stream gradients. *Canadian Journal of Fisheries and Aquatic Sciences*, **55**, 1894–1901.

Taylor, J.N., Courtenay, Jr, W.R. & McCann, J.A. (1984) Known impacts of exotic fishes in the continental United States. In *Distribution, Biology and Management of Exotic Fishes* (ed. W.R. Courtenay, Jr and J.R. Stauffer, Jr), pp. 322–373, Johns Hopkins University Press, Baltimore.

Valley, R.D. & Bremigan, M.T. (2002) Effect of macrophyte bed architecture on largemouth bass foraging: implications of exotic macrophyte invasions. *Transactions of the American Fisheries Society*, **131**, 234–244.

Vanderploeg, A.A., Nalepa, T.F., Jude, J.J., et al. (2002) Dispersal and emerging ecological impacts of Ponto-Caspian species in the Laurentian Great Lakes. *Canadian Journal of Fisheries and Aquatic Sciences*, **59**, 1209–1228.

Vander Zanden, M.J. Casselman, J.M. & Rasmussen, J.B. (1999) Stable isotope evidence for the food web consequences of species invasions in lakes. *Nature*, **401**, 464–467.

Vander Zanden, M.J., Olden, J.D., Thorne, J.H. & Mandrak, N.E. (2004) Predicting the occurrence and impact of bass introductions in north-temperate lakes. *Ecological Applications*, **14**, 132–148.

Vitousek, P.M., D'Antonio, C.M., Loope, L.L., Rejmanek, M. & Westbrooks, R. (1997) Introduced species: a significant component of human-caused global change. *New Zealand Journal of Ecology*, **21**, 1–16.

Vredenburg, V.T. (2004) Reversing introduced species effects: experimental removal of introduced fish leads to rapid recovery of a declining frog. *Proceedings of the National Academy of Sciences of the USA*, **101**, 7646–7650.

Ward, J.M. & Ricciardi, A. (2007) Impacts of *Dreissena* invasions on benthic macroinvertebrate communities: a meta-analysis. *Diversity and Distributions*, **13**, 155–165.

Wilson, S.J. & Ricciardi, A. (2009) Epiphytic macroinvertebrate communities on Eurasian watermilfoil (*Myriophyllum spicatum*) and native milfoils *Myriophyllum sibiricum* and *Myriophyllum alterniflorum* in eastern North America. *Canadian Journal of Fisheries and Aquatic Sciences*, **66**, 18–30.

Witte, F., Goldschmidt, T., Wanink, J., et al. (1992) The destruction of an endemic species flock: quantitative data on the decline of the haplochromine cichlids of Lake Victoria. *Environmental Biology of Fishes*, **34**, 1–28.

Yan, N.D. & Pawson, T.W. (1997) Changes in the crustacean zooplankton community of Harp Lake, Canada, following invasion by *Bythotrephes cederstroemi*. *Freshwater Biology*, **37**, 409–425.

Yule, A.M., Barker, I.K., Austin, J.W. & Moccia, R.D. (2006) Toxicity of *Clostridium botulinum* type E neurotoxin to Great Lakes fish: implications for avian botulism. *Journal of Wildlife Diseases*, **42**, 479–493.

Zaret, T.M. & Paine, R.T. (1973) Species introduction in a tropical lake. *Science*, **182**, 449–455.

Chapter 17

EXPANDING THE PROPAGULE PRESSURE CONCEPT TO UNDERSTAND THE IMPACT OF BIOLOGICAL INVASIONS

Anthony Ricciardi[1], Lisa A. Jones[1,2],
Åsa M. Kestrup[1,2] and Jessica M. Ward[1,2]

[1]Redpath Museum, McGill University, Montreal, Quebec, Canada
[2]Department of Biology, McGill University, Montreal, Quebec Canada

17.1   INTRODUCTION

Most introduced species fail to establish sustainable populations, and many of those that do become established invaders do not cause strong impacts (Williamson 1996; Parker et al. 1999). Others can displace native species, disrupt ecosystem processes, threaten human and animal health and generate large economic costs. Why some introduced species are more successful or more disruptive than others are central questions of invasion ecology. Exploration of these questions has identified propagule pressure – the quantity, richness, or frequency of individuals or life stages of a species released into an area – as the most important determinant of establishment success (Williamson 1996; Lockwood et al. 2005; Simberloff 2009). The basis for this relationship is the greater risk of extinction that small populations suffer because of intrinsic and extrinsic factors including random changes in birth and death rates, small-scale catastrophes like extreme weather events, inbreeding depression and reduced efficiency in mate location and foraging. Indeed, when individuals are introduced in greater numbers or greater frequency they are more likely to form sustainable populations (Sakai et al. 2001).

Although Elton (1958) and others before him recognized the role of the dispersal of propagules in the global distribution of plants and animals, the concept of propagule pressure did not develop until many years later. As Simberloff (2009) noted, propagule pressure was not considered in the key research questions highlighted in the mid-1980s by the international Scientific Committee on Problems of the Environment, whose agenda spurred the growth of invasion ecology during the subsequent two decades (Richardson & Pyšek 2008; Simberloff, this volume). Research interest in the concept began to grow exponentially only during the past decade (Simberloff 2009).

A burgeoning number of studies (reviewed by Lockwood et al. 2005; Simberloff 2009) have shown propagule pressure to be a consistent predictor of establishment and spread of non-indigenous species. Furthermore, recent research points to an overwhelming influence of the pattern of dispersal and non-random variation in propagule supply on the outcome of an invasion (Colautti et al. 2006; Wilson et al. 2009). However, propagule pressure may play an even greater role in the outcome of an invasion than has been previously recognized. Although there has been a surge of experimental and modelling studies that explore the intricacies of the relationship between propagule pressure and establishment success, to our knowledge no study has examined the specific links between propagule pressure and the ecological impact of an invasion. There are no hypotheses that explicitly relate these concepts, beyond the obvious expectation that the frequency and diversity of impacts will increase with invasions (Lockwood et al. 2005; Catford et al. 2009). Here, we consider mechanisms by which propagule pressure can modulate impact and how a predictive understanding of this relationship would be valuable to the management and risk assessment of invasions.

17.2   PROPAGULE PRESSURE AND IMPACT DEFINED

Propagule pressure encompasses variation in the quantity, composition and rate of supply of non-indigenous organisms resulting from the transport conditions and pathways between source and recipient regions. Therefore, measures of propagule pressure generally involve (i) the number of individuals introduced ('propagule abundance'), (ii) the number of taxa or genotypes introduced ('propagule richness') and (iii) the total number or frequency of introduction events (Table 17.1). Lockwood and colleagues (2009) argue for a distinction between the number of introduced species (which they term 'colonization pressure') and other forms of propagule pressure, to better understand the processes that account for variation in non-indigenous species richness. For our purposes, we consider as a single form of propagule pressure the entire richness of introduced propagules, from genotypes to higher taxa. Thus our definition of propagule pressure incorporates colonization pressure as well as genetic (intraspecific) variation among propagules. The importance of genotypic variation to establishment success and post-establishment spread is becoming increasingly recognized (Sexton et al. 2002; Roman 2006; Lavergne & Molofsky 2007), and here we discuss how it may also play a significant role in impact.

A widely used synonym for propagule abundance is 'propagule size' (see, for example, Forsyth & Duncan 2001), which is also used by plant ecologists to des-

Table 17.1 Forms of propagule pressure that are commonly studied. Studies have examined either a single form or proxy variable of propagule pressure, or an aggregate measure of multiple forms. The relative importance of each form to the successful establishment and impact of an introduced species varies among recipient systems.

Form	Definition	Proxy variables	Examples of recent studies
Propagule abundance	The number of individuals of a non-indigenous species introduced to an area	Vector activity (e.g. volume of ballast water released)	Forsyth & Duncan 2001; Ahlroth et al. 2003; Colautti 2005; Verling et al. 2005
Propagule richness	The number of non-indigenous taxa (e.g. species, genera, families) or genotypes introduced to an area	Vector richness; Number of pathways or donor regions; Number of source populations	Ahlroth et al. 2003; Verling et al. 2005; Roman & Darling 2007; Chiron et al. 2009
Propagule frequency	The total number, or rate, of discrete introduction events	Vector activity (e.g. number of ship visits; number of tourists)	Forsyth & Duncan 2001; Colautti 2005; Verling et al. 2005; Drake et al. 2005; Drury et al. 2007; Roman & Darling 2007

cribe the physical dimensions of seeds and other plant reproductive structures, regardless of whether the plant is non-indigenous; therefore, we adopt the former term to avoid confusion. Although propagule abundance involves population-level processes and propagule richness involves community-level processes, these forms are correlated: a greater input of propagules increases both the number of individuals and the number of taxa introduced, in a non-linear manner (Lockwood et al. 2009). Two added dimensions are the physiological condition of the propagules upon release and human cultivation of introduced individuals, both of which can influence establishment success (Mack 2000; Verling et al. 2005) but are not considered here. Finally, it should be noted that herein propagule pressure refers exclusively to the introduction of non-indigenous species to an area, rather than the propagule release (fecundity) associated with invaders after they are established in the area.

We define ecological impact as a measurable change to the properties of organisms, populations, communities or ecosystems. Several factors generate variation in impacts (Parker et al. 1999; Ricciardi 2003). Different impacts arise from different functional roles of the invader within the recipient community; for example, the zebra mussel's effects on pelagic food webs result from its filter-feeding activities, whereas its effects on benthic communities are largely related to its gregarious attachment and structural transformation of submerged surfaces (Ricciardi 2003). An invader's impacts may differ considerably across regions as its abundance and functional role are altered by local abiotic conditions (D'Antonio et al. 2000) and by interactions with resident species (Robinson & Wellborn 1988; Simberloff & Von Holle 1999; Ricciardi 2005). If the invader is a novel organism that uses resources differently than the rest of the community, then resident native species are more likely to be naïve and sensitive to its effects and less likely to offer resistance to its population growth (Ricciardi & Atkinson 2004). Finally, the size of the invaded range determines the absolute scale of the invader's impact; over a larger range, an invader can have a greater impact on biodiversity by affecting a larger number of native species and a greater proportion of the area occupied by those species (Parker et al. 1999). In summary, impact can be described as a function of three principal mediators: the invader's abundance, its functional ecology relative to the recipient community and its range size in the invaded region. Here, we consider how different forms of propagule pressure can influence impact through each of these mediators (Fig. 17.1).

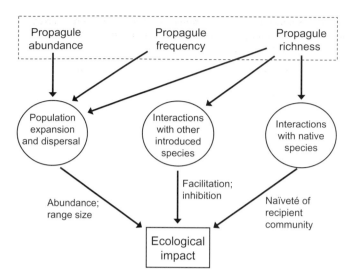

Fig. 17.1 Interactions between three forms of propagule pressure (propagule abundance, frequency and richness) and population-level and community-level processes that mediate the impact of introduced species. Major processes are circled. Specific mediating factors or processes of importance are indicated beside the arrows leading to impact.

17.3 INFLUENCE OF PROPAGULE PRESSURE ON MEDIATORS OF IMPACT

The invader's abundance and population growth

Propagule pressure affects the post-establishment abundance of an introduced species (see Blackburn et al. 2001; Marchetti et al. 2004; Drake et al. 2005; Britton-Simmons & Abbott 2008). Logically, a population will establish and grow more rapidly – and cause impacts sooner – if a large number of individuals are introduced at a given time or if there is a high frequency of introduction events. An increase in propagule abundance can accelerate population growth and establishment by supplying colonists with sufficient numbers to overcome Allee effects (Sakai et al. 2001) or sufficient genetic variation to adapt to local conditions (Ahlroth et al. 2003), thereby reducing the time available for native species to adapt to the stresses imposed by the invader (Schlaepfer et al. 2005). Despite these probable outcomes, the data available for testing correlations between impact and propagule pressure are scarce and yield ambiguous results. For non-indigenous ungulate mammals established in New Zealand (see Forsyth & Duncan 2001) there is no relationship between their ecological impact (reviewed by Lever 1985; Long 2003) and the minimum number of individuals released (logistic regression of ranked high- or low-impact species versus log-transformed propagule abundance, $P > 0.05$; A. Ricciardi, unpublished data), although the most damaging species, red deer, *Cervus elaphus*, had received the greatest introduction effort. In the Great Lakes, invading fishes deemed responsible for native species declines tend to be those that have been introduced frequently through stocking programs (six of nine stocked species versus four of 16 non-stocked species; Fisher's exact test, $P = 0.053$), but this result is confounded by a taxonomic bias: most stocked fishes are salmonids (Crawford 2001).

Frequent introductions over time increase the likelihood that a species will be introduced when abiotic variables are optimal for reaching high abundances (Drake et al. 2006), which ultimately affects the onset and magnitude of impact. Propagule pressure could also restore non-indigenous populations that have suffered a severe decline (Gotelli 1991), and insure that the invader will have an opportunity to recolonize an area after a disturbance. A sufficiently high frequency of releases could allow the non-indigenous species to become established and dominate native species before the latter recover sufficiently to provide resistance (see, for example, Altman & Whitlach 2007). Moreover, in rare instances, propagule pressure can maintain an unsustainable population of a non-indigenous species at a level of abundance sufficient to exert a significant impact. For example, a frequent supply of propagules can erode a population of a closely related native species through hybridization or reproductive interference, as shown by the cumulative negative effect of multiple releases of domestic dogs and cats on the genetic integrity of native canids and wildcats, even where the natives are more abundant (Simberloff 1996; Pierpaoli et al. 2003).

The invader's functional ecology relative to the recipient community

A single transport vector, such as transoceanic shipping (Carlton & Geller 1993), can deliver hundreds of taxa simultaneously, and vectors from different source regions can carry multiple genotypes of a given species (Roman & Darling 2007). Variation in the supply of different taxa or genotypes of the same taxon has several consequences for recipient communities (Fig. 17.2), including the possibility of introducing an aggressive genotype (Fry & Smart 1999) or producing invasive hybrids (Ellstrand & Schierenbeck 2000; Facon et al. 2005). Such events may account for some of the observed long-term changes in the impacts of invaders (Strayer et al. 2006). Given that rapid evolutionary change in invaders is a common phenomenon (Whitney & Gabler 2008), the introduction of additional genotypes could increase functional diversity in populations of an invader (e.g. behaviour, morphology, physiological tolerances) so as to broaden its interactions with, and impacts on, native species by allowing it to invade new habitats (Sexton et al. 2002).

On the other hand, founder populations with low genetic diversity can also evolve rapidly (Dlugosch & Parker 2008) and produce major impacts. A notable example is the Argentine ant *Linepithema humile* introduced to California, where diminished genetic diversity in the invading population has reduced intraspecific aggression, allowing the formation of dense colonies that have expanded rapidly and replaced native species (Tsutsui et al. 2000). These examples underscore the important but poorly understood role of propagule pressure in the post-establishment evolution and impact of invaders.

Theoretical modelling reveals a potentially strong influence of propagule abundance and introduction rate on the impacts of invasive organisms that create or modify habitat (ecosystem engineers). Non-indigenous engineers affect the broadest range of species within the recipient community and can enhance both their own invasion success and impact on resident species by modifying selection pressures in their new environment (Byers 2002; Gonzalez et al. 2008). Their impact on native species will increase with the number of individuals initially released, if their rate of habitat transformation is density dependent. Furthermore, multiple introductions of engineers that fail to establish sustainable populations but persist for a short period of time may nonetheless exert impacts on the native community through cumulative transformations of habitat (Gonzalez et al. 2008).

Finally, an increase in propagule richness raises the probability of introducing a novel predator, competitor or pathogen. Functionally novel organisms are more

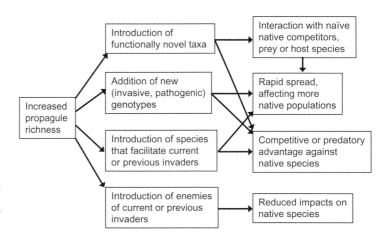

Fig. 17.2 Potential pathways by which an increase in propagule richness (e.g. through an increase in vector or source region diversity) can alter the impact of an invader on native species.

likely to disrupt recipient communities (Ricciardi & Atkinson 2004; Cox & Lima 2006). Owing to ballast water transport from Europe, North American inland waters have been invaded in recent decades by several marine-like animals originating from the freshwater margins of the Black, Azov and Caspian Seas, including several species (e.g. the zebra mussel *Dreissena polymorpha*, quagga mussel *Dreissena bugensis* and round goby *Neogobius melanostomus*) whose unique functional ecology has contributed to strong impacts on native communities in the Great Lakes that lack evolutionary experience with such organisms (Ricciardi & MacIsaac 2000). Before reaching North America, each of these Ponto-Caspian species became abundant at European ports from whence originates the bulk of shipping traffic to the Great Lakes. This example is one of many that show how propagule richness is strongly influenced by the transportation vector activities and pathways between the donor and recipient regions (Wilson et al. 2009).

The invader's range size

Both propagule abundance and frequency of introduction can have positive effects on an invader's range size (see, for example, Duncan et al. 1999). Through the delivery of multiple genotypes, propagule pressure can increase the potential for a non-indigenous species to adapt to local selective pressures (Sexton et al. 2002; Novak & Mack 2005; but see Dlugosch & Parker 2008), and thus governs the range of habitats in which native populations and communities are affected by the invader. Furthermore, a delay in the arrival of these genotypes can cause an invader's impact to change over time (Strayer et al. 2006). Secondary introductions from various sites in Europe raised the adaptive potential of canary grass *Phalaris arundinacea*, allowing it to invade a broader area of North American wetlands (Lavergne & Molofsky 2007). Even a long-established species can suddenly expand its range in the event of an infusion of genetic variation; multiple ballast-water introductions added genetic diversity to European green crab populations on the Atlantic coast of North America, apparently causing them to suddenly spread northwards into colder waters to which they were thought to be intolerant (Roman 2006). Variation in the source of propagules released in a given region may also lead to the creation of competitive hybrids that can rapidly expand their range

(Ellstrand & Schierenbeck 2000; Facon et al. 2005). On the other hand, continuous propagule pressure (gene flow) could also reduce local fitness and impede adaptation (Holt et al. 2005); the recipient habitat could become flooded with an inferior genotype, resulting in inferior hybrids or detrimental genetic swamping. These examples illustrate how frequent introductions of a species from different source populations can alter its invasion potential and the area over which its impacts occur.

17.4 SYNERGISTIC OR ANTAGONISTIC EFFECTS OF PROPAGULE RICHNESS?

Propagule richness, as numbers of introduced genotypes or species, can have varied effects on the recipient community (Fig. 17.3). For a given species, the number of genotypes introduced may have a positive or negative effect on the probability and magnitude of its impact. As noted previously, propagule richness can elevate impact by (i) producing invasive hybrids, (ii) increasing an invader's functional diversity (e.g. physiological tolerances, morphology or behaviour) so as to give it a competitive advantage, or (iii) enabling or accelerating range expansion. However, high genetic diversity might also increase intraspecific aggression or competition, or cause detrimental genetic swamping or the production of inferior hybrids when incoming genotypes are maladaptive (e.g. after rapid adaptation of the invader).

Given that impacts are expected to accumulate with the number of invaders in an area, there should be a positive relationship between propagule richness (as species) and impact, but different situations may arise. An increase in propagule richness provides more opportunities for the introduction of natural enemies of a previously established or future invader, which can constrain the invader's spread and population growth (and hence its impact); but they might instead facilitate the invader if they harm resident species in the invaded range (Colautti et al. 2004). Multiple invaders can interfere with each other in such a way as to attenuate their impacts (Vance-Chalcraft & Soluk 2005; Griffen et al. 2008). Conversely, through a variety of positive interactions, multiple invaders can produce synergistic effects (Simberloff & Von Holle 1999; Richardson et al. 2000; Ricciardi 2005). For example, the establishment of functionally diverse

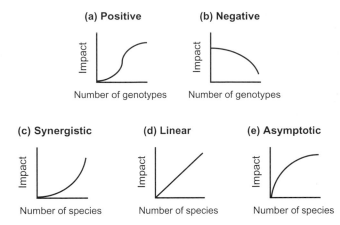

Fig. 17.3 Hypothesized relationships between propagule richness (numbers of introduced genotypes and species) and the probability or magnitude of impact. For a given species, increased numbers of genotypes can elevate impact by enabling or accelerating range expansion (a). However, they may reduce impact when high genetic diversity increases intraspecific competition or genetic swamping with maladaptive genotypes (b). An increase in the number of species can produce a synergistic relationship (c), if the impacts of one or more invaders are amplified in the presence of other invaders. A linear relationship (d) is expected when impacts are additive and could describe a proportional sampling effect. An asymptotic relationship (e) is expected under at least two situations: mutually negative interactions (e.g. competition, interference) between introduced species may diminish their respective impacts; impacts may also attenuate through source pool depletion, if high-impact invaders colonize more rapidly than other species.

predators may increase the extinction risk of resident species through a nonlinear accumulation of deleterious effects (Blackburn et al. 2005). However, in aquatic systems, the number of invaders causing severe declines in native species populations is a linear function of the total number of introduced species in an area, after controlling for species–area effects (Ricciardi & Kipp 2008). This correlation likely reflects proportional sampling and suggests that the probability of receiving a high-impact species increases predictably with propagule richness.

Furthermore, an increase in propagule richness provides opportunities for reassembling co-evolved mutualistic, predator–prey or parasite–host combinations that produce synergistic impacts. In Western Europe, multiple vectors drove sequential invasions that completed the parasitic life cycle of a Ponto-Caspian trematode, *Bucephalus polymorphus*. The introduction of the trematode's first intermediate host (the zebra mussel) and its definitive host (pike-perch *Sander lucioperca*) allowed it to establish and cause high mortality in local populations of cyprinid fishes, which serve as secondary intermediate hosts (Combes & Le Brun 1990).

Finally, an influx of non-indigenous species can alter the potential impacts of a previously established invader, even after a long time period during which it was relatively innocuous. An example is the European weed *Pastinaca sativa* that rapidly evolved an increase in toxicity in response to the introduction of one of its natural herbivores, two centuries after the weed became established in North America (Zangerl & Berenbaum 2005). The consequences of multiple invaders may also depend on the order of their introduction (Robinson & Dickerson 1987). Clearly, the effect of the delivery of multiple invaders to an area is one of the least understood elements of the relationship between propagule pressure and impact.

17.5 IMPLICATIONS FOR THE MANAGEMENT OF INVASIONS

The linkages we have highlighted here demonstrate how vector activity can affect the magnitude and scope of impact. A few predictions follow from these case studies. First, a change in the vector or pathway delivering propagules of a given species to a region can alter the impact potential of the species long after it is established. The delivery of new genetic strains can impede the efficacy of management strategies by supplying potentially adaptive variation to a non-indigenous population that previously may have been benign for many years. Enhanced genetic variation promotes adaptive evolution that reduces vulnerability to enemies (such as biocontrol agents (Burdon & Marshall 1981)) and can trigger rapid growth and spread that make control unfeasible (Roman 2006). Therefore, even after the establishment of a non-indigenous species, efforts should be made to prevent repeated introductions, particularly from new source regions. Second, an increase in the diversity and intensity of vector activity over time will lead to a rise in the number of high-impact invaders (primarily through sampling effects (Ricciardi & Kipp 2008)), which will render ecosystems increasingly difficult to manage. Third, vector activities that transfer species across different biogeographical regions will generate greater impacts than those that transfer species within the same biogeographical region, because the former is more likely to introduce organisms that are functionally novel within the recipient community (Ricciardi & Simberloff 2009). However, this may not apply to freshwater species, because the naïveté of freshwater communities is manifested on a much smaller spatial scale (Cox & Lima 2006). Finally, the risk of receiving highly disruptive invaders will not be proportionally diminished by a simple reduction in propagule abundance, if the number of source regions or vectors is not reduced simultaneously.

The few cases for which long-term data are available do not reveal any straightforward empirical relationship between impact and propagule pressure for a given system. In the Great Lakes, most invasions over the past half century are attributable to shipping (Ricciardi 2006). However, the number of ship-borne invaders that have been implicated in population declines of native species is not correlated with intensity of shipping activity (A. Ricciardi, unpublished data). Most of these invaders are Ponto-Caspian species, which were discovered in the Great Lakes from the mid-1980s to the late-1990s (Ricciardi & MacIsaac 2000), as much as two decades after the peak in shipping activity; it seems unlikely that there was a 10- to 20-year lag between the introduction and detection of these species, given their high reproductive capacity and conspicuous populations. This case suggests that propagule source is at least as important as both prop-

agule abundance and the number of introduction events in influencing the impact of invasion (see also Colautti et al. 2006; Wilson et al. 2009).

17.6 EXPANDING THE CONCEPT OF PROPAGULE PRESSURE: NEW AND EMERGING RESEARCH DIRECTIONS

To advance invasion ecology as a predictive science, we advocate the extension of the propagule pressure concept to unite disparate research foci on the establishment, population dynamics, adaptation, range expansion and impact of invaders. By explicitly including impact, the revised concept would provide the context for addressing questions that have both theoretical and applied value, such as the following.

1 How do different forms of propagule pressure (Table 17.1) vary in importance with respect to the establishment and impact of introduced species? This might be explored through the manipulation of the number, frequency and richness of propagules introduced to experimental mesocosms (see, for example, Drake et al. 2005) and relating these variables to subsequent changes in the structure of replicated recipient communities.

2 Under what conditions does sustained propagule pressure result in a reduction of invasions and their associated impacts versus a rapid accumulation of invaders and their synergistic effects? Frequent introductions of a given species have been observed to overwhelm biotic resistance from the native community to invasion (Von Holle & Simberloff 2005; Hollebone & Hay 2007), but introductions of multiple species could generate resistance (Case 1990). Several hypotheses predict cumulative negative interactions, suggesting an attenuation of impacts among elevated numbers of invaders (Catford et al. 2009). An alternative hypothesis ('invasional meltdown'; Simberloff & Von Holle 1999) predicts an increased frequency of synergistic impacts with additional invaders (Ricciardi 2005).

3 Is the introduction of propagules of co-evolved species more likely to generate synergistic impacts than propagules that do not share an evolutionary history? If co-evolution reduces the intensity of negative interactions (Case & Bolger 1991; Levin et al. 1982) and promotes positive interactions, then invaders with a common evolutionary history may be more facilitative and thus more likely to generate an invasional meltdown.

4 How does propagule pressure interact with other abiotic variables (e.g. disturbance, resource availability, habitat fragmentation) to mediate impact? Anthropogenic disturbance can alter or exacerbate the impacts of introduced species (Byers 2002); habitat fragmentation, for example, can magnify the effects of propagule pressure on hybridization (Simberloff 1996). Complex interactions among the various forms of propagule pressure (Drake et al. 2005; Drury et al. 2007) or between propagule pressure and disturbance (Britton-Simmons & Abbott 2008) pose a challenge to experimental and statistical analysis.

5 How does variation in different forms of propagule pressure affect time lags and long-term changes in the impacts of an invader? What are the mechanisms by which a change in propagule pressure can cause a previously benign invader to become ecologically disruptive? These questions stem from an increasing recognition of temporal variation in the effects of invaders (Strayer et al. 2006).

Such questions offer fertile ground for research that can inform management, even if general predictive models of impact remain elusive. Given its potential to enable a more comprehensive conceptual framework for understanding and managing invasions, the influence of propagule pressure on the short- and long-term effects of established non-indigenous species merits far more attention than it has received.

ACKNOWLEDGEMENTS

We thank Dave Richardson, Hugh MacIsaac and two anonymous referees for commenting on the manuscript. Funding support was provided by NSERC Canada and the Canadian Aquatic Invasive Species Network.

REFERENCES

Ahlroth, P., Alatalo, R.V., Holopainen, A., Kumpulainen, T. & Suhonen, J. (2003) Founder population size and number of source populations enhance colonization success in waterstriders. *Oecologia*, **137**, 617–620.

Altman, S. & Whitlach, R.B. (2007) Effects of small-scale disturbance on invasion success in marine communities. *Journal of Experimental Marine Biology and Ecology*, **342**, 15–29.

Blackburn, T.M., Gaston, K.J. & Duncan, R.P. (2001) Population density and geographic range size in the intro-duced and native passerine faunas of New Zealand. *Diversity and Distributions*, **7**, 209–221.

Blackburn, T.M., Petchey, O.L., Cassey, P. & Gaston, K.J. (2005) Functional diversity of mammalian predators and extinction in island birds. *Ecology*, **86**, 2916–2923.

Britton-Simmons, K.H. & Abbott, K.C. (2008) Short- and long-term effects of disturbance and propagule pressure on a biological invasion. *Journal of Ecology*, **96**, 68–77.

Burdon, J.J. & Marshall, D.R. (1981) Biological control and the reproductive mode of weeds. *Journal of Applied Ecology*, **18**, 649–658.

Byers, J.E. (2002) Impact of non-indigenous species enhanced by anthropogenic alteration of selection regimes. *Oikos*, **97**, 449–458.

Carlton, J.T. & Geller, J.B. (1993) Ecological roulette: the global transport of nonindigenous marine organisms. *Science*, **261**, 78–82.

Case, T.J. (1990) Invasion resistance arises in strongly interacting species-rich model competition communities. *Proceedings of the National Academy of Sciences of the USA*, **87**, 9610–9614.

Case, T.J. & Bolger, D.T. (1991) The role of introduced species in shaping the abundance and distribution of island reptiles. *Evolutionary Ecology*, **5**, 272–290.

Catford, J.A., Jansson, R. & Nilsson, C. (2009) Reducing redundancy in invasion ecology by integrating hypotheses into a single theoretical framework. *Diversity and Distributions*, **15**, 22–40.

Chiron, F., Shirley, S. & Kark, S. (2009) Human-related processes drive the richness of exotic birds in Europe. *Proceedings of the Royal Society B*, **276**, 47–53.

Colautti, R.I. (2005) Are characteristics of introduced salmonid fishes biased by propagule pressure? *Canadian Journal of Fisheries and Aquatic Sciences*, **62**, 950–959.

Colautti, R.I., Grigorovich, I.A. & MacIsaac, H.J. (2006) Propagule pressure: a null model for biological invasions. *Biological Invasions*, **8**, 1023–1037.

Colautti, R.I., Ricciardi, A., Grigorovich, I.A. & MacIsaac, H.J. (2004) Is invasion success explained by the Enemy Release Hypothesis? *Ecology Letters*, **7**, 721–733.

Combes, C. & Le Brun, N. (1990) Invasions by parasites in continental Europe. *Biological invasions in Europe and the Mediterranean Basin* (ed. F. Di Castri, J. Hansen and M. Debussche), pp. 285–296. Kluwer, Dordrecht.

Cox, J.G. & Lima, S.L. (2006) Naiveté and an aquatic-terrestrial dichotomy in the effects of introduced predators. *Trends in Ecology & Evolution*, **21**, 674–680.

Crawford, S.S. (2001) Salmonine introductions to the Laurentian Great lakes: an historical review and evaluation of ecological effects. *Canadian Special Publication of Fisheries and Aquatic Sciences*, **132**, 1–205.

D'Antonio, C.M., Tunison, J.T. & Loh, R.K. (2000) Variation in the impact of exotic grasses on native plant composition in relation to fire across an elevation gradient in Hawaii. *Austral Ecology*, **25**, 507–522.

Dlugosch, K.M. & Parker, I.M. (2008) Founding events in species invasions: genetic variation, adaptive evolution, and the role of multiple introductions. *Molecular Ecology*, **17**, 431–449.

Drake, J.M., Baggenstos, P. & Lodge, D.M. (2005) Propagule pressure and persistence in experimental populations. *Biology Letters*, **1**, 480–483.

Drake, J.M., Drury, K.L.S, Lodge, D.M., Blukacz, A., Yan, N. & Dwyer, G. (2006) Demographic stochasticity, environmental variability, and windows of invasion risk for *Bythotrephes longimanus* in North America. *Biological Invasions*, **8**, 843–861.

Drury, K.L.S., Drake, J.M., Lodge, D.M. & Dwyer, G. (2007) Immigration events dispersed in space and time. *Ecological Modelling*, **206**, 63–78.

Duncan, R.P., Blackburn, T.M. & Veltman, C.J. (1999) Determinants of geographic range sizes: a test using introduced New Zealand birds. *Journal of Animal Ecology*, **68**, 963–975.

Elton, C.S. (1958) *The Ecology of Invasions by Animals and Plants*. Methuen, London.

Ellstrand, N.C. & Schierenbeck, K.A. (2000) Hybridization as a stimulus for the evolution of invasiveness in plants. *Proceedings of the National Academy of Sciences of the USA*, **97**, 7043–7050.

Facon, B., Jarne, P., Pointier, J.P. & David, P. (2005) Hybridization and invasiveness in the freshwater snail *Melanoides tuberculata*: hybrid vigour is more important than increase in genetic variance. *Journal of Evolutionary Biology*, **18**, 524–535.

Forsyth, D.M. & Duncan, R.P. (2001) Propagule size and the relative success of exotic ungulate and bird introductions to New Zealand. *American Naturalist*, **157**, 583–595.

Fry, W.E. & Smart, C.D. (1999) The return of *Phytophthora infestans*, a potato pathogen that just won't quit. *Potato Research*, **42**, 279–282.

Gonzalez, A., Lambert, A. & Ricciardi, A. (2008) When does ecosystem engineering facilitate invasion and species replacement? *Oikos*, **117**, 1247–1257.

Gotelli, N. (1991) Metapopulation models: the rescue effect, the propagule rain, and the core-satellite hypothesis. *American Naturalist*, **138**, 768–776.

Griffen, B.D., Guy, T. & Buck, J.C. (2008) Inhibition between invasives: a newly introduced predator moderates the impacts of a previously established invasive predator. *Journal of Animal Ecology*, **77**, 32–40.

Hollcbone, A.L. & Hay, M.E. (2007) Propagule pressure of an invasive crab overwhelms biotic resistance. *Marine Ecology Progress Series*, **342**, 191–196.

Holt, R.D., Barfield, M. & Gomulkiewicz, R. (2005) Theories of niche conservatism, and evolution: could exotic species be potential tests? *Species Invasions: Insights into Ecology, Evolution, and Biogeography* (ed. D. Sax, J.J. Stachowicz and S.D. Gaines), pp. 259–290. Sinauer, Sunderland, Massachusetts.

Lavergne, S. & Molofsky, J. (2007) Increased genetic variation and evolutionary potential drive the success of an invasive grass. *Proceedings of the National Academy of Sciences of the USA*, **104**, 3883–3888.

Lever, C. (1985) *Naturalized Mammals of the World*. Longman, London.

Levin, B.R., Allison, A.C., Bremmermann, H.J., et al. (1982) Evolution of parasites and hosts. *Population Biology of Infectious Diseases* (ed. R.M. Anderson and R.M. May), pp. 213–243. Springer-Verlag, Berlin.

Lockwood, J.L., Cassey, P. & Blackburn, T. (2005) The role of propagule pressure in explaining species invasions. *Trends Ecology & Evolution*, **20**, 223–228.

Lockwood, J.L., Cassey, P. & Blackburn, T. (2009) The more you introduce the more you get: the role of colonization pressure and propagule pressure in invasion ecology. *Diversity and Distributions*, **15**, 904– 910.

Long, J.L. (2003) *Introduced Mammals of the World*. CABI Publishing, Collingwood.

Mack, R.N. (2000) Cultivation fosters plant naturalization by reducing environmental stochasticity. *Biological Invasions*, **2**, 111–122.

Marchetti, M.P., Moyle, P.B., & Levine, R. (2004) Alien fishes in California watershed: characteristics of successful and failed invaders. *Ecological Applications*, **14**, 587–596.

Novak, S.J. & Mack, R.N. (2005) Genetic bottlenecks in alien plant species: influences of mating systems and introduction dynamics. *Species Invasions: Insights into Ecology, Evolution, and Biogeography* (ed. D. Sax, J.J. Stachowicz and S.D. Gaines), pp. 201–228. Sinauer, Sunderland, Massachusetts.

Parker, I.M., Simberloff, D., Lonsdale, W.M., et al. (1999) Impact: toward a framework for understanding the ecological effects of invaders. *Biological Invasions*, **1**, 3–19.

Pierpaoli, M., Birò, Z.S., Herrmann, M., Hupe, K., Fernandes, M., Ragni, B., Szemethy, L. & Randi, E. (2003) Genetic distinction of wildcat (*Felis silvestris*) populations in Europe, and hybridization with domestic cats in Hungary. *Molecular Ecology*, **12**, 2585–2598.

Ricciardi, A. (2003) Predicting the impacts of an introduced species from its invasion history: an empirical approach applied to zebra mussel invasions. *Freshwater Biology*, **48**, 972–981.

Ricciardi, A. (2005) Facilitation and synergistic interactions among introduced aquatic species. *Invasive Alien Species: A New Synthesis* (ed. H.A. Mooney, R.N. Mack, J. McNeely, L.E. Neville, P.J. Schei, and J.K. Waage), pp. 162–178. Island Press, Washington, DC.

Ricciardi, A. (2006) Patterns of invasion in the Laurentian Great Lakes in relation to changes in vector activity. *Diversity and Distributions*, **12**, 425–433.

Ricciardi, A. & Atkinson, S.K. (2004) Distinctiveness magnifies the impact of biological invaders in aquatic ecosystems. *Ecology Letters*, **7**, 781–784.

Ricciardi, A. & Kipp, R. (2008) Predicting the number of eco-logically harmful exotic species in an aquatic system. *Diversity and Distributions*, **14**, 374–380.

Ricciardi, A. & MacIsaac, H.J. (2000) Recent mass invasion of the North American Great Lakes by Ponto-Caspian species. *Trends in Ecology & Evolution*, **15**, 62–65.

Ricciardi, A. and Simberloff, D. (2009) Assisted colonization is not a viable conservation strategy. *Trends in Ecology & Evolution*, **24**, 248–253.

Richardson, D.M., Allsopp, N., D'Antonio, C., Milton, S.J. & Rejmanek, M. (2000) Plant invasions – the role of mutualisms. *Biological Reviews*, **75**, 65–93.

Richardson, D.M. & Pyšek, P. (2008) Fifty years of invasion ecology – the legacy of Charles Elton. *Diversity and Distributions*, **14**, 161–168.

Robinson, J.F. & Dickerson, J.E. (1987) Does invasion sequence affect community structure? *Ecology*, **68**, 587–595.

Robinson, J.V. & Wellborn, G.A. (1988) Ecological resistance to the invasion of a freshwater clam, *Corbicula fluminea*: fish predation effects. *Oecologia*, **77**, 445–452.

Roman, J. (2006) Diluting the founder effect: cryptic invasions expand a marine invader's range. *Proceedings of the Royal Society B*, **273**, 2453–2459.

Roman, J. & Darling, J.A. (2007) Paradox lost: genetic diversity and the success of aquatic invasions. *Trends in Ecology & Evolution*, **22**, 454–464.

Sakai, A.K., Allendorf, F.W., Holt, J.S., et al. (2001) The population biology of invasive species. *Annual Reviews in Ecology and Systematics*, **32**, 305–332.

Schlaepfer, M.A., Sherman, P.W., Blossey, B. & Runge, M.C. (2005) Introduced species as evolutionary traps. *Ecology Letters*, **8**, 241–146.

Sexton, J.P., McKay, J.K. & Sala. A. (2002) Plasticity and genetic diversity may allow saltcedar to invade cold climates in North America. *Ecological Applications*, **12**, 1652–1660.

Simberloff, D. (1996) Hybridization between native and introduced wildlife species: importance for conservation. *Wildlife Biology*, **2**, 143–150.

Simberloff, D. (2009) The role of propagule pressure in biological invasions. *Annual Reviews in Ecology, Evolution and Systematics*, **40**, 81–102.

Simberloff, D. & Von Holle, B. (1999) Positive interactions of nonindigenous species: invasional meltdown? *Biological Invasions*, **1**, 21–32.

Strayer, D.L., Eviner, V.T., Jesche, J.M. & Pace, M.L. (2006) Understanding the long-term effects of species invasions. *Trends in Ecology & Evolution*, **21**, 645–651.

Tsutsui, N.D., Suarez, A.V., Holway, D.A. & Case, T.J. (2000) Reduced genetic variation and the success of an invasive species. *Proceedings of the National Academy of Sciences of the USA*, **97**, 5948–5953.

Vance-Chalcraft, H.D. & Soluk, D.A. (2005) Multiple predator effects result in risk reduction for prey across multiple prey densities. *Oecologia*, **144**, 471–480.

Verling, E., Ruiz, G.M., Smith, L.D., Galil, B., Miller, A.W. & Murphy, K.R. (2005) Supply-side invasion ecology: characterizing propagule pressure in coastal ecosystems. *Proceedings of the Royal Society B*, **272**, 1249–1257.

Von Holle, B. & Simberloff, D. (2005) Ecological resistance to biological invasion overwhelmed by propagule pressure. *Ecology*, **86**, 3213–3218.

Whitney, K.D. & Gabler, C.A. (2008) Rapid evolution in introduced species, 'invasive traits' and recipient communities: challenges for predicting invasive potential. *Diversity and Distributions*, **14**, 569–580.

Williamson, M. (1996) *Biological Invasions*. Chapman & Hall, New York.

Wilson, J.R.U., Dermott, E.E., Prentis, P.J., Lowe, A.J. & Richardson, D.M. (2009) Something in the way you move: dispersal pathways affect invasion success. *Trends in Ecology & Evolution*, **24**, 136–144.

Zangerl, A.R. & Berenbaum, M.R. (2005) Increase in toxicity of an invasive weed after reassociation with its coevolved herbivore. *Proceedings of the National Academy of Sciences of the USA*, **102**, 15529–15532.

Plate 1 Invasive trees and shrubs have invaded thousands of hectares of natural vegetation in South Africa's Cape Floristic Region. The image shows *Pinus radiata* invading fynbos shrubland in the Langeberg Mountains near Swellendam. (Photograph: D.M. Richardson.) Delegates at an international conference on Mediterranean-type ecosystems in Stellenbosch, South Africa, in 1980 observed that these invasions, into relatively undisturbed ecosystems, went against the notion put forward by Charles Elton in his 1958 book on *The Ecology of Invasions by Animals and Plants* that invasions occurred only in human-modified systems. This was the stimulus for the initiation of a large international programme on the ecology and management of biological invasions in the mid-1980s under the auspices of the Scientific Committee on Problems of the Environment (see Chapter 2).

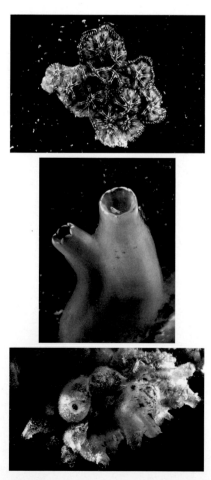

Plate 2 The star-shaped ascidian (sea squirt) *Botryllus schlosseri* (top), the yellow-rimmed ascidian *Ciona intestinalis* (middle) and the ascidian *Molgula manhattensis* (bottom), all thought to be native to the North Atlantic Ocean, have become global invaders, continuing to spread throughout the 20th century, and are representative of the scale of biotic marine globalization that Charles Elton could not have foreseen (see Chapter 3). (Photographs: D. Haydar)

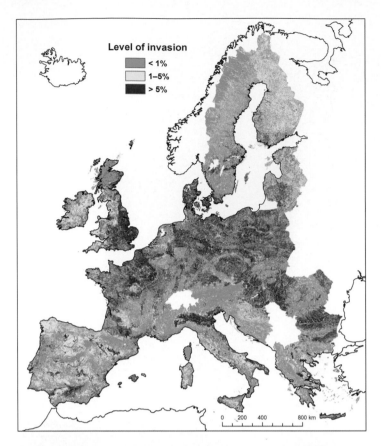

Plate 3 Europe is now probably the most thoroughly and systematically studied region of the world in terms of distribution of alien species and their habitat affiliations (see Chapter 7). This knowledge made it possible to produce this pan-European map of plant invasions based on the occurrence of species in various types of habitat. The estimated level of invasions is based on the mean percentage of neophyte taxa (alien taxa introduced after 1500 AD) in small-scale vegetation plots corresponding to individual land-cover classes as defined in CORINE (Coordination of Information on the Environment, a programme for gathering information on environmental issues for the European Union). Within the mapping limits, areas with non-available land-cover data or insufficient vegetation-plot data are blank. Based on information in Chytrý et al. (2009; *Diversity and Distributions*, **15**, 98–107).

Level of invasion
< 1%
1–5%
> 5%

0   200   400   800 km

Plate 4 Fungal pathogens continue to have a significant impact on both native and non-native forest trees globally (see Chapter 8) (A) Large-scale defoliation of *Pinus radiata* caused by *Phytophthora pinifolia* in Chile. (B) Dothistroma needle blight on *Pinus radiata* in South Africa caused by the introduced *Dothistroma septosporum*. (C) *Leucadendron cordifolium* on Table Mountain, South Africa, dying as a result of infection by the introduced *Phytophthora cinnamomi*. (D) Fusiform rust caused by the native *Cronartium quercuum* on *Pinus elliottii* in the southeastern USA. (E) Quambalaria shoot blight on *Corymbia* sp. caused by the native *Quambalaria pitereka* in northern Queensland. (F) Symptoms associated with infection by *Ceratocytis albifundus*, apparently native to South Africa and that has undergone a host shift to infect non-native plantation-grown *Acacia mearnsii*. (G) Symptoms of *Eucalyptus* rust on *Psidium guajava* and caused by *Puccinia psidii*, native to south and central America on Myrtaceae and that has adapted to infect non-native eucalypts in plantations (Photographs A-E and G: M.J. Wingfield; F: J. Roux).

Plate 5 Many alien plant species become invasive by securing the services of native mutualists, effectively infiltrating naturally occurring networks in the new region. Much work has been done in the past decade on elucidating the roles of seed dispersal agents and pollinators in facilitating invasions (see Chapter 12). The picture shows *Sylvia melanocephala*, a native passerine bird, feeding on the fruits of the alien *Opuntia maxima* in the Canary Islands. Native birds and lizards in these islands also include other alien plants, such as *Lantana camara*, in their diet, playing important roles as dispersal agents. (Photograph: Juan José Hernández.)

Plate 6 Invasive plants of aquatic environments exhibit diverse reproductive systems (see Chapter 15). (a) *Eichhornia paniculata* (Pontederiaceae) infesting a rice field near Camelote, Camagüey, Cuba. The species is an annual relative of water hyacinth native to Brazil and the Caribbean. Most populations from Brazil are outcrossing whereas those in Cuba are predominantly selfing. (b) A population of the perennial wetland invader *Lythrum salicaria* (purple loosestrife; Lythraceae) near Newmarket, Ontario, Canada. The showy floral displays attract diverse insects; populations are self-incompatible and largely outcrossing. (c) Mass flowering of *Eichhornia crassipes* (water hyacinth; Pontederiaceae) in its native range near Salta, Salta province, northwest Argentina. Although predominantly clonal in its introduced range, populations reproduce by both clonal and sexual reproduction in lowland South America, the native range. (Photographs: S.C.H. Barrett).

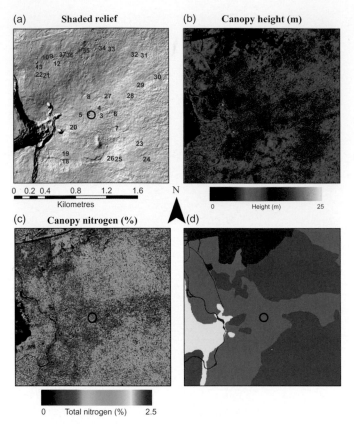

Plate 7 This composite figure illustrates how modern remote sensing technologies can contribute not only to detecting biological invasions, but also to documenting their effects on ecosystems (Chapter 21). It illustrates topographic and canopy properties in a 2 km by 2 km landscape centered on a long-term research site in the Hawaiian archipelago. (a) Topography derived from LiDAR. The long-term research site is indicated by a circle with 50 m radius; points where ground-based information and canopy photographs were obtained are indicated by numbers. Canopy photographs are accessible on-line at http://www.stanford.edu/group/Vitousek/lsaglandscape.htm. (b) Distribution of canopy heights across the Thurston landscape, also obtained via LiDAR. Lower Left. Nitrogen (N) concentration in the forest canopy, measured with High Fidelity Imaging Spectrometer (HiFIS); most of the high nitrogen areas are associated with invasion by *Morella faya*. (c) Cover classes in the Thurston landscape, derived from the remote images and ground observations. Dark green represents native-dominated ecosystems; dark blue represents sites with invasive trees (here predominantly the N-fixer *Morella faya*) at least co-dominant in the canopy; yellow represents volcanic features; and red represents human-disturbed sites. Revised from Vitousek et al. (2010; *Pacific Science*, **64**, 359–366.).

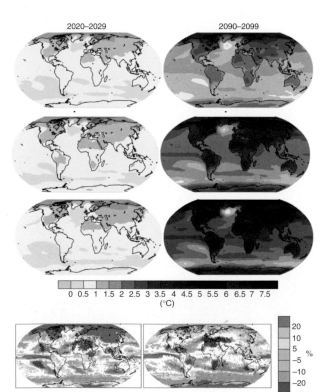

Plate 8 Changes in surface temperatures and precipitation patterns are expected to alter the distributions and impacts of invasive species (see Chapter 26). Climate models suggest that the nature and rate of climate change will vary by region. Top: Projections for surface warming by the periods 2020–2029 and 2090–2099 relative to the period from 1980 to 1999. Projections are derived from averages of several climate models, under three different emissions scenarios with lower (top panels, B1), intermediate (middle panels, A1B) and higher (bottom panels, A2) emission rates. Figure from IPCC (2007). Bottom: Projections for changes in winter (left panel) and summer (right panel) precipitation by the period 2090–2099 relative to 1980–1999 for an intermediate (A1B) emissions scenario. Areas with black stippling indicate greater than 90% agreement among climate models on the direction of change. Figure from IPCC (2007) *Climate Change 2007: The Physical Science Basis. Working Group I Contribution to the Fourth Assessment Report of the Intergovernmental Panel on Climate Change*, figures TS.28 and TS.30. Cambridge University Press.

Part 5

Poster-Child Invaders, Then and Now

ELTON'S INSIGHTS INTO THE ECOLOGY OF ANT INVASIONS: LESSONS LEARNED AND LESSONS STILL TO BE LEARNED

Nathan J. Sanders[1] *and Andrew V. Suarez*[2]

[1]Department of Ecology and Evolutionary Biology, University of Tennessee, Knoxville, TN 37996, USA
[2]Departments of Entomology and Animal Biology, University of Illinois, Urbana, IL 61801, USA

Fifty Years of Invasion Ecology: The Legacy of Charles Elton, 1st edition. Edited by David M. Richardson

18.1   INTRODUCTION

Charles Elton's (1958) classic book, *The Ecology of Invasions by Animals and Plants*, provides numerous case studies of biological invasions. Given the breadth of taxonomic coverage that appears in the book, it is not surprising that Elton would mention several ant species and use them to illuminate several different general points about invasions. With regard to the spread, impact and control (or, more likely, lack thereof) of some invasive ant species, Elton was usually prescient. Much has been learned about the spread and impact of invasive ant species since the publication of *The Ecology of Invasions by Animals and Plants*. In this chapter, we revisit and elaborate on some of the points that Elton raised about ant invasions. Our chapter is loosely organized around the chapter titles and key concepts of Elton's book. In 'The invaders' (section 18.2), we introduce some of the most notorious invasive ant species, with particular focus on the Argentine ant: the ant that Elton most frequently referred to and has since become one of the best studied invasive species (Pysek et al. 2008). In 'New food chains for old' (section 18.3), we discuss some of the diverse ramifications of invasion by the Argentine ant. In 'The balance between populations' (section 18.4), we highlight the effects of the Argentine ant on populations of native species and how those effects can be temporally and spatially dynamic. We also discuss the balance of populations between invasive ants and other ants in the native range of the invasive species. The section 'Ecological resistance and the role of disturbance' (section 18.5) addresses whether there are particular characteristics of communities or ecosystems that make them more or less susceptible to invasion. Finally, in the last section, 'Standing on Elton's shoulders', we suggest some possible avenues of research for the next 50 years of ant ecology.

18.2   THE INVADERS

More than 150 ant species have been introduced outside their native ranges, but not all of these species cause ecological or economic impact (McGlynn 1999). Five ant species – *Anoplolepis gracilipes* (yellow crazy ant), *Linepithema humile* (Argentine ant), *Pheidole megacephala* (big-headed ant), *Solenopsis invicta* (red imported fire ant) and *Wasmannia auropunctata* (little fire ant) – are among the 100 worst invasive species

(Lowe et al. 2000). Additionally, two of these species – *L. humile* at second and *S. invicta* at fourth – are among the four most studied invasive species (Pysek et al. 2008). Of these five notorious invasive ants, Elton mentioned only *L. humile* and *P. megacephala* in his book. However, this probably should not be held against him because at the time much less was known about the other species. E.O. Wilson's first publication on the distribution of *S. invicta* in the southern USA was less than 10 years old (Wilson 1951), and the scope of the ecological problems associated with *S. invicta* was not yet realized (Blu Buhs 2004). Similarly, information on *A. gracilipes* and *W. auropunctata* was largely relegated to the grey literature and regional taxonomic journals until the 1970s (Fabres & Brown 1978; Fluker & Beardsley 1970). *Pheidole megacephala* has continued to receive considerable research attention, especially in Australia (Hoffmann & Parr 2008; Lach & Thomas 2008) and islands in the South Pacific (Savage et al. 2009).

At the time of publication of Elton's book, *L. humile* was relatively well studied in its introduced range. In fact, the first records of *L. humile* are from the Atlantic island of Madeira between 1847 and 1858, pre-dating the type specimen collected in Argentina in 1866 (Wetterer & Wetterer 2006; Wetterer et al. 2009). The native range of the Argentine ant includes the Parana and Uruguay River drainage areas of Argentina, Brazil, Paraguay and Uruguay (Tsutsui et al. 2001; Wild 2004). *Linepithema humile* has become established in mild Mediterranean climates globally (for example California, Chile, the Mediterranean Coast of Europe, coastal Australia, parts of New Zealand, South Africa), and Wetterer et al. (2009) recently documented the occurrence of *L. humile* at over 2100 sites in 95 geographical areas (for example countries, states or islands) throughout the world (Fig. 18.1). Even in the 1950s, however, Elton recognized that the Argentine ant was a global pest:

> 'The Argentine ant has also spread to other countries in an explosive way. In South Africa and Australia there has been the same elimination of native ants. Australia both in the south and west was reached by 1939-1941, and a further bridgehead in New South Wales by 1951.'

Several studies have tracked the rate of invasion or range expansion by Argentine ants in their introduced

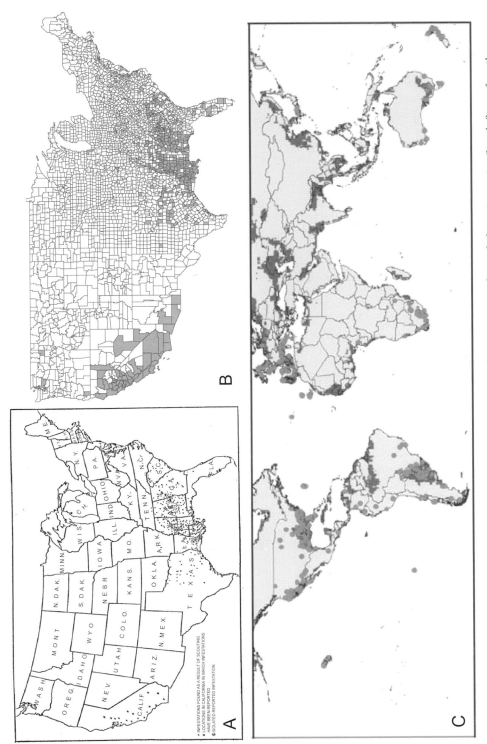

Fig. 18.1 The Argentine ant has become a model organism for spatial analyses of the occurrence and spread of invasive species. Sample figures from the Argentine ant's distribution in the USA (a. Elton 1958; b. Suarez et al. 2001) and globally (c. Roura-Pascual et al. 2004). Whether occurrence data were mapped as points on a map (a), associated with county level data (b) or globally geo-referenced (c), this information has been integral for both predicting regional patterns of spread through multiple dispersal processes (see, for example, Pitt et al. 2009) and for modelling how invasive species will change their distribution under different models of climate change (Roura-Pascual et al. 2004).

range (Suarez et al. 2001). At local spatial scales, the rate of spread of Argentine ants is quite slow (0–250 m yr^{-1}) because colonies disperse by budding rather than by nuptial flights. However, at regional scales, Argentine ants move by human-assisted jump dispersal, resulting in rates of spread of several hundred kilometres per year (Suarez et al. 2001). Elton highlighted this scale-dependent rate of spread in Argentine ants, noting that

> 'The Argentine ant is not, as a matter of fact, a very fast natural invader, for its nuptials take place almost entirely within the nest, and its movements by crawling would not take it more than a few hundred feet a year. It seems that transport in merchandise, especially by railway train, dispersed it so quickly within the United States.'

Long-term monitoring (see, for example, Holway 1998b; Heller et al. 2008b; Sanders et al. 2001) and historical data (Suarez et al. 2001) coupled with elegant experiments (Menke & Holway 2006) suggest that the rate of invasion, and perhaps ultimately the global distribution of Argentine ants, depends strongly on climate.

The impacts of *L. humile* on native biodiversity were also already clear at the time of Elton's book. For example, Elton wrote of the Argentine ant:

> 'Everywhere it multiplied immensely and invaded houses and gardens and orchards, eating food or – out-of-doors – other insects, also farming scale insects and aphids on various trees ... A conspicuous character of this fierce and numerous tropical ant is that it drives out native ants entirely.'

Research over the past 25 years has confirmed that almost everywhere it has become established, the Argentine ant has caused dramatic reductions in the diversity of ants, changed the relative abundance of other arthropods and altered the structure of communities (Ward 1987; Human & Gordon 1997; Holway 1998a; Sanders et al. 2003; Oliveras et al. 2005; Tillberg et al. 2007; Carpintero et al. 2007; Abril & Gomez 2009). For example, at a single site invaded by Argentine ants in southern California, the number of native ant species dropped from 23 to 2 over the course

of several years (Tillberg et al. 2007). Importantly, the impacts of *L. humile* on native diversity can occur almost immediately, and the effects of *L. humile* can cascade across the entire ecosystem by disrupting the co-evolved interactions among species (see Suarez & Case 2002; Lessard et al. 2009; Rodriguez-Cabal et al. 2009), something Elton devoted an entire chapter to called 'New food-chains for old'.

18.3   NEW FOOD CHAINS FOR OLD

> 'On the trees with ants, scale insects were on average five times as common; at the peak of the year 150 times. This was because the Argentine ants killed off many though not all of the natural enemies and parasites of the scales.'

Argentine ants, as well as many other invasive ant species, disrupt food webs and mutualistic interactions with plants and honey excreting hemipterans (see Lach 2003, 2007; O'Dowd et al. 2003; Savage et al. 2009). The ability of Argentine ants to exploit resources more efficiently than many native species may allow them to monopolize an important and relatively stable food resource: honeydew from hemipterans and extrafloral nectar from plants. Argentine ants have long been associated with high densities of honeydew-producing hemipterans, especially in agricultural and riparian systems (Newell & Barber 1913). More generally, there seems to be a growing appreciation that tapping into carbohydrate-rich honeydew fuels ant invasions (O'Dowd et al. 2003; Tillberg et al. 2007; Rowles & Silverman 2009), and Argentine ants are no exception in their ability to exploit hemipterans and extrafloral-nectar-bearing plants. Rowles and Silverman (2009) showed experimentally that access to carbohydrate-rich resources facilitated invasion of natural habitat by Argentine ants in North Carolina. The spread and impacts of other invasive ant species might also be exacerbated by their participation in mutualisms with hemipterans. For example, the yellow crazy ant (*Anoplolepis gracilipes*) has had dramatic effects on the flora and fauna of Christmas Island, in part because of its interactions with scale insects (O'Dowd et al. 2003). Invasion by red imported fire ant (*Solenopsis invicta*) may be facilitated by introduced species of mealybug in some parts of its range (Helms & Vinson 2002).

In addition to providing fuel for a work force, increased access to plant-based carbohydrates may also directly influence the growth and behaviour of invasive ants. In both Argentine ants (Grover et al. 2007) and fire ants (Helms & Vinson 2008) access to carbohydrates can increase colony sizes by up to 50%, primarily through worker longevity. Moreover, access to carbohydrates can increase levels of aggression and forger activity in Argentine ants (Grover et al. 2007). More research needs to be done to understand the role of mutualism in invasion dynamics generally (Traveset & Richardson, this volume). This is especially important for understanding ant invasions, given that ant–hemipteran interactions are ubiquitous in many systems.

Native ants provide numerous functions in ecosystems, and when they are displaced by Argentine ants, those functions, more often than not, are lost. A classic study from the fynbos in South Africa showed the potential consequences of invasion by *L. humile* for the plant species that relied on native ant species to disperse their seeds (Bond & Slingsby 1984). A recent review of the literature by Rodriguez-Cabal et al. (2009) found that, across all studies, sites with *L. humile* contained 92% fewer native seed dispersers than nearby sites without *L. humile*, and that the loss of native ant species had consequences: across all studies, sites with *L. humile* had 47% fewer seeds removed and seedling establishment was 76% lower. The effects of Argentine ants on seed-dispersal mutualisms are not always straightforward, however, and Argentine ants may favour seeds of some species over others (Rowles & O'Dowd 2009). Thus, for at least some plant species, Argentine ants may replace native ants as seed dispersers (Rowles & O'Dowd 2009).

Argentine ants affect more than just ants and the plants whose seeds the native ants disperse. They can also have indirect effects on vertebrate populations. In southern California, the Argentine ant has displaced numerous native ant species (Suarez et al. 1998; Tillberg et al. 2007), and this might have indirectly led to population declines of coastal horned lizards (Suarez et al. 2000; Suarez & Case 2002). Suarez et al. (2000) showed experimentally that when individual horned lizards are offered only *L. humile* as food, their growth rates decline precipitously. One compelling explanation for the decline in horned lizard populations is that *L. humile* displaces the native ants (mostly in the genus *Pogonomyrmex*), and it is these native ants that make up around 90% of the diet of coastal horned lizards. In

addition, continued habitat alteration in southern California has negative impacts on coastal horned lizards but positive effects on *L. humile* by making some sites more susceptible to invasion. Thus it appears that habitat alteration combined with invasion by *L. humile* could be responsible for declines in coastal horned lizard populations. There is also some evidence that Argentine ants affect bird populations as well (Pons et al. 2010).

18.4 THE BALANCE BETWEEN POPULATIONS

Though the impacts of Argentine ants on native species can be immediate, substantial and diverse, they are also temporally dynamic. That is, the impacts might dissipate over time and interact with abiotic conditions such as weather and climate (Heller et al. 2008b). Since 1993, Deborah Gordon's research group has monitored the distribution of the Argentine ant and native species in Jasper Ridge Biological Preserve in northern California. The data illustrate a couple of interesting points. First, and not surprisingly, sites with Argentine ants are species poor, for ants and other invertebrate taxa, relative to sites where Argentine ants have yet to invade (Human & Gordon 1997; Sanders et al. 2001). Second, though the number of ant species in a site can decrease dramatically upon the establishment by *L. humile*, there is a positive correlation between ant species richness and the number of years a site has been invaded (Heller et al. 2008b). That is, in areas where Argentine ants cannot maintain high densities, the negative effects of Argentine ants on diversity might decrease through time. Finally, there seem to be strong seasonal and abiotic effects on the distribution and range expansion of *L. humile* at Jasper Ridge: the distribution of *L. humile* increases in wet years and wet seasons and contracts during hot dry years and seasons (Human et al. 1998; Sanders et al. 2001; Heller et al. 2008b). One of the consequences of this seasonal ebb and flow of Argentine ants is that native species might rapidly colonize, or at least become more active, when Argentine ants retreat from a site (Sanders et al. 2001), something Elton (1958) pointed out:

> 'Just as soon as the Argentine ants begin to disappear, native ants invade the territory, and within a few years are as plentiful as ever.'

Elton suggested that the impact of the Argentine ant on native communities resulted from competitive displacement by interference:

> 'When there is a very clear verdict, as with the Argentine ant, that hard fighting has resulted in regular and catastrophic and general replacement of other species, including other ants, the case is complete: that is replacement through interference.'

Most studies of invasive ants (if not all ants generally) have focused on interference competition among species as the key determinant of community structure (Hölldobler & Wilson 1990; Holway et al. 2002a). However, both interference and exploitative competition occur in ant communities. The outcomes of interference interactions among species are likely determined by worker size (Fellers 1987), which species initiated the interaction (Human & Gordon 1999), or numerical advantages that depend recruitment ability or colony size. Exploitative competition, on the other hand, likely depends on both the ability of individual foragers to quickly and efficiently locate food resources, and the ability of recruits to retrieve the food resources once they are discovered (Fellers 1987; Holway 1999; Human & Gordon 1996).

Argentine ants interfere with native species (Way et al. 1997; Holway 1999; Human & Gordon 1999;). However, detailed behavioural observations on interactions among native ant species and Argentine ants indicate that the case for interference alone as a means of displacement is not as clear as Elton suggested. For example, Human & Gordon (1999) found that, under some conditions, Argentine ants were just as likely to retreat from native ants as native ants were from Argentine ants. Argentine ants were, however, more likely than native ants to behave aggressively toward other species. In a similar experiment, Holway (1999) found that, in head-to-head interactions, single Argentine ant workers experienced mixed success, but Argentine ant colonies were able to displace native ant species more often than not. Holway (1999) noted, 'The discrepancy between worker-level and colony-level interference ability suggests that numerical advantages are key to the Argentine ant's proficiency at interference competition.'

The numerical advantages conferred by a large colony size also increase the Argentine ant's exploita-tive abilities. Several different studies have shown that Argentine ants are more proficient at finding and monopolizing resources than are native species (Human & Gordon 1996; Holway 1999). Taken together, the results of these studies indicate that Argentine ants might not adhere to an exploitation–interference trade-off that many native species are subject to. The basic idea of competitive tradeoffs, in ants at least, was first empirically tested by Fellers (1987). In a classic study, she found that, for a guild of native ant species, there was a negative relationship between interference ability and resource discovery rate. Holway (1999) was the first to show that Argentine ants, at least relative to the suite of native ant species with which they competed most directly, appeared to have broken the exploitation–interference trade-off and are proficient at both interference and exploitation. This competitive double blow, which is partly a product of behaviour, colony size and colony structure, undoubtedly confers a unique advantage to Argentine ants in their introduced range. In the native range of Argentine ants, however, several species are at least as proficient at securing resources as are Argentine ants (LeBrun et al. 2007). This might account for Elton's observation that the Argentine ant 'does not appear to be generally a pest of importance in [Argentina]'. LeBrun et al. (2007) argue that this release from competitive pressure might play an important role in the success of the Argentine ant in its introduced range.

So far, we have documented that the Argentine ant has had substantial impacts on biodiversity and the integrity of ecosystems outside its native range, as do many other invasive ant species (Holway et al. 2002a). But what else have we learned about the success of invasive ants by studying their native ranges? Elton was a strong proponent of studying invasive species in their native range in order to gain insights into what might make them invasive in their introduced ranges. It was perhaps not lost on Elton that most hypotheses suggested for the success of invasive species (for example escape from natural enemies, increased competitive ability, different competitive environment, pre-adapted to disturbance) require some comparison between native and introduced populations.

As pointed out above, Argentine ants experience different competitive environments in their native and introduced populations. This is not the only difference, however; Argentine ants also vary considerably in colony size among populations (Tsutsui et al. 2000;

Pedersen et al. 2006; Heller et al. 2008a; Vogel et al. 2009). Introduced populations of Argentine ants have long been known to have large 'supercolonies' where intraspecific aggression is absent among many spatially separate nests (Newell & Barber 1913). When ants do not exhibit any intraspecific aggression among nests over the scale of an entire population (for example the scale at which the potential for reproduction or competition can occur), that population is considered 'unicolonial' (Bourke & Franks 1995; Hölldobler & Wilson 1977). Nearly all introduced populations of Argentine ants are unicolonial and the only exceptions are a few rare cases where two enormous supercolonies come into contact with one another over large areas (Giraud et al. 2002; Thomas et al. 2006).

However, it was not until relatively recently that the spatial scale of supercolonies was examined in native populations (Suarez et al. 1999). Although Argentine ants also exist in polydomous supercolonies in their native Argentina, they are much smaller and many of them will co-exist in a single population (Pedersen et al. 2006; Tsutsui et al. 2000). Therefore, in contrast to introduced populations where intraspecific aggression almost never occurs, territorial behaviour is relatively common in native populations with dozens of intraspecifically aggressive supercolonies competing for resources.

It has been argued that intraspecific aggression among Argentine ant supercolonies can reduce their ability to monopolize resources, can influence colony growth (Holway et al. 1998) and result in massive mortality of workers at contact zones (Thomas et al. 2006). Unicoloniality may therefore contribute to ecological dominance by enhancing colonization ability, resource exploitation and interference interactions (Helanterä et al. 2009). Indeed, unicolonial colony structures have been documented or inferred for *Anoplolepis gracilipes*, *Linepithema humile*, *Pheidole megacephala*, *Solenopsis invicta* and *Wasmannia auropunctata*. What we do know about unicoloniality is based largely on studies of just a few species, and we still do not know answers to basic questions about the evolution of unicoloniality or how it is maintained in ecological time (Helanterä et al. 2009).

Ultimately, unicoloniality alone cannot account for the success of invasive species in general, or the Argentine ant specifically. As Elton pointed out, the interplay between traits of invaders and characteristics of recipient communities determines invasion success. For example, novel interactions between Argentine ants and resident species in introduced populations may have allowed them to become successful invaders, and characteristics of communities might make them susceptible to invasion. We turn now to the interplay between invader and invaded community.

18.5 ECOLOGICAL RESISTANCE AND THE ROLE OF DISTURBANCE

Although there is little doubt that some characteristics of the Argentine ant make it a successful invader, much less is known about whether there are particular characteristics of ant communities that make them more or less susceptible to invasion. In the literature on non-ant invasive species, two characteristics have received the most research attention: ecological resistance and disturbance.

Elton noted that not all introduced species become highly invasive: 'there are enormously more invasions that never happen, or fail quite soon ... they meet resistance.' Elton also speculated on the nature of this 'resistance':

> '... they will find themselves entering a highly complex community of different populations, they will search for breeding sites and find them occupied, for food that other species are already eating, for cover that other animals are sheltering in ... meeting ecological resistance.'

Ecological resistance is the notion that areas with high species richness should be more difficult to invade than areas with low species richness (Levine et al. 2004). Though this hypothesis (now often called the biotic resistance hypothesis or the diversity–invasibility hypothesis; Fridley, this volume) has been tested both observationally and experimentally, mostly in plant communities (Fridley et al. 2007), it has yet to be tested experimentally in ants. Observational studies of invasion by Argentine ants have not provided direct evidence that native ant diversity prevents or slows the spread of Argentine ants in California (Holway 1998b; Sanders et al. 2003; Carpintero et al. 2007; Roura-Pascual et al. 2009). However, it has been speculated that the species-rich ant communities in Australia, in which competitively dominant native ant species are often abundant, may offer resistance to ant invasions

(Majer 1994). The ideal experimental test of ecological resistance in ant communities would manipulate the number of ant species in a community and then add colonies of invasive species to those communities. To our knowledge, no one has yet tried this experiment, perhaps because of the logistical difficulties of manipulating intact ant communities and because it is ethically questionable to add invasive ant species to areas where they do not yet occur.

Many invasive species, including invasive ant species, often become established in disturbed sites. Most invasive ant species have also spread into relatively undisturbed sites: *Anoplolepis gracilipes* (O'Dowd et al. 2003; Hoffmann & Saul 2010), *Linepithema humile* (Holway 1999; Sanders et al. 2001; Krushelnycky et al. 2005a; Lach 2007), *Pheidole megacephala* (Hoffmann et al. 1999; Vanderwoude et al. 2000), *Solenopsis invicta* (Porter & Savignano 1990; Morris & Steigman 1993; Helms & Vinson 2001; Cook 2003; Stuble et al. 2009) and *Wasmannia auropunctata* (Clark et al. 1982; Walker 2006). Disentangling the combined and relative effects of disturbance and invasive species on native communities has only rarely been tested, even in observational studies Holway (1998b) took advantage of 5 years of monitoring data on the spread of Argentine ants in riparian corridors to assess whether disturbance, native ant species richness (i.e. 'biotic resistance), or stream flow (a measure of moisture availability) best predicted the spread of Argentine ants. Neither biotic resistance nor disturbance accounted for any of the variation in invasion rate. Stream flow alone accounted for 46% of the variation in invasion rate, supporting the notion that physiological tolerance plays an important role in the invasion dynamics of this species. Sanders et al. (2003) also found no relationship between species richness of native ants and probability of invasion by *L. humile* over a study lasting several years. Following up on Holway's observational study, Menke and colleagues (Menke & Holway 2006; Menke et al. 2007) experimentally manipulated moisture availability and the presence of native species. They found that moisture availability was more important than native ant community structure in explaining the invasion dynamics of *L. humile*, again strongly implicating a key role of climate on invasion processes.

Though there is evidence from other non-ant systems in support of the biotic resistance hypothesis (Fridley et al. 2007), there appears to be no evidence that native ant species richness acts as a deterrent to ant invasions. Similarly, Argentine ants appear to invade both disturbed and undisturbed habitats, and no study so far has explicitly tested the hypothesis that disturbed habitats are more susceptible than undisturbed habitats to invasion. This suggests that disturbance alone cannot be used to predict a site's susceptibility to invasion.

Other factors are often correlated with invasion success, namely propagule pressure or propagule size (Simberloff 2010). However, there have been few tests of the role of propagule pressure or size in ant invasions. Recently, Sagata and Lester (2009) showed experimentally that neither propagule size (the size of the 'invading' colony) nor resource availability consistently predicted invasion success by *L. humile*. They went on to suggest that the ability of *L. humile* to modify its behaviour according to local conditions will all but rule out the ability to predict whether a site will be invaded, based simply on resource availability or characteristics of the native community at that site (Sagata & Lester 2009).

To summarize, it seems that three characteristics of Argentine ants, rather than biotic resistance or disturbance, drive invasions by *L. humile*: unicoloniality, their competitive ability and their behavioural flexibility. Those three factors, coupled with appropriate climatic conditions of a site (for example neither too dry nor too cold), seem to account for the success and failure of Argentine ant invasions.

18.6 STANDING ON ELTON'S SHOULDERS

Elton focused on the Argentine ant, because it was (and still is) the best-studied invasive ant species (Pysek et al. 2008). In this chapter, we have also focused on the Argentine ant, but clearly numerous other ant species pose serious threats to biodiversity and the functioning of ecosystems, and much remains to be learned even about those well-studied species like the Argentine ant and the red imported fire ant. Additionally, in the 50 years since the publication of *The Ecology of Invasion by Animals and Plants*, ant ecologists, like all ecologists, have built on what Elton wrote about, in part because of the continued evolution of technological tools that have allowed us to ask questions about the chemical ecology and population genetics of invading populations (Brandt et al. 2009; Vogel et al. 2010), trophic dynamics during invasion

(Tillberg et al. 2007) and the ability to predict global distribution of invasive ants in response to climate change (Roura-Pascual et al. 2004). Here, we highlight what we think are the most substantial advances since the publication of Elton's book and suggest a few areas of future research on the ecology of ant invasions.

Much remains to be learned about the role of positive interactions in invasions (Richardson et al. 2000), and especially for invasive ant species. In some cases, invasive ants may simply replace native ants as mutualists, but in other cases important functions might be lost (Rodriguez-Cabal et al. 2009). And, unfortunately, most studies that have examined whether invasive ant species disrupt mutualisms have focused almost entirely on the Argentine ant (Rodriguez-Cabal et al. 2009). Clearly more work, on more invasive species, needs to be done. Stable isotope analyses (see, for example, Tillberg et al. 2007) will likely play an important role in examining the role of invasive species in new food chains and as mutualists.

Climate clearly influences the distribution of many invasive ant species. This leads to a question that Elton did not raise, but is now at the forefront of considerable research: how will climate change affect the distribution of Argentine ants and other invasive ant species at global scales and perhaps mediate their impact on native biodiversity? In general, environmental niche models suggest that the Argentine ant has the potential for further spread into central Madagascar, Taiwan, southeast Asia, high-elevation Ethiopia and Yemen, and numerous oceanic islands are likely at risk of future establishment (Roura-Pascual et al. 2004; Hartley et al. 2006). Environmental niche modelling can be an important tool for conservation practitioners, but they could be improved by incorporating information on dispersal (see, for example, Roura-Pascual et al. 2009) and ecological interactions with other species, as well as evolutionary responses to changing climates (Fitzpatrick et al. 2007). For example, the outcome of interactions between Argentine ants and native species can depend on environmental temperature (Holway et al. 2002b), but understanding how climatic change will modify interspecific interactions, drive evolutionary responses of both native and invasive species, and ultimately mediate the impact of invasive species is a pressing challenge (Dukes & Mooney 1999; Araújo & Rahbek 2006).

Linepithema humile and *Solenopsis invicta* are the best studied invasive ant species (and perhaps the best studied ant species), and the ever-evolving population genetics toolkit has uncovered many surprising and important characteristics of the population structure of both species in their native and invasive ranges (see, for example, Tsutsui et al. 2000). The other major invasive ant species – *Anoplolepis longicornis*, *Pheidole megacephala* and *Wasmannia auropunctata* – have received some research attention, but not nearly as much as *L. humile* or *S. invicta*. The paucity of information on many other exotic and invasive species is troubling (for example *Doleromyrma darwiniana*, *Lasius neglectus*, *Monomorium pharoensis*, *Monomorium sydneyense*, *Myrmica rubra*, *Pachycondyla chinensis*, *Paratrechina longicornis*, *Paratrechina fulva*, *Pheidole obscurithorax*, *Technomyrmex albipes*, *Tetramorium tsushimae*, *Ochetellus glaber* and several species in the genus *Cardiocondyla*). Unfortunately, what we have learned about *S. invicta* and *L. humile* has been too little too late: they are both responsible for dramatic alterations in native communities. Perhaps by devoting some research effort towards understanding the behaviour and ecology of these other species *before* they become invasive will help minimize or eliminate their potential impacts on biodiversity.

Controlling, or even eradicating, populations of exotic species is a lucrative business and a focus of continued research (Myers et al. 2000). This is especially true for invasive ant species (see Silverman & Brightwell 2008). Though there have been some successful eradications and efforts at controlling spread and impact – *A. gracilipes* on Christmas Island (Green & O'Dowd 2009) and *W. auropunctata* on smaller islands in the Galapagos archipelago (Causton et al. 2005) – much remains to be learned, from both the successes and the colossal failures (for example *S. invicta* in the southeastern USA). For example, what are the traits of species that make them more or less susceptible to eradication and control? Are there particular environmental or ecological conditions that mitigate control efforts? Our view is that ant ecologists studying invasive species should also contribute at least some research effort to improving the ability to control the spread and impact of invasive ant species (see, for example, Krushelnycky et al. 2005b).

Elton was prescient in realizing the importance of studying invaders within native populations to understand what makes them successful. Since Elton's publication, the Argentine ant has received more research attention in its native range than any other invasive ant species, and perhaps more than many other invasive species. The insights gained from these studies are numerous and have highlighted some of the mechanisms that might contribute to its success as an invasive species. However, by comparison, we know very little about the biology of other invasive ants in their native ranges (except perhaps *S. invicta* and more recently *W. auropunctata*). And for most introduced ant species, we still have not even identified from where introduced populations originate. However, for the well-studied species like *L. humile*, recent and ongoing research in nearly all parts of its introduced range offer exciting, but largely untapped, opportunities for comparative studies of the impact of this globally important species.

It is abundantly clear that invasive species are a leading cause of population- and species-level extinctions (Clavero & Garcia-Berthou 2005), and that invasive species can dramatically alter the structure and function of ecosystems (Kurle et al. 2008). In the absence of experimental introductions or removal of invasive species over extensive regions, longitudinal studies of ant communities pre- and post-invasion will help elucidate the consequences of ant invasions on native biodiversity and the functioning of ecosystems (see, for example, Tillberg et al. 2007; Heller et al. 2008b; Hoffmann & Parr 2008). Nevertheless, the consequences of ant invasions, across trophic levels, taxa and for a variety of ecosystem processes are often severe (Holway et al. 2002a). Though the general public's awareness of the impacts of invasive ant species on human health and agriculture has grown substantially in the 50 years since the publication of Elton's book, those who care about the threats that invasive species pose must be vigilant and ensure that, from time to time, we get out of the ivory tower to inform the public and policy makers about the threats that invasive ant species pose, much like Charles Elton did more than 50 years ago.

ACKNOWLEDGEMENTS

We thank Núria Roura-Pascual, Ben Hoffmann and an anonymous reviewer for comments on an earlier draft of this chapter. N.J. Sanders was supported by DOE-PER DE-FG02-08ER64510 and the Center for Macroecology, Evolution and Climate at the University of Copenhagen. A.V. Suarez was supported by DEB 07-16966.

REFERENCES

Abril, S. & Gomez, C. (2009) Ascertaining key factors behind the coexistence of the native ant species *Plagiolepis pygmaea* with the invasive Argentine ant *Linepithema humile* (Hymenoptera: Formicidae). *Sociobiology*, **53**, 559–568.

Araujó, M.B. & Rahbek, C. (2006) How does climate change affect biodiversity? *Science*, **313**, 1396–1397.

Blu Buhs, J. (2004) *The Fire Ant Wars*. University of Chicago Press, Chicago.

Bond, W. & Slingsby, P. (1984) Collapse of an ant–plant mutualism: The Argentine ant (*Iridomyrmex humilis*) and myrmecochorous Proteaceae. *Ecology*, **65**, 1031–1037.

Bourke, A.F.G. & Franks, N.R. (1995) *Social Evolution in Ants*. Princeton University Press, Princeton, New Jersey.

Brandt, M., Van Wilgenburg, E. & Tsutsui, N.D. (2009) Global-scale analsyes of chemical ecology and population genetics in the invasive Argentine ant. *Molecular Ecology*, **18**, 997–1005.

Carpintero, S., Retana, J., Cerda, X., Reyes-Lopez, J. & De Reyna. L.A. (2007) Exploitative strategies of the invasive Argentine ant (*Linepithema humile*) and native ant species in a southern Spanish pine forest. *Environmental Entomology*, **36**, 1100–1111.

Causton, C.E., Sevilla, C.R. & Porter, S.D. (2005) Eradication of the little fire ant, *Wasmannia auropunctata* (Hymenoptera: Formicidae), from Marchena Island, Galapagos: on the edge of success? *Florida Entomologist*, **88**, 159–168.

Clark, D.B., Guayasamin, C. & Pazmino, O. (1982) The tramp ant *Wasmannia auropunctata*: autecology and effects on ant diversity and distribution on Santa Cruz Island, Galapagos. *Biotropica*, **14**, 196–207.

Clavero, M. & Garcia-Berthou, E. (2005) Invasive species are a leading cause of animal extinctions. *Trends in Ecology & Evolution*, **20**, 110–110.

Cook, J.L. (2003) Conservation of biodiversity in an area impacted by the red imported fire ant, *Solenopsis invicta* (Hymenoptera: Formicidae) *Biodiversity and Conservation*, **12**, 187–195.

Dukes, J.S. & Mooney, H.A. (1999) Does global change increase the success of biological invaders? *Trends in Ecology & Evolution*, **14**, 135–139.

Elton, C.S. (1958) *The Ecology of Invasions by Animals and Plants*. Methuen, London.

Fabres, G. & Brown Jr, W.L. (1978) The recent introduction of the pest ant *Wasmannia auropunctata* into New Caledonia. *Journal of the Australian Entomological Society*, **7**, 139–142.

Fellers, J.H. (1987) Interference and exploitation in a guild of woodland ants. *Ecology*, **68**, 1466–1478.

Fitzpatrick, M.C., Weltzin, J.F., Sanders, N.J. & Dunn, R.R. (2007) The biogeography of prediction error: Why doesn't the introduced range of the fire ant predict its native range or vice versa? *Global Ecology and Biogeography*, **15**, 24–33.

Fluker, S.F. & Beardsley, J.W. (1970) Sympatric associations of three ants: *Iridomyrmex humilis, Pheidole megacephala,* and *Anoplolepis longipes* in Hawaii. *Annals of the Entomological Society of America,* **63**, 1291–1296.

Fridley, J.D., Stachowicz, J.J., Naeem, S. et al. (2007) The invasion paradox: reconciling pattern and process in species invasions. *Ecology,* **88**, 3–17.

Giraud, T., Pedersen, J.S. & Keller, L. (2002) Evolution of supercolonies: The Argentine ants of southern Europe. *Proceedings of the National Academy of Sciences of the USA,* **99**, 6075–6079.

Green, P.T. & O'Dowd, D.J. (2009) Management of Invasive Invertebrates: Lessons from the Management of an Invasive Alien Ant. Pages 153–172 in *Invasive Species Management.* M.N. Clout & P.A. Williams editors. Oxford University Press.

Grover, C.D., Kay, A.D., Monson, J.A., Marsh, T.C. & Holway, D.A. (2007) Linking nutrition and behavioural dominance: carbohydrate scarcity limits aggression and activity in Argentine ants. *Proceedings of the Royal Society B,* **274**, 2951–2957.

Hartley, S., Harris, R. & Lester, P.J. (2006) Quantifying uncertainty in the potential distribution of an invasive species: climate and the Argentine ant. *Ecology Letters,* **9**, 1068–1079.

Helanterä, H., Strassmann, J.E., Carrillo, J. & Queller, D.C. (2009) Unicolonial ants: where do they come from, what are they and where are they going? *Trends in Ecology & Evolution,* **24**, 341–349.

Heller, N.E., Ingram, K.K. & Gordon, D.M. (2008a) Nest connectivity and colony structure in unicolonial Argentine ants. *Insectes Sociaux,* **55**, 397–403.

Heller, N.E., Sanders, N.J., Shors, J.W. & Gordon, D.M. (2008b) Rainfall facilitates the spread, and time alters the impact, of the invasive Argentine ant. *Oecologia,* **155**, 385–395.

Helms, K.R. & Vinson, S.B. (2001) Coexistence of native ants with the red imported fire ant, *Solenopsis invicta. Southwestern Naturalist,* **46**, 396–400.

Helms, K.R. & Vinson, S.B. (2002) Widespread association of the invasive ant *Solenopsis invicta* with an invasive mealybug. *Ecology,* **83**, 2425–2438.

Helms, K.R. & Vinson, S.B. (2008) Plant resources and colony growth in an invasive ant: the importance of honeydew-producing Hemiptera in carbohydrate transfer across trophic levels. *Environmental Entomology,* **37**, 487–493.

Hoffmann, B.D., Andersen, A.N. & Hill, G.J.E. (1999) Impact of an introduced ant on native rain forest invertebrates: *Pheidole megacephala* in monsoonal Australia. *Oecologia* **10**, 595–604.

Hoffmann, B.D. & Parr, C.L. (2008) An invasion revisited: the African big-headed ant (*Pheidole megacephala*) in northern Australia. *Biological Invasions,* **10**, 1171–1191.

Hoffmann, B.D. & Saul, W.-C. (2010) Yellow crazy ant (Anoplolepis gracilipes) invasions within undisturbed

mainland Australian habitats: no support for biotic resistance hypothesis. *Biological Invasions,* **9**, 3093–3108.

Hölldobler, B. & Wilson, E.O. (1977) Number of queens – important trait in ant evolution. *Naturwissenschaften,* **64**, 8–15.

Hölldobler, B. & Wilson, E.O. (1990) *The Ants.* The Belknap Press of Harvard University Press, Cambridge, Massachusetts.

Holway, D.A. (1998a) Effect of Argentine ant invasions on ground-dwelling arthropods in northern California riparian woodlands. *Oecologia,* **116**, 252–258.

Holway, D.A. (1998b) Factors governing rate of invasion: a natural experiment using Argentine ants. *Oecologia,* **115**, 206–212.

Holway, D.A. (1999) Competitive mechanisms underlying the displacement of native ants by the invasive Argentine ant. *Ecology,* **80**, 238–251.

Holway, D.A., Lach, L., Suarez, A.V., Tsutsui, N.D. & Case, T.J. (2002a) The causes and consequences of ant invasions. *Annual Review of Ecology and Systematics,* **33**, 181–233.

Holway, D.A., Suarez, A.V. & Case, T.J. (1998) Loss of intraspecific aggression in the success of a widespread invasive social insect. *Science,* **282**, 949–952.

Holway, D.A., Suarez, A.V. & Case, T.J. (2002b) Role of abiotic factors in governing susceptibility to invasion: a test with Argentine ants. *Ecology,* **83**, 1610–1619.

Human, K.G. & Gordon, D.M. (1996) Exploitation and interference competition between the invasive Argentine ant, *Linepithema humile,* and native ant species. *Oecologia,* **105**, 405–412.

Human, K.G. & Gordon, D.M. (1997) Effects of Argentine ants on invertebrate diversity in northern California. *Conservation Biology,* **11**, 1242–1248.

Human, K.G. & Gordon, D.M. (1999) Behavioral interactions of the invasive Argentine ant with native ant species. *Insectes Sociaux,* **46**, 159–163.

Human, K.G., Weiss, S., Weiss, A., Sandler, B. & Gordon, D.M. (1998) The effect of abiotic factors on the local distribution of the invasive Argentine ant (*Linepithema humile*) and native ant species. *Environmental Entomology,* **27**, 822–833.

Krushelnycky, P.D., Joe, S.M., Medeiros, A.C., Daehler, C.C. & Loope, L.L. (2005a) The role of abiotic conditions in shaping the long-term patterns of a high-elevation Argentine ant invasion. *Diversity and Distributions,* **11**, 319–331.

Krushelnycky, P.D., Loope, L.L. & Joe, S.M. (2005b) Limiting spread of a unicolonial invasive insect and characterization of seasonal patterns of range expansion. *Biological Invasions,* **6**, 47–57.

Kurle, C.M., Croll, D.A. & Tershy, B.R. (2008) Introduced rats indirectly change marine rocky intertidal communities from algae- to invertebrate-dominated. *Proceedings of the National Academy of Sciences of the USA,* **105**, 3800–3804.

Lach, L. (2003) Invasive ants: Unwanted partners in ant–plant interactions? *Annals of the Missouri Botanical Garden*, **90**, 91–108.

Lach, L. (2007) A mutualism with a native membracid facilitates pollinator displacement by Argentine ants. *Ecology*, **88**, 1994–2004.

Lach, L. & Thomas, M.L. (2008) Invasive ants in Australia: documented and potential ecological consequences. *Australian Journal of Entomology*, **47**, 275–288.

LeBrun, E.G., Tillberg, C.V., Suarez, A.V., Folgarait, P.J., Smith & Holway, D.A. (2007) An experimental study of competition between fire ants and Argentine ants in their native range. *Ecology*, **88**, 63–75.

Lessard, J.P., Fordyce, J.A., Gotelli, N.J. & Sanders, N.J. (2009) Invasive ants alter the phylogenetic structure of ant communities. *Ecology*, **90**, 2664–2669.

Levine, J.M., Adler, P.B. & Yelenik, S.G. (2004) A meta-analysis of biotic resistance to exotic plant invasions. *Ecology Letters*, **7**, 975–989.

Lowe, S., Browne, M. & Boudjelas, S. (2000) 100 of the world's worst invasive alien species. *Aliens*, **12S**, 1–12.

Majer, J.D. (1994) Spread of Argentine ants (*Linepithema humile*), with special reference to Western Australia. In *Exotic Ants: Biology, Impact, and Control of Introduced Species* (ed. D.F. Williams), pp. 163–173. Westview Press, Boulder, Colorado.

McGlynn, T.P. (1999) The worldwide transport of ants: geographic distribution and ecological invasions. *Journal of Biogeography*, **26**, 535–548.

Menke, S.B., Fisher, R.N., Jetz, W. & Holway, D.A. (2007) Biotic and abiotic controls of Argentine ant invasion success at local and landscape scales. *Ecology*, **88**, 3164–3173.

Menke, S.B. & Holway, D.A. (2006) Abiotic factors control invasion by ants at the community scale. *Journal of Animal Ecology*, **75**, 368–376.

Morris, J.R. & Steigman, K.L. (1993) Effects of polygyne fire ant invasion on native ants of a blackland prairie in Texas. *Southwestern Naturalist*, **38**, 136–140.

Myers, J.H., Simberloff, D., Kuris, A.M. & Carey, J.R. (2000) Eradication revisited: dealing with exotic species. *Trends in Ecology & Evolution*, **15**, 316–320.

Newell, W. & Barber, T.C. (1913) The Argentine ant. *Bulletin of the United States Bureau of Entomology*, **122**, 1–98.

O'Dowd, D.J., Green, P.T. & Lake, P.S. (2003) Invasional 'meltdown' on an oceanic island. *Ecology Letters*, **6**, 812–817.

Oliveras, J., Bas, J.M., Casellas, D. & Gómez, C. (2005) Numerical dominance of the Argentine ant vs native ants and consequences on soil resource searching in Mediterranean cork-oak forests (Hymenoptera: Formicidae) *Sociobiology*, **45**, 643–658.

Pedersen, J.S., Krieger, M.J.B., Vogel, V., Giraud, T. & Keller, L. (2006) Native supercolonies of unrelated individuals in the invasive Argentine ant. *Evolution*, **60**, 782–791.

Pitt, J.P.W., Worner, S.P. & Suarez, A.V. (2009) Predicting Argentine ant spread over the heterogeneous landscape using a spatially-explicit stochastic model. *Ecological Applications*, **19**, 1176–1186.

Pons, P., Bas, J.M. & Estany-Tigerström, D. (2010) Coping with invasive alien species: the Argentine ant and the insectivorous bird assemblage of Mediterranean oak forests. *Biodiversity and Conservation*, **19**, 1711–1723.

Porter, S.D. & Savignano, D.A. (1990) Invasion of polygyne fire ants decimates native ants and disrupts arthropod community. *Ecology*, **71**, 2095–2106.

Pysek, P., Richardson, D.M., Pergl, J., Jarosik, V., Sixtova, Z. & Weber, E. (2008) Geographical and taxonomic biases in invasion ecology. *Trends in Ecology & Evolution*, **23**, 237–244.

Richardson, D.M., Allsopp, N., D'Antonio, C.M., Milton, S.J. & Rejmánek, M. (2000) Plant invasions – the role of mutualisms. *Biological Review of the Cambridge Philosophical Society*, **75**, 65–93.

Rodriguez-Cabal, M., Stuble, K.L., Nuñez, M.A. & Sanders, N.J. (2009) Quantitative analysis of the effects of the exotic Argentine ant on seed dispersal mutualisms. *Biology Letters*, **5**, 499–502.

Roura-Pascual, N., Bas, J.M., Thuiller, W., Hui, C., Krug, R. & Brotons, L. (2009) From introduction to equilibrium: reconstructing the invasive pathways of the Argentine ant in a Mediterranean region. *Global Change Biology*, **15**, 2101–2115.

Roura-Pascual, N., Suarez, A.V., Gomez, C. et al. (2004) Geographical potential of Argentine ants (*Linepithema humile* Mayr) in the face of global climate change. *Proceedings of the Royal Society of London B*, **271**, 2527–2535.

Rowles, A.D. & O'Dowd, D.J. (2009) New mutualism for old: indirect disruption and direct facilitation of seed dispersal following Argentine ant invasion. *Oecologia*, **158**, 709–716.

Rowles, A.D. & Silverman, J. (2009) Carbohydrate supply limits invasion of natural communities by Argentine ants. *Oecologia*, **161**, 161–171.

Sagata, K. & Lester, P.J. (2009) Behavioural plasticity associated with propagule size, resources, and the invasion success of the Argentine ant *Linepithema humile*. *Journal of Applied Ecology*, **46**, 19–27.

Sanders, N.J., Barton, K.E. & Gordon, D.M. (2001) Long-term dynamics of the invasive Argentine ant, *Linepithema humile*, and native ant taxa in northern California. *Oecologia*, **127**, 123–130.

Sanders, N.J., Gotelli, N.J., Heller, N. & Gordon, D.M. (2003) Community disassembly by an invasive species. *Proceedings of the National Academy of Sciences of the USA*, **100**, 2474–2477.

Savage, A.M., Rudgers, J.A. & Whitney, K.D. (2009) Elevated dominance of extrafloral nectary-bearing plants is associated with increased abundances of an invasive ant and reduced native ant richness. *Diversity and Distributions*, **15**, 751–761.

Silverman, J. & Brightwell, R.J. (2008) The Argentine ant: challenges in managing an invasive unicolonial pest. *Annual Review of Entomology*, **53**, 231–252.

Simberloff, D. (2010) The role of propagule pressure in biological invasions *Annual Review of Ecology and Systematics*, **40**, 81–102.

Stuble, K.L., Kirkman, L.K. & Carroll, C.R. (2009) Patterns of abundance of fire ants and native ants in a native ecosystem. *Ecological Entomology*, **34**, 520–526.

Suarez, A.V., Bolger, D.T. & Case, T.J. (1998) Effects of fragmentation and invasion on native ant communities in coastal southern California. *Ecology*, **79**, 2041–2056.

Suarez, A.V. & Case, T.J. (2002) Bottom-up effects on persistence of a specialist predator: ant invasions and horned lizards. *Ecological Applications*, **12**, 291–298.

Suarez, A.V., Holway, D.A. & Case, T.J. (2001) Patterns of spread in biological invasions dominated by long-distance jump dispersal: insights from Argentine ants. *Proceedings of the National Academy of Sciences of the USA*, **98**, 1095–1100.

Suarez, A.V., Richmond, J.Q. & Case, T.J. (2000) Prey selection in horned lizards following the invasion of Argentine ants in southern California. *Ecological Applications*, **10**, 711–725.

Suarez, A.V., Tsutsui, N.D., Holway, D.A. & Case, T.J. (1999) Behavioral and genetic differentiation between native and introduced populations of the Argentine ant. *Biological Invasions*, **1**, 43–53.

Thomas, M.L., Payne-Makrisa, C.M., Suarez, A.V., Tsutsui, N.D. & Holway, D.A. (2006) When supercolonies collide: territorial aggression in an invasive and unicolonial social insect. *Molecular Ecology*, **15**, 4303–4315.

Tillberg, C.V., Holway, D.A., LeBrun, E.G. & Suarez, A.V. (2007) Trophic ecology of invasive Argentine ants in their native and introduced ranges. *Proceedings of the National Academy of Sciences of the USA*, **104**, 20856–20861.

Tsutsui, N.D., Suarez, A.V., Holway, D.A. & Case, T.J. (2000) Reduced genetic variation and the success of an invasive species. *Proceedings of the National Academy of Sciences of the USA*, **97**, 5948–5953.

Tsutsui, N.D., Suarez, A.V., Holway, D.A. & Case, T.J. (2001) Relationships among native and introduced populations of the Argentine ant (*Linepithema humile*) and the source of introduced populations. *Molecular Ecology*, **10**, 2151–2161.

Vanderwoude, C., Lobry de Bruyn, L.A. & House, A.P.N. (2000) Response of an open-forest ant community to invasion by the introduced ant, *Pheidole megacephala*. *Austral Ecology*, **25**, 253–259.

Vogel, V., Pedersen, J.S., d'Ettorre, P., Lehmann, L. & Keller, L. (2009) Dynamics and genetic structure of Argentine ant supercolonies in their native range. *Evolution*, **63**, 1627–1639.

Vogel, V., Pedersen, J.S., Giraud, T., Krieger, M.J.B. & Keller, L. (2010) The worldwide expansion of the Argentine ant. *Diversity and Distributions*, **16**, 170–186.

Walker, K.L. (2006) Impact of little fire ant, *Wasmannia auropunctata*, on native forest ants in Gabon. *Biotropica*, **38**, 666–673.

Ward, P.S. (1987) Distribution of the introduced Argentine ant (*Iridomyrmex humilis*) in natural habitats of the lower Sacramento Valley and its effects on the indigenous ant fauna. *Hilgardia*, **55**, 1–16.

Way, M.J., Cammell, M.E., Paiva, M.R. & Collingwood, C.A. (1997) Distribution and dynamics of the Argentine ant *Linepithema humile* in relation to vegetation, soil conditions, topography and native competitor ants in Portugal. *Insectes Sociaux*, **44**, 415–433.

Wetterer, J.K. & Wetterer, A.L. (2006) A disjunct Argentine ant metacolony in Macaronesia and southwestern Europe. *Biological Invasions*, **8**, 1123–1129.

Wetterer, J.K., Wild, A.L., Suarez, A.V., Roura-Pascual, N. & Espadaler, X. (2009) Worldwide spread of the Argentine ant, *Linepithema humile* (Hymenoptera: Formicidae) *Myrmecological News*, **12**, 187–194.

Wild, A.L. (2004) Taxonomy and distribution of the Argentine ant, *Linepithema humile* (Hymenoptera: Formicidae) *Annals of the Entomological Society of America*, **97**, 1204–1215.

Wilson, E.O. (1951) Variation and adaptation in the imported fire ant. *Evolution*, **5**, 68–79.

Chapter 19

FIFTY YEARS OF 'WAGING WAR ON CHEATGRASS': RESEARCH ADVANCES, WHILE MEANINGFUL CONTROL LANGUISHES

Richard N. Mack

School of Biological Sciences, Washington State University, Pullman, WA 99164, USA

Fifty Years of Invasion Ecology: The Legacy of Charles Elton, 1st edition. Edited by David M. Richardson
© 2011 by Blackwell Publishing Ltd

19.1 INTRODUCTION

Elton's (1958) benchmark book on biological invasions arrived at the mid-point in the history of public
awareness of a cataclysmic biological invasion in
North America: the spread and proliferation of the
Eurasian annual grass *Bromus tectorum* (cheatgrass).
It provides a convenient dividing line by which to
gauge the growth of our knowledge of this invader.
Arguably more important, it also provides the opportunity to evaluate how (or even whether) knowledge

gained in the past 50 years has been translated into
effective control.

Bromus tectorum initially attracted attention early in
the 20th century as a persistent weed in croplands
(Mack 1981). This role persists, but the grass has also
become prolific in the vast arid grasslands in the
Intermountain West of the USA and western Canada.
On these sites, the initial damage stems from its ability
to compete with native species, including the once
dominant perennial grasses (Harris 1967). The
damage cheatgrass causes as a competitor is dwarfed,

Fig. 19.1 Transformation of the arid steppe by *Bromus tectorum* (clockwise from upper right): (a) pristine steppe community
of *Artemisia tridentata* and caespitose grasses; (b) steppe invaded by *B. tectorum*, which is prominent in the understorey;
(c) invaded community is eventually destroyed by cheatgrass-fuelled fires; (d) recurring fires produce a cheatgrass-dominated
landscape; (e) the habit of mature cheatgrass.

however, by the role the vegetative plant plays after death in forming highly combustible fuel in summer. Fire has always been a component of steppe environments, but fires before the dominance of cheatgrass burned slowly and less frequently through the patchily distributed straw of bunchgrasses and perennial dicots (Daubenmire 1970). *Bromus tectorum* now contributes almost all the fuel for recurring catastrophic fires that each can envelop more than 500,000 hectares (O'Driscoll 2007).

Cheatgrass delivers an environmental *coup de grace* the following autumn, when precipitation resumes. The cheatgrass-dominated landscape, having been denuded by fire, becomes dissected by erosion; the resulting sediment soon finds its way into regional waterways. Thus, in less than 120 years, *B. tectorum* has transformed much of the grasslands from central Nevada to southern British Columbia; native regional communities persist only on tiny protected land parcels (Daubenmire 1970) and even these sites are under constant threat from the cheatgrass juggernaut. Although the total cost of this destruction has yet to be tallied, it must nevertheless be horrendous in property damage, human fatalities and the apparently permanent loss in biodiversity.

This chapter has two goals. First, I selectively summarize the voluminous research of the past 50 years that has revealed so much about how and why this grass has become invasive. It is fascinating to assemble the epidemiology of what Elton termed an 'ecological explosion'. However, this research should have another goal: aiding the development of 'best practices' that could curb the invasion, even cause its collapse. Consequently, I also evaluate the success in transferring research results to development of control stratagems and their implementation. Science makes perhaps its greatest contributions when it resolves great issues of societal concern. How well have research results been applied in combating *B. tectorum*?

19.2 HISTORY OF INTRODUCTION AND GENETICS: CHEATGRASS IS NOT A GENETICALLY MONOLITHIC INVADER

Pre-1958 knowledge of the history of *B. tectorum* arrival and spread in North America was based, as it is now, on annotated herbarium records and the somewhat less reliable contemporary accounts of its first appearance in published floras, government botanical surveys and journal articles. Building retrospectively a chronology of any species' range expansion necessitates caveats. Accuracy of the chronology depends totally on accurate species' identifications, a concern that can be alleviated with extant specimens. Furthermore, first detection does not translate directly into first arrival in a new locale. Although knowledge that the grass arrived first in the northeastern USA was widespread by 1958 (Klemmedson & Smith 1964), almost nothing was known about the chronology and likely pattern of its continental spread. This record, bolstered by molecular evidence from extant populations, has been substantially expanded in the past 20 years (Novak & Mack 2001; Valliant et al. 2007; Schachner et al. 2008).

The first report of *B. tectorum* in North America was by Muhlenberg in 1793 at Lancaster, Pennsylvania (Bartlett et al. 2002). I am unaware of any reliable *B. tectorum* collection records for the ensuing 70 years. This long gap in detection has several explanations, including the possibility that the small grass was simply overlooked by the many knowledgeable plant collectors along the East Coast in the early 19th century (Mack 2003, and references therein). Alternatively, Muhlenberg's population may have gone extinct, and later cheatgrass immigrants reestablished the grass in North America. Whatever the explanation, Jackson provided the next oldest known specimen (1859) from West Chester, about 50 km east of Lancaster. In a letter attached to this surviving specimen in the herbarium of the New York Botanical Garden (NY), Jackson reports that the grass, which he had not collected previously, was restricted to a single population (Bartlett et al. 2002).

Whether cheatgrass had been slowly expanding its new range from the late 18th century or had been reintroduced much later, or both, it was collected from 1861 to1880 with increasing frequency along the US Eastern Seaboard from southeastern Pennsylvania to Maine (Bartlett et al. 2002). In the following 20 years (1880–1900), the grass was first detected across much of North America, suggesting rapid spread, coincident with the growth of commerce and transportation. It was collected repeatedly not only west of the Appalachian Range (Ohio, Indiana, Michigan) but also at isolated locations in the mid-continent (Fort Collins, Colorado; Ames, Iowa) (R.N. Mack, unpublished data). Furthermore, the grass was also detected in 1886–1900 at or near major ports in southern Ontario

(Valliant et al. 2007). Spread after 1890 appears to have been along two broad tongues from a central East–West corridor. One tongue spread into the northern Great Plains around 1890; a second avenue developed much later (about 1930) in the southern Great Plains (Oklahoma, New Mexico and the Texas Panhandle) (R.N. Mack, unpublished data).

The late 19th century detection of *B. tectorum* in the Pacific Northwest of the USA and adjacent Canada suggests the grass was also arriving on the West Coast by ship before as well as during its spread from the East (Mack 1981). The grass was detected repeatedly before 1900 in the interior of the Pacific Northwest. In contrast, the only 19th century record of *B. tectorum* in California is inland along its border with Oregon (Mack 1981), despite zealous collecting around San Francisco Bay (see, for example, Brandegee, 1892). However, *Bromus tectorum* was collected with increasing frequency after 1900 near California seaports (e.g. Santa Barbara) (Smith 1952), suggesting direct arrival by sea as well as overland. Today the grass occurs in all US states except Florida, where it has occurred periodically (R.N. Mack, unpublished data), and southern Canadian provinces west of Quebec. Its current status in the Maritime Provinces is problematic (cf. Upadhyaya et al. 1986; Valliant et al. 2007). Ominously, its range has increased in the northern Great Plains (see, for example, Douglas et al. 1990), and it is now found at higher elevations in the Rocky Mountains (greater than 2700 m) (Kao et al. 2008) than previously reported.

Little was known in 1958 about genetic variation in *B. tectorum*. Two taxa had been long recognized (*B. tectorum* and *B. tectorum* var. *glabràtus*, which is denoted by pubescent spikelets), although not universally accepted (cf. Fernald 1950; Hitchcock 1951). Most knowledge before 1958 of its genetic variation stemmed from common gardens trials with populations from the western USA, Connecticut, Kansas, Michigan, North Dakota, plus single populations from Canada and Israel. Results from these trials revealed heritable inter-populational differences in winter hardiness, flowering phenology and senescence (Hulbert 1955). Genetic variation had been detected, but the scope of this variation and its relation to either locales in the native range or admixtures of genotypes in North America was unknown.

Much has emerged in the past 20 years that clearly establishes that *B. tectorum* in North America, despite the grass's almost complete cleistogamy (McKone 1985), is not a monolithic, i.e. genetically uniform, invader. Much of this research examined populations in the native range and North America with starch gel electrophoresis. The genotype that predominates in western European populations is termed simply the Most Common Genotype (MCG); it is also most common among populations along the US Eastern Seaboard, which is not surprising given the trade and human immigration between western Europe and eastern North America for the past 400 years. Nevertheless, many eastern US and southern Ontario populations are admixtures, and some display genotypes found outside western Europe. Paradoxically, one genotype (*Mdh-2b* & *Mdh-3b*) found in the eastern USA is so far known in the native range only from Afghanistan (Bartlett et al. 2002; Valliant et al. 2007).

The case for an East–West spread of *B. tectorum* across the USA is supported by the predominance of the MCG in populations across much of the country from the Appalachians to the Rocky Mountains. However, the current genetic composition of these inland populations is not uniform: many populations contain other genotypes, which have also been detected in Europe. For example, the genotype *Pgm-1a* & *Pgm-2a* has a restricted distribution in Hungary and adjacent Slovakia but is common in populations in the southeastern USA and Great Plains states (Novak & Mack 1993; Schachner et al. 2008; T. Huttanus, unpublished data).

The genetic admixtures in populations in the Pacific Northwest deviate substantially from the general pattern elsewhere on the continent. The MCG is prevalent, and some populations consist largely of this genotype. Strikingly different, however, is the prominence of the *Got-4c* genotype, which has been detected in fewer than 10 populations east of the Rocky Mountains (Schachner et al. 2008; Valliant et al. 2007). Unlike the MCG, *Got-4c* has been detected in Europe only in a narrow belt from Bayreuth, Germany, to Prague (Novak & Mack 1993, 2001). The rarity of this genotype east of the Rockies reduces the likelihood that it spread from the East to the Pacific Northwest. Instead it may have arrived directly by sea along the West Coast. Thus *B. tectorum* may have entered the west as well as east coast of the USA and Canada within the 19th century and then spread inland (Novak & Mack 2001; Valliant et al. 2007).

The rarity of outcrossing in *B. tectorum* has produced North American admixtures of genotypes with detectable native range origins, rather than swarms of genotypes novel to North America. One population in South Dakota is an exception: it consists of four genotypes known in the native range plus the genotype *Pgm-1a & Pgm-2a* and *Mdh-2b & Mdh-3b*, a heterozygote produced after immigration by two genotypes with quite different native ranges (central Europe and Afghanistan, respectively) (Schachner et al. 2008). So far, heterozygotes have been detected is about 10 populations in the USA and Canada (Valliant et al. 2007; Kao et al. 2008; Schachner et al. 2008); others will likely be detected. Whatever the frequency of outcrossing in the new range, the number of novel genotypes appears low. The grass's invasion into so many different habitats appears to be the product of broad phenotypic plasticity, rather than selection among new genotypes (Rice & Mack 1991a,b).

19.3 POPULATION BIOLOGY: THE IMPORTANCE OF SUCCESSIVELY EMERGING COHORTS

Bromus tectorum has traditionally been viewed as a winter annual (see, for example, Stewart & Hull 1949; Hulbert 1955), i.e. it germinates in autumn, overwinters as a basal rosette of leaves and resumes growth in late winter–spring, followed by spring flowering and seed maturation. Additionally, it can germinate in winter or spring, especially if autumn moisture is insufficient to spark germination (Stewart & Hull 1949). Harris (1967) lists a variety of life-histories reported for cheatgrass, including a case of the grass being biennial. Detailed censuses of plant populations in different habitats have quantified the frequency of its modes of persistence (Mack & Pyke 1983).

Much of the persistence of *B. tectorum* in so many habitats is attributable to its capacity to produce seedling cohorts after minimal soil wetting. For example, after an almost rainless summer in eastern Washington, each late summer/early autumn rain induces the emergence of a cohort. This pattern continues even in winter, when precipitation can alternate between snow, hail and rain; germination can occur, provided the soil is neither frozen nor snow-covered. Emergence of cohorts continues well into spring. Given this long annual chronology of emergence, a population by

1 June may consist of cohorts with ages varying from more than 250 days to just a few weeks old (Mack & Pyke 1983).

Successively emerging cohorts effectively buffer the population from seasonally restricted mortality. Between late summer and the following spring, any cheatgrass population faces a gauntlet of risks, from recurring drought in autumn, soil frost-heaving and vole grazing in winter, to smut infestation and further grazing and predation in spring (Mack & Pyke 1984; Pyke 1986, 1987). Any cohort's risk of destruction as well as its fecundity appears directly tied to its age, such that the mixed-age structure populations illustrate a form of bet-hedging that effectively 'covers all bets'. Plants that emerge earliest in autumn and survive until June flowering have reliably the highest fecundity. However, in some years, few if any plants survive from autumn-emerging cohorts, especially those that emerge before the resumption of frequent showers in October or November (Mack & Pyke 1983). Later-emerging cohorts may avoid some hazards (e.g. spring-emerging plants avoid autumn drought, winter frost-damage and predation). Their vegetative growth is restricted, however, to a few spring months, and their seed production is proportionally lower (Mack & Pyke 1983, 1984).

Implications of some individuals reliably surviving, despite an array of environmental hazards, are enormous for the persistence of cheatgrass and consequently for the whole plant community. Calamities can befall a cheatgrass population, yet it remains dominant at a site. Any remaining native species in the community are unlikely to resume their roles even if cheatgrass mortality soars because some cheatgrass cohorts or at least some individuals in different cohorts will survive.

Persistence among populations also has a heritable component that is expressed at several stages in the life cycle. The cessation of precipitation in late spring appears to have selected for plants with genotypes that flower earlier than plants from more mesic and montane environments (Rice & Mack 1991b; Meyer et al. 2004). For example, plants from the arid steppe in central Washington flowered and set seed in a common glasshouse as many as 4 weeks before plants from more mesic sites (Rice & Mack 1991a). Moreover, this earlier flowering was expressed even when plants from the arid steppe were sown in a much cooler, wetter montane environment (Rice & Mack 1991b).

19.4 THE ENVIRONMENTAL ENVELOPE: AN INCOMPLETELY QUANTIFIED MOVING TARGET

The status of cheatgrass in its immense native range (Portugal through Central Asia to Tibet and the northern rim of Africa) is curious, given its abundance in much of arid North America. Aside from its severe restriction under shade (Pierson & Mack 1990a and see below), it remains common but not abundant in much of arid Eurasia. For example, around the Mediterranean Basin it occurs routinely in sites of disturbance and in cereal fields but is not a community dominant. The clear exception to this pattern occurs in Central Asia (e.g. Turkmenistan and Uzbekistan) where it forms the understorey in the arid steppes dominated by *Haloxylon* spp. The physiognomy of these communities most closely resembles the cheatgrass-dominated stands in the Great Basin (Pierson & Mack 1990a; R. N. Mack, unpublished data).

The physical environmental constraints for cheatgrass in North America as known before 1958 were inferred largely from qualitative records in published floras and quantitative vegetation sampling. This view was nevertheless correct in broad outline: a grass with broad, but not unlimited, tolerance of soil texture and salinity and low annual precipitation (more than 200 mm) (Hulbert 1955; Klemmedson & Smith 1964, and references therein). Biotic constraints were characterized anecdotally by the response to competitors, grazers, seed predators and parasites. Although *B. tectorum* can be devastated locally by the smut, *Ustilago bullata* (Fischer 1940), and grazed by livestock as well as native vertebrates (see, for example, Daubenmire 1970; Pyke 1987), none of these mortality agents in themselves destroy a population.

Characterization of the environmental envelope of *B. tectorum* 50 years later remains woefully incomplete. This information is needed, however, if we are to predict cheatgrass responses to global atmospheric change. Cheatgrass is primarily an open-site occupant, and field and glasshouse experiments confirm that shade is a severe restriction. Under prolonged deep shade, *B. tectorum* will die as a seedling or not germinate at all (Pierson & Mack 1990a). This response is most evident in coniferous forests that border the cheatgrass-dominated steppe. However, forest canopies also lower air/soil temperatures, produce thick litter, and provide habitat for mammalian grazers and media for plant parasites, so the restriction of cheatgrass in forests is unlikely to be a response to low light alone (Pierson & Mack 1990a,b; Pierson et al. 1990).

Even though *B. tectorum* in North America and its native range is most prolific in arid, largely treeless communities (Mack 1984), the grass occurs sparsely in deserts in the US Southwest. Smith et al. (1997) hypothesize that cheatgrass may be restricted in deserts by the very warm, dry spring temperatures, which would increase evapo-transpiration for plants several months before they flower. Cheatgrass is tenuously naturalized at high latitude (65° N) in forest openings, such as along pipeline routes in Canada (Cody et al. 2000). The grass is rare or absent at much lower latitudes in an oval-shaped zone in southern Manitoba and adjacent eastern North Dakota. Its absence in this steppe habitat may be caused by the inability of its autumn-emerging cohorts to tolerate bouts of low winter temperatures (below −20° C) on bare ground (Valliant et al. 2007).

Grazers, seed predators and parasites can locally thwart or at least limit the persistence of *B. tectorum*. Cheatgrass seedlings tolerate even repeated bouts of grazing by voles much better than seedlings of native perennial grasses (Pyke 1986). Seed predation, although prevalent, does not apparently become a biotic barrier. Fungal parasitism may become more important in the future as it holds potential in the development of biological control agents (see below).

Consequences of global atmospheric change were unlikely to have been within the realm of concerns for investigators of *B. tectorum* before 1958. The future range and role of this invasive species are now, however, objects of special concern (Bradley et al. 2009). Smith et al. (1997) reported a 50% increase in cheatgrass biomass (i.e. fuel) under a highly elevated ($680\,\mu l\,l^{-1}$) atmospheric CO_2 concentration. More specifically, Ziska et al. (2005) found an increase in biomass of 1.5–2.7 g per plant for every $10\,\mu mol\,mol^{-1}$ increase in CO_2 by 2020. Not only could fuel loads increase, but the grass's range could also be altered in the future. Using bioclimatic envelope modelling, Bradley et al. (2009) predict a dynamic shift in the range of *B. tectorum* with both range contraction and range expansion. Range contraction could occur in the southern part of the Great Basin, whereas major range expansion could occur northward in Wyoming and eastern Montana. These shifts would be driven primarily by changes in precipitation, which would presumably decline at the current southern range boundary.

Less apparent is the driver for northward expansion. The grass is already prominent in the northern Great Plains, and the prediction is based in part on an increase in summer precipitation (Bradley 2008). With seed maturation in late spring and summer seed dormancy that lasts 6–10 weeks, it is not clear how increasing summer precipitation would directly affect the grass's range.

19.5 COMMUNITY/ECOSYSTEM EFFECTS: CHEATGRASS, THE GREAT USURPER

Entry of *B. tectorum* into steppe in the Intermountain West (and to a lesser extent grasslands east of the Rocky Mountains) has produced a concatenation of effects that commonly result in the almost total alteration of the native community; the phrase 'cheatgrass monocultures' (Bradley & Mustard 2005) is only a modest overstatement of the grass's devastating alteration of these communities. Threat of an unfolding cheatgrass invasion was perhaps first noticed in agriculture, as farmers anxiously reported its rapid appearance in wheat fields in south-central Washington before 1920 (Mack 1981, and references therein). The most serious damage was apparent well before 1958: cheatgrass enters steppe communities and eventually proliferates through its severe alteration of the frequency and character of fires (Klemmedson & Smith 1964, and references therein).

The consequences of cheatgrass-induced fire had drawn the attention of ecologists and the public well before 1958 (see, for example, Stewart & Hull 1949), and a vast, disparate and often contradictory literature continues to be assembled. A thorough examination of this topic is, however, beyond the scope of this chapter. I focus here on the consequences identified since 1958 of cheatgrass's direct role in steppe communities, which in many cases are as profound as its role in fire enhancement.

The competition, if not its specific causes, of *B. tectorum* is readily deduced by vegetation sampling of communities before and after its entry. In a few decades at most, cheatgrass replaces native species, even if fire is not directly a factor. Furthermore, succession in a native community is effectively suspended. Once cheatgrass occupies a site, it retains its role indefinitely without extensive control measures (Daubenmire 1970). One major source of competition by *B. tectorum*

stems from its ability to extend its root system downward more than 80 cm during winter, whereas roots of a seedling of a native grass, such as *Agropyron spicatum*, remain inactive. Adult native perennials (e.g. *Festuca idahoensis, Artemisia* spp., *Balsamorhiza sagittata*) are unlikely to be hampered by this root competition, but the effect on native seedlings is severe (Harris 1967). Species replacement nonetheless occurs as adult grasses eventually die and are not replaced.

The root system of *B. tectorum* is also key to profound changes in the cycling of nitrogen, including nitrogen pools, pathways and turnover rates. Aside from available soil moisture, available nitrogen is the chief limitation in native steppe communities. Nitrogen fixation at the soil surface by cyanobacteria and lichens is an important nitrogen contributor. As cheatgrass becomes prominent on a site, nitrogen fixation by these microorganisms and cryptogams declines, one result of the shade cast by the grass, which lowers these species' photosynthesis (Lange et al. 1998); fires sparked by cheatgrass litter also cause their loss or reduction (Anderson et al. 1982; Johansen 2001). As the annual cheatgrass' litter accumulates and soon decomposes in summer, NO_3^- is released and leaches to the subsoil, where it can remain in a nutrient pool until re-acquired by *B. tectorum* roots. With acquisition, NO_3^- is subsequently carried to the plant's above-ground biomass, and the cycle is repeated. Thus, nitrogen cycling has been transformed into a positive feedback loop in which a system driven by nitrogen fixation at the soil surface has been replaced by cheatgrass cycling, which includes pathways for nitrogen from the subsoil (Sperry et al. 2006).

The invasion of *B. tectorum* clearly exerts multifaceted effects on arid ecosystems: for example the availability of soil moisture, nitrogen cycling and, most profoundly, the fire regime. Not surprisingly, practically all other species in the communities are affected either directly or indirectly by cheatgrass. For example, the species richness of much of the soil biota (e.g. fungi, microarthropods, nematodes) declines, even if the ecosystem processes these groups mediate are not substantially affected (Belnap et al. 2005). Loss of native vascular plant species undoubtedly affects native grazers and seed predators, for example the impact of wholesale replacement of native fruits and seeds with the dry caryopses of cheatgrass in summer/autumn. Cheatgrass has permanently altered native communities, hampered agriculture and even caused human fatalities through wild fires (Bonner 2008). Curbing its invasion, or even reducing its range would seem to have warranted a concerted effort to discover and rapidly use the most effective means to combat this horrific plant pest.

19.6 WAGING 'WAR ON CHEATGRASS': FALSE STARTS BUT EMERGING NEW DIRECTIONS

Perhaps the threat of no other invasive species has been reported so elegantly as the menace of cheatgrass: it attracted the attention of America's foremost conservation writer, Aldo Leopold (1949), before his untimely death in 1948. In his essay 'Cheat takes over' Leopold not only saw cheatgrass for the curse it was becoming, he also puzzled over the attentiveness of quarantine services to some pests but not the spread of cheatgrass. Events since 1958 have amply supported Leopold's vision. Based on AVHRR (Advanced Very High Resolution Radiometry), Bradley and Mustard (2005) estimate cheatgrass dominates about 20,000 km^2 in the Great Basin alone. The total area dominated in North America is several-fold larger, given the portions of Idaho, Washington and British Columbia where the grass is equally prominent (Mack 1981). An additional large section of the western Great Plains is also infested and increasing in area (Douglas et al. 1990; Schachner et al. 2008, and references therein). Efforts to control the grass's dominance began well before 1958 (see, for example, Hull & Stewart 1948); the status of control programs since 1958 deserves attention here as well as the more important topic of what research and technology should be encouraged in the future.

In the past 50 years scores of techniques and tools in innumerable combinations have been used in attempting to control cheatgrass in the arid West; singly or in combination, these approaches have failed, often dismally. Detailing these control practices here serves little purpose. To varying degrees, these methods involve alteration (and, often, total destruction) of whatever remained of the native steppe by burning the cheatgrass-infested vegetation, chaining (i.e. mechanically toppling) the sagebrush (*Artemisia* spp.), ploughing, applying herbicides (Monsen et al. 2004a) and then sowing non-native species (e.g. *Agropyron cristatum, Agropyron desertorum, Kochia prostrata*) (Monsen et al. 2004b; Stevens et al. 1985) on these devastated sites. The arsenal of tools and practices

ostensibly hurled against cheatgrass has done little more than destroy remnants of the native steppe and in some cases facilitated cheatgrass. (But see Young & Clements (2009) for a historical account of cheatgrass control, although with a diametrically different assessment of its effectiveness and failure.)

Use of these practices has often devastated, if not eliminated, key members of the native community, while the cheatgrass is at best only temporarily checked. Loss of the essential thin biological soil crust is particularly egregious in this deliberate conversion of badly damaged native communities to communities with *other* non-native species and low species richness. Whether referred to as the cryptogamic or cryptobiotic or simply biological soil crust, these low-stature species (cyanobacteria, green algae, lichens, liverworts and mosses) include indispensable sources of nitrogen fixation (Belnap 2001) that are readily destroyed by burning and mechanical cultivation (Belnap & Eldridge 2001) when the steppe is treated as a farm to be tilled.

These 'restoration' practices, which could have been deemed ineffective by 1958, have only slowly been modified or discouraged outright (Monsen et al. 2004b). For example, use of non-native species is slowly being replaced with dissemination of native species through the Great Basin Native Plant Selection and Increase Project (http://www.fs.fed.us/rm/boise/research/shrub/greatbasin.shtml). The re-introduction of native species (Shaw et al. 2005) needs to be rapidly and substantially expanded in scope, i.e. number of species re-introduced, tonnage of seeds collected and distributed, and the area of cheatgrass-dominated landscapes brought into sustained restoration. Furthermore, at least two lines of research should be greatly expanded (and the resulting lessons implemented) if another 50 years of research on *B. tectorum* is not to remain an untapped reservoir of knowledge.

Search for effective biological control agents for cheatgrass

If mechanical and chemical treatments have failed to curb the invasion of cheatgrass, biological control remains the major tool left. Biological control of grasses presents, however, several unusually high hurdles in the application of host–parasite interaction as well as public perception. Grasses are usually not attacked by specific invertebrate grazers or predators (Evans 1991).

Searching for species-specific invertebrates for bromes may prove fruitless; in North America the genus contains not only native congeners but also introduced species that are valued as forage (e.g. *Bromus inermis*) (Hitchcock 1951). Developing microorganisms for biological control of grasses has been traditionally approached with understandable reluctance: host extension by a microbial control agent to a grain crop would be catastrophic (Goeden & Andres 1999). For example, wheat, oats, barley, corn and even rice are grown within the range of cheatgrass in the USA.

Nevertheless this research has proceeded, although with modest scope and scale. Investigations, so far, have concentrated on two fungi and a bacterium: *Ustilago bullata*, *Pyrenophora semeniperda* and *Pseudomonas flourescens* strain D7. Each has been reported repeatedly to infect *B. tectorum* in North America, so the obstacles in introducing a novel organism to North America through classical biological control are avoided. Pathotypes of *U. bullata* vary in virulence for cheatgrass; least resistance is generally among host lines that co-occur with the head smut pathotypes. Clearly much more research is needed to determine not only the breadth of genetic variation in the smut (Meyer et al. 2001, 2005) but also the full extent of interactions between grass and fungal genotypes. Determination of the full range of susceptible grass taxa other than *B. tectorum* is mandatory. Use of *U. bullata* as a biological control agent could also be restricted by varying intolerance of the smut to temperatures below 25°C, which would restrict the season of infection (Boguena et al. 2007).

The efficacy of the seed bank parasite, *Pyrenophora semeniperda*, may be more promising, as it can attach to developing florets as well as mature caryopses in the soil (Meyer et al. 2008). However, moisture, rather than low temperature, appears to be an environmental constraint to its infectivitiy: caryopses in xeric habitats are more readily infected than caryopses on mesic sites (Beckstead et al. 2007). Other and potentially more serious limitations on its use need to be resolved, including the identification of pathotypes specific to *B. tectorum*, their virulence and the need to use conidia, rather than mycelia, as the infective agents (Medd & Campbell 2005).

Unlike the mode of action by the two proposed fungal agents, the rhizobacterium, *Pseudomonas flourescens* strain D7, retards the root growth of *B. tectorum*. Retardation of root growth averaged 97% for seedlings grown on agar plates with the cell-free supernatant of the bacterium. However, inhibition of root growth was less and varied widely (13–64%) for plants grown in soil under controlled growth conditions (Kennedy et al. 2001). Whether slowing root growth translates into reducing the role of cheatgrass remains problematical, although substantial reductions in cheatgrass have been observed in field sites inoculated with *P. flourescens* strain D7 (A.C. Kennedy, personal communication).

None of these investigations have yet produced an agent for release, but they are illustrative of a research avenue that should be greatly expanded. Not only should the genetic variation in these taxa be comprehensively compared with the genotypes of cheatgrass, but search for other potential microbial agents should also be increased. Sequencing whole genomes of higher plants, such as cheatgrass, and the myriad microbial taxa that may hold potential as biological control agents now seems practical, thanks to recent developments in genomics (Hudson 2008). Admittedly, no agents may be found that meet the high standards of genetic stability and strict specificity to *B. tectorum*. Nevertheless, given the potential benefits, this line of investigation deserves exhaustive exploration.

Restoration and maintenance of the biological soil crust, especially cryptogams

As outlined above, the invasion of *B. tectorum* and especially attempts to control it have been disastrous for the biological soil crust in steppe communities. Any forces that reduce the presence of these organisms (livestock, vehicle tread, fire) (Metting 1991; Johansen 2001) increase the opportunity for cheatgrass entry and persistence (Anderson et al. 1982). In addition to detrimentally affecting nitrogen budgets, the destruction removes what I term here the 'first line of defence' against cheatgrass proliferation (Kaltenecker et al. 1999). Members of the crust, principally the crustose lichen *Diploschistes muscorum* and the moss *Bryum argenteum*, delay or largely thwart the establishment of cheatgrass (Deines et al. 2007; Serpe et al. 2006, 2008). For *D. muscorum*, key in its role is the physical barrier this cryptogam's thalli create to root penetration by seedlings. As a result, *B. tectorum* seedling establishment may be reduced 85% compared with establishment on bare soil (Deines et al. 2007). Furthermore, the water potential of the *D. muscorum* surface is much drier (−4MPa) than the much more mesic water status (−0.6 MPa) on bare soil (Serpe et al. 2008). Admittedly, the biological soil crust in any community is not composed entirely of crustose lichens or *B. argenteum*, or both. Other members of this biotic layer appear to present little or no restriction to cheatgrass (Deines et al. 2007). Nevertheless, the patchy micro-distribution of cryptogams can substantially reduce total cover of cheatgrass.

Practices are needed that not only conserve these fragile organisms but also restore their role. Such restoration is problematic, given the widely varying rates of re-colonization in arid communities. Although cyanobacteria and green algae may re-colonize a surface within 3–5 years, or less (Belnap 1993), colonization by the groups most responsible for thwarting cheatgrass, perennial lichens and mosses, may require many decades (Anderson et al. 1982; Belnap & Eldridge 2001). The rate of re-colonization must then be accelerated, for example by applying an inoculum of the entire cryptogamic community (not simply the species known to hamper *B. tectorum* entry directly).

Research and development has been perplexingly slow for marshalling this important line of defence against cheatgrass in arid communities, despite well-substantiated calls for such action (Belnap 1994) and the availability of technology for producing soil inoculants (see Walter & Paau 1993; Buttars et al. 1998). Large-scale inoculation of mosses and lichens seems more straightforward, i.e. dissemination of mosses as spores and lichens as vegetative fragments (Belnap 1993). Given the clear link between a cryptogamic cover and the status of cheatgrass on a site, a major research initiative seems warranted to develop the technology to apply these organisms at the landscape level.

19.7 CONCLUSIONS

In the 50 years since the publication of Charles Elton's illuminating book we have gained a much better understanding of why and how *Bromus tectorum* has become invasive in North America; unfortunately, knowledge of how to combat this pest has lagged behind. Correcting this situation in the future is challenging but by no means hopeless. Little chance of controlling cheatgrass over such a vast area seems possible without development of effective biological control agents that match the genetic variation in the grass. Developing techniques to mass culture and inoculate

the entire biological soil crusts on different habitats is also essential. But none of those efforts in research and development will be at all meaningful unless deliberate and thorough steps are taken to evaluate and transfer these results into widespread field application. Society rightfully gives its kudos to harbingers of disaster, such as Charles Elton. It reserves its greatest kudos, however, for those who heed the warnings and take concrete steps to turn back disasters, such as the invasion of cheatgrass.

ACKNOWLEDGEMENTS

I thank D.M. Richardson for the opportunity to contribute to this commemorative volume, J. Belnap, T. Huttanus, A. C. Kennedy and S.J. Novak for insights and access to informative references, D. Simberloff and D.M. Richardson for critical reviews of the manuscript, and C. Purdon for diligent copy-editing.

REFERENCES

Anderson, D.C., Harper, K.T. & Rushforth, S.R. (1982) Factors influencing development of cryptogamic soil crusts in Utah deserts. *Journal of Range Management*, **35**, 355–359.

Bartlett, E.A., Novak, S.J. & Mack, R.N. (2002) Genetic variation in *Bromus tectorum* (Poaceae): differentiation in eastern United States. *American Journal of Botany*, **89**, 626–636.

Beckstead, J., Meyer, S.E., Molder, C.J. & Smith, C. (2007) A race for survival: can *Bromus tectorum* seeds escape *Pyrenophora semeniperda*-caused mortality by germinating quickly? *Annals of Botany*, **99**, 907–914.

Belnap, J. (1993) Recovery rates of cryptogamic crusts: inoculant use and assessment methods. *Great Basin Naturalist*, **53**, 89–95.

Belnap, J. (1994) Potential role of cryptobiotic soil crusts in semiarid rangelands. In *Proceedings – Ecology and Management of Annual Rangelands*, pp. 179–185. USFS Intermountain Research Station General Technical Report INT-GTR-313.

Belnap, J. (2001) Factors influencing nitrogen fixation and nitrogen release in biological soil crusts. In *Biological Soil Crusts: Structure, Function and Management* (ed. J. Belnap and O.L. Lange), pp. 241–261. Springer, New York.

Belnap, J. & Eldridge, D.J. (2001) Disturbance and recovery of biological soil crusts. *Biological Soil Crusts: Structure, Function and Management* (ed. J. Belnap and O.L. Lange), pp. 363–383. Springer, New York.

Belnap, J., Phillips, S.L., Sherrod, S.K., & Moldenke, A. (2005) Soil biota can change after exotic plant invasion: does this affect ecosystem processes? *Ecology*, **86**, 3007–3017.

Boguena, T., Meyer, S.E. & Nelson, D.L. (2007) Low temperature during infection limits *Ustilago bullata* (*Ustilaginaceae, Ustilaginales*) disease incidence on *Bromus tectorum* (Poaceae, Cyperales). *Biocontrol Science and Technology*, **17**, 33–52.

Bonner, J.L. (2008) Tests to determine cause of Boise professor's death. *Seattle Times* 28 August 2008. http://seattletimes.nwsource.com/html/localnews/2008143476_boise28.html.

Bradley, B.A. (2008) Regional analysis of the impacts of climate change on cheatgrass invasion shows potential risk and opportunity. *Global Change Biology*, **15**, 196–208.

Bradley, B.A. & Mustard, J.F. (2005) Identifying land cover variability distinct from land cover change: cheatgrass in the Great Basin. *Remote Sensing of Environment*, **94**, 204–213.

Bradley, B.A., Oppenheimer, M. & Wilcove, D.S. (2009) Climate change and plant invasions: restoration opportunities ahead? *Global Change Biology*, **15**, 1511–1521.

Brandegee, K. (1892) Catalogue of the flowering plants and ferns growing spontaneously in the City of San Francisco. *Zoe*, **2**, 334–386.

Buttars, S.M., St. Clair, L.L., Johansen, J.R., et al. (1998) Pelletized cyanobacterial soil amendments: laboratory testing for survival, escapability, and nitrogen fixation. *Arid Soil Research and Rehabilitation*, **12**, 165–178.

Cody, W.J., MacInnes, K.L., Cayouette, J. & Darbyshire, S. (2000) Alien and invasive native vascular plants along the Norman Wells Pipeline, District of Mackenzie, Northwest Territories. *Canadian Field Naturalist*, **114**, 126–137.

Daubenmire, R. (1970) *Steppe Vegetation of Washington*. Washington Agricultural Experiment Station Technical Bulletin 62.

Deines, L., Rosentreter, R., Eldridge, D.J. & Serpe, M.D. (2007) Germination and seedling establishment of two annual grasses on lichen-dominated biological soil crusts. *Plant Soil*, **295**, 23–35.

Douglas, B.J., Thomas, A.G. & Derksen, D.A. (1990) Downy brome (*Bromus tectorum*) invasion into southwestern Saskatchewan. *Canadian Journal of Plant Science*, **70**, 1143–1151.

Elton, C.S. (1958) *The Ecology of Invasions by Animals and Plants*. Methuen, London.

Evans, H. (1991) Biological control of tropical grassy weeds. In *Tropical Grassy Weeds* (ed. F.W.G. Baker and P.J. Terry), pp. 52–71. CAB International, Wallingford, UK.

Fernald, M.L. (1950) *Gray's Manual of Botany*, 8th edn. Van Nostrand, New York.

Fischer, G.W. (1940) Host specialization in the head smut of grasses, *Ustilago bullata*. *Phytopathology*, **30**, 991–1017.

Goeden, R.D. & Andres, L. A. (1999) Biological control of weeds in terrestrial and aquatic environments. In *Handbook of Biological Control: Principles and Applications of Biological Control* (ed. T.S. Bellows and T.W. Fisher), pp. 871–890. Academic Press, London.

Harris, G.A. (1967) Some competitive relationships between *Agropyron spicatum* and *Bromus tectorum*. *Ecological Monographs*, **37**, 89–111.

Hitchcock, A.S. (1951) *Manual of the Grasses of the United States*, 2nd edn. U.S. Government Printing Office, Washington, DC.

Hudson, M. E. (2008) Sequencing breakthroughs for genomic ecology and evolutionary biology. *Molecular Ecology Resources*, **8**, 3–17.

Hulbert, L.C. (1955) Ecological studies of *Bromus tectorum* and other annual bromegrasses. *Ecological Monographs*, **25**, 181–213.

Hull, A.C. & Stewart, G. (1948) Replacing cheatgrass by reseeding with perennial grass on southern Idaho ranges. *Agronomy Journal*, **40**, 694–703.

Johansen, J.R. (2001) Impacts of fire on biological soil crusts. In *Biological Soil Crusts: Structure, Function and Management* (ed. J. Belnap and O.L. Lange), pp. 385–397. Springer, New York.

Kaltenecker, J.H., Wicklow-Howard, M.C. & Pellant, M. (1999) Biological soil crusts: natural barriers to *Bromus tectorum* L. establishment in the northern Great Basin, USA. *People and Rangelands: Building the Future* (ed. D. Eldridge and D. Freudenberger), pp. 109–111. VI International Rangeland Congress, Aitkenvale, Australia.

Kao, R.H., Brown, C.S. & Hufbauer, R.A. (2008) High phenotypic and molecular variation in downy brome (*Bromus tectorum*). *Invasive Plant Science and Management*, **1**, 216–225.

Kennedy, A.C., Johnson, B.N. & Stubbs, T.L. (2001) Host range of a deleterious rhizobacterium for biological control of downy brome. *Weed Science*, **49**, 792–797.

Klemmedson, J.O. & Smith, J.G. (1964) Cheatgrass (*Bromus tectorum* L.). *Botanical Review*, **30**, 226–262.

Lange, O.L., Belnap, J. & Reichenberger, H. (1998) Photosynthesis of the cyanobacterial soil crust lichen *Collema tenax* from arid lands in southern Utah, USA: the role of water content on light and temperature responses. *Functional Ecology*, **12**, 195–202.

Leopold, A. (1949) *A Sand County Almanac and Sketches Here and There*. Oxford University Press, New York.

Mack, R.N. (1981) Invasion of *Bromus tectorum* L. into western North America: an ecological chronicle. *Agro-Ecosystems*, **7**, 145–165.

Mack, R.N. (1984) Invaders at home on the range. *Natural History*, **93**, 40–47.

Mack, R.N. (2003) Plant naturalizations and invasions in the eastern United States: 1634–1860. *Annals of the Missouri Botanical Garden*, **90**, 77–90.

Mack, R.N. & Pyke, D.A. (1983) The demography of *Bromus tectorum* L.: variation in time and space. *Journal of Ecology*, **71**, 69–93.

Mack, R.N. & Pyke, D.A. (1984) The demography of *Bromus tectorum*: the role of microclimate, predation and disease. *Journal of Ecology*, **72**, 731–748.

McKone, M.J. (1985) Reproductive biology of several brome-grasses (*Bromus*): breeding system, pattern of fruit maturation and seed set. *American Journal of Botany*, **72**, 1334–1339.

Medd, R.W. & Campbell, M.A. (2005) Grass seed infection following inundation with *Pyrenophora semeniperda*. *Biocontrol Science and Technology*, **15**, 21–36.

Metting, B. (1991) Biological surface features of semiarid lands and deserts. *Semiarid Lands and Deserts. Soil Resource and Reclamation*. (ed. J. Skujins), pp. 257–293. M. Dekker, New York.

Meyer, S.E., Nelson, D.L. & Clement, S. (2001) Evidence for resistance polymorphism in the *Bromus tectorum* – *Ustilago bullata* pathosystem: implications for biocontrol. *Canadian Journal of Plant Pathology*, **23**, 19–23.

Meyer, S.E., Nelson, D.L. & Carlson, S.L. (2004) Ecological genetics of vernalization response in *Bromus tectorum* L. (*Poaceae*). *Annals of Botany*, **93**, 653–663.

Meyer, S.E., Nelson, D.L., Clement, S., Waters, J., Stevens, M. & Fairbanks, D. (2005) Genetic variation in *Ustilago bullata*: molecular genetic markers and virulence on *Bromus tectorum* host lines. *International Journal of Plant Science*, **166**, 105–115.

Meyer, S.E., Beckstead, J., Allen, P.S. & Smith, D.C. (2008) A seed bank pathogen causes seedborne disease: *Pyrenophora semeniperda* on undispersed grass seeds in western North America. *Canadian Journal of Plant Pathology*, **30**, 525–533.

Monsen, S.B., Stevens, R. & Shaw, N.L. (2004a) *Restoring Western Ranges and Wildlands*. USFS General Technical Report RMRS-GTR-136-vol. 1.

Monsen, S.B., Stevens, R. & Shaw, N. (2004b) Grasses. *Restoring Western Ranges and Wildlands*. pp. 295–424. USFS General Technical Report RMRS-GTR-136-vol. 2.

Novak, S.J. & Mack, R.N. (1993) Genetic variation in *Bromus tectorum* (Poaceae): comparison between native and introduced populations. *Heredity*, **71**, 167–176.

Novak, S.J. & Mack, R.N. (2001) Tracing plant introduction and spread into naturalized ranges: genetic evidence from *Bromus tectorum* (cheatgrass). *BioScience*, **51**, 114–122.

O'Driscoll, P. (2007) Invasive weed a fuel for West's wildfires. http://www.usatoday.com/news/nation/environment/2007-08-29-cheatgrass_N.htm.

Pierson, E.A. & Mack, R.N. (1990a) The population biology of *Bromus tectorum* in forests: distinguishing the opportunity for dispersal from environmental restriction. *Oecologia*, **84**, 519–525.

Pierson, E.A. & Mack, R.N. (1990b) The population biology of *Bromus tectorum* in forests: effect of disturbance, grazing, and litter on seedling establishment and reproduction. *Oecologia*, **84**, 526–533.

Pierson, E.A., Mack, R.N. & Black, R.A. (1990) The effect of shading on photosynthesis, growth, and regrowth following defoliation for *Bromus tectorum*. *Oecologia*, **84**, 534–543.

Pyke, D.A. (1986) Demographic responses of *Bromus tectorum* and seedlings of *Agropyron spicatum* to grazing by small mammals: occurrence and severity of grazing. *Journal of Ecology*, **74**, 739–754.

Pyke, D.A. (1987) Demographic responses of *Bromus tectorum* and seedlings of *Agropyron spicatum* to grazing by small mammals: the influence of grazing frequency and plant age. *Journal of Ecology*, **75**, 825–835.

Rice, K.J. & Mack, R.N. (1991a) Ecological genetics of *Bromus tectorum*: I. A hierarchical analysis of phenotypic variation. *Oecologia*, **88**, 77–83.

Rice, K.J. & Mack, R.N. (1991b) Ecological genetics of *Bromus tectorum*: III. The demography of reciprocally sown populations. *Oecologia*, **88**, 91–101.

Schachner, L., Mack, R.N. & Novak, S.J. (2008) *Bromus tectorum* (Poaceae) in mid-continental United States: population genetic analysis of an ongoing invasion. *American Journal of Botany*, **95**, 1584–1595.

Serpe, M.D., Orm, J.M., Barkes, T., & Rosentreter, R. (2006) Germination and seed water status of four grasses on moss-dominated biological crusts from arid lands. *Plant Ecology*, **185**, 163–178.

Serpe, M.D., Zimmerman, S.J., Deines, L. & Rosentreter, R. (2008) Seed water status and root tip characteristics of two annual grasses on lichen-dominated biological soil crusts. *Plant Soil*, **303**, 191–205.

Shaw, N.L., Lambert, S.M., DeBolt, A.M. & Pellant, M. (2005) Increasing native forb seed supplies for the Great Basin. *USDA Forest Service Proceedings RMRS-P-35*.

Smith, C.F. (1952) *A Flora of Santa Barbara; An Annotated Catalogue of the Native and Naturalized Plants of Santa Barbara, California, and Vicinity*. Santa Barbara Botanic Garden, Santa Barbara, California.

Smith, S.D., Monson, R.K. & Anderson, J.E. (1997) *Physiological Ecology of North American Desert Plants*. Springer, New York.

Sperry, L.J., Belnap, J. & Evans, R.D. (2006) *Bromus tectorum* invasion alters Nitrogen dynamics in an undisturbed arid grassland ecosystem. *Ecology*, **87**, 603–615.

Stevens, R. Jorgensen, K.R., McArthur, E.D. & Davis, J.N. (1985) 'Immigrant' forage kochia. *Rangelands*, **7**, 22–23.

Stewart, G. & Hull, A.C. (1949) Cheatgrass (*Bromus tectorum* L.) – an ecologic intruder in southern Idaho. *Ecology*, **18**, 58–74.

Upadhyaya, M.K., Turkington, R. & McIlverde, D. (1986) The biology of Canadian weeds. 75. *Bromus tectorum* L. *Canadian Journal of Plant Science*, **66**, 689–709.

Valliant, M.T., Mack, R.N. & Novak, S.J. (2007) The introduction history and population genetics of the invasive grass *Bromus tectorum* (Poaceae) in Canada. *American Journal of Botany*, **94**, 1156–1169.

Walter, J.F. & Paau, A.S. (1993) Microbial inoculant production and formulation. In *Soil Microbial Ecology: Applications in Agricultural and Environmental Management*. (ed. F.B. Metting), pp. 579–594. Marcel Dekker, New York.

Young, J.A. & Clements, C.D. (2009) *Cheatgrass. Fire and Forage on the Range*. University of Nevada Press, Reno.

Ziska, L.H., Reeves, J.B. & Blank, B. (2005) The impact of recent increases in atmospheric CO_2 on biomass production and vegetative retention of Cheatgrass (*Bromus tectorum*): implications for fire disturbance. *Global Change Biology*, **11**, 1325–1332.

Part 6

New Directions and Technologies, New Challenges

RESEARCHING INVASIVE SPECIES 50 YEARS AFTER ELTON: A CAUTIONARY TALE

Mark A. Davis

Department of Biology, Macalester College, Saint Paul, MN 55105, USA

Fifty Years of Invasion Ecology: The Legacy of Charles Elton, 1st edition. Edited by David M. Richardson
© 2011 by Blackwell Publishing Ltd

20.1   INTRODUCTION

Rather surprisingly, the publication of Elton's book in 1958 did not instigate a flood of research on species introductions (Simberloff, this volume). There had been a trickle of articles on introduced species before Elton's book (for example Allan 1936; Egler 1942; Baker 1948), which continued during the 1960s and 1970s (for example Sukopp (1962); Jehlík & Slavík (1968); Moyle 1973; Burdon & Chilvers (1977)). However, it was not until the initiation of the SCOPE (Scientific Committee on Problems of the Environment) in 1983, 25 years after the publication of Elton's book, that the modern field of invasion biology really began to take shape (Richardson & Pysek 2006; Simberloff, this volume).

During the past quarter of a century or so, the field has grown enormously (MacIsaac et al., this volume). What had begun as a small group of committed researchers in the 1980s is now a prominent and multi-facetted research field within ecology. Researchers from throughout the world are currently focusing their research on biological invasions and thousands of papers on species introductions, and their spread, impact and management, have been published in the past decade. The result has been a considerable increase in our knowledge in all these areas. At the same time, as is the case with any initiative or organization that grows very quickly, the field has experienced some growing pains. Certain obstacles and pitfalls may have impeded the field's progress. Some of these obstacles are of our own making and can be remedied. Others relate to the inherent properties and dynamics of species introductions and spread and will likely remain as challenges in the future. In this chapter, I briefly describe what I believe have been two missteps by the field: reliance for too long on a niche-based approach to understanding invasions and a tendency to overstate certain conclusions and claims.

20.2   THE NICHE-BASED APPROACH TO UNDERSTANDING INVASIONS: MORE A HINDRANCE THAN A HELP?

Scientific disciplines, like populations, can experience a founder effect. If one looks at the scientists who played a leading role in inaugurating invasion biology through SCOPE in the 1980s, one sees that most were ecologists, particularly community ecologists. As a result, in many ways, invasion biology emerged as a disciplinary offspring of community ecology. At this time, the early 1980s, the field of community ecology was still largely dominated by the niche-based theories of MacArthur and Hutchinson. It is not surprising, then, that the early years of invasion biology were shaped by a perspective that emphasized determinism and local processes. For example, the first two questions originally articulated by the 1983 SCOPE scientific advisory committee on biological invasions, which were intended to focus subsequent research, focused on species traits and local processes.

In invasion biology, the niche-based approach is probably most obvious in the diversity–invasibility hypothesis, which holds that species diverse environments should be more resistant to invasion than species-poor environments (see Fridley, this volume). The essence of this argument has changed little in the 50 years since Elton articulated this line of reasoning, which bore some similarity to Darwin's naturalization hypothesis. Those who have promoted the diversity–invasibility hypothesis, whether invoking species diversity or functional diversity, typically have specifically articulated a niche-limitation argument (see, for example, Fargione & Tilman 2005; Fridley, this volume).

Following in the wake of a larger ship works well as long as the large ship is seaworthy and headed in the right direction. However, in the eyes of many, the ship of community ecology has been foundering, or at least not making much progress, for some time. There has been increasing discontent within community ecology in recent years. Frustrated by what he believed to be lack of progress in community ecology, Lawton (1999) referred to the state of community ecology at that time as 'a mess' and questioned whether community ecology even had a future. Lawton (2000) particularly criticized the localized approach to understanding communities, noting that 'the details and many of the key drivers appear to be different from system to system in virtually every published study ... and we have no means of predicting which processes will be important in which types of system'. Lawton is hardly alone in raising concerns over the lack of progress in the field of community ecology (see, for example, Cuddington & Beisner 2005; Ricklefs 2006). Castle (2005) acknowledged the abundant discussions in community ecology over theory during the past 50 years, but questioned how much new knowledge actually has been acquired.

This general disappointment in progress made in community ecology is mirrored by the souring on the niche-based approach that has been taking place in invasion biology. For example, Williamson (1996) did not believe a niche-based approach to studying invasions held much promise, bluntly concluding, 'it looks as if models of invasion based on niches will be as disappointing as other community studies of niches'. More recently, the use of niche theory in invasion biology has been criticized by Bruno et al. (2005), who charged the field with uncritically accepting the niche-based competition paradigm for several decades. Although small-scale experiments involving constructed environments and communities have often found the invasibility of environments to be inversely correlated with the environment's species richness (see, for example, Naeem et al. 2000; Fargione & Tilman 2005; Stachowicz et al. 2002), most larger-scale studies have found little support for a biotic-resistance model based on species diversity, (see, for example, Stohlgren et al. 1999; Levine et al. 2004; Richardson et al. 2005). In fact, more typically, among naturally occurring environments, species-rich communities have been found to accommodate more introduced species than species-poor communities (see, for example, Stohlgren et al. 2003; Wiser & Allen 2006; Belote et al. 2008).

Certainly, if the impacts of biotic interactions on community assembly are very weak, whether because species respond similarly to one another and to the environment, and/or because regional and stochastic processes normally overwhelm the effects of biotic interactions, then it would seem a niche-based model would be a poor choice to represent community assembly involving recently introduced species. Nevertheless, despite increasing reservations by many about the use of a niche-based and competition approach to understanding invasions, and despite the fact that most data do not support the diversity–invasibility hypothesis, or the related notion of species saturation (Stohlgren et al. 2008), niche-based invasion models have continued to play a major role in invasion theory (see, for example, Shea & Chesson 2002; Fargione et al. 2003; Tilman 2004; Melbourne et al. 2007). Are these models contributing positively to our understanding of invasion dynamics, or are they an example of how some approaches and ideas are able to persist in ecology without strong empirical support, perhaps, as suggested by Graham & Dayton (2002), owing more to other factors, such as the prominence of some of those advocating the ideas?

Levins (1966) observed that good theory rests on three pillars: generality, precision and realism. For biological invasions, niche theory may score high on the first pillar but low on the other two. A question worthy of discussion is whether the localized and niche-based approach to invasions, embodied most clearly in the diversity–invasibility hypothesis, has hindered more than it has helped progress in the field of invasion biology. The localized and niche-based perspective, which, in the early 1980s, likely led the emerging field of invasion biology to focus attention on species traits and local processes, may have delayed the development and emergence of the contemporary view of invasions, which emphasizes history, fluctuating environmental conditions and regional factors (particularly propagule pressure), as well as local processes, in the invasion process (Davis et al. 2000; Lockwood et al. 2005; Rejmánek et al. 2005a; Colautti et al. 2006).

In community ecology in general, and in invasion ecology in particular, it has often seemed difficult for discovery to resolve debates about competing hypotheses and theories. In an unpublished presentation to the British Biological Society in 2004, Peter Grubb expressed this concern, citing the failure of ecologists to reject wrong ideas and faulty interpretations (cited in Grime 2007). Similar concerns were raised by Craine (2005), who noted that some theories in community ecology have persisted in the face of empirical data that have contradicted them. A similar point was made a few years earlier by Graham and Dayton (2002), who argued that without the clarity provided by empirical discovery, some ideas, theories and approaches have dominated the field of ecology despite being supported by little empirical data.

As pointed out by Richardson and Pyšek (2008), although the field of invasion biology has proved itself very prolific in generating new hypotheses and theories, it seems much less inclined, or able, to reject them. This has resulted in a field characterized more by theory accumulation than theory discrimination. This is not a call to halt the formulation of new theory. In fact, recent efforts to develop more overarching and integrative theories (see, for example, Moles et al. 2008; Barney & Whitlow 2008; Catford et al. 2009) are exactly what the field needs. Rather, this is a call to recognize that some theories and hypotheses simply have found little support from empirical data and that, in these instances, it might be best to take these theories off the table. For some reason, this seems a challenge for the field, as illustrated by the Rasputin-like persistence of the diversity–invasibility hypothesis. There is an overwhelming lack of supporting evidence for this hypothesis, beyond support provided from some small-scale constructed communities. In fact, most data from natural systems have shown that increased invasibility tends to be associated with species-rich, not species-poor, communities. Nevertheless, this hypothesis still exhibits considerable vitality in the field, and there is little sign that it is about to be retired any time soon. Why has the field of invasion biology been so hesitant, unwilling or unable to reject this hypothesis?

20.3 THE PROBLEM OF OVERSTATING CLAIMS AND CONCLUSIONS

In any discipline, it is important that preliminary ideas, or tentative conclusions made on the basis of one or a few studies, do not acquire a life of their own, eventually assuming a level of validity and generality that is unjustified on the basis of the actual data. Unfortunately, with common citation practices, it is very easy for this to happen. Lamenting the vitality and longevity of many inflated scientific claims, Gitzen (2007) observed, 'Once bold claims about … a weak result are published, their sins are forgiven and they can be worked into future introductions and discussions at will'.

The process by which preliminary conclusions become inflated generalizations often involves a series of small missteps, each one of which might be regarded as mostly innocuous. For example, when citing a particular finding or conclusion for the first time, authors often take the time to describe the particular context in which the specific finding or conclusion was made. At a later time, the same author may then cite this same finding in another manuscript, or other researchers, without having actually read the original source, may use the information provided by a secondary reference to cite the original work. In both cases, it is common for these subsequent references to leave out the details needed to assess the reliability and generality of the original finding or conclusion. As time goes on, it is not uncommon for the finding or conclusion to be simply stated as fact, with a perfunctory citation of the original author. By now, the original findings or conclusions are often included as boilerplate in introductions and conclusions of articles and proposals. After enough of these iterations, the original finding can become

such an integral part of accepted ecological wisdom that many authors feel comfortable in reporting it without citing any source at all. The general problem is that the more often that preliminary ideas and tentative conclusions are presented as an axiomatic starting point for further discussion and research, the more likely it is that practitioners, particularly young practitioners, begin to regard the statements as factual, believing that they have having been thoroughly and comprehensively empirically confirmed.

A particular striking example of this phenomenon in invasion biology is the conclusion by Wilcove et al. (1998) that non-native species are the second greatest threat to the survival of species in peril. This statement was cited more than 700 times in the decade after its publication, and, no doubt, in many research proposals, management documents and college classes. By the early 2000s, this statement had become common, boilerplate for invasion literature, the conclusion often presented as fact. Given the limitations and some biases in the information used by Wilcove et al. to come to their conclusion, it is difficult to believe that all those who have cited this article actually have read it. First, as the authors were careful to make very clear, little of the information used to support the claim that non-native species were the second greatest extinction threat involved actual data:

> 'We emphasize at the outset that there are some important limitations to the data we used. The attribution of a specific threat to a species is usually based on the judgment of an expert source, such as a USFWS employee who prepares a listing notice or a state Fish and Game employee who monitors endangered species in a given region. Their evaluation of the threats facing that species may not be based on experimental evidence or even on quantitative data. Indeed, such data often do not exist. With respect to species listed under the ESA [(Element Stewardship Abstract, prepared by The Nature Conservancy)], Easter-Pilcher (1996) has shown that many listing notices lack important biological information, including data on past and possible future impacts of habitat destruction, pesticides, and alien species. Depending on the species in question, the absence of information may reflect a lack

of data, an oversight, or a determination by USFWS that a particular threat is not harming the species. The extent to which such limitations on the data influence our results is unknown.'

Second, despite the fact that the article is commonly cited as support for a global claim of extinction threat by non-native species, Wilcove et al. only addressed threats to species in the USA. Third, Wilcove et al.'s findings are dramatically affected by the inclusion of Hawaii, which, although of course part of the USA, clearly has a dramatically different invasion history than does the continental, and substantially majority, portion of the country.

A similar review of extinction threats in Canada found introduced species to be the *least* important of the six categories analysed (habitat loss, overexploitation, pollution, native species interactions, introduced species and natural causes, the last including stochastic events such as storms) (Venter et al. 2006). Venter et al. (2006) reanalysed Wilcove et al.'s data excluding Hawaii and found that the USA and Canada did not differ in the threats posed by introduced species, meaning that the high ranking of non-native species as an extinction threat was due almost entirely to the inclusion of Hawaii. Other studies that have examined species threats over a much larger global area have come to similar conclusions. For example, in their analysis of the causes of species depletions and extinctions in estuaries and coastal marine waters, Lotze et al. (2006) concluded that the threat of non-native species was negligible compared with exploitation and habitat destruction. In Australia, non-native species have been reported to have contributed to the extinctions of some native mammals (see Finlayson 1961; Kinnear et al. 1998). However, the fact that declines in the native species typically began decades before the introductions of species such as cats and foxes (often reputed to be the causes of the extinctions), and the fact that species introductions are usually associated with other types of anthropegenic change that are believed to have contributed to the declines (for example land use change), it is difficult to ascribe extinctions of Australian mammals exclusively to non-native species (Abbott 2002; McKenzie et al. 2007). At the same time, it is likely that non-native species have contributed to some of the Australian extinctions, for example by causing local extinctions of small remnant populations created by drought or land use change (Morton

1990; McKenzie et al. 2007). I am not arguing that non-native species never cause extinctions on continents, just that they are rare and that, on a global scale, they are certainly not the second greatest extinction threat.

Wilcove et al. (1998) cannot be singled out as solely responsible for their preliminary and region-specific conclusion ascending to the status of ecological canon. After all, as shown above, they were careful to describe the limitations of their data. And the title of the article makes it clear that their focus was just regional (the USA), not global. However, the authors must shoulder part of the responsibility. For example, they concluded that 57% of imperilled US plants were threatened by predation or competition from alien species. Because predation is unlikely to be a common threat to plants, one must assume the authors meant to imply that most of the threat to native plants came from non-native plant species. However, it is widely known that the impacts of non-native plants on biodiversity are much less than those of non-native pathogens, herbivores and predators (Rejmánek et al. 2005b). Moreover, when the paper was written there was no evidence that a single native North American plant species had been driven to extinction, or even extirpated within a single US state, by competition from an introduced plant species (John T. Kartesz, Biota of North America Program, University of North Carolina, personal communication). Therefore, concluding, or implying, that non-native plant species threaten a large portion of the US flora with extinction seems quite unjustified. Moreover, the authors certainly would have known that the inclusion of Hawaii in their analyses significantly influenced the results.

Although Wilcove et al. should be held partly accountable for framing their conclusions as they did, the primary responsibility for their conclusion's ascendancy to boilerplate must lie with those who continued to cite the article's region-specific conclusion as a generally accepted global fact, even in the face of considerable and increasing evidence showing that non-native species do not represent a major extinction threat to most species in most environments, with the exception of islands and other insular environments (for example lakes and other freshwater systems), where there are many examples of introduced species causing extinctions. Even in some recent and prominent publications, Wilcove et al. (1998) has been cited as justification for making global statements about non-native species being one of the top two extinction threats, (see, for example, Perrings et al. 2005). In some cases, the statement that non-native species are the second most important cause of species extinctions is made without any citation at all, but simply stated as fact (see, for example, Shine et al. 2005).

There are many documented instances of non-native species causing extinctions in insular environments, particularly oceanic islands and freshwater lakes (Blackburn et al. 2004; Cox & Lima 2006; Sax & Gaines 2008). However, the data collected for terrestrial continental environments and marine systems overwhelmingly show that introduced species seldom drive native species to extinction. Without question, some introduced species in marine and continental terrestrial environments produce other sorts of consequences that may be deemed undesirable, including seriously disrupting ecological services and dramatically reducing the population sizes of some native species. These, and any other impacts that are well documented by actual data, are the effects that should be emphasized. It is time the field puts to rest once and for all any and all general claims that introduced invasive species represent the second greatest global extinction threat to imperilled native species, and instead focuses on the many other sorts of effects that non-native species are having, including impacts on biodiversity that do not involve actual extinctions (Gaertner et al. 2009; Davis 2009).

20.4 CONCLUSIONS

It is the nature of scientific disciplines to face challenges. Disciplines will be most successful if the primary challenges and obstacles they face are those complexities inherent of their subject of study. Disciplines become less effective if they are sidetracked or impeded by obstacles of their own making. The field's very close ties to community ecology when the modern field was emerging in the early 1980s provided opportunities and obstacles. Given the focus of community ecology at that time, the field of invasion biology focused considerable attention on species traits and local and deterministic factors. In particular, a niche-based approach guided much of the field's initial efforts to understand invasion establishment and what makes some environments more invasible than others. Although data frequently have not supported niche-based theories, and despite the fact that Williamson (1996) concluded in his comprehensive review of the field that a niche-based approach to understanding invasions seemed unlikely to offer much of value, this approach has continued to play a prominent role within the field. Unfortunately, this emphasis was, to some extent, at the expense of attention paid to the importance of regional and historical factors in the invasion process, which may have impeded the field's growth. Fortunately, the importance of historical and regional factors has been widely recognized in recent years and invasion biology of the future should be characterized by a more integrated approach, one that recognizes the importance of both local and regional processes.

The problem of preliminary conclusions and tentative statements being transformed into invasion gospel is a challenge the field of invasion biology needs to work hard at preventing. If invasion biology is to be a highly regarded scientific discipline, its primary assertions and conclusions need to be based on comprehensive and thoroughly vetted data sets. When conclusions are preliminary or based on data from a particular region, ecosystem or type of organism, they need to be presented as such, and those that cite these preliminary or limited conclusions need to portray them accurately as such. It is vitally important scientists police themselves in this regard. If we do not, then eventually others will, with a much greater negative and long-lasting impact on the field's credibility.

Fifty years after Elton's book, I believe that the field of invasion biology is undergoing considerable change as it develops into a more nuanced and less intellectually isolated discipline. This is partly due to the influx of many young investigators, who have been attracted to the intellectually rich and socially relevant field developed during the 1980s and 1990s. This bodes very well for the field's future. Disciplines begin to stagnate in the absence of new participants and perspectives (Reiners & Lockwood 2010). The influx of new minds and perspectives into the field is exactly what will ensure the field's vitality in the coming decades. The new investigators and new ideas represent opportunities, not unlike the way that introductions of species can provide the long-term residents with new ecological and evolutionary prospects.

REFERENCES

Abbott, I. (2002) Origin and spread of the cat, *Felis catus*, on mainland Australia, with a discussion on the magnitude of its early impact on native fauna. *Wildlife Research*, **29**, 51–74.

Allan, H.H. (1936) Indigen versus alien in the New Zealand plant world. *Ecology*, **17**, 187–193.

Baker, H.G. (1948) Stages in invasion and replacement demonstrated by species of Melandrium. *Journal of Ecology*, **36**, 96–119.

Barney, J.N. & Whitlow, T.H. (2008) A unifying framework for biological invasions: the state factor model. *Biological Invasions*, **10**, 259–272.

Belote, R.T., Jones, R.H., Hood, S.M. & Wender, B.W. (2008) Diversity–invasibility across an experimental disturbance gradient in Appalachian forests. *Ecology*, **89**, 183–192.

Blackburn, T.M., Cassey, P., Duncan, R.P., Evans K.L. & Gaston K.J. (2004) Avian extinction and mammalian introductions on oceanic islands. *Science*, **305**, 1955–1958.

Bruno, J.F., Fridley, J.D., Bromberg, K. & Bertness, M.D. (2005) Insights into biotic interactions from studies of species invasions. *Insights into Ecology, Evolution, and Biogeography* (ed. D.F. Sax, J.J. Stachowicz and S.D. Gaines), pp. 13–40. Sinauer Associates, Sunderland, Massachusetts.

Burdon, J.J. & Chilvers, G.A. (1977) Preliminary studies on a native Australian eucalypt forest invaded by exotic pines. *Oecologia*, **31**, 1–12.

Castle, D. (2005) Diversity and stability: theories, models, and data. *Ecological Paradigms Lost: Routes of Theory Change* (ed. K. Cuddington and B. Beisner), pp. 201–209. Elsevier, Amsterdam.

Catford, J.A., Jansson, R. & Nilsson, C. (2009) Reducing redundancy in invasion ecology by integrating hypotheses into a single theoretical framework. *Diversity and Distributions*, **15**, 22–40.

Colautti, R.I., Grigorovich, I.A. & MacIsaac, H.J. (2006) Propagule pressure: a null model for biological invasions. *Biological Invasions*, 1023–1037.

Craine, J.M. (2005) Reconciling plant strategy theories of Grime and Tilman. *Journal of Ecology*, **93**, 1041–1052.

Cox, J.G. & Lima, S.L. (2006) Naïveté and an aquatic–terrestrial dichotomy in the effects of introduced predators. *Trends in Ecology & Evolution*, **21**, 674–680.

Cuddington, K. & Beisner, B. (2005) Kuhnian paradigms lost: embracing the pluralism of ecological theory. *Ecological Paradigms Lost: Routes to Theory Change* (ed. K. Cuddington and B. Beisner), pp. 419–426. Elsevier/Academic Press, Oxford.

Davis, M.A. (2009) *Invasion Biology*. Oxford University Press, Oxford.

Davis, M.A., Grime, J.P. & Thompson, K. (2000) Fluctuating resources in plant communities: a general theory of invasibility. *Journal of Ecology*, **88**, 528–536.

Egler, F.E. (1942) Indigene versus alien in the development of arid Hawaiian vegetation. *Ecology*, **23**, 14–23.

Fargione, J., Brown, C.S. & Tilman, D. (2003) Community assembly and invasion: an experimental test of neutral versus niche processes. *Proceeding of the National Academy of Sciences of the USA*, **100**, 8916–8920.

Fargione, J. & Tilman, D. (2005) Diversity decreases invasion via both sampling and complementarity effects. *Ecology Letters*, **8**, 604–611.

Finlayson, H.H. (1961) On central Australian mammals, part IV. The distribution and status of central Australian species. *Records of the South Australian Museum*, **41**, 141–191.

Gaertner, M., Den Breeÿen, A., Hui, C. & Richardson D.M. (2009). Impacts of alien plant invasions on species richness in Mediterranean-type ecosystems: a meta-analysis. *Progress in Physical Geography*, **33**, 319–338.

Gitzen, R.A. (2007) The dangers of advocacy in science. *Science*, **317**, 748.

Graham, M.H. & Dayton, P.K. (2002) On the evolution of ecological ideas: paradigms and scientific progress. *Ecology*, **83**, 1481–1489.

Grime, J.P. (2007) Plant strategy theories: a comment on Craine (2005). *Journal of Ecology*, **95**, 227–230.

Jehlík, V. & Slavík, B. (1968) Beitrag zum Erkennen des Verbreitungscharakters der Art *Bunias orientalis* L. in der Tschechoslowakei. *Preslia*, **40**, 274–293.

Kinnear, J.E., Onus, M.L. & Sumner, N.R. (1998) Fox control and rock-wallaby population dynamics – II. An update. *Wildlife Research*, **25**, 81–88.

Lawton, J.H. (1999) Are there general laws in ecology? *Oikos*, **84**, 177–192.

Lawton, J.H. (2000) Community ecology in a changing world. *Excellence in Ecology Series*, vol. **11** (ed. O. Kinne). International Ecology Institute, Oldendorf/Luhe, Germany.

Levine, J.M., Adler, P.B. & Yelenik, S.G. (2004) A meta-analysis of biotic resistance to exotic plant invasions. *Ecology Letters*, **7**, 975–989.

Levins, R. (1966) The strategy of model building in population biology. *American Scientist*, **54**, 421–431.

Lockwood, J.L., Cassey, P. & Blackburn T.M. (2005) The role of propagule pressure in explaining species invasions. *Trends in Ecology & Evolution*, **20**, 223–228.

Lotze, H.K., Lenihan, H.S., Bourque, B.J., et al. (2006) Depletion, degradation, and recovery potential of estuaries and coastal seas. *Science*, **312**, 1806–1809.

McKenzie, N.L., Burbidge, A.A., Baynes, A., et al. (2007) Analysis of factors implicated in the recent decline of Australia's mammal fauna. *Journal of Biogeography*, **34**, 597–611.

Melbourne, B.A., Cornell, H.V., Davies, K.F., et al. (2007) Invasion in a heterogeneous world: resistance, coexistence or hostile takeover? *Ecology Letters*, **10**, 77–94.

Moles, A.T., Grubber, M.A.M. & Bonser, S.P. (2008) A new framework for predicting invasive plant species. *Journal of Ecology*, **96**, 13–17.

Morton, S.R. (1990) The impact of European settlement on the vertebrate animals of arid Australia: a conceptual model. *Proceedings of the Ecological Society of Australia*, **16**, 210–213.

Moyle, P.B. (1973) Effects of introduced bullfrogs, *Rana catesbeiana*, on the native frogs of the San Joaquin Valley, California. *Copeia*, **1973**, 18–22.

Naeem, S., Knops, J.M.H., Tilman, D., Howe, K.M., Kennedy, T. & Gale, S. (2000) Plant neighbourhood diversity increases plant resistance to invasion in experimental grassland plots. *Oikos*, **91**, 97–108.

Perrings, C., Dalmazzone, S., & Williamson, M. (2005) The economics of biological invasions. In *Invasive Alien Species: A New Synthesis* (ed. H.A. Mooney, R.N. Mack, J.A. McNeely, L E. Neville, P.J. Schei and J.K. Waage), pp. 16–35. Island Press, Washington, DC.

Reiners, W.A. & Lockwood, J.A. (2010) *Philosophical Foundations for the Practices of Ecology.* Cambridge University Press, Cambridge, UK.

Rejmánek, M., Richardson, D.M., Higgins, S.I., Pitcairn, M. J. & Grotkopp. E. (2005a) Ecology of invasive plants: state of the art. In *Invasive Alien Species: A New Synthesis* (ed. H.A. Mooney, R.N. Mack, J.A. McNeely, L.E. Neville, P.J. Schei & J.K. Waage), pp. 104–161. Island Press, Washington, DC.

Rejmánek, M., Richardson, D.M. & Pyšek, P. (2005b) Plant invasions and invasibility of plant communities. *Vegetation Ecology* (ed. E. Van der Maarel), pp. 332–355. Blackwell, Oxford.

Richardson, D.M. & Pyšek, P. (2006) Plant invasions: merging the concepts of species invasiveness and community invasibility. *Progress in Physical Geography*, **30**, 409–431.

Richardson, D.M. & Pyšek, P. (2008) Fifty years of invasion ecology: the legacy of Charles Elton. *Diversity and Distributions*, **14**, 161–168.

Richardson, D.M., Rouget, M., Ralston, S.J., Cowling, R.M., van Rensburg, B.J. & Thuiller, W. (2005) Species richness of alien plants in South Africa: environmental correlates and the relationship with indigenous plant species richness. *Ecoscience*, **12**, 391–402.

Ricklefs, R.E. (2006) Evolutionary diversification and the origin of the diversity–environment relationship. *Ecology*, **87**, S3–S13.

Sax, D.F. & Gaines, S.D. (2008) Species invasions and extinctions: the future of native biodiversity on islands. *Proceedings of the National Academy of Sciences of the USA*, **105**, 11490–11497.

Shea, K. & Chesson, P. (2002) Community ecology theory as a framework for biological invasions. *Trends in Ecology & Evolution*, **17**, 170–176.

Shine, C., Williams, N. & Burhenne-Guilmin, F. (2005) Legal and institutional frameworks for invasive-alien species. In *Invasive Alien Species: A New Synthesis* (ed. H.A. Mooney, R.N. Mack, J.A. McNeely, L.E. Neville, P.J. Schei & J.K. Waage), pp. 233–284. Island Press, Washington, DC.

Stachowicz, J., Fried, H., Osman, R.W. & Whitlatch, R.B. (2002) Biodiversity, invasion resistance, and marine ecosystem function: reconciling pattern and process. *Ecology*, **83**, 2575–2590.

Stohlgren, T.J., Barnett, D.T. & Kartesz, J.T. (2003). The rich get richer: patterns of plant invasions in the United States. *Frontiers in Ecology and the Environment*, **1**, 11–14.

Stohlgren, T.J., Barnett, D., Jarnevich, C.S., Flather, C. & Kartesz, J. (2008) The myth of plant species saturation. *Ecology Letters*, **11**, 313–322.

Stohlgren, T.J., Binkley, D., Chong, G.W., et al. (1999) Exotic plant species invade hot spots of native plant diversity. *Ecological Monographs*, **69**, 25–46.

Sukopp, H. (1962) Neophyten in naturlichen Pflanzengesellschaften Mitteleuropas. *Berichte der Deutschen Botanischen Gesellschaft*, **75**, 193–205.

Tilman, D. (2004) Niche tradeoffs, neutrality, and community structure: a stochastic theory of resource competition, invasion, and community assembly. *Proceedings of the National Academy of Sciences of the USA*, **101**, 10854–10861.

Venter, O., Brodeur, N.N., Nemiroff, L., Belland, B., Dolinsek, I.J. & Grant, J.W.A. (2006) Threats to endangered species in Canada. *Bioscience*, **56**, 903–910.

Wilcove, D.S., Rothstein, D., Dubow, J., Phillips, A. & Losos, E. (1998) Quantifying threats to imperiled species in the United States. *BioScience*, **48**, 607–615.

Williamson, M. (1996) *Biological Invasions*. Chapman & Hall, London.

Wiser, S.K. & Allen, R.B. (2006) What controls invasion of indigenous forests by alien plants? *Biological Invasions in New Zealand* (ed. R.B. Allen and W.G. Lee), pp. 195–209. Springer, Berlin.

Chapter 21

INVASIONS AND ECOSYSTEMS: VULNERABILITIES AND THE CONTRIBUTION OF NEW TECHNOLOGIES

Peter M. Vitousek[1], Carla M. D'Antonio[2] and Gregory P. Asner[3]

[1]Department of Biology, Stanford University, Stanford, CA 94305, USA
[2]Department of Ecology, Evolution and Marine Biology and Program in Environmental Studies, University of California, Santa Barbara, CA 93106, USA
[3]Department of Global Ecology, Carnegie Institution of Washington, Stanford, CA 94305, USA

Fifty Years of Invasion Ecology: The Legacy of Charles Elton, 1st edition. Edited by David M. Richardson
© 2011 by Blackwell Publishing Ltd

21.1   INTRODUCTION

Elton (1958) recognized the potentially transformative nature of biological invasions, referring to the modern acceleration of species introductions as 'a great historical convulsion' that might lead to 'a lost world' of biotic homogenization and lost biological diversity. More recent research has yielded abundant evidence that some introduced species profoundly affect the areas they invade, threatening human health and wealth, threatening the persistence of native species and communities, and altering the structure and functioning of whole ecosystems (Pyšek & Richardson 2010). In these ways, biological invasions represent a component of human-caused global change, one that interacts with other components of change (land use, climate, etc.) (Vitousek et al. 1997).

We focus here on consequences of invasions by plants, in particular for the functioning of ecosystems, and on the development of technologies based on remote sensing that can contribute to detecting, understanding and managing invasions that are capable of transforming ecosystems. This focus on ecosystem transformation postdates Elton (1958); it is a relatively new development in our analyses of invasions and their consequences.

Invasions that transform whole ecosystems are interesting for two major reasons. First, introduced species that affect the structure, hydrology, disturbance regime or biogeochemistry of the ecosystems they invade do more than just compete with or consume native species: they change the rules of existence for all of the species in the ecosystem. In so doing, they often facilitate further invasions or dramatically change the successional trajectory of a site or region. Moreover, by changing the fluxes of energy and materials across ecosystem boundaries, they can affect climate and influence downwind and downstream ecosystems, thereby interacting with other components of human-caused global change (D'Antonio & Vitousek 1992). Identifying the subset of potential invaders that is capable of transforming ecosystems thus should be useful to managers. Second, introduced species that change ecosystems do so as a consequence of physiological, population, community and often evolutionary processes. Understanding such invasions thus contributes to the conceptual and practical integration of ecology, across levels of biological organization (Vitousek 1990).

We recognize that every invasion is an interaction between a potential invader and the community it invades; consideration of both the characteristics that make a species an invader and characteristics that make a community invasible (by that species) have become central to invasion biology and community ecology (Richardson & Pyšek 2006). The pathways by which a subset of invasive species transforms ecosystems – the physiological and morphological properties that drive change, the properties of particular ecosystems that make them vulnerable to that particular change – generally represent an additional interaction beyond the demographic and community processes by which invasions occur.

Several ways in which invasions can alter ecosystem structure and functioning have been identified, including the following.

1 Changing the trophic structure of an ecosystem, most often by adding or eliminating a top predator and thereby initiating a substantial rearrangement of the food web. The invasion of the sea lamprey into the Upper Great Lakes of North America is one example of such a change (Mills et al. 1994).

2 Altering the habitat structure of an ecosystem by physically 'engineering' the substrate. The stabilization of sand dunes by introduced *Ammophila arenaria* on the west coast of the USA is one dramatic example (Wiedemann & Pickart 1996); the increased frequency of landslides in areas invaded by *Miconia calvescens* in the Society Islands may be another (Meyer 1996).

3 Increasing (or decreasing) resource availability within an ecosystem, either by changing the inputs or outputs of resources or by altering the rate at which resources become available. The invasion of young volcanic sites by the symbiotic nitrogen-fixing tree *Morella faya* in Hawai'i (discussed below) is an example of the former; the acceleration of nutrient cycling by the invasive shrub *Berberis thunbergii* in the understorey of deciduous forests in the eastern USA represents the latter (Ehrenfeld et al. 2001).

4 Changing the disturbance regime in an ecosystem, by adding a novel disturbance or altering the frequency or intensity of an existing pathway of disturbance. The invasion of seasonally dry woodlands and shrublands in Hawai'i and many other regions by fire-responsive grasses (discussed below) is one well-studied example (Hughes et al. 1991; D'Antonio et al. 2001; Mack et al. 2001).

In this chapter, we begin by evaluating characteristics that make ecosystems vulnerable to transformation by invasions. We make use of ecosystems of the Hawaiian Islands, where we have studied invasion and ecosystem functioning both separately and together. We then explore the development and application of new technologies in remote sensing that seek not only to detect the presence and distribution of important biological invaders, but also to measure aspects of their effects on ecosystems directly.

21.2 HAWAIIAN ECOSYSTEMS AND THEIR VULNERABILITIES

Because we are interested in invasions that transform ecosystems, we begin by exploring some relevant characteristics of ecosystems across the Hawaiian archipelago. We then ask what species' properties could transform a particular ecosystem, if a species with those properties reached that ecosystem and invaded it. Hawai'i is useful for this analysis because it supports a very wide array of ecosystems that differ in relatively well-understood ways within a small geographical area. Here we consider variation in ecosystems – and attendant vulnerabilities to invasion – along environmental gradients in substrate age and precipitation.

The Hawaiian Islands are being created by a plume of magma rising from Earth's mantle that has been feeding volcanoes for tens of millions of years. The northwesterly movement of the Pacific tectonic plate across this plume has led to a string of volcanoes and islands that increase progressively in age, from currently active Kilauea Volcano in the southeast to the approximately 5 million-year-old island of Kaua'i to the northwest (beyond which lies a much older string of volcanoes that are now reduced to atolls, and ultimately to submerged seamounts). In relatively high-rainfall (approximately $2500\,mm\,yr^{-1}$), mid-elevation (approximately 1200 m) regions along this age sequence, ecosystems on young volcanic substrates (younger than approximately 2000 years) are primary successional forests with developing soils. Plant productivity in these areas is strongly limited by the supply of fixed nitrogen (Vitousek et al. 1993; Harrington et al. 2001), which alone among the major plant nutrients is lacking in the volcanic parent material.

Ecosystems of intermediate-aged substrates (5000–200,000 years) occupy fertile soils with relatively high levels of nitrogen (accumulated from atmospheric deposition and dispersed sources of biological nitrogen fixation) and phosphorus (weathered from primary minerals into biologically accessible forms); these forests are productive, plant tissues are nutrient-rich, and only fertilization with both nitrogen and phosphorus can increase productivity (Crews et al. 1995; Vitousek & Farrington 1997). However, ecosystems on the oldest volcanic substrates (more than 1,000,000 years) occupy soils impoverished by sustained leaching that has depleted almost all of their rock-derived nutrients; their productivity is strongly limited by phosphorus supply (Herbert & Fownes 1995; Harrington et al. 2001), and indeed even their meagre supply of phosphorus depends primarily on long-distance transport of continental dust from central Asia (Chadwick et al. 1999; Kurtz et al. 2001).

Each of the major islands also supports a windward–leeward precipitation gradient, many of which are spectacular. For example, precipitation varies from 180 to $4500\,mm\,yr^{-1}$ on Kohala Volcano on the island of Hawai'i, in a linear distance of only 16 km. The drier leeward soils generally are more fertile than windward soils, due to less-intense leaching (Chadwick et al. 2003). Before the extensive land conversion that followed the Polynesian discovery of Hawai'i and later European colonization, areas with relatively low rainfall ($500–1000\,mm\,yr^{-1}$ on older substrates, extending to wetter areas on young substrates) supported open-canopied woodlands and shrublands (Chadwick et al. 2007), and those woodland areas have persisted in areas where soils are too thin to support agricultural development.

The matrix of ecosystems that we consider here is summarized in Table 21.1; it includes rainforests on young, intermediate-aged and old substrates, and open-canopied woodlands on young substrates. These systems differ substantially in structure and/or biogeochemistry, but they are dominated by the same native species (*Metrosideros polymorpha*; Myrtaceae) and exposed to many of the same potential invasive species. They therefore provide a useful background against which to evaluate invasions that could transform ecosystems. Assuming that demographic properties of the invader and properties of the community allow invasion to take place, what physiological or morphological

Table 21.1 Characteristics of the four Hawaiian ecosystems within which potential transformations resulting from biological invasions were evaluated. The first three occupy a substrate age gradient at constant rainfall (Vitousek 2004), whereas the fourth is a substantially drier woodland on a relatively young substrate (D'Antonio et al. 200).

Site	Substrate age (kyr)	Annual precipitation (mm yr^{-1})	Dominant native tree	Limiting resource	Invader(s) considered
Rainforest	0.3	2500	*Metrosideros*	N	*Morella faya*
Rainforest	20	2500	*Metrosideros*	N and P	*Fraxinus uhdei*
Rainforest	4100	2500	*Metrosideros*	P	*Acacia melanoxylon*
Seasonal woodland	0.5	1500	*Metrosideros*	Water, N	*Schizachyrium, Melinis*

characteristics of an invader could alter particular ecosystems from this array?

Young substrates and biological nitrogen fixers

Native forests developing in young volcanic substrates are strongly limited by nitrogen; the underlying lava supplies all other major plant nutrients, and plants with symbiotic nitrogen-fixing associations are sparse or absent on young substrates. (The absence of native symbiotic nitrogen fixers in these sites raises an interesting but difficult-to-answer why-not question, because actively fixing legumes are present in the native flora and abundant in many forests on older and drier substrates in Hawai'i (Pearson & Vitousek 2002; Baker et al. 2009).) It is reasonable to speculate that nitrogen deficiency makes these systems vulnerable to being transformed by an introduced species that can fix nitrogen symbiotically. Indeed several such invasions have been identified (Vitousek et al. 1987; Hughes & Denslow 2005; Kurten et al. 2008). The invasion of young volcanic sites in Hawai'i Volcanoes National Park by the actinorrhizal nitrogen-fixer *Morella faya* (syn. *Myrica faya*) is well-documented. *Morella* produces abundant bird-dispersed seeds (Vitousek & Walker 1989; Woodward et al. 1990), colonizes early successional open-canopied forests and woodlands and gaps within young primary forests in Hawai'i Volcanoes National Park, and develops nearly monospecific, closed-canopy stands (Vitousek & Walker 1989; Walker & Vitousek 1991). As discussed below, this invasion is detectable by advanced remote sensing technologies, even at its early stages (Asner & Vitousek 2005).

The transformative nature of *Morella* invasion in a young volcanic site is summarized in Fig. 21.1. In the absence of *Morella*, the growth of the native dominant tree is strongly limited by a lack of available nitrogen, as demonstrated by a threefold increase in growth after fertilization with nitrogen (and only nitrogen) (Fig. 21.1a). *Morella*'s fixation of atmospheric nitrogen adds about 20 kg of nitrogen per hectare per year to the site, compared with a total of approximately 10 kg per hectare per year from all other sources of fixed nitrogen (atmospheric deposition and dispersed sources of biological nitrogen fixation in lichens and decaying organic matter); so *Morella* invasion triples inputs of fixed nitrogen (Fig. 21.1b). Invasion by *Morella* increases the biological availability as well as the total quantity of nitrogen: net mineralization increases from near zero to 2.5 mg kg^{-1} per 14 days under *Morella*, meaning that the fixed nitrogen is available to other organisms in the ecosystem (Fig. 21.1c). Finally, invasion by *Morella* influences other species in the site, greatly increasing the abundance of invasive earthworms under *Morella* compared with under native *Metrosideros* (Fig. 21.1d) (Aplet 1990), and favouring colonization by a suite of other introduced plants when *Morella* is removed or dies (Adler et al. 1998; Loh & Daehler 2007). This and other young volcanic sites lack only nitrogen, among the major resources, and by supplying that nitrogen *Morella* transforms ecosystems.

Intermediate-aged substrates and positive feedbacks

Intermediate-aged sites are relatively rich in nutrients, and it would be difficult for introduced species to alter

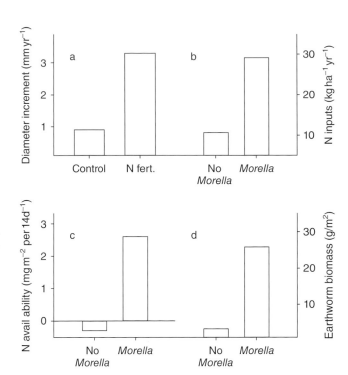

Fig. 21.1 Consequences of invasion by the biological nitrogen fixer *Morella faya* on young volcanic sites in Hawai'i Volcanoes National Park; data from Vitousek and Walker (1989) and Aplet (1990). (a) Fertilization with nitrogen demonstrates that nitrogen supply limits tree growth in young volcanic sites. (b) *Morella* invasion increases inputs of nitrogen threefold. (c) Increased nitrogen mineralization under *Morella* demonstrates that the nitrogen it fixes cycles rapidly into biologically available forms. (d) Populations of earthworms increase under *Morella* compared with native *Metrosideros polymorpha*. Colonization by other invasive plants and fluxes of nitrogen-containing trace gases are also enhanced by *Morella* invasion (Hall & Asner 2007; Loh & Daehler 2007).

these ecosystems by adding still more nutrients. It is possible, however, that an invader could alter these ecosystems by affecting the rate at which nutrients cycle rather than by altering nutrient inputs. The native *Metrosideros polymorpha* trees growing on fertile intermediate-aged sites have higher tissue nutrient concentrations, more rapid leaf turnover and more rapid rates of litter decomposition and nutrient cycling than conspecifics growing on younger (or much older) substrates (Vitousek et al. 1995; Hobbie & Vitousek 2000); they contribute to a positive feedback in which nutrients are not only more abundant, but also cycle more rapidly, on intermediate-ages substrates (Vitousek 2004, 2006). Could an invading species that is adapted to fertile soils in its native range display similar properties to an even greater extent, and so drive a stronger plant–soil–microbial positive feedback than does *Metrosideros*?

The Mexican tree *Fraxinus uhdei* was planted 60 or more years ago in intermediate-aged substrates near one of our long-term research sites in Hawai'i, and it has invaded into the forest surrounding these plant-

ings. *Fraxinus* produces thinner, higher-nutrient litter with lower concentrations of recalcitrant carbon compounds than does *Metrosideros* in comparably rich sites (Rothstein et al. 2004). *Fraxinus* also grows larger, casts deeper shade (as is readily detectable by satellite imagery (unpublished observations) as well as from below the canopy) and supports higher productivity. Finally, *Fraxinus* litter decomposes more rapidly and regenerates available nitrogen and phosphorus more quickly than does *Metrosideros* (Fig. 21.2) (Rothstein et al. 2004), suggesting that *Fraxinus* drives a positive feedback towards more rapid nutrient cycling in intermediate aged, nutrient-rich sites even more effectively than do the *Metrosideros* populations adapted to rich sites.

Similar feedbacks that speed up or slow down rates of nutrient cycling could represent a widespread mechanism by which invaders alter ecosystem functioning, at times in ways that facilitate the success of the invader itself. Baruch and Goldstein (1999) demonstrated that many invaders in Hawai'i have leaf and physiological traits that could speed up nitrogen

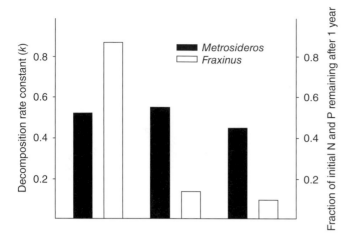

Fig. 21.2 Decomposition and nutrient regeneration from invasive *Fraxinus uhdei* compared with native *Metrosideros polymorpha* trees in a site derived from relatively rich, intermediate-aged substrate on Mauna Kea, Hawai'i. Decomposition rate is summarized by the first-order decay constant k; nutrient retention in decomposing litter is reported as the fraction of the initial nitrogen and phosphorus present in senescent litter that remains within the litter after one year of decomposition. Data from Rothstein et al. (2004).

cycling compared with natives in the same sites, and Allison and Vitousek (2004) documented rapid rates of decomposition and nutrient regeneration in the litter from many of those invaders.

Old substrates and deep roots?

Metrosideros-dominated forests on stable geomorphic surfaces of the oldest islands are profoundly depleted in most nutrients, and their productivity is strongly limited by the supply of phosphorus. Weatherable minerals that contain phosphorus are present at several meters depth in these soils, as indicated by isotopic signatures of rock-derived strontium deep in soil profiles and by a rejuvenation in the supply of rock-derived strontium and phosphorus on eroded slopes and depositional areas within ancient landscapes (Porder et al. 2005; Chadwick et al. 2009). Could a deeply rooted invader obtain access to these nutrients, and bring them to the surface where they would become available to other organisms? Certainly some tropical species can send roots to great depths (more than 15 m) through acidic and aluminium-rich soils (Nepstad et al. 1994); could one of them invade and transform these ecosystems?

Several woody plants have invaded the ancient landscape near our long-term research site on the Island of Kaua'i, Hawai'i, including *Morella faya* and *Psidium cattleianum*. The densest stands of these species occur on erosional and depositional areas that are enriched in rock-derived nutrients. However, the

Australian legume *Acacia melanoxylon* has invaded the stable geomorphic surface near our site. We evaluated whether *Acacia* could obtain its nutrients from deep soil horizons by analysing the $^{87}Sr/^{86}Sr$ ratio in its foliage; atmospheric deposition of marine aerosol and weathering of Hawaiian basalt yield strontium with strong and highly repeatable differences in isotopic ratios (Kennedy et al. 1998). We found that strontium isotope ratios in *Acacia* reflect those in atmospheric deposition, not rock weathering (unpublished data), and concluded that it is not obtaining nutrients from relatively rich sources deep in the soil profile. Perhaps this pathway – deep-soil mining – by which an invader could transform resource supply in ecosystems is not available here; more likely, an invader that could transform ecosystems by this pathway has not (yet) become established in this area.

Seasonally dry environments and the grass-fire feedback

Moderately dry areas across the Hawaiian archipelago support open-canopied woodlands, with an overstorey of trees (often *Metrosideros polymorpha* on young substrates) and an understorey of shrubs and scattered native grasses. Most such areas were converted to agricultural uses by the Polynesian discoverers of Hawai'i, or by later European colonizers, but until recently young substrates with shallow soils have continued to be dominated by native woody plants. These areas now have been invaded by a suite of introduced perennial

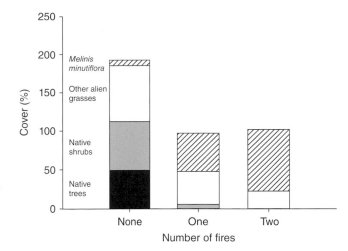

Fig. 21.3 Invasion by alien grasses enhances fire spread and severity in the seasonal submontane zone of Hawai'i Volcanoes National Park, and thereby alters ecosystem structure, function and composition. One post-invasion fire eliminates native trees and reduces the cover of native shrubs; it also favours invasion by the fire-enhancing African grass *Melinis minutiflora*. Subsequent fires eliminate shrubs and reinforce dominance by *Melinis*. Data from Hughes et al. (1991).

grasses, some of which were brought in to support cattle grazing. These grasses create a continuous layer of fine fuel in the understorey of seasonally dry ecosystems – facilitating the spread of fire and the transformation of woodland ecosystems to frequently burned grasslands. These dynamics are by no means unique to the Hawaiian archipelago (D'Antonio & Vitousek 1992), but they are stark and clear there.

We evaluated these dynamics in detail in a seasonal submontane environment in Hawai'i Volcanoes National Park. This area and the adjacent seasonally dry lowlands were invaded by the American perennial grasses *Andropogon virginicus* and *Schizachyrium condensatum* in the late 1960s; before that time, fires were infrequent and small (Tunison et al. 2001). Thereafter, fires became both more frequent and far larger; 7800 hectares burned in the 19 years after grass invasion, compared with 2.3 hectares in the 48 years before invasion. In the submontane seasonal zone, a single grass-fuelled fire sufficed to kill most of the overstorey trees and greatly reduced the cover and diversity of native shrubs (Fig. 21.3). Grass cover was enhanced after fire, and the invasive African grass *Melinis minutiflora* became an important component of the ecosystem, rising to dominance after fire (Hughes et al. 1991). The resulting simplified system is more susceptible to subsequent fires., In turn, the fires are more destructive because *Melinis* is highly competitive against native species, dries out more quickly and ignites more readily than the woodland species it replaces, and the grassland microclimate (especially

higher wind speeds near the homogeneous surface formed by mats of *Melinis*) favours the spread of fire (Freifelder et al. 1998; D'Antonio et al. 2001). Subsequent fires remove the vestiges of native shrubs from the system, and reinforce the dominance of *Melinis* (Fig. 21.3). The resulting grassland differs in many fundamental respects from the woodland it replaces – including its structure, biogeochemistry (Ley & D'Antonio 1998; Mack et al. 2001; Mack & D'Antonio 2003a,b), disturbance dynamics and biodiversity (Hughes et al. 1991). These changes in ecosystem structure and flammability can be detected – and quantified – by advanced remote sensing technologies, as discussed below. This situation represents a clear example of how invasions can transform ecosystems by altering their disturbance regimes.

These same grass invasions and subsequent fires occur in the nearby coastal lowlands, but fires there do not cause the same level of decline in native species cover and richness (D'Antonio et al. 2000). This contrast likely occurs because a few of the native species in the lowlands are tolerant of and responsive to fire. This illustrates the importance of the interplay between invader traits and the recipient ecosystem (including residents and their traits) in determining ecosystem consequences of invasion. In the submontane burned areas, an important part of grass/fire impacts results from the loss of native species, leading to lower productivity, altered nitrogen cycling patterns and greater potential for ecosystem nitrogen loss (Mack et al. 2001).

The grass invasion/fire cycle and subsequent changes in ecosystem structure and functioning occur in continental as well as insular systems (see, for example, D'Antonio & Vitousek 1992; Brooks et al. 2004; Bradley & Mustard 2005). Some continental invaders, such as the annual grass *Bromus tectorum*, alter fire regimes in a range of environments in the western USA (Mack, this volume). Although this species is generally considered to be an ecosystem transformer, its impacts are not the same everywhere (Haubensak et al. 2009). They may be less permanent where fire responsive native species are present or where temporal heterogeneity in resources allows greater opportunity for native species regeneration. Whether these fires are transformative also interacts with post-fire management and land use (see, for example, Eiswerth & Shonkwiler 2006).

Ecosystem vulnerability to transformation by invasion

This Hawai'i-centric discussion of interactions between the properties of ecosystems and those of invaders that are capable of altering ecosystems suggests that the approach has promise for identifying the types of ecosystem that are vulnerable to transformation by particular types of invader. Because many successful invaders have broad ranges in their 'home lands' (Rejmánek 1996), they often invade across a range of habitats where they have been introduced. However, most likely they will only have strong ecological effects in a subset of those habitats, and understanding that subset is critical to targeted control or eradication. Most studies of ecosystem transformation have tended to focus on sites where impacts are large and easy to demonstrate. Although it is critical to document such impacts, a deep and predictive understanding of why certain ecosystems are vulnerable to certain invaders can best be gained by studying a range of invaded sites, and/or by studying the invader within the context of properties of the recipient ecosystem.

21.3 TECHNOLOGIES TO DETECT ECOSYSTEM-TRANSFORMING INVASIONS

Predicting which invaders have the capability to alter particular types of ecosystem is a scientific challenge (and opportunity); however, identifying invasions that are in fact changing ecosystems may be of more practical value to managers, particularly where detection can occur early in the invasion process. Remote sensing offers one promising approach to detecting biological invasions; aircraft- and satellite-based sensors can provide repeated coverage over wider and less accessible areas than can ground-based surveys. However, so far most sensors have provided information that is either too coarse in spatial resolution, or too sparse in information content, to provide detailed information on the early stages of invasions (Huang & Asner 2009).

Nevertheless, a variety of remote-sensing approaches have proved useful in evaluating invasions. These include the following: aerial photography in situations where the brightness or spatial patterning of the invader is distinct from local native species (see Everitt et al. 1995; Müllerová et al. 2005; Fuller 2005); coarse-resolution satellite sensors with frequent temporal coverage (e.g. advanced very high resolution radiometer, moderate resolution imaging spectrometer), where the invader differs in phenology from native species (Bradley & Mustard 2005); active light detection and ranging (LiDAR), where the invader alters structural features of the plant canopy or the substrate (Rosso et al. 2006; Asner et al. 2008a); and, increasingly, hyperspectral imaging spectrometers that are capable of determining detailed spectral profiles of native and introduced plants (Williams & Hunt 2002; Underwood et al. 2003; Pengra et al. 2007).

The combination of LiDAR to assess structure and hyperspectral imaging to evaluate chemistry represents a promising path towards the ability to detect invasions and to analyse their effects on ecosystems remotely. The Carnegie Airborne Observatory (CAO; http://cao.stanford.edu/) is a step on this path; it integrates these sensors with advanced sensor guidance and trajectory sub-systems that provide three-dimensional positioning and attitude data for the entire sensor package onboard the aircraft (Asner et al. 2007).

The CAO has been used actively to detect invasions and to evaluate their effects on ecosystems in the Hawaiian Islands. Hyperspectral data (abbreviated to HiFIS, for high-fidelity imaging spectrometer) were used to estimate upper canopy leaf nitrogen concentrations (percentage nitrogen on a dry-mass basis) and whole canopy water contents (millimetres per unit area), and canopy height profiles (understorey as well as overstorey) were also measured directly with LiDAR

(Asner & Vitousek 2005; Vitousek et al. 2009). This combination of detailed information on canopy height and vertical distribution with direct measurements of foliar nitrogen and canopy water provides a highly resolved sample of important structural and chemical properties of ecosystems across landscapes. Where these attributes are affected by invasion, the remotely derived information not only detects the presence of invasions, but it can also determine at least some of their ecosystem-level consequences, as these are distributed in space and ultimately in time.

In Hawai'i, measurements of canopy nitrogen and water focused initially on the young volcanic sites invaded by *Morella faya*, as discussed above; *Morella* has much higher canopy nitrogen and water content than native *Metrosideros polymorpha*, and the CAO imagery detected the presence of *Morella* and its influence on forest canopies across the landscape surrounding a long-term research site, even at early stages of invasion (Asner & Vitousek 2005) (illustrated for canopy nitrogen content on colour plate 7).

Airborne spectroscopic analysis also revealed areas in which canopy nitrogen was low but water content moderately high. Both LiDAR and ground-based examinations of those areas demonstrated that the introduced large herb Kahili ginger (*Hedychium gardnerianum*) dominated their understories. More interestingly, a careful examination of the HiFIS information demonstrated that upper-canopy (*Metrosideros*) nitrogen concentrations were slightly but significantly lower in the presence of *Hedychium*, an observation that later was confirmed by ground-based sampling (Asner & Vitousek 2005). Although we believe that this effect of *Hedychium* on *Metrosideros* forest primarily represents a competitive rather than an ecosystem-level effect, the ability of high-resolution remote sensing

to detect a previously unsuspected biological interaction (through its consequences) is encouraging.

Subsequent mapping studies focused on detecting invasive species using HiFIS, and then applying LiDAR to measure the effects of these species on the vertical structure of Hawaiian ecosystems (Asner et al. 2008a,b). The invaders analysed included two nitrogen-fixing trees (the legume *Falcataria moluccana* and *Morella faya*), the rich-site tree *Fraxinus uhdei* discussed above, the understorey tree *Psidium cattleianum* and the large herb *Hedychium gardnerianum*. The first three are known to have biogeochemical effects on ecosystems (Vitousek & Walker 1989; Rothstein et al. 2004; Hughes & Denslow 2005), and the LiDAR-based analysis demonstrated that they affect ecosystem structure as well (Asner et al. 2008a). This effect is illustrated for *Falcataria* in Fig. 21.4; *Falcataria* has a taller canopy than the native dominant *Metrosideros*, and one in which most leaves occurs in a single layer; its presence facilitates *Psidium* invasion in the mid-canopy, which in turn decreases light to low levels at the ground level (Hughes & Denslow 2005). Moreover, the very high spatial resolution of LiDAR sampling (approximately 1 m) and its ability to 'see' beneath the canopy means that detecting any invader that differs in structure from native species is straightforward, even at early stages of invasion.

LiDAR also can be used to detect the consequences of grass invasion into the submontane seasonal woodlands, as discussed above (Fig. 21.3). Although a range of remote sensing approaches can detect that grass-invasion-enhanced fires have altered the structure of these ecosystems, LiDAR can provide direct analyses of the fuel loading associated with grass invasion, and so can measure the likelihood of subsequent fires (Varga & Asner 2008).

Fig. 21.4 Vertical structure of the forest canopy along a transect in low-elevation native forest on the Island of Hawai'i, derived from LiDAR measurements by the CAO. Native forest dominated by *Metrosideros polymorpha* is on the left, whereas areas invaded by the taller legume *Falcataria moluccana* are on the right. The distinct mid-canopy layer in the *Falcataria*-invaded forest is the invasive strawberry guava, *Psidium cattleianum*, which is facilitated by *Falcataria* invasion.

Overall, the integration of LiDAR and HiFIS on aircraft is under development; it is not yet widely available. As the technology develops, it is likely to revolutionize how we detect and analyse invasions that can transform ecosystems, with direct value for ecology and cascading value for conservation and management. Already, the CAO allows us to detect invasions and analyse at least some of their ecosystem consequences across the gradients in substrate age and climate discussed above – and it does so on the scale of whole landscapes rather than particular research sites (Elmore & Asner 2006; Martin & Asner 2009; Vitousek et al. 2009; 2010). We suspect that our research questions – our abilities to use this detailed, often process-level landscape information productively – have not yet caught up with developments in the technology.

The practical use of the new aircraft-based imagery is clear as well. During its studies of Hawaiian ecosystems, the CAO has become a source of geospatial information on invasive species to state and federal land-management agencies operating throughout the region. It has detected previously unknown or under-appreciated outbreaks of some of the Hawaiian Islands' most problematic invaders, particularly in remote areas, and its imagery of invasive species has made its way to government decision makers and the public eye, providing a three-dimensional picture of change that most non-scientists can understand. As it becomes more available, the CAO and other forthcoming airborne systems offer the promise that developing technologies will assist us in the detection, analysis and management of ecosystem-transforming invasions, an area in which we need all the help we can get.

ACKNOWLEDGEMENTS

The research discussed here was supported by grants from the National Science Foundation, most recently DEB-0716852, -0715593, -0717382, and -0715674, by NASA Terrestrial Ecology Program-Biodiversity grant NNG-06-GI-87G, by the Andrew W. Mellon Foundation, and by the Carnegie Institution. The Carnegie Airborne Observatory is made possible by the W.M. Keck Foundation and William Hearst III. David Richardson and two anonymous reviewers provided useful comments on an earlier version of this manuscript.

REFERENCES

Adler, P., D'Antonio, C.M. & Tunison, J.T. (1998) Understory succession following a dieback of *Myrica faya* in Hawaii Volcanoes National Park. *Pacific Science*, **52**, 69–78.

Allison, S.D. & Vitousek, P.M. (2004) Rapid nutrient cycling in leaf litter from invasive species in Hawai'i. *Oecologia*, **141**, 612–619.

Aplet, G.H. (1990) Alteration of earthworm community biomass by the alien *Myrica faya* in Hawai'i. *Oecologia*, **82**, 414–416.

Asner, G.P. & Vitousek, P.M. (2005) Remote analysis of biological invasion and biogeochemical change. *Proceedings of the National Academy of Sciences of the USA*, **102**, 4383-4386.

Asner, G.P., Knapp, D.E., Jones, M.O., et al. (2007) Carnegie Airborne Observatory: in-flight fusion of hyperspectral imaging and waveform light detection and ranging (wLiDAR) for three-dimensional studies of ecosystems. *Journal of Applied Remote Sensing*, **1**, doi:10.1117/1.2794018.

Asner, G.P., Hughes, R.F., Vitousek, P.M., et al. (2008a) Invasive plants alter 3-D structure of rainforests. *Proceedings of the National Academy of Sciences of the USA*, **105**, 4519–4523.

Asner, G.P., Knapp, D.E., Kennedy-Bowdoin, T., et al. (2008b) Invasive species detection in Hawaiian rainforests using airborne imaging spectroscopy and LiDAR. *Remote Sensing of Environment*, **112**, 1942–1955.

Baker, P.J., Scowcroft, P.G. & Ewel, J.J. (2009) *Koa (Acacia koa) Ecology and Silviculture*. General Technical Report PSW-GTR-211. US Dept of Agriculture, Forest Service, Pacific Southwest Station, Albany, California.

Baruch, Z. & Goldstein, G. (1999) Leaf construction costs, nutrient concentration and net CO_2 assimilation of native and invasive species in Hawaii. *Oecologia*, **121**, 183–192.

Bradley, B.A. & Mustard, J.F. (2005) Identifying land cover variability distinct from land cover change: cheatgrass in the Great Basin. *Remote Sensing of Environment*, **94**, 204–213.

Brooks, M.L., D'Antonio, C.M., Richardson, D.M., et al. (2004) Effects of invasive alien plants on fire regimes. *Bioscience*, **54**, 677–688.

Chadwick, O.A., Derry, L.A., Vitousek, P.M., Huebert, B.J. & Hedin, L.O. (1999) Changing sources of nutrients during four million years of ecosystem development. *Nature*, **397**, 491–497.

Chadwick, O.A., Gavenda, R.T., Kelly, E.F., et al. (2003) The impact of climate on the biogeochemical functioning of volcanic soils. *Chemical Geology*, **202**, 193–221.

Chadwick, O.A., Kelly, E.F., Hotchkiss, S.C., & Vitousek, P.M. (2007) Precontact vegetation and soil nutrient status in the shadow of Kohala Volcano, Hawaii. *Geomorphology*, **89**, 70–83.

Chadwick, O.A., Derry, L.A., Bern, C.R. & Vitousek, P.M. (2009) Sources of strontium to soil minerals and ecosystems across the Hawaiian Islands. *Chemical Geology*, **267**, 64–76.

Crews, T.E., Kitayama, K., Fownes, J., et al. (1995) Changes in soil phosphorus and ecosystem dynamics across a long soil chronosequence in Hawai'i. *Ecology*, **76**, 1407–1424.

D'Antonio, C.M. & Vitousek, P.M. (1992) Biological invasions by exotic grasses, the grass-fire cycle, and global change. *Annual Review of Ecology and Systematics*, **23**, 63–87.

D'Antonio, C.M, Tunison, J.T. & Loh, R. (2000) Variation in impact of exotic grass fueled fires on species composition across an elevation gradient in Hawai'i. *Austral Ecology*, **25**, 507–522.

D'Antonio, C.M., Hughes, R.F. & Vitousek, P.M. (2001) Factors influencing the dynamics of two invasive C_4 grasses in seasonally dry Hawaiian woodlands. *Ecology*, **82**, 89–104.

Ehrenfeld, J.G., Kourtev, P. & Huang, W. (2001) Changes in soil functions following invasions of exotic understory plants in deciduous forests. *Ecological Applications*, **11**, 1287–1300.

Eiswerth, M.E. & Shonkwiler, J.S. (2006) Examining post-wildfire reseeding on arid rangeland: a multivariate tobit modelling approach. *Ecological Modelling*, **192**, 286–298.

Elmore, A.J. & Asner, G.P. (2006) Effects of deforestation and grazing intensity on soil carbon stocks of Hawaiian dry tropical forest. *Global Change Biology*, **12**, 1761–1772.

Elton, C.S. (1958) *The Ecology of Invasions by Animals and Plants*. Methuen, London.

Everitt, J.H., Anderson, G.L., Escobar, D.E., Davis, M.R., Spencer, N.R. & Andrascik, R.J. (1995) Use of remote sensing for detecting and mapping leafy spurge (*Euphorbia esula*). *Weed Technology*, **9**, 599–609.

Freifelder, R., Vitousek, P.M. & D'Antonio, C.M. (1998) Microclimate change and projected effect on fire following forest-grass conversion in seasonally dry tropical woodland. *Biotropica*, **30**, 286–297.

Fuller, D.O. (2005) Remote detection of invasive melaleuca trees (*Melaleuca quinquenervia*) in South Florida with multispectral IKONOS imagery. *International Journal of Remote Sensing* **26**, 1057–1063.

Hall, S.J. & Asner, G.P. (2007) Biological invasions alter regional N-oxide emissions in Hawaiian rain forests. *Global Change Biology*, **13**, 2143–2160.

Harrington, R.A., Fownes, J.H. & Vitousek, P.M. (2001) Production and resource-use efficiencies in N- and P-limited tropical forest ecosystems. *Ecosystems*, **4**, 646–657.

Haubensak, K.A., D'Antonio, C.M. & Wixon, D. (2009) Effects of fire and environmental variables on plant structure and composition in grazed salt desert shrublands of the Great Basin (USA). *Journal of Arid Environments*, **73**, 643–650.

Herbert, D.A. & Fownes, J.H. (1995) Phosphorus limitation of forest leaf area and net primary productivity on a weathered tropical soil. *Biogeochemistry*, **29**, 223–235.

Hobbie, S.E. & Vitousek, P.M. (2000) Nutrient regulation of decomposition in Hawaiian forests. *Ecology*, **81**, 1867–1877.

Huang, C. & Asner, G.P. (2009) Applications of remote sensing to alien invasive plant studies. *Sensors*, **9**, 4869–4889.

Hughes, R.F., Vitousek, P.M. & Tunison, J.T. (1991) Effects of invasion by fire-enhancing C_4 grasses on native shrubs in Hawaii Volcanoes National Park. *Ecology*, **72**, 743–747.

Hughes, R.F. & Denslow, J.S. (2005) Invasion by an N_2-fixing tree alters function and structure in wet lowland forests of Hawai'i. *Ecological Applications*, **15**, 1615–1628.

Kennedy, M.J., Chadwick, O.A., Vitousek, P.M., Derry, L.A. & Hendricks, D. (1998) Replacement of weathering with atmospheric sources of base cations during ecosystem development, Hawaiian Islands. *Geology*, **26**, 1015–1018.

Kurten, E.L., Snyder, C.P., Iwata, T. & Vitousek, P.M. (2008) *Morella cerifera* invasion and nitrogen cycling on a lowland Hawaiian lava flow. *Biological Invasions*, **10**, 19–24.

Kurtz, A.C., Derry, L.A. & Chadwick, O.A. (2001) Accretion of Asian dust to Hawaiian soils: isotopic, elemental, and mineral mass balances. *Geochimica et Cosmochimica Acta*, **65**, 1971–1983.

Ley, R. & D'Antonio, C.M. (1998) Exotic grasses alter rates of nitrogen fixation in seasonally dry Hawaiian woodlands. *Oecologia*, **113**, 179–187.

Loh, R.K. & Daehler, C.C. (2007) Influence of invasive tree kill rates on native and invasive plant establishment in a Hawaiian forest. *Restoration Ecology*, **15**, 199–211.

Mack, M. & D'Antonio, C.M. (2003a) Exotic grasses alter controls over soil nitrogen dynamics in a Hawaiian woodland. *Ecological Applications*, **13**, 154–166.

Mack, M. & D'Antonio, C.M. (2003b) Direct and indirect effects of introduced C4 grasses on decomposition in a Hawaiian woodland. *Ecosystems*, **6**, 503–523.

Mack, M.C., D'Antonio, C.M. & Ley, R.E. (2001) Pathways through which exotic grasses alter N cycling in a seasonally dry Hawaiian woodland. *Ecological Applications*, **11**, 1323–1335.

Martin, R.E. & Asner, G.P. (2009) Leaf biochemical and optical properties of *Metrosideros polymorpha* across environmental gradients in Hawai'i. *Biotropica*, **41**, 292–301.

Meyer, J.-Y. (1996) Status of *Miconia calvescens* (Melastomataceae), a dominant invasive tree in the Society Islands (French Polynesia). *Pacific Science.* **50**, 66–76.

Mills, E.L., Leach, J.H., Carlton, J.T. & Secor, C.L. (1994) Exotic species and the integrity of the Great Lakes: lessons from the past. *BioScience*, **44**, 666–676.

Müllerová, J., Pyšek, P., Jarošík, V. & Pergl, J. (2005) Aerial photographs as a tool for assessing the regional dynamics of the invasive plant species *Heracleum mantegazzianum*. *Journal of Applied Ecology*, **42**, 1042–1053.

Nepstad, D.C., de Carvalho, C.R., Davidson, E.A., et al. (1994) The role of deep roots in the hydrological and carbon cycles of Amazonian forests and pastures. *Nature*, **372**, 666–669.

Pearson, H.L. & Vitousek, P.M. (2002) Nitrogen and phosphorus dynamics and symbiotic nitrogen fixation across a substrate-age gradient in Hawaii. *Ecosystems*, **5**, 587–596.

Pengra, B.W., Johnston, C.A. & Loveland, T.R. (2007) Mapping an invasive plant, *Phragmites australis*, in coastal wetlands using the EO-1 Hyperion hyperspectral sensor. *Remote Sensing of Environment*, **108**, 74–81.

Porder, S., Paytan, A. & Vitousek, P.M. (2005) Erosion and landscape development affect plant nutrient status in the Hawaiian Islands. *Oecologia*, **142**, 440–449.

Pyšek, P. & Richardson, D.M. (2010) Invasive species, environmental change and management, and ecosystem health. *Annual Review of Environment and Resources*, **35**, doi:10.1146/annurev-environ-033009-095548 (in press).

Rejmánek, M. (1996) A theory of seed plant invasiveness: the first sketch. *Biological Conservation* **78**, 171–181.

Richardson, D.M. & Pyšek, P. (2006) Plant invasions – merging the concepts of species invasiveness and community invasibility. *Progress in Physical Geography*, **30**, 409–431.

Rosso, P.H., Ustin, S.L. & Hastings, A. (2006) Use of lidar to study changes associated with *Spartina* invasion in San Francisco Bay marshes. *Remote Sensing of Environment*, **100**, 295–306.

Rothstein, D.E., Vitousek, P.M. & Simmons, B.L. (2004) An exotic tree accelerates decomposition and nutrient turnover in a Hawaiian montane rainforest. *Ecosystems*, **7**, 805–814.

Tunison, J.T., Loh, R. & D'Antonio, C.M. (2001) Fire, grass invasions and revegetation of burned areas in Hawaii Volcanoes National Park. In *Proceedings of the Invasive Species Workshop: The Role of Fire in the Control and Spread of Invasive Species* (ed. K.E. Galley and T.P. Galley), pp. 122–131. Tall Timbers Research Station Publication No. 11, Allen Press, Lawrence, Kansas.

Underwood, E., Ustin, S. & DiPietro, D. (2003) Mapping non-native plants using hyperspectral imagery. *Remote Sensing of Environment*, **86**, 150–161.

Varga, T.A. & Asner, G.P. (2008) Hyperspectral and LiDAR remote sensing of fire fuels in Hawaii Volcanoes National Park. *Ecological Applications*, **18**, 613–623.

Vitousek, P.M. (1990) Biological invasions and ecosystem processes: towards an integration of population biology and ecosystem studies. *Oikos*, **57**, 7–13.

Vitousek, P.M. (2004) *Nutrient Cycling and Limitation: Hawai'i as a Model System*. Princeton University Press, Princeton.

Vitousek, P.M. (2006) Ecosystem science and human–environment interactions in the Hawaiian Archipelago. *Journal of Ecology*, **94**, 510–521.

Vitousek, P.M. & Farrington, H. (1997) Nutrient limitation and soil development: experimental test of a biogeochemical theory. *Biogeochemistry*, **37**, 63–75.

Vitousek, P.M. & Walker, L.R. (1989) Biological invasion by *Myrica faya* in Hawai'i: plant demography, nitrogen fixation, and ecosystem effects. *Ecological Monographs*, **59**, 247–265.

Vitousek, P.M., Walker, L.R., Whiteaker, L.D., Mueller-Dombois, D. & Matson, P.A. (1987) Biological invasion by *Myrica faya* alters ecosystem development in Hawaii. *Science*, **238**, 802–804.

Vitousek, P.M., Walker, L.R., Whiteaker, L.D. & Matson, P.A. (1993) Nutrient limitation to plant growth during primary succession in Hawaii Volcanoes National Park. *Biogeochemistry*, **23**, 197–215.

Vitousek, P.M., Turner, D.R. & Kitayama, K. (1995) Foliar nutrients during long-term soil development in Hawaiian montane rain forest. *Ecology*, **76**, 712–720.

Vitousek, P.M., D'Antonio, C.M., Loope, L.L., Rejmánek, M. & Westbrooks, R. (1997) Introduced species: a significant component of human-caused global change. *New Zealand Journal of Ecology*, **21**, 1–16.

Vitousek, P.M., Asner, G.P., Chadwick, O.A. & Hotchkiss, S.C. (2009) Landscape-level variation in forest structure and biogeochemistry across a substrate age gradient in Hawai'i. *Ecology*, **90**, 3074–3086.

Vitousek, P.M., Tweiten, M.A., Kellner, J., Hotchkiss, S.C., Chadwick, O.A. & Asner, G.P (2010). Top-down analyses of forest structure and biogeochemistry across Hawaiian landscapes. *Pacific Science*, **64**, 359–366.

Walker, L.R. & Vitousek, P.M. (1991) Interactions of an alien and a native tree during primary succession in Hawaii Volcanoes National Park. *Ecology*, **72**, 1449–1455.

Wiedemann, A.M. & Pickart, A. (1996) The *Ammophila* problem on the northwest coast of North America. *Landscape and Urban Planning*, **34**, 287–299.

Williams, A.P. & Hunt, E.R. Jr (2002) Estimation of leafy spurge cover from hyperspectral imagery using mixture tuned matched filtering. *Remote Sensing of Environment*, **82**, 446–456.

Woodward, S.A., Vitousek, P.M., Matson, K.A., Hughes, R.F., Benvenuto, K. & Matson, P.A. (1990) Avian use of the introduced nitrogen-fixer *Myrica faya* in Hawaii Volcanoes National Park. *Pacific Science*, **44**, 88–93.

Chapter 22

DNA BARCODING OF INVASIVE SPECIES

Hugh B. Cross[1], *Andrew J. Lowe*[1]
and C. Frederico D. Gurgel[1,2]

[1]State Herbarium of South Australia, Science Resource Centre, Department of
Environment and Natural Resources and Australian Centre for Evolutionary Biology
and Biodiversity, School of Earth and Environmental Sciences, University of Adelaide,
North Terrace, SA 5005, Australia
[2]South Australian Research and Development Institute, West Beach, Adelaide, Australia

Fifty Years of Invasion Ecology: The Legacy of Charles Elton, 1st edition. Edited by David M. Richardson

22.1   INTRODUCTION

A new international initiative has the potential to transform research into invasive species. By providing a gigantic resource of universally applicable biological information, DNA barcoding efforts will have the ability to fill in many pieces of the invasive species puzzle. One of the greatest challenges in handling what Elton termed 'ecological explosions' (Elton 1958, p. 15), is assessing the unknown. Even though the native flora and fauna of a particular area may be well documented, when a new invasive species appears, little, if anything, may be known about it. From the moment of detection, information on which species it is, where it has come from or even how long it has been there is usually lacking. Yet all this information is crucial for assessing the threat of an invader and formulating strategies to manage it. Especially critical will be the ability to assess the potential impact on native ecosystems. As Elton wrote, 'we require fundamental knowledge about the balance between populations' (Elton 1958, p. 110) to assess the threat of invasive species properly.

Yet how to obtain this knowledge? In a shrinking world, knowledge is more broadly available and access to biological information about an invasive species' native range is becoming easier. However, despite this increase in communication, research in widely separated places may entail quite different types of data, and even differences in taxonomy, that cannot be easily compared from one place to another. The central goal of DNA barcoding – to use the same genetic regions to obtain unique DNA sequences for each species on Earth – will enable researchers and officials anywhere to compare their native and invasive biota with habitats from all over the world. What is perhaps more important, such a standardized system can be put in the hands of custom agents, park rangers and other officials, in both rich and poor countries, who are on the front lines of the prevention and detection of invasive species. Though these goals still face many challenges, the impact of a DNA-barcoded world will be substantial for research into invasive species, as outlined below.

22.2   AN OVERVIEW OF DNA BARCODING

DNA barcoding is an international initiative that began in 2003 with the idea of developing a standardized method for species identification through the comparative analysis of short DNA sequences (around 650 nucleotides long (Hebert et al. 2003a)). Since then, DNA barcoding has gained momentum around the world. DNA-based identification was not a new technique when DNA barcoding was proposed, as molecular markers to identify and distinguish between species had been developed for many groups of organisms. What DNA barcoding contributes is the idea of having a universal standard molecular marker to identify all species around the world, and a central database that combines complete specimen information with DNA sequence data (Janzen 2004; Hebert & Gregory 2005). There have been attempts to identify a single 'standard' marker for various groups, especially for microorganisms. Microbiologists have long used DNA sequences of the small (16S) and large (23S) subunits of the ribosomal genes as their 'gold standard' for species identification (Sogin et al. 2006; Quince et al. 2009; or see Singh et al. 2009 for review), and mycologists have used the nuclear ribosomal internal transcribed spacer (ITS) as an unofficial standard for identification of fungi (Seifert 2009). The DNA barcoding projects seek to standardize these markers (or agree on a standard among various options). This work has important implications for research into invasive species, because such a system could provide rapid and accurate genetic identification of invasive species and allow comparison across threatened areas and even with the invader's native range (Armstrong & Ball 2005; Darling & Blum 2007; Ivanova et al. 2009).

From initial discussions and preliminary research, two major organizations have arisen that are dedicated to advancing DNA barcoding. The Consortium on the Barcoding of Life (CBOL) (www.barcoding.si.edu), was financed by the Sloane Foundation and the Smithsonian Institution, and consists of over 170 partner organizations around the world. The Consortium has as its primary mission 'to promote the exploration and development of DNA barcoding as a global standard for species identification' and advances this mission by supporting meetings and campaigns around DNA barcoding projects centred on specific groups of taxa or geographic regions. The other major organization, the International Barcoding of Life Project (iBOL, www.ibolproject.org/home.html), based in Guelph, Canada, is committed to assembling and compiling a database of the sequence library for all species, and has received substantial support from Genome Canada to support its mission. Through iBOL's online database system, the Barcoding of Life Database (BOLD) (www.boldsystems.org), over 800,000 barcode records have been assembled for species from every kingdom of life, and from all parts of the world. New projects are begun everyday, including some dedicated to the identification of invasive species.

With coordination between CBOL and iBOL, as well as funding for targeted projects from national and international agencies, DNA barcoding is rapidly developing a large repository of information that has the potential to be an excellent tool for research into invasive species. A current limitation of this tool for full DNA barcode identifications is the limited taxonomic coverage in the BOLD database. New species are continually being added to the database; however, the emphasis on quality, accuracy and completeness of data for each species implies that for some groups it will take some time to have DNA barcodes for every species. In contrast to other publically available DNA sequence databases, such as GenBank, the BOLD database requires complete specimen data, including locality and full taxonomic information (with the possible exception of environmental samples: see below). In addition, DNA barcodes are preferably developed using multiple individuals for each species, which also offers the potential to screen infraspecific variation or tackle taxonomic issues. Thus, although slowing down the accumulation of a complete DNA barcode sequence library, it means that once complete the genetic identification from BOLD will be more accurate.

The ultimate goal of DNA barcoding – and one that may have the most benefit for research into invasive species and monitoring – is the development of a hand-held device, coined the 'DNA barcorder' (Janzen et al. 2005), that will enable on-the-spot, rapid identifications of species. This device is not yet a reality, but technical advancements in molecular biology (see, for example, 'Next-generation sequencing, DNA barcoding and the identification of invasive species' in section 22.6) as well as ongoing research hold promise for the future. As this technology develops, it will prove a boon to monitoring and detecting invasive species. Identification of potential invaders will be easy and fast, not only for scientists but also field agents at borders and airports, inspectors in the horticultural, bushmeat and pet trades, and park rangers. It will provide this at a lower cost in the long run, as little taxonomic expertise will be required for identification. This is especially important for countries with limited budgets for basic scientific research.

22.3   THE SEARCH FOR A UNIVERSAL BARCODE: ONE GENE TO RULE THEM ALL

The original goal of DNA barcoding was to have a single gene region that would provide a unique sequence to differentiate among all species across all life. However, this could not be done as genomes vary considerably across different phyla. The first efforts to develop a standardized system were done on lepidopterans (see Hebert et al. 2003a), using the mitochondrial gene cytochrome oxidase subunit I (COI). This gene has been used as a useful phylogenetic and even phylogeographic tool for a multitude of animal studies. Indeed, it has become the official DNA barcoding marker for most animal groups (see Hebert et al. 2003b). However, COI was found to be unsuitable for other groups, such as plants. In plants, the mitochondrial genome in general and COI in particular have significantly lower mutation rates than in animals (although exceptions exist, see Tomaru et al. (1998)), providing very poor resolution between species. In some fungi large insertions of several hundred base pairs in this gene (Seifert 2009) made COI too difficult to use.

Owing to multiple problems with COI in plants and fungi, other markers were considered that would provide species discrimination across the tree of life (see, for example, Kress et al. 2005, Kress & Erickson 2008). One of the prime candidates for a universal marker was the nuclear ribosomal ITS. As part of a gene region that codes for ribosomes that are fundamental housekeeping genes for all cells, the ITS region is found in all organisms and has highly conserved flanking areas that provide for near universal primers (White et al. 1990). Furthermore as a non-coding spacer region, the ITS is highly variable and has been used as a species-specific marker in a wide array of groups, including plants, fungi, algae and animals (see Kress et al. 2005). However, in many groups ITS has presented problems. The fact that this is not a protein-coding gene, its highly variable nature, the possibility of recombination, removal of information due to concerted evolution, the presence of several copies per genome with occasional intragenomic genetic variation, and the presence of large, sometimes repeated indels (Cho et al. 2009) often makes working with this marker difficult (especially when working with species-rich genera, large phylogenies and comparing evolutionarily distant species). For example, alignment among several phylogenetically distant species within the same genus is often not possible. Nevertheless, the ITS has been successfully applied as either the primary or as an alternative DNA barcode for many groups (particularly fungi and some plant groups see Chen et al. 2010; Seifert 2009).

When it became clear that no single gene was going to work across the vast diversity of life on Earth, it was agreed to have separate multiple DNA barcodes for different groups. The objective of barcoding was still a single gene within each major group, with secondary genes to help discriminate among speciose clades, thus preserving the benefit of having a universal (but group-specific) standard. For animals, COI was retained. For fungi (generally kingdom Fungi and assorted other mycota) two main markers are under consideration: ITS and the adjacent region, the ribosomal 25S long subunit (LSU) (Seifert 2009), both of which have worked well for fungi (although they are not without their problems (Vialle et al. 2009)). Algae comprise an artificial group of phylogenetically unrelated autotrophic organisms. For the different lineages of algae, more specifically the red and brown algae, COI, the chloroplast *rbcL* and others have been considered (Saunders 2005, 2009; Lane et al. 2007). Higher plants have proved to be among the most difficult, as no single locus was found to work across all groups. Several loci were trialled (see, for example, Kress et al. 2007; Plant Working Group 2009), and finally a combination of two chloroplast loci, maturase K (*matK*) and *rbcL*, were found to give the greatest coverage for the most groups of plants (Plant Working Group 2009). Although only about 70% of species could be distinguished using these two loci, additional chloroplast genes did not greatly increase the resolution (but see Fazekas et al. 2008). For plants, the next step will be to establish additional loci that will provide the missing resolution within major groups of plants, as so-called 'local barcodes'.

22.4 A STANDARD REFERENCE FOR CROSS STUDY AND MULTI-REGIONAL COMPARISONS

Creating a DNA barcode library requires matching a unique DNA sequence to a species, which in turn depends on taxonomically well-identified reference specimens. Such requirements emphasize the need for taxonomists to be involved in DNA barcoding projects, especially for megadiverse and taxonomically challenging groups. Only with solid taxonomy can we fully realize the potential of DNA barcoding to act as an identification and discovery tool. As an identification tool, DNA barcodes present researchers with the ability to validate the taxonomy of the species they are working on. As a discovery tool, one of the potentially most powerful applications of DNA barcoding for research on invasive species is the detection and identification of cryptic species, or distinct subspecies lineages (Holland et al. 2004).

DNA barcoding provides a *lingua franca* to identify positively all specimens of a species, and this ensures that the same species is being compared across different regions, no matter what the taxonomic discrepancies. Armstrong and Ball (2005) compared methods for invasive species identification, and found that using DNA barcoding was superior to previous methods involving molecular markers. Not only could DNA barcodes be used to identify the species in other areas, but they also picked up additional taxa that could be then placed in a phylogenetic context. Scheffer et al. (2006) point out, however, that with incomplete sequence datasets for the taxa under study, the number of species present could be overestimated or underestimated. As these gaps in data are filled, this will become less of an issue (Scheffer et al. 2006).

Often a single species will exhibit pronounced morphological variation, and here DNA barcoding can also help resolve questions of whether there are one or several species within and across the range of phenotypes. DNA barcoding, of course, will not resolve all taxonomic issues in difficult groups. However, from the perspective of managing invasive species, identifying taxa based on the barcode identification number (BIN) of BOLD will provide a more precise genetic reference (and thus a more powerful tool) to identify introduced species and invasive and naturalized taxa correctly, and distinguish them from close relatives. DNA barcoding will also allow the identification of invasive species at a lower cost than conventional methods. As an invasive species is by nature from 'somewhere else', there is usually no local taxonomic expertise and it can be quite time consuming and expensive to identify foreign species properly.

22.5 GENETIC IDENTIFICATION ABOVE AND BELOW THE SPECIES LEVEL

For reliable and accurate DNA barcodes, it is crucial that the selected gene regions reflect the species relationships of the groups under study. Of course, the rate of speciation will not always match the degree of genetic differentiation. This is the difficulty of having a 'one size fits all' approach. A locus may either be too slowly evolving (as is common for plant chloroplast markers) or too rapid (leading to homoplasy and generating problems of interpretation). For a genus undergoing rapid radiation, the selected loci may not reflect the speciation pattern as accurately as more slowly evolving taxa. In addition, hybridization and interspecific gene flow will interfere with the genetic boundaries between species and the ability to DNA barcode (especially those based on a single marker). This is particularly a problem for the plant chloroplast genome, which is often exchanged between species as a result of chloroplast capture (Okuyama et al. 2005). These factors can be especially problematic for invasive species that are introduced to areas that contain close relatives with which they can hybridize. In light of these biological realities, it is important that DNA barcoding projects take account of taxonomy, biogeography and population dynamics of species, and for invasive species this applies to related species in both their native and exotic ranges.

For many organisms for which the levels of genetic diversity are high, DNA barcodes are not suitable for phylogenetic studies, as fast-evolving genetic changes will not reliably reflect the relationships among higher levels of taxa. In those cases, contrasts between intraspecific and interspecific genetic variation are often performed (and necessary) so that cut-off levels of genetic distance are determined. In red algae for example, absolute sequence divergence within species are typically below 1.8–2.0% for the large subunit of the ribulose-1,5-bisphosphate carboxylase oxygenase gene (RuBisCO) (Gurgel & Fredericq 2004). As a tool for population genetics, when DNA barcode data are variable enough to highlight information below the species level, it may be possible to identify the source of introduction and population dynamics of the invasive species (see Kang et al. 2007; Prentis et al. 2009; Thomsen et al. 2006). Phylogeographic and population genetic methods open up a wide range of opportunities for ecologists to infer and test recent and past evolutionary hypotheses such as estimates of effective population size (presence of population bottlenecks), size and/or frequency of original introduction(s), comparisons of genetic diversity between invasive and source populations, and synergistic interactions between hosts and pathogens (Wingfield et al., this volume). DNA barcode data can also be used to assess

whether invasive populations are composed of more than one native source (admixed populations) or have hybridized with native species in the new invasive range (although nuclear sequence data or a combination of nuclear and organellar markers will be needed to identify hybrids and in some cases determine distinct genotypes from the native range). Both mechanisms have been linked with 'super-invasibility' (Prentis et al. 2008; Wilson et al. 2009; Dormontt et al., this volume).

22.6 APPLICATIONS OF DNA BARCODING FOR INVASIVE SPECIES RESEARCH

The ability to identify species and potentially source origin using DNA barcoding techniques, even from morphologically compromised samples, highly degraded tissue or fragmentary DNA samples and even organism by-products (e.g. faeces, leather, timber), offers a host of new applications in invasion science.

Early and rapid identification of invasive species

One of the greatest benefits of using DNA barcodes in invasive species programmes is that they offer a means of early and rapid identification of new invasive species. Rapid genetic identification of invasive species, without requirement for any morphologically mature characters, whether vegetative and reproductive, means that invasive species can be detected at any stage of their life cycles.

Morphological identification of species often requires survey of specimens in mature, reproductive or advanced life stages, which are often only available at certain times of year or under specific climatic conditions. Proper identification of plant species (algae and fungi included) usually requires flowering or fruiting specimens, especially in areas where closely related native species occur. In addition, some invasive species of mushrooms are not only restricted by season, but also by climate. The appearance of fruiting bodies will depend on the amount of rain in a given season; many seasons can pass without the appearance of a species. The abundance of invasive mushrooms underground and their effect on native species will sometimes not be apparent for many years. Yet, these species will continue to grow and spread through the soil (Pringle et al. 2009).

Furthermore, the process of morphological identification can be time consuming. Although some species can be distinguished immediately, in some cases experts require several hours or days per specimen to ensure that it is the correct species. This is especially true in the case of algae or fungi, where microscopic preparations are often necessary to observe key cryptic and subtle morphological characters. Moreover, many closely related plant and insect species are also difficult to tell apart. When specific life stages, particular gender or ephemeral body parts are not available, this task is often impossible. In the case of cryptic species, where there are few or no physical differences between species, then genetic identification is the only way to tell them apart.

Early detection offers several advantages. Areas of high occurrence of the invasive can be identified before they reach reproductive maturity, and eradication efforts can begin before the species has become fully established. By the time an invasive plant is detected in its flowering phase, it may have distributed pollen and seed for the next generation, or even begun to hybridize with closely related species. Through genetic identification of agricultural or horticultural seed or vegetative stocks, invasive species can be detected before they are even planted. The speed of identification will increase as technology is improved, such as with the DNA barcorder, as mentioned above.

DNA barcoding has been used successfully to identify the early life stages of invasive species. For example, Armstrong and Ball (2005) found the COI barcode marker to be superior to previous molecular assays in identifying species of moth from eggs found on imported cars. Their work also found that this barcoding marker was able to identify a wider taxonomic range of foreign species than previous methods. In a similar study, Chown et al. (2008) were able to identify specimens of an invasive moth on Marion Island that were in the young, immature life stages. In a study of the invasive plant, *Hydrocotyle ranunculoides*, in the Netherlands, a candidate DNA barcoding marker was able to distinguish between the invasive and closely related native species that are very difficult to tell apart based solely on morphology (van de Wiel et al. 2009).

Data retrieved from historical collections and degraded tissue

The ability to DNA barcode samples where morphological identification is not possible or unreliable frees us from the limitations of morphological identification. Species can be distinguished from historical collections, environmental samples and many cases where the organism has been processed or degraded to such an extent that it is unrecognizable. For research into invasive species, the ability to identify species from older material can add a historical perspective to a study. For example, the timing and distribution of the introduction of invasive species can be important for estimating how fast and from where the species originally spread. Pringle et al. (2009) used DNA barcoding markers on historical herbarium specimens of fungus to identify the first occurrences of the invasive species *Amanita phalloides* in California. Molecular identification was necessary as fungi lose characters once dried, hence morphological distinction of closely related species using archived material becomes unreliable. Their research showed that *A. phalloides* was introduced on imported tree roots as early as the 1930s, on at least two separate occasions (Pringle et al. 2009). Another example of the use of archived material to piece together introduction history was the estimation of oldest introduction of the marine green algal weed *Codium fragile* (Provan et al. 2008). DNA sequences from old herbarium collections were able to assist in determining the oldest date of introduction on record for this marine seaweed in Europe and Australia, which dated back to 18th century European voyages around the world.

The field of ancient DNA has contributed much to these applications of DNA barcoding, both in terms of protocols to obtain good DNA and the precautions necessary when handling compromised material. The main concern for genetic work on degraded tissue of any age is that samples can be easily contaminated. Typically these tissues yield such a small quantity of DNA that any foreign DNA introduced to the mixture can quickly dominate any analysis. Therefore, genetic analysis of degraded tissue requires special decontamination procedures and isolation from other genetic laboratory research (Cooper & Poinar 2000). Another major contribution from ancient DNA research is that degraded tissue will often only yield very short DNA sequences (generally less than 200 base pairs, or in very old or degraded material, less than 100 base pairs), and so protocols have to be adjusted to compensate for this.

As standard DNA barcoding loci are generally longer than 500 base pairs (Hebert et al. 2003a), to DNA barcode degraded tissue it is necessary to redesign standard protocols and primers to sequence either portions of the standard barcoding marker, or obtain the entire length by sequencing from multiple, overlapping polymerase chain reaction (PCR) products. The development of mini-barcodes has proved helpful in this case (Hajibabaei et al. 2006; Meusnier et al. 2008). Obtaining useable sequences will require the redesign of primers, which in many cases will need to be specific to a much narrower range of taxa than the standard DNA barcoding primers. This obviously reduces the 'universality' that is another strength of DNA barcoding protocols, but for specific projects targeting an invasive species and close relatives the information retrieved can still be compared with a reference sequence.

In some cases internal primers have been shown to be nearly as universal as the standard DNA barcoding markers, and have been used for historical and ancient DNA work. Within the *trnL* intron – a commonly used chloroplast marker – primers around the highly variable P6 loop within this locus can be amplified in most plants, and can often distinguish among closely related species. Taberlet et al. (2007) recommended this region as an alternative for DNA barcoding of degraded samples. They tested this region in specific cases of identifying plant species from processed foods, and had generally good results. One issue with the P6 loop is that it is difficult to amplify in some plant groups owing to extensive mononucleotide repeat regions, which can also make sequence reads unreliable and hence the identification suspect. In addition, the region can vary widely in length: from the typical 60–70 base pairs up to almost 200 base pairs in some plant groups (see Taberlet et al. 2007). However, despite these restrictions, these and similar primer combinations are nearly universal across the plant kingdom and have been used in numerous ancient DNA studies. The chloroplast region *rbcL*, which has also been proposed as a plant DNA barcode Plant Working Group 2009), has been used in many studies of ancient plant DNA, including determining a mammoth's diet (van Geel et al. 2008) and ecological studies of palaeoenvironments (Willerslev et al. 2007).

Another region that has potential for producing short barcodes of approximately 200–350 base pairs is the nuclear ribosomal internal transcribed spacer (ITS) region, which is split in two by the highly conserved 5.8S ribosomal gene, providing a natural region for universal internal primers. Often, however, the different segments of ITS on either side of this region (ITS 1 or ITS 2) can be quite long (i.e. more than 500 base pairs). The ITS region is the primary candidate to be the standard DNA barcoding marker for fungi, and has been proposed as an alternate or additional locus for plants (see Kress et al. 2005; Chen et al. 2010).

The central challenge when using smaller fragments of molecular markers is to identify sub-regions of the DNA barcodes that are capable of delivering the same taxonomic resolution using a portion of the DNA sequence. Hajibabaei et al. (2006) trialled portions of the CO1 gene on museum specimens of moths and wasps, and found them to be two to three times more successful at recovering sequences from degraded tissue than full-length barcodes. Furthermore, the mini-barcodes were able to distinguish among as many species as the full-length barcode, with one exception. This is an important point, as a shorter sequence naturally means that there are fewer diagnostic characters with which to make an identification. Although most species could be resolved with the mini-barcodes, this was achieved with less than half as many informative characters, which lowers the confidence of assignment. For the studies of *rbcL* ancient DNA mentioned above (Willerslev et al. 2007; van Geel et al. 2008), resolution was generally only to genus level or above. Nevertheless, for ecological studies and in areas with low species diversity, this level of resolution may be sufficient. In addition, these studies can be bolstered with additional 'mini-barcodes' to increase resolution and support for identification.

Next-generation sequencing, DNA barcoding and the identification of invasive species

For many of the proposed applications of DNA barcoding for invasive species, the amount of DNA sequence data necessary to generate background reference collections will be costly and time consuming to obtain. However, newer sequencing technologies will allow rapid assessment of multiple markers for a fraction of the cost of traditional Sanger sequencing (Margulies et al. 2005; Hall 2007; Rokas & Abbot 2009). The new techniques, collectively called 'next-generation sequencing technologies', will enable whole ecosystems – from all environmental substrates within a region – to be DNA barcoded rapidly and thoroughly. The main technologies currently in use are 'pyrosequencing', or 454 sequencing (454 Life Sciences, Branford, Connecticut, USA), Illumina or Solexa (Solexa, Cambridge, UK) and SOLiD (sequencing by oligonucleotide ligation and detection) by Applied Biosystems (San Francisco, California, USA).

Pyrosequencing and Illumina are the two most commonly used of these technologies. Both work on a principle of massive parallel sequencing, in which fragments of DNA are attached to a substrate and sequenced in a fixed place while a laser reads signals from the independent cells or beads (depending on the particular method) every time a new nucleotide is added to the growing sequence. The resulting sequences are relatively small, but the size is compensated by the immense number of sequence reads generated in a single run. A single 454 run generates over 300,000 sequence reads, averaging about 450 base pairs per read. The reads from the Illumina are much shorter – around 50 base pairs – but a standard run generates substantially more reads, numbering in the tens of millions. Both of these technologies have been further developed, and the length and quantity of reads is expected to increase exponentially in the near future. An even newer technique is being developed by Pacific Biosciences, which will produce even longer reads and allow for single-molecule sequencing, thus bypassing preliminary PCR preparation steps and their associated problems.

The potential applications for next-generation sequencing technologies are enormous; whole genomes can be sequenced very rapidly, and the technology can be applied to scan genomes for polymorphic regions (Swaminathan et al. 2007; van Orsouw et al. 2007; Santana et al. 2009) and other potential gene regions that provide variability at species level (Noor & Feder 2006). Specifically for DNA barcoding and research on invasive species, new markers for species identification and exploration of population genetics can be realized. For groups such as plants there is a great need to develop markers that provide species-level resolution, and for shorter segments of DNA that can be used as mini-barcodes for degraded DNA. Moreover, it is possible to generate genetic information concurrently, not only for the putative invasive species of interest but also for its entire associated microbiota (e.g. plant leaf sample and its associated epiphytic, endophytic and endocytic fungal flora; termites and their associated yeast-related gut flora; seaweeds and their associated epiphytic bacterial flora).

Perhaps the greatest benefit of next generation sequencing is the ability to sequence massive amounts of PCR product in one run, enabling rapid biodiversity assessments of environmental samples and other mixed substrates. Thus multiple barcoding markers from many groups can be sequenced at once, providing not only the identification of invasive species but also an evaluation of the impact of invasives on native macro- and micro- fauna and flora. The amount of data from this type of study has provided a scope of the unseen biodiversity in ecosystems. Many new species have been discovered from such environmental studies, for which there are no physical specimens (Venter et al. 2004; Porter et al. 2008; Quince et al. 2009). Despite the lack of a voucher, it has been argued (T. Bruns, personal communication) that these 'genetic species' should still be catalogued and barcoded, as they can then be compared with data from other ecosystems, again using the barcoding database as a standard reference. The use of DNA barcoding can identify these undescribed species and by clustering them phylogenetically, as is done in the BIN system of BOLD, the threat potential of these species can be better assessed. These ecological studies provide a glimpse of unseen diversity – thanks to high throughput sequencing technology – and can be used to discover invasive taxa that we did not even know about.

22.7   CONCLUSION: DNA BARCODING OFFERS A BRIDGE BETWEEN PHYLOGEOGRAPHY, INVASIVE SPECIES BIOLOGY AND SYSTEMATICS

The intrinsic nature of information contained in highly informative and comprehensive DNA barcode data will provide a bridge between the diverse, yet related, fields of phylogeography, population genetics, taxonomy, systematics, conservation biology and ecology. Comprehensive and multi-marker based DNA barcode datasets contain spatial, historical, taxonomic, evolutionary and temporal information. By providing an international standard reference of data that is cumulative, retrievable, comparable among closely related species, and that holds evolutionary information built up over the past two million to six million years, DNA barcode datasets can generate a more complete picture of the biology of an invasive species not only at their site of introduction but also from their native range.

In addition, more generally we feel that DNA barcode datasets have so far been relatively poorly exploited. DNA barcoding studies often explore one or a few of these dimensions: taxonomy and biodiversity (Hebert et al. 2004), invasive species recognition (Saunders 2009), mating systems (Gomez et al. 2007) or phylogeography (Gurgel et al. 2004). However, the combination of DNA barcoding data with distributional and abiotic data can be used to build predictive models on how, when and where a particular invasive is likely to disperse and invade (see Verbruggen et al. (2009) for a non-invasive species).

The limitations of single DNA barcodes (such as introgression (Rubinoff 2006)) can be overcome by the addition of more data, sampling from different genomes (nuclear, mitochondrial and chloroplast), different gene regions (coding versus non-coding) and using loci with different evolutionary rates. If a single 650 base-pair universal marker has its limitations as described above, freedom from these limitations comes from using two or (far-) more regions with differing levels of genetic variation. The next generation DNA sequencing technologies will be able to produce much more robust datasets (for example, DNA 'barcodes' based on 7–10 genes). So far, the barcoding community have been relatively slow to adopt the new genomics techniques, and the lack of interaction between DNA barcode applications, molecular taxonomy and genomics has been noted (Hajibabaei et al. 2007). Such new genomic information will present new problems due to the daunting task of making sense of massive amounts of new information. However, certain features of genomic organization are inherited thus complementary to a DNA barcode approach. These include gene location (gene order, direction and strand position), gene copy number and overall genomic size (i.e. intraspecific variation in ploidy number). Another potential source of mini-barcode data includes expressed sequence tags, which can allow expressed genes to be probed revealing information on an organism's physiology, environmental adaptations, stress levels, presence of parasites and reproductive status. If we are indeed on the brink of a new genomics era in DNA barcoding, then it is critical that we spend the time now to properly collect reference specimens, curate collections and set up systematic systems that will help progress DNA barcoding methods in a genomics era. As DNA barcoding advances, its contribution to invasive species research, monitoring and detection will increase substantially.

ACKNOWLEDGEMENTS

We thank numerous people whose suggestions and discussions contributed to our understanding of DNA barcoding and research into invasive species, including members of the Australian Barcode Network, the Grass Barcoding of Life Project, the Tree Barcoding of Life Project, the Australian Centre for Ancient DNA, and especially Maria Kuzmina and Paul Hebert and their colleagues at the University of Guelph and the International Barcoding of Life Project. Additionally, we are grateful to Dave Richardson and two anonymous reviewers who provided invaluable suggestions to earlier drafts of this chapter.

REFERENCES

Armstrong, K.F. & Ball, S.L. (2005) DNA barcodes for biosecurity: invasive species identification. *Philosophical Transactions of the Royal Society B*, **360**, 1813–1823.

Chen, S., Yao, H., Han, J., et al. (2010) Validation of the ITS2 region as a novel DNA barcode for identifying medicinal plant species. *PLoS ONE*, **5**, e8613.

Cho, G.Y., Choi, D.W., Kim, M.S. & Boo, S.M. (2009) Sequence repeats enlarge the internal transcribed spacer 1 region of the brown alga *Colpomenia sinuosa* (Scytosiphonaceae, Phaeophyceae). *Phycological Research*, **57**, 242–250.

Chown, S., Sinclair, B. & van Vuuren, B. (2008) DNA barcoding and the documentation of alien species establishment on sub-Antarctic Marion Island. *Polar Biology*, **31**, 651–655.

Cooper, A. & Poinar, H. (2000) Ancient DNA: do it right or not at all. *Science*, **289**, 1139.

Darling, J.A. & Blum, M.J. (2007) DNA-based methods for monitoring invasive species: a review and prospectus. *Biological Invasions*, **9**, 751–765.

Elton, C.S. (1958) *The ecology of invasions by animals and plants.* Methuen, London.

Fazekas, A.J., Burgess, K.S., Kesanakurti, P.R., et al. (2008) Multiple multilocus DNA barcodes from the plastid genome discriminate plant species equally well. *PLoS ONE*, **3**, e2802.

Gomez, A., Wright, P.J., Lunt, D.H., Cancino, J.M., Carvalho, G.R. & Hughes, R.N. (2007) Mating trials validate the use of DNA barcoding to reveal cryptic speciation of a marine bryozoan taxon. *Proceedings of Royal Society London B*, **274**, 199–207.

Gurgel, C.F.D. & Fredericq, S. (2004) Systematics of the Gracilariaceae (Gracilariales, Rhodophyta): a critical assessment based on *rbc*L sequence analysis. *Journal of Phycology*, **40**, 138–159.

Gurgel C.F.D., Fredericq S. & Norris, J.N. (2004) Phylogeography of *Gracilaria tikvahiae* (Gracilariaceae, Rhodophyta): a study of genetic discontinuity in a continuously distributed species. *Journal of Phycology*, **40**, 748–758.

Hajibabaei, M., Singer, G.A., Hebert, P.D. & Hickey, D. (2007) DNA barcoding: how it complements taxonomy, molecular phylogenetics and population genetics. *Trends in Genetics*, **23**, 167–172.

Hajibabaei, M., Smith, M., Janzen, D. H., Rodriguez, J., Whitfield, J. & Hebert, P. (2006) A minimalist barcode can identify a specimen whose DNA is degraded. *Molecular Ecology Notes*, **6**, 959–964.

Hall, N. (2007) Advanced sequencing technologies and their wider impact in microbiology. *Journal of Experimental Biology*, **210**, 1518–1525.

Hebert, P. & Gregory, T. (2005) The promise of DNA barcoding for taxonomy. *Systematic Biology*, **54**, 852–859.

Hebert, P.D.N., Cywinska, A., Ball, S.L. & deWaard, J.R. (2003a) Biological identifications through DNA barcodes. *Proceedings of the Royal Society of London B*, **270**, 313–321.

Hebert, P.D.N., Ratnasingham, S. and deWaard, J.R. (2003b) Barcoding animal life: cytochrome c oxidase subunit 1 divergences among closely related species. *Royal Society of London B*, **270** (Supplement), S96–S99.

Hebert, P.D.N., Penton, E.H., Burns, J.M., Janzen, D.H. & Hallwachs, W. (2004) Ten species in one: DNA barcoding reveals cryptic species in the neotropical skipper butterfly *Astraptes fulgerator*. *Proceedings of the National Academy of Sciences of the USA*, **101**, 14812–14817.

Holland, B.S., Dawson, M.N., Crow, G.L. & Hofmann, D.K. (2004) Global phylogeography of *Cassiopea* (Scyphozoa: Rhizostomeae): molecular evidence for cryptic species and multiple invasions of the Hawaiian Islands. *Marine Biology*, **145**, 1119–1128.

Ivanova, N., Borisenko, A. & Hebert, P.D.N. (2009) Express barcodes: racing from specimen to identification. *Molecular Ecology Resources*, **9**, 35–41.

Janzen, D.H. (2004) Now is the time. *Philosophical transactions of the Royal Society of London B*, **359**, 731–732.

Janzen, D.H., Hajibabaei, M., Burns, J.M., Hallwachs, W., Remigio, E. & Hebert, P.D.N. (2005) Wedding biodiversity inventory of a large and complex Lepidoptera fauna with DNA barcoding. *Philosophical Transactions of the Royal Society B*, **360**, 1835–1845.

Kang, M., Buckley, Y. & Lowe, A.J. (2007) Testing the role of genetic factors across multiple independent invasions of the shrub Scotch broom (*Cytisus scoparius*). *Molecular Ecology*, **16**, 4662–4673.

Kress, W.J., Wurdack, K.J., Zimmer, E.A., Weigt, L.A. & Janzen, D.H. (2005) Use of DNA barcodes to identify flowering plants. *Proceedings of the National Academy of Sciences of the USA*, **102**, 8369–8374.

Kress, W.J. & Erickson, D.L. (2008) DNA barcodes: genes, genomics, and bioinformatics. *Proceedings of the National Academy of Sciences of the USA*, **105**, 2761–2762.

Kress, W., Erickson, D. & Shiu, S. (2007) A two-locus global DNA barcode for land plants: the coding *rbcL* gene complements the non-coding *trnH–psbA* spacer region. *PLoS ONE*, **2**, e508.

Lane, C., Lindstrom, S. & Saunders, G. (2007) A molecular assessment of northeast Pacific *Alaria* species (Laminariales,

Phaeophyceae) with reference to the utility of DNA barcoding. *Molecular Phylogenetics and Evolution*, **44**, 634–648.

Margulies, M., Egholm, M., Altman, W., et al. (2005) Genome sequencing in microfabricated high-density picolitre reactors. *Nature*, **437**, 376–380.

Meusnier, I., Singer, G.A.C., Landry, J.-F., Hickey, D.A., Hebert, P.D.N. & Hajibabaei, M. (2008) A universal DNA mini-barcode for biodiversity analysis. *BMC Genomics*, **9**, 214.

Noor, M. & Feder, J. (2006) Speciation genetics: evolving approaches. *Nature Reviews Genetics*, **7**, 851–861.

Okuyama, Y., Fujii, N., Wakabayashi, M., et al. (2005) Nonuniform concerted evolution and chloroplast capture: heterogeneity of observed introgression patterns in three molecular data partition phylogenies of Asian *Mitella* (Saxifragaceae). *Molecular Biology and Evolution*, **22**, 285–296.

Porter, T., Skillman, J. & Moncalvo, J. (2008) Fruiting body and soil rDNA sampling detects complementary assemblage of Agaricomycotina (Basidiomycota, Fungi) in a hemlock-dominated forest plot in southern Ontario. *Molecular Ecology*, **17**, 3037–3050.

Plant Working Group (2009) A DNA barcode for land plants. *Proceedings of the National Academy of Sciences of the USA*, **106**, 12794–12797.

Prentis, P.J., Sigg, D.P., Raghu, S., Dhileepan, K. & Lowe, A.J. (2009) Understanding invasion history: genetic structure and diversity of two globally invasive plants and implications for their management. *Diversity and Distributions*, **15**, 822–830.

Prentis, P., Wilson, J.R.U., Dormontt, E.E., Richardson, D.M & Lowe, A.J. (2008) Adaptive evolution in invasive species. *Trends in Plant Sciences*, **13**, 288–294.

Pringle, A., Adams, R.I., Cross, H.B. & Bruns, T.D. (2009) The ectomycorrhizal fungus *Amanita phalloides* was introduced and is expanding its range on the west coast of North America. *Molecular Ecology*, **18**, 817–833.

Provan, J., Booth, D., Todd, N.P., Beatty, G.E. & Maggs, C.A. (2008). Tracking biological invasions in space and time: elucidating the invasive history of the green alga *Codium fragile* using old DNA. *Diversity and Distributions*, **14**, 343–354.

Quince, C., Lanzén, A., Curtis, T., et al. (2009) Accurate determination of microbial diversity from 454 pyrosequencing data. *Nature Methods*, 1–6.

Rokas, A. & Abbot, P. (2009) Harnessing genomics for evolutionary insights. *Trends in Ecology & Evolution*, **24**, 192–200.

Rubinoff, D. (2006) Utility of mitochondrial DNA barcodes in species conservation. *Conservation Biology* **20**, 1026–1033.

Santana, Q.C., Coetzee, M., Steenkamp, E.T., Mlonyeni, O.X., Hammond, G.N.A. & Wingfield, B. D., (2009) Microsatellite discovery by deep sequencing of enriched genomic libraries. *BioTechniques*, **46**, 217–223.

Saunders, G.W. (2005) Applying DNA barcoding to red mac-roalgae: a preliminary appraisal holds promise for future applications. *Philosophical Transactions of the Royal Society B*, **360**, 1879–1888.

Saunders, G.W. (2009) Routine DNA barcoding of Canadian Gracilariales (Rhodophyta) reveals the invasive species *Gracilaria vermiculophylla* in British Columbia. *Molecular Ecology Resources*, **9**, 140–150.

Scheffer, S.J., Lewis, M.L. & Joshi, R.C. (2006) DNA barcoding applied to invasive leafminers (Diptera: Agromyzidae) in the Philippines. *Annals of the Entomological Society of America*, **99**, 204–210.

Seifert, K. (2009) Progress towards DNA barcoding of fungi. *Molecular Ecology Resources*, **9**, 83–89.

Singh, J., Behal, A., Singla, N., et al. (2009) Metagenomics: concept, methodology, ecological inference and recent advances. *Biotechnology Journal*, **4**, 480–494.

Sogin, M., Morrison, H., Huber, J., et al. (2006) Microbial diversity in the deep sea and the underexplored 'rare bio-sphere'. *Proceedings of the National Academy of Sciences of the USA*, **103**, 12115–12120.

Swaminathan, K., Varala, K. & Hudson, M. (2007) Global repeat discovery and estimation of genomic copy number in a large, complex genome using a high-throughput 454 sequence survey. *BMC Genomics*, **8**, 132.

Taberlet, P., Coissac, E., Pompanon, F., et al. (2007) Power and limitations of the chloroplast *trn*L (UAA) intron for plant DNA barcoding. *Nucleic Acids Research*, **35**, e14.

Tomaru, N., Takahashi, M., Tsumura, Y., Takahashi, M. & Ohba, K. (1998) Intraspecific variation and phylogeo-graphic patterns of *Fagus crenata* (Fagaceae) mitochondrial DNA. *American Journal of Botany*, **85**, 629–636.

Thomsen, M.S., Gurgel, C.F.D., Fredericq, S. & McGlathery, K.J. (2006) *Gracilaria vermiculophylla* (Rhodophyta, Gracilariales) in Hog Island Bay, Virginia: a cryptic alien invasive macroalga and taxonomic correction. *Journal of Phycology*, **42**, 139–141.

van de Wiel, C., Van Der Schoot, J., Van Valkenburg, J., Duistermaat, H. & Smulders, M. (2009) DNA barcoding discriminates the noxious invasive plant species, floating pennywort (*Hydrocotyle ranunculoides* L.f.), from non-invasive relatives. *Molecular Ecology Resources*, **9**, 1086–1091.

van Orsouw, N., Hogers, R., Janssen, A., et al. (2007) Complexity reduction of polymorphic sequences (CRoPS™): a novel approach for large-scale polymorphism discovery in complex genomes. *PLoS ONE*, **2**, 1–10.

van Geel, B., Aptroot, A., Baittinger, C., et al. (2008) The ecological implications of a Yakutian mammoth's last meal. *Quaternary Research*, **69**, 361–376.

Venter, J.C., Remington, K., Heidelberg, J.F., et al. (2004) Environmental genome shotgun sequencing of the Sargasso Sea. *Science*, **304**, 66–74.

Verbruggen, H., Tyberghein, L., Pauly, K., et al. (2009) Macroecology meets macroevolution: evolutionary niche dynamics in the seaweed *Halimeda*. *Global Ecology and Biogeography*, **18**, 393–405.

Vialle, A., Feau, N., Allaire, M., et al. (2009) Evaluation of mitochondrial genes as DNA barcode for Basidiomycota. *Molecular Ecology Resources*, **9**, 99–113.

White, T., Bruns, T.D., Lee, S. & Taylor, J. (1990) Amplification and direct sequencing of fungal ribosomal RNA genes for phylogenetics. In *PCR Protocols: A Guide to Methods and Applications* (ed. M.D. Innis, D. Gelfand, J. Sninsky and T. White), pp. 315–322. Academic Press, San Diego.

Willerslev, E., Cappellini, E., Boomsma, W., et al. (2007) Ancient biomolecules from deep ice cores reveal a forested southern Greenland. *Science*, **317**, 111–114.

Wilson, J.R.U., Dormontt, E.E., Prentis, P.J., Lowe, A.J. & Richardson, D.M. (2009) Something in the way you move: dispersal pathways affect invasion success. *Trends in Ecology & Evolution*, **24**, 136–144.

Chapter 23

BIOSECURITY: THE CHANGING FACE OF INVASION BIOLOGY

Philip E. Hulme

The Bio-Protection Research Centre, Lincoln University, PO Box 84, Christchurch, New Zealand

Fifty Years of Invasion Ecology: The Legacy of Charles Elton, 1st edition. Edited by David M. Richardson
© 2011 by Blackwell Publishing Ltd

23.1 THE SHAPE OF THINGS TO COME

'A hundred years of faster and bigger transport has kept up and intensified this bombardment of every country by foreign species, brought accidently or on purpose, by vessel and by air, and also overland from places that used to be isolated' Elton (1958)

Charles Elton was clearly well aware of the changes the Industrial Revolution had made to the volume and speed of trade and travel that, following the invention of combustion engines, saw the demise of sail power and the growth of rail, road and air transport. Yet as he wrote those fateful words in *The Ecology of Invasions by Animals and Plants*, even his far-sighted perspective of biological invasions could not have prepared him for the developments that have taken place in the subsequent half-century. Since 1958, the use of private road transport has more than trebled in the UK, while both international as well as domestic air travel has grown by an order of magnitude (DfT 2008). These trends are mirrored globally and taken together with the accelerating rise in container shipping means that the world is more interconnected today than anyone could have imagined in the 1950s (Hulme 2009). As a consequence of these developments, the global perspective of biological invasions, so eloquently described by Charles Elton, has now changed dramatically. This chapter addresses how the study of biological invasions has become increasingly influenced by three major drivers: globalization, environmental change and international legislation, which have not only impacted upon the scale at which invasive species need to be addressed but also shaped how such risks are perceived and managed.

Elton (1958) viewed the increasing transport links across the globe as primarily facilitating the spread and establishment of alien species. This greater global interconnectedness has certainly led to a growth in tourism, passenger and cargo movements, which has increased the risks of alien pest and disease incursions. However, in addition to geographical interconnectedness, the ties between nations have also been strengthened by globalization, which is integrating the world economy and increasing the volume and range of products traded internationally (Hulme 2009). Compared with the 1950s, there are therefore qualitative and quantitative differences in the magnitude of exchanges among the world's biota (Hulme et al. 2008). This includes more subtle, but still poorly understood, consequences of increasing global movements of genetic material as farmers endeavour to increase productivity.

Environmental change, particularly in relation to pollution, land use and climate, is a direct consequence of greater economic and geographical interconnectedness. As economies have grown, owing to increased opportunities for trade and tourism, so human populations have spread into new habitats. This has raised the risk of animal diseases (zoonoses) capable of transmission to human populations as well as human diseases spreading to wildlife (Jones et al. 2008). The intensification of agriculture with concomitant increase in farm and field size, reliance on relatively few cultivars and homogenization of the landscape has increased the potential impact of a pest or disease incursion as well as complicates the ability to contain a pest in a single area. The burning of fossil fuels to support an ever-expanding international transport network contributes one quarter of global CO_2 emissions (Kahn-Ribeiro et al. 2007). Thus, in addition to facilitating the exchange of pests, weeds and diseases, global transport contributes to climate change. Climate change further adds to the spread of pests and diseases by expanding species' distribution ranges or habitats, changing migratory patterns and increasing the probability of weather events that support the spread of disease vectors (Walther et al. 2009; Dukes, this volume).

The risk to agriculture of introduced pests and diseases was widely known at the beginning of the 20th century. The Netherlands Plant Protection Service was established in 1899, the Commonwealth Quarantine Service in Australia came into operation in 1908, whereas in the USA the Federal Horticultural Board was established in 1912 to enforce quarantine measures (Ebbels 2003). Yet as the importance of international trade increased there was a shift from country independence to country interdependence for effective management of such pests. Owing to the profusion of national sanitary and phytosanitary measures often based on quite different and sometimes arbitrary criteria, there was a need to develop harmonized international standards. In 1947, the first global trade agreement, the General Agreement on Trade and Tariffs (GATT), allowed governments to act on trade to protect human, animal or plant life or health, provided they did not discriminate or use this agreement as disguised protectionism. However, it took almost another

50 years before the ground rules for such action could be agreed. The Agreement on the Application of Sanitary and Phytosanitary Measures (the 'SPS Agreement') entered into force with the establishment of the World Trade Organization on 1 January 1995. This binding agreement has had far-reaching implications for the management of pests and diseases worldwide. Over this same period, awareness of the environmental consequences of globalization culminated in the signing of the Convention on Biological Diversity (CBD) wherein 'each Contracting Party shall, as far as possible and appropriate, prevent the introduction of, control or eradicate those alien species which threaten ecosystems, habitats or species' (CBD 1992). Although a non-binding agreement, the CBD has had a major influence on conservation efforts and national policies. Environmental perspectives have also begun to influence phytosanitary regulations established under the auspices of the SPS Agreement with standards for plant pest risk analysis now also addressing risks to non-agronomic ecosystems (Shine 2007).

Although some have criticized Elton for disassociating invasion biology from the rest of ecology (Davis et al. 2001), the reality is that global events have transformed the study of biological invasions. Today, it is naïve to view this discipline as uniquely driven by ecological principles, and the field is increasingly becoming a multidisciplinary subject. The study of biological invasions, particularly when concerned with managing risk, often requires the bringing together of taxonomists and population biologists, statisticians and modellers, economists and social scientists, and its agenda is shaped by politics, legislation and public perceptions. Although its roots may be traced back to Elton's seminal book, the approaches to managing biological invasions have changed irrevocably and have come together to form the new discipline of biosecurity.

23.2 BIOSAFETY, BIOTERRORISM AND BIOSECURITY: A BRIEF BIOGRAPHY

Biosecurity is a relatively new term, only entering the scientific lexicon in the late 1980s and the *Oxford English Dictionary* in 2005 (Fig. 23.1). Although various definitions exist, a growing consensus defines biosecurity as the exclusion, eradication or effective management of risks posed by pests and diseases to the economy, environment and human health (Biosecurity

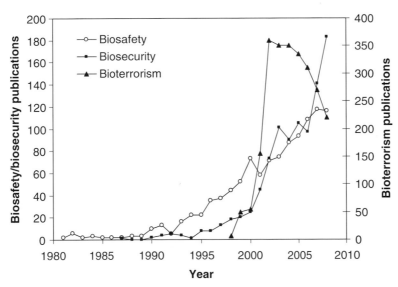

Fig. 23.1 Cumulative number of publications using the terms biosafety, bioterrorism and biosecurity between 1980 and 2008 (from ISI Web of Knowledge, accessed July 2009).

Council 2003; FAO 2007; Beale et al. 2008). In its broadest sense, biosecurity covers all activities aimed at managing the introduction of new species to a particular region and mitigating their impacts should they become established. This includes the regulation of intentional (including illegal) and unintentional introductions and the management of weeds and animal pests by central and local government, industry and other stakeholders. Confusion has arisen over this definition in relation to two sub-areas of biosecurity: biosafety and bioterrorism.

Biosafety covers the containment of new (e.g. genetically modified) and unwanted organisms in laboratories, quarantine facilities and experimental field trials. Biosafety is an older concept that was developed in parallel with recombinant DNA technology in the late 1970s to early 1980s (Fig. 23.1). It addresses the technologies and practices that are implemented to prevent the unintentional exposure to humans or the accidental release of such organisms into the wild whether from the laboratory, field site or during transport. Confusingly, the term 'biosafety' has been used in the Cartagena Protocol on Biosafety specifically in relation to protecting biological diversity from the potential risks posed by genetically modified organisms resulting from modern biotechnology (CBD 2000). Yet there appears to be no scientific basis to treat genetically modified organisms as a different class of biosecurity risk from alien pathogens, pests or weeds (see Conner et al. 2003).

Unfortunately, although Elton (1958) described the role of accidental introductions and misguided deliberate releases, the spectre of bioterrorism involving animal rights extremists or political terrorist organizations is today seen as a further potential risk (Gullino et al. 2008). Bioterrorism is a relatively new term that reached heightened interest after the anthrax attacks in the USA in 2001 (Fig. 23.1). In some sectors, particularly in the USA, bioterrorism is the greatest perceived biosecurity risk, and biosecurity is defined as 'security against the inadvertent, inappropriate, or intentional malicious or malevolent use of potentially dangerous biological agents or biotechnology, including the development, production, stockpiling, or use of biological weapons, as well as natural outbreaks of newly emergent and epidemic diseases' (NRC 2006). However, conceptually bioterrorism is simply another pathway for transmission of unwanted pests and diseases. The intent, however, is quite different and the scale of damage potentially catastrophic.

Irrespective of whether an envelope in the post contains seeds of invasive weeds or anthrax spores, or whether an insect is released for biological control or escapes from a quarantine facility, these threats to the environment and economy can be addressed through a common biosecurity framework. A government will want to know the identity and nature of future alien species risks, and to estimate the nature and magnitude of the hazard, so as to anticipate and allocate resources efficiently. The sheer diversity of potential risks, the difficulty of predicting potential harm for any one species and the speed and stealth of many species in establishing and spreading pose a significant challenge to the implementation of biosecurity strategies. The breadth and scope of these complex challenges is well covered by several recent reviews and syntheses (NRC 2002; FAO 2007; Heather & Hallman 2008). These reviews each cover similar ground and include descriptions of the underlying processes underpinning biological invasions, the policy context, risk assessment and research needs. Rather than repeat this material, the following sections attempt to tease apart the pervasive influence of trade on the assessment of biosecurity risks and address the constraints acting upon existing scientific approaches to deliver new insights in this field.

23.3 TRADING PLACES: UNDERSTANDING THE ROUTES OF BIOSECURITY THREATS

It is intuitive that for a species to establish in a new territory the environmental conditions should be amenable for survival and reproduction. A suitable climate would seem an essential attribute and this fact has resulted in the development of bioclimatic models to predict the vulnerability of new regions to alien pests, weeds and diseases (see, for example, Panetta & Mitchell 1991). A wide range of different statistical models have been applied with varying success to model the potential future ranges of alien organisms (Jeschke & Strayer 2008). However, a suitable climate match, although necessary, may not be a sufficient condition to facilitate establishment. Once a species has breached a region's borders and become introduced, its probability of establishment depends on the life-history and demographic attributes of the species itself as well the suitability of both the abiotic and biotic conditions in the recipient environment.

Yet although these factors may play a role in the likelihood of species' establishment, the magnitude of the introduction effort (or propagule pressure) is increasingly believed to act in concert with climate suitability to determine establishment success of mammals (Bomford et al. 2009a), birds (Duncan et al. 2001), herptiles (Bomford et al. 2009b) and insects (Duncan et al. 2009). The importance of propagule pressure underscores the perils faced by small founder populations early on in the establishment process. Stochasticity in demography, the environment and/or genetics as well as Allee effects will all reduce the probability of establishment of small founder population (NRC 2002; Barrett, this volume). Owing to the constraints on small founder populations, propagule pressure should be particularly important in the success of invasive species yet for most taxa information on introduction effort is often sparse and instead attempts to quantify the likelihood of establishment have used the characteristics of introduction pathways (Hulme et al. 2008). There is increasing evidence that propagule pressure can affect not only the probability of alien species becoming established, but also the magnitude, direction and scope of different impacts through its influence on the abundance, functional ecology and range size of invaders (Ricciardi et al., this volume). For all these reasons, managing propagule pressure is a crucial facet of biosecurity.

In general, pathways describe the processes that result in the movement of alien species from one location to another. Although not using the term 'pathway', Elton (1958) characterized the main means of species arrival in his introductory case histories of invasion and these examples can be shown to map onto a recent framework for invasion pathways (Hulme et al. 2008). Alien species transported as commodities may be introduced as a deliberate release (e.g. the starling *Sturnus vulgaris* in the USA) or as an escape from captivity (e.g. muskrat *Odontra zibethicus* in the Czech Republic). Many species are not intentionally transported by humans but arrive as a contaminant of a commodity as either a host-specific pathogen or pest (e.g. chestnut blight *Cryphonectria parasitica* in the USA). Stowaways are directly associated with human transport but arrive independently of a specific commodity and include organisms transported in ballast water, cargo and airfreight (e.g. the mitten crab *Eriocheir sinensis* in Europe). The corridor pathway highlights the role transport infrastructures play in the introduction of alien species (e.g. the spread of the sea lamprey *Petromyzon marinus* after the opening of the Welland Ship Canal). The unaided pathway describes situations where natural spread results in alien species arriving into a new region from a donor region where it is also alien (e.g. the starling's spread from the USA to Canada or the muskrat spreading across central Europe).

The importance of a particular pathway depends on the taxon being considered. Fish introductions have arisen primarily through deliberate releases for sport fishing, plants as escapes from gardens and horticulture, aquatic invertebrates arrive as stowaways on the hulls or in the ballast water of ships, terrestrial invertebrates enter most frequently as contaminants of commodities whereas pathogenic micro-organisms and fungi are generally introduced as contaminants of their hosts (Hulme et al. 2008). However, even for a single pathway and taxon, probabilities can differ depending on the abundance of the organism in the area of origin, the number and location of arrival points at the destination and whether the life cycle of an organism is of sufficient duration to extend beyond the time in transit (Hulme 2009). In the absence of detailed data on rates of individual species' introductions, accounting for pathways of introduction may be essential to disentangle the role of species and ecosystem traits in biological invasions as well as predict future trends and identify management options. New tools such as DNA barcoding, when combined with distributional data, may play a key role in characterizing invasion pathways and help identify when and where a particular alien species is likely to disperse and invade (Cross et al., this volume).

23.4 QUANTIFYING THREATS TO BIOSECURITY: ESTIMATING THE REAL COSTS OF THE FREE MARKET

In contrast to the detailed description of the redistribution of species across the globe, Elton (1958) was less explicit about the range of impacts introduced species cause in new regions. However, biological invasions impact on the environment, human health, culture and the economy. Environmental impacts include the reduction in population size (in some cases even extinction) of threatened or endemic species, changes to the structure of plant and animal communities, alteration of ecosystem services (e.g. pollination, carbon sequestration) and dynamics (e.g. flood and fire frequency), hybridization and gene flow to native species, and the spread of pathogens (Hulme 2007). Social consequences include impacts on the aesthetics of landscapes, accessibility to habitats and damage to cultural infrastructure (Vilà et al. 2010). Economic and human health consequences comprise both the direct costs arising from impacts on crop and livestock productivity and human diseases, as well indirect costs of higher management inputs to prevent, curtail or eradicate the problem (Pimentel et al. 2001; Perrings, this volume). This is not an exhaustive list of biosecurity hazards but highlights that economic, environmental and social impacts are rarely additive because of the different assessment methods (e.g. monetary, species loss, human values). Species are likely to have consequences for more than one sector and in some cases may have positive impacts in one yet negative consequences in another.

An increasing number of studies have attempted to place a monetary value of the costs of management and lost production arising from established biological invasions (Perrings, this volume). However, prevention or eradication of new problems requires prioritization before a species becomes permanently established within a territory. Such prioritization is usually based on estimates of the potential cost a species would cause should they become established. These estimates are often much larger than the costs incurred in lost production and management identified for established species. For example, the estimate for total current expenditure (from Pimentel et al. 2001) due to all established pests in Australia ($6.24 billion) is less than the estimated total future cost (from Beale et al. 2008) of foot and mouth disease ($6.70 billion). The reason for this disparity is that whereas monetary estimates for established species include the costs due to management and loss of production, they do not account for the impact of changes in market access. In 1996 Karnal bunt (*Tilletia indica*), a fungal pathogen of wheat, was discovered in grain grown in the southwestern USA (Gullino et al. 2008) and subsequently, more than 50 countries adopted phytosanitary trade restrictions against the USA. The result was that although the impact and cost of clean-up measures was limited, the loss of wheat exports amounted to well over $250 million. Two pests of the potato (*Solanum tuberosum*) in the UK further illustrate the importance of accounting for the market. Both the Colorado potato beetle and potato ring rot (*Corynebacterium sepedonicum*) result in yield reductions but the potential annual cost to the UK of the

latter is over ten times higher ($2,992,000 compared with $220,000) because it can easily be spread in tubers and thus has a significant impact on potato seed exports (Waage et al. 2005). These examples highlight that it is unwise to extrapolate from current management costs of established pests and diseases to predictions of costs of future biosecurity threats because most current costs do not include the opportunity cost of lost markets.

Estimating the impact of biological invasions on the environment and human culture is much more complex, particularly if monetary values are sought. Non-market damages often are difficult to quantify because of the complex interactions among species in an ecosystem and a lack of information about the public's preferences across alternative ecological states. As a result, few studies have begun to provide estimates of non-market damages from invasive species (Hoagland & Jin 2006). However, the likelihood is that environmental and social costs of biological invasions will be dwarfed by those of the human health and primary production sectors, even where costs of control and eradication are similar (Smith & Petley 2009). As an example, weeds pose a threat to one-third of all New Zealand nationally threatened plant species, and estimates suggest that without action, weeds could potentially degrade 7% of the conservation estate within a decade, corresponding to a loss of native biodiversity equivalent to $1.3 billion (Williams & Timmins 2002). Although an impressive sum, it is less than a third of the annual potential costs of foot and mouth disease ($4.1 billion) to the New Zealand economy (Biosecurity Council 2003).

23.5 TOOLS OF THE TRADE: HOW BEST TO ASSESS BIOSECURITY RISKS?

The increasing concern of governments with potential, rather than proven, harm has seen a shift in policy focus from the remediation of damage to the prediction of risk. The SPS Agreement is one of the more prominent examples of this trend in that it prescribes scientific risk assessment as a basis for measures dealing with risks to human, animal and plant life or health. Risk assessment is a scientifically based process to identify hazards, characterize their adverse impacts, evaluate the level of exposure of a target to those hazards and estimate the risk (Smith & Petley 2009). In general,

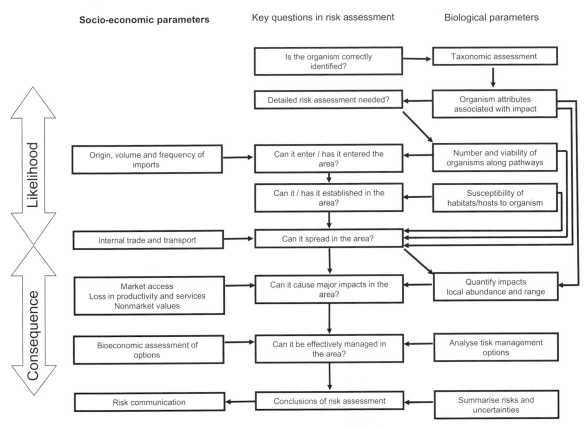

Socio-economic parameters   Key questions in risk assessment   Biological parameters

Fig. 23.2 Key questions required to address the risk of invasion of a newly recorded alien species and examples of some of the biological and socio-economic parameters that are often required to quantify likelihood and consequences (modified from Hulme et al. 2009).

biosecurity risk assessment requires the identification of one or more negative consequences (usually termed a hazard) likely to result from the successful establishment of an alien genotype or species and an estimate of the likelihood of each consequence. The risk is the product of each consequence and likelihood. Uncertainty will be associated with estimates of both consequence and likelihood. Uncertainty represents the limitations in knowledge about the factors that influence risk. For example, there is uncertainty when we assume the behaviour of a species in its native range will be similar in a new region to which it has been introduced, or project future geographical distributions from known locations of a species or when we extrapolate from bioeconomic models of impact to the real world. Uncertainty matters because decision makers and stakeholders need to know the range of possible outcomes and their relative likelihoods (Hart et al. 2006).

A common set of key questions typify risk assessments of alien species (Pheloung et al. 1999; FAO 2004; EPPO 2007) (Fig. 23.2). To answer these questions requires inputs derived from both biological (e.g. species traits, habitat attributes) and socio-economic parameters (e.g. trade volumes, market access). In contrast to the broadly similar key questions, the extent to which the different biological and socio-economic parameters contribute to the overall risk assessment, are actually quantified and/or integrated varies across different protocols. Similarly, the degree to which uncertainty is incorporated in these analyses or communicated is inconsistent.

A wide range of risk assessment approaches exist but in general they fall into one of three categories (Vose 2000): qualitative, semi-quantitative and quantitative. Qualitative approaches rely on expert assessors assembling all available information, weighing up the likelihoods and consequences of a biosecurity risk, and reaching conclusions through consensus (Maguire 2004; Stirling 2008). Although applicable even when data are limited, outcomes can be subjective and dependent on the expertise of the assessors, with the result that the relation between the estimated and true risk is unknown. This approach is currently favoured by several phytosanitary bodies (FAO 2004; EPPO 2007).

Semi-quantitative methods use ordinal scoring to rank the importance of different factors that underpin likelihoods and consequences. Similar scoring approaches have been developed for several taxa including plants (Pheloung et al. 1999), fish (Copp et al. 2009) and terrestrial vertebrates (Bomford 2003). Scoring approaches are more transparent than qualitative approaches, because they make explicit which factors have been considered, how each factor was evaluated and how the factors were combined in the overall assessment. However, assignment of scores is still subjective, the way in which scores are combined is usually a simplified representation of real mechanisms and the relation of the output score to real measures of risk is usually unknown. Nevertheless, semi-quantitative tools have proven remarkably robust (Gordon et al. 2008) and form the basis for the Australian border weed risk assessment (Weber et al. 2009).

Quantitative approaches require measurement and knowledge of the mechanisms involved and attempt to estimate relevant measures of risk in absolute terms, by quantifying key factors determining risk and combining them in ways that are intended to represent real processes. Their advantage over qualitative and scoring approaches is that they attempt an objective model of the risk problem, and provide an estimate of the risk in real terms. However, unless the hazard is specified precisely, for example a transgene release to a native conspecific (Wilkinson et al. 2003), the assessment of biosecurity hazards is often qualitative rather than quantitative. In contrast to consequences, likelihoods can be estimated by the product of a series of sequential probabilities that relate to the introduction,

establishment and spread of an organism (Lambdon & Hulme 2006). Thus quantitative approaches have largely focused on likelihoods rather than consequences. The approaches have been applied to single species (see, for example, Rafoss 2003) or alternatively used to identify general characters of high-risk species.

In most cases, quantitative methods apply deterministic methods that use point estimates for both inputs and outputs. A variety of analytical methods have been applied in this context including logistic regression, discriminant analysis and regression trees that have used species' trait data to distinguish between species that have and have not become invasive (Keller & Drake 2009). In reality both inputs and outputs are variable and probabilistic methods such as interval analysis, fuzzy arithmetic, probability bounds, first- and second-order Monte Carlo methods and Bayesian methods may be more appropriate (Hart et al. 2006). These methods use distributions to represent variability in both the inputs and outputs of the assessment and can provide confidence intervals for the risk estimate, but they are still infrequently used to assess the risks posed by alien species (see, for example, Holt et al. 2006; Batabyal & Nijkamp 2007; Diez et al. 2008).

Owing to the often complex statistical approaches used in deterministic and probabilistic risk assessment methods, these quantitative approaches are rarely used by regulatory bodies. Yet this picture may be about to change and qualitative assessments may no longer be acceptable for establishing quarantine status. For example, the European and Mediterranean Plant Protection Organization (EPPO) has listed the American skunk cabbage (*Lysichiton americanus*) and floating pennywort (*Hydrocotyle ranunculoides*) as quarantine pests because of their threat to ecosystems. It recommends member countries take measures to prevent their introduction or spread, as well as manage established populations (Anon. 2006a,b). These recommendations were based on standard qualitative assessments made by the EPPO expert panel on invasive alien species (EPPO 2007). However, when these qualitative assessments were submitted to the European Food Safety Authority (EFSA), the request to list these species as quarantine pests was declined (EFSA 2007a,b). Although EFSA acknowledged that both species were invasive, to confirm the risk posed and enable further consideration of management options further data were required. This included taking into account the following considerations: (i) the effect of abiotic factors on the establishment, development, reproduction, survival and dispersal of the plant in both the native and introduced range; (ii) the population dynamics of the plant in areas where it is present but not invasive; (iii) the volume of trade in the species entering and moving within Europe; (iv) the nature and occurrence of areas within Europe where conditions are favourable to invasive behaviour of the plant; and (v) the environmental damage caused by the plants in areas where they are invasive. Such decisions set a precedent for increasingly quantitative rather than qualitative risk assessments that depart significantly from most approaches used to assess the risk of invasive species. So can the science deliver?

23.6 SCIENTIFIC HURDLES CANNOT BE USED AS BARRIERS TO TRADE

Although of central importance to biosecurity, compared with other environmental threats, risk assessment for biological invasions is still in its infancy. Initial regulatory discussions on approaches to assess the risks of pests and pathogens occurred as recently as the 1990s, with international standards only appearing in the past decade (Baker et al. 2009). At the same time, several competing frameworks for assessing the risk of biological invasions have been proposed (Andersen et al. 2004; Landis 2004; Andow 2005; Stohlgren & Schnase 2006). Because biosecurity threats are living organisms, risk assessment poses a particular challenge for the reason that the threats can reproduce, disperse and seek out hosts or habitats, as well as evolve. Thus, in the absence of detailed scientific evidence, the 'precautionary principle', which supports taking protective action (e.g. banning particular imports or establishing new quarantine procedures) before there is complete scientific proof of a risk, exists as a fail-safe mechanism. In effect, action should not be delayed simply because full scientific information is lacking. The provisional measure must take into consideration available pertinent information. However, under the SPS Agreement any state adopting precautionary measures must seek to obtain the additional information necessary for a more objective assessment of risk, and must review measures within a reasonable period. Thus precautionary approaches can only be adopted as temporary fail-safe mechanisms and the scientific hurdles faced in undertaking risk assessments must, sooner or later, be overcome to facilitate trade agreements. Although many chal-

lenges exist, three examples that significantly limit the development of a consistent, reproducible framework are described in this section: the ignorance of failures, the complexity of impacts and inconsistent terminology.

The likelihood that, once introduced, a species will become established is small because only a fraction of species introduced to a new region ultimately establish self-replacing populations (Williamson 1996). Understanding the factors that distinguish these successful species from those that are introduced but fail to establish is thus an essential step in risk assessment. Yet for most taxa, this understanding is only based on observations of the successful species because the failures are typically unknown and often unknowable. There are only a few analyses of pest invasions that incorporate information on which species fail to establish (reviewed in Diez et al. 2009). Thus only where this information is available are quantitative assessments of the likelihood of establishment possible. However, omission of the failures risks biasing analyses of the invasion process by conflating the likelihood of introduction with that of subsequent establishment. For example, islands appear more prone to invasion than continental areas if analyses only assess successes, but once failures are accounted for in the analyses such differences disappear (Blackburn & Duncan 2001; Diez et al. 2009).

Many natural hazards can be measured objectively on a scientific scale of magnitude or intensity, for example earthquakes (Richter scale), tornadoes (Fujita scale) or hurricanes (Saffir–Simpson scale). In contrast, the range of potential consequences arising from the introduction of non-native species is complex and manifold. Quantifying biosecurity hazards requires a comparable approach that accounts for the following impact attributes: severity, irreversibility, duration, spatial extent, latency (reflecting the delay between cause and effect) and cumulativeness (where the capacity of a species or gene to accumulate steadily over time leads to a critical threshold being passed whereupon effects become apparent). Such a comparable index is a long way off especially because for many species' introductions information on even the type of consequence, let alone its severity, is unknown (Vilà et al. 2010).

The complexity of quantifying impacts has resulted in most biosecurity risk assessment methods stressing risk in terms of likelihoods rather than consequences. In many cases such risk assessments only identify the probability of a species becoming established. Others go a step further and estimate the probability of a species becoming invasive. Unfortunately, there is no consensus in the scientific community as to what 'invasive' actually means (Falk-Petersen et al. 2006; Colautti & Richardson 2009). Although the policy interpretation of invasive assumes a negative impact (IUCN 2000), many scientists do not agree with this definition and use the term to describe species capable of becoming widespread rather than causing damage (Richardson et al. 2000). In such cases a preferable measure would be actual or potential distribution. However, although such analyses have been undertaken, they differ in both the scale and resolution of this measure, for example range area (Williamson et al. 2009), administrative regions (Bucharova & van Kleunen 2009) or number of islands (Lambdon & Hulme 2006). This limits the comparability of their predictions. Given that performance is likely to vary on a continuous scale, any division into invasive and non-invasive species may be subjective unless transparent criteria are used (see, for example, Dawson et al. 2009). However, even where clear performance measures are presented, the end-point of such analyses is not a full risk assessment in a formal sense because it does not provide an estimate of the probability of harm. Furthermore, range size or distribution may not correlate with the perceived impact species pose and thus may not even be a relevant measure of harm (Ricciardi & Cohen 2007).

23.7 THE SCIENCE OF BIOSECURITY: THE CHALLENGE OF UNCERTAINTY, AMBIGUITY AND IGNORANCE

'*A world that begins to assess its recreation in man-hours probably cares fairly little about the breakdown of Wallace's Realms*'
Elton (1958)

Under the SPS Agreement, risk assessments should be based on scientific evidence and, where this is not forthcoming, provisional assessments can be made but only until additional information has been gained for an objective assessment. The economic sectors that import live alien species (e.g. horticulture, forestry, aquaculture) stand to lose the most from restrictions of trade aimed at limiting biological invasions and have strongly argued that such policies would do more

harm than good (Douglas 2005). There is therefore increasing scrutiny being applied to quarantine decisions and this subsequently places more pressure to deliver quantitative scientific risk assessments.

Ideally, both the likelihood and consequence of a species' introduction would be known perfectly and measured without error, which would allow perfect assessment of potential risks. Different levels in the knowledge of consequences and likelihoods shape the most appropriate strategies to address the threats posed by invasive species (Fig. 23.3). Quantitative risk assessment, by definition, can only proceed where an adequate knowledge of the likelihoods and consequences exists. These conditions occur when decision makers are familiar with the system either because it has been encountered in the past or is sufficiently similar to more familiar systems that decision can easily be translated from one situation to another. Where knowledge of consequences is good but that for likelihoods is poor, decision making faces uncertainty. Uncertainty exists where the available empirical information or analytical models simply do not present a definitive basis for assigning probabilities (Stirling 2008). Under uncertainty the information required to improve decision making is known but insufficient knowledge exists to quantify the probabilities adequately. Semi-quantitative techniques may be applied in which likelihoods are broadly described as coarse probabilities and response scored in terms of both their value as well as level of certainty. In the opposite case, where there is good knowledge of likelihoods but limited understanding of consequences, decision making suffers from ambiguity. Sources of ambiguity may exist in definitions used to describe consequences (e.g. invasive versus non-invasive species), their quantification (e.g. monetary versus non-market values), perceptions (e.g. importer versus exporter) and whether decisions are based solely on costs or also assess the benefits of an introduction. Under these circumstances, qualitative assessments may be the only option but must avoid relying on expert judgment about risk framed in qualitative and value-laden terms, inadvertently mixing the expert's judgment about what is likely to happen with personal preferences (Maguire 2004). Nevertheless, a core part of biosecurity prioritization and threat assessment is expert testimony, i.e. the reliance on expert opinion. Because quantitative data are often lacking there is heavy reliance on expert testimony. This has the potential to inject a range of issues that compromise the integrity

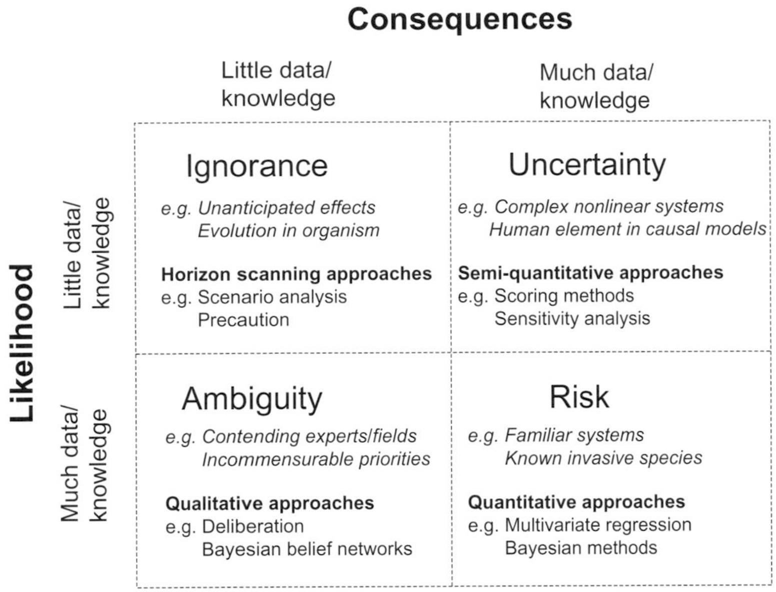

Consequences

	Little data/ knowledge	Much data/ knowledge

Fig. 23.3 Four primary states of knowledge in risk assessment (ignorance, uncertainty, ambiguity and risk) determined by the information available to assess likelihoods and consequences of a biosecurity threat. Schematic examples (in italics) and their implications for the choice of assessment approaches (bold) are presented (adapted from Stirling 2008; Hart et al. 2006) for each of the four states.

of decision making and severely impact consistency and reproducibility (Burgman 2005). Finally, a state of ignorance exists where there is little knowledge of either likelihood or consequence. Here exist the unanticipated effects, unexpected conditions and novel organisms that cannot be predicted *a priori*. Informed decision making under such conditions is almost impossible, even with the most detailed horizon scanning and decision makers should adopt a precautionary approach. Thus, where ignorance about the various implications of a biosecurity threat exists, this in itself should not be used as a reason for postponing or failing to take appropriate eradication, containment and control measures where serious or irreversible environmental damage may occur.

The foregoing sections have highlighted how the shifting landscape of globalization, environmental change and international legislation continue to pose challenges to effective biosecurity. Although scientists may call for further investment to collect more data or develop better analytical methods (Lodge et al. 2006; Keller et al. 2009), these may only have a minor impact on the accuracy of decision making. Obtaining

further data can often reduce uncertainty but different approaches must be adopted where ambiguity is important. However, it is ignorance that is the greatest challenge. Elton (1958) illustrates many examples of how ignorance regarding unanticipated effects led to biological invasions that threaten native species. It might be hoped that 50 years later, science and society have learnt from these mistakes arising from the redistribution of organism among Wallace's realms. Yet there is little evidence of the trend in species' redistribution declining and although information on invasive species is much greater today than in the past (Pyšek & Hulme, this volume) only time will tell if ignorance is less.

REFERENCES

Andersen, M.C., Adams, H., Hope, B. & Powell, M. (2004) Risk analysis for invasive species: general framework and research needs. *Risk Analysis*, **24**, 893–900.

Andow, D.A. (2005) Characterizing ecological risks of introductions and invasions. In *Invasive Alien Species: A New*

Synthesis (ed. H.A. Moody, J.A. McNeely, L. Neville, P.J. Schei and J. Waage), pp. 84–103. Island Press, Washington, DC.

Anon. (2006a) Hydrocotyle ranunculoides. *EPPO Bulletin*, **36**, 3–6.

Anon. (2006b) Lysichiton americanus. *EPPO Bulletin*, **36**, 7–9.

Baker, R., Battisti, A., Bremmer, J., et al. (2009) PRATIQUE: a research project to enhance pest risk analysis techniques in the European Union. *EPPO Bulletin*, **39**, 87–93.

Batabyal, A.A. & Nijkamp, P. (2007) The stochastic arrival of alien species and the number of and the damage from biological invasions. *Ecological Economics*, **62**, 277–280.

Beale R., Fairbrother, J., Inglis, A. & Trebeck, D. (2008) *One biosecurity: A Working Partnership.* Commonwealth of Australia, Barton ACR, Australia.

Biosecurity Council (2003) *Protect New Zealand. The Biosecurity Strategy for New Zealand.* MAF, Wellington, New Zealand.

Blackburn, T.M. & Duncan, R.P. (2001) Determinants of establishment success in introduced birds. *Nature*, **414**, 195–197.

Bomford, M. (2003) *Risk Assessment for the Import and Keeping of Exotic Vertebrates in Australia.* Bureau of Rural Sciences, Canberra.

Bomford, M., Darbyshire, R.O. & Randall, L. (2009a) Determinants of establishment success for introduced exotic mammals. *Wildlife Research*, **36**, 192–202.

Bomford, M., Kraus, F., Barry, S.C. & Lawrence, E. (2009b) Predicting establishment success for alien reptiles and amphibians: a role for climate matching. *Biological Invasions*, **11**, 713–724.

Bucharova, A. & van Kleunen, M. (2009) Introduction history and species characteristics partly explain naturalization success of North American woody species in Europe. *Journal of Ecology*, **97**, 230–238.

Burgman, M. (2005) *Risks and Decisions for Conservation and Environmental Management.* Cambridge University Press, Cambridge, UK.

CBD (1992) *Convention on Biological Diversity.* Secretariat of the Convention on Biological Diversity, Montreal, Canada.

CBD (2000) *Cartagena Protocol on Biosafety to the Convention on Biological Diversity.* Secretariat of the Convention on Biological Diversity, Montreal, Canada.

Colautti, R.I. & Richardson, D.M. (2009) Subjectivity and flexibility in invasion terminology: too much of a good thing? *Biological Invasions*, **11**, 1225–1229.

Conner, A.J., Glare, T.R. & Nap, J.P. (2003) The release of genetically modified crops into the environment – Part II. Overview of ecological risk assessment. *Plant Journal*, **33**, 19–46.

Copp, G.H., Vilizzi, L., Mumford, J., Fenwick, G.V., Godard, M.J. & Gozlan, R.E. (2009) Calibration of FISK, an invasiveness screening tool for non-native freshwater fishes. *Risk Analysis*, **29**, 457–467.

Davis, M., Thompson, K. & Grime, J.P. (2001) Charles S. Elton and the dissociation of invasion ecology from the rest of ecology. *Diversity and Distributions* **7**, 97–102.

Dawson, W., Burslem, D.F.R.P. & Hulme, P.E. (2009) Factors explaining alien plant invasion success in a tropical ecosystem differ at each stage of invasion. *Journal of Ecology*, **97**, 657–665.

DfT (2008) *Transport Statistics Great Britain, 2008 Edition.* Department for Transport, TSO Publications, Norwich, UK.

Diez, J.M., Sullivan, J.J., Hulme, P.E., Edwards, G. & Duncan, R.P. (2008) Darwin's naturalization conundrum: dissecting taxonomic patterns of species invasions. *Ecology Letters*, **11**, 674–681.

Diez, J.M., Williams, P.A., Randall, R.P., Sullivan, J.J., Hulme, P.E. & Duncan, R.P. (2009) Learning from failures: testing broad taxonomic hypotheses about plant naturalization. *Ecology Letters*, **12**, 1174–1183.

Douglas, J. (2005) Exotic plants are the lifeblood of New Zealand: less regulation is needed to allow more new species into this country. *New Zealand Garden Journal*, **8**, 2–6.

Duncan, R.P., Bomford, M., Forsyth, D.M. & Conibear, L. (2001) High predictability in introduction outcomes and the geographical range size of introduced Australian birds: a role for climate. *Journal of Animal Ecology*, **70**, 621–632.

Duncan, R.P., Cassey, P. & Blackburn, T.M. (2009) Do climate envelope models transfer? A manipulative test using dung beetle introductions. *Proceedings of the Royal Society B*, **276**, 1449–1457.

Ebbels, D. L. (2003) *Principles of Plant Health and Quarantine.* CABI Publishing, Wallingford.

EFSA (2007a) Opinion of the EFSA Scientific Panel on Plant Health on request from the Commission on Pest Risk Analysis made by EPPO on *Lysichiton americanus* Hultén & St. John (American or yellow skunk cabbage). *EFSA Journal*, **539**, 1–12 pp.

EFSA (2007b) Opinion of the EFSA Scientific Panel on Plant Health on request from the Commission on pest risk analysis made by EPPO on *Hydrocotyle ranunculoides* L. f. (floating pennywort). *EFSA Journal* **468**, 1–13 pp.

Elton, C.S. (1958) *The Ecology of Invasions by Animals and Plants.* London, Methuen.

EPPO (2007) Guidelines on Pest Risk Analysis: decision-support scheme for quarantine pests EPPO Standard PM 5/3(3). European and Mediterranean Plant Protection Organization, Paris.

Falk-Petersen, J., Bohn, T. & Sandlund, O.T. (2006) On the numerous concepts in invasion biology. *Biological Invasions*, **8**, 1409–1424.

FAO (2004) *Pest risk analysis for quarantine pests including analysis of environmental risks.* International Standards for Phytosanitary Measures Publication No. 11. Rev. 1. Food and Agriculture Organization, Rome.

FAO (2007) *Biosecurity Toolkit.* FAO, Rome.

Gordon, D.R., Onderdonk, D.A., Fox, A.M. & Stocker, R.K. (2008) Consistent accuracy of the Australian weed risk assessment system across varied geographies. *Diversity and Distributions*, **14**, 234–242.

Gullino, M.L., Fletcher, J. & Stack, J.P. (2008) Crop biosecurity: definitions and role in food safety and food security. In *Crop biosecurity – assuring our global food supply.* (ed. M.L. Gullino, J. Fletcher, A. Gamliel and J.P. Stack) pp. 1–11. Springer, Dordrecht.

Hart, A., Roelofs, W., Crocker, J., Murray, N., Boatman, N., Hugo, S., Fitzpatrick, S. & Flari, V. (2006) *Quantitative Approaches to Risk Assessment of GM Crops.* Department for Environment, Food and Rural Affairs, UK.

Heather, N.W. & Hallman, G.J. (2008) *Pest management and phytosanitary trade barriers.* CABI International, Wallingford, UK.

Hoagland, P. & Jin, D. (2006) Science and economics in the management of an invasive species. *Bioscience*, **56**, 931–935.

Holt, J., Black, R. & Abdallah, R. (2006) A rigorous yet simple quantitative risk assessment method for quarantine pests and non-native organisms. *Annals of Applied Biology*, **149**, 167–173.

Hulme, P.E. (2007) Biological Invasions in Europe: Drivers, Pressures, States, Impacts and Responses. In *Biodiversity under Threat* (ed. R. Hester and R.M. Harrison) pp. 56–80 (*Issues in Environmental Science and Technology*, **25**), Royal Society of Chemistry, Cambridge, UK.

Hulme, P.E. (2009) Trade, transport and trouble: managing invasive species pathways in an era of globalization. *Journal of Applied Ecology*, **46**, 10–18.

Hulme, P.E., Bacher, S., Kenis, M., et al. (2008): Grasping at the routes of biological invasions: a framework for integrating pathways into policy. *Journal of Applied Ecology*, **45**, 403–414.

Hulme, P.E., Nentwig, W., Pyšek, P. & Vilà M. (2009) Common market, shared problems: time for a coordinated response to biological invasions in Europe? *Neobiota* **8**, 3–19.

IUCN (2000) *Guidelines for the Prevention of Biodiversity Loss Caused by Alien Invasive Species.* IUCN, Gland, Switzerland.

Jeschke, J.M. & Strayer, D.L. (2008) Usefulness of bioclimatic models for studying climate change and invasive species. *Annals of the New York Academy of Sciences*, **1134**, 1–24.

Jones, K.E., Patel, N.G., Levy, M.A., et al. (2008) Global trends in emerging infectious diseases. *Nature*, **451**, 990–993.

Kahn-Ribeiro, S., Kobayashi S., Beuthe M., et al. (2007) Transport and its infrastructure. In *Climate Change 2007: Mitigation. Contribution of Working Group III to the Fourth Assessment Report of the Intergovernmental Panel on Climate Change* (ed. B. Metz, O.R. Davidson, P.R. Bosch, R. Dave and L.A. Meyer), pp. 323–385. Cambridge University Press, Cambridge, UK.

Keller, R.P. & Drake, J.M. (2009) Trait based risk assessment for invasive species. In *Bioeconomics of Invasive Species: Integrating Ecology, Economics, Policy and Management* (ed. R.P. Keller, D.M. Lodge, M.A. Lewis and J.F. Shogren), pp. 44–62. Oxford University Press, Oxford.

Keller, R.P., Lewis, M.A., Lodge, D.M., Shogren, J.F. & Krkošek, M. (2009) Putting bioeconomic research into practice. In *Bioeconomics of Invasive Species: Integrating Ecology, Economics and Management* (ed. R.P. Keller, D.M. Lodge, M.A. Lewis and J.F. Shogren), pp. 266–284. Oxford University Press, Oxford, UK.

Lambdon, P.W. & Hulme, P.E. (2006) Predicting the invasion success of Mediterranean alien plants from their introduction characteristics. *Ecography*, **29**, 853–865.

Landis, W.G. (2004) Ecological risk assessment conceptual model formulation for nonindigenous species. *Risk Analysis*, **24**, 847–858.

Lodge, D.M., Williams, S., MacIsaac, H.J., et al. (2006) Biological invasions: recommendations for US policy and management. *Ecological Applications*, **16**, 2035–2054.

Maguire, L.A. (2004) What can decision analysis do for invasive species management? *Risk Analysis*, **24**, 859–868.

NRC (2002) *Predicting Invasions of Nonindigenous Plants and Plant Pests.* National Academy Press, Washington, DC.

NRC (2006) *Globalization, Biosecurity, and the Future of the Life Sciences.* National Academies Press, Washington, DC.

Panetta, F.D. & Mitchell, N.D. (1991) Bioclimatic prediction of the potential distributions of some weed species prohibited entry to New Zealand. *New Zealand Journal of Agricultural Research*, **34**, 341–350.

Pheloung, P.C., Williams, P.A. & Halloy, S.R. (1999) A weed risk assessment model for use as a biosecurity tool evaluating plant introductions. *Journal of Environmental Management*, **57**, 239–251.

Pimentel, D., McNair, S., Janecka, J., et al. (2001) Economic and environmental threats of alien plant, animal, and microbe invasions. *Agriculture Ecosystems & Environment*, **84**, 1–20.

Rafoss, T. (2003) Spatial stochastic simulation offers potential as a quantitative method for pest risk analysis. *Risk Analysis*, **23**, 651–661.

Ricciardi, A. & Cohen, J. (2007) The invasiveness of an introduced species does not predict its impact. *Biological Invasions*, **9**, 309–315.

Richardson, D.M., Pyšek, P., Rejmánek, M., Barbour, M.G., Panetta, F.D. & West, C.J. (2000) Naturalization and invasion of alien plants: concepts and definitions. *Diversity and Distributions*, **6**, 93–107.

Shine, C. (2007) Invasive species in an international context: IPPC, CBD, European Strategy on Invasive Alien Species and other legal instruments. *EPPO Bulletin*, **37**, 103–113.

Smith, K. & Petley, D.N. (2009) *Environmental Hazards. Assessing Risk and Reducing Disaster*, 5th edition. Taylor & Francis, Abingdon, UK.

Stirling, A. (2008) Science, precaution and the politics of technological risk: converging implications in evolutionary and social scientific perspectives. *Annals of the New York Academy of Sciences*, **1128**, 95–110.

Stohlgren, T.J. & Schnase, J.L. (2006) Risk analysis for biological hazards: what we need to know about invasive species. *Risk Analysis*, **26**, 163–173.

Vilà, M., Basnou, C., Pyšek, P., et al. (2010) How well do we understand the impacts of alien species on ecosystem services? A pan-European, cross-taxa assessment. *Frontiers in Ecology and the Environment*, **8**, 135–144.

Vose, D.J. (2000) *Risk analysis – A Quantitative Guide*, 2nd edition. John Wiley, London.

Waage, J.K., Fraser, R.W., Mumford, J.D., Cook, D.C. & Wilby, A. (2005) *A New Agenda for Biosecurity*. Horizon Scanning Programme, Department for Environment, Food and Rural Affairs, London.

Walther, G.R., Roques, A., Hulme, P.E., et al. (2009) Alien species in a warmer world: risks and opportunities. *Trends in Ecology & Evolution*, **24**, 686–693.

Weber, J., Panetta, F.D., Virtue, J. & Pheloung, P. (2009) An analysis of assessment outcomes from eight years' operation of the Australian border weed risk assessment system. *Journal of Environmental Management*, **90**, 798–807.

Wilkinson, M.J., Elliott, L.J., Allainguillaume, J., et al. (2003) Hybridization between *Brassica napus* and *B. rapa* on a national scale in the United Kingdom. *Science* **302**, 457–459.

Williams, P.A. & Timmins, S. (2002) Economic impacts of weeds in New Zealand. In *Environmental and Economic Costs of Alien Plant, Animal and Microbe Invasions* (ed. D. Pimentel), pp. 175–184, CRC Press, Boca Raton, Florida.

Williamson, M. (1996) *Biological Invasions*. Chapman and Hall, London.

Williamson, M., Dehnen-Schmutz, K., Kuhn, I., et al. (2009) The distribution of range sizes of native and alien plants in four European countries and the effects of residence time. *Diversity and Distributions*, **15**, 158–166.

Chapter 24

ELTON AND THE ECONOMICS OF BIOLOGICAL INVASIONS

Charles Perrings

School of Life Sciences, Arizona State University, Tempe, AZ 85287, USA

Fifty Years of Invasion Ecology: The Legacy of Charles Elton, 1st edition. Edited by David M. Richardson
© 2011 by Blackwell Publishing Ltd

24.1   BIOINVASIONS AS AN ANTHROPOGENIC PROBLEM

Among ecologists, Elton's (1958) book is seen as a major landmark in our understanding of the development of the ecology of invasive species (Richardson & Pyšek 2007; but see Simberloff, this volume). The book also marks the beginning of a more general appreciation of the anthropogenic dimensions of what McNeely (2001) was later to call 'The great reshuffling'. The origins of the phenomenon that Elton described as 'one of the great historical convulsions in the world's fauna and flora' lie in the economic activities of people. Which species succeed in establishing after introduction depends, at least in part, on traits such as generalism, plasticity and so on (Williamson 1996). However, which systems are most vulnerable to invasion depends on the 'disturbance' caused by agriculture, forestry, urban and industrial development (Pyšek et al. 2010). And which species are introduced depends on patterns of trade, transport and travel (Perrings et al. 2010b). Elton contributed a great deal to our understanding of the invasiveness of species and the invasibility of systems. He contributed rather less to our understanding of the relation between introductions, trade and travel, but he set the stage for later analysis of them.

Elton also made it clear that the consequences of biological invasions go far beyond the impacts that have generally been identified in the ecological literature since that time. The ecological impacts include the effects of bioinvasions on species abundance and richness (Glowka et al. 1994; Williamson 1996, 1999; Rhymer & Simberloff 1996), and consequential changes in ecological functioning (Hawkes et al. 2005; Mack et al. 2000) and resilience (Walker et al. 2004; Kinzig et al. 2006). However, the dominant impacts on human well-being stem from the far more direct effects that invasive species have on human, animal and plant health, and on the production of foods, fuels and fibres (Perrings et al. 2010c). Elton recognized that the most costly invaders are pathogens. Introduced pathogens have historically inflicted more harm on human societies than any other class of invasive species, and they remain the single greatest environmental threat to human well-being. Pathogens are implicated in many of the main ecological impacts of invasive species (Ostfeld et al. 2008). However, they are also an ever-present source of risk in agriculture, forestry and fisheries, and a direct threat to human health and social security (Wingfield et al., this volume). For example, the diseases introduced into Central America during the Columbian exchange – smallpox, measles and typhus – not only decimated local human populations, but destroyed their capacity to resist the human invaders who had vectored the diseases into the area (McNeill 1977; Diamond 1997).

This chapter considers the development of the literature on the anthropogenic dimensions of bioinvasions since the publication of Elton's book. Because people are an integral part of the story, it goes beyond the ecological literature to consider the economic drivers and consequences of invasions, and the way in which ecology, epidemiology and economics have been integrated in models of bioinvasions. There are, however, several parallels with the way in which Elton approached the topic. His emphasis on the effect of invasive species on ecological functioning, and the stress he laid on the intimate connection between invasive species and the simplification of ecosystems maps into the economists' approach to the problem. If we wish to understand the impact of the introduction of any species on human well-being, we need to understand (i) how people depend on the system being affected by the introduced species, (ii) the effect of the introduction on ecosystem functioning and (iii) the implications of a change in ecosystem functioning for the delivery of valued ecosystem services. Invasive species are significant primarily because of their impact on the processes that yield valued services. Human preferences, values and behaviour are critical at each stage.

To understand the way in which people may be expected to prioritize the threats posed by different invasive species, we need to understand the value at risk from their introduction. Whether people care about badgers (or any other species) depends on the impact of badgers on the things that directly affect human well-being. The poster child of invasive species in the USA, the zebra mussel (*Dreissena polymorpha*), is valued because of the damage it inflicts not just on ecosystems but on power plants and water supply facilities. By comparison with the pathogens that have resulted in global pandemics among humans, such as the HIV virus or the Spanish Flu, its impact on human well-being has been trivial. However, compared with many other invasive species that exercise conservationists, the control costs of zebra mussels loom large. In all cases, the potential impact of bioinvasions is a critical part of any calculation of the risk they pose, and hence of the benefit cost calculus involved in undertaking the activities involved. The past 50 years have seen a significant improvement in our understanding of the precise ways in which human behaviour affects the introduction, establishment, spread and impact of invasive species. So the ways in which invasive species risks should be factored into our decisions.

The literature on the economics of invasive species has expanded dramatically since the publication of Elton's book. For surveys see Lovell et al. (2006) on the economics of aquatic invasive species, or Evans (2003) and Eisworth and Johnson (2002) on terrestrial invaders, the last in the context of a paper developing a general model for the management of invasive species. Stutzman et al. (2004) offer an annotated bibliography of economics of invasive plants. For the most part, this literature has involved the application of established principles of conservation to the valuation and optimal control of pests and pathogens. The basic principles of resource conservation had been established by Hotelling (1931), who identified the condition that should hold for people to be indifferent between conserving a set of resources in one state, or allowing the conversion of those resources to some other state. Although Hotelling was thinking of the problem of extracting exhaustible resources, like oil, the principle he identified applies to any set of resources whose value may be expected to change over time, whether because of their increasing scarcity, or to increasing demand (driven by changes in preferences, technology or by consumption growth). He showed that people will be indifferent between the conversion or conservation of some set of resources when the growth in their value in the conserved state is equal to the growth in their value from conversion to some alternative state. If conversion involves the sale of resources, the growth in their value from conversion can be approximated by the rate of return on investment of the proceeds of sale. So the Hotelling principle implies that people will be indifferent between conservation or conversion when the expected growth in value of a conserved resource is equal to the expected rate of return on financial assets (the rate of interest). It follows that conversion of ecosystems to enhance the production of foods, fuels or fibres, or to build roads, dams or cities, occurs because people expect that the value of the converted system will be greater than the value of the unconverted system. However, although conversion increases the value at risk from pests and pathogens, it frequently also increases the vulnerability of the system (the likelihood that introduced pests and pathogens will establish and inflict damage).

The economic literature since Elton has focused on two dimensions of this problem. The first is the calculation of the damage costs of the introduction, establishment and spread of potentially harmful alien species. Because damage costs are generally an externality of international markets (international trade), they are not reflected in the costs of undertaking trade. It follows that development of an appropriate control strategy or control policy requires estimation of the damage to be expected from the introduction of particular species or classes of species, or of trade along particular pathways. The second dimension of the problem addressed by economists is the identification of efficient control strategies or policies. This dimension is complicated by the fact that control of invasive species is a public good at several different levels: national, regional and global. Because it is a public good, it will be undersupplied if left to the market. Because public goods, once provided, are free to all, everyone has an incentive to free-ride on their provision.

This chapter is organized in five sections. The following section describes the way in which the drivers of bioinvasions have been treated outside of the ecological literature. Sections 24.3 and 24.4 then address the two dimensions of the problem that have been the main focus of the economic literature since Elton: estimation of the damage costs of invasions and identification of efficient control strategies. A final section draws some general conclusions.

24.2 THE HUMAN DRIVERS OF BIOINVASIONS

The main factors behind the risks posed by invasive species comprise the characteristics of the source and sink systems: where the introduced species have come from and where they are going to, the vectors involved in the introduction of species, and the characteristics of the introduced species themselves. The last of these include the traits that make species more or less invasive, and are largely (though not entirely) independent of human activities. The most important human drivers of bioinvasions are those affecting (i) the introduction of species through trade and travel and (ii) the vulnerability of host systems as a result of land use change.

Trade and travel

The dependence of the rate of introduction of invasive species on the growth of trade and travel is widely recognized. The opening of new markets or trade routes has resulted in the introduction of new species either as the object of trade or as the unintended consequence of trade, whereas the growth in the volume of trade along existing routes has increased propagule pressure (the frequency with which introductions are repeated), and hence the probability that an introduced species will establish and spread (Cassey et al. 2004; Semmens et al. 2004). We know that the more open economies are, the more vulnerable they are to biological invasions (Dalmazzone 2000; Vila & Pujadas 2001). We also know that the volume and direction of trade are good empirical predictors of introductions (Levine & D'Antonio 2003; Costello et al. 2007). In the same way, the volume of market transactions within a country turn out to be good empirical predictors of the establishment and spread of introduced species. Dehnen-Schmutz et al. (2007) found that the most heavily advertised non-native ornamental species sold in Britain in the 19th century were also the most likely to become invasive.

Whether the introduction of invasive species was deliberate or accidental varies by taxa. Most invasive plants, for example, were introduced as the objects of the agricultural or horticultural trade (Reichard & White 2001). In Australia, for example, 65% of plant species that became established between 1971 and 1995 were introduced as ornamentals (Groves 1998). In Germany, 50% of current alien plants were deliberately introduced, and more than half of these were objects of the horticultural trade (Kuhn & Klotz 2002). Most invasive mammals and fish were similarly deliberately introduced, frequently through the pet trade (Sakai et al. 2001). At the same time, many of the most damaging aquatic species introductions, such as the zebra mussel (*Dreissena polymorpha*) and the Asian clam (*Corbicula fluminea*) were introduced accidentally through ballast water exchange in ships carrying trade goods (Margolis et al. 2005). Most terrestrial pests were introduced in the same way: as 'passengers' on trade goods, or more particularly as 'passengers' on goods shipped between ecosystems, whether traded or not. For example, parthenium weed from Mexico was first detected in Ethiopia in 1988 near food-aid distribution centres, implying that it had accompanied wheat grain distributed as food aid during the drought (Global Invasive Species Program 2004).

The position with pathogens is generally different. Few pathogens have been deliberately introduced as the objects of trade. Most plant and animal pathogens are introduced through trade in infected plants and animals, or in infected plant and animal products. Grey leaf spot (*Circosporda zeae-maydis*), a disease that has had dramatic effects on maize yields in Africa was, for example, thought to have been introduced in US food aid shipments of maize in during the drought years of the 1980s (Ward et al. 1999). There is also strong evidence for the role of trade in the emergence of diseases such as H5N1 (Kilpatrick et al. 2006), West Nile virus (Lanciotti et al. 1999), severe acute respiratory syndrome (SARS) (Guan et al. 2003), and a series of key livestock diseases such as H9N2 avian influenza, bovine spongiform encephalopathy or foot and mouth disease (Rweyemamu & Astudillo 2002; Karesh et al. 2005; Fevre et al. 2006). There is also a growing literature on the disease risks of trade (Perrings et al. 2002, 2005; Hufnagel et al. 2004; Tatem et al. 2006; Smith et al. 2008). The volume and direction of trade turn out to be good predictors of the most likely sources of zoonoses (Pavlin et al. 2009; Smith et al. 2009), as well as the likelihood of transmission of individual human diseases such as West Nile virus, H5N1 avian influenza and 2009 A/H1N1 influenza. Jones et al. (2008) noted that of 335 emerging infectious disease 'events' recorded since the Second World War, 60% involved zoonoses, of which 72% originated in wildlife. They also noted that diseases with this origin were increasing significantly over time, that origins of emerging infectious diseases were significantly correlated with socio-economic, environmental and ecological factors, and that this provided a basis for identifying regions where new emerging infectious diseases were most likely to originate, mostly in tropical areas.

Land use change

The dependence of the rate of establishment on human activity is less widely recognized. This is partly because some dimensions of the vulnerability of host systems are independent of human activities, the bioclimatic distance from the source country for example. However, most depend on human intervention. The existence of predators or competitors of the introduced species, for example, depends not only on the bioclimatic distance between source and host systems but also on the way in which the host system has been transformed by human activities. The simplification of many host systems to enhance their productivity in terms of valued foods, fuels or fibres makes them particularly vulnerable. The fragmentation of systems that have not otherwise been transformed has a similar effect.

The Millennium Ecosystem Assessment concluded that 'human actions are fundamentally, and to a significant extent irreversibly, changing the diversity of life on earth and most of these changes lead to a loss of biodiversity. Changes in important components of biological diversity were more rapid in the past 50 years than at any time in human history' (Millennium Ecosystem Assessment 2005). It also noted that this had largely been due to the conversion of ecosystems for the production of foods, fuels and fibres, the key provisioning services of ecosystems. The rate of conversion has reflected both the increase in human population in the period, and the growth in consumption per person of these services. More land was converted to cropland in the 30 years after 1950 than in the 150 years between 1700 and 1850, with the result that less than 20% of many biomes remained as viable habitat for non-cultivated species.

The trend extends beyond cropland, however. More land has been converted for all forms of human use in the decades since the publication of Elton's book than in the period cited in the Millennium Ecosystem Assessment. Growth of the human population has led to land conversion for the construction of cities. In many cases, this has been arable and pasture land, and its conversion is what has led to the expansion of the agricultural frontier – and the contraction of remaining wildlife refugia. Marine systems have been put under similar pressure. Wild capture fisheries have been systematically extended to meet the demand for protein. The most frequently cited cause of marine stress in pelagic and epipelagic systems, for example, is overfishing (Pauly et al. 2002; Myers & Worm 2003; Hughes et al. 2005), with by-catch (Lewison et al. 2004), loss of habitat (Pandolfi et al. 2003) and climate change being contributory factors (Hughes et al. 2003). The Millennium Ecosystem Assessment (2005) also noted that the number of invasive marine species has been increasing exponentially, and the spread of

pathogens has been identified as a source of stress in marine and terrestrial systems alike. There is evidence that the resilience of coral reefs and kelp forests has been affected in ways similar to many terrestrial systems, and for similar reasons (Steneck et al. 2004; Hughes et al. 2005).

The trend has two important implications for the science of biological invasions. One directly bears on Elton's conclusion that greater species richness makes ecosystems less vulnerable to invasion: the biotic resistance argument. If this argument is correct, simplified agroecosystems may be expected to be more vulnerable to invasions than other ecosystems. Although the evidence for biotic resistance in 'natural' or at least less heavily impacted systems is ambiguous, and depends on the scale at which the analysis is undertaken (Levine & D'Antonio 1999), the evidence for agroecosystems is unambiguous. Simplified agroecosystems are significantly more susceptible to introduced pests and pathogens than other ecosystems, and their susceptibility is greater the fewer the species admitted to the system (Jackson et al. 2007). The transformation of ecosystems through the construction of the built environment – the development of cities – has similarly increased their susceptibility to a range of pests and pathogens. Indeed, the management of most simplified systems is largely about the exclusion of species that either compete with or predate on cultivated or protected species. Whether undertaken by public health officials, farmers, households, pest control agencies, conservation groups or others, the routine of managing simplified systems is the control of potentially invasive species.

The second implication is that conversion of ecosystems to alternative uses, including the production of foods, fuels and fibres and the development of the built environment, changes the value at risk from invasive species, and so the prioritization of control activities. Human intervention in ecosystems is generally designed to enhance their value to people, and this increases the value at risk from introduced pests and pathogens. A crude proxy for this is the price of land. Where land markets are reasonably well developed, the value of the services they yield is reflected in the price that people are willing to pay to acquire (or willing to accept to sell) land. The conversion of land to 'productive' uses is reflected in an increase in its price. The important point here is that the resources that people commit to protect against the pests or pathogens depends on the value under threat – the value at risk. People will commit resources to protect against pests or pathogens up to the point at which the expected benefits of protection (the damage avoided) is just equal to the costs of protection.

24.3 CALCULATING THE DAMAGE COST OF INVASIVE SPECIES

The most widely reported estimates of the damage cost of invasive species are due to David Pimentel. In a series of papers, Pimentel has produced a succession of estimates of the cost of control and damage due to the most common invasive alien pests and pathogens. Before Pimentel's studies, the Office of Technology Assessment of the US Congress (OTA 1993) had reported that the 79 most harmful invasive species recorded in the USA since 1906 had caused damages of around $97 billion. Since then, Pimentel and colleagues have sought to update the OTA estimates and to extend them beyond the USA (Pimentel et al. 2000, 2001, 2005). The second of the Pimentel papers included estimates for three developed (the USA, the UK and Australia) and three developing (South Africa, India and Brazil) countries. Although the exercise is far from complete, it is still the most comprehensive summary of the control costs and lost output associated with invasive species in three productive sectors: agriculture, forestry and fisheries.

The results are summarized in Tables 24.1 and 24.2. They represent a simple sum of various dollar estimates of annual damage costs in the countries concerned made over the preceding decade. There are no estimates of the benefits that accrued from activities leading to the introduction of invasive pests, and the estimates of costs reflect the patchiness of the literature on the topic.

Taking agricultural gross domestic product in 1999 as the numeraire, the estimates reported in Table 24.1 show that invasive species caused damage costs equal to 53% of agricultural gross domestic product in the USA, 31% in the UK and 48% in Australia. By contrast, damage costs in South Africa, India and Brazil were, respectively, 96%, 78% and 112% of agricultural gross domestic product. It is worth repeating, though, that there is considerable uncertainty about the Pimentel estimates given the estimation methods used and the incompleteness of the data. Aside from the Pimentel estimates, there is a growing number of case studies of the damage costs of particular invasive species. Many of these focus on the USA and are summarized in Stutzman et al. (2004). Examples among invasive plants include leafy spurge (Leistritz et al. 2004), tansy ragwort (Coombs et al. 1996), yellow starthistle (Jetter et al. 2003) and tamarisk (Zavaleta 2000). Examples among aquatic species include the zebra mussel, *Dreissena polymorpha* (O'Neill 1997), the green crab, *Carcinus maenas* (Cohen et al. 1995) and lake trout (Mcintosh et al. 2009).

Table 24.1 Economic losses to introduced pests in crops, pastures, and forests in the USA, UK, Australia, South Africa, India and Brazil (billions of (2001) US dollars per year).

Introduced pests	USA	UK	Australia	South Africa	India	Brazil	Total
Weeds							
Crops	27.9	1.4	1.8	1.5	37.8	17.0	87.4
Pastures	6.0	–	0.6	–	0.9	–	7.5
Vertebrates							
Crops	1.0	1.2	0.2	–	–	–	2.4
Arthropods							
Crops	15.9	1.0	0.9	1.0	16.8	8.5	44.1
Forests	2.1	–	–	–	–	–	2.1
Plant pathogens							
Crops	23.5	2.0	2.7	1.8	35.5	17.1	82.6
Forests	2.1	–	–	–	–	–	2.1
Total	78.5	5.6	6.3	4.3	91.0	42.6	228.7

Source: Pimentel et al. (2001).

Table 24.2 Environmental losses to introduced pests in the USA, UK, Australia, South Africa, India and Brazil (billions of (2001) US dollars per year)

Introduced pests	USA	UK	Australia	South Africa	India	Brazil	Total
Plants	0.15	–	–	0.10	–	–	0.18
Mammals							
Rats	19.00	4.10	1.20	2.70	25.00	4.40	56.40
Other	18.12	1.20	4.66	–	–	–	23.96
Birds	1.10	0.27	–	–	–	–	1.37
Reptiles/Amphibians	0.01	–	–	–	–	–	0.01
Fishes	1.00	–	–	–	–	–	1.00
Arthropods	2.14	–	0.23	–	–	–	2.37
Molluscs	1.31	–	–	–	–	–	1.31
Livestock diseases	9.00	–	0.25	0.10	–	–	9.35
Human diseases	6.50	1.00	0.53	0.12	–	2.33	10.47
Total	58.30	6.57	6.87	3.01	25.00	6.73	106.48

Source: Pimentel et al. (2001).

There are fewer examples of pest-specific studies of damage costs outside the USA, although several introduced pests and pathogens are known to have had particularly severe effects on crop yields in sub-Saharan Africa, including witchweed (*Striga hermonthica*), grey leaf spot (*Circosporda zeae-maydis*), the large grain borer (*Prostephanus truncatus*), cassava mealybug (*Phenacoccus manihoti*) and the cassava green mite (*Mononychellus tanajoa*) (Rangi 2004). Among aquatic species, there have been studies of the effects of the comb jelly, *Mnemiopsis leidii*, in the Black Sea (Knowler 2005). Within the African Lakes, Kasulo (2000) analysed the ecological and socio-economic impact of invasive species in African lakes, focusing on introduced fish species and water weeds: the Nile perch (*Lates niloticus*), the Tanganyika sardine (*Limnothrissa miodon*) and water hyacinth (*Eichhornia crassipes*) into Lakes Victoria, Kyoga, Nabugabo, Kariba, Kivu, Itezhi-tezhi and Malawi. There are no estimates of damage costs, but the introduction of Nile perch – which had a major impact on the structure and profitability of fisheries – is believed to have caused the extinction of up to 350 haplochromine cyclid species. Kasulo does, however, offer an estimate of the damage costs associated with the water hyacinth, which has spread through most African lakes obstructing water passages and displacing native plants, fish and invertebrates. His estimate of the annual cost of the hyacinth in terms of its impact on fisheries in this group of lakes was US$71.4 million.

The method generally followed to arrive at estimates of this kind requires an understanding of the 'production functions' for each of the services: the functions that connect the biotic and abiotic components of ecosystems, the ecosystem functions and processes supported by those components, and the supply of the services from ecosystems that people value. It further requires an understanding of the way that those functions are affected by invasive species. If the values of the final services are known, it is possible to derive the values of the marginal impacts of a change in the abundance of the invasive species (Barbier 2000, 2007).

The importance of the approach is that without information on the value of the marginal impact of introduced species, it is not possible to develop an efficient strategy for their control. Although there are many studies of the direct effectiveness of various interventions, almost none base control strategies on the relative costs and benefits of alternative management actions. Because recent developments in research on the value of the services delivered by specific ecosystems make estimation of the marginal value of control actions more reliable, we should expect the costs and benefits of control to be taken into account more frequently in the future. Particularly promising is work that identifies ecosystem services at the landscape scale (Daily & Matson 2008; Nelson et al. 2008, 2009; Polasky et al. 2008), and that is linked to the development of decision-support tools at that same scale (Tallis

& Polasky 2009). Much of this research is spatially explicit, and maps ecosystem services to the landscape in question. It also examines trade-offs between services in particular locations (Chan et al. 2006). Although these studies have not yet been adapted to the valuation and control of invasive species, they have the potential to do so.

24.4   THE CONTROL OF INVASIVE SPECIES

The second aspect of the invasive species problem addressed by economists is the identification of efficient control strategies or policies. Given that the invasive species problem subsumes pest and disease management in many different areas – agriculture, forestry, fisheries, human health – as well as multiple pathways, there is no single modelling strategy that can encompass all cases. Most contributions use some form of dynamic optimization. The most common approaches involve stochastic dynamic programming and optimal control (Eisworth & Johnson 2002; Olson & Roy 2002; Lovell et al. 2006; Olson 2006). The main control options considered include the prevention of introductions (Horan et al. 2002; Sumner et al. 2005), the control of the spread of species (Sharov & Liebhold 1998a,b; Heikkila & Peltola 2004; Cacho et al. 2008), the eradication of an introduced species (Olson & Roy 2002) or the choice between prevention and control (Leung et al. 2002; Finnoff & Tschirhart 2005; Olson & Roy 2005).

Modelling the control of invasive species

Polasky (2010) uses a stochastic dynamic programming approach to model the sequence of management actions – from prevention to control. He shows how the choice between management actions depends on relative costs that, in turn, depend on the state of the invasion process. Port inspections generate the information needed to prevent introductions (generally the least costly management action). Detection of escapees generates information needed to undertake either eradication or control. He establishes the conditions that need to be satisfied for management effort to be allocated optimally between these two actions, and shows how the management options interact.

Specifically, he shows that prevention and control are substitute actions, in the sense that a reduction in the cost or an increase in the effectiveness of one will reduce the resources optimally devoted to the other. However, the information gathering required to support each action, port inspections and detection of escapees respectively, complements that action. A reduction in the cost of port inspections, or an increase in their effectiveness, increases the resources optimally devoted to prevention. Similarly, a reduction in the cost of detecting escapees, or an increase in the effectiveness of the programme for monitoring escapees, increases the resources optimally devoted to control.

To determine the optimal level of alternative management actions, the decision maker needs to know both the costs and the benefits of those actions, where the benefits are the damage avoided by the management actions. As noted above, estimation of the damage avoided through prevention or control requires an understanding of the 'production functions' for the ecosystem services at risk, and the way those production functions are impacted by the introduction of a potentially harmful species. At the core of any economic model of the control of invasive species are the set of functions describing (i) how the production of valued ecosystem services are affected by the spread of an introduced pest or pathogen and (ii) how the control options affect that spread. For example, Perrings et al. (2010b) model the growth of an introduced species as a process that has both density-dependent and density-independent elements. The propagule pressure associated with imports into an area is the density-independent element, whereas the spread of the species once established is density dependent. Prevention acts on the density-independent growth, control acts on the density-dependent growth of the species concerned. The damage associated with an introduced species is then modelled through a production function that captures the impact of the abundance of the introduced pest or pathogen on the value of output, and the management actions are the control variables in an optimal control problem. The solution balances the marginal costs and benefits of the action.

It follows that the optimal level of control of any given pest or pathogen will be sensitive to the value at risk, namely the value of the services being produced that is potentially threatened by the introduction of that pest or pathogen. Understanding the value at risk is essential if we are to calculate the optimal level of the control for any potentially invasive species.

It is interesting to note that although few current bio-invasion control programmes are based on explicit estimates of the value at risk, most reflect that value in some way. The fact that significantly more resources are committed to controlling the spread of human pathogens than plant or animal pathogens, for example, reflects the higher value placed on human well-being, than on the well-being of other species. The link between value at risk and the resources committed to control measures for a single class of trade-transmitted pathogens, animal diseases notified to the World Organisation for Animal Health (OIE) is illustrated by Perrings et al. (2010b), who show that whereas the incidence of disease events is generally increasing in the volume of trade in risk materials – animals and animal products – the incidence of disease events involving pathogens that have the capacity to shut down the market, like foot and mouth disease, is not. For such diseases the value of the agricultural output and merchandise trade at risk is negatively correlated with the frequency of disease events. The greater the value at risk, the greater the effort to control the spread of pathogens.

Despite the economic literature on bioinvasions, this dimension of risk is still largely ignored in the literature on invasive species risk assessment and management. Most research since Elton has focused on the probability that invasive species will spread, rather than the damage they will cause if they do spread. This is problematic for two reasons. First, without an assessment of damage costs, it is not possible to evaluate risk. This is because risk is the product of the probability of an outcome and the value of that outcome. Second, the probability that an introduced species will spread is not independent of the value at risk. This is because (i) the probability of spread depends on human behaviour with respect to the introduced species and (ii) human behaviour with respect to the introduced species depends on the value at risk.

The control of bioinvasions as an international public good

What makes the management of bioinvasions difficult relative to many other environmental issues is both that it is an international problem and that the control of invasive species is a public good. The fact that it is an international problem means that the only admissible solutions are those that can be reached through international agreement. The fact that control of invasive species is a public good means that it will be undersupplied if left to the market. Individual countries are able to take sanitary and phytosanitary measures to reduce disease transmission risks both for imports and exports. They will, however, only take account of the risk they impose on others through exports if it directly affects export earnings. As a general proposition, international transmission rates for particular categories of emerging infectious disease will be higher the more tightly connected countries are through the trade system, the greater the volume of trade and the lower each country's formal responsibility for the disease risks it imposes on others. Because formal responsibilities to other countries are embodied in international agreements, this is a measure of the strength of agreements established to protect the global public good. This dimension of the problem is of much more recent concern than the others, and the literature on it is in an early stage of development. However, it may well turn out to be the most important.

At present, the global governance of invasive species risks is highly fractured, being divided into several areas – human, animal and plant health, conservation, agriculture, forestry and fisheries among others – and spanning a wide range of overlapping international agreements. The most important of the international agreements is the General Agreement on Tariffs and Trade (GATT). The GATT allows actions in restraint of trade where there are demonstrated threats to human, animal or plant health. However, it contains few details on what actions are either mandated or admitted. Admissible strategies are specified in various subsidiary agreements, and especially in the agreements that most directly address trade-related invasive species risks, the Sanitary and Phytosanitary (SPS) Agreement and the International Health Regulations (IHR).

Perrings et al. (2010a) review the options open to countries under these agreements. They note that under the SPS Agreement the default strategy has two parts: border protection and the control of or adaptation to introduced species that have escaped detection at the border. In other words, invasive species policy revolves around unilateral national defensive action as opposed to coordinated international action. By contrast, although the IHR also allows countries to adopt a defensive posture, it also mandates collective action to contain the spread of human pathogens. Perrings et al. (2010c) argue that the restrictions in the SPS

Agreement that prevent similar cooperative action are a major source of inefficiency, and that the solution to the problem requires more proactive global coordination and cooperation in the management of both pathways and sanitary and phytosanitary risks at all scales.

One reason for this is that the control of many invasive species is not just an international public good: it is what is known as a 'weakest link' public good (Sandler 1997, 2004). This means that the benefits of collective efforts to control bioinvasions are only as great as the benefits offered by the least effective country, the 'weakest link' in the chain. This is obvious in the case of infectious diseases. If all but one country were to eradicate the disease, and if the one country that failed to eradicate the disease continued to be integrated into the world system, all other countries would still be exposed. However, it is also true of many other pests and pathogens. Perrings et al. (2010a) claim that controlling the movement and spread of invasive species requires coordination across countries to identify common threats, implement prevention or management measures and ultimately to ensure that new outbreaks do not breach the weakest points in the biosecurity system. Indeed, few countries have the resources, the biogeographical conditions, and the institutional, statutory and regulatory environment to mount effective defences against invasive species. New Zealand and Australia are often cited as the best examples of effective defensive strategies, but even in those countries there are concerns about biosecurity capability. For most bioinvasion problems, cooperative efforts at the regional and global level offer a more cost-effective solution to the management of global risks than the current default strategy of independent national defensive efforts.

24.5 CONCLUSIONS

Although the literature explored in this chapter is largely from the social sciences, of the issues it addresses only the problem of international governance was not presaged in Elton's book. His broad definition of conservation: 'coexistence between man and nature ...' spans the protection of wildlife refugia, the production of foods, fuels and fibres, and the protection of public health. However, it also recognizes that there are trade-offs between many of the objectives of conservation, and that it is not possible to have everything. So the definition continues: '... even if it has to be a modi-

fied kind of man and a modified kind of nature' (Elton 1958, p. 145). This is the starting point for an economic analysis of the problem. Similarly, Elton's identification of the motivations behind conservation – an ethical concern for the rights of other species, an aesthetic and scientific interest in the diversity of life, and the production of the material commodities and conditions that support human well-being – maps neatly into the ecosystem services identified by the Millennium Assessment. It also maps neatly into the way that economists conceptualize the value that people place on nature (Bockstael et al. 2000).

Indeed, much of the research into the economics of bioinvasions has focused on understanding the trade-offs involved in developing strategies to manage invasions, and the potential damage costs associated with successful invasions. It is not as easy to find support for Elton's assertion that the dominant driver of the environmental changes that made ecosystems vulnerable to invasion is human population growth. We have a deeper understanding of the many factors behind the demand for environmental resources now, and even though population growth is a factor, it is a relatively minor one. In the 50 years since Elton's book, the global population has grown two and a half times. In the same period, the global economy has grown eight times, which implies that consumption per person has grown at a much faster rate than the population. More importantly, population growth over the period has been shown to have different impacts in different places (Lopez 1992; Cleaver & Schreiber 1994; Lopez & Scoseria 1996). The link between land conversion and population growth is, for example, tighter in regions where there are few well-defined property rights, such as in parts of sub-Saharan Africa. However, even there, whether population growth has had adverse effects on the environment depends on institutional conditions (Heath & Binswanger 1996).

More significant than changes in the vulnerability of ecosystems to invasion is the change in propagule pressure. In the same 50 years, merchandize exports have grown more than 30 times. Because every container imported to a country presents a sample of the species in the country of export, and because repeated introductions increase the likelihood of establishment, it is not surprising that the growth in trade and travel is the primary source of introductions. Indeed, this turns out to be much more important than growth in the human population. The implication of this is that developing appropriate governance systems for managing international public goods turns is much more important in the solution to the problem of bioinvasions than restraining fertility rates.

The risks of bioinvasions depend both on the likelihood of introduction, establishment and spread and the associated damage. Because these risks depend on the sanitary and phytosanitary capacity of exporting countries, one part of the solution is to build both capacity and commitment in exporting countries to reduce the risks they pose to others. Bilateral trade agreements can be useful in raising sanitary and phytosanitary standards in exporting countries, and the standards emerging out of the work of the three agencies that support the SPS Agreement – the Codex Alimentarius Commission for food safety, the International Office of Epizootics for animal health and the International Plant Protection Convention for plant health – provide a benchmark for that. Where achievement of particular sanitary and phytosanitary standards is a condition on exporters, they have an incentive to commit resources to the problem.

There remains the problem of the costs or benefits of trade imposed on third parties. Given this, Perrings et al. (2010b) argue for bringing the SPS Agreement into conformity with the IHR, as a first step in developing the capacity to respond to the global nature of many trade-related invasive species risks. Although some invasive species are highly localized in their effects, and therefore best dealt with at a local level, many others are not. This is especially the case with infectious diseases spread through international travel. They also argue for global support to individual countries to develop the appropriate capacity, suggesting that the establishment of a mechanism to monitor, evaluate and disseminate information on emerging invasive species risks would be a useful initiative.

In the 50 years since publication of Elton's book a great deal has been learned about the best ways of addressing problems of this kind. The fact that we are still a long way from having an adequate solution is a measure of just how difficult a matter it is. Not only does every country have an incentive to free-ride on the sanitary and phytosanitary measures undertaken by others, they also have an incentive to do nothing to harm their competitive position in the international markets in which they operate. Each exporting country faces a 'prisoner's dilemma': that although every country would be better off if all cooperated in the implementation of sanitary and phytosanitary measures on exports, none can afford to carry the cost of enhanced measures if their trading partners do not. The result is that no country implements such measures.

REFERENCES

Barbier, E.B. (2000) Valuing the environment as input: review of applications to mangrove-fishery linkages. *Ecological Economics*, **35**, 47–61.

Barbier, E.B. (2007) Valuing ecosystem services as productive inputs. *Economic Policy*, 178–229.

Bockstael, N.E., Freeman, A.M., Kopp, R.J., Portney, P.R. & Smith, V.K. (2000) On measuring economic values for nature. *Environmental Science & Technology*, **34**, 1384–1389.

Cacho, O.J., Wise, R.M., Hester, S.M. & Sinden, J.A. (2008) Bioeconomic modeling for control of weeds in natural environments. *Ecological Economics*, **65**, 559–568.

Cassey, P., Blackburn, T.M., Russel, G.J., Jones, K.E. & Lockwood, J.L. (2004) Influences on the transport and establishment of exotic bird species: an analysis of the parrots (Psittaciformes) of the world. *Global Change Biology*, **10**, 417–426.

Chan, K.M.A., Shaw, M.R., Cameron, D.R., Underwood, E.C. & Daily, G.C. (2006) Conservation planning for ecosystem services. *PLoS Biology*, **4**, 2138–2152.

Cleaver, K.M. & Schreiber, A.G. (1994) *Reversing the Spiral: The Population, Agriculture, and Environment Nexus in Sub-Saharan Africa*. World Bank, Washington, DC.

Cohen, A.N., Carlton, J.T. & Fountain, M.C. (1995) Introduction, dispersal and potential impacts of the green crab *Carcinus maenas* in San Francisco Bay, California. *Marine Biology*, **122**, 225–237.

Coombs, E.M., Radtke, H., Isaacson, D.L. & Snyder, S. (1996) Economic and regional benefits from biological control of tansy ragwort, *Senecio jacobaea*, in Oregon. In *International Symposium on Biological Control of Weeds* (ed. V.C. Moran and J.H. Hoffmann), pp. 489–494. University of Cape Town, Stellenbosch.

Costello, C., Springborn, M., Mcausland, C. & Solow, A. (2007) Unintended biological invasions: does risk vary by trading partner? *Journal of Environmental Economics and Management*, **54**, 262–276.

Daily, G.C. & Matson, P.A. (2008) Ecosystem services: from theory to implementation. *Proceedings of the National Academy of Sciences of the USA*, **105**, 9455–9456.

Dalmazzone, S. (2000) Economic factors affecting vulnerability to biological invasions. In *The Economics of Biological Invasions* (ed. C.W. Perrings, M. Williamson and S. Dalmazzone), pp. 17–30. Edward Elgar, Cheltenham, UK.

Dehnen-Schmutz, K., Touza, J., Perrings, C. & Williamson, M. (2007) The horticultural trade and ornamental plant invasions in Britain. *Conservation Biology*, **21**, 224–231.

Diamond, J. (1997) *Guns, Germs, and Steel*. W. W. Norton, New York.

Eisworth, M.E. & Johnson, W.S. (2002) Managing nonindigenous invasive species: insights from dynamic analysis. *Environmental and Resource Economics*, **23**, 319–342.

Elton, C.S. (1958) *The Ecology of Invasions by Animals and Plants*. Methuen, London.

Evans, E.A. (2003) Economic dimensions of invasive species. *Choices*, **June** 5–9.

Fevre, E.M., Bronsvoort, B.M.D.C., Hamilton, K.A. & Cleaveland, S. (2006) Animal movements and the spread of infectious diseases. *Trends in Microbiology*, **14**, 125–131.

Finnoff, D. & Tschirhart, J. (2005) Identifying, preventing, and controlling successful invasive plant speices using their phsyiological traits. *Ecological Economics*, **52**, 397–416.

Global Invasive Species Program (2004) Africa invaded: the growing danger of invasive alien species. Global Invasive Species Program, Cape Town.

Glowka, L., Burhenne-Guilmin, F. & Synge, H.. (1994) *A Guide to the Convention on Biological Diversity*. IUCN, Gland, Switzerland.

Groves, R.H. (1998) Recent incursions of weeds to Australia. *Technical Series* **3**, 1–74. CRC for Weed Management Systems.

Guan, Y., Zheng, B.J., He, Y.Q., et al. (2003) Isolation and characterization of viruses related to the SARS coronavirus from animals in southern China. *Science*, **302**, 276–278.

Hawkes, C.V., Wren, I.F., Herman, D.J. & Firestone, M.K. (2005) Plant invasion alters nitrogen cycling by modifying the soil nitrifying community. *Ecology Letters*, **8**, 976–985.

Heath, J. & Binswanger, H. (1996) Natural resource degradation effects of poverty and population growth are largely policy induce: the case of Colombia. *Environment and Development Economics*, **1**, 65–84.

Heikkila, J. & Peltola, J. (2004) Analysis of the Colorado potato beetle protection system in Finland. *Agricultural Economics*, **31**, 343–352.

Horan, R.D., Perrings, C., Lupi, F. & Bulte, E.H. (2002) Biological pollution prevention strategies under ignorance: the case of invasive species. *American Journal of Agricultural Economics*, **84**, 1303–1310.

Hotelling, H. (1931) The economics of exhaustible resources. *Journal of Political Economy*, **39**, 137–175.

Hufnagel, L., Brockmann, D. & Geisel, T. (2004) Forecast and control of epidemics in a globalized world. *Proceedings of the National Academy of Sciences of the USA*, **101**, 15124–15129.

Hughes, T., Bellwood, D.R., Folke, C., Steneck, R.S. & Wilson, J. (2005) New paradigms for supporting the resilience of marine ecosystems. *Trends in Ecology & Evolution*, **20**, 380–386.

Hughes, T.P., Baird, A.H., Bellwood, D.R., et al. (2003) Climate change, human impacts, and the resilience of coral reefs. *Science*, **301**, 929–933.

Jackson, L.E., Pascual, U., Brussaard, L., De Ruiter, P. & Bawa, K.S. (2007) Biodiversity in agricultural landscapes: Investing without losing interest. *Agriculture Ecosystems & Environment*, **121**, 193–195.

Jetter, K.M., Ditomaso, J.M., Drake, D.J., Klonsky, K.M., Pitcairn, M.J. & Sumner, D.A. (2003) Biological control of yellow starthistle. In *Exotic Pests and Diseases: Biology and Economics for Biosecurity* (ed. D.A. Sumner), pp. 225–241. Iowa State University Press, Ames, Iowa.

Jones, K.E., Patel, N.G., Levy, M.A., et al. (2008) Global trends in emerging infectious diseases. *Nature*, **451**, 990–994.

Karesh, W.B., Cook, R.A., Bennett, E.L. & Newcomb, J. (2005) Wildlife trade and global disease emergence. *Emerging Infectious Diseases*, **11**, 1000–1002.

Kasulo, V. (2000) The impact of invasive species in African lakes. In *The Economics of Biological Invasions* (ed. C. Perrings, M. Williamson and S. Dalmazzone), pp. 183–207. Edward Elgar, Cheltenham.

Kilpatrick, A. M., Chmura, A. A., Gibbons, D. W., Fleischer, R. C., Marra, P. P. & Daszak, P. (2006) Predicting the global spread of H5N1 avian influenza. *Proceedings of the National Academy of Sciences of the USA*, **103**, 19368–19373.

Kinzig, A.P., Ryan, P., Etienne, M., Elmqvist, T., Allison, H. & Walker, B.H. (2006) Resilience and regime shifts: assessing cascading effects. *Ecology and Society*, **11**.

Knowler, D. (2005) Reassessing the costs of biological invasion: Mnemiopsis leidyi in the Black sea. *Ecological Economics*, **52**, 187–199.

Kuhn, I. & Klotz, S. (2002) Floristischer Status und gebietsfremde Arten. *Schriftenreihe Vegetationskunde*, **38**, 47–56.

Lanciotti, R.S., Roehrig, J.T., Deubel, V., et al. (1999) Origin of the West Nile virus responsible for an outbreak of encephalitis in the northeastern United States. *Science*, **286**, 2333–2337.

Leistritz, F.L., Bangsund, D.A. & Hodur, N.M. (2004) Assessing the economic impact of invasive weeds: the case of leafy spurge (*Euphorbia esula*). *Weed Technology*, **18**, 1392–1395.

Leung, B.L., Lodge, D.M., Finnoff, D., Shogren, J.F., Lewis, M.A. & Lamberti, G. (2002) An ounce of prevention or a pound of cure: bioeconomic risk analysis of invasive species. *Proceedings: Biological Sciences*, **269**, 2407–2413.

Levine, J.M. & D'Antonio, C.M. (1999) Elton revisited: a review of evidence linking diversity and invasibility. *Oikos*, **87**, 15–26.

Levine, J.M. & D'Antonio, C.M. (2003) Forecasting biological invasions with increasing international trade. *Conservation Biology*, **17**, 322–326.

Lewison, R.L., Crowder, L.B., Read, A.J. & Freeman, S.A. (2004) Understanding impacts of fisheries bycatch on marine megafauna. *Trends in Ecology & Evolution* **19**, 598–604.

Lopez, R. (1992) Environmental degradation and economic openness in LDCs: the poverty linkage. *American Journal of Agricultural Economics*, **74**, 1138–1145.

Lopez, R. & Scoseria, C. (1996) Environmental sustainability and poverty in Belize: a policy paper. *Environment and Development Economics*, **1**, 289–308.

Lovell, S.J., Stone, S.F. & Fernandez, L. (2006) The economic impacts of aquatic invasive species: a review of the literature. *Review of Agricultural and Resource Economics*, **35**, 195–208.

Mack, R.N., Simberloff, D., Lonsdale, W.M., Evans, H., Clout, M. & Bazzaz, F.A. (2000) Biotic invasions: causes, epidemiology, global consequences, and control. *Ecological Applications*, **10**, 689–710.

Margolis, M., Shogren, J.F. & Fischer, C. (2005) How trade politics affect invasive species control. *Ecological Economics*, **52**, 305–313.

McIntosh, C.R., Finnoff, D.C., Settle, C. & Shogren, J.F. (2009) Economic valuation and invasive species. In *Bioeconomics of invasive species: integrating ecology, economics, policy and management*. (ed. R.P. Keller, D.M. Lodge, M.A. Lewis & J.F. Shogren), pp. 151–179. Oxford University Press, Oxford.

McNeely, J.A. (2001) An introduction to human dimensions of invasive alien species. In *The Great Reshuffling. Human Dimensions of Invasive Alien Species* (ed. J.A. Mcneely), pp. 5–20. IUCN, Gland, Switzerland.

McNeill, W.H. (1977) *Plagues and People*. Anchor Books, New York.

Millennium Ecosystem Assessment (2005) *Ecosystems and Human Well-being: General Synthesis*. Island Press, Washington, DC.

Myers, R.A. & Worm, B. (2003) Rapid worldwide depletion of predatory fish communities. *Nature*, **423**, 280–283.

Nelson, E., Mendoza, G., Regetz, J., et al. (2009) Modeling multiple ecosystem services, biodiversity conservation, commodity production, and tradeoffs at landscape scales. *Frontiers in Ecology and the Environment*, **7**, 4–11.

Nelson, E., Polasky, S., Lewis, D.J., et al. (2008) Efficiency of incentives to jointly increase carbon sequestration and species conservation on a landscape. *Proceedings of the National Academy of Sciences of the USA*, **105**, 9471–9476.

O'Neill, C. (1997) Economic impact of zebra mussels: results of the 1995 Zebra Mussel Information Clearinghouse Study. *Great Lakes Research Review*, **3**, 35–42.

Olson, L.J. (2006) The economics of terrestrial invasive species: a review of the literature. *Review of Agricultural and Resource Economics*, **35**, 178–194.

Olson, L.J. & Roy, S. (2002) The economics of controlling a stochastic biological invasion. *American Journal of Agricultural Economics*, **84**, 1311–1316.

Olson, L.J. & Roy, S. (2005) On prevention and control of an uncertain biological invasion. *Review of Agricultural Economics*, **27**, 491–497.

Ostfeld, R.S., Keesing, F. & Eviner, V.T. (2008) *Infectious Disease Ecology: Effects of Ecosystems on Disease and of Disease on Ecosystems*. Princeton University Press, Princeton, New Jersey.

OTA (1993) Harmful non-indigenous species in the United States. Office of Technology Assessment, US Congress, Washington, DC.

Pandolfi, J.M., Bradbury, R.H., Sala, E., et al. (2003) Global trajectories of the long-term decline of coral reef ecosystems. *Science*, **301**, 955–958.

Pauly, D., Chirstensen, V., Guénette, S., et al. (2002) Towards sustainability in world fisheries. *Nature*, **418**, 689–695.

Pavlin, B., Schloegel, L.M. & Daszak, P. (2009) Risk of importing zoonotic diseases through wildlife trade, United States. *Emerging Infectious Disease* **15**, 1721–1726.

Perrings, C., Burgiel, S., Lonsdale, M., Mooney, H.A. & Williamson, M. (2010a) Globalization and bioinvasions: the international policy problem. In *Globalization and Bioinvasions: Ecology, Economics, Management and Policy* (ed. C. Perrings, H.A. Mooney & M. Williamson), pp. 235–249. Oxford University Press, Oxford.

Perrings, C., Dehnen-Schmutz, K., Touza, J. & Williamson, M. (2005) How to manage biological invasions under globalization. *Trends in Ecology & Evolution* **20**, 212–215.

Perrings, C., Fenichel, E. & Kinzig, A. (2010b) Globalization and invasive alien species: trade, pests and pathogens. In *Globalization and Bioinvasions: Ecology, Economics, Management and Policy* (ed. C. Perrings, H.A. Mooney & M. Williamson), pp. 42–55. Oxford University Press, Oxford.

Perrings, C., Mooney, H.A. & Williamson, M. (2010c) The problem of biological invasions. In *Globalization and Bioinvasions: Ecology, Economics, Management and Policy* (ed. C. Perrings & H.A. Mooney & M. Williamson), pp. 1–18. Oxford University Press, Oxford.

Perrings, C., Williamson, M., Barbier, E.B., et al. (2002) Biological invasion risks and the public good: an economic perspective. *Conservation Ecology*, **6(1)**, 1.

Pimentel, D., Lach, L., Zuniga, R. & Morrison, D. (2000) Environmental and economic costs of nonindigenous species in the United States. *Bioscience*, **50**, 53–56.

Pimentel, D., McNair, S., Janecka, S., et al. (2001) Economic and environmental threats of alien plant, animal and microbe invasions. *Agriculture, Ecosystems* and *Environment*, **84**, 1–20.

Pimentel, D., Zuniga, R. & Morrison, D. (2005) Update on the environmental and economic costs associated with alien-invasive species in the United States. *Ecological Economics*, **52**, 273–288.

Polasky, S. (2010) A model of prevention, detection, and control for invasive species. In *Globalization and Bioinvasions: Ecology, Economics, Management and Policy* (ed. C. Perrings, H. Mooney & M. Williamson), pp. 100–109. Oxford University Press, Oxford.

Polasky, S., Nelson, E., Camm, J., et al. (2008) Where to put things? Spatial land management to sustain biodiversity and economic returns. *Biological Conservation*, **141**, 1505–1524.

Pyšek, P., Chytrý, M. & Jarošík, V. (2010) Habitats and land-use as determinants of plant invasions in the temperate zone of Europe. In *Globalization and Bioinvasions: Ecology, Economics, Management and Policy* (ed. C. Perrings, H. Mooney & M. Williamson), pp. 66–82. Oxford University Press, Oxford.

Rangi, D.K. (2004) Invasive alien species: agriculture and development. In *Global Synthesis Workshop on Biodiversity Loss and Species Extinctions: Managing Risk in a Changing World*. UNEP, Nairobi.

Reichard, S.H. & White, P. (2001) Horticulture as a pathway of invasive plant introductions in the United States. *Bioscience*, **51**, 103–113.

Rhymer, J.M. & Simberloff, D. (1996) Extinction by hybridization and introgression. *Annual Review of Ecology and Systematics*, **27**, 83–109.

Richardson, D.M. & Pyšek, P. (2007) Classics in physical geography revisited: Elton, C.S. 1958: The ecology of invasions by animals and plants. Methuen: London. *Progress in Physical Geography*, **31**, 659–666.

Rweyemamu, M.M. & Astudillo, V.M. (2002) Global perspective for foot and mouth disease control. *Revue Scientifique et Technique de l'Office International Des Epizooties*, **21**, 765–773.

Sakai, A.K., Allendorf, F.W., Holt, J.S., et al. (2001) The population biology of invasive species. *Annual Review of Ecology and Systematics*, **32**, 305–332.

Sandler, T. (1997) *Global Challenges*. Cambridge University Press, Cambridge, UK.

Sandler, T. (2004) *Global Collective Action*. Cambridge University Press, Cambridge, UK.

Semmens, B.X., Buhle, E.R., Salomon, A.K. & Pattengill-Semmens, C.V. (2004) A hotspot of non-native marine fishes: evidence for the aquarium trade as an invasion pathway. *Marine Ecology Progress Series*, **266**, 239–244.

Sharov, A.A. & Liebhold, A.M. (1998a) Bioeconomics of managing the spread of exotic pest species with barrier zones. *Ecological Applications*, **8**, 833–845.

Sharov, A.A. & Liebhold, A.M. (1998b) Model of slowing the spread of gypsy moth (Lepidoptera : Lymantriidae) with a barrier zone. *Ecological Applications*, **8**, 1170–1179.

Smith, K.F., Behrens, M.D., Max, L.M. & Daszak, P. (2008) U.S. drowning in unidentified fishes: scope, implications and regulation of live fish import. *Conservation Letters*, **1**, 103–109.

Smith, R.D., Keogh-Brown, M.R., Barnett, T. & Tait, J. (2009) The economy-wide impact of pandemic influenza on the UK: a computable general equilibrium modelling experiment. *British Medical Journal*, **339**, b4571.

Steneck, R.S., Vavrinec, J. & Leland, A.V. (2004) Accelerating trophic-level dysfunction in kelp forest ecosystems of the Western North Atlantic. *Ecosystems*, **7**, 323–332.

Stutzman, S.K., Jetter, M. & Klonsky, K.M. (2004) *An Annotated Bibliography on the Economics of Invasive Plants*. Agricultural Issues Center, University of California, Davis, California.

Sumner, D.A., Bervejillo, J.E. & Jarvis, L.S. (2005) Public policy, invasive species and animal disease management. *International Food and Agribusiness Management Review* **8**, 78–97.

Tallis, H. & Polasky, S. (2009) Mapping and valuing ecosystem services as an approach for conservation and natural-resource management. *Year in Ecology and Conservation Biology 2009*, **1162**, 265–283.

Tatem, A.J., Rogers, D.J. & Hay, S.I. (2006) Global transport networks and infectious disease spread. *Advances in Parasitology* **62**, 293–343.

Vila, M. & Pujadas, J. (2001) Land-use and socio-economic correlates of plant invasions in European and North African countries. *Biological Conservation*, **100**, 397–401.

Walker, B.H., Holling, C.S., Carpenter, S.R. & Kinzig, A.P. (2004) Resilience, adaptability, and transformability. *Ecology and Society*, **9**, 5.

Ward, J.M.J., Stromberg, E.L., Nowell, D.C. & Nutter, F.W. (1999) Gray leaf spot – a disease of global importance in maize production. *Plant Disease*, **83**, 884–895.

Williamson, M. (1996) *Biological Invasions*. Chapman and Hall, London.

Williamson, M. (1999) Invasions. *Ecography*, **22**, 5–12.

Zavaleta, E. (2000) The economic value of controlling an invasive hrub. *Ambio*, **29**, 462–467.

Chapter 25

MODELLING SPREAD IN INVASION ECOLOGY: A SYNTHESIS

Cang Hui, Rainer M. Krug and David M. Richardson

Centre for Invasion Biology, Department of Botany & Zoology, Stellenbosch University, 7602 Matieland, South Africa

Fifty Years of Invasion Ecology: The Legacy of Charles Elton, 1st edition. Edited by David M. Richardson
© 2011 by Blackwell Publishing Ltd

25.1   INTRODUCTION

Understanding the spatio-temporal dynamics of populations, their expansion and retraction, has always been one of the main pursuits in ecology and biogeography. Spread of invasive species, while posing real and escalating threats to biodiversity conservation and ecosystem functioning, also provides superb natural experiments for unravelling the mechanisms and factors behind the dynamics of species' geographical range. Studies of spread can be traced back to the dawn of mathematical ecology (Fisher 1937; Kolmogorov et al. 1937), with steady development on analysing species' dispersal using partial differential equations and other spatial modelling techniques (Skellam 1951). Spatial modelling currently forms a crucial part of invasion biology. A very wide range of methods, techniques and philosophies underpin modern spatial modelling and it seems worthwhile, five decades after Elton's (1958) pioneering book, to review the current status of the field of spatial modelling relating to biological invasions, to explore the emergence of different approaches and to identify the key challenges.

Organisms have an intrinsic drive to grow and expand. They must constantly be on the move, to alleviate intraspecific competition and inbreeding pressure, and to exploit opportunities provided by disturbances. Such movements of individuals, either through random-walk-like diffusion or directed dispersal, lead to a collective phenomenon of advancing frontiers in the geographic range of species, namely 'spread'. Classic examples of spreading organisms include the natural recolonization of Europe and North America by trees after the ice age, the spread of the European starling in America after its introduction to the region by humans and the expanding wave of the Canadian lynx in response to human-mediated environmental changes (see Hengeveld 1989). Although these examples of spread share some dynamic properties, in this chapter we focus on the spread of post-introduction alien species.

Mathematical modelling has been extremely valuable for uncovering the dynamics behind the spread of introduced species (see, for example, Okubo et al. 1989; van den Bosch et al. 1990; Shigesada & Kawasaki 1997; Caswell et al. 2003). However, most early mathematical ecologists lumped what we now call 'invasive' species (*sensu* Pyšek et al. 2004) with other spreading species, and were essentially interested

in general dispersal and colonization dynamics of organisms. As defined by Richardson et al. (2000a) and Lockwood et al. (2005), the invasion process includes the phases of arrival, establishment and naturalization of an alien species by breaking geographic, environmental and reproductive barriers. Breaching of dispersal barriers allows species to become 'invasive' (see also Wilson et al. 2009a). Spread models need to accommodate all parts of the invasion process.

Our synthesis of advances in spread modelling applied to biological invasions examines three components: the modelling core, environments and methods. Traditional spread models are classified into different categories according to their modelling core (e.g. those using differential equations). Theories about these models are briefly reviewed, and the rate of spread (i.e. the velocity of the travelling waves) is presented analytically where possible. Furthermore, to implement the modelling core in practice, it is necessary to consider other modelling environments and methodologies. Progress in modelling environments and methods (e.g. the neutral landscape model, niche modelling, and individual-based models) have enabled us to incorporate realistic habitat heterogeneity, stochasticity and more powerful algorithms for spatial realization in spread models.

25.2   SPREAD MODELS

A conceptual model

Higgins and Richardson (1996) provided a conceptual framework and a classification of models of alien plant spread. The number and types of models used in invasion ecology (not only for plants) has increased dramatically since this review was published. Models are now not only used to predict spread, but also to explore options for intervention (Jongejans et al. 2008). The essence of spread models has been broadened from a strictly mathematical orientation to a much more cohesive integration of data capture, mathematical analysis and modelling realization. Following these developments, we suggest a conceptual framework representing an 'optimal model' for studying the spatial spread of invasive species (Fig. 25.1).

This conceptual model comprises three compartments: modelling core, environment and method. The modelling core refers to dynamic models used to

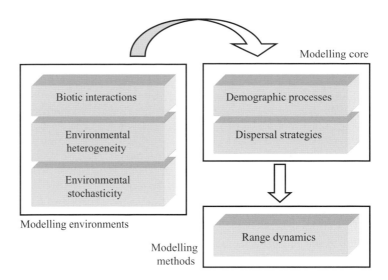

Fig. 25.1 A conceptual model for spread models for use in invasion biology.

describe the demography and spatial dynamics of species (Skellam 1951; Okubo 1980; van den Bosch et al. 1990, 1992). The use of a modelling core, such as differential and integrodifference equations, to analyse range expansions eventually evolved into a multidisciplinary endeavour drawing on dispersal ecology and invasion ecology, and focusing on identifying the ecological and evolutionary determinants and mechanisms of the spatio-temporal dynamics of invasive species, as well as designing the spatially optimal strategies for detection, control and eradication of focal species at different invasion phases.

Modelling environments are the factors that affect the rate of spread and the spatial distributions of invading species. They include biotic interactions, environmental heterogeneity and stochasticity (Hastings et al. 2005). The stage has been set for making real progress towards understanding the effect of environmental heterogeneity on the spread of introduced species, because (i) the potential distribution of a species, as reflected by its realized niche, can now be presented as a heterogeneous map using ecological niche modelling, and (ii) the spatial structure of habitats can now be simulated using, for instance, neutral landscape models.

There have been advances in modelling methods in recent years, especially because of advances in computer technology. Spatial modelling techniques such as individual-based models and cellular automata are fre-

quently incorporated into spatial spread models. Here we review only those methods that are most relevant to spatial spread and that have been most widely applied in invasion biology. Overall, a clear understanding of the assumptions, advantages and drawbacks behind each modelling core, environmental condition and technique enables us to present a clear synthetic view of our current challenges and trends in tackling the spatial spread of invasive species, not only as a theoretical question but also as a backbone for efficient management and control of problematic invasive species.

Modelling core

Ordinary differential equations

We start by considering a simple spatially implicit model that ignores population density: Levins' (1969) patchy occupancy model,

$$\frac{dP}{dt} = cP(1-P) - eP,$$

where P, c and e ($c > e$) denote the proportion of suitable habitat being occupied, the colonization rate and the local extinction rate, respectively. This has been widely applied to the study of metapopulation dynamics in spatial ecology and describes the dynamics of populations balanced by local extinction and recolonization of empty habitat patches. Although the patchy occupancy model is spatially implicit, the occupancy dynamics can be easily interpreted as the spontaneous rate of spread, $v_t = 1/(2\pi^{1/2}P^{1/2})dP/dt$. This model describes the logistic dynamics of species occupancy, with an asymptotic equilibrium determined only by the colonization–extinction process, $\hat{P} = 1 - e/c$. The rate of spread initially depends on the difference between the colonization and extinction rates, as well as the initial range size (P_0), $v_0 \approx (c-e)\sqrt{P_0}/(2\sqrt{\pi})$, then reaches a maximum speed, $v_{max} = (c-e)\sqrt{\hat{P}}/(3\sqrt{3\pi})$, and eventually slows down when the range size approaches its equilibrium.

Though simple, patchy occupancy models nonetheless infer several propositions that have been supported by the invasion biology literature: (i) the early phase of range expansion is often accompanied by an accelerating velocity of spread; (ii) stronger colonizers, indicated by the high colonization rate, are potentially good invaders with a high rate of spread; and (iii) high frequency or intensity of disturbance and a harsh environment result in high rates of local extinction and inhibit rapid spread. In practice, the spatial version of the model should be used and has been applied to reconstruct the spread of Argentine ants in Spain (Roura-Pascual et al. 2009a). One proposition from this model that has not been tested in invasion ecology is that, all else being equal, species with large potential distributions spread faster than those with small potential range. Because potential distribution can be extrapolated from ecological niche modelling (see 'Modelling environments' in section 25.2 for details), it should be possible to test this proposition empirically.

Partial differential equations

Most propositions about the rate of spread are derived using partial differential equations. Consequently, most modelling cores in the literature are based on partial differential equations, because of their advantage of incorporating spatial dimensions (see, for example, Holmes et al. 1994; Turchin 1998; Petrovskii & Li 2006). We focus largely on models with one spatial dimension (i.e. spread along a linear habitat, such as coastline, rivers and roads), although models with two spatial dimensions can easily be implied.

Two classical partial differential equations were first presented by Fisher (1937) and Skellam (1951) as the well-known reaction–diffusion models:

$$\frac{\partial n}{\partial t} = f(n) + D\frac{\partial^2 n}{\partial x^2},$$

where $f(n)$, D and n indicate the population growth rate, the diffusion rate and the population density. For Skellam's (1951) model, the population follows the Malthusian growth $f(n) = r \cdot n$; for Fisher's reaction–diffusion (1937) model, the population follows the logistic growth $f(n) = r \cdot n(1 - n/K)$, where r and K are the intrinsic growth rate and the carrying capacity. The model generates a travelling wave with an asymptotic velocity of $\hat{v} = 2\sqrt{f'(0)D}$. It is worth noting that this is the rate of spread caused by the Brownian-motion diffusion, and thus may not be suitable for representing realistic patterns of animal movement. Seed dispersal by different vectors defies random diffusion and is often studied by incorporating dispersal kernel into the integrodifference equations (Tsoar et al., Chapter 9 of this volume). Furthermore, when the reaction in the model is very fast, the asymptotic velocity can become arbitrarily large, and a solution to such an unrealistic property is to replace the above model by an integrodifferential equation, often known as the generalized Fisher–Kolmogorov–Petrovskii–Piskunov equation (Branco et al. 2007).

It is clear that any effects causing changes to the linear approximation of population growth function at the initial point of invasion $f'(0)$ will have an impact on the rate of spread. For instance, Lewis and Kareiva (1993) examined the rate of spread of the reaction–diffusion model when the population dynamics is subjected to the Allee effect, $f(n) = r \cdot n(n - a)(1 - n)$, and found the asymptotic rate of spread to be sensitive to the intensity of the Allee effect: $\hat{v} = \sqrt{2rD}(0.5 - a)$ if $a < 1/2$ and $\hat{v} = 0$ if $a \geq 1/2$. Furthermore, in a two-dimensional landscape, the expanding wave can only start if the initial radius of the beachhead is greater than a threshold ($\sqrt{D/2r}(0.5 - a)$), i.e. there exists a minimum initial population size for the range expansion (Lewis & Kareiva 1993).

Besides the random-walk diffusion, if species dispersal is biased due to, say, air and water currents, a drift (or convection) term can be added to the reaction–diffusion model; it is then called an advection–diffusion model:

$$\frac{\partial n}{\partial t} = f(n) - v_x\frac{\partial n}{\partial x} + D\frac{\partial^2 n}{\partial x^2},$$

where v_x is the drift velocity along the x-axis direction. The travelling wave of the advection–diffusion model is a simple overlap between the travelling wave of the diffusion model and the convection velocity (van den Bosch et al. 1990).

Another extension to the basic reaction–diffusion model was made because of the realization that animals do not follow strictly random walks in their movements, but tend to move with persistence. This correlated random walk can be depicted by the reaction–telegraph equation (Holmes 1993):

$$\frac{\partial n}{\partial t} = f(n) + \frac{1}{2\lambda}\frac{\partial f}{\partial t} - \frac{1}{2\lambda}\frac{\partial^2 n}{\partial t^2} + \frac{\gamma^2 n}{2\lambda}\frac{\partial^2 n}{\partial x^2},$$

where γ is the velocity of the individual and λ is the rate of changing direction. For logistic growth, $f(n) = r \cdot n(1 - n/K)$, Holmes (1993) found that the rate of spread (i.e. the speed of the travelling wave) is $v = \gamma\sqrt{8r\lambda}/(r+2\lambda)$ if $0 < \sqrt{r/2\lambda} \leq 1$, and $v = \gamma$ if $\sqrt{r/2\lambda} \geq 1$. Furthermore, if we let $D = \gamma^2/2\lambda$, the comparison of the rate of spread from diffusion and telegraph models is made possible, which often predict relatively similar rates of spread in reality (Holmes 1993).

A general density-dependent diffusion model is presented by Okubo (1980) as the crowding-induced diffusion. Aronson (1980) investigated one such general reaction–diffusion (crowding-induced) model:

$$\frac{\partial n}{\partial t} = f(n) + \frac{\partial^2 n^m}{\partial x^2}.$$

where $f(n) = n(1 - n)$. This model has been rescaled so that the equation only has one parameter m. Obviously, if $m = 1$, this model is essentially Fisher's reaction–diffusion model. Individual movement only responds to overcrowding when $m > 1$, and the rate of spread thus significantly slows down. For instance, if $m = 2$, the rate of spread drops by half, and the population density also becomes zero at a certain distance ahead of the wave. Individuals tend to avoid crowded areas when $m < 1$, suggesting a potentially high rate of spread. The crowding-induced diffusion comes from a biased random-walk, instead of a pure or correlated random-walk, and thus differs from the diffusion and the telegraph models (Aronson 1980; Turchin 1998).

Dispersal kernel models

The differential equation models discussed above do not take into account two measurements that are often obtained from ecological surveys: the stage structure of the organism (e.g. age structure or egg–larvae–adult stages of insects) and the different forms of dispersal kernel (i.e. the probability density function with respect to the distance away from the original location after dispersal; a Gaussian dispersal kernel is assumed in diffusion models). A group of models enables us to incorporate both the demographic factors and the dispersal characteristics: the discrete-time dispersal kernel model (also called the *integrodifference equations*) and the continuous-time dispersal kernel model (also called the reproduction-and-dispersal kernel model; *R&D kernel model* for short).

Integrodifference equations are further discretization of integrodifferential models, and can be used to estimate the rate of spread based on any specific dispersal kernels (Weinberger 1982; Kot et al. 1996; Lewis et al. 2006). Let $k(x, y)$ denote the probability density function for the location x to which an individual at y disperses, we have the following integrodifference equation for calculating the population density in locality x at time $t + 1$ (Kot et al. 1996; Neubert & Caswell 2000):

$$n(x,t+1) = \int_{-\infty}^{\infty} k(x,y)b[n(y,t)]n(y,t)\mathrm{d}y,$$

where $b[n]n$ gives the size of the growing population at locality y. The asymptotic rate of spread exists $\hat{v} = \min_{s>0} \ln[b(0)\Phi(s)]$ provided that $b[n]n$ increases monotonically with the population density n (also with no Allee effect) and that the dispersal kernel has a moment-generating function:

$$\Phi(s) = \int_{0}^{\infty} k(z)e^{sz}\mathrm{d}z,$$

where z indicates the distance between x and y. The integrodifference equation can also further incorporate the stage-structured population growth and dispersal, resulting in the stage-structured (matrix) model and enabling us to estimate the elasticity and sensitivity of different demographic stages (Neubert & Caswell 2000). Recently, a special focus of using integrodifference equations, which deserves continuous attention, is to estimate the rate of spread for fat-tailed long-distance dispersal (Higgins & Richardson 1999; Clark et al. 2003; Tsoar et al., Chapter 9 of this volume).

A closely related dispersal kernel model is the continuous-time R&D kernel model (van den Bosch et al. 1992):

$$n(x,t) = \int_{-\infty}^{\infty}\int_{0}^{\infty} k(z,a)l(a)b(x-s,t-a)\mathrm{d}a \cdot \mathrm{d}s.$$

where $l(a)$ is the probability that an individual is still alive at age a. This leads to a rather similar rate of spread as the reaction–diffusion model, suggesting that $\hat{v} = 2\sqrt{f'(0)D}$ is a robust measure of the rate of spread (van den Bosch et al. 1992; Grosholz 1996; Turchin 1998). An important area for future work is to explore the relationship between the rate of spread and different forms of dispersal kernels (e.g. the general power-law function for the long-distance jump dispersal) under different modelling environments (see discussion below).

Modelling environments

Biotic interactions

The relationship between species richness of a community and its invasibility was raised by Elton (1958) and has been one of the most persistent debates in the invasion biology literature ever since (Richardson & Pyšek 2007; Fridley, Chapter 10 of this volume; see also Stohlgren et al. 2003; Tilman 2004). Although recent studies suggest that the impact of biological invasions on native species richness is strongly scale dependent (Gaertner et al. 2009), theoretical studies have confirmed that biotic resistance does have an effect on the rate of spread (Okubo et al. 1989; Dunbar 1983, 1984). For interspecific competition between invaders and native competitors, Okubo et al. (1989) considered a diffusion model of two competing species:

$$\frac{\partial n_1}{\partial t} = D_1 \frac{\partial^2 n_1}{\partial x^2} + r_1 n_1 (1 - \alpha_{11} n_1 - \alpha_{12} n_2)$$

$$\frac{\partial n_2}{\partial t} = D_2 \frac{\partial^2 n_2}{\partial x^2} + r_2 n_2 (1 - \alpha_{21} n_1 - \alpha_{22} n_2)$$

where α is the coefficient of intra- and interspecific competition. Okubo et al. (1989) found that the rate of spread for the invader (species 1) can be slowed down by its native competitor: $\hat{v} = 2\sqrt{r_1 D_1 (1 - \alpha_{12}/\alpha_{11})}$. Dunbar (1983, 1984) examined the effect of predation on the rate of spread in a Lotka–Volterra model:

$$\frac{\partial n_1}{\partial t} = D_1 \frac{\partial^2 n_1}{\partial x^2} + r \cdot n_1 (1 - n_1/K) - a \cdot n_1 n_2$$

$$\frac{\partial n_2}{\partial t} = D_2 \frac{\partial^2 n_2}{\partial x^2} + \gamma \cdot a \cdot n_1 n_2 - \delta \cdot n_2$$

where a, γ and δ are the rate of predation, the conversion rate of captured preys and the death rate of the predators, respectively. If the predator is the invader, then it spreads according to the rate $\hat{v} = 2\sqrt{(\gamma \cdot aK - \delta)D_2}$.

Both studies suggest that biotic interactions such as competition and predation can inhibit the fast spread of invasive species.

Besides the confirmation from such theoretical studies, real evidence is limited. Ferrer et al. (1991) reported that the range expansion of the European starling and the spotless starling in Spain slowed down due to their interspecific competition. Similar evidence has also been reported for other bird species (e.g. the red-whiskered bulbul (Clergeau & Mandon-Dalger 2001)). However, such competitive resistance may not to be important for insects (ants and wasps) (Holway 1998; Walker et al. 2002; Roura-Pascual et al. 2010a), and shows mixed effects on the spread of plants (Higgins et al. 2008). Other interspecific relationships, such as pollination mutualisms and predation, have also been suggested to affect the rate of spread (Lonsdale 1993; Richardson et al. 2000b). Overall, biotic interactions can clearly affect the rate of spread, and such impacts are taxon- and scale-dependent (Traveset & Richardson, Chapter 12 of this volume).

Environmental heterogeneity

Niche theory in ecology posits that the performance of organisms, such as the growth rate and mobility, changes along environmental gradients. Spatial spread of invasive species is also affected by environmental heterogeneity, and this can be incorporated in the modelling core of differential equations and dispersal kernel models using spatially dependent diffusion. Two frequently used methods for this are replacing the diffusion term in the model by the Fokker–Planck equation $\partial n/\partial t = \partial^2(\mu(x)n)/\partial x^2$ or the Fickian diffusion equation $\partial n/\partial t = \partial(D(x)\partial n/\partial x)/\partial x$, where $\mu(x)$ and $D(x)$ are motility and diffusivity at locality x that determine the rate of spread (Turchin 1998). For instance, Shigesada et al. (1986) studied a Fickian diffusion model in a spatial heterogeneous landscape. Clearly, it is difficult to arrive at an analytical solution to the rate of spread. However, numerical simulations can be easily obtained for studying the spread of invasive species in heterogeneous environments. Recently, Dewhirst and Lutscher (2009) have derived an approximate rate of spread in heterogeneous habitats using integrodifference equations, thereby making significant progress in the field (see also Higgins et al. 2003; Kawasaki & Shigesada 2007).

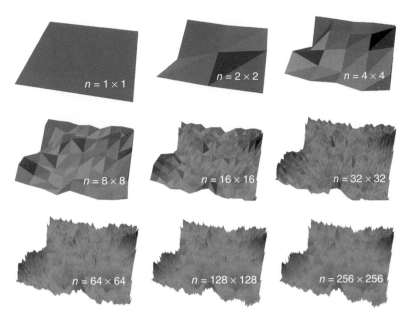

Fig. 25.2 The procedure of generating a spatially heterogeneous fractal landscape using the midpoint displacement algorithm (Fournier et al. 1982).

In theoretical studies of spread in heterogeneous landscapes, spatial heterogeneity is often generated by neutral landscape models, which can be broadly categorized into random, hierarchical and fractal types (With & King 1997). Random neutral landscape is the first-generation model that seeks the critical scales in detecting or comparing ecological processes based on the prediction of percolation theory (see, for example, Gardner et al. 1987; Wiens et al. 1997). The clumped neutral landscape model that incorporates spatial autocorrelation between different localities has been proposed as an improvement (Hiebeler 2000). The second generation of neutral landscape models are hierarchical (Lavorel et al. 1994). Because landscape processes are more or less assembled in a hierarchical structure, such neutral landscape models perform reasonably well in predicting dispersal patterns of species in heterogeneous landscapes (Lavorel et al. 1995). The most realistic neutral landscape models are those of fractal nature (Palmer 1992). Algorithms, such as the midpoint replacement (Fig. 25.2), are often used to generate artificial landscapes for examining species' distributional patterns (Leung et al. 2010). Furthermore, fractal neutral landscape models are clearly consistent with the fractal and self-similar structure of species' distribution and assemblage (With 2002; Hui & McGeoch 2008; Storch et al. 2008), and the effect of the fractal structure of the landscape on the rate of spread is expected to be clarified soon.

For investigating the effect of environmental heterogeneity on spread, ecological niche modelling (also called species distribution modelling or (bio)climatic/ environmental envelope modelling), in its general sense, has become a standard approach for transforming environmental variables into suitability maps that can describe species' fundamental (bottom-up approach) and realized niche (top-down approach) (Fig. 25.3). The bottom-up ecological niche modelling translates species physiological limits into the fundamental niche using fine-scale microclimate data by measuring the survival rate of the organism (see, for example, Kearney & Porter 2009). It thus projects the suitable habitat of focal species in the spatial landscape. The top-down ecological niche modelling, on the other hand, translates the broad-scale presence–absence (and even abundance) records and macroclimate data into the realized niche and, as a result, projects the potential distribution of the species (see, for example, Kriticos et al. 2003; Peterson & Vieglais 2001; Franklin 2010).

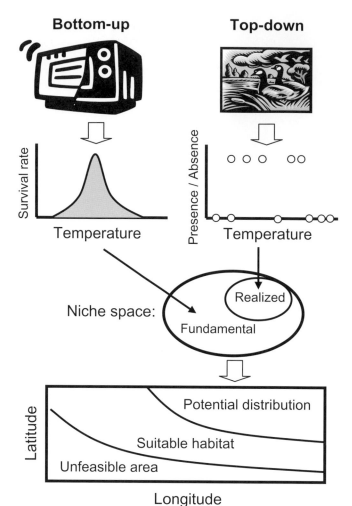

Fig. 25.3 A schematic illustration of the bottom-up and the top-down ecological niche modelling.

Because invasive species experience novel combinations of environmental variables and biotic interactions, the top-down ecological niche modelling for generating the realized niche and projecting the potential distribution of invasive species could be biased (Guisan & Thuiller 2005; Pearson et al. 2006). In contrast, the bottom-up niche modelling can often identify a reliable suitable habitat in the invaded area due to niche conservatism (Wiens & Graham 2005). Furthermore, ecological niche models have been criticized for the low-sensitivity of their prediction (Heikkinen et al. 2006) and the lack of efficiency in the evaluation method, *Area Under Curve* (AUC (Lobo et al.

2008)). Therefore two key challenges in this research are the following: to (i) design (hybrid) models that can encompass population dynamics (see, for example, Gallien et al. 2010) and predict the realized niche shift within its fundamental niche when encountering a new environment; and (ii) design a more sensitive measure than the AUC index to evaluate models that can incorporate spatial information of the input data.

Environmental stochasticity

Even if a spread model is mechanistically correct, its predictive power can still be hindered by the deterministic chaos potentially inherent in the system, and also by the stochasticity of population dynamics. Indeed, no populations can avoid demographic and environmental stochasticity in the real world. Demographic stochasticity refers to the variability in population size caused by the birth–death–dispersal process and could be responsible for the slow rate of spread at initial stage when the population size is small (Hastings 1996). Environmental stochasticity refers to the variability in population size caused by the variations in the life-history parameters from other environmental heterogeneity and disturbances (e.g. the birth rate can be seasonal and locational dependent and further influenced by fire and floods). In general, Mollison (1977, 1991) suggests that stochastic models predict the same rate of spread for density-independent populations as the rate of spread in deterministic models, but could alter the rate of spread in density-dependent populations. Although recent work on modelling the spread of invasive species often includes environmental stochasticity, or uses stochastic models (Schreiber & Lloyd-Smith 2009), further attention needs to be given to clarifying measures of stochasticity and the relationship between the stochasticity and rates of spread. Questions that need to be addressed include the following. (i) Do invasive species experience higher variability in their population dynamics in novel environments? (ii) Do invasive species that encounter high variability spread faster? Clearly, stochasticity and variability require special attention in future studies, not only with reference to rates of spread, but also at all stages of biological invasions.

Spatial modelling methods

Spatially explicit modelling techniques in invasion ecology have advanced gradually (Box 25.1), and involve complex rule-based programming under the scope of agent-based models, often with an intractable analytical solution. The classification of modelling techniques can be based on the way that the population, space and time are considered (e.g. whether discrete or continuous) (Berec 2002).

In general, nearly all techniques for modelling spread belong to the family of individual-based models (IBMs). These simulation techniques treat individuals as unique and discrete entities that often have at least one property that changes during the life cycle (Grimm & Railsback 2005). When modelling spread in a

---

Box 25.1 The development of modelling methods

Methods for the spatial modelling of spread have been developed since the 1950s, with the advance of electronic computers. Cellular automata (CA) were first developed in 1950s by John von Neumann in a failed attempt to prove the emergence of self-reproduction in life by discretizing partial differential equations in lattices. Numerous attempts followed, seeking transition rules in CA to imitate life phenomena (e.g. 'the game of life' (Conway 1970)). The development of CA were also accompanied by the design of parallel computers and digital image processing. In the 1960s, CA began to be used to study neural networks, reaction–diffusion processes of active media as in heart and muscles, and the Ising model in physics (CA with randomness). By the end of 1970s, when computers became widely available, research on CAs accelerated rapidly. In invasion ecology, the spread of invasive species is often simulated by CAs (often called lattice models) with continuous cell state on heterogeneous landscapes (Roura-Pascual et al. 2009a).

The earliest work using individual-based models (IBMs) can be traced back to the 1960s in a doctoral dissertation on the dynamics of Douglas-fir (Newnham 1964). One of earliest papers that brought individual-based models (IBMs) into the ecological literature was by Michael Huston et al. (1988). This visionary work separated IBMs from other types of model and called the development of IBMs 'a self-conscious discipline'. Several reviews have traced the history of IBMs for animals, plants and in general (DeAngelis & Mooij 2005; Grimm & Railsback 2005). IBMs quantify differences between individuals both biologically and spatially (Huston et al. 1988), consistent with the reality in spread models. Given this advantage and increasing computational power, the use of IBMs has grown continuously (DeAngelis & Mooij 2005). Although IBMs are unlikely to lead to any general theories in ecology (DeAngelis & Mooij 2005), they are contributing substantially to predicting case-specific spread in invasion ecology.

spatially explicit context, all individuals need to be assigned a location which may be static in the case of plants, or variable in the case of animals. The 'individual' in IBMs need not be a real individual, but can also represent a group of individuals that share a common entity, for example a colony of insects.

Another class of models are (classical and extended) cellular automata (CA; also termed 'lattice models'). CAs are an idealized 'system in which space and time are discrete' in which the cells have a finite number of possible states (Chopard & Droz 1998). Classical CAs are also deterministic, and the state of a cell at the next time step is defined by the state of the cell and its neighbours in the previous time step (Conway 1970). As these conditions are quite limiting, the stochasticity condition can be relaxed and an infinite number of states for each cell allowed, resulting in the extended CA. CA modelling faces the modifiable areal unit problem (Openshaw 1984) giving that the cell size is often artificially chosen. As a result, the linear dimension of the cell for any spread models should be less than the median of the dispersal kernel.

Given the different units of IBMs and CAs (individuals versus cells), CAs can use IBMs internally by defining the state of a cell as a collaborative measure of all dwelling individuals that are governed by IBM rules. This hybrid is actually the case for spatially and temporally discrete IBMs. Bithell and MacMillan (2006) provide an overview over the problems associated with this approach, and suggest several approaches for the 'escape from the cell'. In contrast to spatially discrete models (CAs), such spatially continuous models do not use an underlying grid to define the location, but store the location as coordinates. This removes the modifiable areal unit problem when using a CA (Hui et al. 2010). Although the spatially continuous model avoids using an artificially chosen cell size, it creates computational challenges: the intensity of interactions between individuals has to be defined as a distance-dependent function (e.g. the dispersal kernel). Furthermore, complications also arise in the construction of these models, as discrete time steps are inherent in the modelling. Even though this is not a problem for species with synchronized life-history events (e.g. breeding and migration), it poses problems for species with no coherence. This problem can be mitigated by either choosing small time steps or using random process to determine the exact time for each event (e.g. the background Poisson process with density dependent rate (Berec 2002)).

In general, IBMs are more computationally intensive than CAs and other formula-based models. Consequently, a major constraint in the use of IBMs is the number of individuals expected, rather than the size of the area simulated. In contrast, CAs are limited by the number of cells simulated (i.e. spatial extent). As a rule of thumb, in cases where large areas need to be simulated, CAs are more suitable than IBMs, unless the number of individuals is expected to be relatively small. However, with recent advances in high-performance computing (e.g. parallel computing), computational requirements have become less restrictive. It is becoming feasible to construct and run advanced simulations that were impossible previously (Bolker et al. 2000).

In cases where the behaviour of the individual is important for predicting the spread of the species, IBMs have to be used, as they can incorporate detailed information about the individual (Jongejans et al. 2008). As outlined by Jongejans et al. (2008), the complexity of spread models varies considerably, depending on (i) the dispersal mode of the species of concern, (ii) the type of model selected and (iii) the required level of realism of the model. CAs are preferred to IBMs when the data related to the spread is limited given that CAs are less data hungry. For programming, we need to specify the initial and boundary conditions and transition rules for CAs and rules governing individual's life-cycle process for IBMs (Grimm & Railsback 2005).

In contrast to data- and computationally intensive dynamic models of spread, other methods exist which can be summarized as non-simulation based static spread models. These models provide no temporal sequence of predictions, but rather use knowledge gained from experts or from statistical models (e.g. regression-trees and multi-criteria decision models) to identify areas likely to be occupied by invading species (see Rouget et al. 2003), prioritize areas for conservation management (Roura-Pascual et al. 2009b, 2010b), or to define integrated strategies to reduce the threat of ecosystem dysfunction and species extinction due to climate- and land use changes and biological invasions (Richardson et al. 2009). These static models are relatively easy to construct and are a good way of capturing expert knowledge, but can be applied only to the area for which they were constructed.

25.3 SYNTHESIS

Although substantial progress has been made in understanding the mechanisms and factors that drive the spatial spread of invasive species in novel environments in the past 50 years, we must acknowledge that the dynamics and management of biological invasions is a multi-stage and multidisciplinary field. From the early work on invasion dynamics (Hengeveld 1989) to the recent emphasis of biological invasions as a multi-stage process, the focus of modelling studies has changed, from an initial preoccupation with only 'spread' to the current interest in parameterizing and simulating dynamics at all stages of the naturalization–invasion continuum. Particular modelling approaches and philosophies are emerging for each stage; these are increasingly linked to management requirements. For instance, human and cargo transportation by ships and aircrafts have been strongly correlated with the number of species introduced to different places across the globe (Drake & Lodge 2004; Tatem 2009). An important theoretical issue at this stage is the estimation of the introduction rate from the discovery records (Solow & Costello 2004). Prevention is clearly the best (most cost-effective) strategy for reducing rates of introduction. Once introduced, only a fraction of alien species establish in the novel environment. The outcome of an introduction event depends on numerous factors associated with the introduction pathway (Wilson et al. 2009b), as well as many other factors, including environmental suitability, resource availability and the ability to adapt and naturalize. The combined influence of many post-introduction factors often causes a time lag before fast spread, during which time eradication may be feasible. Once established, the spatial spread of an invasive species can be modelled in different ways, by stochastic modelling in a heterogeneous landscape, as reviewed in this chapter. Once a species reaches this stage, eradication becomes expensive, and often impractical or impossible, leaving only various options for reducing population sizes or growth rates. The whole process of biological invasion can be incorporated into a probability transition model (see an attempt by Jerde & Lewis (2007)). Advances in the modelling core, environment and methods, as reviewed here, have brought fundamental insights in facilitating the management and control of invasive species, for example identifying crucial demographic stage for fast spread (Neubert & Caswell 2000) and determining optimal control and clearing strategy (Moody & Mack 1988; Moilanen et al. 2009).

Spread dynamics is the collective behaviour of individuals in an expanding population. Obviously, this becomes possible only through dispersal. This leads to a joint focus between dispersal ecology and invasion

biology: the dispersal kernel (Porter & Dooley 1993; Bullock & Clarke 2000; Tsoar et al., Chapter 9 of this volume). The forms of dispersal kernals are diverse, depending on the life forms of the species, dispersal pathways and landscape structure (Greene & Calogeropoulos 2001). Importantly, individuals tend to follow different dispersal types, even within species (Williamson 2001): natal versus breeding dispersal in vertebrates, pollen versus seed dispersal in plants, walking versus flying in insects, within and outside habitat patches, genetic and morphologic difference (e.g. sexual and age difference). Such multi-type dispersal within one species has been suggested to be able to explain the long-distance jump dispersal in fast expanding species (see Clark et al. 1999; Higgins & Richardson 1999). Although a robust conclusion has been reached regarding the rate of spread under the Gaussian dispersal curve (i.e. $\hat{v} = 2\sqrt{f'(0)D}$), a clear deduction of the spreading rate from other forms of dispersal curves is expected to foster a synthesis between invasion biology and dispersal ecology (see, for example, Dewhirst & Lutscher 2009). Furthermore, we have to realize that the evolutionary and genetic identity of a population is constantly changing, causing variations and uncertainties of the rate of spread at different invasion stages (Phillips et al. 2008; Williamson 2009).

Although dispersal kernel and the rate of spread are two main focuses about the spreading dynamics in invasion ecology, it is necessary to link these focuses with other concepts in invasion biology, especially invasiveness and the impact of invasive species. The invasiveness (the ability for alien species to establish in a new environment) has to be stated in the context of the invasibility of the native community and often refers to the ability to cross strong environmental barriers (high stress tolerance and adaptation rate) and suffer no Allee or other founder effect. It is unclear whether species with high invasiveness should have a fat-tail dispersal kernel or a high rate of spread; yet, in general we would expect this, because high dispersability could reflect the life-history strategy for mitigating risks and uncertainty when facing new environments (Trakhtenbrot et al. 2005). However, a fast-expanding species does not necessarily have major impact in the receiving community. In fact, the impacts of such species may be lower, owing to the potential life-history trade-off between 'win stay' and 'lose shift' strategies. This proposition has a clear management implication for prioritizing for eradication of slow-expanding species first. The test of this hypothesis would require a clear understanding of the physiological, genetic and evolutionary background of the species.

Integration within invasion biology is necessary, especially between the application of the type of spread models reviewed in this chapter and the risk assessment framework that is increasingly driving management of biological invasions. Possible applications include the identification of sites threatened by the spread of alien invasive species, testing management strategies and budgets, motivation and justification of budget requirements (Moilanen et al. 2009). In addition, as the need of spread models in management increases, the development of new spread models has to pay more attention to the management demand. This requires new ways to simulate spread and more transparent interpretation of modelling results, as well as problem-driven modelling with targets identified by managers. Web-based technologies have potential to deliver such a user-friendly interface for managers to provide input data and for modellers to deliver the results. In all, the spread of invasive species provides an ideal natural experiment for ecologists to examine the biotic and abiotic factors behind species spatio-temporal dynamics, and poses challenges to managers and modellers for simulating and assessing the trend and impact of the invasion with sufficient accuracy.

ACKNOWLEDGEMENTS

We thank Steven Higgins, Nanako Shigesada, John Wilson and Aziz Ouhinou for helpful comments on the chapter. Our work was largely funded by the DST-NRF Centre of Excellence for Invasion Biology. C.H. acknowledges support from the NRF Blue Sky Programme; R.M.K. and D.M.R. acknowledge support from the Global Environmental Facility (GEF) through the Cape Action for People and the Environment (CAPE) programme.

REFERENCES

Aronson, D.G. (1980) Density-dependent interaction–diffusion systems. *Dynamics and Modelling of Reactive Systems* (ed. W.E. Stewart, W.H. Ray and C.C. Conley), pp. 161–176. Academic Press, New York.

Berec, L. (2002) Techniques of spatially explicit individual-based models: construction, simulation, and mean-field analysis. *Ecological Modelling*, **150**, 55–81.

Bithell, M. & MacMillan, W. (2006) Escape from the cell: Spatially explicit modelling with and without grids. *Ecological Modelling* **200**, 59–78.

Bolker, B.M., Pacala, S.W. & Levin, S.A. (2000) Moment methods for ecological processes in continuous space. *The Geometry of Ecological Interactions: Simplifying Spatial Complexity* (ed. U. Dieckmann, R. Law and J.A.J. Metz), pp. 388–411. Cambridge University Press, Cambridge, UK.

Branco, J.R., Ferreira, J.A. & de Oliveira, P. (2007) Numerical methods for the generalized Fisher–Kolmogorov–Petrovskii–Piskunov equation. *Applied Numerical Mathematics*, **57**, 89–102.

Bullock, J.M. & Clarke, R.T. (2000) Long distance seed dispersal: measuring and modelling the tail of the curve. *Oecologia*, **124**, 506–521.

Caswell, H., Lensink, R. & Neubert, M.G. (2003) Demography and dispersal: life table response experiments for invasion speed. *Ecology*, **84**, 1968–1978.

Chopard, B. & Droz, M. (1998) *Cellular Automata Modeling of Physical Systems*. Cambridge University Press, Cambridge, UK.

Clark, J.S., Lewis, M.A., McLachlan, J.S. & HilleRisLambers, J. (2003) Estimating population spread: what can we forecast and how well? *Ecology*, **84**, 1979–1988.

Clark, J.S., Silman, M., Kern, R., Macklin, E. & HilleRisLambers, J. (1999) Seed dispersal near and far: patterns across temperate and tropical forests. *Ecology*, **80**, 1475–1494.

Clergeau, P. & Mandon-Dalger, I. (2001) Fast colonization of an introduced bird: the case of *Pycnonotus jocosus* on the Mascarene Islands. *Biotropica*, **33**, 542–546.

Conway, J. (1970) The game of life. *Scientific American*, **223**, 120–123.

DeAngelis, D.L. & Mooij, W.M. (2005) Individual-based modeling of ecological and evolutionary processes. *Annual Review of Ecology, Evolution, and Systematics*, **36**, 147–168.

Dewhirst, C. & Lutscher, F. (2009) Dispersal in heterogeneous habitats: thresholds, spatial scales, and approximate rates of spread. *Ecology*, **90**, 1338–1345.

Drake, J.M. & Lodge, D.M. (2004) Global hot spots of biological invasions: evaluating options for ballast-water management. *Proceedings of the Royal Society of London B: Biological Sciences*, **271**, 575–580.

Dunbar, S. (1983) Travelling wave solutions of diffusive Lotka–Volterra equations. *Journal of Mathematical Biology*, **17**, 11–32.

Dunbar, S. (1984) Travelling wave solutions of diffusive Lotka–Volterra equations: a heteroclinic connection in R4. *Transactions of the American Mathematical Society*, **286**, 557–594.

Elton, C.S. (1958) *The Ecology of Invasions by Animals and Plants*. Methuen, London.

Ferrer, X., Motis, A. & Peris, S.J. (1991) Changes in the breeding range of starlings in the Iberian Peninsula during the last 30 years: competition as a limiting factor. *Journal of Biogeography*, **18**, 631–636.

Fisher, R.A. (1937) The wave of advance of advantageous genes. *Annals of Eugenics*, **7**, 355–369.

Fournier, A., Fussel, D. & Carpenter, L. (1982) Computer rendering of stochastic models. *Communications of the ACM*, **25**, 371–384.

Franklin, J. (2010) Moving beyond static species distribution models in support of conservation biogeography. *Diversity and Distributions*, **16**, 321–330.

Gaertner, M., Den Breeyen, A., Hui, C. & Richardson, D.M. (2009) Impacts of alien plant invasions on species richness in Mediterranean-type ecosystems: a meta-analysis. *Progress in Physical Geography*, **33**, 319–338.

Gallien, L., Münkemüller, T., Albert, C.H., Boulangeat, I. & Thuiller, W. (2010) Predicting potential distributions of invasive species: where to go from here? *Diversity and Distributions*, **16**, 331–342.

Gardner, R.H., Milne, B.T., Turner, M.G. & O'Neill, R.V. (1987) Neutral models for the analysis of broad-scale landscape pattern. *Landscape Ecology*, **1**, 19–28.

Greene, D.F. & Calogeropoulos, C. (2001) Measuring and modelling seed dispersal of terrestrial plants. *Dispersal Ecology* (ed. J.M. Bullock, R.E. Kenward and R.S. Hails), pp. 3–23. Blackwell, Oxford.

Grimm, V. & Railsback, S. (2005) *Individual-Based Modeling and Ecology*. Princeton University Press, Princeton, New Jersey.

Grosholz, E.D. (1996) Contrasting rates of spread for introduced species in terrestrial and marine systems. *Ecology*, **77**, 1680–1686.

Guisan, A. & Thuiller, W. (2005) Predicting species distributions: offering more than simple habitat models. *Ecology Letters*, **8**, 993–1009.

Hastings, A. (1996) Models of spatial spread: a synthesis. *Biological Conservation*, **78**, 143–148.

Hastings, A., Cuddington, K., Davies, K.F., et al. (2005) The spatial spread of invasions: new developments in theory and evidence. *Ecology Letters*, **8**, 91–101.

Heikkinen, R.K., Luoto, M., Araújo, M.B., Virkkala, R., Thuiller, W. & Sykes, M.T. (2006) Methods and uncertainties in bioclimatic envelope modelling under climate change. *Progress in Physical Geography*, **30**, 751–777.

Hengeveld, R. (1989) *Dynamics of Biological Invasions*. Chapman and Hall, London.

Hiebeler, D. (2000) Populations on fragmented landscapes with spatially structured heterogeneities: landscape generation and local dispersal. *Ecology*, **81**, 1629–1641.

Higgins, S.I. & Richardson, D.M. (1996) A review of models of alien plant spread. *Ecological Modelling*, **87**, 249–265.

Higgins, S.I. & Richardson, D.M. (1999) Predicting plant migration rates in a changing world: the role of long-distance dispersal. *American Naturalist*, **153**, 464–475.

Higgins, S.I., Lavorel, S. & Revilla, E. (2003) Estimating plant migration rates under habitat loss and fragmentation. *Oikos*, **101**, 354–366.

Higgins S.I., Flores, O. & Schurr, F.M. (2008) Costs of persistence and the spread of competing seeders and sprouters. *Journal of Ecology*, **96**, 679–686.

Holmes, E.E. (1993) Are diffusion model too simple? A comparison with telegraph models of invasion. *American Naturalist*, **142**, 779–795.

Holmes, E.E., Lewis, M.A., Banks, J.E. & Veit, R.R. (1994) Partial differential equations in ecology: spatial interactions and population dynamics. *Ecology*, **75**, 17–29.

Holway, D.A. (1998) Factors governing rate of invasion: a natural experiment using Argentine ants. *Oecologia*, **115**, 206–212.

Hui, C. & McGeoch, M.A. (2008) Does the self-similar species distribution model lead to unrealistic predictions? *Ecology*, **89**, 2946–2952.

Hui, C., Veldtman, R. & McGeoch, M.A. (2010) Measures, perceptions and scaling patterns of aggregated species distributions. *Ecography*, **33**, 95–102.

Huston, M., DeAngelis, D. & Post, W. (1988) New computer models unify ecological theory, *Bioscience*, **38**, 682–691.

Jerde, C.L. & Lewis, M.A. (2007) Waiting for invasions: a framework for the arrival of nonindigenous species. *American Naturalist*, **170**, 1–9.

Jongejans, E., Skarpaas, O. & Shea, K. (2008) Dispersal, demography and spatial population models for conservation and control management. *Perspectives in Plant Ecology, Evolution and Systematics*, **9**, 153–179.

Kawasaki, K. & Shigesada, N. (2007) An integrodifference model for biological invasions in a periodically fragmented environment. *Japan Journal of Industrial and Applied Mathematics*, **24**, 3–15.

Kearney, M. & Porter, W. (2009) Mechanistic niche modelling: combining physiological and spatial data to predict species' ranges. *Ecology Letters*, **12**, 334–350.

Kolmogorov, A.N., Petrovskii, I.G. & Piskunov, N.S. (1937) Studies of the diffusion equation, with the increasing quantity of the substance and its application to a biological problem. *Bull. Moskov. Gos. Univ. Mat. Mekh.*, **1**(6), 1–26.

Kot, M., Lewis, M.A. & van den Driessche, P. (1996) Dispersal data and the spread of invading organisms. *Ecology*, **77**, 2027–2042.

Kriticos, D.J., Sutherst, R.W., Brown, J.R., Adkins, S.W. & Maywald, G.F. (2003) Climate change and the potential distribution of an invasive alien plant: *Acacia nilotica* ssp. *indica* in Australia. *Journal of Applied Ecology*, **40**, 111–124

Lavorel, S., Gardner, R.H. & O'Neill, R.V. (1995) Dispersal of annual plants in hierarchically structured landscapes. *Landscape Ecology*, **10**, 277–289.

Lavorel, S., O'Neill, R.V. & Gardner, R.H. (1994) Spatio-temporal dispersal strategies and annual plant species coexistence in a structured landscape. *Oikos*, **71**, 75–88.

Leung, B., Cacho, O. & Spring, D. (2010) Searching for nonindigenous species: rapidly delimiting the invasion boundary. *Diversity and Distributions*, **16**, 451–460.

Levins, R. (1969) Some demographic and genetic consequences of environmental heterogeneity for biological control. *Bulletin of the Entomological Society of America*, **15**, 237–240.

Lewis, M.A. & Kareiva, P. (1993) Allee dynamics and the spread of invading organisms. *Theoretical Population Biology*, **43**, 141–158.

Lewis, M.A., Neubert, M.G., Caswell, H., Clark, J. & Shea, K. (2006) A guide to calculating discrete-time invasion rates from data. *Conceptual Ecology and Invasion Biology: Reciprocal Approaches to Nature* (ed. M.A. Cadotte, S.M. McMahon and T. Fukami), pp. 162–192. Springer, Berlin.

Lobo, J.M., Jimenez-Valverde, A. & Real, R. (2008) AUC: a misleading measure of the performance of predictive distribution models. *Global Ecology and Biogeography*, **17**, 145–151.

Lockwood, J.L., Cassey, P. & Blackburn, T. (2005) The role of propagule pressure in explaining species invasions. *Trends in Ecology and Evolution*, **20**, 223–228.

Lonsdale, W.M. (1993) Rates of spread of an invading species: *Mimosa pigra* in Northern Australia. *Journal of Ecology*, **81**, 513–521.

Moilanen, A., Wilson, K.A. & Possingham, H. (2009) *Spatial Conservation Prioritization: Quantitative Methods and Computational Tools*. Oxford University Press, Oxford.

Mollison, D. (1977) Spatial contact models for ecological and epidemic spread. *Journal of Royal Statistical Society B*, **39**, 283–326.

Mollison, D. (1991) Dependence of epidemic and population velocities on basic parameters. *Mathematical Biosciences*, **107**, 255–287.

Moody, M.E. & Mack, R.N. (1988) Controlling the spread of plant invasions: the importance of nascent foci. *Journal of Applied Ecology*, **25**, 1009–1021.

Neubert, M.G. & Caswell, H. (2000) Demography and dispersal: calculation and sensitivity analysis of invasion speed for structured populations. *Ecology*, **81**, 1613–1628.

Newnham, R.M. (1964) The development of a stand model for Douglas-fir. PhD thesis, University of British Columbia, Vancouver.

Okubo, A. (1980) *Diffusion and Ecological Problems: Mathematical Models*. Springer, New York.

Okubo, A., Maini, P.K., Williamson, M.H. & Murray, J.D. (1989) On the spatial spread of the grey squirrel in Britain. *Proceedings of the Royal Society of London B*, **238**, 113–125.

Openshaw, S. (1984) *The Modifiable Areal Unit Problem*. GeoBooks, Norwich, UK.

Palmer, M.W. (1992) The coexistence of species in fractal landscapes. *American Naturalist*, **139**, 375–397.

Pearson, R.G., Thuiller, W., Araújo, M.B., et al. (2006) Model-based uncertainty in species range prediction. *Journal of Biogeography*, **33**, 1704–1711.

Peterson, A.T. & Vieglais, D.A. (2001) Predicting species invasion using ecological niche modeling. *BioScience*, **51**, 363–371.

Petrovskii, S.V. & Li, B.L. (2006) *Exactly Solvable Models of Biological Invasion*. CRC Press, London.

Phillips, B.L., Brown, G.P., Travis, J.M.J. & Shine, R. (2008) Reid's paradox revisited: the evolution of dispersal kernels during range expansion. *American Naturalist*, **172**, S34–S38.

Porter, J.H. & Dooley Jr, J.L. (1993) Animal dispersal patterns: a reassessment of simple mathematical models. *Ecology*, **74**, 2436–2443.

Pyšek, P., Richardson, D. M., Rejmánek, M., Webster, G. L., Williamson, M. & Kirschner, J. (2004) Alien plants in checklists and floras: towards better communication between taxonomists and ecologists. *Taxon*, **53**, 131–143.

Richardson, D.M. & Pyšek, P. (2007) Classics in physical geography revisited: Elton, C.S. 1958: The ecology of invasions by animals and plants. Methuen: London. *Progress in Physical Geography*, **31**, 659–666.

Richardson, D.M., Pyšek, P., Rejmánek, M., Barbour, M.G., Panetta, F.D., & West, C.J. (2000a) Naturalization and invasion of alien plants: concepts and definitions. *Diversity and Distributions*, **6**, 93–107.

Richardson, D.M., Allsopp, N., D'Antonio, C.M., Milton, S.J. & Rejmánek, M. (2000b) Plant invasions – the role of mutualisms. *Biological Reviews*, **75**, 65–93.

Richardson, D.M., Hellmann, J.J., McLachlan, et al. (2009) Multidimensional evaluation of managed relocation. *Proceedings of the National Academy of Sciences of the USA*, **106**, 9721–9724.

Rouget, M., Richardson, D., Cowling, R., Lloyd, J. & Lombard, A.T. (2003) Current patterns of habitat transformation and future threats to biodiversity in terrestrial ecosystems of the Cape Floristic Region, South Africa. *Biological Conservation*, **112**, 63–85.

Roura-Pascual, N., Bas, J.M. & Hui, C. (2010a) The spread of the Argentine ant: environmental determinants and impacts on native ant communities. *Biological Invasions*, **12**, 2399–2412.

Roura-Pascual, N., Bas, J.M., Thuiller, W., Hui, C., Krug, R.M. & Brotons, L. (2009a) From introduction to equilibrium: reconstructing the invasive pathways of the Argentine ant in a Mediterranean region. *Global Change Biology*, **15**, 2101–2115.

Roura-Pascual, N., Richardson, D.M., Krug, et al. (2009b) Ecology and management of alien plant invasions in South Africa fynbos: accommodating key complexities in objective decision making. *Biological Conservation*, **142**, 1595–1604.

Roura-Pascual, N., Krug, R.M., Richardson, D.M. & Hui, C. (2010b) Spatially-explicit sensitivity analysis for conservation management: exploring the influence of decisions in invasive alien plant management. *Diversity and Distributions*, **16**, 426–438.

Schreiber, S.J. & Lloyd-Smith, J.O. (2009) Invasion dynamics in spatially heterogeneous environments. *American Naturalist*, **174**, 490–505.

Shigesada, N. & Kawasaki, K. (1997) *Biological Invasions: Theory And Practice*. Oxford University Press, Oxford.

Shigesada, N., Kawasaki, K. & Teramoto, E. (1986) Traveling periodic waves in heterogeneous environments. *Theoretical Population Biology*, **30**, 143–160.

Skellam, J.G. (1951) Random dispersal in theoretical populations. *Biometrika*, **38**, 196–218.

Solow, A.R. & Costello, C.J. (2004) Estimating the rate of species introductions from the discovery record. *Ecology*, **85**, 1822–1825.

Stohlgren, T.J., Barnett, D.T. & Kartesz, J. (2003) The rich get richer: patterns of plant invasions in the United States. *Frontiers in Ecology and the Environment*, **1**, 11–14.

Storch, D., Šizling, A.L., Reif, J., Polechová, J., Šizlingová, E. & Gaston, K.J. (2008) The quest for a null model for macroecological patterns: geometry of species distributions at multiple spatial scales. *Ecology Letters*, **11**, 771–784.

Tatem, A.J. (2009) The worldwide airline network and the dispersal of exotic species: 2007–2010. *Ecography*, **32**, 94–102.

Tilman, D. (2004) Niche tradeoffs, neutrality, and community structure: a stochastic theory of resource competition, invasion, and community assembly. *Proceedings of the National Academy of Sciences of the USA*, **101**, 10854–10861.

Trakhtenbrot, A., Nathan, R., Perry, G. & Richardson, D.M. (2005) The importance of long-distance dispersal in biodiversity conservation. *Diversity and Distributions* **11**, 173–181.

Turchin, P. (1998) *Quantitative Analysis of Movement: Measuring and Modeling Population Redistribution in Animals and Plants*. Sinauer Associates, Sunderland, Massachusetts.

van den Bosch, F., Hengeveld, R. & Metz, J.A.J. (1992) Analyzing the velocity of animal range expansion. *Journal of Biogeography*, **19**, 135–150.

van den Bosch, F., Metz, J.A.F. & Diekmann, O. (1990) The velocity of spatial population expansion. *Journal of Mathematical Biology*, **28**, 529–565.

Walker, P., Leather, S.R. & Crawley, M.J. (2002) Differential rates of invasion in three related alien oak gall wasps (Hymenoptera: Cynipidae). *Diversity and Distributions*, **8**, 335–349.

Weinberger, H.F. 1982. Long-term behavior of a class of biological models. *SIAM Journal on Mathematical Analysis*, **13**, 353–396.

Wiens, J.A., Schooley, R.L. & Weeks, R.D. (1997) Patchy landscapes and animal movements: do beetles percolate? *Oikos*, **78**, 257–264.

Wiens, J.J. & Graham, C.H. (2005) Niche conservatism: integrating evolution, ecology, and conservation biology. *Annual Reviews of Ecology, Evolution, and Systematics*, **36**, 519–539.

Williamson, M. (2001) Overview and synthesis: the tale of the tail. *Dispersal Ecology* (ed. J.M. Bullock, R.E. Kenward and R.S. Hails), pp. 431–443. Blackwell, Oxford.

Williamson, M. (2009) Variation in the rate and pattern of spread in introduced species and its implications. *Bioinvasions and Globalization: Ecology, Economics, Management, and Policy* (ed. C. Perrings, H. Mooney and M. Williamson), pp. 56–65. Oxford University Press, Oxford.

Wilson, J.R.U., Dormontt, E.E., Prentis, P.J., Lowe, A.J. & Richardson, D.M. (2009a) Biogeographic concepts define invasion biology. *Trends in Ecology & Evolution*, **24**, 586.

Wilson, J.R.U., Dormontt, E.E., Prentis, P.J., Lowe, A.J. & Richardson, D.M. (2009b) Something in the way you move: dispersal pathways affect invasion success. *Trends in Ecology & Evolution*, **24**, 136–144.

With, K.A. & King, A.W (1997) The use and misuse of neutral landscape models in ecology. *Oikos*, **79**, 219–229.

With, K.A. (2002) The landscape ecology of invasive spread. *Conservation Biology*, **16**, 1192–1203.

Chapter 26

RESPONSES OF INVASIVE SPECIES TO A CHANGING CLIMATE AND ATMOSPHERE

Jeffrey S. Dukes

Department of Forestry and Natural Resources and Department of Biological Sciences, Purdue University, West Lafayette, IN 47907-2061, USA

Fifty Years of Invasion Ecology: The Legacy of Charles Elton, 1st edition. Edited by David M. Richardson

26.1   INTRODUCTION: A RAPIDLY CHANGING ENVIRONMENT

In 1958, when Charles Elton published *The Ecology of Invasions by Animals and Plants*, an atmospheric chemist named Charles Keeling was busy establishing a research outpost atop Mauna Loa, on the island of Hawaii. In March of that year, he began the first highly accurate series of measurements of atmospheric carbon dioxide, a series that continues to this day. At the time, little was known about how CO_2 concentrations might change from month to month, let alone from year to year. The idea that humans could dramatically influence the composition of the atmosphere seemed far-fetched, and was rarely mentioned in the scientific community, let alone among the general public.

Now, more than 50 years later, the data collected by Keeling's outpost have produced a compelling illustration of our society's power to change the environment. By 2009, CO_2 concentrations at Mauna Loa were 23% above those of Keeling's first measurement, and 39% above those of pre-industrial times (Fig. 26.1), with concentrations increasing more rapidly every year. Researchers around the world have warned that this trend will increasingly affect the climate, and governments around the world are wrestling over how best to reduce CO_2 emissions from fossil-fuel burning and deforestation.

The changing atmosphere directly affects species in many ways. Plants, which depend on CO_2 for growth, respond most directly to the change. Marine organisms face new challenges as atmospheric CO_2 dissolves in the ocean and causes it to become more acidic. And, because the additional CO_2 changes the climate by trapping more infrared radiation in Earth's atmosphere, it indirectly affects the environment for all species. None of these issues would have concerned Elton as he wrote his book, but they concern virtually all ecologists today.

Already, researchers have documented shifts in flowering dates for plants, developmental milestones and migration dates for animals, and the distributions of many taxa in terrestrial and aquatic environments (Parmesan 2006). The directions of these shifts are heavily biased towards the directions expected under recently observed regional changes in climate (Parmesan 2007; Thomas 2010). Some taxa have shifted phenology faster than others, but across taxa, spring phenology has advanced by approximately 2.8 days per decade in the Northern hemisphere (Parmesan 2007), consistent with trends expected in a warming climate.

Greenhouse gas releases have been accelerating, and by the end of the century, we expect Earth's environment to be quite different than it is now. *How* different will depend heavily on societal decisions, but most future scenarios anticipate CO_2 to surpass 600

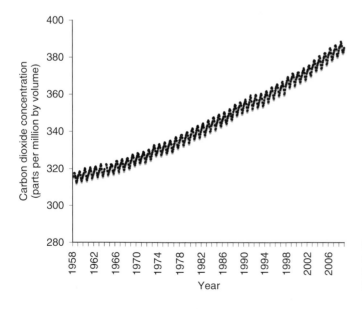

Fig. 26.1 Atmospheric CO_2 concentrations as measured by the station established by Charles Keeling in 1958 atop Mauna Loa. Based on data from Keeling et al. (2009).

parts per million by the end of the century, and envision Earth's average surface temperature rising by 1.1 to 6.4 °C by 2100 (Meehl et al. 2007). The Earth system models that produce these numbers also foresee greater warming over land than over the oceans, and much greater warming as one approaches the higher latitudes in the Northern hemisphere (see Plate 8). Precipitation patterns are also likely to change, with some of the wetter regions of the world receiving more winter precipitation, some of the drier regions of the world becoming drier still, and droughts and deluges becoming more frequent everywhere (see Plate 8). Fossil fuel burning and intensive agriculture also lead to increases in nitrogen deposition in many ecosystems, and increases in low-level ozone.

This cabal of environmental changes will contribute to shifts in species distributions and competitive interactions in ecosystems throughout the world. The results of these shifts, though, will not be simple to predict.

26.2   CLOUDS IN THE CRYSTAL BALL: WHAT CAN BE PREDICTED?

Many types of uncertainty make it challenging to predict responses of ecosystems and communities to environmental change: uncertainty about the rate and nature of the environmental changes (which stem from smaller uncertainties in the science and larger uncertainties about society's future path), uncertainty about how a specific ecosystem functions and how individual species respond to those changes, and uncertainty about the relationships among species within communities, just to name a few (Dukes et al. 2009). Although this uncertainty should make a reader sceptical of the precision of any single prediction for the future, an understanding of the scope of likely changes is valuable for predicting ecological impacts. The scientific community has become increasingly confident in the direction of most environmental changes, even if the rates of some are still uncertain. At the same time, a growing body of experimental research has illuminated the more basic responses of organisms to environmental changes in many simplified ecosystems. Direct responses of many species to climate change can be inferred from elements of their distribution across the landscape. However, species will also respond indirectly to climate change, for instance as climate affects the strength of interspecific interactions. There remains little basis for predicting the net outcome of changes in these indirect responses.

Over the past two decades, invasion biology and climate science have reached the point where conditional predictions can be made that help inform management decisions. Managers, landowners and policy makers must already make planning decisions in the face of an uncertain future, whether the uncertainty is due to economic, political or other factors. Imperfect predictions based on available knowledge, taken with caveats, can help inform decisions with long-term implications and focus attention on potential new problems or opportunities. Indeed, in the face of uncertain climate change impacts, resource managers are already acting, for instance to remove barriers to dispersal in stream habitat in Oregon, USA, allowing native fish access to a wider variety of microsites and stream temperatures (Lawler et al. 2010).

26.3   INVASIVE SPECIES' RESPONSES TO CLIMATE CHANGE

The first discussions of how invasive species might respond to environmental changes (see, for example, Cronk 1995; Huenneke 1997; Dukes & Mooney 1999) had relatively few studies on which to build conclusions. Since then, the topic has received increasing attention, particularly in the form of studies examining potential future distributions of species. A variety of recent publications review potential responses of invasive species to climate change in particular (see, for example, Thuiller et al. 2007; Hellmann et al. 2008; Rahel & Olden 2008; Walther et al. 2009; Bradley et al. 2010; Ziska & Dukes 2011). In 2009, Australia's Invasive Species Council began publishing an eBulletin called *Double Trouble* to focus specifically on potential interactions between climate change and invasive species.

Hellmann et al. (2008) highlighted five ways in which invasive species will respond to climate change. These are as follows.
1 Species will be introduced through new pathways. Pathways of introduction will change as humans try to mitigate or adapt to the changing climate, whether that involves planting new species for bioenergy use (Raghu et al. 2006; Barney & DiTomaso 2008), taking advantage of newly thawed Arctic waterways for transporting goods (Pyke et al. 2008), importing new species for ornamental use or intentionally moving

species of concern to new locations (Richardson et al. 2009).

2 Newly introduced species will experience different biotic and abiotic conditions, affecting the probability that they will become established in the new region. Climate change could alter the probability that a newly arrived species could establish a self-sustaining population through two primary mechanisms. First, it could alter the climatic suitability for the introduced species. Second, it could affect the physiology of resident species directly, potentially altering their competitive abilities or fecundity. Unless native communities rapidly reorganize, and species evolve in response to climate change, existing communities may become more susceptible to invasion by species that arrive pre-adapted to the new conditions. Dukes & Mooney (1999) predicted that ecosystems dominated by long-lived organisms with long juvenile periods and little long-distance seed dispersal will experience the greatest increases in invasibility, because populations of those species will likely be slowest to adapt to the new conditions.

3 Distributions of existing invasives will shift. As environmental conditions change, species distributions will respond. A warming of 3 °C, well within the end-of-century projections from the IPCC (2007), corresponds to isothermal shifts of approximately 250 km in latitude and 500 m in elevation (Gates 1990). Depending on the topography, mean temperatures are expected to migrate horizontally across Earth's land surface at rates of 0.11 to 1.46 km per year (Loarie et al. 2009). Warming will allow many species to extend their ranges on the poleward or uphill sides, where they are currently cold-limited. For instance, in Hawaii, warming may allow invasive mosquitoes to survive at and carry avian malaria to higher altitudes, causing problems for native birds whose last remaining habitats are on the mountain-tops (Benning et al. 2002). Warming may also lead to the retreat of ranges along the warmer edges, through direct physiological responses and/or changes in interactions among species. Changes in precipitation patterns will similarly affect distributions of some terrestrial species. Species' responses can be estimated through climate envelope-type (niche) models, and are further discussed below, in section 26.4.

4 Impacts of existing invasives will change. Impacts of invasive species are likely to change for a variety of reasons, although predicting the nature of these changes will not always be straightforward. Impacts are generally thought of as the product of the range of the invasive species, the per-capita effect of the species on the process or property of interest, and the abundance of the species within its range (Parker et al. 1999). With environmental change, this impact will also depend on any direct response of the process or property to that change. For instance, in South Africa, invasions of woody *Acacia* and *Pinus* species into upland watersheds have decreased streamflow levels (Le Maitre et al. 2000; Dye & Jarmain 2004). Because water is scarce in this region, loss of any water from the streams is a major concern. If less precipitation fell in this region as a consequence of climate change, the relative impact of these invasive species would increase, because each drop of water would be more valuable (Fig. 26.2). Abundances and range sizes of many invasive (and native) species are likely to change in the future (see below), but evidence for changes in species' per-capita impacts is less clear. In

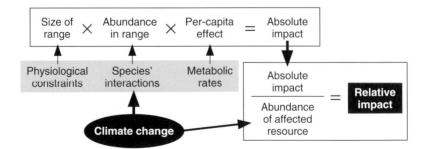

Fig. 26.2 Mechanisms through which climate change could affect the impact of invasive species. Figure adapted from Dukes (2011).

a field experiment in Montana (USA) grassland, Maron and Marler (2008) found little evidence for changes in competitive effects of invasive plant species on natives under increased precipitation. On the other hand, many studies have shown that the outcome of competition can depend on environmental conditions (see, for example, Patterson 1995). For instance, in the Rocky Mountains of North America, native bull trout (*Salvelinus confluentus*) outcompete non-native brook trout (*Salvelinus fontinalis*) in cooler streams, but warmer water would favour the brook trout (Rieman et al. 1997).

5 The effectiveness of strategies for managing and controlling invasive species will change. Environmental changes will alter the effectiveness of management and control strategies by affecting the physiology of invasive species and biocontrol species. Physiological responses of weed species, for instance, can alter their response to herbicides. In an agricultural field, Ziska et al. (2004) found that Canada thistle (*Cirsium arvense*) rebounded from herbicide treatment more rapidly under elevated CO_2 than under ambient CO_2, likely because it had allocated more resources to below-ground structures that were then less completely killed by the glyphosate. The effectiveness of biocontrol techniques also depends on environmental conditions. If distributions of biocontrol species respond differently to climate change than those of their target species, the area in which the biocontrol organism achieves its intended effect will shift.

Much of the research on these issues so far has comprised niche modelling studies and experimental manipulations. I explore results from these studies in more depth below, in sections 26.4–26.6. The subsequent sections concern feedbacks of invasive species to climate change, consequences of societal responses to climate change, future advances and challenges in the field.

26.4 THE ART OF PREDICTION: HOW WILL CLIMATE CHANGE AFFECT INVASIVE SPECIES' DISTRIBUTIONS?

Many studies have tried to predict future distributions of invasive species based on current species ranges, taking a variety of approaches. Niche modelling has been used to estimate future ranges of invasive plants in Australia (see, for example, Kriticos et al. 2003),

Europe (see, for example, Beerling et al. 1995), North America (see, for example, Zavaleta & Royval 2002; Peterson et al. 2003, 2008; Roura-Pascual et al. 2004; Bradley et al. 2009; Jarnevich & Stohlgren 2009) and Africa (see, for example, Richardson et al. 2000; Parker-Allie et al. 2009). Most of these studies base predictions purely on climatic tolerances, and predictions range from dramatic range increases to range decreases, depending on the species and the study. Richardson et al. (2000) found that including soil type in their analysis restricted the potential change in distribution for all five of the invasive plant species they examined, and climate change was predicted to reduce the total area of suitable habitat for all of these.

Similar studies have also examined invasive animals, such as cane toads (*Bufo marinus*) in Australia (Sutherst et al. 1995) and gypsy moth (*Lymantria dispar*) in New Zealand (Pitt et al. 2007). These climate envelope-type studies currently provide the best mechanism for estimating future species ranges, but they have several limitations. For instance, current ranges may not have climatic limits that are relevant for use in predictions. The native ranges of species often occupy more restricted climatic zones than those occupied in the introduced region (Sutherst et al. 1995; Beaumont et al. 2009), and populations in the introduced region may not have had the opportunity to spread to the limits of their climatic tolerances. In addition, models that have a coarse spatial resolution omit topographic variation, and thus underestimate climatic and environmental heterogeneity. Grid cells classified as climatically unsuitable for a species may actually have patches of suitable climate within them (Luoto & Heikkinen 2008; Randin et al. 2009). These factors could lead to an underestimation of the future range of the species. Other factors could lead to overestimation of ranges; limits imposed by competition with other species, by a lack of mutualists or by a dependence on non-climatic environmental conditions such as a specific soil type could restrict ranges to areas much smaller than the zones that models identify as climatically suitable. Beyond this, other environmental changes such as increasing CO_2 concentrations may alter the climatic tolerances of a species in ways that are currently impossible to predict without experimentation.

For plants in particular, *in situ* experimental manipulations can provide important indications of how invasive species may perform in future environments, and could provide useful tests of niche model projections. However, these experiments are expensive, and therefore rare. So far, experimental manipulations have investigated the effects of CO_2, climate change (both warming and precipitation changes) and nutrient deposition on the success of invasive species. These experiments have tested how the abundance of a species may change within its range, as opposed to examining how the range itself may shift. Manipulations that increase resource availability, for instance by adding precipitation (or nutrients; see below), often increase the success of invasive plant species (Dukes & Mooney 1999; Scherer-Lorenzen et al. 2007; Blumenthal et al. 2008). Manipulations of CO_2 and temperature are still too rare to make confident generalizations, but many CO_2 experiments have found positive responses of invasive plant species, at least in some years (see, for example, Smith et al. 2000; and see below). Climatic variability studies are also rare; White et al. (2001) investigated the effect of extreme heating and rainfall events on community invasibility, and found that C_4 plant species may invade C_3 grasslands more successfully as climatic variability increases and climatic extremes become more frequent.

Several observational studies and experiments have examined how the abundance or growth of invasive animal species changes in response to variation in temperature or moisture availability (see, for example, Stachowicz et al. 2002; Chown et al. 2007; Saunders & Metaxas 2007; Heller et al. 2008; Weitere et al. 2009). As with plants, results are largely species specific, although Stachowicz et al. (2002) found increasing recruitment of the three most abundant non-native sea squirts in years with warmer winter waters off the eastern coast of North America. Again, though, few (if any) experiments have examined how climate changes will affect invasive species' distributions.

For the most part, our understanding of how interactions with resident species constrain (or expand) the large-scale distribution of invasive species is poor, and so it is difficult to estimate how climate change might alter these constraints. Few studies have examined how climate change alters community invasibility. Results from several experiments suggest that communities suppress establishment success and growth of some invading species (native and non-native) through resource pre-emption (see, for example, Dukes 2001; Hooper & Dukes 2010). If climate change reduces the ability of resident communities to exploit available resources, then the invasibility of those communities would increase. It seems likely that invasibility of some plant communities, such as those dominated by a few longer-lived species with similar geographic distributions, might increase substantially as the dominants must increasingly tolerate conditions in which they did not evolve. In contrast, invasibility of communities dominated by a large mixture of species with a range of lifespans and largely different distributions might change very little, as shifts in dominance among the resident species could prevent any reduction in resource use by the community. There is some indication that plants entering new ranges experience less impact from pathogens and herbivores than do the resident species, suggesting a potential mechanism for accelerated establishment of new species, and transitions from one vegetation type to another (Engelkes et al. 2008). Climate change seems unlikely to decrease invasibility, except perhaps in extreme environments where conditions become even more extreme, such as deserts that become increasingly water limited. There, the chances of better-adapted species arriving seem slim. Effects of climate change on invasibility of animal communities will similarly depend on the capacity of the extant organisms to tolerate or adapt to new conditions.

Although there is little experimental evidence to describe how climate change may affect the invasibility of communities, there is increasing evidence that, at least for plants, invasive species have a variety of traits that will provide advantages in a changing climate. For instance, within a given community, invasive species are likely to tolerate a wider range of climates than natives, and thus may experience less of a disadvantage. Studies have shown that invasive plant species tolerate a wider range of climates than their non-invasive relatives (see, for example, Rejmánek & Richardson 1996; Qian & Ricklefs 2006) or other, non-invasive resident species (Goodwin et al. 1999). Also, invasive plants in at least one community have adjusted their phenologies to 'keep up' with climate change much more rapidly than competing native plants (Willis et al. 2010). Invasive species have also demonstrated the potential to spread rapidly through the current landscape, either aided or unhindered by roads and human land use. Not all native species will disperse so easily across the patches of fragmented habitat that exist in many parts of the world.

26.5 LESSONS FROM EXPERIMENTAL MANIPULATIONS: INVASIVE SPECIES' RESPONSES TO A CHANGING ATMOSPHERE

It has long been recognized that the increasing atmospheric CO_2 concentration could alter the balance of competition among plants (with implications for the effects of weed species on crop yield, for instance (Patterson 1995), or for competition between C_3 and C_4 plants). Many experiments have examined responses of single plants or single species growing in pots to elevated CO_2. Dukes (2000) suggested that such experiments, collectively, did not indicate that increasing CO_2 favours invasive species over non-invasive species. However, some individual studies of this type have found that invasive species could preferentially benefit. For instance, Song et al. (2009) grew three plant species that are invasive in China along with three similar, but native, co-occurring species in growth chambers at two CO_2 concentrations, and reported stronger biomass responses in the invasives. Ziska (2003) found strong responses of six of North America's more pernicious weed species to both current (relative to pre-industrial) and future CO_2 concentrations.

More realistic tests of this question have featured competition between the invasive species and a native community. Two pot experiments (Bradford et al. 2007; Dukes 2002) collectively show little effect of CO_2 on the overall impact of invasive plants. However, competition experiments in pots still fall short of providing a realistic environment from which to draw strong conclusions.

Field experiments with invasive plant species grown in competitive environments provide the most realistic simulations of future environments. Those conducted so far have found that some invasive species respond quite strongly to elevated CO_2, particularly when other constraints on the species' growth are relaxed (see, for example, Smith et al. 2000; Belote et al. 2003; Hättenschwiler & Körner 2003; Williams et al. 2007). For instance, the annual grass *Bromus madritensis* ssp. *rubens* responded much more positively to elevated CO_2 than native annual species in a very wet year in North America's Mojave Desert (Smith et al. 2000). However, in drier years, *Bromus* failed to germinate or failed to survive and was unaffected by CO_2. Although very few species have been studied in realistic environments, it appears that several invasive species are likely to preferentially benefit from the increase in CO_2, at least in some years or some conditions (Table 26.1).

Although the main byproduct of fossil fuel burning is CO_2, reactive nitrogen molecules are also released during combustion. This nitrogen, along with nitrogen released from agricultural fields, gets deposited on land and in the sea, fertilizing many natural systems. Although this fertilization may have little effect on the abundance of invasive plant species in regions where the native flora already contains many responsive plant species, there are many regions of the world in which the native flora evolved under nutrient-poor conditions. Several lines of evidence suggest that nitrogen deposition increases the prevalence of invasive plants in these regions. Scherer-Lorenzen et al. (2007) recently reviewed this evidence in detail.

Table 26.1 Results from field experiments testing the response of invasive species to elevated CO_2 in a community context.

Species	Community	Response	Reference
Bromus madritensis	Mojave Desert	+/0	Smith et al. 2000
Lonicera japonica	Tennessee forest understorey	+/0	Belote et al. 2003
Microstegium vimineum	Tennessee forest understorey	–/0	Belote et al. 2003
Prunus laurocerasus	Swiss forest understorey	+	Hattenschwiler & Korner 2003
Hypochaeris radicata	Tasmanian grassland	0**	Williams et al. 2007
Leontodon taraxacoides	Tasmanian grassland	0	Williams et al. 2007
Centaurea solstitialis	California valley grassland	+	J.S. Dukes et al., unpublished data

+, Significantly positive effect of CO_2 on biomass production in some years of study; 0, no significant effect of CO_2 on production or population growth rate in some years of study; –, marginally significant negative effect of CO_2 on production in one year of study; **, a marginally significant CO_2-driven increase in population growth rate under ambient temperatures was offset by warming.

26.6 INTERACTING ENVIRONMENTAL CHANGES

The future environment will differ from today's in patterns of human land use, temperature and precipitation, CO_2, nitrogen deposition and runoff, ozone pollution, ocean acidity and many other respects. We might expect some aspects of these environmental changes to counteract each other in some locations. For instance, plants growing under elevated CO_2 often close their stomata partly, leading to water savings. The additional water savings could at least partly counteract increases in water stress due to warming or drought. With so many different environmental changes happening at once, the potential for such interactions is great, and predicting the most consequential interactions presents a challenge.

So far, relatively few studies have examined invasive species' responses to more than one environmental change at a time. Data from Zavaleta et al. (2003), Williams et al. (2007) and Blumenthal et al. (2008) suggest interactions may occasionally be observed, but the frequency with which they will be important is unclear. In North American mixed grass prairie, Blumenthal et al. (2008) found that the success of three invasive forbs depended on addition of supplemental snow. When snow was present, one of the species also responded to nitrogen addition. Results such as this suggest that interactions among environmental factors may be less important than the 'sum' of environmental changes acting on a site. Although the increase in CO_2 will be relatively uniform worldwide, sites that experience dramatic changes in climate and other factors such as nitrogen deposition will move farther from the conditions in which local species evolved than sites in which climatic changes are more subdued and changes in other factors are minimal. I hypothesize that the sites in which environmental conditions depart the farthest from their historical mean will experience the greatest increases in invasibility, as the native species will no longer be well adapted to local conditions. It is unclear how best to measure departures of conditions from a historical mean across multiple factors. An integrating and quantifiable measure in terrestrial plant communities might be aboveground biomass production. Sites in which biomass declines sharply might become more susceptible to invasion of stress-tolerant plant species, whereas those in which biomass increases sharply may be susceptible to invasion by more productive plant species that gain footholds in disturbed microsites. For taxa other than plants, other measures of departures from historical conditions might be more appropriate.

As discussed above, several studies have used niche models to project changes in invasive species' distributions under climate change. However, these models do not incorporate other environmental changes, such as increases in CO_2 and nitrogen deposition. Mechanistic models that simulate a variety of species-level and environmental processes could be used to explore the impacts of a variety of environmental changes, but these models are more difficult to parameterize. Incorporating effects of CO_2, nitrogen and other factors into niche models would not be straightforward, but is worth attempting, considering the evidence that these factors can play important roles in species' success. Such modified niche models could then be used to estimate, at least roughly, the importance of various interactions among environmental change factors for invasive species' success.

26.7 FEEDBACKS FROM INVASIVE SPECIES TO CLIMATE CHANGE

Some plant and animal species, including some invasives, strongly influence disturbance regimes and the cycling of elements in ways that tangibly affect the composition of the atmosphere or the energy balance of the landscape (Dukes & Mooney 2004). Thus some invasive species that respond to an environmental change can create weak to moderate feedbacks to the rate of that environmental change. Species most likely to create perceptible feedbacks are those that dramatically alter the landscape over large regions. Examples include cheatgrass (*Bromus tectorum*) and some insect pests.

Cheatgrass has replaced native shrubland systems in a large region of western North America, altering carbon storage and energy balance in this area. This annual grass, whose tinder allows fires to spread widely, now dominates at least 20,000 km^2 of the Great Basin (Bradley & Mustard 2005), and has the potential to spread across 150,000 km^2 (Suring et al. 2005; Wisdom et al. 2005). Fire-aided conversion of sagebrush shrubland to cheatgrass grassland has transformed parts of western North America from a carbon sink to a carbon source (Bradley & Mustard

2006; Prater et al. 2006), releasing about 8 Tg of carbon so far (Bradley et al. 2006). The invasion also alters albedo (J.S. Dukes, unpublished data) and evapotranspiration (Prater & DeLucia 2006; Prater et al. 2006) across the landscape, with implications for regional climate. Bradley et al. (2009) suggest that the area suitable for cheatgrass may decline under some projected temperature and precipitation regimes. On the other hand, growth chamber studies suggest cheatgrass may have responded positively to increasing atmospheric CO_2 (Ziska et al. 2005), and will continue to respond in the future (Smith et al. 1987; Ziska et al. 2005). Further acceleration of cheatgrass growth would lead to more rapid and greater fine fuel accumulation in the region, likely facilitating more frequent and hotter fires, and greater cheatgrass dominance.

Pathogens and insect pests can also produce feedbacks. Probably the most important feedback mechanism occurs when environmental changes facilitate larger outbreaks of these organisms in forests, leading to widespread tree death. Cold winter temperatures are thought to restrict the range of hemlock woolly adelgid (*Adelgis tsugae*), a non-native pest responsible for the decline of eastern hemlock (*Tsuga canadensis*) in eastern North America. The insect's range could expand rapidly northwards into remaining hemlock stands as annual minimum temperatures rise (Dukes et al. 2009). A similar phenomenon appears to have occurred with the mountain pine beetle (*Dendroctonus ponderosae*), a native insect pest in pine forests of western North America (Kurz et al. 2008). Recent warming has led to more severe outbreaks of mountain pine beetle, and tree deaths from the latest outbreak will release an estimated 270 Mt of carbon over a 20-year period, accelerating climate change. Mountain pine beetle has the potential to create even stronger feedbacks if its range expands to new regions of North America, as would be expected under continued warming.

26.8 THE HUMAN ELEMENT: SOCIETAL RESPONSES TO GLOBAL ENVIRONMENTAL CHANGE, AND CONSEQUENCES FOR INVASIVE SPECIES

We, as a species, will respond to environmental changes ourselves. Some of our attempts to mitigate or adapt to environmental changes will alter the success of invasive species. Examples may include direct introductions, such as those of potentially invasive biofuel species such as *Arundo donax* and *Miscanthus × giganteus* in the USA (Raghu et al. 2006), or of ornamental species that might not have been invasive in a previous climate. We may also have new opportunities to purge invasive species from portions of their ranges that become less hospitable (Bradley et al. 2009). In other cases, threats from invasive species and climate change may lead environmental managers to implement 'managed relocation' of some native species (Richardson et al. 2009). Will any of these species, now transported by humans to a new location, become invasive pests? Similarly, climate change will allow some native species to spread unaided from their historical range into contiguous regions with newly suitable climates. Will these species be considered 'alien' in these new locations (Walther et al. 2009)? Questions such as these, which were rarely discussed before this century, are rapidly becoming more common.

26.9 LOOKING FORWARD

Governments and managers already wage a continuous, demanding battle to intercept, detect and control invasive species. Unfortunately, other global environmental changes complicate this effort. To this point, relatively little guidance has been provided to the invasive species community as to how to prepare for or deal with new challenges arising from other environmental changes. It seems unlikely that shifting environmental conditions will require dramatic changes in invasive species detection or management practices. However, longer-term planning, improved databases, and new predictive tools, such as improved and user-friendly niche models could help prepare managers for likely challenges to come. Resources dedicated to such efforts and tools are likely to provide benefits in any future scenario.

Many researchers are currently working to develop expanded and improved invasive species databases, and to develop useful predictions of invasive species' distributions. A major obstacle to the improvement of predictions is the insularity of data within region-, discipline- or taxon-specific databases, databases that only consider invasive (or native) species, and databases that incorporate climatic or environmental but not biological information. To provide the greatest benefits, data in consolidated, global biological and environmental information systems should be integrated with improved niche models and climate projections in a user-friendly, preferably web-based tool (Lee et al. 2008). Such a major integration project would face several challenging obstacles, and would likely take decades to develop. However, such a system would provide valuable information and predictions for managers around the world.

In *The Ecology of Invasions by Animals and Plants*, Elton (1958) wrote '... provided the native species have their place, I see no reason why the reconstitution of communities to make them rich and interesting and stable should not include a careful selection of exotic forms, especially as many of these are in any case going to arrive in due course and occupy some niche.' Elton clearly recognized that the challenge posed by invasive species was too great to expect to maintain pristine communities. Before we embark as a society on major campaigns to protect ecosystems from new threats from invasive species, we have the responsibility to ask, is it worth it? Are the benefits of any current or future reduction in invasive species' prevalence worth the cost of combating them? Which invasives should we battle, and what is the right amount of money to spend on keeping them in check? At the heart, these are political and philosophical questions, but they are questions that can only be addressed with the help of a strong scientific base of knowledge. Our understanding of biological invasions has grown rapidly in the past half century, and particularly in the past three decades, at about the same time that our awareness and understanding of global environmental changes has blossomed. Although we will never fully describe the ecological, evolutionary or economic consequences of invasive species, many of their consequences are sobering. Other environmental changes have similarly profound consequences. Recent research suggests that in some cases the changes will amplify each other (e.g. through feedbacks), or amplify each other's consequences (e.g. when climate change and an invasive species both decrease availability of the same goods or ecosystem services). Further research will clarify how frequently this is likely to happen. In the meantime, to the extent possible, it will be useful to identify policy and management options that can deal with these problems simultaneously (Pyke et al. 2008).

REFERENCES

Barney, J.N. & DiTomaso, J.M. (2008) Nonnative species and bioenergy: Are we cultivating the next invader? *Bioscience*, **58**, 64–70.

Beaumont, L.J., Gallagher, R.V., Thuiller, W., Downey, P.O., Leishman, M.R. & Hughes, L. (2009) Different climatic envelopes among invasive populations may lead to underestimations of current and future biological invasions. *Diversity and Distributions*, **15**, 409–420.

Beerling, D.J., Huntley, B. & Bailey, J.P. (1995) Climate and the distribution of *Fallopia japonica*: use of an introduced species to test the predictive capacity of response surfaces. *Journal of Vegetation Science*, **6**, 269–282.

Belote, R.T., Weltzin, J.F. & Norby, R.J. (2003) Response of an understory plant community to elevated [CO_2] depends on differential responses of dominant invasive species and is mediated by soil water availability. *New Phytologist*, **161**, 827–835.

Benning, T.L., LaPointe, D., Atkinson, C.T. & Vitousek, P.M. (2002) Interactions of climate change with biological invasions and land use in the Hawaiian islands: modeling the fate of endemic birds using a geographic information system. *Proceedings of the National Academy of Sciences of the USA*, **99**, 14246–14249.

Blumenthal, D., Chimner, R.A., Welker, J.M. & Morgan, J.A. (2008) Increased snow facilitates plant invasion in mixed grass prairie. *New Phytologist*, **179**, 440–448.

Bradford, M.A., Schumacher, H.B., Catovsky, S., Eggers, T., Newington, J.E. & Tordoff, G.M. (2007) Impacts of invasive plant species on riparian plant assemblages: interactions with elevated atmospheric carbon dioxide and nitrogen deposition. *Oecologia*, **152**, 791–803.

Bradley, B.A., Blumenthal, D.M., Wilcove, D.S. & Ziska, L.H. (2010) Predicting plant invasions in an era of global change. *Trends in Ecology & Evolution*, **25**, 310–318.

Bradley, B.A., Houghton, R.A., Mustard, J.F. & Hamburg, S.P. (2006) Invasive grass reduces aboveground carbon stocks in shrublands of the Western US. *Global Change Biology*, **12**, 1815–1822.

Bradley, B.A. & Mustard, J.F. (2005) Identifying land cover variability distinct from land cover change: cheatgrass in the Great Basin. *Remote Sensing of the Environment*, **94**, 204–213.

Bradley, B.A. & Mustard, J.F. (2006) Characterizing the landscape dynamics of an invasive plant and risk of invasion using remote sensing. *Ecological Applications*, **16**, 1132–1147.

Bradley, B.A., Oppenheimer, M. & Wilcove, D.S. (2009) Climate change and plant invasions: restoration opportunities ahead? *Global Change Biology*, **15**, 1511–1521.

Chown, S.L., Slabber, S., McGeoch, M.A., Janion, C. & Leinaas, H.P. (2007) Phenotypic plasticity mediates climate change responses among invasive and indigenous arthropods. *Proceedings of the Royal Society B*, **274**, 2531–2537.

Cronk, Q.C.B. (1995) Changing worlds and changing weeds. In *Weeds in a Changing World* (ed. C. H. Stirton), pp. 3–13. British Crop Protection Council, Farnham, Surrey, UK.

Dukes, J.S. (2000) Will the increasing atmospheric CO_2 concentration affect the success of invasive species? In *Invasive Species in a Changing World* (ed. H.A. Mooney and R.J. Hobbs), pp. 95–113. Island Press, Washington, DC.

Dukes, J.S. (2001) Biodiversity and invasibility in grassland microcosms. *Oecologia*, **126**, 563–568.

Dukes, J.S. (2002) Comparison of the effect of elevated CO_2 on an invasive species (*Centaurea solstitialis*) in monoculture and community settings. *Plant Ecology*, **160**, 225–234.

Dukes, J.S. (2011) Climate change. In *Encyclopedia of Invasive Introduced Species* (ed. D. Simberloff and M. Rejmánek). University of California Press, Berkeley, California (in press).

Dukes, J.S. & Mooney, H.A. (1999) Does global change increase the success of biological invaders? *Trends in Ecology & Evolution*, **14**,135–139.

Dukes, J.S. & Mooney, H.A. (2004) Disruption of ecosystem processes in western North America by invasive species. *Revista Chilena de Historia Natural*, **77**, 411–437.

Dukes, J.S., Pontius, J., Orwig, D., et al. (2009) Responses of insect pests, pathogens, and invasive plant species to climate change in the forests of northeastern North America: What can we predict? *Canadian Journal of Forest Research-Revue Canadienne De Recherche Forestiere*, **39**, 231–248.

Dye, P. & Jarmain, C. (2004) Water use by black wattle (*Acacia mearnsii*): implications for the link between removal of invading trees and catchment streamflow response. *South African Journal of Science*, **100**, 40–44.

Elton, C.S. (1958) *The Ecology of Invasions by Animals and Plants*. Methuen, London.

Engelkes, T., Morrien, E., Verhoeven, K.J.F., et al. (2008) Successful range-expanding plants experience less aboveground and below-ground enemy impact. *Nature*, **456**, 946–948.

Gates, D.M. (1990) Climate change and forests. *Tree Physiology*, **7**, 1–5.

Goodwin, B.J., McAllister, A.J. & Fahrig, L. (1999) Predicting invasiveness of plant species based on biological information. *Conservation Biology*, **13**, 422–426.

Hättenschwiler, S. & Körner, C. (2003) Does elevated CO2 facilitate naturalization of the non-indigenous *Prunus laurocerasus* in Swiss temperate forests? *Functional Ecology*, **17**, 778–785.

Heller, N.E., Sanders, N.J., Shors, J.W. & Gordon, D.M. (2008) Rainfall facilitates the spread, and time alters the impact, of the invasive Argentine ant. *Oecologia*, **155**, 385–395.

Hellmann, J.J., Byers, J.E., Bierwagen, B.G. & Dukes, J.S. (2008) Five potential consequences of climate change for invasive species. *Conservation Biology*, **22**, 534–543.

Hooper, D.U. & Dukes, J.S. (2010) Functional composition controls invasion success in a California serpentine grassland. *Journal of Ecology*, **98**, 764–777.

Huenneke, L.F. (1997) Outlook for plant invasions: interactions with other agents of global change. In *Assessment and Management of Plant Invasions* (ed. J.O. Luken and J.W. Thieret), pp. 95–103. Springer, New York.

IPCC (2007) *Climate Change 2007: The Physical Science Basis*. Cambridge University Press, Cambridge, UK.

Jarnevich, C.S. & Stohlgren, T.J. (2009) Near term climate projections for invasive species distributions. *Biological Invasions*, **11**, 1373–1379.

Keeling, R.F., Piper, S.C., Bollenbacher, A.F. & Walker, J.S. (2009) Atmospheric CO_2 records from sites in the SIO air sampling network. In *TRENDS: A Compendium of Data on Global Change*. Carbon Dioxide Information Analysis Center, Oak Ridge National Laboratory, US Department of Energy, Oak Ridge, Tennessee.

Kriticos, D.J., Sutherst, R.W., Brown, J.R., Adkins, S.W. & Maywald, G.F. (2003) Climate change and the potential distribution of an invasive alien plant: *Acacia nilotica* ssp *indica* in Australia. *Journal of Applied Ecology*, **40**, 111–124.

Kurz, W.A., Dymond, C.C., Stinson, G., et al. (2008) Mountain pine beetle and forest carbon feedback to climate change. *Nature*, **452**, 987–990.

Lawler, J.J., Tear, T.H., Pyke, C., et al. (2010) Resource management in a changing and uncertain climate. *Frontiers in Ecology and the Environment*, **8**, 35–43.

Lee, H., Reusser, D.A., Olden, J.D., et al. (2008) Integrated monitoring and information systems for managing aquatic invasive species in a changing climate. *Conservation Biology*, **22**, 575–584.

Le Maitre, D.C., Versfeld, D.B. & Chapman, R.A. (2000) The impact of invading alien plants on surface water resources in South Africa: a preliminary assessment. *Water SA*, **26**, 397–408.

Loarie, S.R., Duffy, P.B., Hamilton, H., Asner, G.P., Field, C.B. & Ackerly, D.D. (2009) The velocity of climate change. *Nature*, **462**, 1052–1055.

Luoto, M. & Heikkinen, R.K. (2008) Disregarding topographical heterogeneity biases species turnover assessments based on bioclimatic models. *Global Change Biology*, **14**, 483–494.

Maron, J.L. & Marler, M. (2008) Field-based competitive impacts between invaders and natives at varying resource supply. *Journal of Ecology*, **96**, 1187–1197.

Meehl, G.A., Stocker, T.F., Collins, W.D., et al. (2007) Global climate projections. In *Climate Change 2007: The Physical Science Basis. Contribution of Working Group I to the Fourth Assessment Report of the Intergovernmental Panel on Climate Change* (ed. S. Solomon, D. Qin, M. Manning, et al.), pp. 747–845. Cambridge University Press, Cambridge, UK.

Parker, I.M., Simberloff, D., Lonsdale, W.M., et al. (1999) Impact: toward a framework for understanding ecological effects of invaders. *Biological Invasions*, **1**, 3–19.

Parker-Allie, F., Musil, C.F. & Thuiller, W. (2009) Effects of climate warming on the distributions of invasive Eurasian annual grasses: a South African perspective. *Climatic Change*, **94**, 87–103.

Parmesan, C. (2006) Ecological and evolutionary responses to recent climate change. *Annual Review of Ecology and Systematics*, **37**, 637–669.

Parmesan, C. (2007) Influences of species, latitudes and methodologies on estimates of phenological response to global warming. *Global Change Biology*, **13**, 1860–1872.

Patterson, D.T. (1995) Weeds in a changing climate. *Weed Science*, **43**, 685–701.

Peterson, A.T., Papes, M. & Kluza, D.A. (2003) Predicting the potential invasive distributions of four alien plant species in North America. *Weed Science*, **51**, 863–868.

Peterson, A.T., Stewart, A., Mohamed, K.I. & Araújo, M.B. (2008) Shifting global invasive potential of European plants with climate change. *PLoS One*, **3** (6), e2441, doi:10.1371/journal.pone.0002441.

Pitt, J.P.W., Regniere, J. & Worner, S. (2007) Risk assessment of the gypsy moth, *Lymantria dispar* (L), in New Zealand based on phenology modelling. *International Journal of Biometeorology*, **51**, 295–305.

Prater, M.R. & DeLucia, E.H. (2006) Non-native grasses alter evapotranspiration and energy balance in Great Basin sagebrush communities. *Agricultural and Forest Meteorology*, **139**, 154–163.

Prater, M.R., Obrist, D., Arnone, J.A. & DeLucia, E.H. (2006) Net carbon exchange and evapotranspiration in postfire and intact sagebrush communities in the Great Basin. *Oecologia*, **146**, 595–607.

Pyke, C.R., Thomas, R., Porter, R.D., et al. (2008) Current practices and future opportunities for policy on climate change and invasive species. *Conservation Biology*, **22**, 585–592.

Qian, H. & Ricklefs, R.E. (2006) The role of exotic species in homogenizing the North American flora. *Ecology Letters*, **9**, 1293–1298.

Raghu, S., Anderson, R.C., Daehler, C.C., et al. (2006) Adding biofuels to the invasive species fire? *Science*, **313**, 1742–1742.

Rahel, F.J. & Olden, J.D. (2008) Assessing the effects of climate change on aquatic invasive species. *Conservation Biology*, **22**, 521–533.

Randin, C.F., Engler, R., Normand, S., et al. (2009) Climate change and plant distribution: local models predict high-elevation persistence. *Global Change Biology*, **15**, 1557–1569.

Rejmánek, M. & Richardson, D.M. (1996) What attributes make some plant species more invasive? *Ecology*, **77**, 1655–1661.

Richardson, D.M., Bond, W.J., Dean, R.J., et al. (2000) Invasive alien species and global change: a South African perspective. In *Invasive Species in a Changing World* (ed. H.A. Mooney and R.J. Hobbs), pp. 303–349. Island Press, Washington, DC.

Richardson, D.M., Hellmann, J.J., McLachlan, J.S., et al. (2009) Multidimensional evaluation of managed relocation. *Proceedings of the National Academy of Sciences of the USA*, **106**, 9721–9724.

Rieman, B.E., Lee, D.C. & Thurow, R.F. (1997) Distribution, status, and likely future trends of bull trout within the Columbia River and Klamath River basins. *North American Journal of Fisheries Management*, **17**, 1111–1125.

Roura-Pascual, N., Suarez, A.V., Gomez, C., et al. (2004) Geographical potential of Argentine ants (*Linepithema humile* Mayr) in the face of global climate change. *Proceedings of the Royal Society of London B*, **271**, 2527–2534.

Saunders, M. & Metaxas, A. (2007) Temperature explains settlement patterns of the introduced bryozoan *Membranipora membranacea* in Nova Scotia, Canada. *Marine Ecology Progress Series*, **344**, 95–106.

Scherer-Lorenzen, M., Olde Venterink, H. & Buschmann, H. (2007) Nitrogen enrichment and plant invasions: the importance of nitrogen-fixing plants and anthropogenic eutrophication. In *Biological Invasions* (ed. W. Nentwig), pp. 163–180. Springer, Berlin.

Smith, S.D., Huxman, T.E., Zitzer, S.F., et al. (2000) Elevated CO_2 increases productivity and invasive species success in an arid ecosystem. *Nature*, **408**, 79–82.

Smith, S.D., Strain, B.R. & Sharkey, T.D. (1987) Effects of carbon dioxide enrichment on four Great Basin [USA] grasses. *Functional Ecology*, **1**, 139–144.

Song, L.Y., Wu, J.R., Li, C.H., Li, F.R., Peng, S.L. & Chen, B.M. (2009) Different responses of invasive and native species to elevated CO_2 concentration. *Acta Oecologica*, **35**, 128–135.

Stachowicz, J.J., Terwin, J.R., Whitlatch, R.B. & Osman, R.W. (2002) Linking climate change and biological invasions: ocean warming facilitates nonindigenous species invasions. *Proceedings of the National Academy of Sciences of the USA*, **99**, 15497–15500.

Suring, L.H., Wisdom, M.J., Tausch, R.J., et al. (2005) Modeling threats to sagebrush and other shrubland communities. In *Habitat Threats in the Sagebrush Ecosystem: Methods of Regional Assessment and Applications in the Great Basin* (ed. M.J. Wisdom, M.M. Rowland and L.H. Suring), pp. 114–119. Allen Press, Lawrence, Kansas.

Sutherst, R.W., Floyd, R.B. & Maywald, G.F. (1995) The potential geographical distribution of the cane toad, *Bufo marinus* L. in Australia. *Conservation Biology*, **9**, 294–299.

Thomas, C.D. (2010) Climate, climate change, and range boundaries. *Diversity and Distributions*, **16**, 488–945.

Thuiller, W., Richardson, D.M. & Midgley, G.F. (2007) Will climate change promote alien plant invasions? In *Biological Invasions* (ed. W. Nentwig), pp. 197–211. Springer, Berlin.

Walther, G.R., Roques, A., Hulme, P.E., et al. (2009) Alien species in a warmer world: risks and opportunities. *Trends in Ecology & Evolution*, **24**, 686–693.

Weitere, M., Vohmann, A., Schulz, N., Linn, C., Dietrich, D. & Arndt, H. (2009) Linking environmental warming to the fitness of the invasive clam *Corbicula fluminea*. *Global Change Biology*, **15**, 2838–2851.

White, T.A., Campbell, B.D., Kemp, P.D. & Hunt, C.L. (2001) Impacts of extreme climatic events on competition during grassland invasions. *Global Change Biology*, **7**, 1–13.

Williams, A.L., Wills, K.E., Janes, J.K., Schoor, V., Newton, P.C.D. & Hovenden, M. J. (2007) Warming and free-air CO_2 enrichment alter demographics in four co-occurring grassland species. *New Phytologist*, **176**, 365–374.

Willis, C.G., Ruhfel, B.R., Primack, R.B., Miller-Rushing, A.J., Losos, J.B. & Davis, C.C. (2010) Favorable climate change response explains non-native species' success in Thoreau's Woods. *PLoS One*, **5** (1), e8878, doi:10.1371/journal.pone.0008878.

Wisdom, M.J., Rowland, M.M., Suring, L.H., Schueck, L., Meinke, C.W. & Knick, S.T. (2005) Evaluating species of conservation concern at regional scales. In *Habitat Threats in the Sagebrush Ecosystem: Methods of Regional Assessment and Applications in the Great Basin* (ed. M.J. Wisdom, M.M. Rowland and L.H. Suring), pp. 5–74. Allen Press, Lawrence, Kansas.

Zavaleta, E.S. & Royval, J.L. (2002) Climate change and the susceptibility of U.S. ecosystems to biological invasions: two cases of expected range expansion. In *Wildlife Responses to Climate Change* (ed. S.H. Schneider and T.L. Root), pp. 277–341. Island Press, Washington, DC.

Zavaleta, E.S., Shaw, M.R., Chiariello, N.R., et al. (2003) Responses of a California grassland community to three years of elevated temperature, CO_2, precipitation, and N deposition. *Ecological Monographs*, **73**, 585–604.

Ziska, L.H. (2003) Evaluation of the growth response of six invasive species to past, present and future atmospheric carbon dioxide. *Journal of Experimental Botany*, **54**, 395–404.

Ziska, L.H. & Dukes, J. S. (2011) *Weed Biology and Global Climate Change*. Wiley-Blackwell (in press).

Ziska, L.H., Faulkner, S. & Lydon, J. (2004) Changes in biomass and root: shoot ratio of field-grown Canada thistle (*Cirsium arvense*), a noxious, invasive weed, with elevated CO_2: implications for control with glyphosate. *Weed Science*, **52**, 584–588.

Ziska, L.H., Reeves, J.B. & Blank, B. (2005) The impact of recent increases in atmospheric CO_2 on biomass production and vegetative retention of cheatgrass (*Bromus tectorum*): implications for fire disturbance. *Global Change Biology*, **11**, 1325–1332.

Chapter 27

CONCEPTUAL CLARITY, SCIENTIFIC RIGOUR AND 'THE STORIES WE ARE': ENGAGING WITH TWO CHALLENGES TO THE OBJECTIVITY OF INVASION BIOLOGY

Johan Hattingh

Department of Philosophy, Stellenbosch University, South Africa

Fifty Years of Invasion Ecology: The Legacy of Charles Elton, 1st edition. Edited by David M. Richardson
© 2011 by Blackwell Publishing Ltd

27.1 INTRODUCTION

Invasion biology is generally recognized as a serious field of investigation in which researchers use rigorous science to present the facts about, among other things, the damage caused by certain invasive non-native species as objectively as possible to policy makers so that they can decide about appropriate management interventions (Richardson, this volume). Clearly, much is at stake in this advice to policy makers: through their dramatic proliferation and spread, some invasive non-native species can indeed have devastating impacts on ecosystems, often resulting in substantial economic and financial losses. Management of invasions can be an enormously expensive exercise, creating a series of other problems in the context of limited funding for conservation.

For a variety of reasons, it is also sometimes a difficult and expensive political exercise to gain the support of those living in areas affected by eradication or containment programmes. Accordingly, policy makers are highly dependent upon the credibility of the science informing their management interventions. If doubt emerges among policy makers and the public about the scientific rigour of invasion biology, policy interventions may be very difficult, if not impossible, to implement. It thus goes without saying that much depends on maintaining and building public trust in the field of invasion biology.

In ongoing debates about the methodological basis and the socio-political functioning of invasion biology, however, the scientific status of invasion biology has been challenged from different angles by those who argue that it is not possible to find a completely objective basis for claiming that invasive non-native species are bad and therefore require some management intervention. In one prominent version of this challenge, it is argued that there are no objective, non-question-begging definitions of 'invader', 'biodiversity', 'ecosystem integrity', 'stability', 'resilience' or 'harm to the environment' – some of the central concepts of invasion biology – and that these concepts represent value judgements rather than objective reality existing independently from the investigator. Accordingly, it has been argued that invasion biology cannot be a rigorous, mathematical, theory-based science; rather, it should strive to be an inductive, empirical or historical science that works with particular 'case studies' instead of general laws around which theories can be built.

Another version of this challenge suggests that instead of an objective basis, there is rather a narrative basis to the fundamental concepts of invasion biology, and that not only are the policy decisions based on these narratives, but the science of invasion biology itself is fundamentally based on value judgements. Focusing on the language and the dominant metaphors used in invasion biology and in its communication with policy makers and the public, it is argued from this perspective that invasion biology is not a neutral science, but one in which our human values and interests are always firmly entrenched. This is another way of saying that invasive non-native species do not exist separately from humans as purely natural entities; rather, they are the very products of our activities, our decisions and our aspirations. Thus, when non-native invasive species are identified to be problematic, we (humans) are also implicated. Accordingly, the argument goes, when invasion biologists tell the stories of invasive non-native species and what should be done about them, they actually tell the stories of who we are and how we exist in this world, regardless of whether they explicitly acknowledge this or not.

In this chapter, I explore the meaning and implications of these two versions of the challenge to the objectivity of invasion biology. The central question I pose is whether this emphasis on the value basis, or the value dimensions of invasion biology, is indeed undermining the rigour of invasion biology as an objective science. Another way to state the same question is whether, and if so, how it is possible to integrate science and values in invasion biology in a manner that maintains its rigour and reliability, rather than exposing it to bias, prejudice, distortion and ideology. In a third formulation of this question, it can be asked whether it is possible to acknowledge explicitly the functioning of values in invasion biology, and, if so, what difference it would make to the practice of this science and its functioning in society.

Although these questions have been stated and vigorously debated within the different streams of invasion biology since its inception in the 1950s, and even before that (see Davis (2006, 2009) for insightful overviews), I will venture some tentative answers to these questions from the vantage point of recent perspectives in philosophy, ethics and communication theory that, to some extent, stand outside of, or on the margins of, invasion biology. My reasons for choosing this vantage point include that these 'outside' or 'marginal' perspectives have far-reaching implications for the practice of invasion biology and the communication of its results, and that it is important to define clearly the basis of these perspectives, and why it is important to take note of their implications.

For this discussion, I have chosen to focus on exemplars of the two challenges I have distinguished above, instead of giving an overview of their historical unfolding over time and the many direct and indirect responses that have emerged from within the field of invasion biology. Such a historical study would entail a book-length treatment and a review of thousands of publications. In a short chapter like this, a much more focused approach is called for, and a discussion of two exemplars provides the opportunity for a more in-depth and systematic treatment of arguments than would have been possible in a historical overview of limited length.

As a prominent and recent example of a large body of work that predominantly focuses on the assumptions, internal logic and coherence of the central concepts of invasion biology (see, for instance, Brown 1997; Shrader-Frechette 2001; Woods & Moriarty 2001; Aitken 2004; Townsend 2005; Warren 2007), I focus in this chapter on the critique expressed by Mark Sagoff (1999, 2003, 2005, 2006, 2009a,b) against ecological science in general, and invasion biology in particular. In this critique he argues that ecological science and invasion biology alike are unable to define their objects of study (ecosystems and the sense in which they can be identified as 'the same', or 'of a kind'), are unable to test falsifiable hypotheses about changes in the ecosystem, are unable to explain efficient cause for ecosystem structure, pattern, design or function, and are unable to apply their theories to solve real-life problems. Despite this criticism, though, Sagoff is not unsympathetic to the conservation and environmental agenda of invasion biology. In fact, his criticism could be read as an effort to strengthen the cause of invasion biology by pointing out where he thinks its conceptual basis, methodological assumptions and scope of application can be improved.

As a prominent and articulate example of a growing body of work that predominantly focuses on the social functioning of invasion biology, in particular on the manner in which it articulates and disseminates its 'message' to policy makers and the public (see, for example, Harvey 1996; Castree 2001; Hattingh 2001; Hall 2003; Crifasi 2005; Meech 2005; Fall 2005; Mabey 2005; Stokes et al. 2006; Fischer & Van der Wal, 2007; Evans et al. 2008), I also focus on the work of

Brendon Larson (2005, 2007a–c, as well as Larson et al. 2005), on the metaphors used by invasion biologists and the manner in which these metaphors are embedded in wider societal narratives that not only articulate 'who we as humans are' but also co-constitute science in general, and invasion biology in particular. Following a very different line of thinking to that of Sagoff, Larson also understands his work on metaphor, language and values as supportive of invasion biology, and as such it resonates with many of the insights that have been articulated by prominent figures from within invasion biology itself, for instance Jared Diamond, Stephen Jay Gould, Michael Rosenzweig, Steward Pickett, Dov Sax and Mark Davis, to mention a few.

In my discussion I hope to demonstrate that both Sagoff and Larson articulate valuable suggestions about the manner in which invasion biology can hold together science, values, metaphors and narratives in a coherent story about its scientific rigour, credibility and effective social functioning. In this manner, I also hope to shed some light on the question of what it could possibly mean to acknowledge a value or narrative basis for invasion biology, and what the implications of this could be for the scientific rigour of invasion biology.

27.2 MARK SAGOFF'S CHALLENGE

At the outset of this discussion, it is important to note the Sagoff's criticism of invasion biology is not presented here as if it is unique to him, or that he has articulated these ideas for the first time. Important precursors of his ideas about ecology and invasion biology in philosophy and environmental ethics can be found in the work of Shrader-Frechette and McCoy (Shrader-Frechette & McCoy 1990, 1993, 1994a,b; Shrader-Frechette 1996, 2001), as well as Eser (1998). Similarly, much of what Sagoff has to say about ecology and invasion biology has also been articulated earlier from within invasion biology (for example Pimm 1991; Soulé & Lease 1995), or resonates with ideas currently expressed in invasion biology (see, for example, Davis 2006, 2009; Chew & Hamilton, this volume). Rather, Sagoff's work is discussed here as representative of a certain line of philosophical and ethical thinking about invasion biology, and for the systematic articulation that he gives of that line of thinking.

It should also be noted that invasion biology is not presented here as if it were a single, unified field of scientific research with no internal debates between different streams or paradigms within that field. A case in point is the clear distinction that can be made between a more conservation and environmentally focused path in invasion biology that is commonly traced back to the more value-oriented, top down and deductive emphasis of Elton (1958), and a more value-neutral and more strictly scientific path whose first prominent articulations emerged at the first Biological Sciences Symposium of the International Union of Biological Sciences that was held in Asilomar, California, in 1964 (Davis 2006, 2009). The proceedings of this conference was published in *The Genetics of Colonizing Species* (Baker & Stebbins, 1965), which is now also regarded as a classic of invasion biology, similar to Elton's (1958) *The Ecology of Invasions by Animals and Plants*. In the Asilomar stream a more bottom-up and inductive approach is followed, focusing on the study of individual colonizations to better inform general ecological theory (Davis 2006). As I demonstrate later in the chapter, Sagoff's critique of invasion biology is focused on the more dominant Eltonian path, while he articulates a specific preference for a more historical, inductive approach to invasion biology in which the emphasis falls on particular case studies.

A third preliminary point to make is that in all the work that Mark Sagoff has published over the past decade on the topic of environmental issues and ecology in general, and on the topic of invasion biology in particular (Sagoff 1999, 2003, 2005, 2006, 2009a,b), the same basic argument is repeatedly articulated, leading to the same set of conclusions about, and recommendations to, invasion biology. This persistence in following the same line of thought over such a long time is revealing in itself, indicating among other things a strong conviction that there are several serious problems with invasion biology following the Eltonian path, and that it is very important to address these problems properly. In these publications, however, invasion biology is not the only target, for Sagoff argues that the problems of invasion biology are shared by ecology and environmentalism alike, and are also to be found in environmental ethics.

In one of the later articulations of his views, Sagoff argues that the environmental movement of the 1970s found its academic support in an assumption, supported by both ecological science and environmental ethics, that 'natural ecosystems or communities if left undisturbed reached a desirable 'balance' or equilibrium' (Sagoff 2006, p. 148). Linking this assumption back to the Timaeus of Plato and the very long cosmological tradition of the Great Chain of Being, of which ideas of spontaneous natural evolution are, according to Sagoff, only secular versions, this idea of the balance of nature led environmental scientists and ethicists to believe that natural communities or ecosystems 'possess enough order or organization to justify attempts to protect them' (Sagoff 2006, p. 149).

As an example of this belief, Sagoff points out (with reference to the classic study of Stephen Forbes in 1887 on the lake as a microcosm) that 'the beneficent power of natural selection' was usually quoted by ecological scientists to explain the balance and equilibrium of natural communities (Sagoff 2006, p. 148). As such, this balance represented a good for that community, and accordingly it was possible, as it was done up to and through the 1950s in ecological circles, to argue that anything upsetting that natural balance, for instance the interference of man, was unacceptable. This normative concept of natural balance was succinctly captured in the words of Stephen Forbes, who stated that in a natural community 'an equilibrium has been reached and is steadily maintained that actually accomplishes for all the parties involved the greatest good which the circumstances will at all permit' (Forbes 1925, quoted in Sagoff 2006, p. 148; see also Cuddington 2001; and Lodge & Hamlin 2006).

According to Sagoff, however, many ecologists started to lose faith in the idea of the balance of nature or the equilibrium paradigm of ecosystem development. Starting in the 1970s and continuing through the 1990s (see, for instance, Pimm 1991), ecologists started to realize that they had difficulty finding empirical data to support the thesis that general principles or rules of organization could explain ecosystem structure or function (Sagoff 2006, p. 149). Rather, it can be stated that 'the idea that species live in integrated communities is a myth' or that 'local biotic assemblages … [have] never been homeostatic', leading to the conclusion that 'any serious attempt to define the original state of a community or ecosystem leads to a logical and scientific maze' (Soulé & Lease 1995, quoted by Sagoff 2006, p. 149).

To emphasize what he means by this logical and scientific maze, Sagoff quotes five reasons that stand out from many others that cast doubt on the assumption that there is indeed a reality to community structure (2006, pp. 149–151). First, he refers to the 'stunning lack of empirical evidence that ecosystems possess an element of organization'. Second, he mentions that 'no one has been able to explain how natural selection can structure ecosystems'. Third, he draws on the phenomenon of pluralism in ecology to point out that 'most ecologists agree ... they have not arrived at a consensus concerning general principles or rules that may identify and thus help us to observe how ecosystems are organized'. Fourth, he draws attention to the lack of consensus about definitions that 'allow to tell us what kinds of places constitute ecosystems', which implies a fifth point: that it is not possible to distinguish between that which is essential to an ecosystem, and that which is contingent. Accordingly, it is also not possible to tell whether an ecosystem has retained its organizational mode in spite of certain changes, or whether it has collapsed into a different system, or into something that is merely a collection. In short: 'We have no criteria for determining what kinds of changes destroy the system and what changes are consistent with its preservation' (Sagoff 2006, p. 151).

For Sagoff (2006, p. 151), empirical evidence does not support the idea that 'ecosystems are integrated, interconnected systems that conform to organization principles'. He rather argues:

> '... that empirical evidence suggests ... that what are called "communities" or "ecosystems" lack organization, structure or function and that each is idiosyncratic, a law unto itself, a blooming, buzzing confusion of contingency, a temporary even ephemeral outcome of the accidents of history, a Heraclitean flux.' (2006, p. 151)

In the final step of his argument, Sagoff suggests that although there is no empirical basis to prove 'that ecosystems or natural communities are governed by principles of organization' (Sagoff 2006, p. 152), 'compelling evidence' for this supposition can be found in the aesthetic judgements that we form when we experience the most glorious and magnificent productions of the natural world. In his understanding of the aesthetic judgement along Kantian lines (explained in further detail below), it is not the intrinsic unity or the organizational mode of natural beauty that triggers the aesthetic experience in us, together with the moral intuitions often associated with aesthetic experience. It is rather the other way round: the aesthetic experience

brings us to assume that there is a self-organized or otherwise structured unity present in that beautiful phenomenon or place, and this aesthetic judgement, Sagoff argues, often also triggers a moral intuition that this structured unity is worth protecting or demands respect. In his own words, Sagoff states:

> 'Ecosystem scientists may adopt as a research program a commitment to understanding through mathematical modelling the organizational mode of ecosystems because moral intuitions and aesthetic judgements convince them and us that such a design is somehow there to be discovered.' (Sagoff 2006, p. 152).

With this, Sagoff makes the strong point that ecologists will never be able to prove scientifically that there exists a unity or an organizational structure in an ecosystem. If aesthetic judgement is, as Sagoff argues, the primary basis for the assumption of a unity or an organized structure present in an ecosystem, then it is impossible to fall back on the concepts or understandings of science to prove this assumption. The reason for this claim is what Sagoff, following Kant, takes to be the nature of aesthetic experience itself: a perception of the free lawfulness (or the purposiveness without purpose) of the beautiful object. In this mode of perception, Kant argued, the faculty of understanding and the faculty of imagination are in free play in a manner that suggests a concept, but in fact reveals none. Formulated in less technical language, this means that aesthetic experience 'points' to concepts, such as unity or organization, but these concepts are not of the sort that can be tested by empirical science (Sagoff 2006, p. 152).

With this in mind, it is possible for Sagoff to challenge three of the central concepts of environmental ethics: the normative value of biodiversity, intrinsic value and the normative value of the 'natural'. With regard to the problems that he has with the notion of the normative value of biodiversity, Sagoff targets statements like that of Holmes Rolston III to the effect that 'Every individual organism is a distributive increment on a collective good – at least presumably' (Rolston 2006, p. 55; quoted in Sagoff 2006, p. 153). According to Sagoff, this normative understanding of biodiversity is based on the assumption that the abundance and the distribution of plants and animals on Earth were determined by the spontaneous course followed by nature (irrespective of whether this course could be ascribed to evolution or creation). For Sagoff, the problem with this assumption is that it only holds for a distant past. In recent times, however, human activity has come first to influence, and later to determine, which organisms are found where and in which populations. Sagoff's argument in this regard is that accidental or intentional introductions of species to certain areas, as well as advances in breeding practices and biotechnology, have all contributed to what is today referred to as biodiversity (Sagoff 2006, p. 154). As such, the notion of biodiversity seems to be of the same order as that of 'organization', 'structure' or 'unity' in ecology. Therefore, Sagoff argues, only a weak basis is provided to ground any moral injunction to protect and preserve biodiversity. The point that Sagoff makes is that there is no objective or non-arbitrary definition of biodiversity that can be used to base an environmental ethics on; accordingly, any policy advice deduced from this notion of biodiversity will be equally weak.

In environmental ethics, it is also often claimed that the flourishing or well-being of living creatures (living now and in the future) has intrinsic value, and that this intrinsic value provides the rationale to conserve species, ecosystems and habitats. Sagoff, however, poses the question whether this intrinsic value should be respected in all species, ecosystems or habitats (including those that came about as a result of human activity), or only in those that arose as a result of nature's spontaneous course (Sagoff 2006, p. 155). His contention is that it would be drawing a highly arbitrary line if intrinsic value is reserved for only those species, ecosystems and habitats that are 'pristine' and devoid of any human influence. His grounds for this claim have already been alluded to above, namely that there is in the first place 'no evidence that ecosystems possess an organization or design that give them form, structure or function', and in the second place that there is 'certainly no evidence that "natural", pristine or heirloom ecosystems exhibit properties of structure, function or design that differ from those of less "natural" ones' (Sagoff 2006, p. 157). The conclusion that follows from this is that intrinsic value should apply to all species, habitats and ecosystems. However, this conclusion runs up against our aesthetic and moral intuitions that not all species, habitats and ecosystems deserve the same levels of respect and protection. The challenge is therefore, on the one hand, to develop a notion of intrinsic value that is not so wide that it will apply to any- and everything, and thus lose plausibility; and on the other hand, to do so in a manner that does not open the concept up to the challenge that it is applied arbitrarily to certain species, ecosystems and habitats that cannot be distinguished in principle from one another.

In addition to his critique of biodiversity and intrinsic value, Sagoff also turns his attention to the notion of the normative value of the 'natural'. In this regard, he takes issue with the manner in which environmental ethics, as well as the environmental and ecological sciences, have defined the 'natural' in contrast to human activity. The general assumption in these circles, he argues, is that:

> 'Species, habitats and ecosystems may be thought to have intrinsic value insofar as they arise and function within nature's spontaneous course, for example the native wildflowers of the prairie. Species, habitats and ecosystems that result from and function within human-dominated environments, such as Roundup-Ready grain, may have instrumental but not intrinsic value.' (Sagoff 2006, p. 158)

For Sagoff, this idea of species, habitats and ecosystems functioning free from the influence of human activity creates insurmountable problems for environmental ethics and ecological science alike. In the context of ecological science, this distinction forms part of a fundamental methodological commitment. In the practice of ecology, and invasion biology for that matter, this means that human activity is seen as an external disturbance – and this assumption is borne out by the fact that ecologists have traditionally 'sought to study pristine ecosystems to try to get at the workings of nature without the confounding influences of human activity' (Callagher & Carpenter 1997, quoted by Sagoff 2006, p. 158).

In the context of environmental ethics, this separation of the human from the natural creates similar problems. If nature is posited as the norm, it follows that anything that emerges from nature's spontaneous course constitutes ecosystem integrity and health and has intrinsic value, whereas whatever results from human activity is potentially harmful to the natural system. Sagoff, however, seriously questions the assumption, as it has been demonstrated above, that the spontaneous course of nature endows species, habitats and ecosystems with both ecological organization and intrinsic value (Sagoff 2006, p. 159).

With this then, Sagoff has apparently trashed both ecological science and environmental ethics. What Sagoff suggests is that the central ideas of environmental ethics, like respect for intrinsic value and biodiversity, as well as the central ideas of ecological science as discussed above, are nothing but fairy tales (Sagoff 2006, p. 160), or for that matter 'hortatory and aspirational rhetoric' (Sagoff 2006, p. 161). Although this would perhaps not be such a serious charge against environmental ethics that does not claim to be an empirical science per se, but rather a mode of normative reflection, the charge indeed becomes serious to the extent that environmental ethics claims, as it indeed does, to rely on ecological science that is believed to provide an empirical basis for it. Sagoff, however, points to the lack of any empirical basis for the central concepts of ecology. Formulated in his own words, he states that there is 'a stunning lack of evidence that natural communities or ecosystems possess any mode of organization that can serve as the object of environmental protection'. And to this he adds that there is an 'absence of any useful definitions of "community" or "ecosystem" that can identify them through time and change' (Sagoff 2006, p. 157).

There is another side to Sagoff's challenges to invasion biology, though, which lies in the distinction he makes between two different concepts of ecological science:

> 'Ecologists may regard the sites they study either as contingent collections of plants and animals, the relations of which are place-specific and idiosyncratic, or as structured systems and communities that are governed by general rules, forces, or principles. Ecologists who take the first approach rely on observation, induction, and experiment – a case-study or historical method – to determine the causes of particular events. Ecologists who take the second approach, [seek] to explain by inferring events from general patterns or principles ...' (Sagoff 2003, p. 529).

Ecologists who take the second route, Sagoff argues, are faced with insurmountable conceptual obstacles that stem from their efforts to develop a theoretical ecology, turning it into a formal science that studies the mathematical consequences of assumptions without regard for the relationship between these assumptions and the real world (Sagoff 2003, p. 529). These problems have been listed above but can be mentioned again: (i) ecological science is not able to define its objects of study (ecosystems and the sense in which they can be identified as 'the same', or 'of a kind'); (ii) it is not able to test falsifiable hypotheses about changes in the ecosystem; (iii) it is not able to explain efficient cause for ecosystem structure, pattern, design or function; and (iv) it is not able to apply its theories so as to solve real-life problems.

These problems, Sagoff argues, are typically experienced when ecology is taking the second, formal and mathematical route sketched above. It is to ecology and invasion biology practised in terms of this second concept that the brunt of Sagoff's challenge applies. It is at this formal, theoretical ecology (and invasion biology) that he aims his criticism; it is this kind of ecology that he characterizes as ultimately arbitrary (Sagoff 2009b, p. 46) and tautologous (Sagoff 2005, p. 228), because it is developed *in silico*, that is in front of the computer, and not *al fresco*, that is in the great outdoors doing fieldwork (Sagoff 2009b, p. 46).

Sagoff does not write off ecology and invasion biology completely. Rather, he articulates a plea for it to become empirical, to follow an inductive and historical approach. Because it is only by comparing different historical states of ecosystems with one another, or closely observing the interaction between species in real-life settings, that ecologists will arrive at the knowledge on which sensible policy proposals could be based. Indeed, Sagoff explicitly states that he does not deny the fact that certain 'non-native' 'invasive' species can be dangerous to 'native' species, and he also states that he welcomes certain intervention programmes to minimize this danger. What he does not like, however, are some of the ways in which the sources, the nature, the extent and the effects of these dangers are conceptualized, and how the policy responses to them are justified.

In this dislike, Sagoff is not alone. Kristin Shrader-Frechette (2001, p. 507), for instance, following her own analysis of the central concepts of ecology as well as island biogeography, the dominant theory that is used to understand invasibility, came to the same kind of conclusion that Sagoff arrived at by stating that 'there is no comprehensive, predictive theory of invasibility, as part of a larger theory of community structure, that might guide ecological decision making regarding [invasive non-native species]' (see also Shrader-Frechette & McCoy 1994b). Similarly, Brendon Larson at several places in his publications makes observations that clearly resonate with those of Sagoff. Citing Brunson (2000), Hull and Robertson (2000) and Ridder (2007), Larson points out that 'there is no self-evident point at which it is sensible to say that an ecosystem is "native"' (Larson 2007c, p. 996). At another place in the same article, Larson writes in language reminiscent of Sagoff that 'our concerns about invasive species to some extent reflect an image of an enduring and timeless "nature". They imply that nature is static, or even that it should continue on a trajectory present before modern humans arrived. It is difficult to escape this romantic "balance of nature" ideal since it reflects a Christian metaphysic that is part of our everyday worldview: there is a fall from grace, we are in error, and we have a nostalgic wish to return.' (Larson 2007c, p. 995).

My contention is that we should not dismiss Sagoff's sharp attack on a certain version of invasion biology too easily as an attack on a straw-man image of invasion biology, as David Lodge (interviewed by Farrar 2003) and Simberloff (2003) have done. If we followed their lead, then we could very well be overlooking the kernel of truth in Sagoff's claims, which, according to David Lodge, indeed existed. He ironically acknowledged as much in the same sentence that he accused Sagoff of a straw-man attack (Farrar 2003). Opinions may differ on what this kernel of truth could be, but at least two points may not be too controversial to defend: first, that invasion biology would only be able to claim that it is a rigorous science if it followed an inductive, historical approach; and second, that Sagoff's observation that concepts like 'environmental harm' have no basis in biology, but are rather grounded in 'aesthetic, religious, spiritual, historical, cultural, or some other meaning to society' (Sagoff 2009a, p. 84), may not be too mistaken.

As such, Sagoff's challenge to invasion biology is firstly a plea to recognize the conceptual problems that emerge if the value basis of the central concepts of invasion biology are not properly recognized; and secondly it is a plea to move away from an approach where concepts informed by these uncritically accepted and unexplicated value assumptions are used as the basis for investigations that are claimed to have the status of rigorous science. The obvious target of this criticism is clearly what has been identified above as the dominant Eltonian path of invasion biology – in so

far as it serves an uncritically accepted conservation and environmental agenda – that was, and perhaps still is, practised without a self-conscious and critical scrutiny of the value assumptions informing its practice. However, Sagoff's criticism of ecology and invasion biology could equally well apply to a more historical and inductive approach to invasion biology, if exponents of this approach also fail to explicate and critically assess the values informing their research questions, their methodologies, the results of their studies and the manner in which they communicate these results to policy makers and the public.

Although Sagoff gives us no real clues about the manner in which the rigour of inductive invasion biology could be combined with an acknowledgement of the value basis of the central concepts of invasion biology, the question remains how these two apparently exclusive concepts, science and values, could be combined into a single, coherent conceptualization of invasion biology. For an indication of how this could be achieved, we have to turn to the work of Brendon Larson. However, first we have to understand what his challenge to invasion biology entails.

27.3 BRENDON LARSON'S CHALLENGE

As in the case of Mark Sagoff's work discussed previously, Brendon Larson's challenge to invasion biology is not presented here as if it were unique, or articulated for the first time in his work. He also has precursors outside of invasion biology, for instance in ecology (Egler 1942) and in various branches of philosophy. And as in the case of Mark Sagoff, many of his criticisms and suggestions resonate with ideas currently expressed in invasion biology (see, for example, Reise et al. 2006; Davis 2006, 2009). Larson's work is also treated here as an example of a line of thinking critically about invasion biology, and his views are discussed for the insight that they yield in the suggestion that I already alluded to, and is finding more and more support within invasion biology itself: that the values on which it is based should be clearly explicated and critically discussed (Davis 2009, pp. 167–170, 191, 193).

One of the golden threads running through Larson's challenge to invasion biology is his critical discussion of the language and the metaphors that have been, and still are, commonly used in certain circles of invasion biology to conceptualize invasive non-native species and the threats that they pose to ecological systems and the economy. Larson specifically focuses on a set of metaphors that predominantly stems from a militaristic framework. His problem with metaphors in general is that they can easily be used to mobilize the public and policy makers, without a process of rigorous, informed deliberation. His problem with military and similar metaphors in invasion biology is that they give us a restricted understanding of invasive species, and that they narrow the range of possible responses to invasive species that are available to us. Accordingly, Larson argues for an expansion of our understanding of invasive species, and a broadening of our responses to them to include a wider spectrum of options than only 'battle' and 'attack' (Larson 2005; 2007a, p. 132).

One line of thinking that he explores to achieve this broader vision is to scrutinize the assumptions, implications and social functioning of uncritically accepting the narrower set of military metaphors. He also proposes alternative metaphors that would foreground different dimensions of invasive species that are obscured by the 'usual' set of metaphors used to refer to invasive non-native species. Another, perhaps even more fundamental, line of thinking that he explores is his argument that our language and our metaphors reveal as much, and perhaps even more, about ourselves than the invasive species we talk about. His argument in this context is that in telling the stories of invasive species, their dangers and how we should respond to them, we at the same time tell a wider socially embedded story, often an unacknowledged and uncritically accepted story, about who we are, how we are situated in the world and what kind of lives we would like to live.

In his exploration of the assumptions, implications and social functioning of military metaphors in invasion biology, Larson (2007a, pp. 133, 136) points out that the strong association of invasive species with metaphors such as 'invader', 'enemy' or 'terrorist' was established through a long history of repetition that can be traced back to the founding fathers of invasion biology such as Elton (1958). One of the effects of this repetition is that it conjures up a way of looking, or a frame of conceptualization, that precisely because of its repetition seems so 'self-evident and inexorable', that we do not think twice about the framework itself, and almost blindly accept that it provides 'the right way'.

In the case of the metaphor of 'invader', for instance, Larson (2007a, pp. 136–137; 2010, p. 8) argues that a cognitive schema of a 'container' is invoked in terms of which images are formed of a 'borderline' that can be crossed from the 'outside' to the 'inside', threatening the survival of the native species already inside the borderline. Accordingly, it is easy to look at the non-native species as an external 'enemy' that we are at 'war' with. Larson further argues that this cognitive schema is not restricted to invasion biology alone. It is also evident in the way in which we think about national identities located within certain geographical regions, or in the way in which we think of our bodies and how we have to protect them against invasion by diseases. In both contexts, Larson argues, there seems to be a related fear: of the body being invaded by a disease, or a nation being invaded by non-native peoples. As such, the metaphor of 'invader' circulates within different contexts that can, and mostly do, mutually reinforce one another.

A critical scrutiny, however, quickly reveals that the native/non-native dichotomy implied by the 'overly entrenched' and 'too unexamined' (Larson 2007a, p. 136) metaphor of 'invader' does not hold up against the reality that 'invasive species' exist because of human actions and interests. To illustrate this, Larson mentions three considerations. First, invasive species are not intending to be invaders; they have become invaders because we have introduced them to certain areas intentionally or by accident. Second, it can be shown that the 'borderlines' that invasive species are 'crossing' are constantly shifting. And third, it is clear that invaders are regarded as not welcome on the basis of a fear that there is no place for them in the already existing order of native species: the 'inside' is already full. Larson, however, points out that many ecosystems may have 'empty niches' that could happily be filled by non-native species.

Larson goes through a similar kind of analysis of the metaphor of 'terrorist' for a invasive non-native species, showing that it is not only an intensification of the military metaphors discussed above, but also indicates how the discourse of invasion biology can resonate with a certain socio-political context (Larson 2007a, pp. 137–138). With reference to the 9/11 terrorist attacks on the World Trade Center in New York, Larson finds numerous parallels between the discourse used in the fight against invasive species and that of fighting terror in the guise of Al-Qaeda and personified in the figure of Osama bin Laden. Both are characterized by strong calls for a clear definition of the enemy with clear boundaries; both call for grand plans to prevent further attacks/invasions (that include in both cases stricter border control, but also the realization that this will not be 100% effective, which leads in both cases to paranoia that the next attack/invasion can come from anywhere); both entail an anxiety about naturalized individuals; both justify non-target effects to control the enemy; both leave victory ill-defined (turning it into an eternal battle of good against evil); and both do not consider notions of alternative causes for attacks/invasions.

At this point in his argument, it is possible for Larson to observe that the metaphors we use to characterize invasive non-native species are embedded in larger cultural and socio-political patterns (Larson 2007a, p. 138). As such, these metaphors articulate the interests, ambitions, fears and uncertainties prevalent in societies and cultures, and it is exactly because of this that Larson can also argue that the stories articulated in these metaphors are never the only stories, and they are never neutral. On the contrary, there always are a variety of contested stories about invasive non-native species circulating in society (Larson 2007a, p. 133).

Accordingly, Larson introduces several alternative metaphors, in the hope that they can help invasion biology, policy makers and the public to move outside the restrictive box of conventional thinking about invasive non-native species, and to start a creative, new story about them (Larson 2007a, p. 132). In this context, he proposes that we shift our metaphors away from the language of war, and think of non-native species, of which only some may become invasive, as piggy-backers, opportunists, spawn, mirrors, providers, hybrids, tricksters, matrices, transients, founts and teachers.

Drawing on the widely acknowledged insight that invasive species are caused by human introductions, whether intentional or not, the general story-line that emerges from Larson's discussion of these alternative metaphors includes the following.
• Non-native species, of which some may become invasive, draw our attention to human agency, to the fact that the line we draw between non-native and native, between invasive or not, between natural or not is dependent upon our desires and preferences as humans (Larson 2007a, p. 143);
• Invasive non-native species remind us of the way in which we think of ourselves, of our identities and change; how we want to keep things as they are: they

remind us 'that life is characterized by change, and that our concerns derive from trying to keep things as we know them' (Larson 2007a, p. 146);
• Invasive species confront us 'with the radical question whether ecosystems should always comply to our needs: These ecosystems may not serve all of our needs or requirements, but that may not be the only justification for their existence.' (Larson 2007a, p. 148);
• Invasive species confront us with the realization that our self-serving ideas of what nature should be may not coincide with what nature presents us with (Larson 2007a, pp. 148–149).

Having said this, Larson's challenge to invasion biology is in the first place to look at non-native species in a different way, and following from that, to relate and respond to them differently: different invasive species 'require different responses that need to be evaluated in context' (Larson 2007a, p. 149). In this regard his challenge is an appeal for creativity and a greater openness to other dimensions of invasive species than only those captured in conventional military metaphors. What he drives at is a conceptualization of invasive species that is more complex 'than one based in dualities of good-bad, insider-outsider, natural-unnatural' (Larson 2007a, p. 149). What he is looking for is a language and a set of metaphors that are embedded in a narrative that can sustain itself in the long term, that acknowledges the full spectrum of what invasive species in fact are, that overcomes the fear of change and that moves beyond a methodology of hierarchical control (Larson 2007a, p. 132).

As such, Larson's challenge is also an appeal to change the stance from which invasion biology is practised. What he suggests is a shift from the position of disengaged objectivity to that of a contextually engaged and honest knowledge broker (Pielke 2007) that is critically and self-consciously participating in the contested field of socio-political processes in which the 'story-ing' and 're-story-ing' of invasive non-native species and our responses to them take place (Larson 2007a). The metaphor of an honest knowledge broker is not used by Larson himself, but it perfectly captures what he drives at when writing about the role of invasion biology in society. Among other things, it entails an acknowledgement of, and a critical reflection on, the societal values informing the goals and research questions of invasion biology. It also entails an acknowledgement that values play a crucial role in the constitution, the central concepts and the internal operations of science. It furthermore entails critical engagement with narratives about non-native and invasive species that distorts what they talk about by portraying it in a restricted and incomplete fashion, and it appeals for a nuanced response to particular non-native species as they behave differently in different contexts. All in all, a contextually engaged and honest knowledge broker will strive to provide reliable, concrete and contextualized knowledge that can be used in society to help decide on the responses we can justifiably follow for certain invasive species, while acknowledging the gaps and uncertainties in that knowledge and pointing out how this may influence our decisions. However, does this emphasis on values and context not open up invasion biology to subjectivity and relativism, seriously undermining objectivity?

To this counter-challenge, Larson (2007b) answers that invasion biologists, like any other scientist, cannot follow an 'alien' approach to non-native and invasive species. He thus challenges what he calls an inappropriate ideal of objectivity for invasion biology in which the scientist is portrayed as a detached, neutral and value-free observer, as someone who does not have 'strong responses' to their objects of study (Larson 2007b, p. 947), and could just as well have observed the objects of study from the vantage point of an alien from outer space. Larson's argument, however, is that if such an image of science as a totally value-free enterprise would be imposed on invasion biology, invasion biology as a scientific enterprise would become impossible in principle, because its strong emphasis on conservation values would place it in the realm of the subjective, the sphere of emotions and advocacy.

For Larson, however, this implication is unacceptable, because it rests on the assumption that clear-cut distinctions can be made between facts and values, and its correlative: the objective and the subjective. In this regard Larson draws on the numerous studies in the philosophy of science that have pointed out that these distinctions are not as self-evident and clear as some purists would have liked, and that one should rather speak of the value-ladenness of facts and the fact-ladenness of values (Larson 2007b, p. 949). A case in point is the distinction that is often drawn between *contextual values* on the one hand, which may 'colour' the formulation of the research objectives and research questions of scientists from the vantage point of all kinds of prejudices and idiosyncrasies, and which therefore should be kept out of science, and the *constitutive values* of science, like objectivity, on the other hand, that function as the basis for unbiased inquiry.

Following the insights of well-established philosophy of science, however, Larson rightly points out that the desire for objectivity in science is in itself a contextual value, and that contextual values sometimes also play a role in the methodological choices of scientists (Larson 2007b, p. 949).

Accordingly, Larson sets out to demonstrate 'that it is possible for invasion biologists to simultaneously embody their biodiversity values and to be objective' (Larson 2007b, p. 948). Besides the logical difficulties to draw a clear distinction between facts and values already alluded to, Larson does this by pointing out that it is for a variety of reasons impossible for an invasion biologist to 'draw back' into the position of the detached 'alien' observer. One of the reasons for this advanced by Larson is that invasion biology as a field of study is defined and delineated by conservation values. To 'filter out' these values in an effort to become objective (as Colautti & MacIsaac (2004) have tried to do) would therefore entail ignoring the very values on which invasion biology is implicitly based (Larson 2007b, p. 950). Larson finds a second reason in the insights of cognitive science that human perceptions always play a role in the constitution of facts (Larson 2007b, p. 950). Accordingly, there is no such thing as a pure, uninterpreted fact. On the contrary, a fact only becomes one if it is interpreted as such in terms of some human meaning. Elaborating on the latter point, Larson argues in the third place that narrative elements are inseparable from the very observations made in science (Larson 2007b, p. 950). It is therefore not only in the language and metaphors of invasion biology that a story is told of who we are or want to be; it can also be said of the factual observations of invasion biology itself.

So, how is objectivity then reconciled with values in invasion biology without sacrificing its scientific rigour? The first part of Larson's answer to this question is to be found in his observation that there is no such thing as value-free science. Rather, science is an enterprise that unfolds in the tension between different kinds of values. And the mere fact that one holds these values does not necessarily bias one's results. The second part of his answer lies in his observation that 'while objectivity is important, claims to objectivity may also obscure the values that one actually holds ...' (Larson 2007b, p. 950). Because science ultimately functions in social contexts, Larson argues that such obscurity can create a variety of problems. The source of these problems, stemming directly from the view of

science as a value-free enterprise, is an artificial separation of science from society. Such an artificial separation may help some scientists to attain their ideal of
objectivity that is 'uncontaminated' by any values, but
Larson argues that invasion biology practised in such
a way would become a disinterested pursuit of 'interesting ideas' that has little if any relevance to the practical needs of the society that sometimes has to make
hard decisions about invasive species to solve problems
(Larson 2007b, pp. 951–954).

Accordingly, the third part of Larson's answer lies in
his appeal that invasion biology should explicitly
acknowledge that it is a value-laden and a value-
driven enterprise, and that it should openly and critically discuss how its scientific tools and methodologies
are used, could be used and should be used to promote
certain values and not others. As such, Larson suggests that invasion biology should become critically
aware of the story it tells about non-native and invasive species, and how this story articulates what we as
humans have become, or could become in this world.

But is it possible to make room for objectivity in the
sense of a critical reality check in this kind of self-
critically aware science that acknowledges its basis in
metaphors and narratives? A possible, albeit a controversial, response to this conundrum, lies in the fourth
part of what Larson seems to offer as an answer to the
question how values and objectivity can be reconciled
in invasion biology. Following from his contention that
invasion biology should openly acknowledge its value
basis, Larson further suggests that invasion biology
should engage with society in dialogue about its interests and needs, and not withdraw into a mode of
obscure disinterestedness. This engagement with
society can take many forms, but Larson (2007b) suggests that it in essence condenses to actively participating in social processes of joint decision making, where
solutions to concrete problems are sought within practical contexts. According to Larson, the value of this
approach lies in the fact that it constitutes a middle
way between the extreme position where objectivity
functions as an excuse for indifference and inaction,
and the equally extreme opposite position where blind
advocacy dominates. What Larson seems to suggest is
that a moderation of extreme and unrealistic positions
is promoted in this middle-way approach by the reality-
check of actively participating in processes of social
decision making, where different stories are told about
who we are or what we should be. This seems to make

us realize, as ordinary citizens and as invasion biologists, that we have no other option but to engage hon-
estly and openly with others who may differ from us
(Larson 2007a, p. 149) about the 'facts' we present as
basis for policy decisions about invasive non-native
species.

At a first glance, however, this suggestion seems to
be very vague and impractical. How is it possible to
appeal to a dialogue between those who differ from one
another as a channel to promote objectivity in the
sense of a critical assessment of our knowledge? Rather,
is such a dialogical process not a guarantee of sacrificing truth and objectivity to the processes of social consensus building in society? In one of his latest
publications, Larson (2010) argues that it would be a
mistake to conceptualize this problem as a standoff
between a social constructivist position and a realist
position about knowledge formation.

As Larson sees it, the simplistic extremes of social
constructivism and scientific realism can be overcome
by what he refers to as embodied realism. This is a position that, on the one hand, acknowledges that there
are realities independent from the knowing mind that
can be known, such as 'invasions' and 'damage' and
'containment measures', and on the other hand
accepts that our knowledge is always mediated 'by the
form of our embodiment, our particular ways of interacting with the world from specific positions within it'
(Larson 2010, p. 6). Following Hayles (1991), Brown
(2003) and Geertz (1979) in this regard, Larson contends that 'there is no view from nowhere' (2010, p.
6): what we can know of reality is always constrained
and limited by cognitive schemata that are not only
formed by our bodily capacities, but also by our life
experiences and cultural contexts (Larson 2010, p.
13).

From this point of view, Larson draws several far-
reaching conclusions. One of them is that 'every metaphor and schema highlights one aspect of a relation
while hiding others' (Larson 2010, p. 13). Another is
that different people and different cultures may experience the reality of non-native and invasive species in
vastly different ways. The same applies to the way scientists and members of the public may experience non-
native and invasive species (Larson 2010, p. 14).
Accordingly, Larson's argument for embodied realism
is a plea for an in-depth engagement with different perspectives so as to form a richer understanding of the
reasoning, the metaphors and the cognitive schemata

that lead to these different understandings (Larson 2010, p. 15).

What Larson seems to drive at is therefore not so much a guarantee for universal truths about a completely independent, objective reality. Rather, it is to find grounds to engage in open dialogue with others in which we acknowledge from the outset that the fundamental concepts in which we articulate knowledge, including scientific knowledge, about the world are co-constituted by our embodied positions in the world, and that these concepts and the knowledge they articulate, are by definition limited. This is to acknowledge that there are other possibilities of knowing, and that these other ways of knowing can moderate and check what we claim to be knowledge about the reality of non-native and invasive species. What Larson thus argues for is a constant questioning of our own limited understandings of the reality of non-native species so as to form richer and perhaps more complete understandings of it. As he sees it, this cannot be achieved by withdrawing ourselves into a positivistic notion of detached objectivity, but rather by engaging in dialogue with others who differ from us.

Ultimately Larson's argument moves beyond the confines of invasion biology, in that he implores us as humans not only to develop a wider range of responses to invasive non-native species, but also to develop a different kind of self-identity, and through that, a different relationship to that which non-native and invasive species fundamentally are. Larson thus challenges each one of us to situate ourselves differently in the world, in that we as humans do not see ourselves as separate entities apart from nature, pitted in some instances against certain parts of nature as a kind of enemy that we have to overcome or control. Instead, he argues, we should see ourselves as part of nature, shaping it through our activities but also being shaped by its processes. As such, Larson's challenge to invasion biology forms part of a wider call for a new concept of human identity, action and responsibility: it becomes a call for a relational identity, and participative action where human action is not imposed from 'outside', from the heights of culture, on nature, but rather takes place in partnership with nature, participating as co-actor and co-agent in the unfolding of the processes of nature. It becomes an appeal to take responsibility not only for the effects of our actions, but also for the manner in which we shape our knowledge and justify our decisions.

27.4 DISCUSSION AND CONCLUSION

I started this chapter with a reference to the need for rigorous science to inform decision making about our responses to invasive non-native species. A dominant view that is perhaps not applicable to invasion biology maintains that science is an objective enterprise in which that objectivity is guaranteed by 'screening out' all subjectivity, which includes values, metaphors and narratives. Why? Because these subjective factors are perceived as being sources of bias, prejudice and distortion. In our discussion above, it has been demonstrated that both Sagoff and Larson have advanced good reasons why it is important to challenge this view of objective science, as well as those versions of invasion biology that strive to reach this ideal of objectivity (see, for example, Brown & Sax 2004, 2005; Colautti & MacIsaac 2004).

Sagoff's argument in this regard is that the central concepts and distinctions of invasion biology, such as native versus non-native, invasive versus non-invasive, natural versus non-natural, ecosystem structure and harm to the environment, to mention a few, cannot be defined in a manner that corresponds to an 'objective' reality. His argument is that these concepts are fundamentally value laden, and that they can only be justified on aesthetic or religious grounds, or on the basis of contemporary convictions about evolution and the spontaneous course of nature that has replaced religion. Accordingly, he proposes that the only legitimate mode for invasion biology to assume is that of an inductive or historical science, in which non-native and invasive species are studied in a comparative fashion on a case-by-case basis, in specific contexts in which it is possible also to take into account the values on the basis of which certain species are assessed as problematic or not.

In his focus on the language and metaphors of invasion biology, and in his discussion of their social functioning, Larson similarly challenges invasion biology to become self-critically aware of the manner in which our assumptions about ourselves as humans and how we are situated in the world not only influence the manner in which we study non-native and invasive species, but also the stories we tell about them to policy makers and the public. His argument is that in the stories we tell about invasive non-native species, we tell the story of who we are, or who we as society want to be. Because these stories are never neutral, and because they are always contested, he challenges invasion biology to articulate explicitly the story (or stories) that it tells about invasive species/ourselves, to scrutinize critically the social functioning and implications of this story and to seriously engage with others who may tell different stories. As such, Larson prompts invasion biology to become a critical and even a transformative force in society: by opening our eyes to blind spots in the stories we tell about ourselves and invasive non-native species, and by creatively suggesting ways to overcome them (through, among other things, alternative ways of thinking about ourselves, and alternative ways to relate to nature in general and non-native and invasive species in particular).

Above all, Larson challenges invasion biology to acknowledge that its value dimensions are not only to be found in the contextual values that play a role in framing the research questions that it poses. He also challenges invasion biology to acknowledge the cognitive schemas, the metaphors and narrative elements that form part of its *constitutive* values. In this manner he draws attention to the key insight that values do not stand separate and apart from science; rather, he contends that values form part of the very constitution of science as a systematic and methodical study of that which is. Accordingly, he is not trying to develop a kind of hybrid language in which values and science, values and invasion biology, are combined in some way or another in an external relationship to one another (Davis 2009, p. 169). Rather, he tries to find a language in which the values constitutive of invasion biology as a rigorous science can be articulated in terms of their internal relationship to one another. In my view Larson is well on his way to finding and communicating this language, and I think he has made considerable progress in the area of articulating what this language of explicitly acknowledging values could mean in the areas of communicating the results of invasion biology, and thinking about invasion biology as a science. I am not sure, however, whether Larson has been successful thus far in articulating the implications of this language of explicitly acknowledging values in the very constitution of invasion biology for its practice as a rigorous science. I think it will be interesting and important to see how Larson develops his thinking in this regard.

From this overview it is clear that the views of both Sagoff and Larson would not be received well in those circles of invasion biology that would tend to safeguard the rigour of invasion biology with appeals to an ideal of objectivity in which there is no place for values, assumptions, metaphors, narratives, interests and needs. In these versions of invasion biology, it would not be possible to give any place to 'subjective' factors like these, and still claim to be a rigorous science. Sagoff's response to this position would be that invasion biology is by definition a value-based, historical and contextual science, whereas Larson would maintain that invasion biology that strives to realize this kind of objectivity would tend to become disinterested and irrelevant to the needs of society.

It is important to note, however, that much of the substance of the respective challenges of Sagoff and Larson are well taken, or could be well taken, in those circles of invasion biology where it is indeed practised on a case-by-case basis as an inductive or historical study of particular invasions or particular non-native species. In contexts like these, however, the challenge for invasion biology is not so much to acknowledge the relevance of Sagoff's and Larson's perspectives; rather, it is to determine what they actually could mean in concrete terms for its day-to-day practice as rigorous science, and to feed the results of that into the processes of public decision making about non-native and invasive species.

To put it concretely in terms of the concept of 'biotic nativeness', which is arguably one of the most central concepts of invasion biology: although it is possible, and perhaps even courageous, to concede from within invasion biology that it is 'remarkably easy to unravel the conception of biotic nativeness', that 'our habit of preferring natives to aliens is poorly founded' and that 'none of the relationships comprising biotic nativeness is an inevitable, permanent or dependable object supporting a conception of belonging', as Chew and Hamilton have done in Chapter 4 of this volume, it still remains one of the legitimate and highly important conceptual tasks of invasion biology to find working definitions and criteria to distinguish native species from non-native species, or invasive species from non-invasive species in particular biotic contexts (as was done, for example, by Richardson et al. 2000; Klein 2002; Pyšek et al. 2004, 2008; Bean 2007). Following the train of thought of this chapter, it seems fairly obvious to argue that such a search for objective definitions and criteria is important to serve the needs of land managers who have to make timely and justifiable decisions to respond to the challenges of invasive non-native species. On the other hand, though, it does not seem obvious to determine what it could mean in practical terms to acknowledge explicitly the role of values in the very activities that invasion biologists engage in to arrive at their definitions and criteria in a more or less objective, rigorous manner.

I therefore think that it is crucial for invasion biology and its critics, for invasion biologists and philosophers, to continue engaging with one another, policy makers and the public about the central concepts and methodologies used to justify intervention programmes to address the very real problems created by certain invasive species. This engagement, I think, should be actively pursued and sustained, so that invasion biologists and their critics can learn to understand one another better, and to appreciate one another's work better. In this process, the critics of invasion biology may discover that their bright proposals for alternative approaches or methodologies have in fact long been known and practised in invasion biology (as Richardson et al. (2008) recently reminded us); whereas invasion biologists may discover that they also live with some of the questions and uncertainties that their critics write about.

Such an engagement, I submit, could make it possible for us to explicate more fully and scrutinize critically what we actually do when we practise the science of invasion biology and when we choose to respond in a certain manner to non-native and invasive species. If this is done, I submit, we can say with Larson that our studies of, and deliberations about, invasive non-native species 'can help us grow in humanity and in wisdom' (Larson 2007a, p. 149). Among many other things that this wisdom could entail, in relation to invasion biology and invasive non-native species, it should at least enable us to find a balance between the need, on the one hand, sometimes to have a well-managed intervention programme to minimize the damage of an invasive species, and the reality, on the other hand, that we cannot completely prevent or control invasions by non-native species: because we are not able to control completely both nature or human action.

ACKNOWLEDGEMENTS

I thank Mark Davis, Dave Richardson and an anonymous reviewer for their thorough reading of the first draft of this chapter, and for their comments and suggestions, which I greatly appreciate.

REFERENCES

Aitken, G. (2004) *A New Approach to Conservation: The Importance of the Individual Through Wildlife Rehabilitation.* Ashgate, Aldershot, UK.

Baker, H.G. & Stebbins, G.L. (eds) (1965) *The Evolution of Colonizing Species.* Academic Press, New York.

Bean, A.R. (2007) A new system for determining which plant species are indigenous in Australia. *Australian Systematic Botany,* **20**, 1–43.

Brown, N. (1997) Re-defining native woodland. *Forestry,* **70**, 191–198.

Brown, T.L. (2003) *Making Truth: Metaphor in Science.* University of Illinois Press, Chicago.

Brown, J.H. & Sax, D.F. (2004) An essay on some topics concerning invasive species. *Austral Ecology,* **29**, 530–536.

Brown, J.H. & Sax, D.F. reply to Cassey, (2005) Biological invasions and scientific objectivity: et al. *Austral Ecology,* **30**, 481–483.

Brunson, M.W. (2000) Managing naturalness as a continuum: setting limits of acceptable change. In *Restoring Nature: Perspectives from the Social Sciences and Humanities.* (ed. P.H. Gobster and P.B. Hull), pp. 229–244. Island Press, Washington, DC.

Castree, N. (2001) Socializing nature: theory, practice, and politics. In *Social Nature: Theory, Practice, and Politics* (ed. N. Castree, and B. Braun), pp. 1–21. Blackwell, Oxford.

Colautti, R.I. & MacIsaac, H.J. (2004) A neutral terminology to define 'invasive' species. *Diversity and Distributions,* **10**, 135–141.

Crifasi, R.R. (2005) Reflections in a stock pond: are anthropogenically derived freshwater ecosystems natural, artificial, or something else? *Environmental Management,* **36**, 625–639.

Cuddington, K. (2001) The 'balance of nature' metaphor and equilibrium in population ecology. *Biology and Philosophy,* **16**, 463–479.

Davis, M.A. (2006) Invasion biology 1958–2005: the pursuit of science and conservation. In *Conceptual Ecology and Invasions Biology: Reciprocal Approaches to Nature* (ed. M.W. Cadotte, S.M. McMahon and T. Fukami), pp. 35–64. Springer, London.

Davis, M.A. (2009) *Invasion Biology.* Oxford University Press, Oxford.

Egler, F.E. (1942) Indigene versus alien in the development of arid Hawaiian vegetation. *Ecology,* **23**, 14–23.

Elton, C.S. (1958) *The Ecology of Invasions by Animals and Plants.* Methuen, London.

Eser, U. (1998) Assessment of plant invasions: theoretical and philosophical fundamentals. In *Plant Invasions: Ecological Mechanisms and Human Responses* (ed. U. Starfinger, K. Edwards, I. Kowarik and M. Williamson), pp. 95–107. Backhuys Publishers, Leiden, the Netherlands.

Evans, J.M., Wilkie, A.C. & Burkhardt, J. (2008) Adaptive management of nonnative species: moving beyond the 'either–or" through experimental pluralism. *Journal of Agricultural and Environmental Ethics*, **21**, 521–539.

Fall, J. (2005) *Drawing the Line: Nature, Hybridity and Politics in Transboundary Spaces*. Ashgate, Aldershot, UK.

Farrar, S. (2003) Academic blacklisted over threat of invasion. *Times Higher Education Supplement*, 26 September 2003.

Fischer, A. & Van der Wal, R (2007) Invasive plant suppresses charismatic seabird – the construction of attitudes towards biodiversity management options. *Biological Conservation*, **135**, 256–267.

Forbes, S.I. (1925) The lake as a microcosm. *Bulletin of the Illinois State Natural History Survey*, **15**, 537–550.

Gallagher, R. & Carpenter, B. (1997) Human-dominated ecosystems. *Science*, **227**, 485–486.

Geertz, C. (1979) *Local Knowledge*. Basic, New York.

Hall, M. (2003) Editorial: the native, naturalized and exotic – plants and animals in human history. *Landscape Research*, **28**, 5–9.

Harvey, D. (1996) *Justice, Nature and the Geography of Difference*. Blackwell, Oxford.

Hattingh, J. (2001) Human dimensions of invasive alien species in philosophical perspective: towards an ethic of conceptual responsibility. In *The Great Reshuffling: Human Dimensions of Invasive Alien Species* (ed. J.A. McNeely), pp. 183–194. IUCN, Gland, Switzerland.

Hayles, N.K. (1991) Constrained constructivism: locating scientific inquiry in the theater of representation. *New Orleans Review*, **18**, 76–85.

Hull, R.B. & Robertson, D.P. (2000) The language of nature matters: we need a more public ecology. In *Restoring Nature: Perspectives from the Social Sciences and Humanities*. (ed. P.H. Gobster and P.B. Hull), pp. 97–118. Island Press, Washington, DC.

Klein, H. (2002) *Weeds, Alien Plants and Invasive Plants*. PPRI Leaflet Series: Weeds Biocontrol, No.1.1. ARC-Plant Protection Research Institute, Pretoria.

Larson, B.M.H. (2005) The war of the roses: demilitarizing invasion biology. *Frontiers in Ecology and the Environment*, **3**, 495–500.

Larson, B.M.H. (2007a) Thirteen ways of looking at invasive species. In *Invasive Plants: Inventories, Strategies and Action. Topics in Canadian Weed Science*, Volume **5**. (ed. D.R. Clements and S.J. Darbyshire), pp. 131–156. Canadian Weed Science Society, Sainte Anne de Bellevue, Québec.

Larson, B.M.H. (2007b) An alien approach to invasive species: objectivity and society in invasion biology. *Biological Invasions*, **9**, 947–956.

Larson, B.M.H. (2007c) Who's invading what? Systems thinking about invasive species, *Canadian Journal of Plant Science*, **87**, 993–999.

Larson, B.M.H. (2010) Embodied realism and invasive species. In *Philosophy of Ecology and Conservation Biology. Handbook of the Philosophy of Science* (ed. K. de Laplante and K. Peacock), pp. 1–18. Elsevier, Amsterdam.

Larson, B.M.H., Nerlich, B., & Wallis, P. (2005) Metaphors and biorisk: the war on infectious diseases and invasive species. *Science Communication*, **26**, 243–268.

Lodge, D.M. & Hamlin, C. (eds) (2006) *Religion and the New Ecology: Environmental Responsibility in a World in Flux*. University of Notre Dame Press, Notre Dame, Indiana.

Mabey, R. (2005) From corn poppies to eagle owls. *ECOS*, **26** (3/4), 41–46.

Meech, H. (2005) Eradicating non-native mammals from islands: facts and perceptions. *ECOS*, **26** (3/4), 72–80.

Pielke Jr, R.S. (2007) *The Honest Broker: Making Sense of Science in Policy and Politics*. Cambridge University Press, Cambridge, UK.

Pimm, S.L. (1991) *The Balance of Nature? Ecological Issues in the Conservation of Species and Communities*. University of Chicago Press, Chicago.

Pyšek, P., Richardson, D.M., Rejmánek, M., Webster, G.L., Williamson, M. & Kirschner, J. (2004) Alien plants in checklists and floras: towards better communication between taxonomists and ecologists. *Taxon*, **53**, 131–143.

Pyšek, P., Richardson, D.M., Pergl, J., Jaroš, V., Sixtova, Z. & Weber, E. (2008) Geographical and taxonomic biases in invasion ecology. *Trends in Ecology & Evolution*, **23**, 237–244.

Reise, K., Olenin, S. & Thieltges, D.W. (2006) Are aliens threatening aquatic coastal ecosystems? *Helgoland Marine Research*, **60**, 77–83.

Richardson, D.M., Pyšek, P., Rejmánek, M., Barbour, M.G., Panetta, F.D., & West, C.J. (2000) Naturalization and invasion of alien plants: concepts and definitions. *Diversity and Distributions*, **6**, 93–107.

Richardson, D.M., Pyšek, P., Simberloff, D., Rejmánek, M. & Mader, A.D. (2008) Biological invasions – the widening debate: a response to Charles Warren. *Progress in Human Geography*, **32**, 295–298.

Ridder, B. (2007) An exploration of the value of naturalness and wild nature. *Journal of Agricultural and Environmental Ethics*, **20**, 195–213.

Rolston, H. (2006) Intrinsic values on earth: nature and nations. In *Environmental Ethics and International Policy* (ed. H.A.M.J. ten Have), pp. 47–67. UNESCO, Paris.

Sagoff, M. (1999) What's wrong with exotic species? *Report from the Institute for Philosophy and Public Policy*, **19** (4), 16–23.

Sagoff, M. (2003) The plaza and the pendulum: two concepts of ecological science. *Biology and Philosophy*, **18**, 529–552.

Sagoff, M. (2005) Do non-native species threaten the natural environment? *Journal of Agricultural and Environmental Ethics*, **18**, 215–236.

Sagoff, M. (2006) Environmental ethics and environmental science. In *Environmental Ethics and International Policy* (ed. H.A.M.J. ten Have), pp. 145–161. UNESCO, Paris.

Sagoff, M. (2009a) Environmental harm: political not biological. *Journal of Agricultural and Environmental Ethics*, **22**, 81–88.

Sagoff, M. (2009b) Who is the invader? Alien species, property rights, and the police power. *Social Philosophy & Policy*, **26** (2) (summer 2009), 26–52.

Shrader-Frechette, K.S. (1996) Throwing out the bathwater of positivism, keeping the baby of objectivity: relativism and advocacy in conservation biology. *Conservation Biology*, **10**, 912–914.

Shrader-Frechette, K.S. (2001) Non-indigenous species and ecological explanation. *Biology and Philosophy*, **16**, 507–519.

Shrader-Frechette, K.S. & McCoy, E.D. (1990) Theory reduction and explanation in ecology. *Oikos*, **58**, 109–114.

Shrader-Frechette, K.S. & McCoy, E.D. (1993) *Method in Ecology: Strategies for Conservation.* Cambridge University Press, Cambridge, UK.

Shrader-Frechette, K.S. & McCoy, E.D. (1994a) Ecology and environmental problem-solving. *The Environmental Professional*, **16**, 342–347.

Shrader-Frechette, K.S. & McCoy, E.D. (1994b) How the tail wags the dog: how value judgments determine ecological science. *Environmental Values*, **3**, 107–120.

Simberloff, D. (2003) Confronting invasive species: a form of xenophobia? *Biological Invasions*, **5**, 179–192.

Soulé, M.E. & Lease, G. (1995) *Reinventing nature: responses to postmodern deconstruction.* Island Press, Washington, DC.

Stokes, K.E., O'Neill, K.P., Montgomery, W.I., Dick, J.T.A., Maggs, C.A. & McDonald, R.A. (2006) The importance of stakeholder engagement in invasive species management: a cross-jurisdictional perspective in Ireland. *Biodiversity and Conservation*, **15**, 2829–2852.

Townsend, M. (2005) Is the social construction of native species a threat to biodiversity? *ECOS*, **26**, 1–9.

Warren, C.R. (2007) Perspectives on the 'alien' and 'native' species debate: a critique of concepts, language, and practice. *Progress in Human Geography*, **31**, 427–446.

Woods, M. & Moriarty, P.V. (2001) Strangers in a strange land: the problem of exotic species. *Environmental Values*, **10**, 163–191.

Chapter 28

CHANGING PERSPECTIVES ON MANAGING BIOLOGICAL INVASIONS: INSIGHTS FROM SOUTH AFRICA AND THE WORKING FOR WATER PROGRAMME

Brian W. van Wilgen[1], Ahmed Khan[2] and Christo Marais[2]

[1]Centre for Invasion Biology, CSIR Natural Resources and the Environment, P.O. Box 320, Stellenbosch 7599, South Africa
[2]Working for Water Programme, Cape Town, South Africa

Fifty Years of Invasion Ecology: The Legacy of Charles Elton, 1st edition. Edited by David M. Richardson

28.1   INTRODUCTION

Management and control interventions for invasive species have become more effective over the past 50 years, but most focus on particular aspects of the problem, suggesting that invasive species management is based on knowledge and practice that follow reductionist and consequently compartmentalized lines. For example, a recent review (Clout & Williams 2009) divided management approaches into those that addressed particular aspects of the management problem (prevention, risk assessment, detection, eradication or control), particular taxa (terrestrial plants, aquatic plants, invertebrates, fish or mammals) or particular ecosystems (islands, rivers, marine or terrestrial environments). Realization has grown that reductionist approaches alone will be insufficient to deal with the growing problem of biological invasions (McNeely et al. 2001). As a result, national strategies have been developed, including an Australian Pest Animal Strategy and an Australian Weeds Strategy, a National Invasive Species Strategy in the Bahamas, a Biosecurity Strategy in New Zealand and a National Strategy and Implementation Plan for Invasive Species Management in the USA (Pyšek & Richardson 2010). Implementation of national strategies for dealing with biological invasions is, however, a daunting task, and few examples have emerged where this approach has been sustained at a national level, especially in developing countries.

In South Africa, a large inter-departmental programme was initiated in 1995 to address invasive alien plant invasions in a holistic way (van Wilgen et al. 1998, 2002). The programme's name, 'Working for Water', captures the dual goals of conserving an important ecosystem service, while simultaneously providing employment for the rural poor. The programme has received international acclaim, and is often cited as an innovative, holistic and successful approach to the management of biological invasions (Hobbs 2004; Mark & Dickinson 2008; Pejchar & Mooney 2010). We review here the perspectives gained during the establishment and implementation of the Working for Water programme over the past 15 years.

28.2   BIOLOGICAL INVASIONS IN SOUTH AFRICA

South Africa (1.2 million km^2) is topographically and climatically varied and consequently supports many ecosystems. Terrestrial ecosystems include savannas,

grasslands, arid shrublands, Mediterranean-climate shrublands (fynbos), deserts and forests, all of which harbour well-established invasive species. Plants are most important invasive taxon, occupying huge sectors of the country, most conspicuously in the fynbos shrublands, and riparian areas and floodplains in all ecosystems. Terrestrial invasive alien plants and alien freshwater aquatic organisms have the biggest impacts, in contrast to invasive alien mammals, birds, reptiles, amphibians and marine organisms (see van Wilgen & Richardson 2010). In the case of invasive alien invertebrates impacts are as yet insufficiently documented. We acknowledge that invasive alien pathogens have had significant impacts worldwide (Perrings et al. 2010), but these are not considered here as Working for Water did not address this issue.

A total of 234 invasive alien plant species were recognized by 2001 (Henderson 2001), of which 198 were declared as weeds, and the list continues to grow (currently 346). Many ecosystems have been thoroughly transformed by alien trees and shrubs. These invaders include pines (*Pinus* species) and hakeas (*Hakea* species) in fynbos shrublands, Australian wattles (*Acacia* species) and eucalypts (notably *Eucalyptus camaldulensis*) in riparian areas, and mesquite (hybrids of several species of *Prosopis*) in arid areas. Many of these species have been extensively propagated, widely distributed, are predisposed to South Africa's environments and have been in the country for as long as 300 years. Other alien plant species are also widespread and have significant impacts, including lantana (*Lantana camara*) and triffid weed (*Chromolaena odorata*) in savanna, and many cacti (*Opuntia* species) in arid areas and grasslands (Olckers & Hill 1999; Henderson 2001). Perhaps best known in South Africa are the impacts that invading tree species have on water resources, but invasive plant species also exacerbate wildfires, replace palatable plants with unpalatable or poisonous plants, reduce biodiversity and threaten many endemic species, and degrade water bodies (van Wilgen et al. 2008). Given that many invasive species are recent arrivals, the number of invasive species, and therefore the level of impacts, will likely grow.

28.3   HISTORIC APPROACHES TO INVASIVE ALIEN PLANT MANAGEMENT

South Africa has a long history of addressing alien plant invasions. Early botanists, including Peter

MacOwan in 1888 and Rudolf Marloth in 1908, raised concerns that alien plants would replace natural vegetation (Stirton 1978). A landmark publication in 1945 (Wicht 1945) stated that, 'one of the greatest, if not the greatest, threats to which the Cape vegetation is exposed, is suppression through the spread of vigorous exotic plant species'. Most attempts at control in the first half of the 20th century were, however, ad hoc, and records of these efforts are incomplete. Control efforts in the second half of the 20th century were at best uncoordinated and erratic, and did little to stem the spread of invasive alien plants. Although few campaigns were adequately documented, the evidence shows that poor understanding of the ecology of invasive species, as well as a lack of follow-through control after alien plant removal, led to much wasted effort and money. For example, Macdonald et al. (1989) reviewed 47 years of control attempts on the southern Cape Peninsula, and concluded that they were 'almost totally ineffective for the first 35 years'.

In the 1970s and 1980s, coordinated control programmes were introduced in the fynbos (Fenn 1980). These programmes were aimed largely at clearing watershed areas of invasive pines, hakeas and wattles, and they involved mapping invasive plants, and scheduling mechanical clearing in conjunction with prescribed burning. These carefully planned operations had the desired effect of making considerable progress towards achieving clearing targets (Macdonald et al. 1989; van Wilgen 2009).

Biological control for invasive alien plants was developed in parallel with implementation of mechanical clearing, but not necessarily in an explicitly collaborative or coordinated way. The first biological control agents were released in 1913 against the invasive cactus *Opuntia monocantha* (Zimmermann et al. 2004). This release was followed in the 1930s by release of biological control agents against the sweet prickly pear (*Opuntia ficus-indica*). Weed biological control was not, however, effectively organized in South Africa until the early 1970s, resulting in 22 releases of new agents in the 1970s, 30 in the 1980s and 33 in the 1990s. The success of this approach was the result of foresight and leadership by a single individual, Dr D.P. Anneke (Zimmermann et al. 2004).

Coordinated clearing programmes, mainly in the fynbos of South Africa, declined drastically in the late 1980s (van Wilgen 2009) for a combinations of reasons. Increasing restrictions on the use of prescribed burning meant that many prescribed fires, necessary for removal of invasive alien plant seedlings, could not

be done. The government also split the functions of its forestry department (responsible for almost all of the major clearing programmes), with plantation management, which in turn became privatized. Conservation management devolved to unprepared and inexperienced provincial authorities, and the forestry department's research arm was transferred to the Council for Scientific and Industrial Research. This caused fragmentation of responsibilities and loss of capacity and experience. The government of the day, beleaguered by anti-apartheid sanctions, had to cut funding, resulting in further losses of capacity. As a result, programmes to control invasive alien plants lagged, and cleared areas were under threat of re-invasion (van Wilgen et al. 1990, 1997).

28.4 A NEW APPROACH: THE BIRTH OF WORKING FOR WATER

The Working for Water programme was established by South Africa's newly elected democratic government in 1995. South Africa's remarkable and relatively peaceful transition to democracy in 1994 was accompanied by the election of a cadre of new politicians and a fierce public desire to change national priorities and practices, and to change them for the better. This single event was arguably the biggest turning point in the country's history, and was accompanied by new-found optimism and a widespread willingness to accept change in many approaches and activities. Against this background a group of ecologists argued to the government that (i) invasive alien plants were a large and growing national problem, (ii) these invasions were specifically a serious threat to water resources (Le Maitre et al. 1996; van Wilgen et al. 1996) that needed to be addressed and (iii) by forcefully addressing these problems, important employment opportunities would arise for poor people in underdeveloped rural areas.

A proposal was presented to Professor Kader Asmal, the first Minister of Water Affairs and Forestry in President Nelson Mandela's Cabinet, on 2 June 1995. Asmal immediately sought and received funding of R25 million (1US$ ≈ R7.5) for the last six months of the 1995/6 financial year. Project managers were appointed, workers were recruited and clearing projects for invasive species began almost immediately. Further funding followed the next year, and the Working for Water programme rapidly became the flagship of the government's poverty-relief programmes (van Wilgen et al. 2002).

28.5   APPROACHES ADOPTED BY WORKING FOR WATER

The Working for Water programme has been characterized by the encompassing goal to create invasive plant interventions that have broad support, to foster collaboration between stakeholders, to leverage the opportunities to broaden and maximize benefits, to be un-swayed by vested interests and to be sustainable. The unique socio-political conditions in South Africa in the mid-1990s, outlined above, made it possible to develop effective approaches rapidly. We describe below important aspects of the programme.

Adopting a strategic approach at national level

The Working for Water programme was initiated as a poverty relief programme of the national government for all provinces. Although the programme has been managed within the government's Department of Water Affairs, it was recognized from the outset, at least by the programme's leaders, that invasive species affected more than water. Environmental management was (and still is) fragmented in South Africa, with key and often overlapping responsibilities falling under departments responsible for the environment, forestry, water and agriculture. Working for Water actively and deliberately sought to engage these departments in its goals. Engagement was also needed to access the capacity to manage the programme's funding. The necessary capacity was spread across departments, and implementation of the programme required use of existing capacity, mainly in government conservation agencies, science councils and in the forestry sector.

At first, Working for Water had no formal strategy, and the initial focus of the programme, understandably, was on implementation. This focus proved fortuitous, because the government was under pressure to deliver on election promises. Most of the departments eligible for poverty-relief funds were simply not structured to receive and rapidly deploy the funds. Given Working for Water's goals, it received a generous share of the poverty-relief funding, providing in return a success story for its sponsors. Although the need for a strategy was recognized at the outset, its development was only possible after the programme had overcome the challenges of initial establishment and rapid growth. The programme's current strategy requires a partnership of the departments responsible for implementing the fragmented policy and legislation pertaining to invasive alien species.

The programme's mission statement has two broad goals, both to be achieved through the management of invasive alien plants. The first seeks to enhance the sustainable use and conservation of South Africa's natural resources. The second promotes socio-economic development within the Government's Expanded Public Works programme. The strategy calls for the first goal to be achieved through the prevention of new invasive alien plant problems, the reduction of the impacts arising from existing invasive alien plant problems and the enhancement of capacity and commitment to solve problems caused by invasive alien plant. The second goal is to be achieved through creating sustainable jobs in the natural-resource management market. The strategy provides the basis for coordination and collaboration with the management of biological invasions in South Africa, and the approaches described below are all aligned with the broad strategy.

Maintaining a focus on ecosystem services

In any nation, the competition for public funds is fierce, and requires that any claim on these funds must be well motivated and based on defensible estimates. Working for Water was initiated with a focus on a particular ecosystem service (water) and it broadened that focus over the next 15 years. Assessment of the impacts of invasive alien plants on water resources (Le Maitre et al. 1996; van Wilgen et al. 1996) was later expanded through research either funded directly by Working for Water, or performed on its behalf (Prinsloo & Scott 1999; Le Maitre et al. 2000, 2002; Dye & Jarmain 2004). In addition, further research examined a wider range of ecosystem services, including ecotourism, harvested products, the preservation of genetic diversity and pollination services (Higgins et al. 1997; Turpie et al. 2003). The impacts of invasions on these services was then quantified in economic terms, providing monetary estimates of the returns on investment from control operations for invasive species (see Hosking & du Preez 1999; Turpie & Heydendrych 2000; De Wit et al. 2003).

Finally, the potential impacts on ecosystem services have been assessed in a hypothetical situation where invasive alien plants occupy all of the land that could be potentially invaded (van Wilgen et al. 2008) and expressed in monetary terms (De Lange & van Wilgen, 2011). The first study estimates the impact of invasive alien plants on water resources, livestock production and biodiversity in five terrestrial biomes. Potential reductions in water resources would be more than eight times greater than current impacts if invasive alien plants were to occupy their full potential range. Reductions in grazing capacity under current levels of invasion are estimated to be approximately 1% of the potential number of livestock that could be supported, but future impacts could decrease livestock production by as much as 71%. A 'biodiversity intactness index' (the remaining proportion of pre-modern populations) ranged from 89% to 71% for the five biomes. With the exception of the fynbos biome, current invasions were estimated to have had almost no impact on the intactness of biodiversity. Under future levels of invasion, however, these intactness values decrease to around 30% for the savanna, fynbos and grassland biomes, but to even lower values (13% and 4%) for the two karoo biomes. Thus the current impacts of invasive alien plants may be relatively low (with the exception of those on surface water runoff), but the future impacts could be very high. Although these estimates likely include substantial errors, the predicted impacts are sufficiently large to make a serious case for concern. In monetary terms, the estimated economic losses under current levels of impact total R6.5 billion per annum (about 0.3% of South Africa's gross domestic product of around R2000 billion in 2009). The bulk of losses are attributable to water (R5.8 billion) with the balance to grazing (R300 million) and other biodiversity-related value (R400 million). Estimates of possible future impacts (van Wilgen et al. 2008) suggest that economic losses could reach 5.2% of gross domestic product for these three ecosystem services if invasive plants are allowed to reach their full potential.

Developing systems of payment for ecosystem services

Quantification of the value of ecosystem services, especially water, has allowed the development of a voluntary payment system for ecosystem services in South Africa, based on service delivery (Turpie et al. 2008). Such systems are especially important for supporting conservation efforts, as government funding for conservation is shrinking globally, while threats to the integrity of ecosystems rise simultaneously. Early studies funded by Working for Water showed, for example, that clearing the invasive alien plants from the catchments of existing dams could deliver water at a fraction of the cost of constructing new dams (van Wilgen et al. 1997). This conclusion led, in turn, to adding the cost of alien plant control to levies on water tariffs by the national Department of Water Affairs (Blignaut et al. 2007). As the hydrological benefits of these projects became more widely appreciated, water utilities and municipalities contracted Working for Water to control invasive alien plants in their water catchments. This payment system for ecosystem services is unique in that the service providers are previously unemployed people who compete for contracts to restore public or private lands, rather than the landowners themselves (Turpie et al. 2008). Furthermore, by protecting an 'umbrella' ecosystem service such as water yield, ecosystems and their component biodiversity are also conserved, and will continue to deliver additional services that may be less easily quantified, and consequently more difficult to justify (for example, biodiversity is notoriously difficult to quantify in economic terms (Turpie 2004)).

Influencing new national legislation

Inauguration of South Africa's first democratically elected government was followed by an enthusiastic overhaul of the country's laws, including many with direct relevance for the management of invasive alien species (Richardson et al. 2003). Working for Water engaged in the revision of legislation, and sometimes initiated revision. The first intervention by Working for Water was the revision of the list of declared invasive species under the Conservation of Agricultural Resources Act. Originally, this act listed 'declared weeds' in a single category, and required them to be controlled. As a result of interventions by Working for Water, the regulations were broadened in 2001 to cover three categories and many more species of weeds: the first category (weeds that must be controlled); the second category (invasive alien plants that have commercial value and can be grown on demarcated areas and traded, provided steps are taken to control their spread); the third category (invasive alien plants that have ornamental uses and can be grown on

---

Box 28.1 Features of South Africa's draft regulations to govern the management of invasive species

New draft regulations to govern the management of invasive species have been drafted in terms of South Africa's legislation that governs biodiversity management. These regulations have several innovative features, including the following:
• Creation of one overarching set of regulations covering multi-departmental interests, and incorporating various regulations from other acts relating to invasive alien species.
• The incorporation of best practices from across the world.
• Balancing the need for a precautionary principle to apply to species entering the country for the first time ('guilty-until-proven-innocent'), with an 'innocent-until-proven-guilty' approach for alien species within the country.
• Taking a pragmatic approach, allowing for tolerance of invasive species with positive attributes (such as

trout, jacarandas or indigenous but extra-limital species).
• Allowing for control of accidental or illegal introductions through listed invasion pathways, as well as having an onus of proof that an alien species is legally in the country in terms of risk assessment requirements.
• Introducing differentiated duty of care requirements for listed invasive species, with different categories to cater for a range of circumstances.
• The inclusion of comprehensive lists of species, across taxa (previous approaches included only plants).
• Including strong measures that focus on accountability, including for propagule pollution, the use of rates as incentives and disincentives, and the ability to forbid the sale or transfer of property that contains prioritized invasive species.

---

demarcated areas, but not traded or replanted) (Henderson 2001). More recently, Working for Water, in collaboration with local biologists and ecologists, has made a substantial contribution to the development of regulations for the management of invasive alien species under the country's new act governing biodiversity-related issues. The draft regulations (currently undergoing public review) are arguably the most comprehensive anywhere (Box 28.1).

Involving important stakeholder groups

Working for Water sought to actively engage stakeholders whose activities led to, or could lead to, the introduction and spread of invasive species, in order to identify solutions to these problems that would accommodate the strategic interests of both parties. Foremost among these stakeholders was the nursery industry (mainly people who traded in ornamental plants) and the plantation forest industry (a significant source of invasive trees). Engagement with the nursery industry started with surveys of species offered for sale, and the identification of invasive or potentially invasive species. Members of the nursery industry were also involved in

workshops to identify species for listing in different categories in regulations, and in proposals to develop a 'green label' for plants that had been cleared by Working for Water as non-invasive.

With the forest industry, Working for Water took a strong stand over the 'polluter pays' principle. Staff members from the forest industry were seconded to Working for Water, where they collaborated on the development of mutually acceptable approaches to managing both escaping plantation trees as well as invasive alien species on forest estates. The Forestry Stewardship Council, which certifies wood-based products as being produced in a sustainable way, was also influential in shaping policy toward invasive species in forestry.

Promoting the effective use of biological control

The biological control of invasive alien plants is recognized by the Working for Water programme as essential to the sustainable management of invasive alien plants (Moran et al. 2005). South Africa has been fortunate in having had significant capacity in place in this field on which to build, and Working for Water provides significant support to the biological control research effort (about R10 million annually). Under the auspices of the Working for Water programme, innovative approaches have been developed (see below).

A national approach

Biological control in South Africa is coordinated at a national level, and the available research capacity is focused on priority problems. All members of the South African biological control research community work closely together to identify target weed species and review research progress, in collaboration with Working for Water managers. Annual research meetings have been held for over two decades, and regular assessments of progress are published (see, for example, Olckers & Hill 1999). As a result, high levels of consensus have emerged about the use of biological control agents.

Addressing emerging weeds

Biological control against emerging weeds (plants in an early stage of invasion) has not yet been widely practised, largely because limited budgets are directed at invasions that have already become detrimental. In 2003, the Working for Water programme allocated funds for biological control programmes against emerging weed species (Olckers 2004), giving formal recognition, for the first time, to the rationale of targeting incipient weeds. Species selected for initial investigation include pompom weed (*Campuloclinium macrocephalum*), American bramble (*Rubus cuneifolius*), balloon vine (*Cardiospermum grandiflorum*), parthenium weed (*Parthenium hysterophorus*) and yellow bells (*Tecoma stans*).

Building capacity in biological control

Working for Water has recognized the risks arising from retirements and other losses in the ranks of biological researchers and their employers, and has set aside funds to build capacity. These funds have been used variously for postgraduate bursaries in biological control research, the appointment and development of new recruits, and the development and presentation of short courses for alien plant control managers.

Supporting and expediting implementation

Working for Water has allocated funds to enhance application of biological control in the field. It has appointed field staff specifically to ensure that biological control is used to its full potential in the regions for which they are responsible, and it has established four stations at which biological control agents are mass-reared and released (Zimmermann et al. 2004).

Demonstrating returns on investment

Working for Water commissioned holistic economic evaluations of South African endeavours to manage invasive alien plants using biological control (van Wilgen et al. 2004; De Lange & van Wilgen 2011). De Lange & van Wilgen's (2011) work focuses on the delivery of ecosystem services from habitats that are invaded by groups of weeds, rather than by single species. They conclude that the value of lost ecosystem services would have amounted to an estimated additional R41.7 billion had no control been performed, and that 5–75% (depending on the group of weeds) of this protection was due to biological control. The benefit:cost ratios arising from biological control research range from 50:1 for invasive subtropical shrubs to 3726:1 for invasive Australian trees. Although this research invokes some assumptions by necessity, the conclusion that biological control has brought about a huge level of protection of ecosystem services remains robust, even after subjecting the assumptions to sensitivity analysis.

Despite this progress, risk assessment and management of biological control agents of weeds remains in a quandary, compared with other potentially invasive alien species. Bio-control agents are extensively investigated for risks of non-targeted impacts (with an average testing period of between one and eight years, costing over R1 million per species (Moran et al. 2005)), are aimed at combating biological invasions and have an impeccable safety record (Moran et al. 2005). Yet frivolous introductions, for marginal profit or pleasure, have been allowed into the country with comparatively perfunctory assessments. Permits for the release of biological control agents have been delayed by a lack of capacity to process applications, poor understanding of the risk assessments, and fundamentalist opposition to the use of biological control. These delays have absorbed scarce funds within Working for Water.

Expanding the social benefits

Working for Water's second broad goal is to promote socio-economic development as part of the Government's Expanded Public Works programme (see 'Adopting a strategic approach at national level' in section 28.5). Consequently, the programme set out to maximize employment opportunities. During its first year of operations (from October 1995 to March 1996), the programme employed 6163 people. This figure rose significantly in 1997, and, since then, levels of employment has been between 25,000 and 30,000 employees (Magadlela & Mdzeke 2004). The varied aspects of the employment programme have been extremely important in obtaining and maintaining broad political support.

A focus on the rural poor

Since inception, the programme has strived to provide employment to previously disadvantaged people living in under-developed rural areas. Working for Water also emphasizes addressing gender imbalances; it seeks to ensure that at least 60% of the wages are earned by women.

The provision of training

All staff in the programme are provided with training, both to equip them for their immediate work assignments as well as to provide some life and development skills. Training consequently had three components: training in work-related activities (skills in machine and herbicide use, and worker safety), training in health (with a focus on HIV/AIDS) and contractor development. The last of these components of training allowed some employees to develop entrepreneurial skills.

The development of entrepreneurial skills

The programme initially introduced a contractor scheme, which sought progressively to advance people from a daily wage approach to work, initially through piece work (in which workers are paid for pre-defined 'pieces' of work, such as an area to be cleared), then closed contracts (where contractors are hired without tendering on the open market) to the final stage of independent contractor, which allowed people to apply for contract work. All work is now done under the auspices of the programme and is performed through such contracts with service suppliers.

Re-integration of ex-offenders and military veterans

South Africa has a large prison population and former offenders have difficulty finding work. The programme gave employment to ex-offenders, to address the inability of the prison system to re-socialize former offenders. In addition, military veterans are being offered employment in the clearing of invasive alien plants from bases and training grounds.

Developing a hierarchical system of prioritization

Working for Water has recognized the need for transparent and objective approaches to prioritizing its operations. Despite the programme's relatively high levels of funding, resources are insufficient to deal with the entire invasive species problem at once and choices must be made. These choices should be guided by a set of priorities that are transparent, defensible, logical and flexible. Working for Water has thus embarked on the development of a hierarchical approach to prioritizing its current clearing operations. At the highest level, South Africa's major terrestrial biomes (fynbos, grasslands, arid savannas, moist savannas, thicket, forest, desert, Indian Ocean coastal belt, succulent karoo and nama karoo) were ranked, followed by primary catchments within each biome and finally quaternary catchments (i.e. nested subdivisions of primary, secondary and tertiary catchments) within primary catchments. South Africa has 1911 quaternary catchments; they represent the level at which control projects for alien plants are established, and where competition for limited funding is most acute.

The prioritization exercise used the analytic hierarchy process (Saaty 1990), a multi-criteria decision-making tool for setting priorities when both qualitative and quantitative aspects of a decision need to be considered. The process requires identification of criteria on which to base the prioritization and the assignment of relative importance to each criterion. The areas to be prioritized are then systematically compared with each other for each criterion to arrive at a final ranking. Criteria and their importance were decided in workshops involving stakeholders from participating gov-

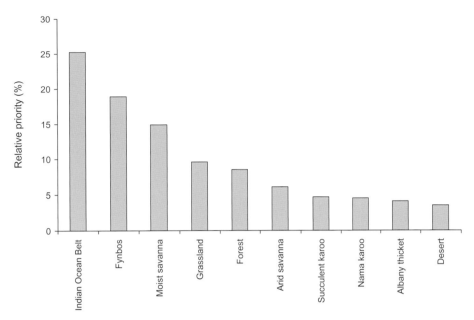

Fig. 28.1 South Africa's terrestrial biomes ranked according to their relative importance for controlling invasive alien plants. Figure from van Wilgen et al. (2010).

ernment departments, and local experts. For some criteria, where data were available, comparisons could be made quantitatively. The criteria include the presence and density of priority invasive alien plants (from available maps); the relative conservation importance of the area (from conservation planning exercises); impacts on water resources and grazing (from estimates of surface runoff and livestock carrying capacity); and whether the area fell within an identified priority poverty node. Other criteria required subjective estimates by experts or those with local knowledge. These criteria included those related to the value of harvested products, the impacts of invasion on fire hazard, the impact of removal of alien species that carried some benefit and the capacity to hold any gains made in initial clearing operations.

The hierarchical prioritization process is not yet complete, but results are available for the higher-level (biome and primary catchment) prioritizations. The priorities derived in this way (Fig. 28.1) were compared with the budgets allocated to the different biomes (Fig. 28.2). The Indian Ocean coastal belt received little funding compared with its priority as determined in the exercise; this indicates that the level of funding

may need to be adjusted. However, it is at the lowest level (quaternary catchments) where competition for funding is likely to be most acute, and the technique promises to be very useful in this regard.

Fostering relevant research

The Working for Water programme, with the provision of limited funding for research, has facilitated an expansion in research in invasion ecology and management in South Africa. Although the Working for Water programme is not primarily a research-funding organization, it has wielded significant influence and promoted relevant research. This has been done in part through strategic research partnerships with a view to accessing co-funding for projects of direct interest to Working for Water, or of influencing the direction of research (Table 28.1). The Working for Water programme has established a research advisory panel that assists in the identification of priority research questions and monitors the quality of research outputs that are directly funded by Working for Water. Most directly funded research has been in

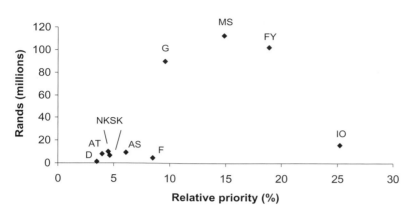

Fig. 28.2 The 2009/2010 budget for invasive alien plant clearing projects in the main terrestrial biomes of South Africa in relation to identified priorities (see Fig. 28.1). The biomes are as follows: F, forest; AT, Albany thicket; AS, Arid savanna; NK, Nama karoo; SK, Succulent karoo; D, desert; MS, Moist savanna; G, Grassland; FY, Fynbos; IO, Indian Ocean Coastal Belt. Figure from van Wilgen et al. (2010).

Table 28.1 Research partner organizations that have provided Working for Water with research and related additional support.

Research partner organization	Research focus	Additional support to Working for Water
Council for Scientific and Industrial Research	Hydrological effects of invasions. Economic impacts of invasions. Prioritization studies.	Management capacity in initial stages. Scientific advisory service.
Water Research Commission	Water use by invasive alien plants. The economics of clearing operations. The inclusion of control considerations for alien plants in water-resource planning.	None.
Agricultural Research Council	Biological control. Remote sensing for mapping the extent of invasions.	Revision of regulations relating to alien species management.
The Department of Science and Technology – National Research Foundation's Centre of Excellence for Invasion Biology	Genetic techniques to improve understanding of invasions and their management. Effects of control and rehabilitation projects. Effects of invasive eucalypts on riparian areas.	Revision of regulations relating to alien species management.
South African National Biodiversity Institute	Emerging invaders.	None.

biological control (see 'Promoting the effective use of biological control' in section 28.5), but further limited funding has addressed landscape hydrology, the ecology and control of invasive plants, resource economics, social aspects and the development of operational solutions to management problems. The questions asked often include whether predictions of significant benefits arising from the control of invasive alien plants can be substantiated by good science, and whether solutions can be developed to significantly reduce the threat of invasive alien plants, or increase the efficiency of their control.

Dealing with emerging invaders

Although Working for Water's primary focus has been on clearing existing infestations of invading alien plants, the programme has also recognized the imperative to identify and deal effectively with new invasions. It has thus recently (2008) specifically set aside funding to establish an early detection and rapid response programme. Building and maintaining capacity to address this component of invasive alien species management is critical. The programme for Early Detection and Rapid Response for Invasive Alien Plants is housed within the South African National Institute for Biodiversity. The programme's objectives include inter-organization coordination, surveillance, capacity development, rapid response, information management, research, awareness-raising, and monitoring and evaluation. The programme has already established projects to address emergent weed species (see, for example, Zenni et al. 2009; Le Roux et al. 2010), and more are being added.

28.6 FACTORS PROMOTING SUCCESS

Many factors contributed to the success of the Working for Water programme, and if any one of them had not been in place, the outcome could have been quite different. These factors are discussed briefly below.

A unique opportunity

The unique opportunity offered by the new democratic government, and the climate of an acceptance of change, was critically important in facilitating the initiation of the project. It is unlikely that it would have been possible to secure the resources necessary to initiate the programme under different circumstances.

Political vision and backing

Political backing in the form of funding and cabinet-level support, and the visionary political leadership that came from the programme's sponsor, Professor Kader and his leadership, was a critical ingredient for success. Without the high-level political support (both of Asmal and successive ministers), it is doubtful that the programme would have been commanded the status it subsequently enjoyed.

Providing employment

Although studies had shown that the programme would have had significant benefits even if job creation had been very small, the programme's ability to create many jobs, and therefore directly affect the lives of many people, was fundamental to securing funding and sustaining political support. Linking the programme to job creation, and aligning it with the new government's Reconstruction and Development Programme, provided the novel strategy that made the package eminently saleable.

Addressing a serious problem

Many people, especially environmentalists, were acutely aware of the serious environmental threat posed by invasive plants, and of the need to deal with the problem. This high level of awareness arose, in part, through efforts of the SCOPE (Scientific Committee on Problems of the Environment) programme on biological invasions (Macdonald et al. 1986). The resulting widespread concern about invasive species led to rapid endorsement of the Working for Water programme by ecologists and environmentalists.

Committed managers

A crucial factor in the programme's success was the availability of a core of dedicated people. This relatively small group realized the programme's value and its aims, after having become increasingly frustrated with the reality of declining funding and an inability to deal with the problem over many years. When the new funding did become available, these dedicated conservationists, who included the management and communications teams, the project managers and employees of several conservation agencies, were committed to apply the additional effort to initiate the programme.

Selling the story

Creation of a communications programme added an important dimension. Informative promotional brochures and newsletters were produced (Fig. 28.3), promotional 'open days' were held and valuable contacts were forged with key reporters, who carried several articles in influential national and local newspapers. Without the conscious allocation of resources to this

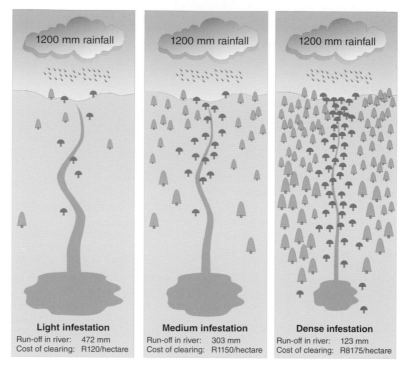

Light infestation | Medium infestation | Dense infestation
Run-off in river: 472 mm | Run-off in river: 303 mm | Run-off in river: 123 mm
Cost of clearing: R120/hectare | Cost of clearing: R1150/hectare | Cost of clearing: R8175/hectare

Fig. 28.3 Working for Water used simple yet effective depictions of the plausible impacts on water runoff that would arise if alien plant invasions were not addressed. This example from a typical information brochure shows that surface runoff from a catchment with annual rainfall of 1200 mm would decrease by 74%, and would be accompanied by a 68-fold increase in clearing costs, if invasive plants were allowed to spread over 50 years.

function, the good work done elsewhere may have gone largely unnoticed.

28.7   REMAINING PROBLEMS

The Working for Water programme has made considerable progress and much has been learnt over the past 15 years. The programme has initiated novel and innovative approaches, and it has received recognition worldwide, in part because it illustrates the dividends of a comprehensive and collaborative undertaking with widespread support. However, as could be expected, difficult issues remain.

Conflicts of interest

Invasive alien species that also have commercial, ornamental or other uses give rise to difficult prob-

lems. Controlling these species, especially in natural (rather than agricultural) environments gives rise to 'public good' benefits, where the individual marginal benefit (the amount of benefit gained by any one person) is small. Where individual marginal benefits are small, people tend not to voice strong concern. Stakeholders who stand to lose, on the other hand, often vigorously oppose the control of these species. For example, pines (*Pinus* species) and wattles (especially *Acacia mearnsii*) underpin the plantation industry in South Africa but are aggressively invasive. Biological control using seed-feeding agents offers hope for reducing the invasive tendencies of these species, while still allowing for their cultivation. Much progress has been made in controlling wattles in this way, although it took decades before the wattle industry accepted biological control. In the case of pines, however, research on a seed-feeding beetle (Moran et al. 2000) was abandoned for fear the insect would allow the ingress of pitch canker (Lennox et al. 2009). Proposals to clear eucalypts invading along rivers brought a storm of protest from bee-keepers, whose charges are highly dependent on eucalypts. Bee-keepers pointed to the value of bees as pollinators of deciduous fruit orchards, and to the potential drop in annual fruit production without ample bee pollination. Mesquite trees (*Prosopis* species) that provide fodder and firewood, but impact negatively on groundwater and pastures, are hotly debated by farmers. The listing of jacarandas (*Jacaranda mimosaefolia*, the iconic street tree of South Africa's capital city) as a declared invader led to fears that they would be removed, a possibility regarded by many with disbelief, leading to negative publicity for Working for Water. In similar vein, the citizens of Cape Town often protest vigorously when conservation authorities clear pines from the slopes of Table Mountain. These protests persist despite studies that clearly demonstrate that, overall, the costs of invasive trees far exceed any benefits they may bring (see, for example, De Wit et al. 2003), and the resultant opposition seriously retards progress.

Invasive alien species other than plants

Working for Water has until now only addressed the problem of invasive alien plants. There is clear a need to expand its scope to address other taxa effectively. In South Africa, freshwater ecosystems are particularly vulnerable to invasion, especially by alien fish. In addition, very little is known of the impacts and potential for control of alien invertebrates and disease organisms that could affect human health (van Wilgen & Richardson 2010).

Monitoring progress

Whether control attempts are having substantial impacts on alien plant invasions is an important issue. The Working for Water programme was initially put forward as a 20-year activity (van Wilgen et al. 1998), but clearing major infestations within that timeframe will not be possible. Marais et al., (2004) estimate that, at the rates of clearing prevailing at the time, infestations of several important species would only be cleared within 30–85 years. They warned, however, that these estimates were unrealistic and that, at prevailing rates of management, the problem will not be contained. These predictions are proving correct. For example, in the case of pines, Working for Water's

clearing records indicate that a greater area than was estimated to be under pines in 1995 had already been cleared by 2009 (74,519 hectares cleared compared with an estimate of 65,000 hectares covered in 1996 (Le Maitre et al. 2000)); yet invasive pines still dominate much of the landscape. Either the original estimate was far too low, or pines are spreading faster than they can be cleared. Either way, it illustrates the difficulties associated with assessing progress. The same problem arises with demonstrating benefits; most estimates of benefits are based on models rather than actual field monitoring (see, for example, Le Maitre et al. 2002; van Wilgen et al. 2008).

Stabilizing institutional arrangements

Working for Water, as a poverty-relief programme, lacks an explicit mandate and an independent institutional identity. Working for Water addresses an issue that cuts across the interests of many government departments. Initially, the programme was given the scope to develop on the fringes of government bureaucracy, enabling it to initiate many activities. Predictably, bureaucracy returned as the new government became more settled. Officials in the Department of Water Affairs took over the responsibility for management of the programme, and were often reluctant to implement new ideas. This impasse has persisted for some time, and it retards the programme's ability to be truly effective.

28.8 CONCLUSIONS

Working for Water has been shaped by a range of environmental, historical and social factors (Table 28.2). Natural forests in South Africa cover less than 1% of the land surface, and this drove early colonists to import and establish many alien trees. As a result, a large proportion of South Africa's alien flora are trees, and, because of their water use, the programme has a significant focus on woody invasive alien plants. Invasive alien plants that are spread by fire, particularly those with long-range dispersal abilities (such as pines and hakeas), have invaded large areas in inaccessible and rugged mountains, where they remain beyond the reach of most clearing teams. The major focus of the programme has been on relatively accessible lowland areas, where teams clear tress mainly

Table 28.2 Features and outcomes associated with selected environmental, historical and social factors that have shaped the Working for Water programme.

Type of shaping factor	Feature	Outcome
Environmental	Largely treeless landscape.	Widespread introduction of alien trees for a range of purposes, leading to invasion by many species.
	Overall water scarcity, with large proportion of arid areas.	Strong motivation for clearing plants that use excessive amounts of water.
	Fire-prone ecosystems.	Selection for fire-adapted invasive trees and shrubs.
	A wide diversity of landscapes, and a high proportion of remote and inaccessible terrain.	A wide variety of invasive species manage to establish.
		Invasions in remote areas remain largely uncontrolled.
Historical	Long (350 years) period of colonization.	Many species have had centuries to establish and become invasive, especially in the south.
	Relatively high historical investment in research.	High level of understanding of the problem.
		Capacity to develop motivation for programme establishment and management solutions.
		High levels of 'boundary management' (*sensu* Kueffer & Hirsch Hadorn 2008), in which researchers from different fields have collaborated across traditional boundaries.
Social	Unique set of political circumstances.	Willingness to accept change and adopt new approaches.
	High levels of poverty and inequality.	Strong motivation to establish labour-intensive clearing approaches, with insufficient effort spent on prevention and detection.
		Additional considerations for prioritization of clearing operations.

along riparian areas. Working for Water has established specially trained 'high altitude teams', whose task it is to clear isolated outbreaks of invading trees, but with limited success. Invasive pines seem to be increasing, notably in the rugged areas, and this remains a major challenge.

A historical investment in research capacity by South Africa's pre-democracy government, although it excluded the country's black majority, did create the capacity to raise awareness of the problem of invasions, and to support the processes of establishing Working for Water. Kueffer and Hirsch Hadorn (2008) discuss the importance of 'boundary management' in research aimed at creating effective solutions to environmental problems. Boundary management is a process in which experts and stakeholders clarify

appropriate approaches to problems through active deliberation, and there are many examples of where this has helped to shape Working for Water. The initial research that developed the rationale for the programme required collaboration across traditional research boundaries such as forest hydrology, applied ecology, water systems engineering and resource economics (see papers in a special issue of the *South African Journal of Science* for a review (van Wilgen 2004)). Managers and researchers also collaborated actively in the design of research initiatives (for example biological control (see 'Promoting the effective use of biological control' in section 28.5) and prioritization exercises (see 'Developing a hierarchical system of prioritization' in section 28.5)), whereas researchers often strive to address topics of practical importance, such as the impacts of invasive alien plants on riparian areas (see Esler et al. 2008).

Finally, it was the unique set of political circumstances, and a widespread willingness to change, that had the largest effect on shaping Working for Water. The pressing need for poverty relief and employment creation added an unusual set of criteria for the establishment of clearing projects, and arguably drew attention away from other important aspects of invasion management, such as prevention and early detection.

The factors that shaped Working for Water have resulted in several strengths and weaknesses. The unusual levels of political support (both in terms of funding and the revision of legislation), and the levels of boundary management achieved by collaborating managers, researchers and other stakeholders, are definite strengths. The weaknesses include an ineffective institutional model, a disproportionate focus on plants and an inability to deal with invasions in inaccessible areas.

Whether the Working for Water model can be replicated elsewhere depends on several factors. The special circumstances that allowed the programme to be established are unlikely to arise easily elsewhere, and the focus on poverty relief would probably not be universally attractive. However, the use of payments for ecosystem services (see 'Developing systems of payment for ecosystem services' in section 28.5) meshes well with a wider interest in using instruments of that kind, and could well be pursued elsewhere. In addition, Working for Water's positive experience in involving stakeholders across traditional divides suggests that the promotion of similar approaches elsewhere may be advantageous.

ACKNOWLEDGEMENTS

We thank Guy Preston, Dick Mack, Charles Perrings and Dave Richardson for valuable reviews of an earlier draft.

REFERENCES

Blignaut, J.N., Marais, C. & Turpie, J.K. (2007) Determining a charge for the clearing of invasive alien plant species (IAPs) to augment water supply in South Africa. *Water SA*, **33**, 27–34.

Clout, M. & Williams, P.A. (2009) *Invasive Species Management: A Handbook of Principles and Techniques*. Oxford University Press, Oxford.

De Lange, W.J. & van Wilgen, B.W. (2011) An economic assessment of the contribution of weed biological control to the management of invasive alien plants and to the protection of ecosystem services in South Africa. *Biological Invasions*, doi:10.1007/s10530-010-9811-y (in press).

De Wit, M.P., Crookes, D.J. & van Wilgen, B.W. (2003) Conflicts of interest in environmental management: estimating the costs and benefits of black wattle (*Acacia mearnsii*) in South Africa. *Biological Invasions*, **3**, 167–178.

Dye, P. & Jarmain, C. (2004) Water use by black wattle (*Acacia mearnsii*): implications for the link between removal of invading trees and catchment streamflow response. *South African Journal of Science*, **100**, 40–44.

Esler, K.J., Holmes, P.M., Richardson, D.M. & Witkowski, E.T.F. (2008) Riparian vegetation management in landscapes invaded by alien plants: insights from South Africa. *South African Journal of Botany*, **74**, 397–400.

Fenn, J.A. (1980) Control of hakea in the western Cape. In *Proceedings of the Third National Weeds Conference of South Africa* (ed. S. Neser and A.L.P. Cairns), pp. 167–173. Balkema, Cape Town.

Henderson, L. (2001) *Alien Weeds and Invasive Plants: A Complete Guide to Declared Weeds and Invaders in South Africa*. Plant Protection Research Institute, Pretoria.

Higgins, S.I., Azorin, E.J., Cowling, R.M. & Morris, M.H. (1997) A dynamic ecological–economic model as a tool for conflict resolution in an invasive alien-plant, biological control and native-plant scenario. *Ecological Economics*, **22**, 141–154.

Hobbs, R.J. (2004) The Working for Water programme in South Africa: the science behind the success. *Diversity and Distributions*, **10**, 501–503.

Hosking, S.G. & du Preez, M. (1999) A cost benefit analysis of removing alien trees in the Tsitsikamma mountain catchment. *South African Journal of Science*, **95**, 442–448.

Kueffer, C. & Hirsch Hadorn, G. (2008) How to achieve effectiveness in problem-oriented landscape research: the example of research on biotic invasions. *Living Review of Landscape Research*, **2**, http://www.livingreviews.org/lrlr-2008-2.

Le Maitre, D.C., van Wilgen, B.W., Chapman, R.A. & McKelly, D. (1996) Invasive plants and water resources in the Western Cape Province, South Africa: modelling the consequences of a lack of management. *Journal of Applied Ecology*, **33**, 161–172.

Le Maitre, D.C., Versfeld, D.B. & Chapman, R.A. (2000) The impact of invading alien plants on surface water resources in South Africa: a preliminary assessment. *Water SA*, **26**, 397–408.

Le Maitre, D.C., van Wilgen, B.W., Gelderblom, C.M., Bailey, C., Chapman, R.A. & Nel, J.A. (2002) Invasive alien trees and water resources in South Africa: case studies of the costs and benefits of management. *Forest Ecology and Management*, **160**, 143–159.

Lennox, C.L., Hoffmann, J.H., Coutinho, T.A. & Roques, A. (2009) A threat of exacerbating the spread of pitch canker precludes further consideration of a cone weevil, *Pissodes validirostris*, for biological control of invasive pines in South Africa. *Biological Control*, **50**, 179–184.

Le Roux, J.J., Geerts, S., Ivey, P., et al. (2010) Molecular systematics and ecology of invasive kangaroo paws in South Africa: management implications for a horticulturally important genus. *Biological Invasions* (in press).

Macdonald, I.A.W., Kruger, F.J. & Ferrar, A.A. (eds) (1986) *The Ecology and Control of Biological Invasions in South Africa*. Oxford University Press, Cape Town.

Macdonald, I.A.W., Clark, D.L. & Taylor, H.C. (1989) The history and effects of alien plant control in the Cape of Good Hope Nature Reserve, 1941–1947. *South African Journal of Botany*, **55**, 56–75.

Magadlela, D. & Mdzeke, N. (2004) Social benefits in the Working for Water programme as a public works initiative. *South African Journal of Science*, **100**, 94–96.

Marais, C., van Wilgen, B.W. & Stevens, D. (2004) The clearing of invasive alien plants in South Africa: a preliminary assessment of costs and progress. *South African Journal of Science*, **100**, 97–103.

Mark, A.F. & Dickinson, K.J.M. (2008) Maximizing water yield with indigenous non-forest vegetation: a New Zealand perspective. *Frontiers in Ecology and the Environment*, **6**, 25–34.

McNeely, J.A., Mooney, H.A., Neville, L.E., Schei, P. & Waage, J.K. (2001) *A Global Strategy on Invasive Alien Species*. IUCN, Gland, Switzerland.

Moran, V.C., Hoffmann, J.H., Donnelly, D., Zimmermann, H.G. & van Wilgen, B.W. (2000) Biological control of alien invasive pine trees (*Pinus* species) in South Africa. In *Proceedings of the Xth International Symposium on Biological Control of Weeds* (ed by Neal R. Spencer), pp. 941–953. Montana State University, Bozeman.

Moran, V.C., Hoffmann, J.H. & Zimmermann, H.G. (2005) Biological control of invasive alien plants in South Africa: necessity, circumspection, and success. *Frontiers in Ecology and the Environment*, **3**, 77–83.

Olckers, T. & Hill, M.P. (1999) *Biological Control of Weeds in South Africa*. (*African Entomology* Memoir No. 1.) Entomological Society of Southern Africa.

Olckers, T. (2004) Targeting emerging weeds for biological control in South Africa: the benefits of halting the spread of alien plants at an early stage of their invasion. *South African Journal of Science*, **100**, 64–68.

Pejchar, L. & Mooney, H. (2010) The impact of invasive alien species on ecosystem services and human well-being. In *Bioinvasions and Globalization: Ecology, Economics, Management and Policy* (ed. C. Perrings, H.A. Mooney and M. Williamson), pp. 161–182. Oxford University Press, Oxford.

Perrings, C., Mooney, H. & Williamson, M. (2010) The problem of biological invasions. In *Bioinvasions and Globalization: Ecology, Economics, Management and Policy* (ed. C. Perrings, H.A. Mooney and M. Williamson), pp. 1–16. Oxford University Press, Oxford.

Prinsloo, F.W. & Scott, D.F. (1999) Streamflow responses to the clearing of alien invasive trees from riparian zones at three sites in the Western Cape Province. *South African Forestry Journal*, **185**, 1–7.

Pyšek, P. & Richardson, D.M. (2010) Invasive species, environmental change and management, and health. *Annual Review of Environment and Resources*, **35**, doi:10.1146/annurev-environ-033009-095548 (in press).

Richardson, D.M., Cambray, J.A., Chapman, R.A., et al. (2003) Vectors and pathways of biological invasions in South Africa – past, future and present. In *Invasive Species: Vectors and Management Strategies* (ed. G. Ruiz and J.T. Carlton), pp. 292–349. Island Press, Washington, DC.

Saaty, T.L. (1990) How to make a decision: the analytic hierarchy process. *European Journal of Operational Research*, **48**, 9–26.

Stirton, C.H. (1978) *Plant Invaders: Beautiful, but Dangerous*. Department of Nature and Environmental Conservation, Cape Town.

Turpie, J. (2004) The role of resource economics in the control of invasive alien plants in South Africa. *South African Journal of Science*, **100**, 87–93.

Turpie, J.K. & Heydendrych, B.J. (2000) Economic consequences of alien infestation of the Cape Floral Kingdom's Fynbos vegetation. In *The Economics of Biological Invasions* (ed. C. Perrings, C.M. Williamson and S. Dalmazzone), pp. 152–182. Edward Elgar, Cheltenham, UK.

Turpie, J.K., Heydenrych, B.J. & Lamberth, S.J. (2003) Economic value of terrestrial and marine biodiversity in the Cape Floristic Region: implications for defining effective and socially optimal conservation strategies. *Biological Conservation*, **112**, 233–251.

Turpie, J.K., Marais, C. & Blignaut, J.N. (2008) The Working for Water programme: evolution of a payments for ecosystem services mechanism that addresses both poverty and ecosystem service delivery in South Africa. *Ecological Economics*, **65**, 788–798.

van Wilgen, B.W. (2004) Scientific challenges in the field of invasive alien plant management. *South African Journal of Science*, **100**, 19–20.

van Wilgen, B.W. (2009) The evolution of fire and invasive alien plant management practices in fynbos. *South African Journal of Science*, **105**, 335–342.

van Wilgen, B.W., Cowling, R.M. & Burgers, C.J. (1996) Valuation of ecosystem services: a case study from the fynbos, South Africa. *BioScience*, **46**, 184–189.

van Wilgen, B.W., de Wit, M.P., Anderson, H.J., et al. (2004) Costs and benefits of biological control of invasive alien plants: case studies from South Africa. *South African Journal of Science*, **100**, 113–122.

van Wilgen, B.W., Everson, C.S. & Trollope, W.S.W. (1990) Fire management in southern Africa: some examples of current objectives, practices and problems. In *Fire in the Tropical Biota: Ecosystem Processes and Global Challenges* (ed. J.G. Goldammer), pp. 179–209. Springer, Berlin.

van Wilgen, B.W., Le Maitre, D.C. & Cowling R.M. (1998) Ecosystem services, efficiency, sustainability and equity: South Africa's Working for Water programme. *Trends in Ecology & Evolution*, **13**, 378.

van Wilgen, B.W., Le Maitre, D.C., Forsyth, G.G. & O'Farrell, P. (2010) *The Prioritization of Terrestrial Biomes for Invasive Alien Plant Control in South Africa*. Report CSIR/NRE/ECO/ER/2010/0004/B, CSIR, Stellenbosch.

van Wilgen, B.W., Little, P.R., Chapman, R.A., Görgens, A.H.M., Willems, T. & Marais, C. (1997) The sustainable development of water resources: history, financial costs and benefits of alien plant control programmes. *South African Journal of Science*, **93**, 404–411.

van Wilgen, B.W., Marais, C., Magadlela, D., Jezile, N. & Stevens, D. (2002) Win–win–win: South Africa's Working for Water programme. In *Mainstreaming Biodiversity in Development: Case Studies from South Africa* (ed. S.M. Pierce, R.M. Cowling, T. Sandwith and K. MacKinnon), pp. 5–20. World Bank, Washington, DC.

van Wilgen, B.W., Reyers, B., Le Maitre, D.C., Richardson, D.M. & Schonegevel, L. (2008) A biome-scale assessment of the impact of invasive alien plants on ecosystem services in South Africa. *Journal of Environmental Management*, **89**, 336–349.

van Wilgen, B.W. & Richardson, D.M. (2010) Current and future consequences of invasion by alien species: a case study from South Africa. In *Bioinvasions and Globalization* (ed. C. Perrings, H.A. Mooney and M. Williamson), pp. 183–201. Oxford University Press, Oxford.

Wicht, C.L. (1945) *Report of the Committee on the Preservation of the Vegetation of the South Western Cape*. Royal Society of South Africa, Cape Town.

Zenni, R.D., Wilson, J.R.U., LeRoux, J.J. & Richardson, D.M. (2009) Evaluating the invasiveness of *Acacia* paradoxa in South Africa. *South African Journal of Botany*, **75**, 485–496.

Zimmermann, H.G., Moran, V.C. & Hoffmann, J.H. (2004) Biological control of invasive alien plants in South Africa, and the role of the Working for Water programme. *South African Journal of Science*, **100**, 34–40.

Part 7

Conclusions

Chapter 29

INVASION SCIENCE: THE ROADS TRAVELLED AND THE ROADS AHEAD

David M. Richardson

Centre for Invasion Biology, Department of Botany & Zoology, Stellenbosch University, 7602 Matieland, South Africa

Fifty Years of Invasion Ecology: The Legacy of Charles Elton, 1st edition. Edited by David M. Richardson

29.1 INTRODUCTION

Charles Elton (1958) has been acknowledged for pulling together insights and concepts from previously disparate fields of enquiry to sow the seeds for the emergence of 'invasion ecology' as a discrete field (Richardson & Pyšek 2007, 2008). After a lag phase and rather hesitant beginnings, the systematic investigation of invasions has exploded as a field of study, and hundreds of publications appear every year on an increasingly broad range of themes (Ricciardi & MacIsaac 2008; Richardson & Pyšek 2008, MacIsaac et al., this volume; Simberloff, this volume). The chapters in this book have reviewed the very substantial progress made in some of the disciplines mentioned by Elton, the emergence of many new avenues of research, increasing synergies between disciplines, and some of the challenges that face researchers working on various aspects of biological invasions.

This chapter considers the dimensions of research that is currently being conducted on non-native species and the phenomenon of biological invasions, and discusses some priorities for the future. To guide the assessment of current research foci, I assembled a large sample of research outputs published in the year (2008) in which the 50th anniversary of the publication of Elton's (1958) book was commemorated. My sample comprises 500 papers covering the full spectrum of research on non-native species and issues pertaining to them (see Appendix 29.1 for details). I used this selection of outputs to provide a snapshot of the current research agenda in 'invasion science' (see Fig. 29.1 for elucidation of the term). My framework for discussing the dimensions of research in this area follows that proposed by Kueffer and Hirsch Hadorn (2008). They suggest that research efforts in a problem-orientated field such as invasion science typically cluster around three focal areas: systems knowledge (the analysis of casual relationships), target knowledge (essentially clarification of conflicts of interest and values) and transformation knowledge (the quest for appropriate actions for management). They also suggest that another feature of problem-orientated research involves deliberations between experts and stakeholders about the framing of appropriate research questions about processes, values and practices for effective problem solving; they term this type of research 'boundary management' (Fig. 29.1). My treatment builds on their valuable analysis, and seeks further clarification and a preliminary quantification of the key research thrusts within these areas with reference to the sample of papers from 2008 and the conclusions provided in the chapters of this book. I also suggest key areas where further research is needed.

29.2 THE STUDY OF BIOLOGICAL INVASIONS 50 YEARS AFTER ELTON

Examination of the 500 papers published in 2008 suggests that research related to biological invasions is dominated by studies that fall within the 'systems knowledge' category of Kueffer and Hirsch Hadorn (2008); 74% of published papers were thus classified (some papers fitted into more than one category). This research explores the 'nuts and bolts' of biological invasions, focusing largely on research in the fields of community ecology (including reproduction and pollination biology: 119 papers), biogeography/macroecology (30), population biology (28), evolutionary biology (26) and molecular ecology (20). These are very broad classifications, and some papers were difficult to place in such traditional groupings. Nonetheless, I suggest that this provides a reasonable cross-section of the global research agenda on fundamental issues that mediate invasions. The main issues that were dealt with in 'systems' papers (as captured in the primary keywords assigned) were impacts (49), invasibility (37), invasiveness (23) and introduction/invasion history (21). Others with 10 or more papers were dispersal ecology, pollination and pathways. Those with five or more entries were competition, distribution, drivers of invasion, homogenization, hybridization, inventory, phylogeography and spread.

Papers categorized as aiming to produce 'target knowledge' comprised 14% of the sample and were mostly grouped in the fields of community ecology (30), biogeography (9), risk analysis (9), ecosystem ecology (4) and resource economics (4). Together, studies in these fields account for 78% of the 'target knowledge' sample. Almost half the studies in this category dealt with the assessment of impacts, five papers dealt with issues relating to the debate on the native/alien status of taxa, whereas the rest dealt with a wide range of management-orientated issues. Papers aiming to produce 'transformation knowledge' were grouped as follows: 'management science' (24), restoration ecology (13), risk analysis (8), policy development (7), and vector science (5); together these groups make up 69% of the 83 papers in this category. The best repre-

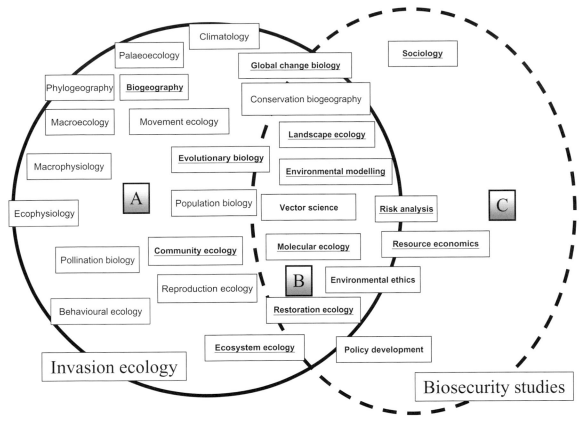

Fig. 29.1 Fields of research on issues relating to biological invasions, as shown in an analysis of 500 papers published in 2008 (Appendix 29.1) (main fields are underlined). Research in fields at the left of the diagram (zone A) are largely those that produce 'systems knowledge'; those closer to and within zone B, 'target knowledge'; and those within zone B and into zone C, largely 'transformation knowledge'. 'Boundary management' occurs towards the right of the diagram (see text for details). Zones A, B and C together define the domain of 'invasion science'.

sented primary keywords were restoration (12), management (9), climate change (6) and pathways (5). Although many papers mentioned the potential benefits of 'boundary management', involving cross-sectoral collaboration or transdisciplinarity, only nine papers from the sample of 500 (1.8%) provide any concrete contributions in this area. These papers cover a wide of disciplines and key issues: biogeography/macroecology, conservation biogeography, horizon scanning, policy development, resource economics and risk analysis.

The groups 'systems-', 'target-' and 'transformational knowledge' and 'boundary management' as discussed by Kueffer and Hirsch Hadorn (2008) provide a convenient framework for classifying the very diverse research activity on various aspects of biological invasions. It is clear that research on the 'nuts and bolts' of invasions is flourishing and that this work is infiltrating every conceivable area of biology and ecology. This work is using a range of approaches, from biogeographical/macroecological assessments using natural experiments to increasingly ambitious, innovative and sophisticated manipulative experiments at smaller spatial and temporal scales. Zone A in Fig. 29.1 shows the main broad fields in which this research is taking place (although many studies straddle these fields). All the main areas of biogeography, ecology and evolutionary biology are represented, and new spin-off areas of specialization such as macrophysiology (Chown & Gaston 2008) and movement ecology (Nathan et al. 2008) are increasing in prominence as important fields of research that provide new lenses through which to view biological invasions. Increasingly, insights from biogeographical and ecological research are also being applied to addressing management issues (zone B in Fig. 29.1). Relatively well established disciplines such as global change biology (see, for example, Van der Wal et al. 2008), landscape ecology (see, for example, Gillson et al. 2008) and restoration ecology (see, for example, Shafroth & Briggs 2008) are becoming increasingly infiltrated by studies and perspectives relating to biological invasions. Environmental modelling (including species distribution modelling (Pyron et al. 2008)) and molecular ecology (see, for example, Dlugosch & Parker 2008; Suarez & Tsutsui 2008) are fertile disciplines for research on biological invasions, and invasion-related work is driving new developments in these fields. New disciplines such as conservation biogeography have emerged recently in an attempt to bridge the gap between the science of biogeography and the need for practical information to guide the management of biodiversity; increasingly, this includes the need to manage biological invasions (Richardson & Whittaker 2010). As discussed by Hobbs and Richardson (this volume), invasion ecology and restoration ecology are becoming increasingly intermeshed, as invasions become more pervasive and therefore increasingly important drivers of degradation. The study of the vectors and pathways of movement of species around the world ('vector science' *sensu* Carlton & Ruiz (2005)) has grown considerably in scope and sophistication (Areal et al. 2008; Barry et al. 2008; Carrete & Tella 2008; Fernandez 2008; Hulme et al. 2008). The words 'vector' and 'pathway' appeared in 27% and 24% of the 500 sample papers from 2008, respectively. The notion of propagule pressure as a fundamental driver of invasions, and the obvious implications of this for management and policy has exploded in the research agenda of invasion science; the term appears in 22% of the 500 sample papers. Research in zones A and B falls within the realms of 'invasion ecology' as traditionally defined. Although it still forms a relatively small part of the overall research tied to biological invasions, there is increasing interest and a growing research agenda in zone C of Fig. 29.1: areas where insights from biology and ecology are relevant and important, but where socio-political dimensions are generally of overriding

importance. Such research initiatives, although sometimes lumped within the ambit 'invasion ecology', are diverging from, and are increasingly clustered together outside, the domain of this field (Fig. 29.1). The field of biosecurity is a very rapidly growing area, where synergies and linkages are being formed between disciplines that have traditionally been underrepresented in published work on issues relating to invasions (Brasier 2008; Hulme, this volume).

29.3   THE STUDY OF BIOLOGICAL INVASIONS: INTO THE FUTURE

When deliberating priorities for research on invasions in the future, one must consider scenarios for the changing context of biological invasion within the larger set of environment issues. First, there is no sign of the rate of invasions slowing down, despite several decades of efforts in some regions to prevent, contain and otherwise manage alien species. Even if prevention measures were to become dramatically more successful soon, the large number of alien species already present in different regions, many still in the lag phase and yet to spread widely, represents a major 'invasion debt' that will drive invasions for decades and centuries to come. Invasions will become ever more pervasive, and very few ecosystems will be spared impacts from invasive species. For this reason alone, researchers working on any aspect of biodiversity or conservation biology will increasingly be confronted by non-native species. Invasion ecology will therefore become increasingly intermeshed with all other sub-disciplines of ecology, biogeography and environmental management. Secondly, interactions between invasions and other environmental issues will undoubtedly become increasingly complex. Invasive species are already a crucial ingredient of the 'lethal cocktail' of factors driving many forms of environmental degradation, and are sure to become implicated in others in the future as non-native species infiltrate more networks. This will demand departures, sometimes radical, from current approaches and philosophies for environmental management. Climate change is certain to change the dynamics of invasions profoundly (Thuiller et al. 2007; Bradley et al. 2010; Dukes, this volume). Conservation biologists, restoration ecologists and others will, in some cases, need to revise purist notions and consider more pragmatic ways of dealing with increasingly ubiquitous alien species. The increasing interest in 'novel ecosystems' (Hobbs et al. 2006) and the intensifying debate on new conservation approaches such as managed relocation (Richardson et al. 2009) heralds the dawning of a new era in environmental management and conservation. With these points in mind, I now discuss some issues and themes that I believe will feature prominently in research agendas for invasion science in the future.

Macroecological perspectives: rocking in the real world

Increasingly sophisticated and far-reaching big-picture analyses, using recently assembled global, national and regional databases and modern approaches from biogeography and macroecology are starting to emerge. Results from certain regions (see, for example, Pyšek & Hulme, this volume and references therein for insights from Europe) and for certain groups, for example pests and pathogens of forest trees (Wingfield et al., this volume) and mycorrhizal fungi (Vellinga et al. 2009), are yielding crucial new insights, generating new research questions and providing key information for improved management and policy formulation. The extent to which alien species and alien-dominated assemblages conform with emerging 'ecogeographical rules' (Gaston et al. 2008) is a fascinating question that remains to be properly addressed.

Besides the opportunities for gaining fundamental insights on invasion ecology from macroecological studies, there is also much potential, largely untapped, for using macroecological approaches to inform management strategies. Combined perspectives of invasions resulting from forestry operations over several centuries in Australia, New Zealand and South Africa have been used to formulate objective scenarios and to predict the performance of different taxa and cultivation strategies in South America (which has a much shorter history of major plantings and where invasions are much less widespread) (Richardson et al. 2008; Simberloff et al. 2010).

Bird watchers use the term 'rocking' when they zoom in and out gently by tweaking the focus of their binoculars to gain an improved perspective of difficult-to-identify birds. I suggest that invasion science would profit by applying more 'rocking' by seeking perspectives of a wide range of phenomena relating to invasions by examining these at multiple spatial and temporal scales.

Networks as a fundamental framework for invasion ecology

The conceptualization of linkages between species, including alien species, in the form of networks has enjoyed much research attention in the past decade in particular, especially with reference to pollination biology, seed dispersal and food webs (Tylianakis 2008). This research is shedding new light on how introduced species can be integrated into ecosystems, and how this can generate and regulate impacts. Importantly, such frameworks are also yielding information on options for management, for example for elucidating, predicting and potentially preventing or mitigating trophic cascades and secondary invasions in the wake of management interventions (Pyšek & Richardson 2010). This is a fertile area for further research (Cumming et al. 2010; Traveset & Richardson, this volume).

Adding more evolutionary spice to the invasion science stew

Exciting advances are being made in the understanding of evolutionary processes during invasions (Barrett et al. 2008; Dlugosh & Parker 2008; Suarez & Tsutsui 2008; Whitney & Gabler 2008; Dormontt et al., this volume). Molecular ecology is revolutionizing insights on introduction and invasion histories (Darling et al. 2008; Facon et al. 2008; Hufbauer & Sforza 2008). These developments are radically improving our knowledge of the mechanisms of invasion and will enhance our ability to predict the outcome of introductions and manage invasions (Le Roux & Wieczorek 2008). This is an area of intense research effort at present, and results from this work are sure to enrich the theoretical foundation of invasion ecology, advancing our understanding of invasiveness, invasibility and other dimensions of invasion ecology.

Drivers of invasion: towards the effective linking of propagule pressure and other factors in models

Several years ago I wrote that 'Gaining a more robust and practical understanding of propagule pressure is probably the biggest challenge facing invasion ecologists. Understanding propagule pressure is the new frontier in invasion ecology' (Richardson 2004). This has been borne out by the large output of research on diverse aspects relating to propagule pressure in the ensuing 6 years. I agree with Colautti et al. (2006), who called for propagule pressure to be considered a null model when inferring process from patterns of invasion. Propagule pressure is now well accepted as a fundamental driver of invasions; its importance has been verified in experimental studies (see, for example, Britton-Simmons & Abbott 2008) and it is starting to be incorporated into synthetic models (Barney & Whitlow 2008; Catford et al. 2009; Davis 2009). Efforts are also being made to incorporate insights on its driving role into improved management strategies (see, for example, Funk et al. 2008). Further experimental verification of the role of propagule pressure as a component driver of invasions of different taxa and in different systems, and the incorporation of such insights into different types of models, is sure to remain a crucial focus of research in invasion science. Linking such insights with emerging perspectives on the roles of facilitation and inhibition (Brooker et al. 2008; Griffen et al. 2008) is an obvious area where invasion ecologists and community ecologists in general could forge productive partnerships.

Changing climate, changing directions and priorities for invasion science

As discussed above, priorities for invasion science must be considered within the larger picture of changing environmental concerns. Among the facets of global change, rapid climate change probably has the most potential to influence research direction in invasion science. Given the limited knowledge of exactly how climate affects range limits of species in different taxonomic groups (see, for example, Thomas 2010) and how changes could influence community composition and invasibility, the research challenges in this area are daunting. Given the multiple linkages and complex feedback and feed-forward loops implicated, innovative deductions from natural experiments are required to elucidate the nature and potential magnitude of effects that are likely or possible. Manipulative experiments on carefully selected taxa and systems are also needed to shed light on crucial factors to fine-tune our understanding and our ability to model invasions

(Thuiller et al. 2007). Hellmann et al. (2008) proposed five crucial consequences of climate change for biological invasions: (i) altered mechanisms of transport and introduction; (ii) altered climatic constraints on invasive species; (iii) altered distribution of existing invasive species; (iv) altered impacts of existing invasive species; and (v) altered effectiveness of management strategies. A key challenge is to define and initiate management strategies that are sufficiently flexible to allow for the rapid integration of new research results from disparate fields.

Vectors and pathways: understanding the dynamics of human-mediated introductions

The definition and elucidation of the dynamics of introduction pathways for biological invasions has received considerable attention in the past decade. This work has focused on developing general models (Hulme et al. 2008; Wilson et al. 2009 and references therein), and on exploring links between trade dynamics and pathways, in general terms (Hulme 2009), with special reference to transport hubs (Floerl et al. 2009), and for specific taxa, for example plants (see Thuiller et al. 2005) and spiders (see Kobelt & Nentwig 2008). An important development has been the clarification of the dimensions and dynamics of pathways for those sectors of human endeavour known to be crucial drivers of species introductions. Sectors for which important advances have been made in this regard include forestry (Wilson et al. 2009; Richardson 2011) and timber importation (Piel et al. 2008; Skarpaas & Økland 2009), the pet trade (Carette & Tella 2008; Van Wilgen et al. 2010), the cut flower trade (Areal et al. 2008) and shipping, in general terms (Fernandez 2008), as well as for different facets thereof, notably introductions through ballast water (Barry et al. 2008). The emerging biofuel industry, and the potential of the introduction and cultivation of alien feedstock species for generating invasions, is an important and growing area for research (Raghu et al. 2006). Many of the sector-focused assessments of vectors and pathways have been formulated within a formal risk-assessment framework. This is a crucial area where much additional work is needed to provide objective guidelines to inform policy and legislation.

Identification, detection and mapping: taking the technology train

Rapid technological advances are opening new doors for the identification of alien species, the detection of new incursions, and for mapping the extent of invasions. Cross et al. (this volume) reviewed the emergence of DNA barcoding and the many applications of this technology in invasion science. Other high-tech diagnostic tools have been developed for detecting even small numbers of micro-organisms (reviewed in Pyšek & Richardson 2010). Radical advances in remote sensing technology (see Vitousek et al., this volume) allow for the assessment of the extent of many types of invasive species, and increasingly for the quantification of impacts. Such tools have great value for setting priorities for management, for monitoring the success of management actions and for gaining insights on processes that have not been quantifiable over large spatial scales. Together, technological advances such as those discussed above promise to improve the accuracy and completeness of alien species inventories, especially for micro-organisms. Linked with advances in modelling techniques (Hui et al., this volume), there are endless opportunities for harnessing these technologies to improve our ability to manage invasive species. There are countless research opportunities and priorities in this area.

Impacts: towards objective criteria and a common currency

An urgent priority for invasion science is to develop and fine-tune objective and practical methods for quantifying the full range of effects of invasive species on recipient ecosystems. Research is needed to quantify impacts at multiple scales and multiple levels of organization, and to synthesize available data on different response variables. Richardson et al. (this volume) discuss the multiple facets of impact and allude to some of the research challenges on that front.

Linking research results with management: an overarching theme

For every area discussed above, frameworks need to be put in place to allow for research results to be used in management. Environmental management interventions are increasingly taking place within a risk-analysis framework. Many national and international policies and legal instruments now call for formal risk assessments for many decisions relating to the importation and use of alien species. There has been a rapid growth in studies of risk assessment and management for biosecurity. These studies include pre-border assessments to provide objective criteria for screening in a wide range of situations, sector-specific analyses (see above), as well as risks of post-border dissemination and spread and the generation of impacts (see Pyšek & Richardson 2010 for a recent review). Research is needed to provide the means for objectively assigning risk to specific stages and phases in the naturalization–invasion continuum and to different activities associated with introduction pathways that potentially determine invasiveness or invasibility.

Effective linking of research results with management will require a major investment in what Kueffer and Hirsch Hadorn (2008) term 'boundary management', involving deliberations among experts and various stakeholders on the framing of adequate research questions about processes, values and practices for effective problem-solving. Van Wilgen et al. (this volume) describe perspectives after 15 years of the well-publicized Working for Water Programme in South Africa, which seeks to merge effective management of invasive species with various socio-political imperatives. Experiments in applied ecology along these lines are urgently needed in other parts of the world if real progress it to be made in dealing with the escalating problems associated with biological invasions.

Philosophical foundations and linkages with other disciplines

Much research remains to be done on the philosophy and history of science with reference to the emergence of fields like invasion ecology, restoration ecology, conservation biology and global change biology, on the linkages and contrasts between these fields, and on the various ways in which these lines of enquiry have sought to bridge the divide between academic/theoretical issues and the provision of objective information to guide management. The role of normative thinking in shaping the research agenda for invasion ecology needs to be thoroughly deconstructed. Research is needed on the anatomy of the expanding network of specialist areas to forge more effective synergies. There was clear evidence among the sample of 500 papers published in 2008 of encouraging activity in this direction. For example, productive links between invasion ecology and the following disciplines were noted: conservation biology (Bradshaw et al. 2008; Jeschke & Strayer 2008), epidemiology (Jones et al. 2008), hydrology (Gurnell et al. 2008), palaeoecology (Gillson et al. 2008), population biology (Liebhold & Tobin 2008; Ramula et al. 2008), resource economics (Keller et al. 2008) and restoration ecology (Funk et al. 2008; see also Hobbs & Richardson, this volume). Given the increasing extent of invasions worldwide and the increasing complexity of interactions with other environmental stressors and the need for objective guidelines for management, the strengthening of links with these and other related disciplines is an urgent priority (see, for example, van Kleunen & Richardson 2007).

ACKNOWLEDGEMENTS

Corlia Richardson provided valuable assistance in the compilation of the database of 500 sample papers from 2008. I acknowledge support from the DST-NRF Centre of Excellence for Invasion Biology and the Hans Sigrist Foundation.

REFERENCES

Areal, F.J., Touzab, J., MacLeod, A., et al. (2008) Integrating drivers influencing the detection of plant pests carried in the international cut flower trade. *Journal of Environmental Management*, **89**, 300–307.

Barrett, S.C.H., Colautti, R.I. & Eckert, C.G. (2008) Plant reproductive systems and evolution during biological invasion. *Molecular Ecology*, **17**, 373–383.

Barney, J.N. & Whitlow, T.H. (2008) A unifying framework for biological invasions: the state factor model. *Biological Invasions*, **10**, 259–272.

Barry, S.C., Hayes, K.R., Hewitt, C.L., Behrens, H.L., Dragsund, E. & Bakke, S.M. (2008) Ballast water risk assessment: principles, processes, and methods. *ICES Journal of Marine Science*, **65**, 121–131.

Bradley, B.A., Blumenthal, D.M., Wilcove, D.S. & Ziska, L.H. (2010) Predicting plant invasions in an era of global change. *Trends in Ecology & Evolution*, **25**, 310–318.

Bradshaw, C.J.A., Giam, X., Tan, H.T.W., Brook, B.W. & Sodhi, N.S. (2008) Threat or invasive status in legumes is related to opposite extremes of the same ecological and life-history attributes. *Journal of Ecology*, **96**, 869–883.

Brasier, C.M. (2008) The biosecurity threat to the UK and global environment from international trade in plants. *Plant Pathology*, **57**, 792–808.

Britton-Simmons, K.H. & Abbott, K.C. (2008) Short- and long-term effects of disturbance and propagule pressure on a biological invasion. *Journal of Ecology*, **96**, 68–77.

Brooker, R.W., Maestre, F.T., Callaway, R.M. et al. (2008) Facilitation in plant communities: the past, the present, and the future. *Journal of Ecology*, **96**, 18–34.

Carlton, J.T. & Ruiz, G.M. (2005) Vector science and integrated vector management on bioinvasion ecology: Conceptual frameworks. *Invasive alien species. A new synthesis* (ed. H.A. Mooney, R.N. Mack, J.A. McNeely, L.E. Neville, P.J. Schei & J.K. Waage), pp. 36–58. Island Press, Washington, DC.

Carrete, M. & Tella, J.L. (2008) Wild-bird trade and exotic invasions: a new link of conservation concern? *Frontiers in Ecology and the Environment*, **6**, 207–211.

Catford, J.A., Jansson, R. & Nilsson, C. (2009) Reducing redundancy in invasion ecology by integrating hypotheses into a single theoretical framework. *Diversity and Distributions*, **15**, 22–40.

Chown, S.L. & Gaston, K.J. (2008) Macrophysiology for a changing world. *Proceedings of the Royal Society B*, **275**, 1469–1478.

Colautti, R.I., Grigorovich, I.A. & MacIsaac, H.J. (2006) Propagule pressure: a null model for biological invasions. *Biological Invasions*, **8**, 1023–1037.

Cumming, G.S., Bodin, O., Ernstson, H. & Elmqvist, T. (2010) Network analysis in conservation biogeography: challenges and opportunities. *Diversity and Distributions*, **16**, 414–425.

Darling, J.A., Bagley, M.J., Roman, J., Teplot, C.K. & Geller, J.B. (2008) Genetic patterns across multiple introductions of the globally invasive crab genus *Carcinus*. *Molecular Ecology*, **17**, 4992–5007.

Davis, M.A. (2009) *Invasion Biology*. Oxford University Press, Oxford.

Dlugosch, K.M. & Parker, I.M. (2008) Founding events in species invasions: genetic variation, adaptive evolution, and the role of multiple introductions. *Molecular Ecology*, **17**, 431–449.

Elton, C.S. (1958) *The Ecology of Invasions by Animals and Plants*. Methuen, London.

Facon, B., Pointier, J.-P., Jarne, P., Sarda, V. & David, P. (2008) High genetic variance in life-history strategies within invasive populations by way of multiple introductions. *Current Biology*, **18**, 363–367.

Fernandez, L. (2008) NAFTA and member country strategies for maritime trade and marine invasive species. *Journal of Environmental Management*, **89**, 308–321.

Floerl, O., Inglis, G.J., Dey, K. & Smith, A. (2009) The importance of transport hubs in stepping-stone invasions. *Journal of Applied Ecology*, **46**, 37–45.

Funk, J.L., Cleland, E.E., Suding, K.N. & Zavaleta, E.S. (2008) Restoration through reassembly: plant traits and invasion resistance. *Trends in Ecology & Evolution*, **23**, 695–703.

Gaston, K.J., Chown, S.L. & Evans, K.L (2008) Ecogeographical rules: elements of a synthesis. *Journal of Biogeography*, **35**, 483–500.

Gillson, L., Ekblom, A., Willis, K.J. & Froyd, C. (2008) Holocene palaeo-invasions: the link between pattern, process and scale in invasion ecology? *Landscape Ecology*, **23**, 757–769.

Griffen, B.D., Guy, T. & Buck, J.C. (2008) Inhibition between invasives: a newly introduced predator moderates the impacts of a previously established invasive predator. *Journal of Animal Ecology*, **77**, 32–40.

Gurnell, A., Thompson, K., Goodson, J. & Moggridge, H. (2008) Propagule deposition along river margins: linking hydrology and ecology. *Journal of Ecology*, **96**, 553–565.

Hellmann, J.J., Byers, J.E., Bierwagen, B.G. & Dukes, J.S. (2008) Five potential consequences of climate change for invasive species. *Conservation Biology*, **22**, 534–543.

Hobbs, R.J., Arico, S., Aronson, J., et al. (2006) Novel ecosystems: theoretical and management aspects of the new ecological world order. *Global Ecology and Biogeography*, **15**, 1–7.

Hufbauer, R.A. & Sforza, R. (2008) Multiple introductions of two invasive *Centaurea* taxa inferred from cpDNA haplotypes. *Diversity and Distributions*, **14**, 252–261.

Hulme, P.E. (2009) Trade, transport and trouble: managing invasive species pathways in an era of globalization. *Journal of Applied Ecology*, **46**, 10–18.

Hulme, P.E., Bacher, S., Kenis, M. et al. (2008) Grasping at the routes of biological invasions: a framework for integrating pathways into policy. *Journal of Applied Ecology*, **45**, 403–414.

Jeschke, J.M. & Strayer, D.L. (2008) Are threat status and invasion success two sides of the same coin? *Ecography*, **31**, 124–130.

Jones, K. E., Patel, N.G., Levy, M.A., Storeygard, A., Balk, D., Gittleman, J.L. & Daszak, P. (2008) Global trends in emerging infectious diseases. *Nature*, **451**, 990–993.

Keller, R.P., Frang, K. & Lodge, D.M. (2008) Preventing the spread of invasive species: Economic benefits of intervention guided by ecological predictions. *Conservation Biology*, **22**, 80–88.

Kobelt, M. & Nentwig, W. (2008) Alien spider introductions to Europe supported by global trade. *Diversity and Distributions*, **14**, 273–280.

Kueffer, C. & Hirsch Hadorn, G. (2008) How to achieve effectiveness in problem-oriented landscape research: The example of research on biotic invasions. *Living Reviews in Landscape Research*, **2**, http://www.livingreviews.org/lrlr-2008-2.

Le Roux, J.J. & Wieczorek, A.M. (2008) Molecular systematics and population genetics of biological invasions: towards a better understanding of invasive species management. *Annals of Applied Biology*, **154**, 1–17.

Liebhold, A.M. & Tobin, P.C. (2008) Population ecology of insect invasions and their management. *Annual Review of Entomology*, **53**, 387–408.

Nathan, R., Getz, W.M., Revilla, E., et al. (2008) A movement ecology paradigm for unifying organismal movement research. *Proceedings of the National Academy of Sciences of the USA*, **105**, 19052–19059.

Piel, F., Gilbert, M., De Cannière, C. & Grégoire, J.-C. (2008) Coniferous round wood imports from Russia and Baltic countries to Belgium. A pathway analysis for assessing risks of exotic pest insect introductions. *Diversity and Distributions*, **14**, 318–328.

Pyron, R.A., Burbrink, F.T. & Guiher, T.J. (2008) Claims of potential expansion throughout the U.S. by invasive python species are contradicted by ecological niche models. *PloS One*, **3** (8), e2931, doi:10.1371/journal.pone.0002931.

Pyšek, P. & Richardson, D.M. (2010) Invasive species, environmental change and management, and ecosystem health. *Annual Review of Environment and Resources*, **35**, doi:10.1146/annurev-environ-033009-095548 (in press).

Pyšek, P., Richardson, D.M., Pergl, J., Jarošík, V., Sixtová, Z. & Weber, E. (2008) Geographical and taxonomical biases in invasion ecology. *Trends in Ecology & Evolution*, **23**, 237–244.

Raghu, S., Anderson, R.C., Daehler, C.C., et al. (2006) Adding biofuels to the invasive species fire? *Science*, **313**, 1742.

Ramula, S., Knight, T.M., Burns, J.H. & Buckley, Y.M. (2008) General guidelines for invasive plant management based on comparative demography of invasive and native plant populations. *Journal of Applied Ecology*, **45**, 1124–1133.

Ricciardi, A. & MacIsaac, H.J. (2008) The book that began invasion ecology. *Nature*, **452**, 34.

Richardson, D.M. (2004) Plant invasion ecology – dispatches from the front line. *Diversity and Distributions*, **10**, 315–319.

Richardson, D.M. (2011) Forestry and agroforestry. In *Encyclopedia of Biological Invasions* (ed. D. Simberloff and M. Rejmánek). University of California Press, Berkeley (in press).

Richardson, D.M., Hellmann, J.J., McLachlan, J. et al. (2009) Multidimensional evaluation of managed relocation. *Proceedings of the National Academy of Sciences of USA*, **106**, 9721–9724.

Richardson, D.M. & Pyšek, P. (2007) Classics in physical geography revisited: Elton, C.S. 1958: The ecology of invasions by animals and plants. Methuen: London. *Progress in Physical Geography*, **31**, 659–666.

Richardson, D.M. & Pyšek, P. (2008) Fifty years of invasion ecology – the legacy of Charles Elton. *Diversity and Distributions*, **14**, 161–168.

Richardson, D.M., van Wilgen, B.W. & Nunez, M. (2008) Alien conifer invasions in South America – short fuse burning? *Biological Invasions*, **10**, 573–577.

Richardson, D.M. & Whittaker, R.J. (2010) Conservation biogeography – foundations, concepts and challenges. *Diversity and Distributions*, **16**, 313–320.

Shafroth, P.B. & Briggs, M.K. (2008) Restoration ecology and invasive riparian plants: an introduction to the special section on *Tamarix* spp. in Western North America. *Restoration Ecology*, **16**, 94–96.

Simberloff, D., Nuñez, M., Ledgard, N.J. et al. (2010) Spread and impact of introduced conifers in South America: Lessons from other southern hemisphere regions. *Austral Ecology*, **35**, 489–504.

Skarpaas, O. & Økland, B. (2009) Timber import and the risk of forest pest introductions. *Journal of Applied Ecology*, **46**, 55–63.

Suarez, A.V. & Tsutsui, N.D. (2008) The evolutionary consequences of biological invasions. *Molecular Ecology*, **17**, 351–360.

Thomas, C.D. (2010) Climate, climate change and range boundaries. *Diversity and Distributions*, **16**, 488–495.

Thuiller, W., Richardson, D.M., Pyšek, P., Midgley, G.F., Hughes, G. & Rouget, M. (2005) Niche-based modelling as a tool for predicting the risk of alien plant invasions at a global scale. *Global Change Biology*, **11**, 2234–2250.

Thuiller, W., Richardson, D.M. & Midgley, G.F. (2007) Will climate change promote alien plant invasions? In *Biological Invasions* (ed. W. Nentwig), pp. 197–211. Springer, Berlin.

Tylianakis, J.M. (2008) Understanding the web of life: the birds, the bees, and sex with aliens. *PLoS Biology*, **6** (2), e47, doi:10.1371/journal.pbio.0060047.

Van der Wal, R., Truscott, A.-M., Pearce, I.S.K., Cole, L., Harris, M.P. & Wanless, S. (2008) Multiple anthropogenic changes cause biodiversity loss through plant invasion. *Global Change Biology*, **14**, 1428–1436.

Van Kleunen, M. & Richardson, D.M. (2007) Invasion biology and conservation biology: time to join forces to explore the links between species traits and extinction risk and invasiveness. *Progress in Physical Geography*, **31**, 447–450.

Van Wilgen, N.J., Elith, J., Wilson, J.R.U., Wintle, B. & Richardson, D.M. (2010) Alien invaders and reptile traders: What drives pet imports and dissemination in South Africa? *Animal Conservation*, **13**, doi:10.1111/j.1469-1795.2009.00298.x (in press).

Vellinga, E.C., Wolfe, B.E. & Pringle, A. (2009) Global patterns of ectomycorrhizal introductions. *New Phytologist*, **181**, 960–973.

Whitney, K.D. & Gabler, C.A. (2008) Rapid evolution in introduced species, 'invasive traits' and recipient communities: challenges for predicting invasive potential. *Diversity and Distributions*, **14**, 569–580.

Wilson, J.R.U., Dormontt, E.E., Prentis, P.J., Lowe, A.J. & Richardson, D.M. (2009) Something in the way you move: dispersal pathways affect invasion success. *Trends in Ecology & Evolution*, **24**, 136–144.

APPENDIX 29.1 DETAILS OF 500 PAPERS SELECTED TO GIVE A SNAPSHOT OF THE CURRENT DIMENSIONS OF RESEARCH ON BIOLOGICAL INVASIONS.

The sample comprises only papers published in peer-reviewed journals during 2008. No books, book chapters, contributions on the web or articles in the 'grey literature' were considered. Papers were collected in two main ways: (i) the foundation collection arose through the accumulation of material for the production of this book, in the course of my own research, post-graduate teaching and as a reviewer and editor of journal papers; (ii) searching of the published literature (especially on ISI Web of Knowledge, the Internet, annual reports of research institutions, reference lists of recent books and review papers) to add papers from all the main journals known to publish research on alien species. All the best-cited papers dealing with alien species from the top ecology/biogeography/invasions journals for 2008 are included. The sample is considered an unbiased cross-section of research in the field. One hundred and forty-two journals are represented in the sample, and 10 journals were represented with more 10 papers or more (in order: *Diversity and Distributions*; *Biological Invasions*; *Journal of Ecology*; *Conservation Biology*; *South African Journal of Botany*; *Molecular Ecology*; *Plant Ecology*; *Ecology*; *Journal of Applied Ecology*; and *Frontiers in Ecology and the Environment*). One journal is a surprising inclusion in this list (*South African Journal of Botany*; owing to a large special issue on plant invasions in 2008) but the others regularly carry many papers on aspects of biological invasions (together these top-ten journals carried 41% of the sample of 500 papers). The bias in favour of plants in the sample (59% of papers had plants as main or co-main focus group) is in line with the overall bias in favour of this group in invasion ecology (Pyšek et al. 2008). Invertebrates (9%), amphibians and fishes (each 4%), mammals and seaweeds (2%) and birds (1%) were the next best represented groups (10% of papers considered multiple groups). Each paper was examined to determine its main focus area: 'systems', 'target', 'transformation' or 'boundary' research as defined by Kueffer and Hirsh Haddon (2008), or a combination of these areas (see text for details). Each paper was also assigned values in each of the following fields: 'primary taxon/ taxa', 'primary field of enquiry', and 'primary keyword' (indicating the primary contribution). Additional keywords were also listed. For further analysis, all 500 papers were also searched for key words/phrases using Copernic Desktop Search 3 (www.copernic.com).

Chapter 30

A COMPENDIUM OF ESSENTIAL CONCEPTS AND TERMINOLOGY IN INVASION ECOLOGY

David M. Richardson[1], Petr Pyšek[2] and James T. Carlton[3]

[1]Centre for Invasion Biology, Department of Botany & Zoology, Stellenbosch University, 7602 Matieland, South Africa
[2]Institute of Botany, Academy of Sciences of the Czech Republic, CZ-252 43 Průhonice, Czech Republic; and Department of Ecology, Faculty of Science, Charles University, Viničná 7, CZ-128 01 Praha 2, Czech Republic
[3]Maritime Studies Program, Williams College-Mystic Seaport, Mystic, CT 06355, USA

30.1 INTRODUCTION

This chapter provides a list of definitions of selected concepts and terms used in invasion ecology. Many of these have been used in different ways by different authors. The uncritical use of terms and concepts is hampering conceptual advances in some parts of the field and is impeding the smooth flow of research results into management and policy arenas. There are geographical and historical differences in the usage of terminology, some of which are attributable to the origin of terms in languages other than English. Differences also exist in the use of terms for different taxonomic groups and between terrestrial, freshwater and marine systems. The schemes proposed so far are usually restricted taxonomically, or refer specifically to either terrestrial or aquatic environments. The list provided here is not exhaustive, but includes important terms and concepts used in this book and elsewhere in the current literature that may be unfamiliar to all readers and for which a clearer understanding is needed. Key references are given where appropriate. We are grateful to many colleagues for inputs and useful discussions on the evolving list, and especially to Spencer Barrett and Phil Hulme for valuable comments on a near-final version of the compendium.

30.2 THE WAY AHEAD

Quine (1936) remarked that, 'the less a science has advanced the more its terminology tends to rest upon an uncritical assumption of mutual understanding.' As much of the present volume emphasizes, invasion science is a young discipline with comparatively shallow roots, and with literally thousands of papers appearing only in the past 25 years. The result has been both a welcome cornucopia of questions, hypotheses and insights, accompanied by an often opaque panoply of definitions that focus on taxon- and habitat-specific phenomena that have not strived to seek the more fundamental ecological and evolutionary threads that bind these elements together. As such, considerable 'license and creativity' (see Carlton 2002) now accompany the terminology of the science, as reflected in the many different definitions excavated by Falk-Petersen et al. (2006) and as discussed below. It is our view that a lack of stabilization of fundamental concepts impedes both the science and management of alien species.

In 1958 a famous international symposium was held in Italy on the 'Classification of Brackish Waters'. This meeting produced one of the rare international conventions ('The Venice System for the Classification of Marine Waters') that for many decades led to an increased level of clarity and a reduced level of misunderstanding among those studying estuaries around the world. As with all such conventions, it was not without a difficult birth, and it has been continually critiqued and refined in the past 50 years. While the glossary presented in this chapter was not precipitated by such an international convention seeking consensus, we believe that we reflect here much (but certainly not all) of the general agreements and disagreements among many leading workers in the field. We propose that the definitions presented here should act as stage-setters from which we can now proceed. Generating a uniform, broadly accepted and acceptable set of terms and concepts for invasion science, which while acknowledging the debates within and conflicting perceptions (often generated by differing spatial and temporal scales) universal to all science, will, we argue, profoundly advance the discipline. Without a common language on the *science* side, translating the critical aspects of why concerns about the prevention and management of alien species are fundamental to both the environment and ecosystem services on the *policy* side will often remain confused and confusing to the public, the press and the political world.

The way ahead is to seek both consensus and concession among the broadest possible realm of invasion scientists.

30.3 OVERVIEW OF CONCEPTS

The concepts are presented in the form of a glossary. Terms in bold type indicate cross-references to other concepts in the list.

Alien species (synonyms: adventive, exotic, foreign, introduced, non-indigenous, non-native) – Those whose presence in a region is attributable to human actions that enabled them to overcome fundamental biogeographical barriers (i.e. human-mediated extra-range dispersal). Some **alien species** (a small proportion) form self-replacing populations in the new region (see **tens rule**). Of these, a subset has the capacity to spread over substantial distances from **introduction** sites. Depending on their status within the **naturalization–invasion continuum**, **alien species** may be objectively classified as **casual**, **naturalized** or **invasive** (Richardson et al. 2000b; Pyšek et al. 2004). Note: designation of a species as alien should include a statement about the region under discussion; depending on the scale of observation, a species can be alien to a country but native to the continent (see discussion in Lambdon et al. 2008).

Baker's rule (also called Baker's law) – For plants and some animals, the notion that organisms capable of uniparental reproduction are more likely to establish populations after **long-distance dispersal** than are organisms that require mates because they are obligately outcrossing. This originated with the following statement by Baker (1955): '*With self-compatible individuals a single propagule is sufficient to start a sexually-reproducing colony, making its establishment much more likely than if the chance growth of two self-incompatible yet cross-compatible individuals sufficiently close together spatially and temporally is required.*' (see Barrett, this volume).

Biological invasions (synonyms: bioinvasions, biotic invasions, species invasions) – The phenomenon of, and suite of processes involved in determining, the following: (i) the transport of organisms, through human activity (intentionally or accidentally, through **introduction pathways**) to areas outside the potential range of those organisms as defined by their natural dispersal mechanisms and biogeographical barriers; and (ii) the fate of such organisms in their new ranges, including their ability to survive, establish, reproduce, disperse, spread, proliferate, interact with **resident** biota and exert influence in many ways on and in invaded ecosystems. There is a school of thought that advocates that the concept of **biological invasions** should more broadly embrace both **range expansions** (involving no obvious human mediation), because the fundamental processes (except, critically, the means of negotiating a major biogeographical barrier (Wilson et al. 2009a)) are the same (both involve the movement of individuals from a donor community into a recipient community (Sorte et al. 2010)) see discussion under **dispersal pathway**. Cf. **range expansions**.

Biosecurity – The management of risks posed by organisms to the economy, environment and human health through exclusion (the prevention of initial **introduction** of a species), mitigation, adaptation, control and **eradication** (Hulme, this volume).

Biotic acceptance (synonym: 'The rich get richer' concept) – A notion that argues that the dominant general pattern in invasion ecology at multiple spatial scales is one where natural ecosystems tend to accommodate the establishment and coexistence of **alien species** despite the presence and abundance of **native species** (i.e. the opposite of what we would expect from the **biotic resistance** hypothesis). At large spatial scales, the same abiotic conditions that promote high diversity of **native species** (energy, substrate and habitat heterogeneity, etc.) also support diverse floras of **alien species**. The patterns of **invasibility** may be more closely related to the degree of resources available in **native** plant communities, independent of species richness. (Stohlgren et al. 1999, 2006). Most treatments discuss only species richness within the same trophic level as the **alien species**. However, because cross-taxon facilitation and inhibition are crucial mediators of **invasibility**, a broader consideration of biodiversity is appropriate (Richardson et al. 2000a). Much remains to be understood about these hypotheses across habitats and taxa; for example, see Leprieur et al. (2008), who found no support for either **biotic acceptance** or **biotic resistance** among global freshwater fish invasions. Cf. **biotic resistance**.

Biotic homogenization – The addition to, and often the partial if not extensive replacement of, local biotas by **alien species** (see McKinney & Lockwood 1999), which can result in decreased compositional turnover (β-diversity) of species between distant areas, both in terms of taxonomic and phylogenetic similarity (Winter et al. 2009).

Biotic resistance – Resistance by **resident species** to the establishment (or post-establishment survival, proliferation and spread) of **alien species**. A classic hypothesis, first articulated by Charles Elton (1958) (the **diversity–invasibility hypothesis**), is that **biotic resistance** is greater in more diverse communities. Most evidence for **biotic resistance** comes from experimental work using synthetic assemblages that vary in diversity, and from modelling (Tilman 1999). Empirical tests of the effects of species richness on **invasibility** have produced unambiguous results (Levine & D'Antonio 1999). The hypothesis is usually tested by exploring the relationship between the numbers of **native** and **alien species**, which appears negative (supporting **biotic resistance**) at very small spatial scales but positive at larger scales (more **alien species** in areas with high richness of **native species**) where it led to the formulation of a '**biotic acceptance**' concept (Stohlgren et al. 2006). This discrepancy, termed the '**invasion paradox**' by Fridley et al. (2007), is largely explained by the spatial scale of observation (Fridley et al. 2004; Herben et al. 2004) and by covarying external factors (Shea & Chesson 2002).

Casual species – Those **alien species** that do not form self-replacing populations in the invaded region and whose persistence depends on repeated **introductions** of propagules (Richardson et al. 2000b; Pyšek et al. 2004). The term is generally used for plants.

Colonization pressure – A recent variant of the concept of **propagule pressure**; defined as the number of species introduced or released to a single location, some of which will go on to establish a self-sustaining population and some of which will not. Lockwood et al. (2009) argue that **colonization pressure** should serve as a null hypothesis for understanding temporal or spatial differences in **alien species** richness, as the more species are introduced, the more we should expect to establish. They show that **propagule pressure** is related to **colonization pressure**, but in a nonlinear manner (see also Blackburn et al., this volume).

Competitive release hypothesis – A hypothesis that predicts that **alien species** may be released from competition in habitats with novel competitors or no competitors (Sorte et al. 2010); part of several competition models in **invasion ecology** also related to the **evolution of increased competitive ability hypothesis** (**EICA**), which predicts increased competitive ability through the relaxation of herbivore pressure.

Corridor – As used in **invasion ecology**, a dispersal route (a physical connection of suitable habitats) linking previously unconnected regions (Hulme et al. 2008; Wilson et al. 2009b) (see **dispersal pathway**; **introduction pathway**; **vector**).

Cryptogenic species – Species of unknown biogeographical history which cannot be ascribed as being **native** or **alien** (Carlton 1996a; see also Carlton 2009 for a discussion of the misapplication of the concept). Species can be recognized as clearly alien (based upon palaeontological, archaeological, historical, biogeographic, vector, genetic and other evidence), although their geographic origin may be unknown; these are not **cryptogenic species**.

Darwin's naturalization hypothesis – The notion than **alien** species with close **native** relatives in their introduced range may have reduced chances of establishment and **invasion**; based on ideas formulated by Charles Darwin (1859) in chapter 3 of *The Origin of Species*, borrowing ideas from Alphonse de Candolle, in the context of his discussion on the 'struggle for existence' between similar organisms: '*As species of the same genus have usually, though by no means invariably, some similarity in habits and constitution, and always in structure, the struggle will generally be more severe between species of the same genus, when they come into competition with each other, than between species of distinct genera*' (Daehler 2001a; Proches et al. 2008; Thuiller et al. 2010).

Dispersal pathway – The combination of processes and opportunities resulting in the movement of propagules from one area to another, including aspects of the **vectors** involved, features of the original and recipient environments, and the nature and timing of what exactly is moved. The definition thus combines phenomenological and mechanistic aspects. Wilson et al. (2009b) define six types of **dispersal pathway**: leading edge; corridor; jump dispersal; extreme **long-distance dispersal**; mass dispersal; and cultivation. Human mediation is only essential in the last two of these categories (which form a sub-group: **introduction pathways**). This definition emphasizes that the distinction between an **invasion** and a **range expansion** is not absolute and that dispersal events are best considered as points on a continuum. Note that Carlton and Ruiz (2005) argue that the term 'pathway', as currently used in the invasion literature, means three distinctly different things: the cause of **invasion**, the geographic route and the **vector** itself. Cf. **vectors**.

Diversity–invasibility hypothesis – The proposition that more biologically diverse communities are less susceptible to **invasion** by novel species or genotypes (related terms and concepts include: biotic-resistance hypothesis; diversity-resistance hypothesis; species-richness hypothesis) (Fridley, this volume). See also **biotic acceptance**, **biotic resistance**, **invasional meltdown**.

EICA (the **evolution of increased competitive ability hypothesis**) – Predicts that plants introduced to an environment that lacks their usual herbivores (or disease agents) will experience selection favouring individuals that allocate less energy to defence and more to growth and reproduction (Blossey & Nötzold 1995).

Enemy release hypothesis (ERH) – Proposes that **alien species** have a better chance of establishing and becoming dominant when released from the negative effects of natural enemies that, in their native range, lead to high mortality rates and reduced productivity (Keane & Crawley 2002). Colautti et al. (2004) argue that the **ERH** is often accepted without recognizing that all **alien species** will lose at least some natural enemies owing to bottlenecks during transport. See also Dormontt et al. (this volume).

Eradication – The extirpation of an entire population of an **alien species** within a designated management unit. When a species can be declared eradicated (that is, how long a period of time after the management intervention) depends on the species and the situation and must take into account factors such as seed-bank longevity (for plants). The probability that a species should be quantified?

Feral species – **naturalized species** that have reverted to the wild from domesticated stock, i.e. have undergone some change in phenotype, genotype and/or behaviour owing to artificial selection in captivity.

Fluctuating resources theory of invasibility – A theory, formulated for plants by Davis et al. (2000), that predicts that pronounced fluctuations in resource availability enhances community **invasibility** if coinciding with the availability of sufficient propagules to initiate an **invasion**. It is based on the assumption that an invading species must have access to available resources (e.g. light, nutrients, water for plants, food, shelter, space, mates for animals) and that a species will be more successful in invading a community if it does not encounter intense competition for these resources from **resident species**. An increase in resource availability can arise from several phenomena: the rate at which resources are supplied from external sources is faster than the rate at which the **resident biota** can use them, the **resident biota's** use of resources declines or the resources themselves become more available within the community (part of patch-dynamic theory, which includes the novel creation of often large open spaces owing to abrupt physical or biological disturbance, which may eliminate all or most of the previous biota). A short-term pulse in the availability of resources can have long-term conse-

quences once the invading species is established in the community.

Foreign – see **alien species**.

Genetically modified organism (GMO, synonym: living modified organism) – An organism that possesses a novel combination of genetic material engineered through recombinant DNA technology, and which may have adverse effects on the conservation and sustainable use of biodiversity, owing to the risk of the organism becoming **invasive**, effects on human health and other factors (CBD 2000).

Hub-and-spoke model – The concept that **alien species** expand on a local, regional or global scale owing to the continued establishment of multiple loci, which form new population epicentres (hubs) that then interface with novel dispersal routes (spokes) (Carlton 1996b). A global example would be a species being carried from one seaport (visited by a certain set of ships and shipping routes) to another seaport on a different continent (frequented by ships on different routes).

Impact – The description or quantification of how an **alien species** affects the physical, chemical and biological environment. Parker et al. (1999) proposed that **impact** should be conceptualized as the product of the range size of the invader, its average abundance per unit area across that range and the effect per individual or per biomass unit of the invader. Lockwood et al. (2007) list the following categories of **impacts** associated with **biological invasions**: genetic, individual, population, community, ecosystem, and landscape, regional and global. Another approach, used by the Millennium Ecosystem Assessment, assesses **impacts** relative to specific types of ecosystem services: supporting, regulating, provisioning and cultural (Vilà et al. 2010). Major issues relating to **impacts** of **invasive species** include their perception and recognition with reference to human value systems (Richardson et al. 2008), and the quest for a common and objective currency, including the means for translating **impacts** into financial and other costs (Pyšek & Richardson 2010; Vilà et al. 2010). A fundamental construct of properly quantifying **impact** is experimental science, rather than deductions based on assumptions or correlations (such as a **native species** declining and an **alien species** increasing, perhaps for unrelated reasons). Equally crucial is to recognize that '**impact**' is a scaled and gradational phenomenon requiring careful, replicated quantification; and that it is not a concept that can conveniently be divided into simple dichotomous bins of '**impact**' and 'no impact'.

Introduced – see **alien species**.

Introduction – Movement of a species, intentionally or accidentally, owing to human activity, from an area where it is **native** to a region outside that range ('introduced' is synonymous with **alien**). The act of an **introduction** (inoculation of propagules) may or may not lead to **invasion**.

Introduction pathway – Describes the processes that result in the **introduction** of **alien species** from one geographical location to another. Hulme et al. (2008) suggested a universal framework applicable to a wide range of taxonomic groups in terrestrial and aquatic ecosystems. **Alien species** may arrive through three broad mechanisms: importation of a commodity; arrival of a transport vector; natural spread from a neighbouring region where the species is itself alien. These three mechanisms result in six principal pathway classes: release, escape, contaminant, stowaway, corridor and unaided. **Introduction pathways** form a subset of **dispersal pathways** – those that are mediated by human activities.

Invasion – The multi-stage process whereby an **alien** organism negotiates a series of potential barriers in the **naturalization–invasion continuum** (Richardson et al. 2000b) (cf. **range expansion**).

Invasion cliff – A construct integrating community **invasibility** and **propagule pressure**, which together constitute **invasion pressure**, defined as the probability that an environment will experience an **invasion** within a specified period. The theoretical model shows that changes in **invasion pressure** can alternatively be very sensitive or very insensitive to changes in **invasibility** and/or **propagule pressure**, depending on the magnitude of the two variables as well as on their relative values. The relationship between **invasion pressure** and its two primary components is nonlinear; in a three-dimensional graph of the three variables this sensitivity is reflected by a cliff-like feature connecting areas of unlikely **invasion** with those where invasion is almost certain. **Invasion pressure** is thus best described by two relatively stable states, separated by a tipping point (Davis 2009, and this volume). This concept is important for management, because when a system is not near the **invasion cliff**, even substantial changes in invasibility

and/or **propagule pressure** due to management interventions have little potential effect on the probability of **invasion**. Alternatively, relatively minor changes could dramatically influence invasion probability if the system is positioned near the **invasion cliff** (Richardson 2009).

Invasion complex – A situation where one **invasive species** facilitates, directly or indirectly, the establishment of one or more 'secondary' **alien species**, potentially with impacts greater than the sum of the individual species. An example of *direct* facilitation is an **alien** frugivorous bird promoting the spread of an alien fruit-bearing tree, as occurred in the Hawaiian Islands when **introduced** birds promoted spread of the alien tree *Morella faya* by eating its fruits and dispersing its seeds. *Indirect* facilitation involves an **alien species** modifying environmental conditions or disturbance regimes in a manner that promotes the establishment of subsequent invaders, for example soil disturbance from rooting by **alien** wild boar promotes establishment of alien plants in several ecosystems (for examples, see Richardson et al. (2000a)) (see also **invasional meltdown**) (D'Antonio 1990).

Invasion debt – A concept that posits that even if **introductions** cease (and/or other drivers of **invasion** are relaxed, e.g. **propagule pressure** is reduced), new **invasions** will continue to emerge, **naturalized species** that are present will enter the **invasion** stage and already-**invasive species** will continue to spread and cause potentially greater **impacts**, because large numbers of **alien species** are already present, many of them in a **lag phase** (Richardson, this volume).

Invasion ecology – The study of the causes and consequences of the **introduction** of organisms to areas outside their native range as governed by their dispersal mechanisms and biogeographical barriers. The field deals with all aspects relating to the **introduction** of organisms, their ability to establish, **naturalize** and **invade** in the target region, their interactions with **resident organisms** in their new location, and the consideration of costs and benefits of their presence and abundance with reference to human value systems (Richardson & van Wilgen 2004). This term is often used interchangeably with 'invasion biology' in the literature; see also **invasion science**.

Invasion paradox – A term used in at least two broad contexts in the recent literature. The most widely used meaning relates to contrasting lines of support for both negative and positive relationships between **native** biodiversity and various measures of 'success' of **alien species** (Fridley et al. 2007; see also **biotic resistance**). Sax and Brown (2000) also used the term to describe biological invasions in general, in particular: *'why are exotic organisms, which come from distant locations and have had no opportunity to adapt to the local environment, able to become established and sometimes to displace native species, which have had a long period of history in which to adapt to local conditions?'*.

Invasion pressure – The probability that an environment will experience an **invasion** within a specified period (Davis 2009). Cf. **invasion cliff**.

Invasion science (synonym: invasion research) – A term used to describe the full spectrum of fields of enquiry that address issues pertaining to **alien species** and **biological invasions**. The field embraces **invasion ecology**, but increasingly involves non-biological lines of enquiry, including economics, ethics, sociology, and inter- and transdisciplinary studies (Richardson, this volume).

Invasibility – The properties of a community, habitat or ecosystem that determine its inherent vulnerability to **invasion** (Lonsdale 1999). Early studies tended to use the concept deterministically (particular systems were deemed either invasible or not), but **invasibility** is more appropriately considered probabilistically, and the degree of **invasibility** may change markedly over time owing to, for instance, changes in biotic or abiotic features of the ecosystem. **Invasibility** is ideally measured as the survival rate of **alien species** introduced to the system, thus accounting for losses due to competition with **resident biota**, effects of enemies, chance events and other factors (Lonsdale 1999). **Invasibility** differs from the **level of invasion**, which integrates the effects of invasibility, **propagule pressure** and climate (Chytrý et al. 2008). (see also **biotic acceptance**, **biotic resistance**, **colonization pressure**, **fluctuating resources theory of Invasibility**, **invasion cliff**, **invasion complex**, **invasion pressure**, **invasiveness**, **lag phase**, **level of invasion**).

Invasional meltdown – A phenomenon whereby **alien species** facilitate one another's establishment, spread and **impacts** (Simberloff & Von Holle 1999; see Simberloff (2006) for examples and conceptual discussion).

Invasive species – **Alien** species that sustain self-replacing populations over several life cycles, produce reproductive offspring, often in very large numbers at considerable distances from the parent and/or site of **introduction**, and have the potential to spread over long distances (Richardson et al. 2000b; Occhipinti-Ambrogi & Galil 2004; Pyšek et al. 2004). **Invasive species** are a subset of **naturalized species**; not all **naturalized species** become invasive. This definition explicitly excludes any connotation of **impact**, and is based exclusively on ecological and biogeographical criteria (for discussion, see Daehler 2001b; Rejmánek et al. 2002; Ricciardi & Cohen 2007). It should be noted that the definition supported by the World Conservation Union (IUCN), the Convention on Biological Diversity and the World Trade Organization explicitly assumes that **invasive species** cause **impacts** to the economy, environment or health (see IUCN 2000). This important difference has implications for **risk analyses** of **invasive species** (Hulme, this volume). Consequently, it is crucial for **risk assessment** protocols to assign dimensions of risk separately for elements of **invasion** and **impact**. Note: designation of a species as **invasive** should include a statement about the region under discussion; for example a species **alien** to a state can be **native** to a continent (see discussion in Lambdon et al. (2008)).

Invasiveness – The features of an **alien** organism, such as their life-history traits and modes of reproduction that define their capacity to invade, i.e. to overcome various barriers to **invasion**. The level of **invasiveness** of a species can change over time owing to, for example, changes in genetic diversity through hybridization, introgression or the continued arrival of new propagules of the same species that is already established in a region, but from new and different (meta)populations, such that genetic diversity may increase. This last concept is important in management strategies, which sometimes assume that less concern needs to be paid to the continued **introduction** of species (the continued arrival of propagules, whether accidental or intentional) that are *already* well-established in a region, overlooking the critical potential for elevated **invasiveness** over time.

Jump dispersal – A category of **long-distance dispersal**, sometimes over substantial scales, whereby connection (gene flow) between the new and original ranges is maintained. Cf. **dispersal pathway**.

Lag phase (synonym: latency period) – the time between when an **alien** species arrives in a new area and the onset of the phase of exponential increase. Multiple factors are frequently implicated in the persistence or dissolution of the **lag phase** in **invasions**, including an initial shortage of **invasible** sites, the absence or shortage of essential mutualists, inadequate genetic diversity and the relaxation of competition or predation (owing to other alterations in the **resident biota**). However, Aikio et al. (2010) show that **lag phases** may equally be the result of statistical or sampling artefacts commonly found in time series of records of **alien species**.

Lag time – The broad set of lag (the period of time from one event to another) phenomena across the entire **invasion** sequence, which may include the following: (i) the apparent long-term failure of species to **invade** successfully from potential donor regions to potential recipient regions (until they do, owing to, for example, changes in the environments of donor and/or recipient regions, to changes in vectors or to changes in other phenomena); (ii) lags in population increase (see **lag phase**); and (iii) lags in geographic expansion, whereby a species may appear to remain resident in one relatively small and restricted region for a long period of time, but then begin to suddenly expand (owing, in part, to the fact that spread increases exponentially once multiple foci have had time to establish).

Level of invasion – Actual number or proportion of **alien species** in a community, habitat or region, resulting from an interplay of its **invasibility**, **propagule pressure** and climate (Hierro et al. 2005, Chytrý et al. 2008). The **level of invasion** is determined by the product of the number of **alien species** **introduced** to the system (**propagule pressure**) and their survival rate, which differs in individual habitats based on their **invasibility** (Lonsdale 1999). Relatively resistant communities can be invaded to a high level if exposed to high **propagule pressure**. Even relatively vulnerable communities will experience low-level invasions if **propagule pressure** is low.

Long-distance dispersal – Dispersal of propagules over a long distance, defined either by the absolute distance travelled, or by a set proportion of all propagules that disperse the farthest. **Long-distance dispersal** occurs at various scales; extremely, propagules may move beyond the disper-

sal range seen over ecological timescales (Wilson et al. 2009b).

Managed relocation (synonym: assisted migration, translocation, transplantation) – A form of management intervention aimed at reducing the negative effects of global change (especially rapid climate change) on defined biological units such as populations, species or ecosystems. It involves the intentional movement of biological units from their current areas of occupancy to locations where the probability of future persistence is predicted to be higher (Richardson et al. 2009). Such movements may include **introduction** of the species to areas outside their current or known historic range; as such it potentially represents an important **introduction pathway** and its potential for causing new **invasions** is an important criticism advanced by opponents of this strategy (e.g. Ricciardi & Simberloff 2009).

Native species (synonym: indigenous species) Species that have evolved in a given area or that arrived there by natural means (through **range expansion**), without the intentional or accidental intervention of humans from an area where they are native (see Pyšek et al. 2004).

Naturalized species (synonym: established species) – Those **alien species** that sustain self-replacing populations for several life cycles or a given period of time (10 years is advocated for plants) without direct intervention by people, or despite human intervention (Richardson et al. 2000b; Pyšek et al. 2004). The term is currently mainly used with reference to terrestrial plant invasions, although it was previously widely used for mammals.

Naturalization–invasion continuum – A conceptualization of the progression of stages and phases in the status of an **alien** organism in a new environment which posits that the organism must negotiate a series of barriers. The extent to which a species is able to negotiate sequential barriers (which is mediated by **propagule pressure** and **residence time**, and which frequently involves a **lag phase**) determines the organism's status as an **alien**: **casual**, **naturalized** or **invasive species** (Richardson et al. 2000b).

Non-indigenous (**nonindigenous**) and **non-native** (**nonnative**) – see **alien species** (students should note that web searches with and without the hyphen will yield different results).

Novel ecosystems – Those comprising species that occur in combinations and relative abundances that have not occurred previously at a given location or biome. Such ecosystems result from either the degradation or **invasion** of natural ecosystems (those dominated by **native species**) or the abandonment of intensively managed systems (Hobbs et al. 2006).

Pests – A cultural term often applied to animals (not necessarily alien) that live in places where they are not wanted and which have detectable economic or environmental impact or both (Pyšek et al. 2009). Cf. **weeds**.

Propagule pressure – A concept that encompasses variation in the quantity, quality, composition and rate of supply of **alien** organisms resulting from the transport conditions and pathways between source and recipient regions (see also **colonization pressure**) (Simberloff 2009). **Propagule pressure** has emerged as a fundamental determinant of the **level of invasion**; Colautti et al. (2006) suggest that it should serve as the basis of a null model for studies of **biological invasions** when inferring process from patterns of **invasion**.

Range expansion – The process whereby a species spreads into new areas (usually new regions, rather than local-scale movements) owing to natural or human-mediated dispersal; such expansion may be assisted or primarily driven by human-mediated changes to the environment. Differs from **invasion** in that human-mediated extra-range dispersal (i.e. across a biogeographical barrier) is not implicated; the concept can be applied to both **native** and **alien species**.

Residence time – The time since the **introduction** of a species to a region; because the **introduction** date is usually derived from post-hoc records and is likely inaccurate, the term *minimum residence time* has been suggested (Rejmánek 2000). The extent of invasion of **alien species** generally increases with increasing **residence time** as species have more time to fill their potential ranges (Wilson et al. 2007; Williamson et al. 2009).

Resident biota/organisms – Species that are present in a community, habitat or region at the time of **introduction** of an **alien species**. The pool of resident species includes both **native species** and **alien species** introduced previously. (See also **biotic resistance, novel ecosystems**.)

Resource-enemy release hypothesis – Fast-growing plant species adapted to high resource availability have less constitutive defences against enemies, and therefore incur relatively large costs when enemies are present. These fast-growing species benefit most from **enemy release**, and the two mechanisms can act in concert to cause **invasion**; this could explain both the strong effects of resource availability on **invasion** and the extraordinary success of some **alien species** (see Blumenthal 2006; Blumenthal et al. 2009).

Risk assessment – The estimation of the quantitative or qualitative value of risk (the likelihood of an event occurring within a specified time frame and the consequences if it occurs). In the context of **invasion science**, **risk assessment** is undertaken to evaluate the likelihood of the entry, establishment and spread of a species (intentionally or accidentally) in a given region, negotiating given barriers in the **naturalization–invasion continuum**, and the extent and severity of ecological, social and economic **impacts** (see Hulme, this volume).

Tens rule – A probabilistic assessment of the proportion of species that reach particular stages in the **naturalization–invasion continuum**. It predicts that 10% of imported species (species brought in for cultivation or held in captivity) become **casual**, 10% of casuals become **naturalized** and 10% of **naturalized species** become pests (Williamson & Brown 1986; Williamson & Fitter 1996). The rule was developed from European plant data, but the general principle that **invasions** are rare, and that achievement of this status depends on **propagule pressure**, biology and location, holds worldwide and across all taxonomic groups, although the 10% is probably an artefact of the history of biological invasions worldwide and is likely to increase with increasing **residence times** of **alien species** in floras (Richardson & Pyšek 2006). Caley et al. (2008) point out that the **tens rule** refers to the distribution for the probability of an invader reaching a stage in the **naturalization–invasion continuum**; the point estimate 0.1 is a measure of central tendency, although it is frequently misinterpreted as a rule describing point estimates for the transition probabilities for each stage. The **tens rule** is thus not meant to be interpreted as meaning or predicting that 10% is a standard or fixed outcome of invasion probabilities. Interpretation of the **tens**

rule is also dependent on the definition and perception of *pest* species: in many cases, for example, 0% of imported species may become casual and 0% of casual species may become naturalized, or, conversely, a much larger proportion than 10% of naturalized species may be considered nuisance species, depending on the value systems assigned.

Transformers – Invasive species that change the character, condition, form or nature of ecosystems (Richardson et al. 2000b).

Vectors – A broadly defined phenomenon involving dispersal mechanisms that can be both non-human mediated (wind, water, birds, mammals, amphibians, etc.) and human mediated. Carlton and Ruiz (2005) propose a classification framework for the human-mediated movement of organisms that includes six elements: *cause* (why a species is transported; that is, whether accidentally or deliberately); *route* (the geographic path over which a species is transported from the origin to the destination, which they synonymize with passageway, course and corridor); *vector* (how a species is transported – that is, the physical means or agent, such as ballast, clothing, commercial oyster movement, animal feeds or vehicles; **vector** is synonymized with mode, transport mechanism, carrier, and bearer); *vector tempo* (how a given **vector** operates through time, in terms of size and rate, speed and timing: size and rate are defined as the frequency with which the **vector** operates to deliver propagules to the target region, measured as the quantity of the **vector** (in units appropriate to the **vector**) expressed per unit time (for example, gallons of ballast water per day, number of container boxes per month, etc.)); *vector biota* (quantitative and/or qualitative description of the all of the living organisms being transferred by a given **vector**, in terms of diversity, density and condition; see **propagule pressure**); and *vector strength* (the relative number or rate of established invasions that result within a specific period from a given **vector** in a particular geographic region). They note that in the invasion literature the term **pathway** can thus have very different meanings, including cause, route (corridor) and the **vector** itself. Cf. discussion at **dispersal pathway**.

Weeds – A plant is a **weed** '*if, in any specified geographical area, its populations grow entirely or predominantly in situations markedly disturbed by man (without, of course, being deliberately cultivated plants)*' (Baker 1965); in cultural terms, **weeds** are plants (not nec-

essarily **alien**) that grow in sites where they are not wanted and that have detectable economic or environmental **impacts** (Pyšek et al. 2004).

REFERENCES

Aikio, S., Duncan, R.P. & Hulme, P.E. (2010) Time lags in alien plant invasions: separating the facts from the artefacts. *Oikos*, **119**, 370–378.

Baker, H.G. (1955) Self-compatibility and establishment after 'long-distance' dispersal. *Evolution*, **9**, 347–349.

Baker, H.G. (1965) Characteristics and modes of origins of weeds. *The Genetics of Colonizing Species* (ed. H. G. Baker and G. L. Stebbins), pp. 147–172. Academic Press, New York.

Blossey, B. & Nötzold, R. (1995) Evolution of increased competitive ability in invasive nonindigenous plants: a hypothesis. *Journal of Ecology*, **83**, 887–889.

Blumenthal, D.M. (2006) Interactions between resource availability and enemy release in plant invasion. *Ecology Letters*, **9**, 887–895.

Blumenthal, D., Mitchell, C.E., Pyšek, P. & Jarošík, V. (2009) Synergy between pathogen release and resource availability in plant invasion. *Proceedings of the National Academy of Sciences of the USA*, **106**, 7899–7904.

Caley, P., Groves, R.H. & Barker, R. (2008) Estimating the invasion success of introduced plants. *Diversity and Distributions*, **14**, 196–203.

Carlton, J.T. (1996a) Biological invasions and cryptogenic species. *Ecology*, **77**, 1653–1655.

Carlton, J.T. (1996b) Pattern, process, and prediction in marine invasion ecology. *Biological Conservation* **78**, 97–106.

Carlton, J.T. (2002) Bioinvasion ecology: assessing invasion impact and scale. In *Invasive Aquatic Species of Europe. Distribution, Impacts, and Management* (ed. E. Leppäkoski, S. Gollasch and S. Olenin) pp. 7–19. Kluwer Academic Publishers, Dordrecht, the Netherlands.

Carlton, J.T. (2009) Deep invasion ecology and the assembly of communities in historical time. In *Biological Invasions in Marine Ecosystems* (ed. G. Rilov and J.A. Crooks), pp. 13–56. Springer, Berlin and Heidelberg.

Carlton, J.T. & Ruiz, G. M. (2005) Vector science and integrated vector management in bioinvasion ecology: conceptual frameworks. In *Invasive Alien Species: A New Synthesis* (ed. H.A. Mooney, R.N. Mack, J.A. McNeely, L.E. Neville, P.J. Schei and J.K. Waage), pp. 36–58. Island Press, Washington, DC.

CBD (2000) *Cartagena Protocol on Biosafety to the Convention on Biological Diversity*. Secretariat of the Convention on Biological Diversity, Montreal, Canada.

Chytrý, M., Jarošík, V., Pyšek, P., et al. (2008) Separating habitat invasibility by alien plants from the actual level of invasion. *Ecology*, **89**, 1541–1553.

Colautti, R.I., Grigorovich, I.A. & MacIsaac, H.J. (2006) Propagule pressure: a null model for biological invasions. *Biological Invasions*, **8**, 1023–1037.

Colautti, R.I., Ricciardi, A., Grigorovich, I.A. & MacIsaac, H.J. (2004) Is invasion success explained by the enemy release hypothesis? *Ecology Letters*, **7**, 721–733.

Daehler, C.C. (2001a) Darwin's naturalization hypothesis revisited. *American Naturalist*, **158**, 324–330.

Daehler, C.C. (2001b) Two ways to be an invader, but one is more suitable for ecology. *Bulletin of the Ecological Society of America*, **82**, 101–102.

Darwin, C. (1859) *On the Origin of Species by Means of Natural Selection or the Preservation of Favoured Races in the Struggle for Life*. John Murray, London.

D'Antonio, C.M. (1990) Seed production and dispersal in the non-native, invasive succulent *Carpobrotus edulis* (Aizoaceae) in coastal strand communities of central California. *Journal of Applied Ecology*, **27**, 693–702.

Davis M.A. (2009) *Invasion Biology*. Oxford University Press, Oxford.

Davis, M.A., Grime, J.P. & Thompson, K. (2000) Fluctuating resources in plant communities: a general theory of invasibility. *Journal of Ecology*, **88**, 528–534.

Elton, C.S. (1958) *The Ecology of Invasions by Animals and Plants*. Methuen, London.

Falk-Petersen, J., Bøhn, T., & Sandlund, O.T. (2006) On the numerous concepts in invasion biology. *Biological Invasions*, **8**, 1409–1424.

Fridley, J.D., Brown, R.L. & Bruno, J.F. (2004) Null models of exotic invasion and scale-dependent patterns of native and exotic species richness. *Ecology*, **85**, 3215–3222.

Fridley, J.D., Stachowicz, J.J., Naeem, S., et al. (2007) The invasion paradox: reconciling pattern and process in species invasions. *Ecology*, **88**, 3–17.

Herben, T., Mandák, B., Bímová, K. & Münzbergová, Z. (2004) Invasibility and species richness of a community: a neutral model and a survey of published data. *Ecology*, **85**, 3223–3233.

Hierro, J.L., Maron, J.L. & Callaway, R.M. (2005) A biogeographical approach to plant invasions: the importance of studying exotics in their introduced and native range. *Journal of Ecology*, **93**, 5–15.

Hobbs, R.J., Arico, S., Aronson, J., et al. (2006) Novel ecosystems: theoretical and management aspects of the new ecological world order. *Global Ecology and Biogeography*, **15**, 1–7.

Hulme, P.E., Bacher, S., Kenis, M., et al. (2008) Grasping at the routes of biological invasions: a framework for integrating pathways into policy. *Journal of Applied Ecology*, **45**, 403–414.

IUCN (2000) *Guidelines for the Prevention of Biodiversity Loss Caused by Alien Invasive Species*. IUCN, Gland, Switzerland.

Keane, R.M. & Crawley, M.J. (2002) Exotic plant invasions and the enemy release hypothesis. *Trends in Ecology & Evolution*, **17**, 164–170.

Leprieur, F., Beauchard, O., Blanchet, S., Oberdorff, T. & Brosse, S. (2008) Fish invasions in the world's river systems: when natural processes are blurred by human activities. *PLoS Biology*, **6**, 404–410.

Levine, J.M. & D'Antonio, C.M. (1999) Elton revisited: a review of evidence linking diversity and invasibility. *Oikos*, **87**, 15–26.

Lambdon, P.W., Pyšek, P., Basnou, C., et al. (2008) Alien flora of Europe: species diversity, temporal trends, geographical patterns and research needs. *Preslia*, **80**, 101–149.

Lockwood, J.L., Cassey, P. & Blackburn, T.M. (2009) The more you introduce the more you get: the role of colonization and propagule pressure in invasion ecology. *Diversity and Distributions*, **15**, 904–910.

Lockwood, J.L., Hoopes, M.F. & Marchetti, M.P. (2007) *Invasion Ecology*. Blackwell, Oxford.

Lonsdale, M. (1999) Global patterns of plant invasions and the concept of invasibility. *Ecology*, **80**, 1522–1536.

McKinney, M.L. & Lockwood, J.L. (1999) Biotic homogenization: a few winners replacing many losers in the next mass extinction. *Trends in Ecology & Evolution*, **14**, 450–453.

Occhipinti-Ambrogi, A. & Galil, B.S. (2004) A uniform terminology on bioinvasions: a chimera or an operative tool? *Marine Pollution Bulletin*, **49**, 688–694.

Parker, I.M., Simberloff, D., Lonsdale, W.M., et al. (1999) Impact: toward a framework for understanding the ecological effect of invaders. *Biological Invasions*, **1**, 3–19.

Procheş, Ş., Wilson, J.R.U., Richardson, D.M. & Rejmánek, M. (2008) Searching for phylogenetic pattern in biological invasions. *Global Ecology and Biogeography*, **17**, 5–10.

Pyšek, P., Hulme, P.E. & Nentwig, W. (2009) Glossary of the main technical terms used in the handbook. *Handbook of Alien Species in Europe* (ed. DAISIE), pp. 375–379. Springer, Berlin.

Pyšek, P. & Richardson, D.M. (2010) Invasive species, environmental change and management, and health. *Annual Review of Environment and Resources*, **35**, doi:10.1146/annurev-environ-033009-095548 (in press).

Pyšek, P., Richardson, D.M., Rejmánek, M., Webster, G.L., Williamson, M. & Kirschner, J. (2004) Alien plants in checklists and floras: towards better communication between taxonomists and ecologists. *Taxon*, **53**, 131–143.

Quine, W.V. (1936) Truth by convention. In *Philosphical Essays for Alfred North Whitehead*, pp. 90–124. Longmans, Green & Co., New York.

Rejmánek, M. (2000) Invasive plants: approaches and predictions. *Austral Ecology*, **25**, 497–506.

Rejmánek, M., Richardson, D.M., Barbour, M.G., et al. (2002) Biological invasions: politics and discontinuity of ecological terminology. *Bulletin of the Ecological Society of America*, **83**, 131–133.

Ricciardi, A. & Cohen, J. (2007) The invasiveness of an introduced species does not predict its impact. *Biological Invasions*, **9**, 309–315.

Ricciardi, A. & Simberloff, D. (2009) Assisted colonization is not a viable conservation strategy. *Trends in Ecology & Evolution*, **24**, 248–253.

Richardson, D.M. (2009) Invasion biology deconstructed. *Trends in Ecology & Evolution*, **24**, 258–259.

Richardson, D.M., Allsopp, N., D'Antonio, C.M., Milton, S.J. & Rejmánek, M. (2000a) Plant invasions: the role of mutualisms. *Biological Reviews*, **75**, 65–93.

Richardson, D.M., Hellmann, J.J., McLachlan, J., et al. (2009) Multidimensional evaluation of managed relocation. *Proceedings of the National Academy of Sciences of the USA*, **106**, 9721–9724.

Richardson, D.M. & Pyšek, P. (2006) Plant invasions – merging the concepts of species invasiveness and community invasibility. *Progress in Physical Geography*, **30**, 409–431.

Richardson, D.M., Pyšek, P., Simberloff, D., Rejmánek, M. & Mader, A.D. (2008) Biological invasions – the widening debate: a response to Charles Warren. *Progress in Human Geography*, **32**, 295–298.

Richardson, D.M., Pyšek, P., Rejmánek, M., Barbour, M.G., Panetta, D.F. & West, C.J. (2000b) Naturalization and invasion of alien plants: concepts and definitions. *Diversity and Distributions*, **6**, 93–107.

Richardson, D.M. & van Wilgen, B.W. (2004) Invasive alien plants in South Africa: how well do we understand the ecological impacts? *South African Journal of Science*, **100**, 45–52.

Sax, D.F. & Brown, J.H. (2000) The paradox of invasion. *Global Ecology and Biogeography*, **9**, 363–372.

Shea, K. & Chesson, P. (2002) Community ecology theory as a framework for biological invasions. *Trends in Ecology & Evolution*, **17**, 70–76.

Simberloff, D. (2006) Invasional meltdown 6 years later: important phenomenon, unfortunate metaphor, or both? *Ecology Letters* **9**, 912–919.

Simberloff, D. (2009) The role of propagule pressure in biological invasions. *Annual Review of Ecology, Evolution, and Systematics*, **40**, 81–102.

Simberloff, D. & Von Holle, B. (1999) Positive interactions of nonindigenous species: invasional meltdown? *Biological Invasions*, **1**, 1–32.

Sorte, C.J.B., Williams, S.L. & Carlton, J.T. (2010) Marine range shifts and species introductions: comparative spread rates and community impacts. *Global Ecology and Biogeography* **19**, 303–316.

Stohlgren, T.J., Binkley, D., Chong, G.W., et al. (1999) Exotic plant species invade hot spots of native plant diversity. *Ecological Monographs*, **69**, 25–46.

Stohlgren, T., Jarnevich, C., Chong, G.W. & Evangelista, P.H. (2006) Scale and plant invasions: a theory of biotic acceptance. *Preslia*, **78**, 405–426.

Thuiller, W., Gallien, L., Boulangeat, I., et al. (2010) Resolving Darwin's naturalization conundrum: a quest for evidence. *Diversity and Distributions*, **16**, 461–475.

Tilman, D. (1999) The ecological consequences of changes in biodiversity: a search for general principles. *Ecology*, **80**, 1455–1474.

Vilà, M., Basnou, C., Pyšek, P., et al. & DAISIE partners (2010) How well do we understand the impacts of alien species on ecological services? A pan-European cross-taxa assessment. *Frontiers in Ecology and the Environment*, **8**, 135–144.

Williamson, M., Dehnen-Schmutz, K., Kühn, I., et al. (2009) The distribution of range sizes of native and alien plants in four European countries and the effects of residence time. *Diversity and Distributions*, **15**, 158–166.

Williamson, M. & Brown, K.C. (1986) The analysis and modelling of British invasions. *Philosophical Transactions of the Royal Society of London B*, **314**, 505–522.

Williamson M. & Fitter A. (1996) The varying success of invaders. *Ecology*, **77**, 1661–1666.

Wilson, J.R.U., Dormontt, E.E., Prentis, P.J., Lowe, A.J. & Richardson, D.M. (2009a) Biogeographic concepts define invasion biology. *Trends in Ecology & Evolution*, **24**, 586.

Wilson, J.R.U., Dormontt, E.E., Prentis, P.J., Lowe, A.J. & Richardson, D.M. (2009b) Something in the way you move: dispersal pathways affect invasion success. *Trends in Ecology & Evolution*, **24**, 136–144.

Wilson, J.R.U., Richardson, D.M., Rouget, M., et al. (2007) Residence time and potential range: crucial considerations in modelling plant invasions. *Diversity and Distributions*, **13**, 11–22.

Winter, M., Schweiger, O., Klotz, S., et al. (2009) Plant extinctions and introductions lead to phylogenetic and taxonomic homogenization of the European flora. *Proceedings of the National Academy of Sciences of the USA*, **106**, 21721–21725.

TAXONOMIC INDEX

Page numbers in italics indicate charts, graphs, photographs or text boxes
(common names are listed only where these are widely used)

Acacia mearnsii (black wattle) *Plate 4F*, 93, 383
Acacia melanoxylon (Australian blackwood) 282
Acer negundo (box elder) 136
Acer platanoides (Norway maple) 136
Acer pseudoplatanus (sycamore) 76, *76*, *77*
Achillea millefolium (common yarrow) 136
Acridotheres tristis (Indian myna) 54, 169
Adelgis tsugae (hemlock woolly adelgid) 353
Aegilops triuncialis (barbed goatgrass) 137, 178
Ageratina adenophora (crofton weed) 136
Agropyron cristatum (crested wheatgrass) 136, 260
Agropyron desertorum (desert wheatgrass) 260
Agropyron spicatum (bluebunch wheatgrass) 259
Ailanthus altissima (tree of heaven) 13
Alliaria petiolata (garlic mustard) 139, 178, 183
Alosa pseudoharengus (alewife) 216
Alosa sapidissima (American shad) *28*
Alstroemeria aurea (Lily of the Incas) 152
Amanita phalloides (death cap) 294
Ambrosia artemisiifolia (common ragweed) 178, 203
Ammophilia arenaria (marram grass) 133–134, 148, 182, 178
Amphibalanus eburneus 28
Amphibalanus improvisus 28
Amylostereum areolatum 94
Andropogon virginicus (broomsedge bluestem) 283
Anopheles gambiae (malaria mosquito) 13

Anoplolepis gracilipes (yellow crazy ant) 54, 58, 153, 240, 242, 245, 246
Anoplolepis longicornis (black crazy ant) 247
Anoplophora glabripennis (Asian longhorn beetle) 54
Apalopteron familiare 169
Apera spica-venti (silky bentgrass) *178*
Apiaceae 77, *180*
Apis mellifera (European honey bee) 149
Ardisia elliptica (shoebutton ardisia) 54
Aristida meridionalis 137
Artemisia tridentata 254
Arundo donax (giant reed) 353
Asparagopsis armata 28
Asteraceae 178,*178*, 203, 204
Atlantoxerus getulus 150
Austrominius modestus 28
Avena barbata (slender oat) *178*

Balanus eburneus (ivory barnacle) *28*
Balanus improvisus (bay barnacle) *28*
Balsamorhiza sagittata (arrowleaf balsamroot) 259
Batrachochytrium dendrobatidis (frog chytrid fungus) 54, 56
Berberis thunbergii (Japanese barberry) 278
Biddulphia sinensis 28
Bidens pilosa (Spanish needle) 136
Boiga irregularis (brown tree snake) 54, 57
Bombus dahlbomii 152
Bombus ruderatus (bumblebee) 152
Boraginaceae *178*
Botryllus schlosseri (ascidian) *Plate 2*
Botryosphaeriaceae 92, 94
Brachypodium sylvaticum (false brome) *178*
Brassicaceae 139, *178*, *180*, 183

Bromus inermis (smooth brome) 136, 261
Bromus madritensis (compact brome) 351, *351*
Bromus mollis (soft brome) *178*
Bromus tectorum (cheatgrass) 21, 138, *178*, 253–265, *254*, 352
Bryophyllum delagoense (mother of millions) 178
Bryum argenteum 261
Bucephalus polymorphus (trematode) 217, 231
Bufo marinus (cane toad) 54, 57, 349
Bulbulcus ibis (cattle egret) 41
Bursaphelenchus xylophilus (pine wood nematode) 91
Bythotrephes longimanus (spiny waterflea) 52

Cameraria ohridella horse chestnut (leaf-miner) 74, *75*
Campanulaceae 152
Campuloclinium macrocephalum (pompom weed) 383
Capsella bursa-pastoris (shepherd's purse) *178*
Carcinus maenas (European green crab) 54, 230
Cardiospermum grandiflorum (balloon vine) 383
Carnegiea gigantea (saguaro) 43
Carpobrotus edulis (Hottentot fig) 77
Carpobrotus species 154
Caryophyllaceae *180*
Castanea sativa (European chestnut) *42*
Caulerpa taxifolia (green alga/caulerpa seaweed) 54, 57, 58, 77
Cenchrus biflorus (Indian sandbur) 137
Centaurea diffusa (diffuse knapweed) 146, *178*

Fifty Years of Invasion Ecology: The Legacy of Charles Elton, 1st edition. Edited by David M. Richardson
© 2011 by Blackwell Publishing Ltd

Centaurea maculosa (spotted knapweed) 146

Centaurea solstitialis (yellow starthistle) *351*

Centaurea stoebe micranthos (spotted knapweed) 135–136, *178*

Ceratocystis albifundus Plate 4F, 93

Ceratonia siliqua (carob tree) *113*

Cercopagis pengoi (fishhook waterflea) 54, 58

Cervus elaphus (red deer) 228

Channa argus (northern snakehead) 42

Chenopodiaceae 138

Chondrilla juncea (skeleton weed) *178*

Chromolaena odorata (triffid weed) 138, 378

Cinara cupressi (cypress aphid) 54

Cinchona pubescens (quinine tree) 54

Ciona intestinalis (ascidian) *Plate 2*

Circosporda zeae-maydis (grey leaf spot) 318, 321

Circus approximans (swamp harrier) 169

Cirsium arvense (Canada thistle) 349

Clidemia hirta (soapbush; Koster's curse) 54, 133, *178*, 182

Clusiaceae *180*

Cneorum tricoccon (spurge olive) 152

Codium fragile (dead man's fingers) 294

Corbicula fluminea (Asian clam) 216, 318

Corbula amurensis (marine clam) 57

Corvus splendens (house crow) 168

Corynebacterium sependonicum 306

Craspedacusta sowerbyi (freshwater jellyfish) 217

Crassostrea gigas 28

Crassulaceae 154, *178*

Crataegus monogyna (oneseed hawthorn) 146

Crepidula fornicata 28

Cronartium quercuum (fusiform rust) *Plate 4D*, 92

Cryphonectria cubensis 91

Cryphonectria parasitica (chestnut blight) 54, 56, 90–91, 305

Cryphonectriaceae 92, 93

Cyprinus carpio (common carp) 41, 54, 57

Cytisus scoparius (Scotch broom) 147, 149, *178*

Dendroctonus ponderosae (mountain pine beetle) 353

Depressaria pastinacella (parsnip webworm) 183

Doleromyrma darwiniana 247

Diplodia pinea 94

Diploschistes muscorum 261

Diplotaxis erucoides 152

Dorosoma petenense (threadfin shad) 216

Dothistroma septosporum 92, 94

Dreissena polymorpha (zebra mussel) 54, 56, 57, 212, 230, 316, 318, 320

Dreissena rostriformus bugensis (quagga mussel) 216, 230

Drosophila 185

Ducula aurorae (Tahiti imperial pigeon) 169

Echium plantagineum (Paterson's curse) *178*

Eichhornia crassipes (water hyacinth) 54, 57, 197, 204, *205*, 321, *Plate 6c*

Eichhornia paniculata (Brazilian water hyacinth) 184–185, 197, 198–199, *199*, 200, 205, *Plate 6a*

Elminius modestus 28

Elodea canadensis (Canadian pondweed) 196

Epipactis helleborine (broad leaved helleborine) *178*

Eragrostis lehmanniana (Lehmann lovegrass) 137

Erigeron annuus (daisy fleabane) *180*

Eriocheir sinensis (Chinese mitten crab) 12, 26, *28*, 56, 305

Esox lucius (northern pike) 216

Eucalyptus camaldulensis (river red gum) 378

Euglandina rosea (rosy wolf snail) 54

Euphorbia esula (leafy spurge) 136

Fabaceae 147, *178*, *180*

Falcataria moluccana 285, *285*

Fallopia japonica (Japanese knotweed) 58, 204

Festuca idahoensis (Idaho fescue) 136, 259

Festuca ovina (blue fescue) 135–136

Ficopomatus enigmaticus 28

Ficus spp. (figs) 147, 149, 150

Ficus carica (common fig) 112, *113*

Ficus sycomorus (sycamore fig) *113*

Fraxinus uhdei (evergreen ash) 281, *282*, 285

Gambusia amistadensis (Amistad gambusia) 214

Gopherus agassizi (desert tortoise) 43

Hakea spp. (hakeas) 378

Halogeton glomeratus 138

Hedychium gardnerianum (Kahili ginger) 54, 285

Helianthus debilis (beach sunflower) 186

Heracleum mantegazzianum (giant hogweed) 77, *180*

Heracleum persicum (Persian hogweed) *180*

Heracleum sosnowskyi (Sosnowskyi hogweed) *180*

Herpestes javanicus (small Indian mongoose) 54

Hieracium aurantiacum (orange hawkweed) *204*

Hieracium pilocella (European hawkweed) *204*

Hiptage benghalensis (hiptage) 54

Hirschfeldia incana (hoary mustard) *180*

Hydrocotyle ranunculoides (floating pennywort) 308

Hydrilla verticillata (hydrilla) 217

Hylaeus volcanica 153

Hypericum canariense (Canary Island St. John's wort) *180*

Hypericum perforatum (St. John's wort) *180*, 184

Hypochaeris radicata (cat's ear) *351*

Imperata cylindrica (cogon grass) 57

Iridaceae 203

Jacaranda mimosaefolia (jacaranda) 389

Kalanchoe pinnata (air plant) 154

Kochia prostrata (forage kochia) 260

Koeleria cristata 136

Lamium album (white dead-nettle) 14

Lantana camara (lantana) 57, 378

Larrea tridentata (creosote bush) 43

Lasius neglectus 247

Lasthenia californica 137

Lates niloticus (Nile perch) 54, 57, 58–59, 212, 321

Leander adspersus 28

Leontodon taraxacoides 351

Lepomis macrochirus (bluegill sunfish) 216

Leptinotarsa decemlineata (Colorado beetle) 74, 75

Leucadendron cordifolia Plate 4C

Leucadendron spp. 91

Leucaena leucocephala (leucaena) 54

Leucospermum spp. 91

Ligustrum robustum (tree privet) *180*

Limnothrissa miodon (Tanganyika sardine) 321

Linepithema humile (Argentine ant) 153, 229, 240–248, *241*

Linum lewisii 136

Lolium perenne (perennial ryegrass) 137

Lonicera japonica (Japanese honeysuckle) *351*

Lymantria dispar (gypsy moth) 54, 57, 90, 349

Lysichiton americanus (American skunk cabbage) 308
Lythraceae 201, *Plate 6b*
Lythrum salicaria (purple loosestrife) 146, 183, 185, 201, *202*, *Plate 6b*

Martes martes (pine marten) 152
Megachile rotundata 149
Melastomataceae 54, 93, 133, *178*, 182
Melia azedarach (Chinaberry tree) *113*
Melinus minutiflora (molasses grass) 283, *283*
Mercierella enigmatica 28
Meterosideros polymorpha (Ohia) 94, 279, 280, 281–282
Metridium senile (large white sea anemone) 27
Micropterus dolomieu (smallmouth bass) 216
Micropterus salmoides (largemouth bass 215
Microstegium vimineum (Japanese stiltgrass) 351
Mikania micrantha (mile-a-minute weed) 54
Mimosa pigra (mimosa / giant sensitive tree) 57
Miscanthus X giganteus 353
Mnemiopsis leidyi (comb jelly) 52, 54, 74, 321
Molgula manhattensis (ascidian) *Plate 2*
Monomorium pharoensis 247
Monomorium sydneyense 247
Mononychellus tanajoa (cassava green mite) 321
Morella faya (fire tree) Plate 7, 54, 278, 280–281, *280*, *281*, 285, 414
Morone saxatilis 28
Morus alba (white mulberry) 112, *113*
Mugil cephalus 28
Mus musculus (house mouse) 54
Mya arenaria (North Atlantic softshell clam) 26, 28
Mycosphaerella nubilosa 94
Myrax fugax 28
Myrica faya see Morella faya
Myriophyllum spicatum (Eurasian watermilfoil) 216, 217
Myrmica rubra 247
Myrtaceae 93, 279
Mysis diluviana (formerly *M. relicta*) (opossum shrimp) 215
Mysis relicta see Mysis diluviana
Mytilaster lineatus 28
Mytilus galloprovincialis (Mediterranean mussel) 56

Neanthes succinea 28
Neofusicoccum australe (= *Botryosphaeria australis*) 92
Neogobius melanostomus (Eurasian round goby) 217, 230
Neptunus pelagicus 28
Neptunus sanguinolentus 28
Nereis succinea 28
Nesocodon mauritianus 152
Nicotiana glauca (wild tobacco) 149
Nymphoides peltata 204
Nitzschia sturionis 28
Nucifraga columbiana Clark's nutcracker) 150

Ocinebrina inornata 28
Ochetellus glaber 247
Odontella sinensis 28
Odontra zibethicus see Ondatra zibethicus
Oleaceae *113*, *180*
Olea europaea cuspidata (wild olive) *113*, *180*
Olea europaea europaea (cultivated olive) *113*, *180*
Onchorhynchus kisutch (Coho salmon) 28
Onchorhynchus mykiss (rainbow trout) 54, *57*, 216
Onchorhynchus nerka (landlocked salmon) 28, 215
Onchorhynchus tschawytscha (chinook salmon) 28
Ondatra zibethicus (muskrat) 41, 74, 75, 305
Ophiostoma ulmi (Dutch elm disease) 54, 56
Opuntia maxima Plate 5, 148, 148, 150
Orchidaceae *178* see also orchids
Oreochromis mossambicus (Mozambique tilapia) 57
Oryctolagus cuniculus (rabbit) 41, 54
Ostrea edulis 28
Ostrea gigas 28
Oxalis pes-caprae (Bermuda buttercup) 152

Pachycondyla chinensis 247
Palaemon adspersus 28
Paphia philippinarum 28
Paratrechina longicornis 247
Paratrechina fulva 247
Parthenium hysterophorus (parthenium weed) 383
Pastinaca sativa (wild parsnip) 183, 231
Pelophylax (formerly *Rana*) *lessonae* 41–42
Pennisetum setaceum (fountain grass) *180*
Petromyzon marinus (sea lamprey) 212, 305

Phalaris arundinacea (reed canarygrass) *180*, 230
Pheidole megacephala (big-headed ant) 240, 245, 246, 247
Pheidole obscurithorax 247
Phenacoccus manihoti (cassava mealybug) 321
Phoenix dactylifera (true date palm) *113*
Phyla canescens (lippia) *180*, 184, 187
Phytophthora cinnamomi (phytophthora root rot) *Plate 4C*, 54, 56, 91, 95
Phytophthora pinifolia *Plate 2*, 90
Phytophthora ramorum 93
Pinctada radiata 28
Pinctada vulgaris 28
Pinus albicaulis (whitebark pine) 150
Pinus elliottii (slash pine) *Plate 4D*, 92
Pinus radiata (Monterey pine) *Plate 1*, *Plate 4*, 90–91
Plantago erecta (dotseed plantain) 137
Plasmodium relictum (avian malaria) 54, 56
Poaceae 138, *178*, *180*, 254–265
Pomarea nigra (Tahiti flycatcher) 169
Pontederiaceae 198, 204, *Plate 6*
Potamocorbula amurensis see Corbula amurensis
Potamogeton cripsus (curly pondweed) 216
Processa aequimana (Red Sea shrimp) 28
Prunus laurocerasus (cherry laurel) *351*
Prunus serotina (black cherry) 135, 136
Pseudomonas flourescens 261
Psidium cattleianum (strawberry guava) 282, 285, 285
Psidium guajava (guava) *Plate 4G*, 147
Psittacula krameri (ring-necked parakeet, rose-ringed parakeet) 74, 75, 76, 167
Pteropodidae 111–112
Puccinia psidii *Plate 4G*, 93, 94
Pueraria lobata (kudzu) 19, 54, *180*
Pycnonotus cafer (red-vented bulbul) 54
Pycnonotus jocosus (red-whiskered bulbul) 152
Pyrenophora semenipera 261

Quambalaria eucalypti 92
Quambalaria pitereka Plate 4E, 92

Rana catesbeiana (bullfrog) 54
Rana lessonae see Pelophylax lessonae
Rattus rattus (ship rat) 54, 57
Rhithropanopeus harrisii 28
Rhododendron ponticum (common rhododendron) 76–77, *76*, 124
Roccus saxatilis 28\
Rosaceae *180*

Rosa multiflora (multiflora rose) 76–77, 76

Rousettus aegyptiacus (fruit bat) 109–115, 110–111

Rubus alceifolius (giant bramble) *180*

Rubus cuneifolius (American bramble) 383

Rubus ellipticus yellow (Himalayan raspberry) 54

Ruditapes philippinarum 28

Saccharomyces spp. 92

Salmo salar (Atlantic salmon) 28, 217

Salmo trutta (European brown trout) 57, 214

Salvelinus confluentus (bull trout) 349

Salvelinus fontinalis (brook trout) 349

Salvelinus namaycush (lake trout) 214

Sander lucioperca (pike-perch 217, 231

Sander vitreus (walleye) 216

Sapium sebiferum (Chinese tallow tree) 182

Saussurea deltoidea 136

Senecio jacobaea (tansy ragwort 182, 183, 320

Schinus terebinthifolius (Brazilian pepper tree) 54, *180*

Schizachyrium condensatum (bush beardgrass) 283

Silene latifolia (white campion) *180*

Silene vulgaris (bladder campion) *180*

Sirex noctilio (sirex wood wasp) 94

Solanum tuberosum (potato) 306

Solenopsis invicta (red imported fire ant; RIFA) 240, 242, 245–247

Solidago gigantea (giant goldenrod) 183

Sorghum halepense (Johnsongrass) 139

Spartina alterniflora (smooth cordgrass) *180*

Spartina townsendii (hybrid cordgrass 12, 28, 76, *76*, 196

Spathodea campanulata (African tulip tree) 54

Sciurus carolinensis (grey squirrel) – see grey squirrel

Sciurus vulgaris (red squirrel) – see red squirrel

Striga hermonthica (witchweed) 321

Sturnus vulgaris (common starling) 54, 57, 168, 305, 330

Sus scrofa (pig) 54, 57

Sylvia melanocephala, Plate 5

Tamarix aphylla (Athel tamarisk) 43

Tamarix chinensis (five-stamen tamarisk) 43

Tamarix ramosissima (saltcedar) 43

Tarpon atlanticus (tarpon) 28

Technomyrmex albipes 153, 247

Tecoma stans (yellow bells) 383

Teratosphaeria nubilosa 92

Tetramorium tsushimae 247

Thenus orientalis 28

Tibouchina spp. 93

Tilletia indica 306

Triadica sebifera see *Sapium sebiferum*

Trifolium hirtum (rose clover) *180*

Trifolium pratense (red clover) 147

Trifolium repens (white cover) 137

Tritonalia japonica (Japanese oyster drill) 28

Tsuga canadensis (eastern hemlock) *353*

Urosalpinx cinerea 28

Ustilago bullata 258, 261

Verbenaceae *180*, 184, 187

Wasmannia auropunctata (little fire ant) 54, 153, 240, 245, 246, 247

Zea mays ssp. *parviglumis* (wild maize) 188

Zea mays ssp. *mays* (maize) 188

Zosterops japonicas (Japanese white-eye) 152, 169

Zosterops lateralis (western silver-eye) 152

GENERAL INDEX

Page numbers in italics indicate charts, graphs, photographs or text boxes

acacias 92, 93, 282, 348, 378 see also *Acacia* spp.
adaptation, local *see* local adaptation
adaptive evolution *see* evolution, adaptive
Adelaide symposium (1977) *see* 'Exotic species – their establishment and success'
aerial surveys 284 *see also* remote sensing (in detection of invasion)
adventive species *see* alien species
African Lakes, the 321 *see also* Lake Victoria
Agreement on the Application of Sanitary and Phytosanitary Measures (SPS Agreement) 303, 306, 310, 323, 324
agriculture *see* cultivation; livestock
ALARM *see* Assessing Large Scale Risks for Biodiversity using Tested Methods
alien species *see also* invasive species
  and biotic acceptance 411
  control of 322–324, 339
  definition 38, 64–65, 410
  economic effects of 316–325, *320*, *321*, 323
  five criteria for determining (Lindroth) 31
  impacts of 65, 413–414
  management approaches 378
    see also risk assessment
  publications on 52–59, *55*, 398–407
  and socio-political values 66, 360–373
  spread, modelling of 329–343
Allee effects 201, 203–204, 228
allometric relationships (and animal seed dispersal) *110–111*
American slipper-limpet 12
anemones (Metridium) 26, 27

Animal Ecology (Charles Elton, 1927) 4, 12, 20, *63*, 162
ant invasions 239–250
anthrax 304
anthropogenic dimensions of biological invasions 315–328
anthropogenic replacement 43
  see also human-assisted species introductions
aquatic plants, introduced 215–216
aquatic predators, introduced 214–216, *215*
Argentine ant 229, 240–248 *see also Linepithema humile*
asexual reproduction, invasion through 204
Asian longhorn beetle *see Anoplophora glabripennis*
Asilomar conference (California, 1964) 15, 196, 362
Asmal, Kader 379, 387
assembly rules (in restoration ecology) 62
Assessing Large Scale Risks for Biodiversity using Tested Methods (ALARM) 79, 85
assisted migration 206 *see also* managed relocation
Aurelia (jellyfish) 27
Australia
  ant communities in 240–242, 245–246
  Codium fragile in 294
  deliberate plant introductions 318
  economic losses due to introduced pests 306, 320–321, *320*, *321*
  impact of introduced species 15, 273
  invasive biology research 18
  management strategies for invasive species 378
  Pheidole megacephala in 240
avian malaria *see Plasmodium relictum*

'back introduction' (in tree pathogens) 93
Baker, Herbert G. *xiv*, 196, 362
  see also Genetics of Colonising Species, The
Baker's law *see* Baker's rule
Baker's rule 147, 177, 203, 411
ballast, water in ships
  role in biological invasions in 26, 28, 30–31, 52, *227*, 230, 305, 318, 402, 418
banana bunchy top virus 56
Banks and Solander Flora (1770) 40–41
BAP *see* Bureau of Animal Population
Barcoding of Life Database (BOLD) 291, 296
Bates, Marston 27
bees, displacement of 153
beetles, transmission over North Atlantic 30–31
Belnap, Jayne 138
Bever, Jim *132*, 134
big-headed ant *see Pheidole megacephala*
biofuels 353, 402
biogeography of invasions 134–135
bioinvasions *see* biological invasions
biological invasions
  definition 411
  economics of 315–325, *320*, *321*
  financial costs of 84, 305–306, 320–322
  global control of 323–324
  stages of 104–105
biosafety 303–304, *303*
biosecurity 65, 301–314, *303*, 399, 411
bioterrorism 303–304, *303*
biotic acceptance 411, 412
biotic homogenization 411
biotic invasions *see* biological invasions
biotic nativeness 35–47, 372
biotic resistance 144, 319, 411

Fifty Years of Invasion Ecology: The Legacy of Charles Elton, 1st edition. Edited by David M. Richardson
© 2011 by Blackwell Publishing Ltd

biotic resistance hypothesis *see* diversity – invasibility hypothesis
bird invasions 161–171
black bass 12
Blackstone, Sir William 38
body mass, role in animal seed dispersal *110–111*
BOLD *see* Barcoding of Life Database
Bonin Islands 149, 153, 169
botulism bacteria 217
'boundary management' *390*, 390, 398, 399, *399*, 403
Brazil
　Anopheles gambiae in 13
　economic losses due to introduced pests 320–321, *320*, *321*
　Eichhornia paniculata in 198, *199*, 200
Britain
　Acer pseudoplatanus in 76
　Chinese mitten crab in 12, 26
　early invasive species 12–13
　government conservation model 6–7
　grey squirrel in, 12–13
　Lamium album in *14*
　muskrat in 12, 13
　nursery trade and 318
　Pelophylax (formerly *Rana*) *lessonae* in 41
　Rhododendron ponticum in 76, 124
British Ecological Society 20
Brookhaven Symposium (1969) 16
brown tree snake *see Boiga irregularis*
brown trout 214–215
　see also Salmo trutta
brush fires, role of cheatgrass in 255, 259, 284
bullfrog (*Rana catesbeiana*) 54
bumblebee *see Bombus* species
Bureau of Animal Population (BAP) (University of Oxford) 4–5, 7, 8, 13, 14–15

CA *see* cellular automata
California
　Aegilops triuncialis in *178*
　Amanita phalloides in 294
　Ammophilia in 133
　Avena barbata in *178*
　Bromus tectorum in 256
　contamination of harbour fauna 26
　Cytisus scoparius in *178*
　Dorosoma petenense in 216
　Dreissena polymorpha in 56
　freshwater invasions 213, 219
　Linepithema humile (Argentine ant) in 229, 240, 242, 243
　Pennisetum setaceum in *180*
　Trifolium hirtum in *180*

Callaway, Ragan *132*, 135, 136
Canada
　Apera spica-venti in *178*
　Bromus tectorum in 254, 256, 258
　Epipactis helleborine in *178*
　extinction threats in 273
　Lythrum salicaria in Plate 6b
Canada thistle *see Cirsium arvense*
Canadian lynx 330
Candolle, Alphonse de *xiv*, 38–39, 41, 412
cane toad *see Bufo marinus*
canopy nitrogen measurement 284–285, *285*
CAO *see* Carnegie Airborne Observatory
Cape Floristic Region, pathogens in 91
carbon dioxide *see* CO_2
Carnegie Airborne Observatory (CAO) 284–285, 286
Caspian Sea 28, *28*, 29, 230
'casual' (classification category for alien species) 40, *146*, 412, 417–418
CBD *see* Convention on Biological Diversity
CBOL *see* Consortium on the Barcoding of Life
cellular automata 331, *337*, 338
cheatgrass 253–265, 352–353 *see also Bromus tectorum*
chestnut blight *see Cryphonectria parasitica*
Chew, Matthew 6–7, *63*
Chinese mitten crab (*Eriocheir sinensis*) 12, 26
Chitty, Dennis 20
Christmas Island 242, 247
Chromolaena 138
Chrysoporthe species 93
Chrzanowski, Thomas H. *132*
clams (*Mya*) 28
Clark's nutcracker *see Nucifraga columbiana*
climate change
　effect on biological invasions 66, 81, 95–96, 164, 198, *241*, 247, 302, 347–349
　effect on invasive species' distribution 349–350
　interactions with biological invasions 31, 52, 319
　invasive species' responses to 347–349, *348*, 351, *351*
　research on 59, 402
climate envelope-type studies 335, 348, 349 *see also* niche modelling
Codex Alimentarius Commission 325
cogon grass *see Imperata cylindrica*
COI (cytochorome oxidase subunit I) 291

CO_2 (in atmosphere) 346–347, *346*, 349, 350–352, *351*
'colonist', Watson definition 37
colonization pressure 163, 226, 412
　see also propagule pressure
colonization – extinction cycles 198
Colorado potato beetle 306 *see also Leptinotarsa decemlineata*
Columbian exchange 316
comb jelly *see Mnemiopsis leidyi*
'Committee of the British Ecological Society' (1944) report 20
Commonwealth Quarantine Service 302
community ecology 7–9, 18, 63, 122, 125, 132, 270–272, 278, 398, 399, 402
conservation biogeography 399, 400
Consortium on the Barcoding of Life (CBOL) 290–291
competitive release hypothesis 412
control of invasive species *see* invasive species, control of
Convention on Biological Diversity (CBD) 303
corridor, definition 412
crayfish 12, 212
creosote bush *see Larrea tridentata*
crop monocultures 132
crop rotation 132
cryptogenic species, definition 412
cultivation, effect on invasions 167–168, 302, 319
Cybele Britannica (H.C. Watson, 1859) 39

DAISIE *see* Delivering Alien Invasive Species Inventories for Europe
damage cost of invasive species 320–322, *320*, *321 see also* economics of biological invasions
Darlington, P.J. 127–128
Darwin, Charles
　on evolutionary significance of biodiversity 122, *123*, 128
　on invasive species xiv, 14
　naturalization hypothesis 126, 145, 196, 270, 412
　Origin of Species (1859) 8, 14, 39, 127–128, 144, 145, 196, 412
'Darwin's naturalization hypothesis' *see* Darwin, Charles
de Candolle, Alphonse *see* Candolle, Alphonse de
deliberate introductions 28, 29, 31, 165
Delivering Alien Invasive Species Inventories for Europe (DAISIE) 77–78, 79, 85
'denizen', Watson definition 37
DeWalt, Saara *132*, 133

DIH *see* diversity – invasibility hypothesis

dispersal curves 339

dispersal kernel 105–107, 111–113, 115, 332–334, 338–339

dispersal pathway, definition 412 *see also* introduction pathway

direct facilitation (of invasion) 414

disturbance (as factor in invasions) 16, 79, 245–246

diversity – invasibility hypothesis (DIH)
  in animal invasions 127
  ants 245–246
  and biotic resistance 411
  contrary evidence 125
  definition 122–123, 245, 412
  development of *123*
  Elton and MacArthur on 127–128, 162–163
  and niche-based approach 270272

diversity – stability relationship (in ecosystems) 16

DNA barcoding 289–299, 403

Dobzhansky, Theodosius 128

Dothistroma needle blight *Plate 4B*, 92

Double Trouble (e-bulletin) 347

Dutch elm disease *(Ophiostoma ulmi)* 54, 91

early detection 294, 387, 391

Early Detection and Rapid Response for Invasive Alien Plants programme (South Africa) 387

ecological impact, definition 227

'ecological resistance' to invasion
  arguments for (Elton) 123–124
  Argentine ant 245–246

ecological succession (in restoration ecology) 62

'Ecological and Genetic Implications of Fish Introductions' (1990 international symposium) 212

Ecology of Animals (Charles Elton, 1933) 12

Ecology of Invasions by Animals and Plants, The (Charles Elton, 1958)
  animal bias in 74–76, *76*
  ant invasions 240–248
  biodiversity 122, 123–124
  bird invasions 161–171
  context 5–7
  genetic and evolutionary issues in 196
  human-induced change 302–303, 315–325
  marine invasions 26–33
  overview 3–10, 12
  plant taxa in 76
  restoration ecology 62

soil biota *132*

species dispersal 104, 114

tree pests 90

economics of biological invasions 315–325, *320, 321*
  see also biological invasions, financial costs of 305–306

ecosystem degradation, study of 64

ecosystem engineers 154, 229

ecosystem services 6, 84, 92, 305, 316, 319–324, 354, 378, 380–383

ecosystem transformation, invasions and 278–288

Egler, Frank *xiv*, 20

EIAP see Ecology of Invasions by Animals and Plants, The

EICA see evolution of increased competitive ability hypothesis

Elton, Charles Sutherland (1900–1991) *xiv*
  academic milieu 4–10, 12
    on ant invasions 239–250
  biographical details 4–9, 12
  citation history *xiv*, 52
  on dispersal processes 104–105
  European biological invasions 74–77
  introduced species, views on 12–14
  relationship with Aldo Leopold 6, 9, 21, *63*, 124
  on marine invasions 25–33
  publication overview 4
  radio broadcasts 13, 26
  influence on modern biology 14–21
  see also Ecology of Animals, The; Ecology of Invasions by Animals and Plants; The; Animal Ecology; Exploring the Animal World; Land Ethic, The; Pattern of Animal Communities, The; Bureau of Animal Population (BAP)

'Elton symposium' *see* 'Fifty years of invasion ecology – the legacy of Charles Elton'

'Elton's book' *see Ecology of Invasions by Animals and Plants, The*

enemy escape 92, *132*, 134–135

enemy release hypothesis (ERH) 124, 135–136, 144, 167, 177–183, 413

English Nature *see* Nature Conservancy (UK)

environmental heterogeneity, modelling 334–336

environmental stochasticity *331*, 336–337

Eppinga, Maarten 138

EPPO *see* European and Mediterranean Plant Protection Organization

Eradication 43, 284, 294, 303, 306, 311, 322, 331, 339, 360, 378, 411
  Anopheles gambiae Brazil 13
  ants 243, 247
  definition 413

ERH *see* enemy release hypothesis

ethics, environmental 361–365

Eucalyptus 20, 91, 92, 93, 94

Eucalyptus rust 93

Europe
  as donor of alien species 74
  alien species in 73–85, 134–135, 217
  key studies in 79–82, *80–81*
  management of alien species 83–85

European Alien Species Database *see* Delivering Alien Invasive Species Inventories for Europe (DAISIE)

European and Mediterranean Plant Protection Organization (EPPO) 308

European Expertise Registry *see* Delivering Alien Invasive Species Inventories for Europe (DAISIE)

European Invasive Alien Species Information System *see* Delivering Alien Invasive Species Inventories for Europe (DAISIE)

evolution, adaptive
  and enemy release 177–183
  future directions in research 186–188, 401
  genetic constraints on 201
  and invasion success 175–193
  and propagule pressure 176–177

evolution of increased competitive ability hypothesis (EICA) 182–183, 412–413

exotic species *see* alien species

'Exotic species – their establishment and success' (symposium, Adelaide, 1977) 18

Exploring the Animal World (Charles Elton, 1933c) 12

454 sequencing 295–296

farming, effect on invasions
  see crop rotation; crop monocultures; cultivation; livestock

Faunal Connections between Europe and North America, The (C.H. Lindroth, 1957) 30–31

Federal Horticultural Board 302

feral species, definition 413

fig (*Ficus* species) 147
  see also Ficus carica, Ficus sycomorus, Ficus spp.

financial costs of invasive alien species *see*
biological invasions, financial costs
of
'Fifty years of invasion ecology – the
legacy of Charles Elton'
(symposium, Stellenbosch, Nov.
2008) xi, xiii, *xiv*, 12,
fire tree *see Morella faya*
fish introduction symposium (1990) *see*
'Ecological and Genetic
Implications of Fish Introductions'
fisheries 31, 52
fishhook waterflea *see Cercopagis pengoi*
Flathead Lake (Montana) 215
floating pennywort *see Hydrocotyle
ranunculoides*
Flora Europaea (Tutin et al.,
1964–1980) 77
floral trade, role in spread of
pathogens 94
fluctuating resources theory of
invasibility 413
Forbes, Stephen 362
foreign species *see* alien species
forest pests *see* tree pests
forests, invasion patterns in 92–93
freshwater ecosystems, impacts of
biological invasions on 211–224,
213, 219
freshwater invasions *see* freshwater
ecosystems
freshwater jellyfish *see Craspedacusta
sowerbyi*
fruit bat, and seed dispersal 103–115 *see
also Rousettus aegyptiacus*
Fusarium 138
fynbos vegetation *Plate 1*, 16, 91, 150,
243, 378–379, 381, *385–386 see
also* Cape Floristic Region

Galapagos Islands 147, 154, 247
garlic mustard *see Alliaria petiolata*
GATT *see* General Agreement on Tariffs
and Trade
Gatun Lake 28
Gause, Georgii Frantsevich 123
Gaussian dispersal curve 339
GDP, relationship with number of alien
species, Europe 78, *78*
General Agreement on Tariffs and Trade
(GATT) 302, 323
genetic diversity
ecological consequences of 205–207
in invasive plant species *178–181*
and reproductive systems 198–199
standing variation vs. new
mutation 184–185
genetic identification *see* DNA barcoding

genetic mixing 185–186
genetic variation *see* genetic diversity
genetically modified organism (GMO) 413
Genetics of Colonising Species, The (eds
Baker and Stebbins) 196, 362
genome scans 187, 188
Geographical Ecology (R.H. MacArthur,
1972) *123*, 125–128
giant goldenrod *see Solidago gigantea*
global warming 348, 350 *see also* climate
change
globalization, role in spread of alien
species 94, 302–303
goats 154
Great Basin Native Plant Selection and
Increase Project 260
Great Lakes
Alosa pseudoharengus in 216
high- vs. low-impact invasions *215, 219*
interspecies interactions 216, 217, 220
invasion by sea lamprey 212, 214,
216, 278
invasion by zebra mussel 56, 212
shipping and invasions 54, 56, 230,
232
stocked fish 228
waterfowl die-offs 217, 218
greenhouse gases 346 *see also* CO_2 (in
atmosphere)
grey squirrel (*Sciurus carolinensis*) as
invader in Britain 12–13, 21
Grinnell, Joseph *xiv*, 4
gypsy moth *see Lymantria dispar*

Hawaii
Acacia in 282
Clidemia hirta in 133, *178*, 182
Ficus species in 150
fire-responsive grasses 278, 283–284,
283
Fraxinus uhdei in 281, *282*
Keeling data collection on 346
Melinis minutiflora in 283, *283*
Metrosideros polymorpha in 279, *280,
281, 282*, 282
Morella faya in *Plate 7*, 278, *280, 281*,
pollinators in 153
Puccinia psidii in 94
vulnerable ecosystems 150, 278–284,
280
wild olive in *178*
Zosterops japonica in 152
Henslow, John 37, 40, 44–45
HiFIS 284, 286
hogweed see *Heracleum* species
honeybee 149, 152–153, 155 see also
Apis mellifera
Hooker, Joseph xiv

horse chestnut leaf-miner *see Cameraria
ohridella*
horticultural trade *see* nursery trade
host-specific pathogens (in trees) 91, 92,
93
Hotelling, Harold 317
house mouse *see Mus musculus*
hub-and-spoke model 413
Hudson River 215
Huffaker, Carl *xiv*
human agency *see* human-assisted species
introductions
human-assisted species introductions
Darwin's discussion of 196
definition 40–42
economic costs of 320–322
C. Elton on 302–303
impact of 104, 302
in lakes and rivers 212–219
methods of 317–320
hundred worst invasive species list (Lowe
et al. 2004) 52–59, *57*, 240
Huxley, Julian 5, 12
hybridization
and bees 153
role in diversity 200
in freshwater fishes 214
as a stimulus for invasion 185–186
technology to detect 293
in tree management 92
hybrids *see* hybridization
hyperspectral data *see* (HiFIS) 284, 286

IBMs *see* individual-based models
iBOL *see* International Barcoding of Life
Project
IHR *see* International Health Regulations
impact (of alien species)
definition 413
in freshwater systems 211–224
'incognita', Watson definition of 38
India, economic losses due to introduced
pests 320–321, *320, 321*
indirect facilitation (of invasion) 414
individual-based models (IBMs) 330,
331, 337, *337*
Industrial Revolution 302
infectious diseases, control of 324
International Barcoding of Life Project
(iBOL) 290–291
International Health Regulations'
(IHR) 323, 325
International Office of Epizootics 325
International Plant Protection
Convention 325
International Union of Biological Sciences
symposium 1964 *see* Asilomar
conference

introduced species *see* alien species
introduction
     definition 414
introduction pathways 412, 414
invasibility
     definition 415
     effect of species richness on 411–12
     mapping 84
     predicting 64
invasion (of alien species), definition 414
invasion biology *see* invasion ecology
invasion cliff 414
invasion complex 149, 414
invasion debt 400
     definition 414
invasion ecology
     definition 414
     history of studies in 51–60
     relationship with restoration
          ecology 62–66
invasion, level of 52, 415, 416
invasion paradox, definition 414–415
invasion pressure, definition 414, 415
invasion process, stages of 330
invasion science 397–407, 415
invasional meltdown (Simberloff & von
          Holle 1999) 144, 149–151, 412,
          414–415
invasive species
     control of 322–324
     cumulative publication rate on 52–59
     definition 310, 415
     detection of 293–294
     spread of 329–343
     see also alien species
invasiveness, definition 415
invasive casual, naturalized or invasive
          classification 411
islands
     and invasion complexes 149–150
     effects of invasions 153–154, 165
     invasibility of 169
     see also entries for individual islands
ITS *see* nuclear ribosomal ITS
IUCN list of 100 worst invasive species *see*
          hundred worst invasive species list

Janzen – Connell hypothesis *132*, 133
Japanese knotweed *see Fallopica japonica*
jellyfish, freshwater *see Craspedacusta
          sowerbyi*
jump dispersal, definition 415

Kalanchoe 149, 154
Kaphra beetle 53
Keeling, Charles 346, *346*
Kitching, Roger L. 15, 17–18
Klamath Mountains (California) 216

Klironomos, John *132*, 134
Kofoid, Charles Atwood 27–28
kokanee *see Onchorhynchus nerka*
Krigia 134
Kruger, Fred 16
kudzu *see Pueraria lobata*
Kulmatiski, Andrew *132*, 137

lag phase, definition 416
lag time, definition 416
Lake Biwa (Japan) 215
Lake Erie 56 *see also* Great Lakes
Lake Malawi 321
Lake Victoria 56, 212, 214 *see also*
          African Lakes
lakes, alien species in 211–224
'land ethic' (Aldo Leopold) 63, 124
largemouth bass *see Micropterus salmoides*
Larson, Brendon 361, 366–373
latency period *see* lag phase
'lattice models' *see* cellular automata
Lawton, John H. 270
LDD *see* long-distance dispersal
leafy spurge 320 see also *Euphorbia esula*
legislation and biological invasions xvii,
          302–303, 311, 380–382, 391,
          402
Leopold, Aldo 6, 20–21, 62, 63, 260
          see also 'land ethic', *Sand County
          Almanac*
Liao, Chengzhang *132*
LiDAR *see* light detection and ranging
light detection and ranging (LiDAR) 284,
          285, 286
limiting similarity, concept of xv, 19, 64,
          122–*123*, 125–128
Lindroth, Carl Hildebrand 19, *27*, 30–31
little fire ant *see Wasmannia auropunctata*
livestock (and biological invasions) 84,
          306, 319, 321, 381
local adaptation, rapid evolution
          of 199–201
Lodge, David M. 366
long-distance dispersal 104–108, 111,
          113, 115, 145, 147, 198, 201,
          339, 348, 411–412, 415, 416
longhorn beetle, Asian *see Anoplophora
          glabripennis*
Lotka – Volterra predator – prey
          model *123*, 125–126, 334
Lyell, Charles xiv

MacArthur, Robert H. 7, *123*, 125–128,
          163, 270
MacOwan, Peter 378–379
maize *see Zea mays* ssp. *mays*
malaria mosquito *see* mosquito, malaria;
          Anopheles gambiae

managed relocation 153, 155, 353, 401,
          416
management of invasive species *see* alien
          species, control of
*Man's Role in Changing the Face of the
          Earth* (Thomas, W.L., 1956) 27
marine invasions 25–33, 55–56, 319
Marloth, Rudolf 379
Matamek Conference on Biological Cycles
          (Labrador, 1931) 6, 13, 63
Mayr, Ernst 5, 15
Mediterranean Basin
     as donor of alien species 74
     as receiver of alien species 78
     rats and rabbits in 150
Mediterranean mussel *see Mytilus
          galloprovincialis*
metaphors (martial/militaristic) in
          invasion ecology 12–14, 65,
          366–367
microbes, role in plant
          invasions 138–139
mile-a-minute weed *see Mikania
          micrantha*
Millennium Ecosystem Assessment 319,
          324, 413
minimum residence time *see* residence
          time
mitten crab *see* Chinese mitten crab
molecular markers *178–181*, 185
molecular signature *see* molecular
          markers
monocultures *see* crop monocultures
Mooney, Harold A. 16
mosquito, malaria (*Anopheles gambiae*) 13
Most Common Genotype (MCG)
          (cheatgrass) 256
mountain pine beetle 95, 353 *see also
          Dendroctonus ponderosae*
Murdoch, William 14–15
muskrat 12–13, 19–20, 305 *see also
          Ondatra zibethicus*
mutualisms 143–159 *see also* seed
          dispersal
Mycosphaerella leaf blotch 92
myna 168 *see also Acridotheres tristis*
mysid shrimp 212

Natural Environment Research Council
          see Nature Conservancy (UK)
national management strategies 378
          see also entries for individual
          countries
native-alien dichotomy 36–37, 367
nativeness, biotic 35–47
native species
     and biotic acceptance 411
     diagnosing 40–42, 90

definition 416
effects of Argentine ant on 240–245
natural enemies, absence of 144
naturalized (discussion of classification
        category for alien species) 38,
        105, *146*, 411, 413–418
naturalized species, definition 416
naturalization hypothesis *see* Darwin's
        naturalization hypothesis
naturalization – invasion
        continuum 145–147, *146*, 416,
        417
naturalized species vs. invasive
        species 415
Nature Conservancy (UK) 6–7
Nature Conservance Act (1949) 7
Netherlands Plant Protection Service 302
New Zealand
    bees in 147, 149–150
    Biosecurity Strategy 378
    birds in 157, 165, 168
    Hieracium pilosella in 204
    impact of alien plants in 74, 76, 306
    Linepithema humile in 240
    non-indigenous ungulate
        mammals 228
    Salmo trutta in 214–215
    Thomson work on invasions 19, *27*
next-generation sequencing 295–296
niche-based approach to invasions 270–
        272, 274
niche-driven community assembly 126
niche modelling 247, 331, 335–336,
        336, 349
niche width 149
Nile perch *see Lates niloticus*
non-indigenous species *see* alien species
non-native species *see* alien species
North America
    Adelgis tsugae in 353
    beetles in 31
    Brachypodium sylvaticum in 178
    Capsella bursa-pastoris in 178
    Centaurea diffusa in 178
    Centaurea stoebe micranthos in 178
    cheatgrass (*Bromus tectorum*)
        invasion 253–265, 352–353
    climate change, effects on weed
        species 351
    Dendroctonus ponderosae (mountain pine
        beetle) in 95, 353
    Dreissena polymorpha in 56
    extinctions in 273
    forest pests 353
    freshwater invasions in 214, *215*
    Gambusia amistadensis in 214
    non-native birds 165–166, *166*
    non-native plants 134–135

Pastinaca sativa in 231
Phalaris arundinacea in 230
Rana catesbeiana in 216
trout 349
See also Great Lakes
novel ecosystems 66, 401, 417
nuclear ribosomal ITS 291–292
nursery trade, role in movement of alien
        species 93–94, 318, 382, 402

oceans, invasions in *Plate 2*, 25–33, 56,
        78, *80*, 84, 125, 127, 154, *213*,
        218, *219*, 273–274, 294, 319
Old World vs. New World invasive
        plants 182
Oosting, Henry J. 19–20
orchids 149 see also Orchidaceae
Origin of Species see Darwin, Charles
ornamental plants 37, 318, 347–348,
        353, 381–382, 388 see also
        nursery *trade, role in movement of
        alien species*
Ottowa Assembly (SCOPE, 1982) 16
overfishing 319
Oxford University Bureau of Animal
        Population *see* Bureau of Animal
        Population
oyster culture 26, 28–29
oyster-tingles 12 *see also* drills

Panama Canal 28, 30
parthenium weed 318 *see also Parthenium
        hysterophorus*
partial differential equations (for rate of
        spread) 332–333
Pattern of Animal Communities, The
        (Charles Elton, 1966) 13, 15
pests, definition 417
Pew Ocean Commission 31
phytophthora root rot *see Phytophthora
        cinnamomi*
phytophthora disease of alder 91
pig *see Sus scrofa*
Pimentel, David 15, 320
pines 90–91, 147, 378 see also *Pinus*
        spp.
place, as argument for nativeness 36, 38,
        42–43
plantations *see* forests
plant – animal mutualisms 144–159
        (*see also* seed dispersal)
plant – flower visitor network *148*
plant – soil feedbacks 134, 135–137
pollination (and alien species) 66, 146–
        155, 334, 380, 398, *399*, 401
pool frogs *see Pelophylax lessonae*
population ecology *xiv*, 7, 74
population growth (human) 324

positive interactions (in invasions) 144–
        150, 247 see also mutualisms
Potomac River *215*, 216–217
prairies (US Midwest), restoration of 62
predictability (of invasion impact)
        217–218
prevention of invasive species spread *see*
        alien species, control of
'prisoner's dilemma' 325
propagule abundance 226–227, *227*
propagule frequency 226–227, *227*
propagule pressure
    and rapid adaptive evolution 176–77
    ant invasions 246, 305
    and biosecurity 305, 324
    definition 163, 226–227, 417
    driver of survival of alien species 21,
        80–81, 163, 165, 167
    and invasion cliff 414
    as mediator of impact 225–251, 305
    emergence as a concept 271, 400
    interaction with other factors 82, 126,
        163, 165, 169, 177, 305, 401–
        402, 414, 416, 417
    research on *213*, 232–233, 401–402
    role in pathways 82, 318, 322, 324,
        418
    in understanding impact of biological
        invasions 225–235, *227*,
        228
    see also colonization pressure
propagule richness 226, 227, *227*, *229*,
        229–231
purple loosestrife *see Lythrum salicaria*
pyrosequencing 295, 296
Pythium 134, 135

quagga mussel *see Dreissena rostriformus
        bugensis*
quantitative trait loci mapping *see* QTL
        mapping
quarantine measures (to prevent
        invasions) 93, 94, 302
QTL mapping 187–188

rabbit *see Oryctolagus cuniculus*
RAD tags 187–188
ragweed *see Ambrosia artemisiifolia*
rainbow trout *see Onchorhynchus mykiss*
range expansions 230–231, *231*, 411–
        412, 417
red imported fire ant (RIFA) *see Solenopsis
        invicta*
red squirrel 13
red-vented bulbul *see Pycnonotus cafer*
red-whiskered bulbul 334 see also
        Pycnonotus jocosus
Reinhart, Kurt *132*, 135, 136

remote sensing (in detection of invasion) 278, 279, 283–285, 403
reproductive diversity, influence on genetic variation (plants) 196–197, *197*
reproductive systems
  and genetic diversity 198–199
  and invasion biology 195–210
residence time 82, 176, 416–417
resident biota 411, 417
resident organisms 411
resource-enemy release hypothesis 417
restoration ecology 61–69, 398, *399*, 400
Rhine River *215*
the 'rich get richer' concept *see* biotic acceptance
Rinderpest virus 56
risk analysis *see* risk assessment
risk assessment (for biosecurity risks) 84–85, 306–311, *307, 311*, 403, 417
riverine ecosystems *see* freshwater ecosystems, impacts of biological invasions on
Rout, Marnie E. *132*
'rule of ten' 4 *see also* tens rule

Sagoff, Mark 361–366, 371
Salisbury, Sir Edward 77
salmon 29 *see also Salmo* species
San Diego Bay 26
Sand County Almanac (Aldo Leopold, 1949) 20–21, *63*, 124
Sanitary and Phytosanitary (SPS) Agreement *see* Agreement on the Application of Sanitary and Phytosanitary Measures
satellite tracking 106, 284 *see also* remote sensing (in detection of invasion)
SCOPE (Scientific Committee on Problems of the Environment) programme on biological invasions xi, Plate 1, 16–19, 21, 270, 387
Science and Policy Working Group (Ecological Society of America) 64
Scientific Committee on Problems of the Environment *see* SCOPE
sea lamprey *see Petromyzon marinus*
'secondary invasions' 66
seed dispersal and plant invasions 103–115, 145–146, *146*, 147
shad 29
Shelford, Victor 4
shipping, as vector for invasion 28, 30–32, 56

in Black, Caspian and Azov seas 56, 229, 232
on American Pacific Coast 26–27
ship rat *see Rattus rattus* 54
Shrader-Frechette, Kristin 365–366
Simberloff, Daniel 366
Sirex wood wasp *see Sirex noctilio*
slipper limpet, American *see* American slipper-limpet
smallmouth bass *see Micropterus dolomieu*
snakehead, northern *see Channa argus*
Society for Ecological Restoration International 64
soil biota (and plant invasions) 131–142, *132*
soil-borne disease *see* soil biota
soil pathogens *see* soil biota
SOLiD (sequencing by oligonucleotide ligation and detection) 295–296
South Africa
  alien plant management 378–393
  biological invasions in 378
  bird dispersal in 147
  draft regulations on invasive species management *382*
  economic losses due to introduced pests 320–321, *320, 321*, 381
  introduced birds 165
  Phytophthora invasions in 91
  Working for Water Programme 377–393, 403
  see also Cape Floristic Region, fynbos vegetation
Southampton Conference (1984) 16
Southwood, T. Richard E 8
spatial modelling 330–340, *337*
species distribution modelling 335, 400
  see also niche modelling
species packing 126, 128
species removal 66
spiny waterflea *see Bythotrephes longimanus*
spread, modelling of 329–343, *331*
SPS Agreement *see* Agreement on the Application of Sanitary and Phytosanitary Measures
St. John's wort *see Hypericum perforatum*
starling *see Sturnus vulgaris*
Stebbins, G. Ledyard (plant evolutionist) 196, 204, 362
  see also Genetics of Colonising Species, The
Stellenbosch Conference (1980) *see* Third International Conference on Mediterranean-Type Ecosystems
Stellenbosch symposium (2008) *see* 'Fifty years of invasion ecology – the legacy of Charles Elton'

steppe, North American *254*
stochastic dynamic programming 322
succession ecology 12, 14
sudden oak death 93–94 *see also Phytophthora ramorum*
Suez Canal *28*, 105
supercolonies 245
'systems knowledge' research 398, *399*

Tahiti imperial pigeon *see Ducula aurorae*
'Tamarisk coalition' 44
Tamarix (in USA) 42–44, 320 see also *Tamarix* spp.
Tanganyika sardine *see Limnothrissa miodon*
'target knowledge' research 398, *399*
tens rule 417
Teosinte *see Zea mays* ssp. *parviglumis*
terminological debates
  restoration ecology/invasion biology 65
  see also martial metaphors (for introduced species)
'terrorist' metaphor (for invasive species) 367 *see also* metaphors
The Ecology of Invasions by Animals and Plants (Charles Elton) *see Ecology of Invasions by Animals and Plants, The*
The Genetics of Colonising Species (eds Baker and Stebbins) *see Genetics of Colonising Species, The*
Third International Conference on Mediterranean-Type Ecosystems (Stellenbosch, 1980) 16
Thomas, William L. 27
threshold dynamics (in restoration ecology) 62
trade, as means of species spread 52, 93–94, 302–303, 316, 318, 323
transformers 418
'transformation knowledge' research 398, *399*
translocation *see* managed relocation
transplantation *see* managed relocation
transport links and species spread 302, 305, 318 *see also* trade
travel *see* transport links
tree endophytes 94
tree pathogens *see* tree pests
tree pests 89–99, 401
trematode *see Bucephalus polymorphus*
triffid weed *see Chromolaena odorata*
tropics, invasibility of 165–167

UK, economic losses due to introduced pests 320–321, *320, 321* see also Britain

unicoloniality (in Argentine ants) 245
United States Commission on Ocean
        Policy 31
USA, economic losses due to introduced
        pests 320–321, *320, 321 see also*
        North America

value basis of invasion biology 360–373
van der Putten, Wim *132*, 133,
        136–137
vectors
    definition 418
    in marine invasions 30
    studies of 398, 400, 402
'vector-centred' approach (in seed
        dispersal) 105
'vector science' 398, *399, 400 see also*
        vectors
vicariance 5
von Neumann, John *337*

Wallace, Alfred Russel 4, 7, 14
walleye *see Sander vitreus*
wattles 388 *see also* acacias, *Acacia* spp.
warming, global see global warming
Warwick, Tom 13, 14
water hyacinth *see Eichhornia crassipes,
        E. paniculata*
Watson, Hewett C. 37–38, 39–40
Watson's civil model of biotic
        nativeness 37–38, 39
weed, definition 418
Weeds and Aliens (E. Salisbury, 1961) 77
Wegener (continental drift theory) 5
white pine blister rust 91
Wilcove et al. (on extinction
        threats) 272–273
wind, seed dispersal by 108
witchweed *see Striga hermonthica*
wood packaging, role in movement of
        pests 93

Working for Water Programme 377–
        393, *390*
Wytham Ecological Survey 8, 13
Wytham Woods estate 8
    see also Bureau of Animal Population

yellow crazy ant *see Anoplolepis gracilipes*
yellow starthistle 320

zebra mussel
    in ballast water 52, 56, 318
    economic costs of 316
    in Great Lakes 212, 214, 216, 230
    studies on 54, 56–58 *57*
    see also Dreissena polymorpha
zoonoses 318